Operator Theory: Advances and Applications
Vol. 160

Recent Advances in Operator Theory and its Applications

The Israel Gohberg Anniversary Volume

International Workshop on Operator Theory and its Applications
IWOTA 2003, Cagliari, Italy

Marinus A. Kaashoek
Sebastiano Seatzu
Cornelis van der Mee
Editors

Birkhäuser Verlag
Basel · Boston · Berlin

Editors:

Marinus A. Kaashoek
Department of Mathematics, FEW
Vrije Universiteit
De Boelelaan 1081A
1081 HV Amsterdam
The Netherlands
e-mail: m.a.kaashoek@few.vu.nl

Sebastiano Seatzu
Cornelis van der Mee
Dipartimento di Matematica
Università di Cagliari
Viale Merello 92
09123 Cagliari
Italy
e-mail: seatzu@unica.it
cornelis@krein.unica.it

2000 Mathematics Subject Classification 34, 35, 45, 47, 65, 93

A CIP catalogue record for this book is available from the
Library of Congress, Washington D.C., USA

Bibliographic information published by Die Deutsche Bibliothek
Die Deutsche Bibliothek lists this publication in the Deutsche Nationalbibliografie; detailed bibliographic data is available in the Internet at <http://dnb.ddb.de>.

ISBN 3-7643-7290-7 Birkhäuser Verlag, Basel – Boston – Berlin

Member of the BertelsmannSpringer Publishing Group
Printed on acid-free paper produced from chlorine-free pulp. TCF ∞
Cover design: Heinz Hiltbrunner, Basel
Printed in Germany
ISBN 10: 3-7643-7290-7
ISBN 13: 978-3-7643-7290-3
e-ISBN: 3-7643-7398-9

9 8 7 6 5 4 3 2 1

www.birkhauser.ch

Contents

Editorial Preface

This volume contains a selection of papers in modern operator theory and its applications. Most of them are directly related to lectures presented at the Fourteenth International Workshop on Operator Theory and its Applications (IWOTA 2003) held at the University of Cagliari, Italy, in the period of June 24–27, 2003.

The workshop, which was attended by 108 mathematicians – including a number of PhD and postdoctoral students – from 22 countries, presented eight special sessions on

1) control theory,
2) interpolation theory,
3) inverse scattering,
4) numerical estimates for operators,
5) numerical treatment of integral equations,
6) pseudodifferential operators,
7) realizations and transformations of analytic functions and indefinite inner product spaces, and
8) structured matrices.

The program consisted of 19 plenary lectures of 45 minutes and 78 lectures of 30 minutes in four parallel sessions.

The present volume reflects the wide range and rich variety of topics presented and discussed at the workshop, both within and outside the special sessions. The papers deal with inverse scattering, numerical ranges, pseudodifferential operators, numerical analysis, interpolation theory, multidimensional system theory, indefinite inner products, spectral factorization, and stationary processes.

Since in the period that the proceedings of IWOTA 2003 were being prepared, *Israel Gohberg*, the president of the IWOTA steering committee, reached the age of 75, we decided to dedicate these proceedings to *Israel Gohberg on the occasion of his 75th birthday.* All of the authors of these proceedings have joined the editors and dedicated their papers to him as well.

The Editors

Israel Gohberg, the president
of the IWOTA steering committee

Operator Theory:
Advances and Applications, Vol. 160, 1–20

Inverse Scattering with Rational Scattering Coefficients and Wave Propagation in Nonhomogeneous Media

Tuncay Aktosun, Michael H. Borkowski, Alyssa J. Cramer and Lance C. Pittman

Dedicated to Israel Gohberg on the occasion of his 75th birthday

Abstract. The inverse scattering problem for the one-dimensional Schrödinger equation is considered when the potential is real valued and integrable and has a finite first-moment and no bound states. Corresponding to such potentials, for rational reflection coefficients with only simple poles in the upper half complex plane, a method is presented to recover the potential and the scattering solutions explicitly. A numerical implementation of the method is developed. For such rational reflection coefficients, the scattering wave solutions to the plasma-wave equation are constructed explicitly. The discontinuities in these wave solutions and in their spatial derivatives are expressed explicitly in terms of the potential.

Mathematics Subject Classification (2000). Primary 34A55; Secondary 34L40 35L05 47E05 81U40.

Keywords. Inverse scattering, Schrödinger equation, Rational scattering coefficients, Wave propagation, Plasma-wave equation.

1. Introduction

Consider the Schrödinger equation

$$\psi''(k,x) + k^2\psi(k,x) = V(x)\,\psi(k,x), \qquad x \in \mathbf{R}, \tag{1.1}$$

The research leading to this paper was supported by the National Science Foundation under grants DMS-0243673 and DMS-0204436 and by the Department of Energy under grant DE-FG02-01ER45951.

where the prime denotes the x-derivative, and the potential V is assumed to have no bound states and to belong to the Faddeev class. The bound states of (1.1) correspond to its square-integrable solutions. By the Faddeev class we mean the set of real-valued and measurable potentials for which $\int_{-\infty}^{\infty} dx\,(1+|x|)|V(x)|$ is finite.

Via the Fourier transformation

$$u(x,t) = \frac{1}{2\pi}\int_{-\infty}^{\infty} dk\,\psi(k,x)\,e^{-ikt},$$

we can transform (1.1) into the plasma-wave equation

$$\frac{\partial^2 u(x,t)}{\partial x^2} - \frac{\partial^2 u(x,t)}{\partial t^2} = V(x)\,u(x,t), \qquad x,t\in\mathbf{R}. \tag{1.2}$$

In the absence of bound states, (1.1) does not have any bounded solutions for $k^2<0$. The solutions for $k^2>0$ are known as the scattering solutions. Each scattering solution can be expressed as a linear combination of the two (linearly-independent) Jost solutions from the left and the right, denoted by f_{l} and f_{r}, respectively, satisfying the respective asymptotic conditions

$$f_{\mathrm{l}}(k,x) = e^{ikx}\,[1+o(1)]\,, \quad f_{\mathrm{l}}'(k,x) = ik\,e^{ikx}\,[1+o(1)]\,, \qquad x\to+\infty,$$

$$f_{\mathrm{r}}(k,x) = e^{-ikx}\,[1+o(1)]\,, \quad f_{\mathrm{r}}'(k,x) = -ik\,e^{-ikx}\,[1+o(1)]\,, \qquad x\to-\infty.$$

We have

$$f_{\mathrm{l}}(k,x) = \frac{1}{T(k)}\,e^{ikx} + \frac{L(k)}{T(k)}\,e^{-ikx} + o(1), \qquad x\to-\infty,$$

$$f_{\mathrm{r}}(k,x) = \frac{1}{T(k)}\,e^{-ikx} + \frac{R(k)}{T(k)}\,e^{ikx} + o(1), \qquad x\to+\infty,$$

where L and R are the left and right reflection coefficients, respectively, and T is the transmission coefficient.

The solutions to (1.1) for $k=0$ require special attention. Generically, $f_{\mathrm{l}}(0,x)$ and $f_{\mathrm{r}}(0,x)$ are linearly independent on $\mathbf{R}$, and we have

$$T(0)=0, \quad R(0)=L(0)=-1.$$

In the exceptional case, $f_{\mathrm{l}}(0,x)$ and $f_{\mathrm{r}}(0,x)$ are linearly dependent on $\mathbf{R}$ and we have

$$T(0)=\sqrt{1-R(0)^2}>0, \quad -1<R(0)=-L(0)<1.$$

When V belongs to the Faddeev class and has no bound states, it is known [1–5] that either one of the reflection coefficients R and L contains the appropriate information to construct the other reflection coefficient, the transmission coefficient T, the potential V, and the Jost solutions f_{l} and f_{r}. Our aim in this paper is to present explicit formulas for such a construction when the reflection coefficients are rational functions of k with simple poles on the upper half complex plane $\mathbf{C}^+$. We will use $\mathbf{C}^-$ to denote the lower half complex plane and let $\overline{\mathbf{C}^+} := \mathbf{C}^+\cup\mathbf{R}$ and $\overline{\mathbf{C}^-} := \mathbf{C}^-\cup\mathbf{R}$.

The recovery of V from a reflection coefficient constitutes the inverse scattering problem for (1.1). There has been a substantial amount of previous work [2,

6–14] done on the inverse scattering problem with rational reflection coefficients. The solution to this inverse problem can, for example, be obtained by solving the Marchenko integral equation [1–5]. Another way to solve this inverse problem is to use Sabatier's method [2, 12–14] utilizing transformations resembling Darboux transformations [1–3]. Dolveck-Guilpart developed [7] a numerical implementation of Sabatier's method. Yet another method is based on the Wiener-Hopf factorization of a 2×2 matrix [6] related to the scattering matrix for (1.1). It is also possible to use [15, 16] a minimal realization of a rational reflection coefficient and to recover the potential explicitly. The method discussed in our paper is closely related to that given in [6]. Here, we are able to write down the Jost solutions explicitly in terms of the poles in $\mathbf{C}^+$ and the corresponding residues of the reflection coefficients. This also enables us to construct explicitly certain solutions to (1.2), which we call the Jost waves.

Our paper is organized as follows. In Section 2 we present the preliminary material needed for later sections, including an outline of the construction of T and L from the right reflection coefficient R. In Section 3 we present the explicit construction of the potential and the Jost solutions for $x > 0$ in terms of the poles in $\mathbf{C}^+$ and the corresponding residues of R. Having constructed the left reflection coefficient L in terms of R, in Section 4 we present the explicit construction of the potential and the Jost solutions for $x < 0$ in terms of the poles in $\mathbf{C}^+$ and the corresponding residues of L. In Section 5 we turn our attention to (1.2) and explicitly construct its solutions by using the Fourier transforms of the Jost solutions to (1.1). In Section 6 we analyze the discontinuities in such wave solutions and in their x-derivatives at each fixed t. Finally, in Section 7 we remark on the numerical implementation of our method.

2. Preliminaries

For convenience, we introduce the Faddeev functions from the left and right, denoted by m_{l} and m_{r}, respectively, defined as

$$m_{\mathrm{l}}(k,x) := e^{-ikx} f_{\mathrm{l}}(k,x), \qquad m_{\mathrm{r}}(k,x) := e^{ikx} f_{\mathrm{r}}(k,x). \tag{2.1}$$

From (1.1) and (2.1) it follows that

$$m_{\mathrm{l}}''(k,x) + 2ik\, m_{\mathrm{l}}'(k,x) = V(x)\, m_{\mathrm{l}}(k,x), \qquad x \in \mathbf{R}, \tag{2.2}$$

$$m_{\mathrm{r}}''(k,x) - 2ik\, m_{\mathrm{r}}'(k,x) = V(x)\, m_{\mathrm{r}}(k,x), \qquad x \in \mathbf{R}.$$

It is known [1–5] that

$$f_{\mathrm{l}}(-k,x) = -R(k)\, f_{\mathrm{l}}(k,x) + T(k)\, f_{\mathrm{r}}(k,x), \qquad k \in \mathbf{R}, \tag{2.3}$$

$$f_{\mathrm{r}}(-k,x) = T(k)\, f_{\mathrm{l}}(k,x) - L(k)\, f_{\mathrm{r}}(k,x), \qquad k \in \mathbf{R}, \tag{2.4}$$

or equivalently

$$m_{\mathrm{l}}(-k,x) = -R(k)\, e^{2ikx} m_{\mathrm{l}}(k,x) + T(k)\, m_{\mathrm{r}}(k,x), \qquad k \in \mathbf{R}, \tag{2.5}$$

$$m_{\mathrm{r}}(-k,x) = T(k)\, m_{\mathrm{l}}(k,x) - L(k)\, e^{-2ikx} m_{\mathrm{r}}(k,x), \qquad k \in \mathbf{R}. \tag{2.6}$$

When R and L are rational functions of k, their domains can be extended meromorphically from $\mathbf{R}$ to the entire complex plane $\mathbf{C}$. Similarly, the Jost solutions and the Faddeev functions have extensions that are analytic in $\mathbf{C}^+$ and meromorphic in $\mathbf{C}^-$. We have

$$f_{\rm l}(-k^*,x)=f_{\rm l}(k,x)^*,\quad f_{\rm r}(-k^*,x)=f_{\rm r}(k,x)^*,\qquad k\in\mathbf{C},$$
$$T(-k^*)=T(k)^*,\quad R(-k^*)=R(k)^*,\quad L(-k^*)=L(k)^*,\qquad k\in\mathbf{C},$$

where the asterisk denotes complex conjugation. Note, in particular, that

$$|R(k)|^2=R(k)\,R(-k),\qquad k\in\mathbf{R}. \tag{2.7}$$

The scattering coefficients satisfy

$$T(k)\,T(-k)+R(k)\,R(-k)=1,\qquad k\in\mathbf{R}, \tag{2.8}$$
$$T(k)\,T(-k)+L(k)\,L(-k)=1,\qquad k\in\mathbf{R},$$
$$L(k)\,T(-k)+R(-k)\,T(k)=0,\qquad k\in\mathbf{R}, \tag{2.9}$$

with appropriate meromorphic extensions to $\mathbf{C}$.

In the rest of this section we outline the construction of T and L from R. From (2.7) and (2.8) we get

$$T(k)=\left[1-|R(k)|^2\right]\frac{1}{T(-k)},\qquad k\in\mathbf{R}. \tag{2.10}$$

If $R(k)$ is a rational function of $k\in\mathbf{R}$, so is $1-|R(k)|^2$. When V belongs to the Faddeev class and has no bound states, it is known [1–5] that $T(k)$ is analytic in $\mathbf{C}^+$, continuous in $\overline{\mathbf{C}^+}$, nonzero in $\overline{\mathbf{C}^+}\setminus\{0\}$, and $1+O(1/k)$ as $k\to\infty$ in $\overline{\mathbf{C}^+}$. Generically $T(k)$ has a simple zero at $k=0$, and $T(0)\neq 0$ in the exceptional case. We have $R(k)=o(1/k)$ as $k\to\pm\infty$ in $\mathbf{R}$, and hence the rationality of R implies that $R(k)=O(1/k^2)$ as $k\to\infty$ in $\mathbf{C}$. With the help of (2.10), by factoring both the numerator and the denominator of $1-|R(k)|^2$, we can obtain $T(k)$ by separating the zeros and poles of $1-|R(k)|^2$ in $\mathbf{C}^+$ and in $\mathbf{C}^-$.

In the exceptional case it is known [1–5] that $|R(k)|<1$ for $k\in\mathbf{R}$. Thus, in that case we get

$$1-|R(k)|^2=\frac{\prod(k-k_a^+)\prod(k-k_b^-)}{\prod(k-k_m^+)\prod(k-k_n^-)},\qquad k\in\mathbf{R}, \tag{2.11}$$

where $k_a^+,k_m^+\in\mathbf{C}^+$ and $k_b^-,k_n^-\in\mathbf{C}^-$. Note that the left-hand side in (2.11) is an even function of k and it converges to 1 as $k\to\pm\infty$. As a result we find that the degrees of the four polynomials $\prod(k-k_a^+)$, $\prod(k-k_b^-)$, $\prod(k-k_m^+)$, and $\prod(k-k_n^-)$ are all the same. Hence, from (2.10) and (2.11) we get

$$T(k)\,\frac{\prod(k-k_n^-)}{\prod(k-k_b^-)}=\frac{1}{T(-k)}\,\frac{\prod(k-k_a^+)}{\prod(k-k_m^+)}, \tag{2.12}$$

where the left-hand side is analytic in $\mathbf{C}^+$, continuous in $\overline{\mathbf{C}^+}$, and $1+O(1/k)$ as $k\to\infty$ in $\overline{\mathbf{C}^+}$. Similarly, the right-hand side of (2.12) is analytic in $\mathbf{C}^-$, continuous in $\overline{\mathbf{C}^-}$, and $1+O(1/k)$ as $k\to\infty$ in $\overline{\mathbf{C}^-}$. With the help of Morera's theorem, we conclude that each side of (2.12) must be equal to an entire function on $\mathbf{C}$ that converges to 1 at ∞. By Liouville's theorem both sides must then be equal to 1. Therefore, we obtain

$$T(k) = \frac{\prod(k - k_b^-)}{\prod(k - k_n^-)}, \qquad k \in \mathbf{C}. \tag{2.13}$$

The argument given above can easily be adapted to the generic case. In the generic case, it is known [1–5] that $T(k) = O(k)$ as $k \to 0$, and the construction of T from R is similarly obtained by replacing exactly one of k_b^- with zero and exactly one of k_a^+ with zero in the above argument.

Let us write the expression in (2.13) for $T(k)$ in a slightly different notation which will be useful in Section 5:

$$T(k) = \begin{cases} \dfrac{k\prod_{j=1}^{n_z}(k+z_j)}{\prod_{j=1}^{n_z+1}(k+p_j)}, & T(0)=0, \\[2ex] \dfrac{\prod_{j=1}^{n_z}(k+z_j)}{\prod_{j=1}^{n_z}(k+p_j)}, & T(0)\neq 0, \end{cases} \tag{2.14}$$

where the z_j for $1 \le j \le n_z$ correspond to the zeros of $T(-k)$ in $\mathbf{C}^+$ and the p_j correspond to the poles there. Thus, the poles of $1/T(-k)$ in $\mathbf{C}^+$ occur at $k = z_j$, and let us use τ_j to denote the residues there:

$$\tau_j := \operatorname{Res}\left(\frac{1}{T(-k)}, z_j\right), \qquad j = 1, \dots, n_z. \tag{2.15}$$

Once $T(k)$ is constructed, with the help of (2.9) we obtain

$$L(k) = -\frac{R(-k)\,T(k)}{T(-k)}.$$

3. Construction on the positive half-line

In this section, when $x > 0$ we explicitly construct the Jost solutions and the potential in terms of the poles and residues of $R(k)$ in $\mathbf{C}^+$. We use n_{l} to denote the number of poles of $R(k)$ in $\mathbf{C}^+$, assume that such poles are simple and occur at $k = k_{\mathrm{l}j}$ in $\mathbf{C}^+$, and use $\rho_{\mathrm{l}j}$ to denote the corresponding residues.

We define

$$B_{\mathrm{l}}(x,\alpha) := \frac{1}{2\pi}\int_{-\infty}^{\infty} dk\, [m_{\mathrm{l}}(k,x) - 1]\, e^{-ik\alpha}. \tag{3.1}$$

When $\alpha < 0$, we have $B_{\mathrm{l}}(x,\alpha) = 0$ due to, for each fixed x, the analyticity of $m_{\mathrm{l}}(k,x)$ in $\mathbf{C}^+$, the continuity of $m_{\mathrm{l}}(k,x)$ in $\overline{\mathbf{C}^+}$, and the fact that $m_{\mathrm{l}}(k,x) =$

$1 + O(1/k)$ as $k \to \infty$ in $\overline{\mathbf{C}^+}$. From (2.5) we get

$$\begin{aligned}\frac{1}{2\pi}\int_{-\infty}^{\infty} dk\, [m_{\rm l}(-k,x) - 1]\, e^{ik\alpha} = &-\frac{1}{2\pi}\int_{-\infty}^{\infty} dk\, R(k)\, m_{\rm l}(k,x) e^{ik(2x+\alpha)} \\ &+ \frac{1}{2\pi}\int_{-\infty}^{\infty} dk\, [T(k)\, m_{\rm r}(k,x) - 1]\, e^{ik\alpha}.\end{aligned} \tag{3.2}$$

The second integral on the right-hand side of (3.2) vanishes when $\alpha > 0$ due to the fact that $T(k)$ and $m_{\rm r}(k,x)$ are analytic for $k \in \mathbf{C}^+$ and continuous for $k \in \overline{\mathbf{C}^+}$, and $T(k)\, m_{\rm r}(k,x) = 1 + O(1/k)$ as $k \to \infty$ in $\overline{\mathbf{C}^+}$. Thus, from (3.1) and (3.2) we obtain

$$B_{\rm l}(x,\alpha) = -\frac{1}{2\pi}\int_{-\infty}^{\infty} dk\, R(k)\, e^{2ikx+ik\alpha} m_{\rm l}(k,x), \qquad \alpha > 0. \tag{3.3}$$

From (3.1) and the fact that $B_{\rm l}(x,\alpha) = 0$ for $\alpha < 0$, we have

$$m_{\rm l}(k,x) = 1 + \int_0^{\infty} d\alpha\, B_{\rm l}(x,\alpha)\, e^{ik\alpha}. \tag{3.4}$$

When $2x + \alpha > 0$, the integral in (3.3) can be evaluated as a contour integral along the boundary of $\mathbf{C}^+$, to which the only contribution comes from the poles of $R(k)$ in $\mathbf{C}^+$. Since such poles are assumed to be simple, we get

$$B_{\rm l}(x,\alpha) = -i\sum_{j=1}^{n_{\rm l}} \rho_{{\rm l}j}\, e^{2ik_{{\rm l}j}x + ik_{{\rm l}j}\alpha} m_{\rm l}(k_{{\rm l}j},x), \qquad 2x+\alpha > 0, \quad \alpha > 0. \tag{3.5}$$

Using (3.5) in (3.4), with the help of

$$\int_0^{\infty} d\alpha\, e^{i(k+k_{{\rm l}j})\alpha} = \frac{i}{k + k_{{\rm l}j}}, \qquad k \in \overline{\mathbf{C}^+},$$

we get

$$m_{\rm l}(k,x) = 1 + \sum_{j=1}^{n_{\rm l}} \frac{\rho_{{\rm l}j}\, e^{2ik_{{\rm l}j}x}}{k + k_{{\rm l}j}}\, m_{\rm l}(k_{{\rm l}j},x), \qquad x \ge 0. \tag{3.6}$$

We are interested in determining $m_{\rm l}(k_{{\rm l}j},x)$ appearing in (3.5) and (3.6). To do so, we put $k = k_{{\rm l}p}$ in (3.6) for $1 \le p \le n_{\rm l}$. Then, for $x \ge 0$ we obtain

$$m_{\rm l}(k_{{\rm l}p},x) = 1 + \sum_{j=1}^{n_{\rm l}} \frac{\rho_{{\rm l}j}\, e^{2ik_{{\rm l}j}x}}{k_{{\rm l}p} + k_{{\rm l}j}} m_{\rm l}(k_{{\rm l}j},x), \qquad p = 1,\dots,n_{\rm l}. \tag{3.7}$$

Notice that (3.7) is a linear algebraic system and can be written as

$$\mathbf{M}_{\rm l}(x) \begin{bmatrix} m_{\rm l}(k_{{\rm l}1},x) \\ \vdots \\ m_{\rm l}(k_{{\rm l}n_{\rm l}},x) \end{bmatrix} = \begin{bmatrix} 1 \\ \vdots \\ 1 \end{bmatrix}, \tag{3.8}$$

where $\mathbf{M}_{\rm l}(x)$ is the $n_{\rm l} \times n_{\rm l}$ matrix-valued function whose (p,q)-entry is given by

$$[\mathbf{M}_{\rm l}(x)]_{pq} := \delta_{pq} - \frac{\rho_{{\rm l}q}\, e^{2ik_{{\rm l}q}x}}{k_{{\rm l}p} + k_{{\rm l}q}}, \tag{3.9}$$

with δ_{pq} denoting the Kronecker delta. The unique solvability of the linear system in (3.8) and hence the invertibility of $\mathbf{M}_{\mathrm{l}}(x)$ follow from Corollary 4.2 of [6]. Using (3.8) in (3.6) we get for $x \geq 0$

$$m_{\mathrm{l}}(k,x) = 1 + \begin{bmatrix} \dfrac{\rho_{\mathrm{l}1}\, e^{2ik_{\mathrm{l}1}x}}{k+k_{\mathrm{l}1}} & \cdots & \dfrac{\rho_{\mathrm{l}n_{\mathrm{l}}}\, e^{2ik_{\mathrm{l}n_{\mathrm{l}}}x}}{k+k_{\mathrm{l}n_{\mathrm{l}}}} \end{bmatrix} \mathbf{M}_{\mathrm{l}}(x)^{-1} \begin{bmatrix} 1 \\ \vdots \\ 1 \end{bmatrix}. \tag{3.10}$$

We can simplify (cf. p. 12 of [17]) the bilinear form in (3.10) and obtain for $x \geq 0$

$$m_{\mathrm{l}}(k,x) = 1 - \frac{1}{\det \mathbf{M}_{\mathrm{l}}(x)} \begin{vmatrix} 0 & \dfrac{\rho_{\mathrm{l}1}\, e^{2ik_{\mathrm{l}1}x}}{k+k_{\mathrm{l}1}} & \cdots & \dfrac{\rho_{\mathrm{l}n_{\mathrm{l}}}\, e^{2ik_{\mathrm{l}n_{\mathrm{l}}}x}}{k+k_{\mathrm{l}n_{\mathrm{l}}}} \\ 1 & & & \\ \vdots & & \mathbf{M}_{\mathrm{l}}(x) & \\ 1 & & & \end{vmatrix}, \tag{3.11}$$

and hence we have written $m_{\mathrm{l}}(k,x) - 1$ as the ratio of two determinants that are constructed solely in terms of the $k_{\mathrm{l}j}$ and $\rho_{\mathrm{l}j}$ with $1 \leq j \leq n_{\mathrm{l}}$.

Similarly, from (3.5) and (3.8) we get

$$B_{\mathrm{l}}(x,\alpha) = -i \begin{bmatrix} \rho_{\mathrm{l}1}\, e^{ik_{\mathrm{l}1}(2x+\alpha)} & \cdots & \rho_{\mathrm{l}n_{\mathrm{l}}}\, e^{ik_{\mathrm{l}n_{\mathrm{l}}}(2x+\alpha)} \end{bmatrix} \mathbf{M}_{\mathrm{l}}(x)^{-1} \begin{bmatrix} 1 \\ \vdots \\ 1 \end{bmatrix},$$

or equivalently

$$B_{\mathrm{l}}(x,\alpha) = \frac{\det \Gamma_{\mathrm{l}}(x,\alpha)}{\det \mathbf{M}_{\mathrm{l}}(x)}, \qquad x \geq 0, \quad \alpha > 0, \tag{3.12}$$

where $\Gamma_{\mathrm{l}}(x,\alpha)$ is the $(n_{\mathrm{l}}+1) \times (n_{\mathrm{l}}+1)$ matrix defined as

$$\Gamma_{\mathrm{l}}(x,\alpha) := \begin{bmatrix} 0 & i\rho_{\mathrm{l}1}\, e^{2ik_{\mathrm{l}1}x + ik_{\mathrm{l}1}\alpha} & \cdots & i\rho_{\mathrm{l}n_{\mathrm{l}}}\, e^{2ik_{\mathrm{l}n_{\mathrm{l}}}x + ik_{\mathrm{l}n_{\mathrm{l}}}\alpha} \\ 1 & & & \\ \vdots & & \mathbf{M}_{\mathrm{l}}(x) & \\ 1 & & & \end{bmatrix}. \tag{3.13}$$

It is pleasantly surprising that we have

$$\det \Gamma_{\mathrm{l}}(x, 0^+) = \frac{d}{dx} \det \mathbf{M}_{\mathrm{l}}(x). \tag{3.14}$$

The proof of (3.14) is somehow involved, and we briefly describe the basic steps in the proof. First, in the matrix $\Gamma_{\mathrm{l}}(x,0^+)$, multiply the $(j+1)$st column by $e^{-ik_{\mathrm{l}j}x}$ and the $(j+1)$st row by $e^{ik_{\mathrm{l}j}x}$ for all $1 \leq j \leq n_{\mathrm{l}}$. The determinant remains unchanged. Then, use the cofactor expansion of the resulting determinant with respect to the first column and get

$$\det \Gamma_{\mathrm{l}}(x,0^+) = 0\,|\cdot| - e^{ik_{\mathrm{l}1}x}\,|\cdot| + e^{ik_{\mathrm{l}2}x}\,|\cdot| - \cdots + (-1)^{n_{\mathrm{l}}} e^{ik_{\mathrm{l}n_{\mathrm{l}}}x}\,|\cdot|, \tag{3.15}$$

where $|\cdot|$ denotes the appropriate subdeterminant. Next, put each coefficient $e^{ik_{\mathrm{l}j}x}$ on the right-hand side of (3.15) into the first row of the corresponding subdeterminant. We need to show that the resulting quantity is equal to the x-derivative of $\det \mathbf{M}_{\mathrm{l}}(x)$. In order to do so, in the matrix $\mathbf{M}_{\mathrm{l}}(x)$ multiply the jth row by $e^{ik_{\mathrm{l}j}x}$ and the jth column by $e^{-ik_{\mathrm{l}j}x}$ for $1 \le j \le n_{\mathrm{l}}$, which results in no change in $\det \mathbf{M}_{\mathrm{l}}(x)$. Now take the x-derivative of the resulting determinant and write it as a sum where the jth term is the determinant of a matrix obtained by taking the x-derivative of the jth row of $\mathbf{M}_{\mathrm{l}}(x)$. Then rewrite each term in the summation so that the row whose derivative has been evaluated is moved to the first row while the remaining rows are left in the same order. By comparison, we then conclude (3.14).

Using (3.14) in (3.12) we get

$$B_{\mathrm{l}}(x, 0^+) = \frac{\frac{d}{dx}\det \mathbf{M}_{\mathrm{l}}(x)}{\det \mathbf{M}_{\mathrm{l}}(x)}, \qquad x \ge 0. \tag{3.16}$$

It is known [1–5] that

$$V(x) = -2\frac{d}{dx}B_{\mathrm{l}}(x, 0^+), \qquad x \in \mathbf{R}. \tag{3.17}$$

Therefore, using (3.16) in (3.17) we get

$$V(x) = -2\frac{d}{dx}\left[\frac{\frac{d}{dx}\det \mathbf{M}_{\mathrm{l}}(x)}{\det \mathbf{M}_{\mathrm{l}}(x)}\right], \qquad x > 0. \tag{3.18}$$

From (3.9) we see that $\mathbf{M}_{\mathrm{l}}(x)$ is uniquely constructed in terms of the poles and residues of $R(k)$ in $\mathbf{C}^+$. Thus, in (3.18) we have expressed $V(x)$ for $x > 0$ in terms of the $(2n_{\mathrm{l}})$ constants $k_{\mathrm{l}j}$ and $\rho_{\mathrm{l}j}$ alone.

Alternatively, having constructed $m_{\mathrm{l}}(k,x)$ for $x \ge 0$ as in (3.11), we can use (2.2) and obtain the potential for $x > 0$ as

$$V(x) = \frac{m_{\mathrm{l}}''(k,x) + 2ik\, m_{\mathrm{l}}'(k,x)}{m_{\mathrm{l}}(k,x)}. \tag{3.19}$$

Note that even though the parameter k appears in the individual terms on the right-hand side in (3.19), it is absent from the right-hand side as a whole. In particular, using $k = 0$ in (3.19) we can evaluate $V(x)$ for $x > 0$ as

$$V(x) = \frac{m_{\mathrm{l}}''(0,x)}{m_{\mathrm{l}}(0,x)}. \tag{3.20}$$

We have shown that, starting with a rational right reflection coefficient R, one can explicitly construct $m_{\mathrm{l}}(k,x)$ for $x \ge 0$ and $V(x)$ for $x > 0$. We can then obtain $f_{\mathrm{l}}(k,x)$ via (2.1). This means that we also have $f_{\mathrm{l}}(-k,x)$ in hand. Then, using (2.3) we also get $f_{\mathrm{r}}(k,x)$ for $x \ge 0$ as

$$f_{\mathrm{r}}(k,x) = \frac{f_{\mathrm{l}}(-k,x) + R(k)\, f_{\mathrm{l}}(k,x)}{T(k)},$$

with $T(k)$ as in (2.14).

4. Construction on the negative half-line

In this section we present the explicit construction of the Jost solutions and the potential when $x < 0$. Finding $m_{\mathrm{r}}(k,x)$ and $V(x)$ for $x < 0$ in terms of $L(k)$ is similar to the construction of $m_{\mathrm{l}}(k,x)$ and $V(x)$ for $x > 0$ from R as outlined in Section 3. As we shall see in (4.6) and (4.10), the explicit formulas for $m_{\mathrm{r}}(k,x)$ and $V(x)$ for $x < 0$ are written in terms of the poles and residues of $L(k)$ in $\mathbf{C}^+$. We let n_{r} denote the number of poles of $L(k)$ in $\mathbf{C}^+$, use $k_{\mathrm{r}j}$ and $\rho_{\mathrm{r}j}$ to denote those (simple) poles and the corresponding residues, respectively.

Let

$$B_{\mathrm{r}}(x,\alpha) := \frac{1}{2\pi}\int_{-\infty}^{\infty} dk\,[m_{\mathrm{r}}(k,x)-1]\,e^{-ik\alpha}\,dk. \tag{4.1}$$

When $\alpha < 0$, we get $B_{\mathrm{r}}(x,\alpha) = 0$ because, for each fixed x, $m_{\mathrm{r}}(k,x)$ is analytic in $\mathbf{C}^+$, continuous in $\overline{\mathbf{C}^+}$, and $1 + O(1/k)$ as $k \to \infty$ in $\overline{\mathbf{C}^+}$. Starting with (2.6) we show [cf. (3.3)] that

$$B_{\mathrm{r}}(x,\alpha) = -\frac{1}{2\pi}\int_{-\infty}^{\infty} dk\,L(k)\,e^{-2ikx+ik\alpha}m_{\mathrm{r}}(k,x), \qquad \alpha > 0, \tag{4.2}$$

and thus, [cf. (3.4)] we obtain

$$m_{\mathrm{r}}(k,x) = 1 + \int_0^{\infty} d\alpha\,B_{\mathrm{r}}(x,\alpha)\,e^{ik\alpha}. \tag{4.3}$$

In order to evaluate the integral in (4.2), we use a contour integration along the infinite semicircle which is the boundary of $\mathbf{C}^+$. Since the poles of $L(k)$ in $\mathbf{C}^+$ are assumed to be simple, we obtain [cf. (3.5)]

$$B_{\mathrm{r}}(x,\alpha) = -i\sum_{j=1}^{n_{\mathrm{r}}} \rho_{\mathrm{r}j}\,e^{-2ik_{\mathrm{r}j}x+ik_{\mathrm{r}j}\alpha}m_{\mathrm{r}}(k_{\mathrm{r}j},x), \qquad -2x+\alpha > 0, \quad \alpha > 0. \tag{4.4}$$

Using (4.4) in (4.3) we get [cf. (3.6)]

$$m_{\mathrm{r}}(k,x) = 1 + \sum_{j=1}^{n_{\mathrm{r}}} \frac{\rho_{\mathrm{r}j}\,e^{-2ik_{\mathrm{r}j}x}}{k+k_{\mathrm{r}j}}\,m_{\mathrm{r}}(k_{\mathrm{r}j},x), \qquad x \le 0. \tag{4.5}$$

Proceeding as in Section 3 leading to (3.10), we get for $x \le 0$

$$m_{\mathrm{r}}(k,x) = 1 + \begin{bmatrix} \dfrac{\rho_{\mathrm{r}1}\,e^{-2ik_{\mathrm{r}1}x}}{k+k_{\mathrm{r}1}} & \dots & \dfrac{\rho_{\mathrm{r}n_{\mathrm{r}}}\,e^{-2ik_{\mathrm{r}n_{\mathrm{r}}}x}}{k+k_{\mathrm{r}n_{\mathrm{r}}}} \end{bmatrix} \mathbf{M}_{\mathrm{r}}(x)^{-1} \begin{bmatrix} 1 \\ \vdots \\ 1 \end{bmatrix},$$

or equivalently

$$m_{\mathrm{r}}(k,x) = 1 - \frac{1}{\det \mathbf{M}_{\mathrm{r}}(x)} \begin{vmatrix} 0 & \dfrac{\rho_{\mathrm{r}1}\,e^{-2ik_{\mathrm{r}1}x}}{k+k_{\mathrm{r}1}} & \dots & \dfrac{\rho_{\mathrm{r}n_{\mathrm{r}}}\,e^{-2ik_{\mathrm{r}n_{\mathrm{r}}}x}}{k+k_{\mathrm{r}n_{\mathrm{r}}}} \\ 1 & & & \\ \vdots & & \mathbf{M}_{\mathrm{r}}(x) & \\ 1 & & & \end{vmatrix}, \tag{4.6}$$

where $\mathbf{M}_{\mathrm{r}}(x)$ is the $n_{\mathrm{r}} \times n_{\mathrm{r}}$ matrix-valued function whose (p,q)-entry is given by

$$[\mathbf{M}_{\mathrm{r}}(x)]_{pq} := \delta_{pq} - \frac{\rho_{\mathrm{r}q}\, e^{-2ik_{\mathrm{r}q}x}}{k_{\mathrm{r}p} + k_{\mathrm{r}q}}. \tag{4.7}$$

Let us remark that the invertibility of $\mathbf{M}_{\mathrm{r}}(x)$ follows from Corollary 4.2 of [6]. Then [cf. (3.12)]

$$B_{\mathrm{r}}(x,\alpha) = \frac{\det \Gamma_{\mathrm{r}}(x,\alpha)}{\det \mathbf{M}_{\mathrm{r}}(x)}, \qquad x \le 0, \quad \alpha > 0, \tag{4.8}$$

where $\Gamma_{\mathrm{r}}(x,\alpha)$ is the $(n_{\mathrm{r}}+1) \times (n_{\mathrm{r}}+1)$ matrix defined as

$$\Gamma_{\mathrm{r}}(x,\alpha) := \begin{bmatrix} 0 & i\rho_{\mathrm{r}1}\, e^{-2ik_{\mathrm{r}1}x + ik_{\mathrm{r}1}\alpha} & \dots & i\rho_{\mathrm{r}n_{\mathrm{r}}}\, e^{-2ik_{\mathrm{r}n_{\mathrm{r}}}x + ik_{\mathrm{r}n_{\mathrm{r}}}\alpha} \\ 1 & & & \\ \vdots & & \mathbf{M}_{\mathrm{r}}(x) & \\ 1 & & & \end{bmatrix}. \tag{4.9}$$

Similarly as in the proof of (3.14), it can be shown that

$$\det \Gamma_{\mathrm{r}}(x,0^+) = -\frac{d}{dx} \det \mathbf{M}_{\mathrm{r}}(x),$$

and hence (4.8) implies

$$B_{\mathrm{r}}(x,0^+) = \frac{-\frac{d}{dx} \det \mathbf{M}_{\mathrm{r}}(x)}{\det \mathbf{M}_{\mathrm{r}}(x)}, \qquad x \le 0.$$

It is known [1–5] that

$$V(x) = 2\frac{d}{dx} B_{\mathrm{r}}(x,0^+), \qquad x \in \mathbf{R},$$

and hence we obtain

$$V(x) = -2\frac{d}{dx}\left[\frac{\frac{d}{dx}\det \mathbf{M}_{\mathrm{r}}(x)}{\det \mathbf{M}_{\mathrm{r}}(x)}\right], \qquad x < 0. \tag{4.10}$$

Alternatively, having constructed $m_{\mathrm{r}}(k,x)$ as in (4.6) for $x \le 0$, we can evaluate $V(x)$ for $x < 0$ via [cf. (3.19) and (3.20)]

$$V(x) = \frac{m_{\mathrm{r}}''(k,x) - 2ik\, m_{\mathrm{r}}'(k,x)}{m_{\mathrm{r}}(k,x)}, \tag{4.11}$$

$$V(x) = \frac{m_{\mathrm{r}}''(0,x)}{m_{\mathrm{r}}(0,x)}. \tag{4.12}$$

Note that the right-hand side in (4.11) is independent of k as a whole.

If we start with R, we can construct T and L as in Section 2. Then, using the poles and residues of $L(k)$ in $\mathbf{C}^+$, we can construct $m_{\mathrm{r}}(k,x)$ for $x \le 0$ and $V(x)$ for $x < 0$. We then obtain $f_{\mathrm{r}}(k,x)$ with the help of (2.1) and (4.6). Having $f_{\mathrm{r}}(k,x)$ and $f_{\mathrm{r}}(-k,x)$ in hand, via (2.4) we construct $f_{\mathrm{l}}(k,x)$ for $x \le 0$ by using

$$f_{\mathrm{l}}(k,x) = \frac{f_{\mathrm{r}}(-k,x) + L(k)\, f_{\mathrm{r}}(k,x)}{T(k)}.$$

5. Wave propagation and Jost waves

We now wish to analyze certain solutions to the plasma-wave equation (1.2) when V belongs to the Faddeev class, there are no bound states, and the corresponding reflection coefficients are rational functions of k with simple poles in $\mathbf{C}^+$.

We define the Jost wave from the left as

$$J_l(x,t) := \frac{1}{2\pi}\int_{-\infty}^{\infty} dk\, f_l(k,x)\, e^{-ikt}. \tag{5.1}$$

Using (2.1) in (5.1), we get

$$J_l(x,t) = \frac{1}{2\pi}\int_{-\infty}^{\infty} dk\, e^{ik(x-t)} + \frac{1}{2\pi}\int_{-\infty}^{\infty} dk\, [m_l(k,x)-1]\, e^{ik(x-t)}. \tag{5.2}$$

From (3.1), (5.2), and the fact that $\int_{-\infty}^{\infty} dk\, e^{ik\alpha} = 2\pi\,\delta(\alpha)$, we obtain

$$J_l(x,t) = \delta(x-t) + B_l(x,t-x), \qquad x,t\in\mathbf{R}, \tag{5.3}$$

where $\delta(x)$ denotes the Dirac delta distribution. Note that $B_l(x,t-x)=0$ if $t-x<0$ because $B_l(x,\alpha)=0$ for $\alpha<0$, as we have seen in Section 3. Thus,

$$J_l(x,t) = 0, \qquad x-t>0. \tag{5.4}$$

Comparing (5.3) with (3.12), we see that when $x\geq 0$ and $t-x>0$, we can express $B_l(x,t-x)$ as the ratio of two determinants that are constructed explicitly with the help of (3.9) and (3.13). Hence, we have

$$J_l(x,t) = \frac{\det \Gamma_l(x,t-x)}{\det \mathbf{M}_l(x)}, \qquad x\geq 0, \quad t-x>0. \tag{5.5}$$

We also need $J_l(x,t)$ in the region with $x-t<0$ and $x+t<0$. Towards that goal, we can use (2.6) in (5.2) and get

$$J_l(x,t) = \frac{1}{2\pi}\int_{-\infty}^{\infty} dk\, e^{ik(x-t)} + \frac{1}{2\pi}\int_{-\infty}^{\infty} dk \left[\frac{m_r(-k,x)}{T(k)}-1\right] e^{ik(x-t)}$$
$$+ \frac{1}{2\pi}\int_{-\infty}^{\infty} dk\, \frac{L(k)}{T(k)}\, m_r(k,x)\, e^{-ik(x+t)},$$

or equivalently

$$\begin{aligned} J_l(x,t) = \delta(x-t) &+ \frac{1}{2\pi}\int_{-\infty}^{\infty} dk \left[\frac{m_r(k,x)}{T(-k)}-1\right] e^{-ik(x-t)} \\ &+ \frac{1}{2\pi}\int_{-\infty}^{\infty} dk\, \frac{L(k)}{T(k)}\, m_r(k,x)\, e^{-ik(x+t)}. \end{aligned} \tag{5.6}$$

This separation causes each of the two integrands in (5.6) to have a simple pole at $k=0$ in the generic case. The zeros of $T(-k)$ in $\mathbf{C}^+$ contribute to the first integral in (5.6). The poles of $L(k)$ in $\mathbf{C}^+$ contribute to the second integral in (5.6). We find that the contribution from $k=0$ to the two integrals on the right-hand side of (5.6) is given by

$$i\,\theta(-x)\,[\theta(x+t)-\theta(t-x)]\; m_{\rm r}(0,x)\,\text{Res}\,(1/T(k),0)\,, \tag{5.7}$$

where $\theta(x)$ is the Heaviside function defined as

$$\theta(x):=\begin{cases}1, & x>0,\\ 0, & x<0.\end{cases}$$

Note that (5.7) can be evaluated by using the fact that $L(0)=-1$ in the generic case and that $T(k)$ is analytic and nonzero in $\mathbf{C}^+$.

As in Section 4, let us use $k_{{\rm r}j}$ to denote the poles of $L(k)$ in $\mathbf{C}^+$ and $\rho_{{\rm r}j}$ the residues there. Similarly, as in (2.14) and (2.15) let us use z_j for the poles of $1/T(-k)$ in $\mathbf{C}^+$ and τ_j for the corresponding residues for $1\le j\le n_z$. The contributions to the right-hand side of (5.6) from the zeros of $T(-k)$ and the poles of $L(k)$ in $\mathbf{C}^+$ can be evaluated by using a contour integration along the infinite semicircle enclosing $\mathbf{C}^+$. Hence, in the region with $t-x>0$ and $x+t<0$, that contribution is given by

$$i\sum_{j=1}^{n_{\rm z}}\tau_j\, m_{\rm r}(z_j,x)\, e^{-iz_j(x-t)}+i\sum_{j=1}^{n_{\rm r}}\frac{\rho_{{\rm r}j}\, e^{-ik_{{\rm r}j}(x+t)}}{T(k_{{\rm r}j})}\, m_{\rm r}(k_{{\rm r}j},x). \tag{5.8}$$

From (5.7) and (5.8) we see that, in the region with $t-x>0$ and $x+t<0$, the Jost wave from the left is given by

$$\begin{aligned}J_{\rm l}(x,t)=-i\, m_{\rm r}(0,x)\,\text{Res}\,(1/T(k),0)&+i\sum_{j=1}^{n_{\rm z}}\tau_j\, m_{\rm r}(z_j,x)\, e^{-iz_j(x-t)}\\ &+i\sum_{j=1}^{n_{\rm r}}\frac{\rho_{{\rm r}j}\, e^{-ik_{{\rm r}j}(x+t)}}{T(k_{{\rm r}j})}\, m_{\rm r}(k_{{\rm r}j},x).\end{aligned} \tag{5.9}$$

Note that $m_{\rm r}(z_j,x)$ and $m_{\rm r}(k_{{\rm r}j},x)$ can be evaluated explicitly by using (4.6) and $T(k_{{\rm r}j})$ by using (2.14).

Finally, we will evaluate $J_{\rm l}(x,t)$ in the region with $x<0$ and $x+t>0$. In that region, the contribution to the first integral in (5.6) from the zeros of $T(-k)$ in $\mathbf{C}^+$ is evaluated with the help of a contour integration along the boundary of $\mathbf{C}^+$ and we get the first summation term in (5.8). However, the second integral in (5.6) needs to be evaluated as a contour integral along the boundary of $\mathbf{C}^-$ due to the presence of the exponential term $e^{-ik(x+t)}$ in the integrand. With the help of (2.9) we write that integral as

$$\frac{1}{2\pi}\int_{-\infty}^{\infty}dk\,\frac{L(k)}{T(k)}\, m_{\rm r}(k,x)\, e^{-ik(x+t)}=-\frac{1}{2\pi}\int_{-\infty}^{\infty}dk\,\frac{R(k)}{T(k)}\, m_{\rm r}(-k,x)\, e^{ik(x+t)}, \tag{5.10}$$

where the right-hand side can now be evaluated as a contour integral along the boundary of $\mathbf{C}^+$. Let us now evaluate the contribution to that integral coming from the poles of $R(k)$ in $\mathbf{C}^+$ and also the poles of $m_{\rm r}(-k,x)$ in $\mathbf{C}^+$. From Section 3 and (4.5) we see that the former poles occur at $k = k_{lj}$ and the latter occur at $k = k_{rj}$. As a result, such contributions to the right-hand side of (5.10) can be explicitly evaluated. For example, when the sets $\{k_{lj}\}_{j=1}^{n_l}$ and $\{k_{rj}\}_{j=1}^{n_r}$ do not intersect, we get

$$-i\sum_{j=1}^{n_l} \frac{\rho_{lj}\, e^{ik_{lj}(x+t)}}{T(k_{lj})}\, m_r(-k_{lj},x) - i\sum_{j=1}^{n_r} \frac{R(k_{rj})\, e^{ik_{rj}(x+t)}}{T(k_{rj})} \mathrm{Res}\,(m_r(-k,x),k_{rj})\,. \tag{5.11}$$

From (4.5) we see that

$$\mathrm{Res}\,(m_r(-k,x),k_{rj}) = -\rho_{rj}\, e^{-2ik_{rj}x}\, m_r(k_{rj},x),$$

and thus, in the region with $x < 0$ and $x + t > 0$, with the help of (5.6)–(5.11) we obtain

$$\begin{aligned} J_l(x,t) = i\sum_{j=1}^{n_z} \tau_j\, m_r(z_j,x)\, e^{-iz_j(x-t)} &- i\sum_{j=1}^{n_l} \frac{\rho_{lj}\, e^{ik_{lj}(x+t)}}{T(k_{lj})}\, m_r(-k_{lj},x) \\ &+ i\sum_{j=1}^{n_r} \frac{\rho_{rj}\, R(k_{rj})\, e^{-ik_{rj}(x-t)}}{T(k_{rj})}\, m_r(k_{rj},x), \end{aligned} \tag{5.12}$$

where we note that there is no contribution to $J_l(x,t)$ from the poles at $k = 0$ [cf. (5.7)] in the region with $x < 0$ and $x+t > 0$. In case the sets $\{k_{lj}\}_{j=1}^{n_l}$ and $\{k_{rj}\}_{j=1}^{n_r}$ partially or wholly overlap, the integral on the right-hand side of (5.10) can be evaluated explicitly in a similar way as a contour integral along the boundary of $\mathbf{C}^+$ and the result in (5.12) can be modified appropriately.

We can write the Jost wave $J_l(x,t)$ by combining (5.3)–(5.5), (5.9), and (5.12) as

$$\begin{aligned} J_l(x,t) =& \delta(x-t) + \theta(x)\,\theta(t-x)\frac{\det \Gamma_l(x,t-x)}{\det \mathbf{M}_l(x)} \\ &+ i\,\theta(-x)\,\theta(t-x)\sum_{j=1}^{n_z} \tau_j\, m_r(z_j,x)\, e^{-iz_j(x-t)} \\ &- i\,\theta(t-x)\,\theta(-x-t)\, m_r(0,x)\,\mathrm{Res}\,(1/T(k),0) \\ &+ i\,\theta(t-x)\,\theta(-x-t)\sum_{j=1}^{n_r} \frac{\rho_{rj}\, e^{-ik_{rj}(x+t)}}{T(k_{rj})}\, m_r(k_{rj},x) \\ &- i\,\theta(-x)\,\theta(x+t)\sum_{j=1}^{n_l} \frac{\rho_{lj}\, e^{ik_{lj}(x+t)}}{T(k_{lj})}\, m_r(-k_{lj},x) \\ &+ i\,\theta(-x)\,\theta(x+t)\sum_{j=1}^{n_r} \frac{\rho_{rj}\, R(k_{rj})\, e^{-ik_{rj}(x-t)}}{T(k_{rj})}\, m_r(k_{rj},x), \end{aligned}$$

and hence, in terms of the quantities that have been constructed explicitly once the rational reflection coefficient R is known, we have constructed $J_{\rm l}(x,t)$ for all $x,t\in\mathbf{R}$ except when $x=t$ and also when $x=-t$ with $x<0$. In Section 6, we will see that $J_{\rm l}(x,t)$ may have jump discontinuities when $x=t$ and also when $x=-t$ with $x<0$, and we will evaluate those discontinuities.

We define the Jost wave from the right in a similar manner, by letting

$$J_{\rm r}(x,t):=\frac{1}{2\pi}\int_{-\infty}^{\infty}dk\,f_{\rm r}(k,x)\,e^{-ikt}. \tag{5.13}$$

Using (2.1) in (5.13), we get [cf. (5.2)]

$$J_{\rm r}(x,t)=\frac{1}{2\pi}\int_{-\infty}^{\infty}dk\,e^{-ik(x+t)}+\frac{1}{2\pi}\int_{-\infty}^{\infty}dk\,[m_{\rm r}(k,x)-1]\,e^{-ik(x+t)}, \tag{5.14}$$

which can be written as [cf. (4.1)]

$$J_{\rm r}(x,t)=\delta(x+t)+B_{\rm r}(x,x+t). \tag{5.15}$$

Note that $B_{\rm r}(x,x+t)=0$ if $x+t<0$ due to, for each fixed x, the analyticity in $\mathbf{C}^+$, the continuity in $\overline{\mathbf{C}^+}$, and the $O(1/k)$-behavior as $k\to\infty$ in $\overline{\mathbf{C}^+}$ of $m_{\rm r}(k,x)-1$. Thus,

$$J_{\rm r}(x,t)=0,\qquad x+t<0. \tag{5.16}$$

Comparing (4.8) and (5.15), when $x<0$ and $x+t>0$, we see that we can write $B_{\rm r}(x,x+t)$ as the ratio of two determinants and obtain [cf. (5.5)]

$$J_{\rm r}(x,t)=\frac{\det\Gamma_{\rm r}(x,x+t)}{\det\mathbf{M}_{\rm r}(x)},\qquad x\le 0,\quad x+t>0, \tag{5.17}$$

where $\mathbf{M}_{\rm r}(x)$ and $\Gamma_{\rm r}(x,\alpha)$ are as in (4.7) and (4.9), respectively.

Next, we will obtain $J_{\rm r}(x,t)$ in the region with $x+t>0$ and $x-t>0$. Using (2.5) in (5.14), we get [cf. (5.6)]

$$\begin{aligned}J_{\rm r}(x,t)=\delta(x+t)&+\frac{1}{2\pi}\int_{-\infty}^{\infty}dk\left[\frac{m_{\rm l}(k,x)}{T(-k)}-1\right]e^{ik(x+t)}\\&+\frac{1}{2\pi}\int_{-\infty}^{\infty}dk\,\frac{R(k)}{T(k)}\,m_{\rm l}(k,x)\,e^{ik(x-t)}.\end{aligned} \tag{5.18}$$

The zeros of $T(-k)$ in $\mathbf{C}^+$ contribute to the first integral on the right-hand side of (5.18). As in (5.6), we evaluate that integral as a contour integral along the boundary of $\mathbf{C}^+$. The poles of $R(k)$ in $\mathbf{C}^+$ contribute to the second integral in (5.18). In the generic case, each of the two integrands in (5.18) has a simple pole at $k=0$ because of the simple zero of $T(k)$ there; the contribution from $k=0$ in the two integrals in (5.18) can be evaluated as in (5.7) and we get

$$i\,\theta(x)\,[\theta(t-x)-\theta(x+t)]\,m_{\rm l}(0,x)\,\mathrm{Res}\,(1/T(k),0)\,. \tag{5.19}$$

Recalling that $\{k_{\mathrm{l}j}\}$ is the set poles of $R(k)$ in $\mathbf{C}^+$ and $\{z_j\}$ is the set of zeros of $T(-k)$ in $\mathbf{C}^+$, in the region with $x+t>0$ and $x-t>0$ we get [cf. (5.9)]

$$\begin{aligned} J_{\mathrm{r}}(x,t) = & -i\, m_{\mathrm{l}}(0,x)\,\mathrm{Res}\,(1/T(k),0) \\ & + i\sum_{j=1}^{n_{\mathrm{z}}} \tau_j\, m_{\mathrm{l}}(z_j,x)\, e^{iz_j(x+t)} \\ & + i\sum_{j=1}^{n_{\mathrm{l}}} \frac{\rho_{\mathrm{l}j}\, e^{ik_{\mathrm{l}j}(x-t)}}{T(k_{\mathrm{l}j})}\, m_{\mathrm{l}}(k_{\mathrm{l}j},x), \end{aligned} \tag{5.20}$$

where the first term on the right-hand side is the contribution from (5.19). Note that $m_{\mathrm{l}}(z_j,x)$ and $m_{\mathrm{l}}(k_{\mathrm{l}j},x)$ can be evaluated explicitly from (3.11) and $T(k_{\mathrm{l}j})$ from (2.14).

Finally, let us evaluate $J_{\mathrm{r}}(x,t)$ in the region with $x>0$ and $t-x>0$. From (5.19) we see that the contribution from the pole at $k=0$ to $J_{\mathrm{r}}(x,t)$ is nil when $x>0$ and $t-x>0$. To obtain $J_{\mathrm{r}}(x,t)$ when $x>0$ and $t-x>0$, we can use (5.18) and evaluate the first integral there via a contour integration along the boundary of $\mathbf{C}^+$ to get the first summation term in (5.20). To evaluate the second integral in (5.18), since $t-x>0$, with the help of (2.9) we get

$$\frac{1}{2\pi}\int_{-\infty}^{\infty} dk\, \frac{R(k)}{T(k)} m_{\mathrm{l}}(k,x)\, e^{ik(x-t)} = -\frac{1}{2\pi}\int_{-\infty}^{\infty} dk\, \frac{L(k)}{T(k)} m_{\mathrm{l}}(-k,x)\, e^{-ik(x-t)},$$

where the right-hand side is to be evaluated as a contour integral along the boundary of $\mathbf{C}^+$ with the contributions coming from the poles $k_{\mathrm{r}j}$ of $L(k)$ in $\mathbf{C}^+$ and the poles of $m_{\mathrm{l}}(-k,x)$ in $\mathbf{C}^+$. From (3.6) or (3.10) we see that the latter poles occur at exactly $k=k_{\mathrm{l}j}$, which are the poles of $R(k)$ in $\mathbf{C}^+$. If the sets $\{k_{\mathrm{l}j}\}_{j=1}^{n_{\mathrm{l}}}$ and $\{k_{\mathrm{r}j}\}_{j=1}^{n_{\mathrm{r}}}$ do not intersect, we get [cf. (5.11)]

$$\begin{aligned} & -\frac{1}{2\pi}\int_{-\infty}^{\infty} dk\, \frac{L(k)}{T(k)}\, m_{\mathrm{l}}(-k,x)\, e^{-ik(x-t)} \\ & = -i\sum_{j=1}^{n_{\mathrm{r}}} \frac{\rho_{\mathrm{r}j}\, e^{-ik_{\mathrm{r}j}(x-t)}}{T(k_{\mathrm{r}j})}\, m_{\mathrm{l}}(-k_{\mathrm{r}j},x) \\ & \quad - i\sum_{j=1}^{n_{\mathrm{l}}} \frac{L(k_{\mathrm{l}j})\, e^{-ik_{\mathrm{l}j}(x-t)}}{T(k_{\mathrm{l}j})} \mathrm{Res}\,(m_{\mathrm{l}}(-k,x),k_{\mathrm{l}j}). \end{aligned} \tag{5.21}$$

From (3.6) we see that

$$\mathrm{Res}\,(m_{\mathrm{l}}(-k,x),k_{\mathrm{l}j}) = -\rho_{\mathrm{l}j}\, e^{2ik_{\mathrm{l}j}x}\, m_{\mathrm{l}}(k_{\mathrm{l}j},x). \tag{5.22}$$

In case the sets $\{k_{\mathrm{l}j}\}_{j=1}^{n_{\mathrm{l}}}$ and $\{k_{\mathrm{r}j}\}_{j=1}^{n_{\mathrm{r}}}$ overlap, the result in (5.21) can appropriately be modified.

By combining (5.15)–(5.22), we can write the Jost wave $J_r(x,t)$ for any $x,t \in \mathbf{R}$ as

$$\begin{aligned} J_r(x,t) =& \delta(x+t) + \theta(-x)\,\theta(x+t)\,\frac{\det \Gamma_r(x,x+t)}{\det \mathbf{M}_r(x)} \\ &+ i\,\theta(x)\,\theta(x+t) \sum_{j=1}^{n_z} \tau_j\, m_l(z_j,x)\, e^{iz_j(x+t)} \\ &- i\,\theta(x+t)\,\theta(x-t)\, m_l(0,x)\, \mathrm{Res}\,(1/T(k),0) \\ &+ i\,\theta(x+t)\,\theta(x-t) \sum_{j=1}^{n_l} \frac{\rho_{lj}\, e^{ik_{lj}(x-t)}}{T(k_{lj})}\, m_l(k_{lj},x) \\ &- i\,\theta(x)\,\theta(t-x) \sum_{j=1}^{n_r} \frac{\rho_{rj}\, e^{-ik_{rj}(x-t)}}{T(k_{rj})}\, m_l(-k_{rj},x) \\ &+ i\,\theta(x)\,\theta(t-x) \sum_{j=1}^{n_l} \frac{\rho_{lj}\, L(k_{lj})\, e^{ik_{lj}(x+t)}}{T(k_{lj})}\, m_l(k_{lj},x). \end{aligned}$$

Thus, in terms of the quantities that have been constructed explicitly once the rational reflection coefficient $R(k)$ is known, we have obtained $J_r(x,t)$ for all $x,t \in \mathbf{R}$ except when $x=-t$ and also when $x=t$ with $x>0$. In the next section, we will see that $J_r(x,t)$ may have jump discontinuities when $x=-t$ and also when $x=t$ with $x>0$, and we will evaluate those discontinuities.

6. Discontinuities in the Jost waves

For each fixed $t \in \mathbf{R}$, let us now analyze the discontinuities in the Jost waves $J_l(x,t)$ and $J_r(x,t)$ and in their x-derivatives. From (3.9) and (3.18), we see that for $x>0$ the potential V is the ratio of linear combinations of various exponential functions of x and that ratio decays exponentially as $x \to +\infty$. Similarly, for $x<0$, from (4.7) and (4.10) we see that V is the ratio of linear combinations of various exponential functions and that ratio decays exponentially as $x \to -\infty$. In the absence of bound states, it is known [3] that $m_l(0,x)>0$ and $m_r(0,x)>0$ for $x \in \mathbf{R}$. Hence, from (3.20) and (4.12) we see that the only discontinuities in V and its derivatives can occur at $x=0$.

From (7.5)–(7.7) of [18], as $k \to \infty$ in $\overline{\mathbf{C}^+}$ we have

$$m_l(k,x) = 1 - \frac{\gamma_l(x)}{2ik} - \frac{1}{8k^2}\left[\gamma_l(x)^2 - 2\,q_l(k,x)\right] + O(1/k^3), \tag{6.1}$$

$$m_r(k,x) = 1 - \frac{\gamma_r(x)}{2ik} - \frac{1}{8k^2}\left[\gamma_r(x)^2 + 2\,q_r(k,x)\right] + O(1/k^3), \tag{6.2}$$

$$m_l'(k,x) = \frac{\gamma_l(x)}{2ik} + O(1/k^2), \quad m_r'(k,x) = \frac{\gamma_r(x)}{2ik} + O(1/k^2), \tag{6.3}$$

with

$$\gamma_\mathrm{l}(x) := \int_x^\infty dy\, V(y), \quad \gamma_\mathrm{r}(x) := \int_{-\infty}^x dy\, V(y),$$

$$q_\mathrm{l}(k,x) := V(x) + \theta(-x)\left[V(0^+) - V(0^-)\right] e^{-2ikx},$$

$$q_\mathrm{r}(k,x) := -V(x) + \theta(x)\left[V(0^+) - V(0^-)\right] e^{2ikx}.$$

Note that $q_\mathrm{l}(k,x)$ and $q_\mathrm{r}(k,x)$ are continuous at $x = 0$, and we have

$$q_\mathrm{l}(k,0) = V(0^+), \quad q_\mathrm{r}(k,0) = -V(0^-).$$

With the help of (5.2) and (5.3), let us define

$$U_\mathrm{l}(x,t) := J_\mathrm{l}(x,t) - \delta(x-t). \tag{6.4}$$

We will refer to $U_\mathrm{l}(x,t)$ as the tail of the Jost wave $J_\mathrm{l}(x,t)$. From (5.2) we see that the discontinuity in the tail $U_\mathrm{l}(x,t)$ is caused by the $(1/k)$-term in the expansion of $m_\mathrm{l}(k,x) - 1$ as $k \to \infty$ in $\overline{\mathbf{C}^+}$. By using

$$\frac{1}{2\pi i}\int_{-\infty}^\infty dk\, \frac{e^{ik\xi}}{k + i0^+} = -\theta(-\xi), \tag{6.5}$$

we evaluate that contribution as $(1/2)\,\gamma_\mathrm{l}(x)\,\theta(t-x)$, and hence the only discontinuity in the tail $U_\mathrm{l}(x,t)$ occurs at the wavefront $x = t$ and is given by

$$U_\mathrm{l}(t+0^+,t) - U_\mathrm{l}(t-0^+,t) = -\frac{1}{2}\int_t^\infty dy\, V(y).$$

In other words,

$$U_\mathrm{l}(x,t) = \begin{cases} 0, & x > t, \\ \dfrac{1}{2}\displaystyle\int_t^\infty dy\, V(y), & x = t - 0^+. \end{cases}$$

Next, let us analyze the discontinuities in $\partial U_\mathrm{l}(x,t)/\partial x$. From (5.2) and (6.4), we see that

$$\frac{\partial U_\mathrm{l}(x,t)}{\partial x} = \frac{1}{2\pi}\int_{-\infty}^\infty dk\, \{m_\mathrm{l}'(k,x) + ik\,[m_\mathrm{l}(k,x) - 1]\}\, e^{ik(x-t)}, \tag{6.6}$$

and for each fixed $t \in \mathbf{R}$, the discontinuity in $\partial U_\mathrm{l}(x,t)/\partial x$ is caused by the $(1/k)$-term in the expansion of the integrand of (6.6) as $k \to \infty$ in $\overline{\mathbf{C}^+}$. Using (6.1), (6.3), and (6.5) in (6.6) we see that there are exactly two such discontinuities. The first discontinuity occurs at the wavefront $x = t$, and the second occurs at $x = -t$. At the wavefront, we get

$$\frac{\partial U_\mathrm{l}(x,t)}{\partial x} = \begin{cases} 0, & x > t, \\ \dfrac{V(x)}{4} - \dfrac{1}{2}\displaystyle\int_t^\infty dy\, V(y) - \dfrac{1}{8}\left[\int_t^\infty dy\, V(y)\right]^2, & x = t - 0^+. \end{cases}$$

The contribution to the discontinuity at $x = -t$ is obtained as

$$\frac{\partial U_\mathrm{l}(t+0^+,t)}{\partial x} - \frac{\partial U_\mathrm{l}(t-0^+,t)}{\partial x} = -\frac{1}{4}\,\theta(-x)\left[V(0^+) - V(0^-)\right].$$

In analogy to (6.4), with the help of (5.14) let us define

$$U_{\mathrm{r}}(x,t) := J_{\mathrm{r}}(x,t) - \delta(x+t). \tag{6.7}$$

We will refer to $U_{\mathrm{r}}(x,t)$ as the tail of the Jost wave $J_{\mathrm{r}}(x,t)$. Using (6.2) and (6.5) in (6.7), we see that for each fixed $t \in \mathbf{R}$, the only discontinuity in $U_{\mathrm{r}}(x,t)$ occurs at the wavefront $x = -t$ and is described by

$$U_{\mathrm{r}}(x,t) = \begin{cases} 0, & x < -t, \\ \dfrac{1}{2}\displaystyle\int_{-\infty}^{t} dy\, V(y), & x = -t + 0^+. \end{cases}$$

To determine the discontinuities in $\partial U_{\mathrm{r}}(x,t)/\partial x$, first from (5.14) we obtain

$$\frac{\partial U_{\mathrm{r}}(x,t)}{\partial x} = \frac{1}{2\pi}\int_{-\infty}^{\infty} dk\, \{m_{\mathrm{r}}'(k,x) - ik\,[m_{\mathrm{r}}(k,x) - 1]\}\, e^{-ik(x+t)}. \tag{6.8}$$

Using (6.2), (6.3), and (6.5) in (6.8), for each fixed $t \in \mathbf{R}$, we see that the discontinuities in $\partial U_{\mathrm{r}}(x,t)/\partial x$ may occur only at $x = -t$ and at $x = t$. The former occurs at the wavefront and is described by

$$\frac{\partial U_{\mathrm{r}}(x,t)}{\partial x} = \begin{cases} 0, & x < -t, \\ -\dfrac{V(x)}{4} - \dfrac{1}{2}\displaystyle\int_{-\infty}^{t} dy\, V(y) + \dfrac{1}{8}\left[\int_{-\infty}^{t} dy\, V(y)\right]^2, & x = -t + 0^+. \end{cases}$$

Finally, the discontinuity at $x = t$ is given by

$$\frac{\partial U_{\mathrm{r}}(t+0^+,t)}{\partial x} - \frac{\partial U_{\mathrm{r}}(t-0^+,t)}{\partial x} = \frac{1}{4}\,\theta(x)\left[V(0^+) - V(0^-)\right].$$

7. Numerical implementation

One of the authors (Borkowski) has implemented the theoretical method described in this paper as a *Mathematica 4.2* notebook. The user inputs a rational function for $R(k)$ and instructs *Mathematica* to evaluate the notebook. *Mathematica* then calculates all of the quantities relevant to (1.1); namely, the Faddeev functions $m_{\mathrm{l}}(k,x)$ and $m_{\mathrm{r}}(k,x)$, Jost solutions $f_{\mathrm{l}}(k,x)$ and $f_{\mathrm{r}}(k,x)$, the potential $V(x)$, the scattering coefficients $T(k)$ and $L(k)$, and the quantities $B_{\mathrm{l}}(x,\alpha)$ and $B_{\mathrm{r}}(x,\alpha)$ given in (3.1) and (4.1), respectively.

The implemented program first reduces $R(k)$, then calculates $T(k)$ and $L(k)$ as described in Section 2, and then reduces those to cancel common factors appearing both in the numerator and in the denominator. The reduction in each scattering coefficient is achieved by computing all the zeros and poles in $\mathbf{C}^+$, comparing them within a chosen numerical precision, and cancelling the common factors appearing in the numerator and in the denominator. This reduction is necessary because *Mathematica* cannot usually cancel the terms by itself or cannot simplify enough in certain circumstances. Finally, the Faddeev functions and the

Jost solutions, along with the potential are determined for $x > 0$, and then for $x < 0$.

Our program has been able to duplicate the numerical results in [9]. However, we have not been able to duplicate the two numerical examples in [7] given by Dolveck-Guilpart. We have also verified that the result of our program agrees with the analytical example of Sabatier [14]. Prof. Paul Sacks of Iowa State University has used his *Matlab* program for the solution of the inverse scattering problem based on transforming the relevant inverse problem into an equivalent time-domain problem and solving it by a time-domain method [19]; he also was able to duplicate the results in [9], but not in [7], and he has confirmed to us that our results are in complete agreement with his as far as the two examples in [7] are concerned. Prof. Sabatier later has informed us that the two numerical examples given in [7] by Dolveck-Guilpart were indeed incorrect, and the rational reflection coefficients used in those two examples were outside the domain of the applicability of the method of [14].

References

[1] T. Aktosun and M. Klaus, *Chapter 2.2.4, Inverse theory: problem on the line,* in: E.R. Pike and P.C. Sabatier (eds.), *Scattering,* Academic Press, London, 2001, pp. 770–785.

[2] K. Chadan and P.C. Sabatier, *Inverse problems in quantum scattering theory,* 2nd ed., Springer, New York, 1989.

[3] P. Deift and E. Trubowitz, *Inverse scattering on the line,* Comm. Pure Appl. Math. **32** (1979), 121–251.

[4] L.D. Faddeev, *Properties of the S-matrix of the one-dimensional Schrödinger equation,* Am. Math. Soc. Transl. (ser. 2) **65** (1967), 139–166.

[5] V.A. Marchenko, *Sturm-Liouville operators and applications,* Birkhäuser, Basel, 1986.

[6] T. Aktosun, M. Klaus, and C. van der Mee, *Explicit Wiener-Hopf factorization for certain nonrational matrix functions,* Integral Equations Operator Theory **15** (1992), 879–900.

[7] B. Dolveck-Guilpart, *Practical construction of potentials corresponding to exact rational reflection coefficients,* in: P.C. Sabatier (ed.), *Some topics on inverse problems,* World Sci. Publ., Singapore, 1988, pp. 341–368.

[8] I. Kay, *The inverse scattering problem when the reflection coefficient is a rational function,* Comm. Pure Appl. Math. **13** (1960), 371–393.

[9] K.R. Pechenick and J.M. Cohen, *Inverse scattering – exact solution of the Gel'fand-Levitan equation,* J. Math. Phys. **22** (1981), 1513–1516.

[10] K.R. Pechenick and J.M. Cohen, *Exact solutions to the valley problem in inverse scattering,* J. Math. Phys. **24** (1983), 406–409.

[11] R.T. Prosser, *On the solutions of the Gel'fand-Levitan equation,* J. Math. Phys. **25** (1984), 1924–1929.

[12] P.C. Sabatier, *Rational reflection coefficients in one-dimensional inverse scattering and applications,* in: J.B. Bednar et al. (eds.), *Conference on inverse scattering: theory and application*, SIAM, Philadelphia, 1983, pp. 75–99.

[13] P.C. Sabatier, *Rational reflection coefficients and inverse scattering on the line,* Nuovo Cimento B **78** (1983), 235–248.

[14] P.C. Sabatier, *Critical analysis of the mathematical methods used in electromagnetic inverse theories: a quest for new routes in the space of parameters,* in: W. M. Boerner et al. (eds.), *Inverse methods in electromagnetic imaging*, Reidel Publ., Dordrecht, Netherlands, 1985, pp. 43–64.

[15] D. Alpay and I. Gohberg, *Inverse problem for Sturm-Liouville operators with rational reflection coefficient,* Integral Equations Operator Theory **30** (1998), 317–325.

[16] C. van der Mee, *Exact solution of the Marchenko equation relevant to inverse scattering on the line,* in: V.M. Adamyan et al. (eds.), *Differential operators and related topics,* Vol. I, Birkhäuser, Basel, 2000, pp. 239–259.

[17] R. Courant and D. Hilbert, *Methods of mathematical physics,* Vol. I, Interscience Publ., New York, 1953.

[18] T. Aktosun and J.H. Rose, *Wave focusing on the line,* J. Math. Phys. **43** (2002), 3717–3745.

[19] P.E. Sacks, *Reconstruction of steplike potentials,* Wave Motion **18** (1993), 21–30.

Acknowledgment

We have benefited from discussions with Profs. Pierre C. Sabatier, Paul Sacks, and Robert C. Smith.

Tuncay Aktosun, Michael H. Borkowski, Alyssa J. Cramer and Lance C. Pittman
Department of Mathematics and Statistics
Mississippi State University
Mississippi State, MS 39762, USA
e-mail: aktosun@math.msstate.edu

Operator Theory:
Advances and Applications, Vol. 160, 21–39

Aluthge Transforms and the Convex Hull of the Spectrum of a Hilbert Space Operator

Tsuyoshi Ando

Dedicated to Professor Israel Gohberg on the occasion of his 75th birthday

Abstract. For a bounded linear operator T on a Hilbert space its Aluthge transform $\Delta(T)$ is defined as $\Delta(T) = |T|^{\frac{1}{2}}U|T|^{\frac{1}{2}}$ with the help of a polar representation $T = U|T|$. In recent years usefulness of the Aluthge transform has been shown in several directions. In this paper we will use the Aluthge transform to study when the closure of the numerical range $W(T)$ of T coincides with the convex hull of its spectrum. In fact, we will prove that it is the case if and only if the closure of $W(T)$ coincides with that of $W\Big(\Delta(T)\Big)$. As a consequence we will show also that for any operator T the convex hull of its spectrum is written as the intersection of the closures of the numerical ranges of all iterated Aluthge transforms $\Delta^n(T)$.

Mathematics Subject Classification (2000). Primary 47A12; Secondary 47A10.

Keywords. Aluthge transform; Numerical range; Convex hull of spectrum; Convexoid operator; Norm inequalities.

1. Introduction

Among familiar quantities related to a (bounded linear) operator T on a Hilbert space are the (operator) *norm* $||T||$ and the *spectral radius* $r(T)$. Also among familiar sets related to T are the *spectrum* $\sigma(T)$ and the *numerical range* $W(T)$, defined as

$$W(T) \stackrel{\text{def}}{=} \{\langle Tx, x\rangle : ||x|| = 1\}.$$

The quantity

$$w(T) \stackrel{\text{def}}{=} \sup\{|\langle Tx, x\rangle| : ||x|| = 1\}$$

is called the *numerical radius*. Obviously

$$||T|| \geq w(T) \geq r(T).$$

Consider a *polar representation* $T = U|T|$ where $|T|$ is the positive semi-definite square root of T^*T and U is a *partial isometry* with $U^*U|T| = |T|$. Aluthge [2] assigned to T an operator $\tilde{T}$ defined by

$$\tilde{T} \equiv |T|^{\frac{1}{2}}U|T|^{\frac{1}{2}}.$$

It is easy to see that $\tilde{T}$ does not depend on a choice of a partial isometry U with $U^*U|T| = |T|$. We will call the correspondence $T \longmapsto \tilde{T}$ the *Aluthge transform* and use the notation

$$T \longmapsto \Delta(T) \equiv |T|^{\frac{1}{2}}U|T|^{\frac{1}{2}}. \tag{1.1}$$

With $\Delta^0(T) \equiv T$, the nth iterate of the Aluthge transform will be denoted by $\Delta^n(T)$

$$\Delta^n(T) \stackrel{\text{def}}{=} \Delta^{n-1}\Big(\Delta(T)\Big) \qquad (n = 1, 2, \ldots).$$

If $T = U|T|$ is normal, U can be chosen to commute with $|T|$ (hence $|T|^{\frac{1}{2}}$), so that $T = \Delta(T)$. But this relation does not characterize normality.

Since $||T|| = ||\ |T|\ || = ||\ |T|^{\frac{1}{2}}||^2$, the following inequality is obvious

$$||T|| \geq ||\Delta(T)||. \tag{1.2}$$

According to the Toeplitz-Hausdorff theorem (see [8] p. 113) $W(T)$ is a convex set of the complex plane. Using the representation theorem of a closed convex set (of the complex plane) as the intersection of all closed half-planes containing it, Hildebrandt [9] proved

$$\overline{W(T)} = \bigcap_{\zeta\in\mathbb{C}}\{\xi : |\xi - \zeta| \leq ||T - \zeta I||\}. \tag{1.3}$$

As a corollary we have

$$\overline{W(T)} = \bigcap_{\zeta\in\mathbb{C}}\{\xi : |\xi - \zeta| \leq w(T - \zeta I)\}. \tag{1.4}$$

In Section 2, generalizing (1.2) we will establish the inequality (Theorem 2)

$$||T - \zeta I|| \geq ||\Delta(T) - \zeta I|| \qquad (\zeta \in \mathbb{C}),$$

which proves via (1.3) a known result (Corollary 3)

$$\overline{W(T)} \supset \overline{W\Big(\Delta(T)\Big)}. \tag{1.5}$$

It is well known (see [8] p. 43) that for any pair of operators A, B

$$\sigma(AB) \cup \{0\} = \sigma(BA) \cup \{0\}.$$

Since $0 \in \sigma(T) \Longleftrightarrow 0 \in \sigma(|T|^{\frac{1}{2}}) \cup \sigma(U)$, we can see that $0 \in \sigma(T)$ is equivalent to $0 \in \sigma\Big(\Delta(T)\Big)$. Therefore by (1.1) we arrive at the following known relation

$$\sigma(T) = \sigma\Big(\Delta(T)\Big). \tag{1.6}$$

It is well known (see [8] p. 115) that $\sigma(T)$ is contained in the closure $\overline{W(T)}$, so that by convexity of $\overline{W(T)}$ we have

$$\overline{W(T)} \supset \operatorname{conv}\Big(\sigma(T)\Big),$$

where conv$(\cdot)$ denotes the convex hull. When combined with (1.5) and (1.6), this yields

$$\overline{W(T)} \supset \overline{W\Big(\Delta(T)\Big)} \supset \overline{W\Big(\Delta^2(T)\Big)} \supset \cdots \supset \operatorname{conv}\Big(\sigma(T)\Big). \tag{1.7}$$

As a consequence of the spectral representation, it is known (see [8] p. 116) that if T is *normal* the convex hull of $\sigma(T)$ coincides with the closure of $W(T)$. But the converse is not true in general.

Let us call an operator T *convexoid* if

$$\overline{W(T)} = \operatorname{conv}\Big(\sigma(T)\Big).$$

It follows from (1.7) that if T is convexoid then

$$(*) \qquad \overline{W(T)} = \overline{W\Big(\Delta(T)\Big)}.$$

In Section 3 we will prove (Theorem 6) that the condition $(*)$ completely characterizes convexoidity. Then by (1.4) we can show (Corollary 7) that T is convexoid if and only if

$$w\big(\Delta(T) - \zeta I\big) = w(T - \zeta I) \qquad (\zeta \in \mathbb{C}).$$

In Section 4 we will establish a representation result (Theorem 9) for the convex hull of $\sigma(T)$;

$$\operatorname{conv}\Big(\sigma(T)\Big) = \bigcap_{n=1}^{\infty} \overline{W\Big(\Delta^n(T)\Big)},$$

and derive, as a related result, a known result (Theorem 10)

$$\lim_{n\to\infty} ||\Delta^n(T)|| = r(T).$$

Theorem 6 and Theorem 9 for a (finite-dimensional) matrix were established in an earlier paper [3].

2. Norm inequalities

Let T be an operator on a Hilbert space $\mathcal{H}$ with polar representation $T = U|T|$. For notational simplicity, write

$$P \equiv |T|$$

so that

$$T = UP \quad \text{and} \quad \Delta(T) = P^{\frac{1}{2}}UP^{\frac{1}{2}}.$$

When T is not invertible, the polar part U may not be unitary. But we can reduce many problems to the case with unitary U by the following lemma.

Lemma 1. *Suppose that T is a bounded linear operator on a Hilbert space $\mathcal{H}$ with polar representation $T = UP$. When T is not invertible, define an operator $\hat{T}$ on the direct sum $\mathcal{H} \oplus \mathcal{H}$ as $\hat{T} = \hat{U}\hat{P}$, where*

$$\hat{U} \stackrel{\text{def}}{=} \begin{bmatrix} U & (I - UU^*)^{\frac{1}{2}} \\ (I - U^*U)^{\frac{1}{2}} & -U^* \end{bmatrix} \quad \text{and} \quad \hat{P} \stackrel{\text{def}}{=} \begin{bmatrix} P & 0 \\ 0 & 0 \end{bmatrix}.$$

Then $\hat{U}$ is unitary and $\hat{P} \geq 0$, and $\hat{T}$ satisfies the following properties:

(a) $\sigma(T) = \sigma(\hat{T})$ *and* $\sigma\Big(\Delta(T)\Big) = \sigma\Big(\Delta(\hat{T})\Big)$,

(b) $W(T) \subset W(\hat{T}) \subset \overline{W(T)}$ *and* $W\Big(\Delta(T)\Big) \subset W\Big(\Delta(\hat{T})\Big) \subset \overline{W\Big(\Delta(T)\Big)}$,

(c) $||T - \zeta I|| = ||\hat{T} - \zeta\hat{I}||$ *and* $||\Delta(T) - \zeta I|| = ||\Delta(\hat{T}) - \zeta\hat{I}|| \qquad (\zeta \in \mathbb{C})$,

(d) $||(T - \zeta I)^{-1}|| = ||(\hat{T} - \zeta\hat{I})^{-1}||$ *and*

$$||\Big(\Delta(T) - \zeta I\Big)^{-1}|| = ||\Big(\Delta(\hat{T}) - \zeta\hat{I}\Big)^{-1}|| \qquad (\zeta \notin \sigma(T)),$$

where $\hat{I}$ is the identity operator on $\mathcal{H}\oplus\mathcal{H}$. In particular, if $W(T)$ (resp. $W\Big(\Delta(T)\Big)$) is closed, so is $W(\hat{T})$ (resp. $W\Big(\Delta(\hat{T})\Big)$).

Proof. The operator $\hat{U}$ is the so-called *unitary dilation* of U. Since $U^*UP = P$ by definition, we have

$$\hat{T} = \begin{bmatrix} UP & 0 \\ 0 & 0 \end{bmatrix} = \begin{bmatrix} T & 0 \\ 0 & 0 \end{bmatrix} \quad \text{and} \quad \Delta(\hat{T}) = \begin{bmatrix} \Delta(T) & 0 \\ 0 & 0 \end{bmatrix}. \tag{2.1}$$

By (2.1) we have

$$\sigma(\hat{T}) = \sigma(T) \cup \{0\} \quad \text{and} \quad \sigma\Big(\Delta(\hat{T})\Big) = \sigma\Big(\Delta(T)\Big) \cup \{0\}.$$

But since neither T or $\Delta(T)$ is invertible, by (1.6) we have

$$0 \in \sigma(T) = \sigma\Big(\Delta(T)\Big), \tag{2.2}$$

and we can conclude (a).

Again by (2.1) we can see by definition of numerical range

$$W(\hat{T}) = \mathrm{conv}\Big(W(T), 0\Big) \quad \text{and} \quad W\Big(\Delta(\hat{T})\Big) = \mathrm{conv}\Big(W\Big(\Delta(T)\Big), 0\Big).$$

On the other hand, since by (2.2)

$$0 \in \sigma(T) \subset \overline{W(T)} \quad \text{and} \quad 0 \in \sigma\Big(\Delta(T)\Big) \subset \overline{W\Big(\Delta(T)\Big)},$$

we can conclude (b) by convexity of $\overline{W(T)}$ and $\overline{W\Big(\Delta(T)\Big)}$.

Again it follows from (2.1) that

$$||\hat{T} - \zeta\hat{I}|| = \max\left\{||T - \zeta I||, |\zeta|\right\} \quad \text{and} \quad ||\Delta(\hat{T}) - \zeta\hat{I}|| = \max\left\{||\Delta(T) - \zeta I||, |\zeta|\right\}.$$

Therefore for (c) it suffices to prove that

$$||T - \zeta I|| \geq |\zeta| \quad \text{and} \quad ||\Delta(T) - \zeta I|| \geq |\zeta| \qquad (\zeta \in \mathbb{C}). \tag{2.3}$$

By (2.2) there is a sequence of unit vectors x_n $(n = 1, 2, \ldots)$ such that

$$\lim_{n\to\infty} Tx_n = 0 \quad \text{or} \quad \lim_{n\to\infty} T^* x_n = 0. \tag{2.4}$$

In the former case of (2.4), we have

$$||T - \zeta I|| \geq \lim_{n\to\infty} ||(T - \zeta I)x_n|| = |\zeta|$$

while in the latter case

$$||T - \zeta I|| = ||T^* - \bar{\zeta} I|| \geq \lim_{n\to\infty} ||(T^* - \bar{\zeta} I)x_n|| = |\zeta|,$$

which prove the first part of (2.3).

By (2.2) the same arguments prove the second part of (2.3). These establish (c). The proof of (d) is quite similar and is omitted. □

Theorem 2. *Let T be a Hilbert space operator with polar representation $T = UP$. Then*

$$||T - \zeta I|| \geq ||\Delta(T) - \zeta I|| \qquad (\ \zeta \in \mathbb{C}\ ), \tag{2.5}$$

and

$$||(T - \zeta I)^{-1}|| \geq ||\Big(\Delta(T) - \zeta I\Big)^{-1}|| \qquad (\ \zeta \notin \sigma(T)\ ). \tag{2.6}$$

Proof. If U is not unitary, consider $\hat{T}$ in Lemma 1. Therefore we may assume that U is unitary. For $\epsilon > 0$ define

$$T_\epsilon \stackrel{\text{def}}{=} U(P + \epsilon I).$$

Then all T_ϵ are invertible, and

$$\lim_{\epsilon\downarrow 0} ||T_\epsilon - \zeta I|| = ||T - \zeta I|| \qquad (\zeta \in \mathbb{C}),$$

and

$$\lim_{\epsilon\downarrow 0} ||\Delta(T_\epsilon) - \zeta I|| = ||\Delta(T) - \zeta I|| \qquad (\zeta \in \mathbb{C}).$$

In a similar way

$$\lim_{\epsilon\downarrow 0} ||(T_\epsilon - \zeta I)^{-1}|| = ||(T - \zeta I)^{-1}|| \qquad \Big(\zeta \notin \sigma(T)\Big),$$

and

$$\lim_{\epsilon\downarrow 0} ||\Big(\Delta(T_\epsilon) - \zeta I\Big)^{-1}|| = ||\Big(\Delta(T) - \zeta I\Big)^{-1}|| \qquad \Big(\zeta \notin \sigma(T)\Big).$$

Therefore to prove (2.5) and (2.6) we may assume further that T is invertible.

First we claim a general assertion that if an operator S *commutes* with $T = UP$ then

$$(\flat) \qquad ||P^{\frac{1}{2}}SUP^{\frac{1}{2}}|| \leq ||SUP||.$$

In fact, since for a selfadjoint operator the norm coincides with the spectral radius, we can see

$$\begin{aligned} ||P^{\frac{1}{2}}SUP^{\frac{1}{2}}||^2 &= ||P^{\frac{1}{2}}SUPU^*S^*P^{\frac{1}{2}}|| \\ &= r(P^{\frac{1}{2}}SUPU^*S^*P^{\frac{1}{2}}) = r(PSUPU^*S^*) \\ &\leq ||U^* \cdot UPS \cdot U \cdot (SUP)^*|| \leq ||UPS|| \cdot ||(SUP)^*|| \\ &= ||SUP||^2, \end{aligned}$$

proving ($\flat$).

Now, to see (2.5), take in ($\flat$)

$$S \equiv (T - \zeta I)T^{-1} = (UP - \zeta I)(UP)^{-1}.$$

Then since

$$\begin{aligned} P^{\frac{1}{2}}SUP^{\frac{1}{2}} &= P^{\frac{1}{2}}(UP - \zeta I)P^{-1}U^*UP^{\frac{1}{2}} \\ &= (P^{\frac{1}{2}}UP^{\frac{1}{2}} - \zeta I)P^{-\frac{1}{2}}U^*UP^{\frac{1}{2}} = \Delta(T) - \zeta I \end{aligned}$$

and

$$SUP = T - \zeta I,$$

it follows from ($\flat$) that

$$||\Delta(T) - \zeta I|| \leq ||T - \zeta I||,$$

proving (2.5).

To see (2.6), take

$$S \equiv (T - \zeta I)^{-1}T^{-1} = (UP - \zeta I)^{-1}(UP)^{-1}.$$

Then since

$$\begin{aligned} P^{\frac{1}{2}}SUP^{\frac{1}{2}} &= P^{\frac{1}{2}}\Big(UP - \zeta I\Big)^{-1}P^{-1}U^*UP^{\frac{1}{2}} \\ &= (P^{\frac{1}{2}}UP^{\frac{1}{2}} - \zeta I)^{-1}P^{-\frac{1}{2}}U^*UP^{\frac{1}{2}} = \Big(\Delta(T) - \zeta I\Big)^{-1} \end{aligned}$$

and

$$SUP = (T - \zeta I)^{-1},$$

it follows from ($\flat$) that

$$||\Big(\Delta(T) - \zeta I\Big)^{-1}|| \leq ||(T - \zeta I)^{-1}||,$$

proving (2.6). This completes the proof. □

It should be mentioned that in a recent paper [7] (see also [10]) general inequalities

$$||f\big(\Delta(T)\big)|| \;\le\; ||f(T)|| \quad \forall\, f(\zeta) \text{ analytic on } \sigma(T)$$

were established.

Corollary 3. (Yamazaki [14], Wu [13]) *For a Hilbert space operator T and its Aluthge transform $\Delta(T)$ the following holds*

$$\overline{W(T)} \;\supset\; \overline{W\big(\Delta(T)\big)}.$$

Proof. According to the formula (1.3) it follows from Theorem 2

$$\begin{aligned}\overline{W(T)} \;&=\; \bigcap_{\zeta\in\mathbb{C}}\{\xi :\ |\xi-\zeta|\le ||T-\zeta I||\}\\ &\supset\; \bigcap_{\zeta\in\mathbb{C}}\{\xi :\ |\xi-\zeta|\le ||\Delta(T)-\zeta I||\} \;=\; \overline{W\big(\Delta(T)\big)}.\end{aligned}$$

This completes the proof. □

3. Convexoid operators

Let T be an operator on a Hilbert space $\mathcal{H}$ with polar representation $T = UP$, and let $\Delta(T) \equiv P^{\frac{1}{2}}UP^{\frac{1}{2}}$ be its Aluthge transform.

The numerical range $W(T)$ is not closed in general. But for the problems under discussion we may assume the closedness of $W(T)$ as seen in Lemma 4 below. For this, let us start with a general setup following the idea of Berberian [4].

Consider the Banach space l^∞ of bounded complex sequences (ζ_n). Remark that the indexing is from 0 to ∞. For a sequence

$$(\zeta_n) \;=\; (\zeta_0, \zeta_1, \zeta_2, \ldots)$$

the (backward) shifted sequence (ζ_{n+1}) is defined as

$$(\zeta_{n+1}) \;\equiv\; (\zeta_1, \zeta_2, \zeta_3, \ldots).$$

We consider a *Banach limit* Lim (ζ_n) (see [6] pp. 84–86) as a unital positive linear functional on the commutative C^*-algebra l^∞ (see [6], p. 256). Here *unitalness* means

$$\mathrm{Lim}(\zeta_n) = 1 \quad \text{when} \quad \zeta_n = 1 \;\; \forall\, n$$

while *positivity* means

$$\mathrm{Lim}(\zeta_n) \ge 0 \quad \text{whenever } \zeta_n \ge 0 \;\; \forall\, n.$$

The characteristic property of a Banach limit is

(*shift invariance*) $$\mathrm{Lim}\,(\zeta_n) \;=\; \mathrm{Lim}\,(\zeta_{n+1}).$$

By linearity, positive unitalness and shift invariance we can see

$$\varliminf_{n\to\infty} |\zeta_n| \;\le\; |\mathrm{Lim}\,(\zeta_n)| \;\le\; \varlimsup_{n\to\infty} |\zeta_n| \tag{3.1}$$

and

$$\mathrm{Lim}\,(\zeta_n) \;=\; \lim_{n\to\infty} \zeta_n \qquad \text{(if } \zeta_n \text{ converges).} \tag{3.2}$$

Next consider the space $l^\infty(\mathcal{H})$ of bounded sequences $\mathbf{x} = (x_n)$ of a Hilbert space $\mathcal{H}$, and define a sesquilinear form

$$\langle \mathbf{x}, \mathbf{y} \rangle \;\stackrel{\text{def}}{=}\; \mathrm{Lim}\Big(\langle x_n, y_n \rangle\Big)$$

and the corresponding Hilbertian seminorm

$$||\mathbf{x}|| \;\stackrel{\text{def}}{=}\; \sqrt{\langle \mathbf{x}, \mathbf{x} \rangle} \;=\; \sqrt{\mathrm{Lim}\,(||x_n||^2)}.$$

$l^\infty(\mathcal{H})$ becomes a pre-Hilbert space. The associated Hilbert space, that is, the completion of $l^\infty(\mathcal{H})/\{\mathbf{x} : ||\mathbf{x}|| = 0\}$ will be called the *ultra-sum* (based on the Banach limit) of $\mathcal{H}$ and denoted by $\mathcal{K}$.

The map

$$x \;\longmapsto\; \mathbf{x} \;=\; (x_n) \quad (\text{with } x_n = x \;\forall\; n)$$

gives a canonical isometric embedding of $\mathcal{H}$ into $\mathcal{K}$. This embedding will be simply written as

$$\mathbf{x} \;=\; (x).$$

When (A_n) is a bounded sequence of operators on $\mathcal{H}$, define a linear map $\mathbf{A}$ on $l^\infty(\mathcal{H})$ by

$$\mathbf{A}\mathbf{x} \;\stackrel{\text{def}}{=}\; (A_n x_n) \quad \text{for } \mathbf{x} \;=\; (x_n).$$

The operator $\mathbf{A}$ can be canonically lifted to an operator on $\mathcal{K}$, which will be denoted by the same $\mathbf{A}$. We will denote this relation by

$$\mathbf{A} \;=\; (A_n).$$

It is immediate from definition that for $\mathbf{A} = (A_n)$, $\mathbf{B} = (B_n)$ and $\alpha, \beta \in \mathbb{C}$

$$\mathbf{A}\cdot\mathbf{B} \;=\; (A_n B_n), \quad \alpha\mathbf{A} + \beta\mathbf{B} \;=\; (\alpha A_n + \beta B_n) \quad \text{and} \quad \mathbf{A}^* \;=\; (A_n^*). \tag{3.3}$$

Therefore there is a canonical C^*-embedding of $\mathcal{B}(\mathcal{H})$ into $\mathcal{B}(\mathcal{K})$:

$$A \;\longmapsto\; \mathbf{A} \;=\; (A_n) \qquad (\text{with } A_n = A \;\;\forall\; n).$$

This $\mathbf{A}$ will be written as

$$\mathbf{A} \;=\; (A).$$

Just as in [4] we can prove the following Lemma.

Lemma 4. *For any operator T on a Hilbert space $\mathcal{H}$ the operator $\mathbf{T} = (T)$ on the ultra-sum $\mathcal{K}$ of $\mathcal{H}$ possesses the following properties;*

$$\sigma(\mathbf{T}) \;=\; \sigma(T),$$

and

$$W(\mathbf{T}) \;=\; \overline{W(T)} \quad \text{and} \quad W\Big(\Delta(\mathbf{T})\Big) \;=\; \overline{W\Big(\Delta(T)\Big)}.$$

Though the proof of Lemma 4 does not use shift-invariance of the Banach limit, the proof of the following lemma, which gives generalizations of (3.1) and (3.2), is based on the shift invariance.

Lemma 5. *For any bounded sequence* (A_n) *of operators on a Hilbert space* $\mathcal{H}$ *the norm of the operator* $\mathbf{A} = (A_n)$ *on the ultra-sum* $\mathcal{K}$ *satisfies the following inequalities:*

$$\varliminf_{n\to\infty} ||A_n|| \;\le\; ||\mathbf{A}|| \;\le\; \varlimsup_{n\to\infty} ||A_n||,$$

so that

$$||\mathbf{A}|| \;=\; \lim_{n\to\infty} ||A_n|| \quad (if\ ||A_n||\ converges).$$

Proof. For any bounded sequence $\mathbf{x} = (x_n)$ in $\mathcal{H}$ and $k = 1, 2, \ldots$, by shift invariance and positivity of the Banach limit

$$\begin{aligned} ||\mathbf{A}\mathbf{x}|| &= \operatorname{Lim} ||A_{n+k}x_{n+k}|| \;\le\; \operatorname{Lim} ||A_{n+k}|| \cdot ||x_{n+k}|| \\ &\le \{\sup_n ||A_{n+k}||\} \cdot \operatorname{Lim} ||x_{n+k}|| \;=\; \{\sup_n ||A_{n+k}||\} ||\mathbf{x}||, \end{aligned}$$

which implies

$$||\mathbf{A}\mathbf{x}|| \;\le\; \varlimsup_{n\to\infty} ||A_n|| \cdot ||\mathbf{x}||,$$

so that

$$||\mathbf{A}|| \;\le\; \varlimsup_{n\to\infty} ||A_n||.$$

Conversely, for any $\epsilon > 0$ and $n = 0, 1, 2, \ldots$ take $x_n \in \mathcal{H}$ such that

$$||x_n|| = 1 \quad \text{and} \quad ||A_n x_n|| \ge ||A_n|| - \epsilon.$$

Then again by shift invariance and positivity we have for any $k = 1, 2, \ldots$

$$\begin{aligned} ||\mathbf{A}\mathbf{x}|| &= \operatorname{Lim} ||(A_{n+k}x_{n+k})|| \;\ge\; \operatorname{Lim} (||A_{n+k}|| - \epsilon) \\ &\ge \inf_n ||A_{n+k}|| - \epsilon, \end{aligned}$$

which implies

$$||\mathbf{A}|| \;\ge\; ||\mathbf{A}\mathbf{x}|| \ge \varliminf_{n\to\infty} ||A_n|| - \epsilon.$$

Since $\epsilon > 0$ is arbitrary, we can conclude

$$||\mathbf{A}|| \;\ge\; \varliminf_{n\to\infty} ||A_n||.$$

This completes the proof. □

Recall that an operator T is said to be *convexoid* when

$$\overline{W(T)} \;=\; \operatorname{conv}\Big(\sigma(T)\Big).$$

Theorem 6. *A Hilbert space operator T is convexoid if and only if*

$$\overline{W(T)} = \overline{W\Big(\Delta(T)\Big)}.$$

Proof. The "only if" part was already pointed out in Introduction. To prove the "if" part, considering $\mathbf{T}$ in Lemma 4 and further $\hat{\mathbf{T}}$ in Lemma 1 if necessary, we may assume that $W(T)$ and $W\Big(\Delta(T)\Big)$ are closed and $T = UP$ with unitary U.

There fore we have to prove that if $W(T)$ is closed and U is unitary then

$$W(T) = W\Big(\Delta(T)\Big) \quad \Longrightarrow \quad W(T) = \mathrm{conv}\Big(\sigma(T)\Big).$$

Since $W(T)$ is a compact convex set containing $\mathrm{conv}\Big(\sigma(T)\Big)$, by the separation theorem for a compact convex set (see [12] p. 40), with the help of a rotation of the complex plane (around the origin) if necessary, for the proof it suffices to show, under the condition

$$(*) \qquad \overline{W(T)} = W(T) = W\Big(\Delta(T)\Big),$$

that for any $\tau \in \mathbb{R}$

$$(\sharp) \qquad W(T) \subset \{\zeta : \mathrm{Re}(\zeta) \leq \tau\} \quad \text{and} \quad W(T) \cap \{\zeta : \mathrm{Re}(\zeta) = \tau\} \neq \emptyset$$
$$\Longrightarrow \quad \sigma(T) \cap \{\zeta : \mathrm{Re}(\zeta) = \tau\} \neq \emptyset.$$

When $\tau = 0$ and T is not invertible, we have obviously

$$0 \in \sigma(T) \cap \{\zeta : \mathrm{Re}(\zeta) = \tau\} \neq \emptyset.$$

Therefore in the following we have to consider two cases; the first is the case of $\tau = 0$ with invertible T and the second is the case of $\tau \neq 0$.

Write $P \equiv |T|$ for simplicity, so that

$$T = UP \quad \text{and} \quad \Delta(T) = P^{\frac{1}{2}} U P^{\frac{1}{2}}. \tag{3.4}$$

Denote by Q the orthoprojection onto the closure of the range of P. Then according to the decomposition $I = Q \oplus (I - Q)$, write the positive semi-definite operator P as $P = P_1 \oplus 0$ where P_1 is a positive definite operator on $\mathrm{ran}(Q)$, the range of Q.

Notice that if $\tau < 0$ the (closed) numerical range $W(T)$ is contained in the open left half-plane and hence T is invertible, so that $Q = I$.

Since U is unitary, the first of the assumptions in $(\sharp)$ implies

$$\mathrm{Re}(UP) \leq \tau I \quad \text{and} \quad \mathrm{Re}(PU) = U^* \cdot \mathrm{Re}(UP) \cdot U \leq \tau I, \tag{3.5}$$

where $\mathrm{Re}(\cdot)$ denotes the selfadjoint part: $\mathrm{Re}(T) = \frac{1}{2}(T + T^*)$.

Now let us consider an operator-valued (strongly) continuous function $\Phi(\zeta)$ on the strip $\{\zeta : |\mathrm{Re}(\zeta)| \leq \frac{1}{2}\}$ defined by

$$\Phi(\zeta) \stackrel{\text{def}}{=} P^{\frac{1}{2} - \zeta} U P^{\frac{1}{2} + \zeta} \quad \text{with} \quad P^{\zeta} \stackrel{\text{def}}{=} \exp(\zeta \log P_1) \oplus 0. \tag{3.6}$$

Take the resolution of the identity $E(\lambda)$ $(0 \leq \lambda < \infty)$ for the operator P_1 (see [1] p. 249). Then

$$P_1 = \int_0^\infty \lambda dE(\lambda) \tag{3.7}$$

and

$$P_1^\zeta = \exp(\zeta \log P_1) = \int_0^\infty \lambda^\zeta dE(\lambda). \tag{3.8}$$

Notice that by definition (3.6) we have

$$\Phi(\tfrac{1}{2}) = QUP = QT, \ \Phi(0) = \Delta(T) \quad \text{and} \quad \Phi(-\tfrac{1}{2}) = PUQ. \tag{3.9}$$

Now $\zeta \longmapsto \Phi(\zeta)$ is analytic in the interior of the strip, and on the boundary $\{\pm\frac{1}{2} + it : t \in \mathbb{R}\}$ we have

$$\langle \Phi(\tfrac{1}{2} + it)x, x\rangle = \langle UP(P^{it}x), P^{it}x\rangle \tag{3.10}$$

and

$$\langle \Phi(-\tfrac{1}{2} + it)x, x\rangle = \langle PU(P^{it}x), P^{it}x\rangle. \tag{3.11}$$

Since

$$||P^{it}x|| \leq ||x|| \quad \text{(with equality if } \tau < 0),$$

we can conclude from (3.5)

$$\mathrm{Re}\Big(\Phi(\zeta)\Big) \leq \tau I \qquad (|\mathrm{Re}(\zeta)| = \tfrac{1}{2}). \tag{3.12}$$

Then it follows from (3.12) via a variant of the *three lines theorem* in complex analysis (see [5] Chap. VI, Sect. 3) that

$$\mathrm{Re}\Big(\Phi(\zeta)\Big) \leq \tau I \qquad (|\mathrm{Re}(\zeta)| \leq \tfrac{1}{2}). \tag{3.13}$$

Since by $(*)$ the second of the assumptions in $(\sharp)$ implies

$$\{\zeta : \mathrm{Re}(\zeta) = \tau\} \cap W\Big(\Phi(0)\Big) = \{\zeta : \mathrm{Re}(\zeta) = \tau\} \cap W\Big(\Delta(T)\Big) \neq \emptyset,$$

there exists a vector u such that

$$||u|| = 1 \quad \text{and} \quad \langle \mathrm{Re}\Big(\Phi(0)\Big)u, u\rangle = \tau. \tag{3.14}$$

Fix this u and consider the numerical analytic function

$$\varphi(\zeta) \stackrel{\text{def}}{=} \langle \Phi(\zeta)u, u\rangle \qquad (|\mathrm{Re}(\zeta)| \leq \tfrac{1}{2}).$$

Since by (3.13) and (3.14)

$$\mathrm{Re}\Big(\varphi(\zeta)\Big) \leq \tau \quad (|\mathrm{Re}(\zeta)| \leq \tfrac{1}{2}) \quad \text{and} \quad \mathrm{Re}\Big(\varphi(0)\Big) = \tau,$$

it follows from a variant of the *maximum principle* for an analytic function ([5] p. 128) that

$$\mathrm{Re}\Big(\varphi(\zeta)\Big) = \tau \qquad (|\mathrm{Re}(\zeta)| \leq \tfrac{1}{2}). \tag{3.15}$$

Then from (3.13) and (3.15) by an inequality of Cauchy–Schwarz type we can see (see [8] p. 75)

$$\Big[\tau I - \mathrm{Re}\Big(\Phi(\zeta)\Big)\Big]u \;=\; 0 \qquad (|\mathrm{Re}(\zeta)| \le \tfrac{1}{2}).$$

In particular,

$$\Big[\tau I - \mathrm{Re}\Big(\Phi(0)\Big)\Big]u \;=\; 0.$$

These considerations show that

$$\mathcal{M} \overset{\text{def}}{=} \ker\Big[\tau I - \mathrm{Re}\Big(\Phi(0)\Big)\Big] \tag{3.16}$$

is a non-trivial closed subspace and

$$\Big[\tau I - \mathrm{Re}\Big(\Phi(\zeta)\Big)\Big]x \;=\; 0 \qquad (x \in \mathcal{M},\ (|\mathrm{Re}(\zeta)| \le \tfrac{1}{2}). \tag{3.17}$$

Now we are in position to prove ($\sharp$) under the condition $\mathcal{M} \neq \emptyset$. First let us show that $\mathcal{M}$ is included in $\mathrm{ran}(Q)$. This is immediate when T is invertible, because then P is invertible with $Q = I$ so that $\mathrm{ran}(Q)$ is the whole space. As mentioned in the beginning of the proof of ($\sharp$), it remains to treat the case $\tau \neq 0$. When $\tau \neq 0$, writing

$$\mathrm{Re}\Big(\Phi(0)\Big) \;=\; P^{\frac{1}{2}}\mathrm{Re}(U)P^{\frac{1}{2}} \;=\; Q\cdot\mathrm{Re}\Big(\Phi(0)\Big)\cdot Q,$$

we have by definition (3.16) that

$$\tau(I-Q)x \oplus \Big[\tau I - \mathrm{Re}\Big(\Phi(0)\Big)\Big]Qx \;=\; 0 \quad (x \in \mathcal{M}).$$

This implies $(I - Q)x = 0$, that is, $Qx = x$, hence $\mathcal{M} \subset \mathrm{ran}(Q)$.

Now since $Q = P^{it}$ with $t = 0$ and $Qx = x$ for $x \in \mathcal{M}$, we can derive as before from (3.5), (3.9), (3.10) and (3.11) together with (3.17)

$$\Big[\tau I - \mathrm{Re}(UP)\Big]x = 0 \quad \text{and} \quad \Big[\tau I - \mathrm{Re}(PU)\Big]x = 0 \quad (x \in \mathcal{M}). \tag{3.18}$$

Then since

$$\begin{aligned} 2\tau I - 2\mathrm{Re}(UP) &= 2\tau I - UP - PU^* \\ &= \{2\tau I - Q\mathrm{Re}(U)QP - PQ\mathrm{Re}(U)\} - (I-Q)\mathrm{Re}(U)P \\ &\qquad\qquad -i\{\mathrm{Im}(U)P - P\mathrm{Im}(U)\} \end{aligned}$$

where $\mathrm{Im}(U) = \frac{1}{2i}(U - U^*)$, and similarly

$$\begin{aligned} 2\tau I - 2\mathrm{Re}(PU) &= 2\tau I - PU - U^*P \\ &= \{2\tau I - Q\mathrm{Re}(U)QP - PQ\mathrm{Re}(U)\} - (I-Q)\mathrm{Re}(U)P \\ &\qquad\qquad +i\{\mathrm{Im}(U)P - P\mathrm{Im}(U)\}, \end{aligned}$$

we can conclude from (3.18)

$$[2\tau I - Q\mathrm{Re}(U)QP - PQ\mathrm{Re}(U)]x - (I-Q)\mathrm{Re}(U)Px = 0 \quad (x \in \mathcal{M}), \tag{3.19}$$

and

$$[\mathrm{Im}(U)P - P\mathrm{Im}(U)]x = 0 \quad (x \in \mathcal{M}). \tag{3.20}$$

With $\zeta = it$ ($t \in \mathbb{R}$) in (3.17) we can see

$$\Big[\tau I - \mathrm{Re}\big(\Phi(0)\big)\Big] P^{it} x \;=\; 0 \quad (x \in \mathcal{M};\; t \in \mathbb{R}).$$

This shows that $\mathcal{M}$ is invariant for P^{it} ($t \in \mathbb{R}$). Let Q_0 be the orthoprojection onto the subspace $\mathcal{M}$ of $\mathrm{ran}(Q)$. Since P^{it} is unitary on $\mathrm{ran}(Q)$, it follows from the invariance of $\mathcal{M}$ for P^{it} ($t \in \mathbb{R}$) that

$$Q_0 P^{it} \;=\; P^{it} Q_0 \quad (t \in \mathbb{R}). \tag{3.21}$$

We claim further that

$$(\natural) \qquad Q_0 P = P Q_0.$$

To see this, notice that it follows from (3.8) and (3.21) that for any $x, y \in \mathrm{ran}(Q)$ and $t \in \mathbb{R}$

$$\begin{aligned} \int_0^\infty \lambda^{it} \langle d\langle E(\lambda) Q_0 x, y\rangle &= \langle P^{it} Q_0 x, y\rangle \\ &= \langle P^{it} x, Q_0 y\rangle = \int_0^\infty \lambda^{it} d\langle E(\lambda) x, Q_0 y\rangle. \end{aligned}$$

Now by the injectivity of the Fourier transform on the set of measures on the real line (see [11] p. 134) we can conclude that for any $x, y \in \mathrm{ran}(Q)$

$$\langle E(\lambda) Q_0 x, y\rangle \;=\; \langle E(\lambda) x, Q_0 y\rangle \quad (0 \le \lambda < \infty),$$

which implies that Q_0 commutes with all $E(\lambda)$, and hence with P by (3.7), establishing the claim ($\natural$).

Now it follows from definition (3.16) via ($\natural$) that the subspace $\mathcal{M}$ is invariant for any function $f(P)$ of P, that is,

$$P^{1/2}\mathrm{Re}(U)P^{1/2} \cdot f(P)x \;=\; \tau f(P)x \quad (x \in \mathcal{M}). \tag{3.22}$$

With $f(t) = \sqrt{t}$, we can see from (3.22) that

$$P^{\frac{1}{2}}\mathrm{Re}(U)Px \;=\; \tau \cdot P^{\frac{1}{2}}x \qquad (x \in \mathcal{M}),$$

and hence

$$Q\mathrm{Re}(U)Q \cdot Px = \tau x \qquad (x \in \mathcal{M})$$

because $P^{1/2}$ is injective on $\mathcal{M}$. Again since $P(\mathcal{M})$ is dense in $\mathcal{M}$, this implies further that $\mathcal{M}$ is invariant for $Q\mathrm{Re}(U)Q$ too, and

$$P \cdot Q\mathrm{Re}(U)Qx = Q\mathrm{Re}(U)Q \cdot Px = \tau x \qquad (x \in \mathcal{M}). \tag{3.23}$$

Therefore the positive semidefinite operator P and the selfadjoint contraction $Q\mathrm{Re}(U)Q$ on $\mathcal{M}$ have common approximate (unit) eigenvectors, say v_n ($n = 1, 2, \ldots$) in $\mathcal{M}$, that is, for some $\lambda \ge 0$ and $0 \le \theta \le \pi$

$$\lim_{n\to\infty} (\lambda I - P)v_n = 0 \quad \text{and} \quad \lim_{n\to\infty} \Big[\cos\theta \cdot I - Q\mathrm{Re}(U)\Big] v_n = 0. \tag{3.24}$$

It follows from (3.23) and (3.24) that

$$\lambda \cos\theta = \tau. \tag{3.25}$$

Here $\lambda > 0$, because we are treating the case $\tau \neq 0$

Then it follows from (3.19), (3.20), (3.23) and (3.24) that

$$\lim_{n\to\infty} (I - Q)\mathrm{Re}(U)v_n \;=\; 0 \quad \text{and} \quad \lim_{n\to\infty} (\lambda I - P)\mathrm{Im}(U)v_n \;=\; 0, \tag{3.26}$$

which implies again by (3.24) that

$$\lim_{n\to\infty} \Big[\cos\theta \cdot I - \mathrm{Re}(U)\Big] v_n \;=\; 0. \tag{3.27}$$

If $|\cos\theta| = 1$, by (3.27) and the fact that U is unitary we have

$$\lim_{n\to\infty} (\cos\theta \cdot I - U)v_n = 0.$$

Then by (3.24) and (3.25) we have

$$U \cdot \lim_{n\to\infty} (\tau I - PU)v_n = \lim_{n\to\infty} (\tau I - T)Uv_n = 0,$$

and hence

$$\tau \in \sigma(T) \cap \{\zeta : \mathrm{Re}(\zeta) = \tau\}.$$

Finally suppose that $|\cos\theta| < 1$ and hence $\sin\theta \neq 0$. Then by the unitarity of U and (3.27), each v_n can be written in the form

$$v_n = x_n + y_n \quad (n = 1, 2, \ldots),$$

such that

$$\lim_{n\to\infty} (e^{i\theta} I - U)x_n = 0 \quad \text{and} \quad \lim_{n\to\infty} (e^{-i\theta} I - U)y_n \;=0. \tag{3.28}$$

Then it follows from (3.26) and (3.28) that

$$\sin\theta \cdot \lim_{n\to\infty} (\lambda I - P)(x_n - y_n) = \lim_{n\to\infty} (\lambda I - P)\mathrm{Im}(U)v_n = 0. \tag{3.29}$$

Since $\sin\theta \neq 0$, by (3.24) and (3.29) we have

$$\lim_{n\to\infty} (\lambda I - P)(x_n + y_n) = 0 \quad \text{and} \quad \lim_{n\to\infty} (\lambda I - P)(x_n - y_n) = 0$$

so that

$$\lim_{n\to\infty} (\lambda I - P)x_n = 0 \quad \text{and} \quad \lim_{n\to\infty} (\lambda I - P)y_n \;=\; 0,$$

and hence

$$\lim_{n\to\infty} (\lambda e^{i\theta} I - T)x_n = 0 \quad \text{and} \quad \lim_{n\to\infty} (\lambda e^{-i\theta} I - T)y_n \;=\; 0.$$

Now by (3.25) $\lambda e^{i\theta}$ or $\lambda e^{-i\theta}$ is an approximate eigenvalue of T with real part $\lambda\cos\theta = \tau$, so that

$$\sigma(T) \cap \{\zeta\ ;\ \mathrm{Re}(\zeta) = \tau\} \;\neq\; \emptyset.$$

This completes the proof of ($\sharp$). □

Corollary 7. *A Hilbert space operator T is convexoid if and only if*

$$w(T-\zeta I) = w\Big(\Delta(T)-\zeta I\Big) \qquad (\zeta\in\mathbb{C}).$$

Proof. First assume that T is convexoid. Then by Theorem 6

$$\overline{W(T)} = \overline{W\Big(\Delta(T)\Big)}$$

which implies, by definition of numerical radius,

$$\begin{aligned} w(T-\zeta I) &= \sup\{|\xi-\zeta| : \xi\in\overline{W(T)}\} \\ &= \sup\{|\xi-\zeta| : \xi\in\overline{W\Big(\Delta(T)\Big)}\} = w\Big(\Delta(T)-\zeta I\Big), \end{aligned}$$

proving the relation.

Conversely, if the relation is valid, by the general formula (1.4) we have

$$\begin{aligned} \overline{W(T)} &= \bigcap_{\zeta\in\mathbb{C}}\{\xi : |\xi-\zeta|\le w(T-\zeta I)\} \\ &= \bigcap_{\zeta\in\mathbb{C}}\{\xi : |\xi-\zeta|\le w\Big(\Delta(T)-\zeta I\Big)\} = \overline{W\Big(\Delta(T)\Big)}. \end{aligned}$$

Now the assertion follows again from Theorem 6. □

4. Convex hull of spectrum

When an operator T is not convexoid, it is an interesting question how to represent $\operatorname{conv}\Big(\sigma(T)\Big)$ in terms of numerical ranges related to T.

To answer this question, we need an operator, different from that in Lemma 4, on the ultra-sum.

Lemma 8. *Suppose that T is a bounded linear operator on a Hilbert space $\mathcal{H}$. Then the sequence of operators $\mathbf{T} = \Big(\Delta^n(T)\Big)$ determines a bounded linear operator on the ultra-sum $\mathcal{K}$ of $\mathcal{H}$ with the following properties:*

$$\sigma(\mathbf{T}) \subset \sigma(T) \quad \textit{and} \quad \overline{W(\mathbf{T})} = \overline{W\Big(\Delta(\mathbf{T})\Big)} = \bigcap_{n=1}^{\infty}\overline{W\Big(\Delta^n(T)\Big)}.$$

Proof. The operator $\mathbf{T}$ is bounded, because by (1.2)

$$||T|| \ge ||\Delta(T)|| \ge ||\Delta^2(T)|| \ge \cdots.$$

Then it is easy to see

$$\Delta(\mathbf{T}) = \Big(\Delta^{n+1}(T)\Big). \tag{4.1}$$

Take $\zeta\notin\sigma(T)$. Then since by Theorem 2

$$||(T-\zeta I)^{-1}|| \ge ||\Big(\Delta(T)-\zeta I\Big)^{-1}|| \ge ||\Big(\Delta^2(T)-\zeta I\Big)^{-1}|| \ge \cdots$$

the operator

$$\mathbf{S} \stackrel{\text{def}}{=} \Big((\Delta^n(T) - \zeta I)^{-1}\Big)$$

is bounded on $\mathcal{K}$ and becomes the inverse of $\mathbf{T} - \zeta\ \mathbf{I}$. Therefore $\zeta \notin \sigma(\mathbf{T})$, hence

$$\sigma(\mathbf{T}) \subset \sigma(T).$$

Since by (4.1)

$$\Delta^k(\mathbf{T}) - \zeta\ \mathbf{I} = \Big(\Delta^{n+k}(T) - \zeta I\Big)$$

and by Theorem 2

$$||T - \zeta I|| \geq ||\Delta(T) - \zeta I|| \geq ||\Delta^2(T) - \zeta I|| \geq \cdots,$$

we can conclude from Lemma 5 that

$$||\Delta^k(\mathbf{T}) - \zeta\ \mathbf{I}|| = \inf_n ||\Delta^n(T) - \zeta I|| = ||\mathbf{T} - \zeta\ \mathbf{I}|| \quad (\zeta \in \mathbb{C}). \tag{4.2}$$

Now using (4.2) for $k = 1$, by the general formula (1.3) we have

$$\begin{aligned} \overline{W(\mathbf{T})} &= \bigcap_{\zeta\in\mathbb{C}} \Big\{\xi : |\xi - \zeta| \leq ||\mathbf{T} - \zeta\ \mathbf{I}||\Big\} \\ &= \bigcap_{\zeta\in\mathbb{C}} \Big\{\xi : |\xi - \zeta| \leq ||\Delta(\mathbf{T}) - \zeta\ \mathbf{I}||\Big\} = \overline{W\Big(\Delta(\mathbf{T})\Big)}. \end{aligned}$$

In a similar way we have by (4.2)

$$\begin{aligned} \overline{W(\mathbf{T})} &= \bigcap_{\zeta\in\mathbb{C}} \Big\{\xi : |\xi - \zeta| \leq ||\mathbf{T} - \zeta\ \mathbf{I}||\Big\} \\ &= \bigcap_{\zeta\in\mathbb{C}} \bigcap_{n=1}^{\infty} \Big\{\xi : |\xi - \zeta| \leq ||\Delta^n(T) - \zeta I||\Big\} = \bigcap_{n=1}^{\infty} \overline{W\Big(\Delta^n(T)\Big)}. \end{aligned}$$

This completes the proof. □

Theorem 9. *For any Hilbert space operator T the convex hull of its spectrum is represented as*

$$\operatorname{conv}\Big(\sigma(T)\Big) = \bigcap_{n=1}^{\infty} \overline{W\Big(\Delta^n(T)\Big)}.$$

Proof. Consider the operator $\mathbf{T}$ in Lemma 8 such that

$$\overline{W(\mathbf{T})} = \overline{W\Big(\Delta(\mathbf{T})\Big)} = \bigcap_{n=1}^{\infty} \overline{W\Big(\Delta^n(T)\Big)} \quad \text{and} \quad \sigma(\mathbf{T}) \subset \sigma(T).$$

Then by Theorem 6 $\mathbf{T}$ is convexoid, so that

$$\operatorname{conv}\Big(\sigma(\mathbf{T})\Big) = \overline{W(\mathbf{T})} = \bigcap_{n=1}^{\infty} \overline{W\Big(\Delta^n(T)\Big)}.$$

Finally since $\sigma(T) \supset \sigma(\mathbf{T})$, this implies

$$\operatorname{conv}\Big(\sigma(T)\Big) \supset \bigcap_{n=1}^{\infty} \overline{W\Big(\Delta^n(T)\Big)}.$$

The reverse inclusion is obvious by (1.7). This completes the proof. □

Theorem 10. (Yamazaki [15]) *The spectral radius of a Hilbert space operator T is represented as*

$$r(T) = \lim_{n\to\infty} ||\Delta^n(T)|| = \inf_{n=1,2,\dots} ||\Delta^n(T)||.$$

Proof. Since by (1.6)

$$||\Delta^n(T)|| \geq r\Big(\Delta^n(T)\Big) = r(T),$$

the inequality

$$\inf_{n=1,2,\dots} ||\Delta^n(T)|| \geq r(T)$$

is immediate. To see the reverse inequality, we may assume

$$\inf_{n=1,2,\dots} ||\Delta^n(T)|| = 1, \tag{4.3}$$

and have to prove $r(T) \geq 1$.

To this end, consider the operator $\mathbf{T}$ on the ultra-sum $\mathcal{K}$ in Lemma 8. Then by (4.2) and (4.3) we have

$$||\mathbf{T}|| = \inf_{n=1,2,\dots} ||\Delta^n(T)|| = 1.$$

Choose unit vectors $x_n \in \mathcal{H}$ such that

$$\lim_{n\to\infty} ||\Delta^n(T)x_n|| = 1,$$

and let $\mathbf{x} = (x_n) \in \mathcal{K}$. For any $k \geq 1$ let

$$\mathbf{x}_k \stackrel{\text{def}}{=} (x_{n+k})$$

Then we have, by definition (4.1) of $\mathbf{T}$, for every $k = 1, 2, \dots$

$$||\mathbf{x}_k|| = 1 \quad \text{and} \quad ||\Delta^k(\mathbf{T})\mathbf{x}_k|| = \lim_{n\to\infty} ||\Delta^{n+k}(T)x_{n+k}|| = 1. \tag{4.4}$$

We claim, for an operator S with polar representation $S = V|S|$ and $||S|| = 1$, that for any $k \geq 0$

$$(\dagger) \qquad ||\Delta^k(S)u|| = ||u|| = 1 \implies ||S^{k+1}u|| = 1.$$

Since $\Delta^0(S) = S$ by definition, (†) for $k = 0$ is immediate. Suppose that (†) for some $k \geq 0$ is true in general, and suppose

$$||\Delta^{k+1}(S)u|| = ||u|| = 1.$$

Then since by (1.2)

$$1 = ||S|| \geq ||\Delta(S)|| \geq ||\Delta^{k+1}(S)|| \geq ||\Delta^{k+1}(S)u|| = 1,$$

we have $||\Delta(S)|| = 1$. Since

$$1 = ||\Delta^{k+1}(S)u|| = ||\Delta^k\Big(\Delta(S)\Big)u||,$$

apply the induction assumption to $\Delta(S)$ instead of S to conclude

$$||\Delta(S)^{k+1}u|| = 1. \tag{4.5}$$

Since

$$\Delta(S)^{k+1} = |S|^{\frac{1}{2}}(V|S|)^k V|S|^{\frac{1}{2}} \quad \text{and} \quad 0 \leq |S|^{\frac{1}{2}} \leq I,$$

we have by (4.5)

$$1 = ||\Delta(S)^{k+1}u|| \leq ||(V|S|)^k V|S|^{\frac{1}{2}}u|| \leq ||\ |S|^{\frac{1}{2}}u|| \leq ||u|| = 1$$

and hence

$$||(V|S|)^k V|S|^{\frac{1}{2}}u|| = ||\ |S|^{\frac{1}{2}}u|| = 1. \tag{4.6}$$

Recall the following well-known result (see [8] p.75);

(‡) $\quad 0 \leq A \leq I,\ ||x|| \leq 1,\ ||Ax|| = 1 \quad \Longrightarrow \quad A^2x = Ax = x.$

Now consider the vector $v \equiv (V|S|)^k V|S|^{\frac{1}{2}}u$. Then by (4.6) and (4.5) we have $||v|| = 1$ and

$$||\ |S|^{\frac{1}{2}}v|| = ||\Delta(S)^{k+1}u|| = 1.$$

Now we can apply (‡) to $|S|^{\frac{1}{2}}$ instead of A and to u and v instead of x to get

$$\begin{aligned} ||S^{k+2}u|| &= ||V|S|\cdot(V|S|)^k V|S|u|| = ||(|S|^{\frac{1}{2}})^2(V|S|)^k V|S|^{\frac{1}{2}}|S|^{\frac{1}{2}}u|| \\ &= ||(|S|^{\frac{1}{2}})^2(V|S|)^k V|S|^{\frac{1}{2}}u|| = ||(|S|^{\frac{1}{2}})^2 v|| = 1. \end{aligned}$$

This completes induction for (†).

Now applying (†) to (4.4) we can conclude

$$||\mathbf{T}^{k+1}|| \geq ||\mathbf{T}^{k+1}\mathbf{x}_k|| = 1 \quad (k = 0, 1, 2, \ldots).$$

Then the Gelfand formula (see [8] p.48) yields

$$r(\mathbf{T}) = \lim_{k\to\infty} ||\mathbf{T}^k||^{\frac{1}{k}} \geq 1. \tag{4.7}$$

Finally since $\sigma(\mathbf{T}) \subset \sigma(T)$ implies $r(\mathbf{T}) \leq r(T)$, by (4.7) we arrive at $r(T) \geq 1$. This completes the proof. □

References

[1] N.I. Akhiezer and I.M. Glazman, *Theory of Linear Operators in Hilbert Space*, (English Translation) Pitman Pub., Boston, 1981.

[2] A. Aluthge, *On p-hyponorma l operators for* $0 < p < 1$, Integral Equations Operator Theory **13** (1990), 307–315.

[3] T. Ando, *Aluthge transforms and the convex hull of the eigenvalues of a matrix*, Linear Multilinear Algebra, **52**, 281–292.

[4] S.K. Berberian, *Approximate proper vectors*, Proc. Amer. Math. Soc. **13** (1962), 111–114.

[5] J.B. Conway, *Functions of One Complex Variable I*, Springer, New York, 1978.

[6] J.B. Conway, *A Course in Functional Analysis*, Springer, New York, 1985.

[7] C. Foiaş, I.B. Jung, E. Ko and C. Pearcy, *Complete contractivity of maps associated with the Aluthge and Duggal transforms*, Pacific J. Math. **209** (2003), 249–259.

[8] P. Halmos, *A Hilbert Space Problem Book*, Springer, New York, 1985.

[9] S. Hildebrandt, *Über den numerischen Wertebereich eines Operators*, Math. Ann. **163** (1966), 230–247.

[10] I.B. Jung, E. Ko, and C. Pearcy, *Aluthge transforms of operators*, Integral Equations Operator Theory **37** (2000), 437–448.

[11] Y. Katznelson, *An Introduction to Harmonic Analysis* (Second corrected edition), Dover, New York, 1976.

[12] S.R. Lay, *Convex Sets and Their Applications*, Wiley, New York, 1982.

[13] P.Y. Wu, *Numerical range of Aluthge transform of operators*, Linear Algebra Appl. **357**(2002), 295–298.

[14] T. Yamazaki, *On numerical range of the Aluthge transformation*, Linear Alg. Appl. **341**(2002), 111–117.

[15] T. Yamazaki, *An expression of spectral radius via Aluthge transformation*, Proc. Amer. Math. Soc. **130** (2002), 1131–1137.

Tsuyoshi Ando
Shiroishi-ku, Hongo-dori 9
Minami 4-10-805
Sapporo 003-0024, Japan
e-mail: ando@es.hokudai.ac.jp

Operator Theory:
Advances and Applications, Vol. 160, 41–52

Maximal Nevanlinna-Pick Interpolation for Points in the Open Unit Disc

W. Bhosri, A.E. Frazho and B. Yagci

Dedicated to Israel Gohberg on the occasion of his seventy-fifth birthday

Abstract. This note uses a modification of the classical optimization problem in prediction theory to derive a maximal solution for the Nevanlinna-Pick interpolation problem for each point in the open unit disc. This optimization problem is also used to show that the maximal solution is unique. A state space realization for the maximal solution is given.

1. A positive interpolation problem

In this note we will use a modification of the classical optimization problem in prediction theory [14, 15] to derive a special set of solutions for the Nevanlinna-Pick interpolation or covariance problem in [7, 8, 10, 16]. For each α in the open unit disc, we compute a state space solution to the Nevanlinna-Pick interpolation problem that uniquely satisfies a maximum principle. For $\alpha = 0$ our maximal solution reduces to the central or maximal entropy solution in [7, 8, 10, 16].

To introduce our Nevanlinna-Pick interpolation problem, let $\mathcal{U}$ be Hilbert space and T on $\ell^2_+(\mathcal{U})$ be the strictly positive Toeplitz operator matrix given by

$$T = \begin{bmatrix} R_0 & R_1 & R_2 & \cdots \\ R_{-1} & R_0 & R_1 & \cdots \\ R_{-2} & R_{-1} & R_0 & \cdots \\ \vdots & \vdots & \vdots & \ddots \end{bmatrix} \text{ on } \ell^2_+(\mathcal{U}) . \tag{1.1}$$

(An operator P strictly positive if P is an invertible positive operator.) Now let A be a stable operator on a Hilbert space $\mathcal{X}$. By stable we mean that the spectrum of A is contained in the open unit disc $\mathbb{D}$, that is, $r_{\text{spec}}(A) < 1$. Let C be an operator mapping $\mathcal{X}$ onto the whole space $\mathcal{U}$. Let W be the observability operator mapping

$\mathcal{X}$ into $\ell_+^2(\mathcal{U})$ defined by

$$W = \begin{bmatrix} C \\ CA \\ CA^2 \\ \vdots \end{bmatrix} : \mathcal{X} \to \ell_+^2(\mathcal{U})\,. \tag{1.2}$$

Throughout we assume that the pair $\{C, A\}$ is observable. In other words, we assume that W is left invertible, or equivalently, W^*W is invertible. Hence $\Lambda = W^*TW$ is a strictly positive operator on $\mathcal{X}$.

The operator $\Lambda = W^*TW$ is a solution to a Lyapunov equation of the form

$$\Lambda = A^*\Lambda A + C^*\widetilde{C} + \widetilde{C}^*C. \tag{1.3}$$

Here $\widetilde{C}$ is an operator from $\mathcal{X}$ into $\mathcal{U}$. To obtain this Lyapunov equation, let $\widetilde{C}$ be the operator from $\mathcal{X}$ into $\mathcal{U}$ defined by

$$\widetilde{C} = \frac{1}{2}R_0C + \sum_{j=1}^{\infty} R_jCA^j\,. \tag{1.4}$$

Now let S be the standard forward shift on $\ell_+^2(\mathcal{U})$. By employing $S^*W = WA$, we obtain

$$\begin{aligned} \Lambda - A^*\Lambda A &= W^*TW - W^*STS^*W = W^*\left(T - STS^*\right)W \\ &= W^* \begin{bmatrix} R_0 & R_1 & R_2 & \cdots \\ R_{-1} & 0 & 0 & \cdots \\ R_{-2} & 0 & 0 & \cdots \\ \vdots & \vdots & \vdots & \ddots \end{bmatrix} W \\ &= C^*R_0C + \sum_{j=1}^{\infty} C^*R_jCA^j + \sum_{j=1}^{\infty} A^{*j}C^*R_{-j}C = C^*\widetilde{C} + \widetilde{C}^*C\,. \end{aligned}$$

Therefore Λ is a solution to the Lyapunov equation in (1.3).

This naturally leads to a Nevanlinna-Pick interpolation problem. The data set is a triple of operators $\{A, C, \Lambda\}$ where $\{C, A\}$ is a stable, observable pair and C is onto. Moreover, Λ is a strictly positive operator on $\mathcal{X}$ satisfying a Lyapunov equation of the form (1.3). Finally, we assume that $\mathcal{U}$ is finite dimensional. Then our Nevanlinna-Pick interpolation problem is to find the set of all strictly positive Toeplitz operators T on $\ell_+^2(\mathcal{U})$ satisfying $\Lambda = W^*TW$. The set of all solutions to this problem is given in [7, 8, 16]. It turns out that the Nevanlinna-Pick interpolation problem is equivalent to a state covariance problem arising in linear systems [10, 11]. Reference [10] uses J expansive functions to derive the set of all solutions. Here we will use some optimization problems from classical prediction theory [14, 15], to present an elementary derivation of a special set of solutions to this interpolation problem. Our set of solutions is parameterized by the open unit disc $\mathbb{D}$. For each α in $\mathbb{D}$, we obtain a solution to the Nevanlinna-Pick interpolation problem which uniquely satisfies a maximal principle. For $\alpha = 0$, our solution turns out to be the central solution to the Nevanlinna-Pick interpolation problem presented in [7, 8, 10, 16]. Finally, it is noted that we do not need to obtain $\widetilde{C}$ to compute a solution to our Nevanlinna-Pick interpolation problem. All we need to know is that Λ is a solution to a Lyapunov equation of the form (1.3).

This interpolation problem encompasses the classical Carathéodory interpolation problem. To see this assume that A is the upper shift on $\mathcal{X} = \oplus_1^n \mathcal{U}$ given by

$$\begin{aligned} A &= \begin{bmatrix} 0 & I & 0 & \cdots & 0 \\ 0 & 0 & I & \cdots & 0 \\ \vdots & \vdots & \vdots & \ddots & \vdots \\ 0 & 0 & 0 & \cdots & I \\ 0 & 0 & 0 & 0 & 0 \end{bmatrix} \\ C &= \begin{bmatrix} I & 0 & 0 & \cdots & 0 \end{bmatrix}. \end{aligned} \tag{1.5}$$

Notice that the state space $\mathcal{X}$ is simply n orthogonal copies of $\mathcal{U}$. In this setting the observability operator W in (1.2) is given by

$$W = \begin{bmatrix} I \\ 0 \end{bmatrix} : \oplus_1^n \mathcal{U} \to \ell_+^2(\mathcal{U}).$$

In other words, W embeds $\mathcal{X} = \oplus_1^n \mathcal{U}$ into the first n components of $\ell_+^2(\mathcal{U})$. Now assume that T is the strictly positive Toeplitz operator given in (1.1). Then $\Lambda = W^*TW$ is the strictly positive $n \times n$ Toeplitz matrix contained in the upper left-hand corner of T, that is,

$$\Lambda = W^*TW = \begin{bmatrix} R_0 & R_1 & R_2 & \cdots & R_{n-1} \\ R_{-1} & R_0 & R_1 & \cdots & R_{n-2} \\ R_{-2} & R_{-1} & R_0 & \cdots & R_{n-3} \\ \vdots & \vdots & \vdots & \ddots & \vdots \\ R_{1-n} & R_{2-n} & R_{3-n} & \cdots & R_0 \end{bmatrix}. \tag{1.6}$$

Now assume that Λ is a strictly positive Toeplitz matrix of the form (1.6). Let $\widetilde{C}$ be the operator given by

$$\widetilde{C} = \begin{bmatrix} R_0/2 & R_1 & R_2 & \cdots & R_{n-1} \end{bmatrix}.$$

Then Λ is a solution to the Lyapunov equation in (1.3). Therefore $\{A, C, \Lambda\}$ with $\{C, A\}$ in (1.5) and Λ in (1.6) strictly positive is a data set for our Nevanlinna-Pick interpolation problem. In this setting, our Nevanlinna-Pick interpolation problem is to find the set of all strictly positive Toeplitz operators T on $\ell_+^2(\mathcal{U})$ such that Λ is contained in the $n \times n$ upper left-hand corner of T. This is precisely the classical Carathéodory interpolation problem.

To introduce our solution to the Nevanlinna-Pick interpolation problem, we need some additional notation. Let Θ be a function in $H^\infty(\mathcal{L}(\mathcal{U}))$. (Here $H^\infty(\mathcal{L}(\mathcal{U}))$ is the Hardy space consisting of the set of all uniformly bounded analytic functions in the open unit disc whose values are bounded operators on $\mathcal{U}$.) Then T_Θ is the lower triangular Toeplitz operator on $\ell^2_+(\mathcal{U})$ defined by

$$T_\Theta = \begin{bmatrix} \Theta_0 & 0 & 0 & \cdots \\ \Theta_1 & \Theta_0 & 0 & \cdots \\ \Theta_2 & \Theta_1 & \Theta_0 & \cdots \\ \vdots & \vdots & \vdots & \ddots \end{bmatrix} \quad \text{on } \ell^2_+(\mathcal{U}) \tag{1.7}$$

where $\Theta(\lambda) = \sum_0^\infty \Theta_n \lambda^n$ is the Taylor series expansion for Θ. We say that Θ is an *invertible outer function* if both Θ and Θ^{-1} are function in $H^\infty(\mathcal{L}(\mathcal{U}))$. A function Θ in $H^\infty(\mathcal{L}(\mathcal{U}))$ is an invertible outer function if and only if T_Θ is an invertible operator on $\ell^2_+(\mathcal{U})$. We say that Θ is an *outer spectral factor* for the Toeplitz operator T in (1.1) if Θ is an outer function in $H^\infty(\mathcal{L}(\mathcal{U}))$ satisfying $T = T_\Theta^* T_\Theta$. It is well known that the outer spectral factor is unique up to a unitary constant on the left. Finally, T is a strictly positive Toeplitz operator if and only if T admits an invertible outer spectral factor; see Theorem 1.1 page 534 in [6].

Throughout $\mathcal{F}$ is the standard Fourier transform mapping $\ell^2_+(\mathcal{U})$ onto $H^2(\mathcal{U})$. Here $H^2(\mathcal{U})$ is the Hardy space formed by the set of all analytic functions in the open unit disc with values in $\mathcal{U}$ whose Taylor coefficients are square summable. If W is the observability operator in (1.2), then $\mathcal{F}Wx = C(I - \lambda A)^{-1}x$ where x is in $\mathcal{X}$.

Now assume that T is a solution to the Nevanlinna-Pick interpolation problem for the data set $\{A, C, \Lambda\}$. In other words, assume that T is a strictly positive Toeplitz operator such that $\Lambda = W^*TW$. Then T admits a unique invertible outer spectral factor Θ. Therefore $\Lambda = W^*T_\Theta^*T_\Theta W$. Clearly, there is a one to one correspondence between the set of all solutions to our Nevanlinna-Pick interpolation problem and the set of all invertible outer functions Θ satisfying $\Lambda = W^*T_\Theta^*T_\Theta W$. Motivated by this we say that Θ is a *spectral interpolant* for the data $\{A, C, \Lambda\}$ if Θ is an invertible outer function in $H^\infty(\mathcal{L}(\mathcal{U}))$ satisfying $\Lambda = W^*T_\Theta^*T_\Theta W$.

2. Two optimization problems

As before, let Θ be an invertible outer function in $H^\infty(\mathcal{L}(\mathcal{U}))$. Let α be a fixed scalar in the open unit disc $\mathbb{D}$. Consider the classical optimization problem

$$\mu(y, \alpha) = \inf\{\|\Theta h\|^2 : h \in H^2(\mathcal{U}) \text{ and } h(\alpha) = y\}\,. \tag{2.1}$$

In optimal control theory, the error $\mu(y, \alpha)$ in the optimization problem is referred to as the "cost". The idea is to design a controller with the lowest cost possible. Motivated by control theory we will refer to $\mu(y, \alpha)$ as the cost. If $\alpha = 0$, this is precisely the classical optimization problem arising in prediction theory [14, 15, 18], which also played a role in the maximal entropy solution to the state covariance problem in [10]. Now let φ_α be the reproducing kernel in H^2 defined by $\varphi_\alpha(\lambda) = 1/(1 - \overline{\alpha}\lambda)$. Set $d_\alpha = (1 - |\alpha|^2)^{1/2}$. The optimal solution h_{opt} to the optimization problem in (2.1) is unique and given by

$$h_{\text{opt}}(\lambda) = d_\alpha^2 \varphi_\alpha(\lambda)\Theta(\lambda)^{-1}\Theta(\alpha)y \quad \text{and} \quad \mu(y, \alpha) = d_\alpha^2 \|\Theta(\alpha)y\|^2\,. \tag{2.2}$$

To show that $h_{\rm opt}$ is the optimal solution, first observe that $h_{\rm opt}(\alpha) = y$. Let h be any function in $H^2(\mathcal{U})$ satisfying $h(\alpha) = y$. Recall that if g is a function in $H^2(\mathcal{U})$ and u is in $\mathcal{U}$, then $(g(\alpha), u) = (g, \varphi_\alpha u)$. Using this reproducing property of φ_α, we obtain

$$\begin{aligned}
\|\Theta h\|^2 &= \|\Theta(h - h_{\rm opt}) + \Theta h_{\rm opt}\|^2 \\
&= \|\Theta(h - h_{\rm opt})\|^2 + 2\Re(\Theta(h - h_{\rm opt}), d_\alpha^2 \varphi_\alpha \Theta(\alpha)y) + \|\Theta h_{\rm opt}\|^2 \\
&= \|\Theta(h - h_{\rm opt})\|^2 + 2\Re(\Theta(\alpha)(y - y), d_\alpha^2 \Theta(\alpha)y) + \|\Theta h_{\rm opt}\|^2 \\
&= \|\Theta(h - h_{\rm opt})\|^2 + \|\Theta h_{\rm opt}\|^2 \geq \|\Theta h_{\rm opt}\|^2 .
\end{aligned}$$

This readily implies that

$$\|\Theta h_{\rm opt}\|^2 \leq \|\Theta(h - h_{\rm opt})\|^2 + \|\Theta h_{\rm opt}\|^2 = \|\Theta h\|^2 . \tag{2.3}$$

Hence $h_{\rm opt}$ is an optimal solution to the optimization problem in (2.1). The inequality in (2.3) shows that $\|\Theta h_{\rm opt}\| = \|\Theta h\|$ if and only if $\|\Theta(h - h_{\rm opt})\| = 0$, or equivalently, $h = h_{\rm opt}$. In other words, $h_{\rm opt}$ is the unique solution to the optimization problem in (2.1). Since $d_\alpha \varphi_\alpha$ is a unit vector, the cost $\mu(y, \alpha) = \|\Theta h_{\rm opt}\|^2 = d_\alpha^2 \|\Theta(\alpha)y\|^2$.

Now let us introduce the second optimization problem, which plays a fundamental role in our approach to the Nevanlinna-Pick interpolation problem. Recall that operator $\Lambda = W^*TW$ is the solution to the Lyapunov equation in 1.3. Let

$$\nu(y, \alpha) = \inf\{(\Lambda x, x) : C(I - \alpha A)^{-1}x = y\} . \tag{2.4}$$

Here α is a fixed scalar in $\mathbb{D}$, and y is a fixed vector in $\mathcal{U}$. This is a standard least squares optimization problem whose solution is unique and given by

$$\begin{aligned}
x_{\rm opt} &= \Lambda^{-1}(I - \overline{\alpha}A^*)^{-1}C^*\Delta y \quad \text{and} \quad \nu(y, \alpha) = (\Delta y, y) \\
\Delta &= \left(C(I - \alpha A)^{-1}\Lambda^{-1}(I - \overline{\alpha}A^*)^{-1}C^*\right)^{-1} .
\end{aligned} \tag{2.5}$$

To develop a connection between (2.4) and the Nevanlinna-Pick interpolation problem, assume that Θ is a spectral interpolant for the data $\{A, C, \Lambda\}$, that is, Θ is an invertible outer function satisfying $\Lambda = W^*T_\Theta^*T_\Theta W$. If x is in $\mathcal{X}$, then $(\mathcal{F}T_\Theta W)(\lambda) = \Theta(\lambda)C(I - \lambda A)^{-1}x$ where $\lambda \in \mathbb{D}$. Hence

$$(\Lambda x, x) = (W^*T_\Theta^*T_\Theta Wx, x) = \|T_\Theta Wx\|^2 = \|\Theta C(I - \lambda A)^{-1}x\|_{H^2}^2 .$$

This readily implies that

$$\begin{aligned}
(\Delta y, y) &= \inf\{(\Lambda x, x) : C(I - \alpha A)^{-1}x = y\} \\
&= \inf\{\|\Theta C(I - \lambda A)^{-1}x\|^2 : C(I - \alpha A)^{-1}x = y\} .
\end{aligned} \tag{2.6}$$

Notice that the solution $x_{\rm opt}$ to this optimization problem is independent of the spectral interpolant Θ. We claim that

$$\Delta \geq d_\alpha^2 \Theta(\alpha)^*\Theta(\alpha). \tag{2.7}$$

If x is a vector in $\mathcal{X}$ satisfying $C(I - \alpha A)^{-1}x = y$, then $h(\lambda) = C(I - \lambda A)^{-1}x$ is a function in $H^2(\mathcal{U})$ satisfying $h(\alpha) = y$. In other words, the optimization problem

$$\nu(y, \alpha) = \inf\{\|\Theta C(I - \lambda A)^{-1}x\|^2 : C(I - \alpha A)^{-1}x = y\} \tag{2.8}$$

searches over a smaller set than the optimization problem in (2.1). This readily implies that the cost $\nu(y,\alpha) \geq \mu(y,\alpha)$. By virtue of $\nu(y,\alpha) = (\Delta y, y)$ and $\mu(y,\alpha) = d_\alpha^2 \|\Theta(\alpha)y\|^2$, we arrive at the inequality in (2.7).

The previous analysis yields the following maximal principle: *If Θ is any spectral interpolant for $\{A, C, \Lambda\}$, then $\Delta \geq d_\alpha^2 \Theta(\alpha)^*\Theta(\alpha)$.* We say that Θ is an α-*maximal spectral interpolant* for $\{A, C, \Lambda\}$ if Θ is a spectral interpolant satisfying $\Delta = d_\alpha^2 \Theta(\alpha)^*\Theta(\alpha)$.

Assume that Θ is an α-maximal spectral interpolant. This means that two optimization problems in (2.1) and (2.8) have the same cost, that is, $\mu(y,\alpha) = \nu(y,\alpha)$ for all y in $\mathcal{U}$. Recall that the optimization problem in (2.8) searches over a smaller set than the optimization problem in (2.1). Because the solution to these two optimization problems are unique, we must have $h_{\text{opt}}(\lambda) = C(I-\lambda A)^{-1}x_{\text{opt}}$ for all y in $\mathcal{U}$. By consulting (2.2) and (2.5) this readily implies that

$$d_\alpha^2 \varphi_\alpha(\lambda)\Theta(\lambda)^{-1}\Theta(\alpha) \quad = \quad C(I-\lambda A)^{-1}\Lambda^{-1}(I-\overline{\alpha}A^*)^{-1}C^*\Delta\,. \tag{2.9}$$

Since $\Delta = d_\alpha^2 \Theta(\alpha)^*\Theta(\alpha)$, without loss of generality we can assume that $d_\alpha \Theta(\alpha) = \Delta^{1/2}$. Then equation (2.9) implies that

$$\Theta(\lambda) \quad = \quad d_\alpha \left((1-\overline{\alpha}\lambda)C(I-\lambda A)^{-1}\Lambda^{-1}(I-\overline{\alpha}A^*)^{-1}C^*\Delta^{1/2}\right)^{-1}. \tag{2.10}$$

Observe that Θ is uniquely determined by the formula in (2.10). In other words, if there exists a spectral interpolant Θ satisfying $\Delta = d_\alpha^2 \Theta(\alpha)^*\Theta(\alpha)$, then Θ is uniquely given by (2.10). (Here we do not distinguish between two outer functions which are equal up to a unitary constant on the left.) So far we have shown that if there exists an α-maximal spectral interpolant for $\{A, C, \Lambda\}$, then Θ is unique and given by (2.10). The following result shows that there exists a unique α-maximal spectral interpolant for any data set.

Theorem 2.1. *Let $\{A, C, \Lambda\}$ be the data set for a Nevanlinna-Pick interpolation problem. Moreover, let Ω be the function in $H^\infty(\mathcal{L}(\mathcal{U}))$ defined by*

$$\begin{aligned} \Omega(\lambda) &= d_\alpha^{-1}(1-\overline{\alpha}\lambda)C(I-\lambda A)^{-1}\Lambda^{-1}(I-\overline{\alpha}A^*)^{-1}C^*\Delta^{1/2} \\ \Delta &= \left(C(I-\alpha A)^{-1}\Lambda^{-1}(I-\overline{\alpha}A^*)^{-1}C^*\right)^{-1}. \end{aligned} \tag{2.11}$$

Then the following holds.

(i) *The inverse $\Theta(\lambda) = \Omega(\lambda)^{-1}$ is the unique α-maximal spectral factor for $\{A, C, \Lambda\}$. In particular, Ω is an invertible outer function.*

(ii) *A realization for Θ is given by*

$$\begin{aligned} \Theta(\lambda) &= D - \lambda DC(I-\lambda J)^{-1}(A-\overline{\alpha}I)B(CB)^{-1} \\ B &= \Lambda^{-1}(I-\overline{\alpha}A^*)^{-1}C^* \\ J &= A-(A-\overline{\alpha}I)B(CB)^{-1}C \\ D &= d_\alpha \Delta^{-1/2}(CB)^{-1}. \end{aligned} \tag{2.12}$$

Finally, the operator J is stable, that is, $r_{\text{spec}}(J) < 1$.

Remark 2.1. *If $\alpha = 0$, then the 0-maximal solution in Theorem 2.1 is given by $\Theta = \Omega^{-1}$ where*

$$\Omega(\lambda) = C(I - \lambda A)^{-1}\Lambda^{-1}C^*\Delta^{1/2} \quad \text{and} \quad \Delta = \left(C\Lambda^{-1}C^*\right)^{-1}.$$

This Θ is precisely the central solution to the Nevanlinna-Pick interpolation problem presented in [7, 8, 10, 16].

Remark 2.2. *Let Λ be the strictly positive Toeplitz operator on $\mathcal{X} = \oplus_1^n \mathcal{U}$ given in (1.6), and A and C the operators on $\mathcal{X}$ defined in (1.5). In this case, the Nevanlinna-Pick interpolation problem for $\{A, C, \Lambda\}$ reduces to the Carathéodory interpolation problem. Notice that*

$$C(I - \lambda A)^{-1} = \begin{bmatrix} I & \lambda I & \lambda^2 I & \cdots & \lambda^{n-1} I \end{bmatrix} := \Psi(\lambda). \tag{2.13}$$

By consulting Theorem 2.1, we see that the α-maximal solution to the Carathéodory interpolation problem is given by $\Theta = \Omega^{-1}$ where Ω is the polynomial determined by

$$\begin{aligned} \Omega(\lambda) &= d_\alpha^{-1}(1 - \overline{\alpha}\lambda)\Psi(\lambda)\Lambda^{-1}\Psi(\alpha)^*\Delta^{1/2} \\ \Delta &= \left(\Psi(\alpha)\Lambda^{-1}\Psi(\alpha)^*\right)^{-1}. \end{aligned} \tag{2.14}$$

If $\alpha = 0$, then $\Omega = \begin{bmatrix} I & \lambda I & \cdots & \lambda^{n-1} I \end{bmatrix} \Lambda^{-1}C^*\Delta^{1/2}$ and $\Delta = (C\Lambda^{-1}C^*)^{-1}$. In this case, $\Theta = \Omega^{-1}$ is the outer spectral factor computed from the Levinson filter.

Proof of Theorem 2.1. Recall that if F admits a state space realization of the form

$$F(\lambda) = N + \lambda C(I - \lambda A)^{-1}E \tag{2.15}$$

where N is invertible, then the inverse of F exists in some neighborhood of the origin and is given by

$$F(\lambda)^{-1} = N^{-1} - \lambda N^{-1}C\left(I - \lambda(A - EN^{-1}C)\right)^{-1}EN^{-1}. \tag{2.16}$$

Using $(I - \lambda A)^{-1} = I + \lambda(I - \lambda A)^{-1}A$, it follows that Ω in (2.11) admits a state space realization of the form

$$d_\alpha\Omega(\lambda)\Delta^{-1/2} = CB + \lambda C(I - \lambda A)^{-1}(A - \overline{\alpha}I)B. \tag{2.17}$$

Lemma 2.1 below shows that CB is invertible. By consulting (2.15) and (2.16), we see that $\Theta = \Omega^{-1}$ is given by the state space realization in (2.12).

We claim that J satisfies the Lyapunov equation

$$\Lambda = J^*\Lambda J + C^*D^*DC. \tag{2.18}$$

To obtain the Lyapunov equation in (2.18), set

$$P = I - B(CB)^{-1}C \quad \text{and} \quad Q = B(CB)^{-1}C.$$

Using $P + Q = I$ with x in $\mathcal{X}$, we obtain

$$(\Lambda x, x) = (\Lambda(P+Q)x, (P+Q)x) = (\Lambda Px, Px) + 2\Re(\Lambda Px, Qx) + (\Lambda Qx, Qx). \tag{2.19}$$

By employing $AP = J - \overline{\alpha}Q$, we obtain

$$\begin{aligned}\|\Lambda^{1/2}APx\|^2 &= \|\Lambda^{1/2}Jx - \overline{\alpha}\Lambda^{1/2}Qx\|^2 \\ &= \|\Lambda^{1/2}Jx\|^2 - 2\Re(\Lambda(APx + \overline{\alpha}Qx), \overline{\alpha}Qx) + |\alpha|^2\|\Lambda^{1/2}Qx\|^2 \\ &= (\Lambda Jx, Jx) - 2\Re(Px, \overline{\alpha}A^*\Lambda Qx) - |\alpha|^2(\Lambda Qx, Qx)\,. \qquad (2.20)\end{aligned}$$

Notice that $CP = 0$. By applying P^* to the left and P to the right of the Lyapunov equation in (1.3), we obtain $P^*\Lambda P = P^*A^*\Lambda AP$. This with (2.19) and (2.20) yields

$$\begin{aligned}(\Lambda x, x) - (\Lambda Jx, Jx) &= (\Lambda Px, Px) + 2\Re(Px, \Lambda Qx) + (\Lambda Qx, Qx) \\ &\quad -(\Lambda APx, APx) - 2\Re(Px, \overline{\alpha}A^*\Lambda Qx) - |\alpha|^2(\Lambda Qx, Qx) \\ &= 2\Re(Px, (I - \overline{\alpha}A^*)\Lambda Qx) + d_\alpha^2(\Lambda Qx, Qx) \\ &= 2\Re(CPx, (CB)^{-1}Cx) + d_\alpha^2(\Lambda Qx, Qx) \\ &= d_\alpha^2(B^*\Lambda B(CB)^{-1}Cx, (CB)^{-1}Cx) = (C^*D^*DCx, x)\,.\end{aligned}$$

The last equation follows from the fact that $B^*\Lambda B = \Delta^{-1}$. Therefore the Lyapunov equation in (2.18) holds.

Now let us show that J is stable. Set $L = (A - \overline{\alpha}I)B(CB)^{-1}$, and let k be any positive integer. Using $J = A - LC$, we obtain

$$\begin{bmatrix} C \\ CA \\ CA^2 \\ \vdots \\ CA^k \end{bmatrix} = \begin{bmatrix} I & 0 & 0 & \cdots & 0 \\ CL & I & 0 & \cdots & 0 \\ CAL & CL & I & \cdots & 0 \\ \vdots & \vdots & \vdots & \ddots & \vdots \\ CA^{k-1}L & CA^{k-2}L & CA^{k-3}L & \cdots & I \end{bmatrix} \begin{bmatrix} C \\ CJ \\ CJ^2 \\ \vdots \\ CJ^k \end{bmatrix}. \qquad (2.21)$$

The square operator matrix in (2.21) is a lower triangular Toeplitz matrix with the identity on the main diagonal. In particular, this matrix is invertible. For the moment assume that $\mathcal{X}$ is finite dimensional. Because $\{C, A\}$ is observable, equation (2.21) implies that $\{C, J\}$ is observable. Notice that D is invertible, and thus, $\{DC, J\}$ is also observable. Since Λ is a strictly positive solution to the Lyapunov equation in (2.18) and $\{DC, J\}$ is observable, it follows that J is stable. The stability of A and J imply that Ω and Ω^{-1} are both function in $H^\infty(\mathcal{L}(\mathcal{U}))$. In other words, $\Theta = \Omega^{-1}$ is an invertible outer function.

Now assume that $\mathcal{X}$ is infinite dimensional. Let W_k and V_k be the operators mapping $\mathcal{X}$ into $\oplus_0^k\mathcal{U}$ defined by

$$W_k = \begin{bmatrix} C \\ CA \\ \vdots \\ CA^k \end{bmatrix} \quad \text{and} \quad V_k = \begin{bmatrix} C \\ CJ \\ \vdots \\ CJ^k \end{bmatrix}.$$

Because A is stable, the operator W_k converges to W in the operator topology as k tends to infinity. Recall that W is left invertible. So there exists an integer k such that W_k is left invertible. Since the lower triangular matrix in (2.21) is invertible, V_k is also left invertible. Recall that J satisfies the Lyapunov equation in (2.18).

By consulting Lemma 5.4 in [7], we see that J is stable. Therefore $\Theta = \Omega^{-1}$ is an invertible outer function.

We claim that

$$\Theta(\lambda)C(I-\lambda A)^{-1} = DC(I-\lambda J)^{-1}. \tag{2.22}$$

Recall that $J = AP + \overline{\alpha}Q$ and $P + Q = I$. Using this we obtain

$$\begin{aligned}\Theta(\lambda)C(I-\lambda A)^{-1} &= \left(D - \lambda DC(I-\lambda J)^{-1}(A-\overline{\alpha}I)B(CB)^{-1}\right)C(I-\lambda A)^{-1}\\ &= DC\left(I - \lambda(I-\lambda J)^{-1}(A-\overline{\alpha}I)Q\right)(I-\lambda A)^{-1}\\ &= DC(I-\lambda J)^{-1}\left((I-\lambda J) - \lambda(A-\overline{\alpha}I)Q\right)(I-\lambda A)^{-1}\\ &= DC(I-\lambda J)^{-1}\left(I - \lambda A(P+Q)\right)(I-\lambda A)^{-1}\\ &= DC(I-\lambda J)^{-1}.\end{aligned}$$

Therefore (2.22) holds.

Recall that $\Lambda = \sum_0^\infty J^{*n}C^*D^*DCJ^n$ is the unique solution to the Lyapunov equation in (2.18). By employing (2.22) with x in $\mathcal{X}$, we obtain

$$\begin{aligned}\|T_\Theta Wx\|^2 &= \|\Theta C(I-\lambda A)^{-1}x\|_{H^2}^2 = \|DC(I-\lambda J)^{-1}x\|_{H^2}^2\\ &= \|\sum_{n=0}^\infty \lambda^n DCJ^n x\|^2 = \sum_{n=0}^\infty \|DCJ^n x\|^2\\ &= \sum_{n=0}^\infty (J^{*n}C^*D^*DCJ^n x, x) = (\Lambda x, x).\end{aligned}$$

Thus $(W^*T_\Theta^*T_\Theta Wx, x) = \|T_\Theta Wx\|^2 = (\Lambda x, x)$ for all x in $\mathcal{X}$. Hence Θ is a spectral interpolant for the data $\{A, C, \Lambda\}$. Notice that $\Theta(\alpha)^{-1} = \Omega(\alpha) = d_\alpha \Delta^{-1/2}$. In other words, $d_\alpha\Theta(\alpha) = \Delta^{1/2}$. This readily implies that $\Delta = d_\alpha^2\Theta(\alpha)^*\Theta(\alpha)$. Therefore Θ is the unique α-maximal spectral interpolant for $\{A, C, \Lambda\}$. This completes the proof. □

Lemma 2.1. *Let $\{A, C, \Lambda\}$ be a data set, and $B = \Lambda^{-1}(I-\overline{\alpha}A^*)^{-1}C^*$ where $\alpha \in \mathbb{D}$. Then the operator CB is invertible.*

Proof. If $\alpha = 0$, then $CB = C\Lambda^{-1}C^*$. In this case, CB is strictly positive. Therefore the Lemma holds when $\alpha = 0$.

Now assume that α is nonzero, and recall that $\mathcal{U}$ is finite-dimensional. Let us proceed by contradiction and assume that $CBu = 0$ for some nonzero u in $\mathcal{U}$. This implies that Bu is in the kernel of C. Recall that

$$x_{\text{opt}} = \Lambda^{-1}(I-\overline{\alpha}A^*)^{-1}C^*\Delta y = B\Delta y$$

is the unique solution to the optimization problem in (2.4). So if we set $y = \Delta^{-1}u$, then the optimal solution $x_{\text{opt}} = Bu$ is in the kernel of C. Moreover, $\nu(y,\alpha)$ is nonzero. By employing the Lyapunov equation in (1.3), we see that $(\Lambda x, x) =$

$(\Lambda Ax, Ax)$ for all x in the kernel of C. Using this we obtain

$$\begin{aligned}
\nu(y,\alpha) &= (\Lambda x_{\mathrm{opt}}, x_{\mathrm{opt}}) \geq \inf\{(\Lambda x, x) : Cx = 0 \text{ and } C(I-\alpha A)^{-1}x = y\} \\
&= \inf\{(\Lambda x, x) : C\alpha x = 0 \text{ and } C(I-\alpha A)^{-1}A\alpha x = y\} \\
&= |\alpha|^{-2}\inf\{(\Lambda x, x) : Cx = 0 \text{ and } C(I-\alpha A)^{-1}Ax = y\} \\
&= |\alpha|^{-2}\inf\{(\Lambda Ax, Ax) : Cx = 0 \text{ and } C(I-\alpha A)^{-1}Ax = y\} \\
&\geq |\alpha|^{-2}\inf\{(\Lambda x, x) : C(I-\alpha A)^{-1}x = y\} \\
&= |\alpha|^{-2}\nu(y,\alpha).
\end{aligned}$$

Hence $|\alpha|^2\nu \geq \nu \neq 0$. Thus $|\alpha| \geq 1$. Since $\alpha \in \mathbb{D}$, we arrive at a contradiction. Therefore CB is invertible. This completes the proof. □

3. An approximation result

As before, let T be a specified strictly positive Toeplitz matrix on $\ell^2_+(\mathcal{U})$, and $\{A, C, \Lambda\}$ the corresponding data set where $\Lambda = W^*TW$. In applications one is trying to estimate T from the data set $\{A, C, \Lambda\}$. Recall that $T = T_\Phi^* T_\Phi$ where Φ is an invertible outer function in $H^\infty(\mathcal{L}(\mathcal{U}))$. So in practice one is interested in determining how close the α-maximal spectral interpolant Ω^{-1} is to a specified spectral interpolant Φ. The following result shows that if Δ is approximately equal to $d_\alpha^2\Phi(\alpha)^*\Phi(\alpha)$, then Ω^{-1} is approximately equal to Φ.

Proposition 3.1. *Let Φ be a spectral interpolant for the data set $\{A, C, \Lambda\}$, and Ω^{-1} the α-maximal spectral interpolant. Then the following equality holds*

$$\left\| \left(\Phi\Omega\Delta^{1/2} - d_\alpha\Phi(\alpha)\right) d_\alpha\varphi_\alpha y\right\|^2 = (\Delta y, y) - d_\alpha^2\|\Phi(\alpha)y\|^2 . \tag{3.1}$$

In particular, if $\alpha = 0$, then we have

$$\|\Phi\Omega\Delta^{1/2}y - \Phi(0)y\|^2 = (\Delta y, y) - \|\Phi(0)y\|^2 . \tag{3.2}$$

Proof. Notice that $C(I-\lambda A)^{-1}x_{\mathrm{opt}} = d_\alpha\varphi_\alpha\Omega\Delta^{1/2}y$. Using the fact that φ_α is a reproducing kernel for H^2 with $C(I-\alpha A)^{-1}x_{\mathrm{opt}} = y$, we obtain

$$\begin{aligned}
\left\| \left(\Phi\Omega\Delta^{1/2} - d_\alpha\Phi(\alpha)\right) d_\alpha\varphi_\alpha y\right\|^2 &= \|\Phi C(I-\lambda A)^{-1}x_{\mathrm{opt}} - d_\alpha^2\varphi_\alpha\Phi(\alpha)y\|^2 \\
&= \|\Phi C(I-\lambda A)^{-1}x_{\mathrm{opt}}\|^2 + d_\alpha^2\|\Phi(\alpha)y\|^2 \\
&\quad -2\Re(\Phi C(I-\lambda A)^{-1}x_{\mathrm{opt}}, d_\alpha^2\varphi_\alpha\Phi(\alpha)y) \\
&= (\Lambda x_{\mathrm{opt}}, x_{\mathrm{opt}}) + d_\alpha^2\|\Phi(\alpha)y\|^2 \\
&\quad -2d_\alpha^2\Re(\Phi(\alpha)C(I-\alpha A)^{-1}x_{\mathrm{opt}}, \Phi(\alpha)y) \\
&= (\Delta y, y) - d_\alpha^2\|\Phi(\alpha)y\|^2 .
\end{aligned}$$

Therefore (3.1) holds. This completes the proof. □

References

[1] C.I. Byrnes, T.T. Georgiou, and A. Lindquist, A generalized entropy criterion for Nevanlinna-Pick interpolation: A convex optimization approach to certain problems in systems and control, *IEEE Transactions on Automatic Control*, **42** (2001) pp. 822–839.

[2] P.E. Caines, *Linear Stochastic Systems*, Wiley, New York, 1988.

[3] M.J. Corless and A.E. Frazho, *Linear Systems and Control; An Operator Perspective,* Marcel Decker, New York, 2003.

[4] R.L. Ellis, I. Gohberg and D.C. Lay, Extensions with positive real part, a new version of the abstract band method with applications, *Integral Equations and Operator Theory*, **16** (1993) pp. 360–384.

[5] C. Foias and A.E. Frazho, *The Commutant Lifting Approach to Interpolation Problems,* Operator Theory: Advances and Applications, vol. 44, Birkhäuser, 1990.

[6] C. Foias, A.E. Frazho, I. Gohberg and M. A. Kaashoek, *Metric Constrained Interpolation, Commutant Lifting and Systems,* Operator Theory: Advances and Applications, vol. 100, Birkhäuser, 1998.

[7] A.E. Frazho and M.A. Kaashoek, A band method approach to a positive expansion problem in a unitary dilation setting, *Integral Equations and Operator Theory*, **42** (2002) pp. 311–371.

[8] A.E. Frazho and M.A. Kaashoek, A Naimark dilation perspective of Nevanlinna-Pick interpolation, *Integral Equations and Operator Theory*, to appear.

[9] T.T. Georgiou, Spectral estimation via selective harmonic amplification, *IEEE Transactions on Automatic Control*, **46** (2001) pp. 29–42.

[10] T.T. Georgiou, Spectral analysis based on the state covariance: the maximum entropy spectrum and linear fractional parameterization, *IEEE Transactions on Automatic Control*, **47**, (2002) pp. 1811–1823.

[11] T.T. Georgiou, The structure of state covariances and its relation to the power spectrum of the input, *IEEE Transactions on Automatic Control*, **47**, (2002) pp. 1056–1066.

[12] I. Gohberg, S. Goldberg and M.A. Kaashoek, *Classes of Linear Operators I,* Operator Theory: Advances and Applications, **49**, Birkhäuser Verlag, Basel, 1990.

[13] I. Gohberg, S. Goldberg and M.A. Kaashoek, *Classes of Linear Operators II*, Operator Theory: Advances and Applications, **63**, Birkhäuser Verlag, Basel, 1993.

[14] H. Helson and D. Lowdenslager, Prediction theory and Fourier series in several variables, *Acta Math.*, **99** (1958), pp. 165–202.

[15] H. Helson and D. Lowdenslager, Prediction theory and Fourier series in several variables II, *Acta Math.*, **106** (1961), pp. 175–213.

[16] M.A. Kaashoek and C.G. Zeinstra, The band method and generalized Carathéodory-Toeplitz interpolation at operator points, *Integral Equations and Operator Theory*, **33** (1999) pp. 175–210.

[17] T. Kailath, *Linear Systems*, Englewood Cliffs: Prentice Hall, New Jersey, 1980.

[18] B. Sz.-Nagy and C. Foias, *Harmonic Analysis of Operators on Hilbert Space,* North Holland Publishing Co., Amsterdam-Budapest, 1970.

W. Bhosri
School of Aeronautics and Astronautics
Purdue University
West Lafayette, IN 47907-1282, USA
e-mail: wbhosri@ecn.purdue.edu

A.E. Frazho
School of Aeronautics and Astronautics
Purdue University
West Lafayette, IN 47907-1282, USA
e-mail: frazho@ecn.purdue.edu

B. Yagci
School of Aeronautics and Astronautics
Purdue University
West Lafayette, IN 47907-1282, USA
e-mail: byagci@ecn.purdue.edu

Operator Theory:
Advances and Applications, Vol. 160, 53–79

On the Numerical Solution of a Nonlinear Integral Equation of Prandtl's Type

M.R. Capobianco, G. Criscuolo and P. Junghanns

Dedicated to Professor Israel Gohberg on the Occasion of his 75th Birthday

Abstract. We discuss solvability properties of a nonlinear hypersingular integral equation of Prandtl's type using monotonicity arguments together with different collocation iteration schemes for the numerical solution of such equations.

Mathematics Subject Classification (2000). Primary 65R20; Secondary 45G05.

Keywords. Nonlinear hypersingular integral equation, Collocation method.

1. Introduction

We are interested in the numerical solution of integral equations of the form

$$-\frac{\varepsilon}{\pi}\int_{-1}^{1}\frac{g(y)}{(y-x)^2}dy+\gamma(x,g(x))=f(x)\,,\quad |x|<1\,, \tag{1.1}$$

where $0<\varepsilon\le 1$ and the unknown function g satisfies the boundary conditions

$$g(\pm 1)=0\,. \tag{1.2}$$

The integral has to be understood as the "finite part" of the strongly singular integral in the sense of Hadamard, who introduced this concept in relation to the Cauchy principal value.

This type of strongly singular integral equations can be used effectively to model many problems in fracture mechanics (see [6, 7, 10, 11] and the references given there). Denote by D the linear Cauchy singular integral operator

$$(Dg)(x)=\frac{1}{\pi}\int_{-1}^{1}\frac{g(y)}{y-x}dy,\quad |x|<1\,.$$

Using the boundary conditions (1.2), we can write

$$(Lg)(x) := \frac{1}{\pi}\int_{-1}^{1}\frac{g(y)}{(y-x)^2}dy = \frac{d}{dx}(Dg)(x) = (Dg')(x)\,. \tag{1.3}$$

In a two-dimensional crack problem g is the crack opening displacement defined by the density of the distributed dislocations $v(x)$ as

$$g(x) = -\int_{-1}^{x} v(y)dy\,.$$

If we suppose that the nondimensional half crack length is equal to 1, the parameter ε in (1.1) corresponds to the inverse of the normalized crack length, measured in terms of a physical length parameter which is small relative to the physical crack length. The stress field at a crack tip has a square-root singularity with respect to the distance measured from the crack tip. This requires that the dislocation density $v(x)$ is similarly singular, and it turns out that the Cauchy singular integral remains bounded at the crack tip. Thus, we suppose that

$$g(x) = \varphi(x)u(x), \qquad \varphi(x) = \sqrt{1-x^2}\,. \tag{1.4}$$

Then, by relations (1.3) and (1.4), we can rewrite equation (1.1) as

$$-\frac{\varepsilon}{\pi}\frac{d}{dx}\int_{-1}^{1}\frac{\varphi(y)u(y)}{y-x}dy + \gamma(x,\varphi(x)u(x)) = f(x)\,, \quad |x|<1\,. \tag{1.5}$$

Moreover, it can be supposed that the functions $f(x)$ and $\gamma(x,g)$ will both be nonnegative by physical reasons. This happens since f and γ represent the applied tensile tractions that pull the crack surfaces apart and the stiffness of the reinforcing fibres that resist crack opening, respectively (for more details the reader is referred to [10]). But, in the present paper we will not make use of such nonnegativity assumptions. We are particularly interested in the class of problems for which $\gamma(x,g)$ is a monotone function with respect to g, i.e.,

$$[g_1-g_2][\gamma(x;g_1)-\gamma(x;g_2)] \geq 0, \qquad |x|\leq 1,\ g_1,g_2\in\mathbb{R}.$$

As examples, in the literature $\gamma(x,g)$ is chosen as follows

$$\gamma(x,g) = \Gamma(x)g, \qquad |x|\leq 1,\ g\in\mathbb{R}, \tag{1.6}$$

and

$$\gamma(x,g) = \Gamma(x)\sqrt{|g|}\,\mathrm{sgn}\,g, \qquad |x|\leq 1,\ g\in\mathbb{R}, \tag{1.7}$$

where $\Gamma(x)>0\,,\ |x|\leq 1\,.$ Both cases occur in the analysis of a relatively long crack in unidirectionally reinforced ceramics (see [10]). The case corresponding to (1.6) is extensively treated in [2, 3].

The paper is organized as follows. In Section 2 we study the solvability of (1.5) and smoothness properties of the solutions. In Section 3 the convergence of a collocation method is proved and iteration methods for the solution of the collocation equations are investigated. In the foundation of such iteration methods for nonlinear operator equations the Lipschitz continuity plays an important role. But, this Lipschitz continuity is not satisfied in example (1.7). Hence, in Section 4 we use a transformation of the unknown function and study a collocation method for the transformed equation together with an iteration method for solving the respective collocation equations. In the last section we present and discuss the results of some numerical experiments.

2. Solvability and regularity properties of the solution

By $p_n^\varphi(x)$ we denote the normalized Chebyshev polynomial of the second kind

$$p_n^\varphi(\cos s) = \sqrt{\frac{2}{\pi}}\,\frac{\sin(n+1)s}{\sin s}, \qquad n = 0, 1, \dots ,$$

and by $\mathbf{L}_\varphi^2$ the real Hilbert space of all square integrable functions $u : (-1, 1) \longrightarrow \mathbb{R}$ with respect to the weight $\varphi(x)$ equipped with the inner product

$$\int_{-1}^{1} u(x)v(x)\varphi(x)\,dx = \sum_{n=0}^{\infty} \langle u, p_n^\varphi\rangle_\varphi \langle v, p_n^\varphi\rangle_\varphi \,.$$

We consider equation (1.1) in the pair $\mathbf{X} \longrightarrow \mathbf{X}^*$ of the real Banach space $\mathbf{X} = \mathbf{L}_\varphi^{2,\frac{1}{2}}$ and its dual space $\mathbf{X}^* = \mathbf{L}_\varphi^{2,-\frac{1}{2}}$ with respect to the dual product

$$\langle u, v\rangle_\varphi = \sum_{n=0}^{\infty} \langle u, p_n^\varphi\rangle_\varphi \langle v, p_n^\varphi\rangle_\varphi, \qquad u \in \mathbf{X}^*,\ v \in \mathbf{X}\,,$$

where, for $s \geq 0$, $\mathbf{L}_\varphi^{2,s}$ is the subspace of $\mathbf{L}_\varphi^2$ of all $u \in \mathbf{L}_\varphi^2$ for which

$$\|u\|_{\varphi,s} := \sqrt{\sum_{n=0}^{\infty} (n+1)^{2s} \left|\langle u, p_n^\varphi\rangle_\varphi\right|^2} < \infty$$

and $\mathbf{L}_\varphi^{2,-s} := \left(\mathbf{L}_\varphi^{2,s}\right)^*$. Write (1.1) in the form

$$A(u) := \varepsilon V u + F(u) = f\,, \tag{2.1}$$

where $V : \mathbf{L}_\varphi^{2,\frac{1}{2}} \longrightarrow \mathbf{L}_\varphi^{2,-\frac{1}{2}}$ is an isometrical isomorphism given by

$$(Vu)(x) = -\frac{d}{dx}\frac{1}{\pi}\int_{-1}^{1} \frac{\varphi(y)u(y)}{y-x}\,dy, \qquad |x| < 1,$$

or, which is the same, by

$$Vu = \sum_{n=0}^{\infty} (n+1)\,\langle u, p_n^\varphi\rangle_\varphi\, p_n^\varphi \,. \tag{2.2}$$

For $u, v \in \mathbf{L}_\varphi^{2,s}$, consider the inner product

$$\langle u, v\rangle_{\varphi,s} = \sum_{n=0}^{\infty} (1+n)^{2s} \langle u, p_n^\varphi\rangle_\varphi \langle v, p_n^\varphi\rangle_\varphi \,.$$

Note that

$$\left|\langle u,v\rangle_{\varphi,s}\right| \le \|u\|_{\varphi,s-t}\,\|v\|_{\varphi,s+t}\,, \quad u\in\mathbf{L}_\varphi^{2,s-t}\,,\ v\in\mathbf{L}_\varphi^{2,s+t}\,. \tag{2.3}$$

Moreover, for the operator $V:\mathbf{L}_\varphi^{2,s+\frac12}\longrightarrow\mathbf{L}_\varphi^{2,s-\frac12}$ defined by (2.2), we have

$$\langle Vu,u\rangle_{\varphi,s} = \|u\|^2_{\varphi,s+\frac12}\,, \quad u\in\mathbf{L}_\varphi^{2,s+\frac12}\,. \tag{2.4}$$

(See [2] for a more detailed analysis of the operator V.) The operator $F:\mathbf{X}\to\mathbf{X}^*$ is defined by

$$\Big(F(u)\Big)(x) = \gamma(x,\varphi(x)u(x)).$$

With respect to the function $\gamma:[-1,1]\times\mathbb{R}\to\mathbb{R}$ we can make different assumptions, for example

(A) $(g_1-g_2)[\gamma(x,g_1)-\gamma(x,g_2)]\ge 0\,,\ x\in[-1,1]\,,\ g_1,g_2\in\mathbb{R}\,,$

and

(B) $|\gamma(x,g_1)-\gamma(x,g_2)|\le\lambda(x)\,|g_1-g_2|^\alpha\,,\ x\in[-1,1],\ g_1,g_2\in\mathbb{R}$, for some $0<\alpha\le 1$, where

$$c_\alpha := \left\{\begin{array}{lcc} \displaystyle\int_{-1}^1 [\lambda(x)]^{\frac{2}{1-\alpha}}\,[\varphi(x)]^{\frac{1+\alpha}{1-\alpha}}\,dx & : & 0<\alpha<1 \\ \sup\{\lambda(x)\varphi(x): -1\le x\le 1\} & : & \alpha=1 \end{array}\right\} < \infty$$

and $\gamma(.,0)\in\mathbf{L}_\varphi^2$.

In any case we assume that $y\mapsto\gamma(x,y)$ is continuous on $\mathbb{R}$ for almost all $x\in[-1,1]$ and that $x\mapsto\gamma(x,y)$ is measurable for all $y\in\mathbb{R}$. The following definitions are taken from [18].

Definition 2.1. An operator $A:\mathbf{X}\to\mathbf{X}^*$ is called

- **hemicontinuous**, if the function $s\mapsto\langle A(u+sv),w\rangle$ is continuous on $[0,1]$ for any fixed $u,v,w\in\mathbf{X}$;
- **strictly monotone**, if $\langle A(u)-A(v),u-v\rangle>0$ for all $u,v\in\mathbf{X}$ with $u\ne v$;
- **strongly monotone**, if there exists a constant $m>0$ such that

$$\langle A(u)-A(v),u-v\rangle \ge m\,\|u-v\|^2_{\mathbf{X}}$$

  for all $u,v\in X$;
- **coercive**, if there exists a function $\rho:[0,\infty)\to\mathbb{R}$ satisfying

$$\lim_{s\to\infty}\rho(s)=\infty \quad\text{and}\quad \langle A(u),u\rangle\ge\rho\left(\|u\|_{\mathbf{X}}\right)\|u\|_{\mathbf{X}}$$

  for all $u\in\mathbf{X}$.

Lemma 2.2. *If (A) is fulfilled and if F maps $\mathbf{X}$ into $\mathbf{X}^*$, then the operator $A:\mathbf{X}\longrightarrow\mathbf{X}^*$ in (2.1) is strongly monotone (with $m=\varepsilon$) for each $\varepsilon>0$.*

Proof. Let $u, v \in \mathbf{X}$. In view of (A) and (2.2), we have

$$\langle A(u) - A(v), u - v \rangle_{\varphi} \geq \varepsilon \sum_{n=0}^{\infty} (n+1) |\langle u - v, p_n^{\varphi} \rangle_{\varphi}|^2 = \varepsilon \, ||u - v||^2_{\varphi, \frac{1}{2}},$$

which proves the lemma. □

Lemma 2.3. *If (B) is fulfilled, then the operator F maps $\mathbf{L}^2_{\varphi}$ into $\mathbf{L}^2_{\varphi}$, where $F : \mathbf{L}^2_{\varphi} \longrightarrow \mathbf{L}^2_{\varphi}$ is Hölder continuous with exponent α.*

Proof. For $u, v \in \mathbf{L}^2_{\varphi}$, we get

$$\begin{aligned} \|F(u) - F(v)\|^2_{\varphi} &\leq \int_{-1}^{1} [\lambda(x)]^2 [\varphi(x)]^{1+2\alpha} |u(x) - v(x)|^{2\alpha} \, dx \\ &\leq c_{\alpha}^{1-\alpha} \, \|u - v\|^{2\alpha}_{\varphi} \end{aligned}$$

in case $0 < \alpha < 1$ and

$$\|F(u) - F(v)\|^2_{\varphi} \leq c_1^2 \, \|u - v\|^2_{\varphi}$$

in case $\alpha = 1$. In particular, for $v = 0$, we have

$$\|F(u) - F(0)\|_{\varphi} \leq \text{const} \, \|u\|^{\alpha}_{\varphi} \, ,$$

which together with $\gamma(.,0) \in \mathbf{L}^2_{\varphi}$ implies $F(u) \in \mathbf{L}^2_{\varphi}$ for all $u \in \mathbf{L}^2_{\varphi}$. The Lemma is proved. □

Corollary 2.4 ([18], Theorem 26.A). *If the assumptions (A) and (B) are fulfilled, then the operator A is also coercive and equation (2.1) has a unique solution in $\mathbf{X}$ for each $f \in \mathbf{X}^*$ and $\varepsilon > 0$.*

Corollary 2.5. *Let the assumptions (A) and (B) be fulfilled. Moreover, let $f \in \mathbf{L}^{2,s}_{\varphi}$ for some $s > 0$. If $u \in \mathbf{L}^{2,1}_{\varphi}$ implies $F(u) \in \mathbf{L}^{2,s}_{\varphi}$, then the unique solution $u^* \in \mathbf{X}$ of equation (2.1) belongs to $\mathbf{L}^{2,s+1}_{\varphi}$.*

Proof. Since $u^* \in \mathbf{X}$ and $f \in \mathbf{L}^{2,s}_{\varphi} \supset \mathbf{L}^2_{\varphi}$ we have, due to $F(u) \in \mathbf{L}^2_{\varphi}$ (see Lemma 2.3), also $Vu^* \in \mathbf{L}^2_{\varphi}$, which implies $u^* \in \mathbf{L}^{2,1}_{\varphi}$. Consequently, $F(u^*) \in \mathbf{L}^{2,s}_{\varphi}$ and so $Vu^* \in \mathbf{L}^{2,s}_{\varphi}$ and $u^* \in \mathbf{L}^{2,s+1}_{\varphi}$. □

The previous corollary can be generalized in the following way.

Corollary 2.6. *Let the assumptions (A) and (B) be fulfilled. Moreover, let $f \in \mathbf{L}^{2,s}_{\varphi}$ for some $s > 1$. If there is a $t_0 \in (0,1]$ such that $u \in \mathbf{L}^{2,1}_{\varphi}$ implies $F(u) \in \mathbf{L}^{2,t_0}_{\varphi}$ and such that $u \in \mathbf{L}^{2,t_0+r}_{\varphi}$ implies $F(u) \in \mathbf{L}^{2,\min\{t_0+r,s\}}_{\varphi}$ for $r = 1, 2, \dots$, then the solution u^* of equation (2.1) belongs to $\mathbf{L}^{2,s+1}_{\varphi}$.*

Proof. As in the proof of Cor. 2.5 we get $u \in \mathbf{L}^{2,1}_{\varphi} \supset \mathbf{L}^{2,t_0}_{\varphi}$. This implies $Vu^* \in \mathbf{L}^{2,t_0}_{\varphi}$ and $u^* \in \mathbf{L}^{2,t_0+1}_{\varphi}$, and so on. □

Remark 2.7. If $\gamma(x,g)$ fulfils assumption (B) and if the partial derivatives $\gamma_x(x,g)$ and $\gamma_g(x,g)$ are continuous functions on $(-1,1)\times\mathbb{R}$ satisfying

$$|\gamma_x(x,g)|^2 \le \frac{\mathrm{const}}{(1-x^2)^2}\left[(1-x^2)^{\delta-\frac{1}{2}}+g^2\right] \quad \text{and} \quad |\gamma_g(x,g)| \le \mathrm{const}$$

for some $\delta>0$, then $u\in\mathbf{L}^{2,1}_\varphi$ implies $F(u)\in\mathbf{L}^{2,1}_\varphi$.

Proof. As a result of [1] (see pp. 196,197) we have that the condition $u\in\mathbf{L}^{2,1}_\varphi$ is equivalent to $u\in\mathbf{L}^2_\varphi$ and $u'\in\mathbf{L}^2_{\varphi^3}$. Due to Lemma 2.3 it remains to show that $u\in\mathbf{L}^{2,1}_\varphi$ implies $h'\in\mathbf{L}^2_{\varphi^3}$, where

$$h(x)=\gamma\left(x,u(x)\sqrt{1-x^2}\right).$$

By our assumptions it follows

$$\begin{aligned}
&|h'(x)|^2(1-x^2)^{\frac{3}{2}}\\
&\le \left|\gamma_x\left(x,u(x)\sqrt{1-x^2}\right)\right|^2(1-x^2)^{\frac{3}{2}}\\
&\quad+\left|\gamma_g\left(x,u(x)\sqrt{1-x^2}\right)\right|^2\left(|u'(x)|^2(1-x^2)+\frac{|x\,u(x)|^2}{1-x^2}\right)(1-x^2)^{\frac{3}{2}}\\
&\le \mathrm{const}\left[(1-x^2)^{\delta-1}+|u(x)|^2(1-x^2)^{\frac{1}{2}}+|u'(x)|^2(1-x^2)^{\frac{5}{2}}\right],
\end{aligned}$$

thus $|h'(x)|^2(1-x^2)^{\frac{3}{2}}$ is summable. □

Remark 2.8. Based on some results of [17], in [13] it is shown that, for any $f\in\mathbf{L}^q_\mu$, problem (1.1), (1.2) has a unique solution $g\in\mathbf{L}^p_\tau$ with $g'\in\mathbf{L}^q_\mu$ if $\gamma(x,g)$ is non-decreasing in $g\in\mathbb{R}$ for almost all $t\in(-1,1)$, if

$$|\gamma(x,g)|\le A(x)+B\,\sigma(x)|g|^{p-1},$$

where $p>1$, $A\in\mathbf{L}^q_\mu$, $B>0$, and, if $p>2$,

$$g\,\gamma(x,g)\ge C\,\sigma(x)|g|^p-D(x),$$

where $D\in\mathbf{L}^1$ and $C>0$.

Here, the norm in $\mathbf{L}^p_\psi$ is defined by

$$\|g\|_{\mathbf{L}^p_\psi}=\left(\int_{-1}^1|g(x)|^p\psi(x)\,dx\right)^{\frac{1}{p}},$$

$p^{-1}+q^{-1}=1$, and the weight functions are chosen as

$$\sigma(x)=\left(1-x^2\right)^{-\frac{1}{2}},\quad \mu(x)=\left(1-x^2\right)^{\frac{q-1}{2}}.$$

Note that, by means of the substitution $x=\cos\tau$, problem (1.1), (1.2) can be written equivalently as

$$-\frac{\varepsilon\sin\tau}{\pi}\int_0^\pi\frac{\widetilde{g}'(\tau)\,d\sigma}{\cos\tau-\cos\sigma}+\widetilde{\gamma}(\tau,\widetilde{g}(\tau))=\widetilde{f}(\tau),\ 0<\tau<\pi, \tag{2.5}$$

with $\widetilde{g}(0) = \widetilde{g}(\pi) = 0$, where $\widetilde{g}(\tau) = g(\cos\tau)$, $\widetilde{\gamma}(\tau,\widetilde{g}) = \gamma(\cos\tau,\widetilde{g})\sin\tau$, and $\widetilde{f}(\tau) = f(\cos\tau)\sin\tau$.

Corollary 2.9 ([13], Concl. 1). *Let $\widetilde{\gamma}(x,\widetilde{g})$ be a monotone Carathéodory function satisfying*

$$|\widetilde{\gamma}(\tau,\widetilde{g})| \le a(\tau) + B\,|\widetilde{g}|^{p-1} \quad \textit{and, if } p>1, \quad \widetilde{g}\,\widetilde{\gamma}(\tau,\widetilde{g}) \ge C\,|\widetilde{g}|^{p} - d(\tau)$$

with some $a \in \mathbf{L}^q$, $d \in \mathbf{L}^1$, and constants $B, C > 0$. Then, for any $\widetilde{f} \in \mathbf{L}^q$, problem (2.5) has a unique solution $\widetilde{g} \in \mathbf{L}^p$ with $\widetilde{g}' \in \mathbf{L}^q$, where $p > 1$, $p^{-1} + q^{-1} = 1$.

3. A collocation method

Denote by

$$x^{\varphi}_{nk} = \cos\frac{k\pi}{n+1}, \qquad k = 1,\dots,n,$$

the zeros of the nth orthonormal polynomial $p^{\varphi}_n(x)$. Let $\mathbf{X}_n$ denote the space of all algebraic polynomials of degree less than n and let L^{φ}_n be the Lagrange interpolation operator onto $\mathbf{X}_n$ with respect to the nodes x^{φ}_{nk}, $k = 1,\dots,n$. We recall that L^{φ}_n is defined by

$$L^{\varphi}_n(f;x) = \sum_{k=1}^{n} f(x^{\varphi}_{nk})\ell^{\varphi}_{nk}(x), \quad \ell^{\varphi}_{nk}(x) = \prod_{r=1,r\ne k}^{n} \frac{x - x^{\varphi}_{nr}}{x^{\varphi}_{nk} - x^{\varphi}_{nr}}.$$

Moreover, in order to prove the convergence of the collocation method, we recall a well-known result on the Lagrange interpolation.

Lemma 3.1 (see [1, 4, 8]). *For $s > \frac{1}{2}$ and for all $f \in \mathbf{L}^{2,s}_{\varphi}$, we have*

$$\lim_{n\to\infty} \|f - L^{\varphi}_n f\|_{\varphi,s} = 0$$

and

$$\|f - L^{\varphi}_n f\|_{\varphi,r} \le \text{const}\, n^{r-s}\|f\|_{\varphi,s}, \quad 0 \le r \le s.$$

We look for an approximate solution $u_n \in \mathbf{X}_n$ to the solution of equation (2.1) by solving the collocation equations

$$A_n(u_n) := \varepsilon V u_n + F_n(u_n) = L^{\varphi}_n f, \quad u_n \in \mathbf{X}_n, \tag{3.1}$$

where $F_n(u_n) := L^{\varphi}_n F(u_n)$.

Theorem 3.2. *Consider equation (2.1) for a function $f : (-1,1) \longrightarrow \mathbb{C}$. Assume that the conditions (A) and (B) are satisfied. Then the equations (3.1) have a unique solution $u^*_n \in \mathbf{X}_n$. If the solution $u^* \in \mathbf{X}$ of (2.1) belongs to $\mathbf{L}^{2,s+1}$ for some $s > \frac{1}{2}$, then the solutions u^*_n converge in $\mathbf{X}$ to u^*, where*

$$\|u^*_n - u^*\|_{\varphi,\frac{1}{2}} \le \text{const}\, n^{-s}\,\|u^*\|_{\varphi,s+1}$$

and the constant does not depend on n, ε, and u^*.

Proof. At first we observe that, for $u_n, v_n \in \mathbf{X}_n$,

$$\begin{aligned}&\langle F_n(u_n) - F_n(v_n), u_n - v_n\rangle_\varphi \\ &\quad = \sum_{k=1}^{n} \lambda_{nk}^\varphi \left[\gamma\left(x_{nk}^\varphi, \varphi(x_{nk}^\varphi)\xi_k\right) - \gamma\left(x_{nk}^\varphi, \varphi(x_{nk}^\varphi)\eta_k\right)\right](\xi_k - \eta_k) \ge 0,\end{aligned} \tag{3.2}$$

where $\lambda_{nk}^\varphi = \pi[\varphi(x_{nk}^\varphi)]^2/(n+1)$ are the Christoffel numbers with respect to the weight function $\varphi(x)$, and $\xi_k = u_n(x_{nk}^\varphi)$, $\eta_k = v_n(x_{nk}^\varphi)$. This implies the strong monotonicity of the operator $\varepsilon V + F_n : \mathbf{X}_n \subset \mathbf{X} \longrightarrow \mathbf{X}_n \subset \mathbf{X}^*$. Moreover, the estimation (here let $0 < \alpha < 1$; for the case $\alpha = 1$ see the proof of Cor. 3.4)

$$\begin{aligned}\|F_n(u_n) - F_n(v_n)\|_\varphi^2 &= \sum_{k=1}^{n} \lambda_{nk}^\varphi \left|\gamma\left(x_{nk}^\varphi, \varphi(x_{nk}^\varphi)\xi_k\right) - \gamma\left(x_{nk}^\varphi, \varphi(x_{nk}^\varphi)\eta_k\right)\right|^2 \\ &\le \sum_{k=1}^{n} \lambda_{nk}^\varphi [\lambda(x_{nk}^\varphi)]^2 [\varphi(x_{nk}^\varphi)]^{2\alpha} |\xi_k - \eta_k|^{2\alpha} \\ &\le \left(\sum_{k=1}^{n} \left((\lambda_{nk}^\varphi)^{1-\alpha} [\lambda(x_{nk}^\varphi)]^2 [\varphi(x_{nk}^\varphi)]^{2\alpha}\right)^{\frac{1}{1-\alpha}}\right)^{1-\alpha} \left(\sum_{k=1}^{n} \lambda_{nk}^\varphi |\xi_k - \eta_k|^2\right)^{\alpha} \\ &=: c_{n\alpha} \|u_n - v_n\|_\varphi^{2\alpha}\end{aligned}$$

gives the Hölder continuity of $F_n : \mathbf{X}_n \subset \mathbf{L}_\varphi^2 \longrightarrow \mathbf{X}_n \subset \mathbf{L}_\varphi^2$. Thus, equation (3.1) is uniquely solvable (comp. Cor. 2.4). With the help of

$$L_n^\varphi F(u^*) = L_n^\varphi F(L_n^\varphi u^*) = F_n(L_n^\varphi u^*)$$

we can estimate

$$\begin{aligned}&\varepsilon \|u_n^* - L_n^\varphi u^*\|_{\varphi,\frac{1}{2}}^2 \\ &\quad\le \langle \varepsilon V u_n^* + F_n(u_n^*) - \varepsilon V L_n^\varphi u^* - F_n(L_n^\varphi u^*), u_n^* - L_n^\varphi u^*\rangle_\varphi \\ &\quad= \langle L_n^\varphi f - \varepsilon V L_n^\varphi u^* - L_n^\varphi F(u^*), u_n^* - L_n^\varphi u^*\rangle_\varphi \\ &\quad= \varepsilon \langle L_n^\varphi V u^* - V L_n^\varphi u^*, u_n^* - L_n^\varphi u^*\rangle_\varphi \\ &\quad\le \varepsilon \left(\|L_n^\varphi V u^* - V u^*\|_{\varphi,-\frac{1}{2}} + \|u^* - L_n^\varphi u^*\|_{\varphi,\frac{1}{2}}\right) \|u_n^* - L_n^\varphi u^*\|_{\varphi,\frac{1}{2}}.\end{aligned}$$

Hence, taking into account $\|L_n^\varphi V u^* - V u^*\|_{\varphi,-\frac{1}{2}} \le \|L_n^\varphi V u^* - V u^*\|_\varphi$ and Lemma 3.1, we obtain

$$\|u_n^* - L_n^\varphi u^*\|_{\varphi,\frac{1}{2}} \le \text{const}\left(n^{-s} \|Vu^*\|_{\varphi,s} + n^{-s-\frac{1}{2}} \|u^*\|_{\varphi,s+1}\right),$$

and the theorem is proved. □

Corollary 3.3. *Additionally to the assumptions of Theorem 3.2 assume that there is an $r \ge \frac{1}{2}$ such that*

$$\langle F_n(u_n) - F_n(v_n), u_n - v_n\rangle_{\varphi,r} \ge 0, \quad u_n, v_n \in \mathbf{X}_n, \ n \ge n_0, \tag{3.3}$$

and such that $s \geq r - \frac{1}{2}$. *Then*

$$\|u_n^* - u^*\|_{\varphi, r+\frac{1}{2}} \leq \text{const}\, n^{r-\frac{1}{2}-s} \|u^*\|_{\varphi, s+1}, \tag{3.4}$$

where the constant does not depend on n, ε, *and* u^*.

Proof. Using (2.4), (2.3), and Lemma 3.1 we get, analogously to the proof of Theorem 3.2,

$$\begin{aligned}
&\|u_n^* - L_n^\varphi u^*\|_{\varphi, r+\frac{1}{2}}^2 \\
&\leq \langle L_n^\varphi V u^* - V L_n^\varphi u^*, u_n^* - L_n^\varphi u^* \rangle_{\varphi, r} \\
&\leq \left(\|L_n^\varphi V u^* - V u^*\|_{\varphi, r-\frac{1}{2}} + \|V(u^* - L_n^\varphi u^*\|_{\varphi, r-\frac{1}{2}} \right) \|u_n^* - L_n^\varphi u^*\|_{\varphi, r+\frac{1}{2}} \\
&\leq \text{const} \left(n^{r-\frac{1}{2}-s} \|V u^*\|_{\varphi, s} + n^{r+\frac{1}{2}-s-1} \|u^*\|_{\varphi, s+1} \right) \|u_n^* - L_n^\varphi u^*\|_{\varphi, r+\frac{1}{2}}.
\end{aligned}$$

Since, again due to Lemma 3.1, $\|u^* - L_n^\varphi u^*\|_{\varphi, r+\frac{1}{2}} \leq \text{const}\, n^{r-\frac{1}{2}-s} \|u^*\|_{\varphi, s+1}$ the assertion is proved. □

Let us discuss the question how we can solve the collocation equations (3.1). Although Theorem 3.2 holds for all $0 < \alpha \leq 1$, for the following we need to assume $\alpha = 1$.

Corollary 3.4. *If condition (B) with* $\alpha = 1$ *is fulfilled, then the operator* $A : \mathbf{X} \longrightarrow \mathbf{X}^*$ *as well as the operator* $A_n : \mathbf{X}_n \subset \mathbf{X} \longrightarrow \mathbf{X}_n \subset \mathbf{X}^*$ *are Lipschitz continuous with constant* $c_1 + \varepsilon$.

Proof. We give the proof for the operator A_n. The proof for A is analogous. Let $u_n^1, u_n^2, u_n \in \mathbf{X}_n$. Then

$$\begin{aligned}
&\left| \left\langle F_n(u_n^1) - F_n(u_n^2), u_n \right\rangle_\varphi \right| \\
&\leq \sum_{k=1}^n \lambda_{nk}^\varphi \left| \gamma\left(x_{nk}^\varphi, \varphi(x_{nk}^\varphi) u_n^1(x_{nk}^\varphi)\right) - \gamma\left(x_{nk}^\varphi, \varphi(x_{nk}^\varphi) u_n^2(x_{nk}^\varphi)\right) \right| |u_n(x_{nk}^\varphi)| \\
&\leq \sum_{k=1}^n \lambda_{nk}^\varphi \lambda(x_{nk}^\varphi) \varphi(x_{nk}^\varphi) \left| u_n^1(x_{nk}^\varphi) - u_n^2(x_{nk}^\varphi) \right| |u_n(x_{nk}^\varphi)| \\
&\leq c_1 \left\| u_n^1 - u_n^2 \right\|_\varphi \|u_n\|_\varphi \leq c_1 \left\| u_n^1 - u_n^2 \right\|_{\varphi, \frac{1}{2}} \|u_n\|_{\varphi, \frac{1}{2}},
\end{aligned}$$

and we are done. □

Remark that V is just the dual mapping between the spaces $\mathbf{X}$ and $\mathbf{X}^*$ as well as between $\mathbf{X}_n \subset \mathbf{X}$ and $\mathbf{X}_n \subset \mathbf{X}^*$. Hence, we consider (for some fixed $t > 0$) the equations

$$u_n = u_n - tV^{-1}\left[\varepsilon V u_n + F_n(u_n) - L_n^\varphi f\right] =: B_n(u_n), \tag{3.5}$$

which are equivalent to (3.1). If we choose $t \in (0, t_\varepsilon)$ with $t_\varepsilon = 2\varepsilon/(c_1^2 + \varepsilon^2)$ then the operator $B_n : \mathbf{X}_n \subset \mathbf{X} \longrightarrow \mathbf{X}_n \subset \mathbf{X}$ is a k_ε-contractive mapping with $k_\varepsilon = \sqrt{(1-t\varepsilon)^2 + t^2 c_1^2} < 1$, i.e.,

$$\left\|B_n(u_n^1) - B_n(u_n^2)\right\|_{\varphi,\frac{1}{2}} \le k_\varepsilon \left\|u_n^1 - u_n^2\right\|_{\varphi,\frac{1}{2}} . \tag{3.6}$$

This follows from

$$\begin{aligned}
&\left\|B_n(u_n^1) - B_n(u_n^2)\right\|_{\varphi,\frac{1}{2}}^2 \\
&= (1-t\varepsilon)^2 \left\|u_n^1 - u_n^2\right\|_{\varphi,\frac{1}{2}}^2 - 2t\left\langle V^{-1}[F(u_n^1) - F_n(u_n^2)], u_n^1 - u_n^2\right\rangle_{\varphi,\frac{1}{2}} \\
&\quad + t^2 \left\|V^{-1}[F(u_n^1) - F_n(u_n^2)]\right\|_{\varphi,\frac{1}{2}}^2 \\
&= (1-t\varepsilon)^2 \left\|u_n^1 - u_n^2\right\|_{\varphi,\frac{1}{2}}^2 - 2t\left\langle F(u_n^1) - F_n(u_n^2), u_n^1 - u_n^2\right\rangle_{\varphi} \\
&\quad + t^2 \left\|V^{-1}[F(u_n^1) - F_n(u_n^2)]\right\|_{\varphi,\frac{1}{2}}^2 \\
&\le k_\varepsilon^2 \left\|u_n^1 - u_n^2\right\|_{\varphi,\frac{1}{2}}^2
\end{aligned}$$

taking into account (3.2). Consequently, under the assumptions (A) and (B) with $\alpha = 1$ the collocation equations (3.1) can be solved by applying the method of successive approximation to the fixpoint equation (3.5). The smallest possible k_ε for given ε and c_1 is equal to

$$k_\varepsilon^* = \frac{c_1}{\sqrt{\varepsilon^2 + c_1^2}} \quad \left(\Leftrightarrow \quad t = t_\varepsilon^* = \frac{\varepsilon}{\varepsilon^2 + c_1^2} \right). \tag{3.7}$$

Remark that, if we directly apply the formulas of [18, Sect. 25.4] to the fixed point equation (3.5) we obtain, for $t \in (0, \widetilde{t}_\varepsilon)$ with $\widetilde{t}_\varepsilon = 2\varepsilon/(\varepsilon + c_1)^2$, the contraction constant $\widetilde{k}_\varepsilon = \sqrt{1 - 2t\varepsilon + t^2(\varepsilon + c_1)^2}$, whose minimal value is

$$\widetilde{k}_\varepsilon^* = \sqrt{1 - \left(\frac{\varepsilon}{\varepsilon + c_1}\right)^2} \quad \left(\Leftrightarrow \quad t = \widetilde{t}_\varepsilon^* = \frac{\varepsilon}{(\varepsilon + c_1)^2} \right).$$

If we seek the solution of (3.1) in the form

$$u_n(x) = \sum_{k=1}^{n} \xi_{nk} \ell_{nk}^{\varphi}(x) ,$$

then (3.1) can be written as

$$\varepsilon\, \mathbf{V}_n \mathbf{\Lambda}_n \xi_n + \mathbf{F}_n(\xi_n) = \eta_n , \quad \xi_n = [\xi_{nk}]_{k=1}^{n} , \tag{3.8}$$

where $\eta_n = [f(x_{nk}^{\varphi})]_{k=1}^{n}$, $\mathbf{V}_n = \mathbf{U}_n^T \mathbf{D}_n \mathbf{U}_n$, $\mathbf{U}_n = [p_j^{\varphi}(x_{nk}^{\varphi})]_{j=0,k=1}^{n-1,\,n}$, and $\mathbf{D}_n = \operatorname{diag}[1, \dots, n]$, $\mathbf{\Lambda}_n = \operatorname{diag}[\lambda_{n1}^{\varphi}, \dots, \lambda_{nn}^{\varphi}]$, $\mathbf{F}_n(\xi_n) = [\gamma(x_{nk}^{\varphi}, \varphi(x_{nk}^{\varphi})\xi_{nk})]_{k=1}^{n}$ (see [2, (4.12)]). We recall that, due to the orthogonality relations of p_j^{φ}, we have

$$\mathbf{U}_n \mathbf{\Lambda}_n \mathbf{U}_n^T = \mathbf{I}_n =: [\delta_{jk}]_{j,k=1}^{n} . \tag{3.9}$$

Thus, the fixed point iteration for (3.5) takes the form

$$\xi_n^{(m+1)} = (1-t\varepsilon)\xi_n^{(m)} - t\mathbf{\Lambda}_n^{-1}\mathbf{V}_n^{-1}\left[\mathbf{F}_n(\xi_n^{(m)}) - \eta_n\right], \quad m = 0,1,\dots, \tag{3.10}$$

where, taking into account (3.9),

$$\mathbf{\Lambda}_n^{-1}\mathbf{V}_n^{-1} = \mathbf{\Lambda}_n^{-1}\mathbf{U}_n^{-1}\mathbf{D}_n^{-1}(\mathbf{U}_n^T)^{-1} = \mathbf{U}_n^T\mathbf{D}_n^{-1}\mathbf{U}_n\mathbf{\Lambda}_n\,.$$

Remark that the matrix $\mathbf{U}_n$ can be written as

$$\mathbf{U}_n = \widetilde{\mathbf{U}}_n\widetilde{\mathbf{D}}_n^{-1}, \quad \widetilde{\mathbf{U}}_n = \sqrt{\frac{2}{\pi}}\left[\sin\frac{jk\pi}{n+1}\right]_{j,k=1}^n, \quad \widetilde{\mathbf{D}}_n = \operatorname{diag}\left[\sin\frac{k\pi}{n+1}\right]_{k=1}^n,$$

and that the matrix $\widetilde{\mathbf{U}}_n$ can be applied to a vector with $O(n\log n)$ computational complexity (see [15, 16]).

Remark 3.5. In [13] the approximate solution is represented in the form

$$\widetilde{v}_n(\tau) = \varphi(\cos\tau)\,u_n(\cos\tau) = \sin\tau\,u_n(\cos\tau)\,.$$

With $\widetilde{\zeta}_{nk} = \widetilde{v}_n\left(\frac{k\pi}{n+1}\right)$ we get $\widetilde{\zeta}_n = \widetilde{\mathbf{D}}_n\xi_n$, such that (3.8) (in case $\varepsilon = 1$) is equivalent to (use $\mathbf{\Lambda}_n\widetilde{\mathbf{D}}_n^{-1} = \frac{\pi}{n+1}\widetilde{\mathbf{D}}_n$)

$$\varepsilon\mathbf{A}_n\widetilde{\zeta}_n + \mathbf{\Phi}_n(\widetilde{\zeta}_n) = \widetilde{\eta}_n\,, \tag{3.11}$$

where

$$\mathbf{A}_n = [\alpha_{jk}^n]_{j,k=1}^n = \frac{\pi}{n+1}\widetilde{\mathbf{U}}_n^T\mathbf{D}_n\widetilde{\mathbf{U}}_n = \frac{2}{n+1}\left[\sum_{\ell=1}^n \ell\sin\frac{j\ell\pi}{n+1}\sin\frac{k\ell\pi}{n+1}\right]_{j,k=1}^n,$$

$$\mathbf{\Phi}_n(\widetilde{\zeta}_n) = \left[\mathbf{\Phi}_{nj}(\widetilde{\zeta}_{nj})\right]_{j=1}^n = \left[\varphi(x_{nj}^\varphi)\,\gamma\left(x_{nj}^\varphi, \frac{\widetilde{\zeta}_{nj}}{\varphi(x_{nj}^\varphi)}\right)\right]_{j=1}^n, \quad \widetilde{\eta}_n = \widetilde{\mathbf{D}}_n\eta_n\,.$$

In [13] it is shown that the sequence $\left\{\widetilde{\zeta}_n^m\right\}_{m=1}^\infty$ defined by the nonlinear Jacobi iteration

1. solve $\quad \alpha_{jj}^n\widetilde{\zeta}_{nj} + \mathbf{\Phi}_{nj}(\widetilde{\zeta}_{nj}) = \widetilde{\eta}_{nj} - \sum\limits_{k=1,k\neq j}^n \alpha_{jk}^n\widetilde{\zeta}_{nk}^m\,,\ j = 1,\dots,n\,,$
2. set $\quad \widetilde{\zeta}_j^{m+1} = \widetilde{\zeta}_j^m + \omega(\widetilde{\zeta}_j - \widetilde{\zeta}_j^m)\,,\ j = 1,\dots,n\,,$

where ω is taken from the interval $(0,1]$, converges to the unique solution $\widetilde{\zeta}_n^*$ of (3.11) for any $\widetilde{\zeta}_n^0 \in \mathbb{R}^n$, if $\gamma(x,g)$ fulfills condition (A), is continuous in $g \in \mathbb{R}$ for all $x \in [-1,1]$ and bounded in $x \in [-1,1]$ for all $g \in \mathbb{R}$, and if $f(x)$ is bounded in $x \in [-1,1]$.

Instead of (3.1), (3.5) let us consider the collocation-iteration scheme

$$\widetilde{u}_m = L_{n_m}^\varphi\left[\widetilde{u}_{m-1} - tV^{-1}\left(\varepsilon V\widetilde{u}_{m-1} + F(\widetilde{u}_{m-1}) - f\right)\right], \quad m = 1,2,\dots,$$

where $1 < n_0 < n_1 < n_2 < \cdots$ and $\widetilde{u}_0 \equiv 0$. This is equivalent to

$$\widetilde{u}_m = \widetilde{u}_{m-1} - tV^{-1}\left[\varepsilon V\widetilde{u}_{m-1} + F_{n_m}(\widetilde{u}_{m-1}) - L_{n_m}^\varphi f\right] =: T_m(\widetilde{u}_{m-1}) \tag{3.12}$$

with $T_m u = B_{n_m}(P_{n_m} u)$, where $P_n : \mathbf{L}_\varphi^{2,\frac{1}{2}} \longrightarrow \mathbf{L}_\varphi^{2,\frac{1}{2}}$ denotes the projection

$$P_n u = \sum_{k=0}^{n-1} \langle u, p_k^\varphi \rangle_\varphi \, p_k^\varphi .$$

Due to $\|P_n\|_{\mathbf{L}_\varphi^{2,\frac{1}{2}} \to \mathbf{L}_\varphi^{2,\frac{1}{2}}} = 1$ and (3.6) the operator $T_m : \mathbf{L}_\varphi^{2,\frac{1}{2}} \longrightarrow \mathbf{L}_\varphi^{2,\frac{1}{2}}$ is a k_ε-contractive operator. Thus, the equation

$$v_m = T_m(v_m) \,, \quad v_m \in \mathbf{L}_\varphi^{2,\frac{1}{2}} \,,$$

has a unique solution v_m^*, which is nothing else then the solution of the collocation method (3.1) for $n = n_m$. Hence, under the assumptions of Theorem 3.2 with (B) in the case $\alpha = 1$,

$$\|v_m^* - u^*\|_{\varphi, \frac{1}{2}} = O\left(n_m^{-s}\right) . \tag{3.13}$$

Now, for $m = 2, 3, \dots$,

$$\begin{aligned} \|\widetilde{u}_m - v_m^*\| &\leq \|T_m(\widetilde{u}_{m-1}) - T_m(v_m^*)\| \\ &\leq k_\varepsilon \|\widetilde{u}_{m-1} - v_m^*\| \\ &\leq k_\varepsilon \left(\left\|\widetilde{u}_{m-1} - v_{m-1}^*\right\| + \left\|v_{m-1}^* - v_m^*\right\|\right) \end{aligned}$$

and, by repeating this,

$$\|\widetilde{u}_m - v_m^*\| \leq k_\varepsilon^{m-1} \|\widetilde{u}_1 - v_1^*\| + \sum_{\ell=1}^{m-1} k_\varepsilon^{m-\ell} \left\|v_\ell^* - v_{\ell+1}^*\right\| .$$

Taking into account the Toeplitz convergence theorem (comp. [18, Prop. 25.1, Problem 25.1]) we get $\|\widetilde{u}_m - v_m^*\| \longrightarrow 0$, which implies together with (3.13) the convergence of $\widetilde{u}_m$ to u^*. Of course, these ideas apply also if we use the iteration $\widetilde{u}_m = T_m^r(\widetilde{u}_{m-1})$ for some fixed integer $r > 1$ instead of (3.12). (For more details concerning projection-iteration methods, we refer the reader to [18, Sect.s 25.1, 25.2].) Practically we can write the collocation-iteration (3.12) in the form

$$\widetilde{\xi}_m = (1 - t\,\varepsilon)\mathbf{E}_m \widetilde{\xi}_{m-1} - t\mathbf{\Lambda}_{n_m}^{-1} \mathbf{V}_{n_m}^{-1} \left[\mathbf{F}_{n_m}(\mathbf{E}_m \widetilde{\xi}_{m-1}) - \eta_{n_m}\right] , \quad \widetilde{\xi}_0 = 0 \,, \tag{3.14}$$

where $\mathbf{E}_m = \mathbf{U}_{n_m}^T \mathbf{P}_{n_m n_{m-1}} \mathbf{U}_{n_{m-1}} \mathbf{\Lambda}_{n_{m-1}}$ and $\mathbf{P}_{nm} = [\delta_{jk}]_{j=1,k=1}^{n\ ,\ m}$. Since the fast transformations $\mathbf{U}_n$ can be realized most effectively for particular n, for example $n = 2^r - 1$, $r \in \mathbb{N}$ (comp. the examples in Section 5), an appropriate way to use the idea of (3.14) is to combine it with the method (3.10):

1. Choose a finite sequence $n_1 < n_2 < \cdots < n_{M+1}$ of natural numbers and a natural number K.
2. For $m = 1, \dots, M$ do K iterations of the form (3.10) on level $n = n_m$ and use (3.14) to get a good initial approximation $u_{n_{m+1}}^{(0)}$ for (3.10) on level n_{m+1}.
3. Apply (3.10) with the initial approximation $u_{n_{M+1}}^{(0)}$ till the desired accuracy is achieved.

4. Collocation for the case (1.7)

We can try to apply the collocation method described in Section 3 to example (1.7) (comp. Theorem 3.2). Of course, condition (A) is fulfilled. Moreover, if

$$\int_{-1}^{1} [\Gamma(x)]^4 [\varphi(x)]^3 \, dx < \infty$$

then (B) is satisfied with $\alpha = \frac{1}{2}$ and $u \in \mathbf{L}^2_\varphi$ implies $F(u) \in \mathbf{L}^2_\varphi$ (see Lemma 2.3). But, the conditions of Remark 2.7 are not satisfied such that the convergence rate established in Theorem 3.2 cannot be proved.

Let us consider a little bit more general example

$$\gamma(x,g) = \Gamma(x)\,|g|^\alpha \operatorname{sgn} g\,, \quad |x| \le 1,\, g \in \mathbb{R}\,, \tag{4.1}$$

with $0 < \alpha < 1$ and a continuous function $\Gamma : [-1,1] \longrightarrow (0,\infty)$. The problem in the application of the collocation method (3.1) together with an iteration scheme (3.5) or (3.12) is that condition (B) with $\alpha = 1$ is not satisfied. To overcome this difficulty we transform the respective equation (1.1) with $\gamma(x,g)$ defined in (4.1) as follows: Define a new unknown function $\widetilde{g}(x) = [\Gamma(x)]^{\delta\alpha} |g(x)|^\alpha \operatorname{sgn} g(x)$, where $\delta = (1+\alpha)^{-1}$. Then, equation (1.1) with (4.1) is equivalent to

$$-\frac{\varepsilon}{\pi} \int_{-1}^{1} \frac{|\widetilde{g}(y)|^\beta \widetilde{g}(y)}{[\Gamma(y)]^\delta (y-x)^2} \, dy + [\Gamma(x)]^\delta \widetilde{g}(x) = f(x)\,, \ |x| < 1\,, \quad \widetilde{g}(\pm 1) = 0\,, \tag{4.2}$$

where $\beta := \alpha^{-1} - 1 > 0$. In [14, Sect. 1]) there is proved that the solution of

$$-(L\,g)(x) = -(D\,g')(x) = f(x)\,, \ -1 < x < 1\,, \quad g(\pm 1) = 0$$

(comp. (1.3)) is given by the formula

$$g(x) = \frac{1}{\pi} \int_{-1}^{1} h(x,y) f(y) \, dy\,,$$

where

$$h(x,y) = \ln \frac{1 - xy + \sqrt{(1-x^2)(1-y^2)}}{|y-x|}\,.$$

Thus, equation (4.2) can be written in the form

$$\varepsilon |\widetilde{g}(x)|^\beta \widetilde{g}(x) + \frac{[\Gamma(x)]^\delta}{\pi} \int_{-1}^{1} h(x,y) [\Gamma(y)]^\delta \widetilde{g}(y)] \, dy = \widetilde{f}(x)\,, \tag{4.3}$$

with

$$\widetilde{f}(x) = \frac{[\Gamma(x)]^\delta}{\pi} \int_{-1}^{1} h(x,y) f(y) \, dy\,.$$

Taking into account, for $x, y \in (-1,1)$ and $x \neq y$,

$$\begin{aligned}\frac{\partial h(x,y)}{\partial y} &= -\frac{1}{y-x} + \frac{-x\sqrt{1-y^2} - y\sqrt{1-x^2}}{\sqrt{1-y^2}\left(1 - xy - \sqrt{(1-x^2)(1-y^2)}\right)} \\ &= -\frac{\sqrt{1-y^2} - x^2\sqrt{1-y^2} - xy\sqrt{1-x^2} + \sqrt{1-x^2}}{(y-x)\sqrt{1-y^2}\left(1 - xy - \sqrt{(1-x^2)(1-y^2)}\right)} \\ &= -\frac{\sqrt{1-x^2}}{(y-x)\sqrt{1-y^2}},\end{aligned}$$

by partial integration we get

$$\begin{aligned}(Hf)(x) &:= \frac{1}{\pi}\int_{-1}^{1} h(x,y) f(y)\, dy = \frac{\sqrt{1-x^2}}{\pi}\int_{-1}^{1} \frac{\int_{-1}^{y} f(t)\, dt}{\sqrt{1-y^2}(y-x)}\, dy \\ &= \frac{\sqrt{1-x^2}}{\pi}\int_{-1}^{1} \frac{\int_{-1}^{y} f(t)\, dt - \frac{1+y}{2}\int_{-1}^{1} f(t)\, dt}{\sqrt{1-y^2}\,(y-x)}\, dy \\ &\qquad + \frac{\sqrt{1-x^2}}{2\pi}\int_{-1}^{1} \frac{(1+y)\, dy}{\sqrt{1-y^2}\,(y-x)} \int_{-1}^{1} f(t)\, dt \\ &= \frac{\sqrt{1-x^2}}{\pi}\int_{-1}^{1} \frac{\int_{-1}^{y} f(t)\, dt - \frac{1+y}{2}\int_{-1}^{1} f(t)\, dt}{\sqrt{1-y^2}\,(y-x)}\, dy \\ &\qquad + \frac{1}{2}\sqrt{1-x^2}\int_{-1}^{1} f(t)\, dt.\end{aligned}$$

For $0 < \mu < \frac{1}{2}$, define the Banach space $\mathcal{H}_0^\mu$ of all functions $u : (-1,1) \longrightarrow \mathbb{R}$ such that φu is Hölder continuous on $[-1,1]$ with exponent μ and $(\varphi u)(\pm 1) = 0$, where the norm in $\mathcal{H}_0^\mu$ is given by

$$\|u\|_{\mathcal{H}_0^\mu} = \|\varphi u\|_{\mathcal{H}^\mu}$$

with

$$\|f\|_{\mathcal{H}^\mu} := \|f\|_\infty + \sup\left\{\frac{|f(x_1) - f(x_2)|}{|x_1 - x_2|^\mu} : -1 \le x_1 < x_2 \le 1\right\},$$

$\|f\|_\infty = \sup\{|f(x)| : -1 \le x \le 1\}$. The Cauchy singular integral operator $D : \mathcal{H}_0^\mu \longrightarrow \mathcal{H}_0^\mu$ is bounded (see [5, Sect. 1.6]). Hence, we have

$$\|Hf\|_{\mathcal{H}^\mu} \le \text{const}\, \|f\|_{\mathbf{L}^{\frac{1}{1-\mu}}} \tag{4.4}$$

since $f \in L^{\frac{1}{1-\mu}}$ implies $\|F\|_{\mathcal{H}^\mu} \le \text{const}\, \|f\|_{\mathbf{L}^{\frac{1}{1-\mu}}}$ for $F(x) = \displaystyle\int_{-1}^{x} f(t)\, dt$. Since $\mathcal{H}^\mu$ is compactly embedded into the space of continuous functions, the operator $H : \mathbf{L}^p \longrightarrow \mathbf{L}^{q_0}$ is linear and compact for all $p > 1$ and $q_0 \ge 1$.

Due to [18, Prop. 26.7] the operator $G_\beta : \mathbf{L}^p \longrightarrow \mathbf{L}^q$ defined by

$$\Big(G_\beta(g)\Big)(x) = |g(x)|^\beta g(x)\,, \quad p = \beta + 2\,,\ p^{-1} + q^{-1} = 1\,,$$

is continuous and bounded with

$$\|G_\beta(g)\|_{\mathbf{L}^q} \le \|g\|_{\mathbf{L}^p}^{p-1}\,,$$

strictly monotone as well as coercive with

$$\langle G_\beta(g), g\rangle \ge \|g\|_{\mathbf{L}^p}^p\,.$$

Let f belong to $\mathbf{L}^p$ for some $p \ge 2$ and let $v = Hf$. Then $f \in \mathbf{L}^2_\varphi$ and $v = \varphi u = \varphi V^{-1} f$. Using (2.2), we obtain, for $f \ne 0$,

$$\left\langle \Gamma^\delta H \Gamma^\delta f, f\right\rangle = \left\langle V^{-1}\Gamma^\delta f, \Gamma^\delta f\right\rangle_\varphi = \sum_{n=0}^\infty \frac{1}{n+1}\left|\left\langle \Gamma^\delta f, p_n^\varphi\right\rangle_\varphi\right|^2 > 0\,. \tag{4.5}$$

Hence, defining $B : \mathbf{L}^p \longrightarrow \mathbf{L}^q$, $p = \beta + 2$, $p^{-1} + q^{-1} = 1$, by $B(g) = \varepsilon G_\beta(g) + \Gamma^\delta H \Gamma^\delta g$, we conclude that B is strictly monotone and coercive with

$$\langle B(g), g\rangle \ge \varepsilon\, \|g\|_{\mathbf{L}^p}^p\,.$$

Consequently, due to [18, Theorem 26.A] we have the following.

Theorem 4.1. *For each* $f \in \mathcal{L}^1 := \bigcup_{p>1} \mathbf{L}^p$, *there is a unique solution* $\widetilde{g} \in \mathbf{L}^{\beta+2}$ *of equation* (4.3)*.*

Moreover, the operator $G_\beta : \mathbf{L}^p \longrightarrow \mathbf{L}^q$, $p = \beta + 2$, $p^{-1} + q^{-1} = 1$, is uniformly monotone. Indeed, using $\beta > 0$ and the inequalities

$$\frac{x^{\beta+1} - 1}{(x-1)^{\beta+1}} > 1\,, \quad 1 < x < \infty\,,$$

and

$$\frac{x^{\beta+1} + 1}{(x+1)^{\beta+1}} \ge \min\left\{1, 2^{1-\beta}\right\} =: d_\beta\,, \quad 0 \le x < \infty\,,$$

we get

$$\left(|x|^\beta x - |y|^\beta y\right)(x - y) \ge d_\beta |x - y|^{\beta+2}\,, \quad x, y \in \mathbb{R}\,. \tag{4.6}$$

This implies

$$\langle G_\beta(g) - G_\beta(f), g - f\rangle \ge d_\beta\, \|g - f\|_{\mathbf{L}^p}^{\beta+2} = a\left(\|g - f\|_{\mathbf{L}^p}\right) \|g - f\|_{\mathbf{L}^p}$$

with $a(s) = d_\beta s^{\beta+1}$.

To solve equation (4.3) or, which is the same, the equation

$$B(\widetilde{g}) = \Gamma^\delta H f\,, \quad \widetilde{g} \in \mathbf{L}^{\beta+2}\,, \tag{4.7}$$

numerically, we consider the collocation method

$$B_n(\widetilde{g}_n) := \varepsilon G_{\beta,n}(\widetilde{g}_n) + H_n \widetilde{g}_n = M_n^\varphi \Gamma^\delta H L_n^\varphi f\,, \quad \widetilde{g}_n \in \mathbf{X}_n\,, \tag{4.8}$$

where

$$G_{\beta,n}(\widetilde{g}_n) = M_n^\varphi G_\beta(\widetilde{g}_n)\,, \quad H_n \widetilde{g}_n = M_n^\varphi \Gamma^\delta H L_n^\varphi \Gamma^\delta \widetilde{g}_n\,, \quad M_n^\varphi = \varphi L_n^\varphi \varphi^{-1} I\,.$$

In what follows, let $p = \beta + 2$, $p^{-1} + q^{-1} = 1$, and let $\mathbf{X}_n^p$ denote the space $\mathbf{X}_n$ of polynomials of degree less than n equipped with the $\mathbf{L}^p$-norm.

Lemma 4.2. *The operator $H_n : \mathbf{X}_n^p \longrightarrow \mathbf{X}_n^q$ is strictly monotone.*

Proof. Using the Gaussian rule w.r.t. the weight $\varphi(x)$ and the fact that $V^{-1}\widetilde{g}_n \in \mathbf{X}_n$ if $\widetilde{g}_n \in \mathbf{X}_n$, we get, for all $\widetilde{g}_n \in \mathbf{X}_n^p \setminus \{\Theta\}$,

$$\begin{aligned}\langle H_n\widetilde{g}_n, \widetilde{g}_n\rangle &= \left\langle L_n^\varphi \Gamma^\delta \varphi^{-1} H L_n^\varphi \Gamma^\delta \widetilde{g}_n, \widetilde{g}_n\right\rangle_\varphi \\ &= \sum_{k=1}^n \lambda_{nk}^\varphi \Gamma^\delta(x_{nk}^\varphi)\left(V^{-1}L_n^\varphi\Gamma^\delta\widetilde{g}_n\right)(x_{nk}^\varphi)\,\widetilde{g}_n(x_{nk}^\varphi) \\ &= \left\langle V^{-1}L_n^\varphi\Gamma^\delta\widetilde{g}_n, L_n^\varphi\Gamma^\delta\widetilde{g}_n\right\rangle_\varphi > 0\end{aligned}$$

taking into account (4.5). □

In what follows we need the existence of a constant $M_p > 1$ such that ([8, Theorems 2.7])

$$\|p_n\sigma\|_{\mathbf{L}^p} \le M_p \left(\sum_{k=1}^n \lambda_n(\sigma^p; x_{nk}^\varphi)|p_n(x_{nk}^\varphi)|^p\right)^{\frac{1}{p}} \tag{4.9}$$

for all $p_n \in \mathbf{X}_n$, where $\lambda_n(\sigma^p; x)$ denotes the Christoffel function w.r.t. the Jacobi weight $\sigma^p(x) = (1-x^2)^\eta$. Recall that this relation holds if and only if $\frac{\sigma}{\varphi} \in \mathbf{L}^p$, and $\frac{\varphi}{\sigma} \in \mathbf{L}^q$, i.e., $\frac{\beta}{2} < \eta < \frac{3\beta}{2} + 2$. Recall that ([12, Theorem 6.3.28])

$$\lambda_n(\sigma^p; x) \sim \frac{1}{n}\left(\sqrt{1-x} + \frac{1}{n}\right)^{2\eta+1}\left(\sqrt{1+x} + \frac{1}{n}\right)^{2\eta+1}. \tag{4.10}$$

Lemma 4.3. *Let $\frac{\beta}{2} < \eta < \frac{3\beta}{2} + 2$. Then the operators $G_{\beta,n} : \mathbf{X}_n^p \longrightarrow \mathbf{X}_n^q$ and $B_n : \mathbf{X}_n^p \longrightarrow \mathbf{X}_n^q$ are uniformly monotone with*

$$\left\langle G_{\beta,n}(\widetilde{g}_n) - G_{\beta,n}(\widetilde{f}_n), \widetilde{g}_n - \widetilde{f}_n\right\rangle \ge b\left(\left\|\widetilde{g}_n - \widetilde{f}_n\right\|_{\mathbf{L}^p}\right)\left\|\widetilde{g}_n - \widetilde{f}_n\right\|_{\mathbf{L}^p}$$

and

$$\left\langle B_n(\widetilde{g}_n) - B_n(\widetilde{f}_n), \widetilde{g}_n - \widetilde{f}_n\right\rangle \ge \varepsilon\, b\left(\left\|\widetilde{g}_n - \widetilde{f}_n\right\|_{\mathbf{L}^p}\right)\left\|\widetilde{g}_n - \widetilde{f}_n\right\|_{\mathbf{L}^p}, \quad \widetilde{f}_n, \widetilde{g}_n \in \mathbf{X}_n^p,$$

where $b(s) = \mathrm{const}\, n^{-2\eta} s^{\beta+1}$.

Proof. From (4.10) we obtain

$$\frac{\lambda_{nk}^\varphi}{\varphi(x_{nk}^\varphi)\lambda_n(\sigma^p; x_{nk}^\varphi)} \sim (1 - {x_{nk}^\varphi}^2)^{-\eta} > 1. \tag{4.11}$$

Thus, taking into account (4.6), (4.9) and (4.11),

$$\begin{aligned}
&\left\langle G_{\beta,n}(\widetilde{g}_n) - G_{\beta,n}(\widetilde{f}_n), \widetilde{g}_n - \widetilde{f}_n \right\rangle \\
&= \left\langle L_n^{\varphi}\varphi^{-1}\left[G_\beta(\widetilde{g}_n) - G_\beta(\widetilde{f}_n)\right], \widetilde{g}_n - \widetilde{f}_n \right\rangle_{\varphi} \\
&= \sum_{k=1}^{n} \frac{\lambda_{nk}^{\varphi}}{\varphi(x_{nk}^{\varphi})} \left[|\widetilde{g}_n(x_{nk}^{\varphi})|^{\beta}\widetilde{g}_n(x_{nk}^{\varphi}) - |\widetilde{f}_n(x_{nk}^{\varphi})|^{\beta}\widetilde{f}_n(x_{nk}^{\varphi})\right]\left[\widetilde{g}_n(x_{nk}^{\varphi}) - \widetilde{f}_n(x_{nk}^{\varphi})\right] \\
&\geq d_\beta \sum_{k=1}^{n} \frac{\lambda_{nk}^{\varphi}}{\varphi(x_{nk}^{\varphi})} \left|\widetilde{g}_n(x_{nk}^{\varphi}) - \widetilde{f}_n(x_{nk}^{\varphi})\right|^{\beta+2} \geq \mathrm{const} \left\|(\widetilde{g}_n - \widetilde{f}_n)\sigma\right\|_{\mathbf{L}^p}^{p} .
\end{aligned}$$

On the other hand, due to [12, Cor. 6.3.15] we have

$$\left\|(\widetilde{g}_n - \widetilde{f}_n)\sigma\right\|_{\mathbf{L}^p}^{p} \geq \mathrm{const}\, n^{-2\eta} \left\|\widetilde{g}_n - \widetilde{f}_n\right\|_{\mathbf{L}^p}^{p} .$$

Together with Lemma 4.2 we get the assertions. □

Corollary 4.4. *The operators* $G_{\beta,n} : \mathbf{X}_n^p \longrightarrow \mathbf{X}_n^q$ *and* $B_n : \mathbf{X}_n^p \longrightarrow \mathbf{X}_n^q$ *are coercive with*

$$\langle G_{\beta,n}(\widetilde{g}_n), \widetilde{g}_n\rangle \geq \mathrm{const}\, n^{-2\eta} \|\widetilde{g}_n\|_{\mathbf{L}^p}^{p} \quad \text{and} \quad \langle B_n(\widetilde{g}_n), \widetilde{g}_n\rangle \geq \mathrm{const}\, \varepsilon\, n^{-2\eta} \|\widetilde{g}_n\|_{\mathbf{L}^p}^{p} ,$$

$\widetilde{g}_n \in \mathbf{X}_n^p$.

Lemma 4.2, Lemma 4.3, and Cor. 4.4 state that we can preserve the essential properties of the operator of equation (4.7) for the operator of the collocation method (4.8).

Lemma 4.5. *For each function* $h : [-1,1] \longrightarrow \mathbb{R}$ *with* $\left\|\varphi^{\frac{1}{q}} h\right\|_{\infty} < \infty$ *and each polynomial* $\widetilde{g}_n \in \mathbf{X}_n$ *, we have*

$$\langle M_n^{\varphi} h, \widetilde{g}_n\rangle \leq \mathrm{const} \left\|\varphi^{\frac{1}{q}} h\right\|_{\infty} \|\widetilde{g}_n\|_{\mathbf{L}^p} ,$$

where the constant does not depend on h *,* $\widetilde{g}_n$ *, and* n *.*

Proof. Using the Gaussian rule as well as ([12, Theorem 9.25]), we can estimate

$$\begin{aligned}
\langle M_n^{\varphi} h, \widetilde{g}_n\rangle &= \left\langle L_n^{\varphi}\varphi^{-1} h, \widetilde{g}_n\right\rangle_{\varphi} = \sum_{k=1}^{n} \frac{\lambda_{nk}^{\varphi}}{\varphi(x_{nk}^{\varphi})} h(x_{nk}^{\varphi})\widetilde{g}_n(x_{nk}^{\varphi}) \\
&\leq \left(\sum_{k=1}^{n} \frac{\lambda_{nk}^{\varphi}}{\varphi^2(x_{nk}^{\varphi})} \left|\varphi^{\frac{1}{q}}(x_{nk}^{\varphi}) h(x_{nk}^{\varphi})\right|^{q}\right)^{\frac{1}{q}} \left(\sum_{k=1}^{n} \frac{\lambda_{nk}^{\varphi}}{\varphi(x_{nk}^{\varphi})} |\widetilde{g}_n(x_{nk}^{\varphi})|^{p}\right)^{\frac{1}{p}} \\
&\leq \mathrm{const} \left\|\varphi^{\frac{1}{q}} h\right\|_{\infty} \|\widetilde{g}_n\|_{\mathbf{L}^p} ,
\end{aligned}$$

and the lemma is proved. □

Theorem 4.6. *For each $n \in \mathbb{N}$, the collocation equation (4.8) has a unique solution $\widetilde{g}_n^* \in \mathbf{X}_n$. If $\widetilde{g} \in \mathbf{L}^p$ is the unique solution of (4.2) (comp. Theorem 4.1) and if $L_n^\varphi \widetilde{g}$ converges in $\mathbf{L}^p$ to $\widetilde{g}$, and if, for some $r > 1$ and $\eta \in \left(\frac{\beta}{2}, \frac{3\beta}{2} + 2\right)$,*

$$\lim_{n\to\infty} \left(\left\| L_n^\varphi \Gamma^\delta \widetilde{g} - \Gamma^\delta \widetilde{g} \right\|_{\mathbf{L}^r} + \| L_n^\varphi f - f \|_{\mathbf{L}^r} \right) n^{2\eta} = 0,$$

then $\widetilde{g}_n^*$ converges to $\widetilde{g}$, where

$$\|\widetilde{g}_n^* - L_n^\varphi \widetilde{g}\|_{\mathbf{L}^p} \le \left[\operatorname{const} \varepsilon^{-1} \left(\left\| L_n^\varphi \Gamma^\delta \widetilde{g} - \Gamma^\delta \widetilde{g} \right\|_{\mathbf{L}^r} + \| L_n^\varphi f - f \|_{\mathbf{L}^r} \right) n^{2\eta} \right]^{\frac{1}{p-1}} .$$

Proof. The unique solvability of (4.2) follows from [18, Theorem 26.A] and Lemma 4.3. Furthermore, with the help of Lemma 4.5 as well as the relations $L_n^\varphi \Gamma^\delta L_n^\varphi \widetilde{g} = L_n^\varphi \Gamma^\delta \widetilde{g}$ and $G_{\beta,n}(L_n^\varphi \widetilde{g}) = M_n^\varphi G_\beta(\widetilde{g})$ we get

$$\begin{aligned}
&\varepsilon\, b\,(\|\widetilde{g}_n^* - L_n^\varphi \widetilde{g}\|_{\mathbf{L}^p})\, \|\widetilde{g}_n^* - L_n^\varphi \widetilde{g}\|_{\mathbf{L}^p} \\
&\quad \le \langle B_n(\widetilde{g}_n^*) - B_n(L_n^\varphi \widetilde{g}), \widetilde{g}_n^* - L_n^\varphi \widetilde{g} \rangle \\
&\quad = \left\langle M_n^\varphi \Gamma^\delta H L_n^\varphi f - \varepsilon\, G_{\beta,n}(L_n^\varphi \widetilde{g}) - H_n L_n^\varphi \widetilde{g}, \widetilde{g}_n^* - L_n^\varphi \widetilde{g} \right\rangle \\
&\quad = \left\langle M_n^\varphi \Gamma^\delta H f - \varepsilon\, M_n^\varphi G_\beta(\widetilde{g}) - H_n L_n^\varphi \widetilde{g} + M_n^\varphi \Gamma^\delta H(L_n^\varphi f - f), \widetilde{g}_n^* - L_n^\varphi \widetilde{g} \right\rangle \\
&\quad = \left\langle M_n^\varphi \Gamma^\delta H(\Gamma^\delta \widetilde{g} - L_n^\varphi \Gamma^\delta \widetilde{g}) + M_n^\varphi \Gamma^\delta H(L_n^\varphi f - f), \widetilde{g}_n^* - L_n^\varphi \widetilde{g} \right\rangle \\
&\quad \le \operatorname{const} \left(\left\| H(\Gamma^\delta \widetilde{g} - L_n^\varphi \Gamma^\delta \widetilde{g}) \right\|_\infty + \| H(L_n^\varphi f - f) \|_\infty \right) \|\widetilde{g}_n^* - L_n^\varphi \widetilde{g}\|_{\mathbf{L}^p} .
\end{aligned}$$

Now, apply (4.4) for some $\mu \in (0, \frac{1}{2})$ and use $b(s) = \operatorname{const} n^{-2\eta} s^{p-1}$. □

Let us investigate the iteration method

$$\widetilde{g}_n^{(m+1)} = \widetilde{g}_n^{(m)} - t\, R_n(\widetilde{g}_n^{(m)}), \quad \widetilde{g}_n^{(0)} \equiv 0, \; t > 0, \tag{4.12}$$

where $R_n(\widetilde{g}_n) = \varepsilon \widetilde{G}_{\beta,n}(\widetilde{g}_n) + \widetilde{H}_n \widetilde{g}_n - \widetilde{f}_n$ and

$$\widetilde{G}_{\beta,n}(\widetilde{g}_n) = L_n^\varphi \varphi^{-1} G_\beta(\widetilde{g}_n), \quad \widetilde{H}_n \widetilde{g}_n = L_n^\varphi \varphi^{-1} \Gamma^\delta H L_n^\varphi \Gamma^\delta \widetilde{g}_n, \quad \widetilde{f}_n = L_n^\varphi \varphi^{-1} \Gamma^\delta H L_n^\varphi f,$$

for solving the collocation equation (4.8). For this, denote by $\mathbf{Y}_n^\varphi$ the space $\mathbf{X}_n$ of polynomials of degree less than n (with real coefficients) equipped with the inner product

$$\left\langle \widetilde{g}_n, \widetilde{f}_n \right\rangle_\varphi = \int_{-1}^1 \varphi(x) \widetilde{g}_n(x) \widetilde{f}_n(x)\, dx = \sum_{k=1}^n \lambda_{nk}^\varphi \widetilde{g}_n(x_{nk}^\varphi) \widetilde{f}_n(x_{nk}^\varphi) .$$

Since $\Gamma : [-1,1] \longrightarrow (0,\infty)$ is assumed to be continuous, there exist constants $\gamma_0, \gamma_1 \in \mathbb{R}$ such that

$$\gamma_1 \ge \Gamma(x) \ge \gamma_0 > 0, \quad -1 \le x \le 1 . \tag{4.13}$$

Then, for $\widetilde{g}_n, \widetilde{f}_n \in \mathbf{Y}_n^\varphi$,

$$\left\langle G_{\beta,n}(\widetilde{g}_n) - G_{\beta,n}(\widetilde{f}_n), \widetilde{g}_n - \widetilde{f}_n \right\rangle_\varphi \ge 0$$

and (see (4.5) and the proof of Lemma 4.2)

$$\left\langle \widetilde{H}_n\widetilde{g}_n, \widetilde{g}_n \right\rangle_\varphi = \sum_{k=0}^{n-1} \frac{1}{k+1} \left| \langle L_n^\varphi \Gamma^\delta \widetilde{g}_n, p_k^\varphi \rangle_\varphi \right|^2$$

$$\geq \frac{1}{n} \left\| L_n^\varphi \Gamma^\delta \widetilde{g}_n \right\|_\varphi^2 = \frac{1}{n} \sum_{k=1}^{n} \lambda_{nk}^\varphi \left| \gamma^\delta(x_{nk}^\varphi) \widetilde{g}_n(x_{nk}^\varphi) \right|^2 \geq \frac{\gamma_0^{2\delta}}{n} \left\| \widetilde{g}_n \right\|_\varphi^2 .$$

Furthermore,

$$\left\| \widetilde{H}_n \widetilde{g}_n \right\|_\varphi = \sqrt{\sum_{k=1}^{n} \lambda_{nk}^\varphi \left| \Gamma^\delta(x_{nk}^\varphi)(V^{-1} L_n^\varphi \Gamma^\delta \widetilde{g}_n)(x_{nk}^\varphi) \right|^2}$$

$$\leq \gamma_1^\delta \left\| V^{-1} L_n^\varphi \Gamma^\delta \widetilde{g}_n \right\|_\varphi \leq \gamma_1^{2\delta} \left\| \widetilde{g}_n \right\|_\varphi .$$

Taking into account that $\|\widetilde{g}_n\|_\varphi \leq r$ implies $|\widetilde{g}_n(x_{nk}^\varphi)| \leq \frac{r}{\sqrt{\lambda_{nk}^\varphi}}$, $k = 1, \ldots, n$, and that $\lambda_{nk}^\varphi = \dfrac{\pi \sin^2 \frac{k\pi}{n+1}}{n+1} \geq \dfrac{4\pi}{(n+1)^3}$, we get, for $\|\widetilde{g}_n\|_\varphi , \left\|\widetilde{f}_n\right\|_\varphi \leq r$,

$$\left\| \widetilde{G}_{\beta,n}(\widetilde{g}_n) - \widetilde{G}_{\beta,n}(\widetilde{f}_n) \right\|_\varphi^2$$

$$= \frac{\pi}{n+1} \sum_{k=1}^{n} \left| |\widetilde{g}_n(x_{nk}^\varphi)|^\beta \widetilde{g}_n(x_{nk}^\varphi) - |\widetilde{f}_n(x_{nk}^\varphi)|^\beta \widetilde{f}_n(x_{nk}^\varphi) \right|^2$$

$$\leq \frac{\pi[(\beta+1)r^\beta]^2}{n+1} \sum_{k=1}^{n} \frac{1}{(\lambda_{nk}^\varphi)^\beta} |\widetilde{g}_n(x_{nk}^\varphi) - \widetilde{f}_n(x_{nk}^\varphi)|^2$$

$$\leq \frac{\pi[(\beta+1)r^\beta]^2}{(4\pi)^{\beta+1}} (n+1)^{3\beta+2} \left\| \widetilde{g}_n - \widetilde{f}_n \right\|_\varphi^2 .$$

Consequently, the operator $R_n : \mathbf{Y}_n^\varphi \longrightarrow \mathbf{Y}_n^\varphi$ is strongly monotone and locally Lipschitz continuous, which implies that, for sufficiently small $t > 0$, the iteration method (4.12) converges in $\mathbf{Y}_n^\varphi$ to the solution $\widetilde{g}_n \in \mathbf{Y}_n^\varphi$ of (4.8) (see [18, Prop. 26.8, Theorem 26.B]).

Remark that the collocation equation (4.8) can be written as

$$\varepsilon \mathbf{G}_n(\widetilde{\xi}_n) + \mathbf{H}_n \widetilde{\xi}_n = \mathbf{H}_n \mathbf{\Gamma}_n^{-1} \eta_n ,$$

where $\eta_n = [f(x_{nk}^\varphi)]_{k=1}^n$ and $\widetilde{g}_n = \sum_{k=1}^{n} \widetilde{\xi}_{nk} \ell_{nk}^\varphi$, i.e., $\widetilde{\xi}_n = [\widetilde{\xi}_{nk}]_{k=1}^n = [\widetilde{g}_n(x_{nk}^\varphi)]_{k=1}^n$,

$$\mathbf{G}_n(\widetilde{\xi}_n) = \left[|\widetilde{\xi}_{nk}|^\beta \widetilde{\xi}_{nk} \right]_{k=1}^n , \ \mathbf{\Gamma}_n = \mathrm{diag}[\Gamma^\delta(x_{nk}^\varphi)]_{k=1}^n , \ \mathbf{H}_n = \mathbf{\Gamma}_n \widetilde{\mathbf{D}}_n \mathbf{U}_n^T \mathbf{D}_n^{-1} \mathbf{U}_n \mathbf{\Lambda}_n \mathbf{\Gamma}_n$$

(comp. the definitions associated with (3.8)). Thus, the iteration equation (4.12) is equivalent to

$$\widetilde{\xi}_n^{(m+1)} = \widetilde{\xi}_n^{(m)} - t \left[\varepsilon \mathbf{G}_n(\widetilde{\xi}_n^{(m)}) + \mathbf{H}_n \widetilde{\xi}_n^{(m)} - \mathbf{H}_n \mathbf{\Gamma}_n^{-1} \eta_n \right] . \qquad (4.14)$$

5. Numerical examples

Let us consider equation (1.5) for different functions $f(x)$ and $\gamma(x,g)$.

Example 1. We solve the (linear) equation (1.5) with $\gamma(x,g) = (1-x^2)^{-1/2}g$ and

(a) $f(x) = f_a(x) = x|x| - \frac{\varepsilon x}{\pi}\left(\frac{2-3x^2}{\sqrt{1-x^2}}\ln\frac{1+\sqrt{1-x^2}}{1-\sqrt{1-x^2}} - 6\right)$

as well as with

(b) $f(x) = f_b(x) = x|x|$

by the collocation method (3.1) together with the iteration method (3.10) (with $u_n^{(0)} \equiv 0$) as well as with the combination of (3.10) and (3.14) as described at the end of Section 3.

In case of the method (3.10) the iteration is stopped if the $\mathbf{L}_\varphi^{2,\frac{1}{2}}$-norm of the difference of two consecutive iterations is smaller than *toll*, i.e.,

$$\left\| u_n^{(N)} - u_n^{(N-1)} \right\|_{\varphi,\frac{1}{2}} < toll\,, \tag{5.1}$$

where $u_n^{(m)} = \sum_{k=1}^{n} \xi_{nk}^{(m)} \ell_{nk}^{\varphi}$. For the combination of (3.10) and (3.14) we choose the sequence

$$n_1 = 7 < \cdots < n_j = 2^{j+2} - 1 < \cdots < n = n_{M+1} = 2^{M+3} - 1$$

and the number K of iterations realized on the levels $n_1, \ldots, n_M$. The number of iterations needed on the last level n_{M+1} to get the same accuracy (5.1) is denoted by N_K. Condition (3.3) is satisfied for all r.

(a) In this case the solution is given by $u^*(x) = x|x|$ (independent of ε). We have, for $n = 2k-1$,

$$a_{2k-1}^* := \left\langle u^*, p_{2k-1}^{\varphi} \right\rangle_{\varphi} = \frac{4\sqrt{2}\,(-1)^k (n+1)}{\sqrt{\pi}(n^2-4)n(n+4)}, \qquad k = 1, 2, \ldots,$$

such that $u^* \in \mathbf{L}_\varphi^{2,2.5-\delta}$ for all $\delta > 0$. Thus, the convergence rate predicted by (3.4) for $t = 0.5$ is

$$\|u_n^* - u^*\|_{\varphi,1} = O\left(n^{\delta-1.5}\right), \quad \delta > 0 \text{ arbitrarily small,}$$

which is confirmed by the numerical results presented in the following table, in which the values

$$d_s = \left\| u_n^{(N)} - P_n u^* \right\|_{\varphi,s} = \sqrt{\sum_{k=0}^{n-1} (k+1)^{2s} \left[\left\langle u_n^{(N)}, p_k^{\varphi} \right\rangle_{\varphi} - a_k^* \right]^2}$$

are presented for $s = 0.5$ and $s = 1$. To compute these values we use the relation

$$\left[\left\langle u_n^{(N)}, p_k^{\varphi} \right\rangle_{\varphi} \right]_{k=0}^{n-1} = \mathbf{U}_n \mathbf{\Lambda}_n \xi_n^{(N)}.$$

n	N	N_1	N_3	$d_{1/2}$	d_1	$n^{2.0}d_{1/2}$	$n^{1.5}d_1$
31	16			1.13e-03	5.43e-03	1.090	0.938
63	16			2.93e-04	2.00e-03	1.164	0.999
127	16			7.46e-05	7.21e-04	1.204	1.032
255	16			1.88e-05	2.58e-04	1.225	1.049
511	16			4.73e-06	9.15e-05	1.235	1.057
1023	16	13	12	1.19e-06	3.24e-05	1.241	1.062
2047	16	12	11	2.97e-07	1.15e-05	1.243	1.064
4095	16	11	10	7.42e-08	4.06e-06	1.245	1.065
8191	16	10	10	1.86e-08	1.44e-06	1.246	1.066
16383	16	9	9	4.64e-09	5.08e-07	1.246	1.066
32767	16	9	8	1.16e-09	1.80e-07	1.246	1.066
65535	16	8	7	2.92e-10	6.41e-08	1.256	1.076

Example 1, (a): $\varepsilon = 1.0$, $t = 0.75$, $toll = 10^{-12}$

Here the value t is equal to $1.5\,t_\varepsilon^*$ with t_ε^* from (3.7), where $c_1 = 1$. For $t = t_\varepsilon^*$, $N = 31$ iteration steps are needed in (3.10) to get the same accuracy in (5.1).

The next table presents the respective results for $\varepsilon = 0.2$. We observe the same convergence rates and that the numbers in the last two columns do not depend on ε as predicted by Cor. 3.3. Otherwise, the number of iterations is much higher than for $\varepsilon = 1.0$ as expected by (3.7). Here t is about $9\,t_\varepsilon^*$. For $t = t_\varepsilon^*$ $N = 410$ iterations are needed in (3.10) to fulfil (5.1).

n	N	N_3	N_5	N_{10}	$d_{1/2}$	d_1	$n^{2.0}d_{1/2}$	$n^{1.5}d_1$
31	40				1.06e-03	5.17e-03	1.018	0.892
63	42				2.82e-04	1.94e-03	1.119	0.972
127	43				7.30e-05	7.11e-04	1.178	1.017
255	44				1.86e-05	2.56e-04	1.210	1.041
511	44				4.70e-06	9.12e-05	1.228	1.053
1023	44	37	36	36	1.18e-06	3.24e-05	1.237	1.060
2047	44	35	33	33	2.96e-07	1.15e-05	1.241	1.063
4095	44	32	30	30	7.42e-08	4.06e-06	1.244	1.065
8191	44	30	27	26	1.86e-08	1.44e-06	1.245	1.065
16383	44	27	24	23	4.64e-09	5.08e-07	1.246	1.066
32767	44	24	20	20	1.16e-09	1.80e-07	1.246	1.066
65535	44	21	17	16	2.92e-10	6.41e-08	1.256	1.076

Example 1, (a): $\varepsilon = 0.2$, $t = 1.7$, $toll = 10^{-12}$

(b) In this case the solution u^* is unknown. For this reason we compare the approximate solutions $u_n^{(N)}$ with $u_{65535}^{(N)}$. We have $f_b \in \mathbf{L}_\varphi^{2,2.5-\delta}$ for all $\delta > 0$. Thus, by Cor. 2.6 and by (3.4), we get

$$\|u_n^* - u^*\|_{\varphi,1} = O\left(n^{\delta-2.5}\right) , \quad \delta > 0 \text{ arbitrarily small.}$$

n	N	N_1	N_3	$D_{1/2}$	D_1	$n^{3.0}D_{1/2}$	$n^{2.5}D_1$
31	14			2.90e-05	1.43e-04	0.865	0.763
63	14			3.83e-06	2.68e-05	0.958	0.845
127	14			4.93e-07	4.90e-06	1.009	0.890
255	14			6.25e-08	8.80e-07	1.036	0.914
511	14			7.87e-09	1.57e-07	1.050	0.926
1023	14	10	8	9.87e-10	2.78e-08	1.057	0.932
2047	14	9	6	1.24e-10	4.93e-09	1.060	0.935
4095	14	8	5	1.55e-11	8.71e-10	1.061	0.935
8191	14	7	3	2.00e-12	1.60e-10	1.100	0.974

Example 1, (b): $\varepsilon = 1.0\,,\ t = 0.75\,,\ toll = 10^{-12}$

The numerical results presented in the tables confirm these theoretical estimate. Here we observe the values of the norms $D_s = \|u_n^{(N)} - u_{65535}^{(N)}\|_{\varphi,s}$ for $s = 0.5$ and $s = 1$. Of course, here the values of the last two columns depend on ε since the exact solution u^* and so $\|u^*\|_{\varphi,3.5}$ in (3.4) depends on ε .

n	N	N_3	N_5	N_{10}	$D_{1/2}$	D_1	$n^{3.0}D_{1/2}$	$n^{2.5}D_1$
31	37				1.24e-04	6.17e-04	3.697	3.303
63	38				1.76e-05	1.24e-04	4.405	3.919
127	39				2.36e-06	2.36e-05	4.829	4.281
255	39				3.05e-07	4.31e-06	5.064	4.479
511	39				3.89e-08	7.76e-07	5.189	4.582
1023	39	29	26	25	4.91e-09	1.38e-07	5.253	4.635
2047	39	26	22	20	6.16e-10	2.46e-08	5.286	4.662
4095	39	23	17	15	7.72e-11	4.35e-09	5.300	4.673
8191	39	20	13	10	9.77e-12	7.80e-10	5.368	4.738

Example 1, (b): $\varepsilon = 0.2\,,\ t = 1.7\,,\ toll = 10^{-12}$

Example 2. We solve the equation (1.5) with $\gamma(x,g) = |g|\arctan(g)$ and $f(x)$ equal to

$$x^2\sqrt{1-x^2}\arctan\left(x|x|\sqrt{1-x^2}\right) - \frac{\varepsilon x}{\pi}\left(\frac{2-3\,x^2}{\sqrt{1-x^2}}\ln\frac{1+\sqrt{1-x^2}}{1-\sqrt{1-x^2}} - 6\right)$$

by the collocation method (3.1) together with the iteration method (3.10) (with $u_n^{(0)} \equiv 0$) as well as with the combination of (3.10) and (3.14) as described at the end of Section 3.

Condition (B) is fulfilled with $\alpha = 1$ and $c_1 = \pi/2$. The solution is given by $u^*(x) = x|x|$ (independent of ε). In the following tables (where $\varepsilon = 1.0$ and $\varepsilon = 0.2$) we can observe the convergence rate which is predicted by (3.4). (The notations from the previous example are used.)

n	N	N_3	$d_{1/2}$	d_1	$n^{2.0}d_{1/2}$	$n^{1.5}d_1$
31	12		1.16e-03	5.52e-03	1.116	0.953
63	12		2.97e-04	2.01e-03	1.178	1.007
127	12		7.51e-05	7.24e-04	1.212	1.036
255	12		1.89e-05	2.58e-04	1.229	1.051
511	12		4.74e-06	9.16e-05	1.237	1.058
1023	12	8	1.19e-06	3.25e-05	1.242	1.062
2047	12	8	2.97e-07	1.15e-05	1.244	1.064
4095	12	7	7.43e-08	4.06e-06	1.245	1.065
8191	12	6	1.86e-08	1.44e-06	1.246	1.066
16383	12	6	4.64e-09	5.08e-07	1.246	1.066
32767	12	5	1.16e-09	1.80e-07	1.246	1.066
65535	12	5	2.92e-10	6.41e-08	1.256	1.076

Example 2: $\varepsilon = 1.0$, $t = 0.9$, $toll = 10^{-12}$

n	N	N_3	N_5	N_{10}	$d_{1/2}$	d_1	$n^{2.0}d_{1/2}$	$n^{1.5}d_1$
31	22				1.16e-03	5.51e-03	1.113	0.951
63	22				2.97e-04	2.01e-03	1.178	1.007
127	22				7.51e-05	7.24e-04	1.211	1.036
255	22				1.89e-05	2.58e-04	1.229	1.051
511	22				4.74e-06	9.16e-05	1.237	1.058
1023	22	15	15	15	1.19e-06	3.25e-05	1.242	1.062
2047	22	14	14	14	2.97e-07	1.15e-05	1.244	1.064
4095	22	12	12	12	7.43e-08	4.06e-06	1.245	1.065
8191	22	11	11	11	1.86e-08	1.44e-06	1.246	1.066
16383	22	10	10	10	4.64e-09	5.08e-07	1.246	1.066
32767	22	9	9	9	1.16e-09	1.80e-07	1.246	1.066
65535	22	8	7	7	2.92e-10	6.41e-08	1.256	1.076

Example 2: $\varepsilon = 0.2$, $t = 3.4$, $toll = 10^{-12}$

Example 3. We solve equation (1.7) with $\Gamma(x) = (1-x^2)^{\frac{1}{4}}$ and

$$f(x) = x\left[\sqrt{1-x^2} - \frac{\varepsilon}{\pi}\left(\frac{2-3\,x^2}{\sqrt{1-x^2}}\ln\frac{1+\sqrt{1-x^2}}{1-\sqrt{1-x^2}} - 6\right)\right]$$

(a) by the collocation method (3.1) together with the iteration method (3.10) (with $u_n^{(0)} \equiv 0$), although condition (B) is not satisfied for $\alpha = 1$, and

(b) by the collocation method (4.8) together with the iteration method (4.14).

We observe that the iteration method (3.10) can converge (although the condition on the Lipschitz continuity is not satisfied), but does usually not converge for greater n if the iteration parameter t is not small enough. On the other hand, in the iteration method (4.14) the number of iterations essentially depends on the discretization parameter n and increases strongly with n. Thus, a more effective and stable way to solve equation (1.7) seems to be the application of a Newton iteration method to the collocation equations (4.8) or a combination of a Newton iteration with the method (4.14). We will discuss this in a forthcoming paper.

(a) The solution is given by $g^*(x) = \sqrt{1-x^2}u^*(x)$ with $u^*(x) = x|x|$. In the following tables we use the same notations as in the previous examples.

n	N	$d_{1/2}$	d_1	$n^{2.0}d_{1/2}$	$n^{1.5}d_1$
31	68	1.06e-03	5.16e-03	1.017	0.891
63	68	2.72e-04	1.89e-03	1.078	0.946
127	68	6.89e-05	6.81e-04	1.111	0.975
255	68	1.73e-05	2.43e-04	1.128	0.990
511	68	4.35e-06	8.64e-05	1.136	0.998
1023	20000	2.90e-04	1.13e-03		
2047	20000	2.93e-04	1.13e-03		
4095	20000	2.91e-04	1.13e-03		
8191	20000	2.91e-04	1.12e-03		
16383	20000	2.91e-04	1.12e-03		
32767	20000	2.91e-04	1.12e-03		
65535	20000	2.91e-04	1.12e-03		

Example 3, (a): $\varepsilon = 1.0$, $t = 0.3$, $toll = 10^{-12}$

n	N	$d_{1/2}$	d_1	$n^{2.0}d_{1/2}$	$n^{1.5}d_1$
31	365	1.06e-03	5.16e-03	1.017	0.891
63	365	2.72e-04	1.89e-03	1.078	0.946
127	365	6.89e-05	6.81e-04	1.111	0.975
255	365	1.73e-05	2.43e-04	1.128	0.990
511	365	4.35e-06	8.64e-05	1.136	0.998
1023	365	1.09e-06	3.06e-05	1.141	1.002
2047	365	2.73e-07	1.08e-05	1.143	1.004
4095	365	6.82e-08	3.83e-06	1.144	1.005
8191	365	1.71e-08	1.36e-06	1.145	1.005
16383	365	4.27e-09	4.80e-07	1.145	1.006
32767	365	1.07e-09	1.70e-07	1.145	1.006
65535	365	2.70e-10	6.06e-08	1.158	1.017

Example 3, (a): $\varepsilon = 1.0$, $t = 0.06$, $toll = 10^{-12}$

(b) The iteration is stopped if the $\mathbf{L}^2_\varphi$-norm of the difference of two consecutive iterations is smaller than *toll*, i.e.,

$$\left\| \widetilde{g}_n^{(N)} - \widetilde{g}_n^{(N-1)} \right\|_\varphi < toll\,, \quad \text{where} \quad \widetilde{g}_n^{(m)} = \sum_{k=1}^n \widetilde{\xi}_{nk}^{(m)} \ell_{nk}^\varphi\,.$$

Then, the transformation

$$\widetilde{\xi}_{nk}^{(N)} \mapsto \xi_{nk}^{(N)} = \frac{\left|\widetilde{\xi}_{nk}^{(N)}\right|^{\beta} \widetilde{\xi}_{nk}^{(N)}}{[\Gamma(x_{nk}^{\varphi})]^{\delta}\varphi(x_{nk}^{\varphi})}$$

is applied and the approximations $\widetilde{u}_n = \sum_{k=1}^{n} \xi_{nk}^{(N)} \ell_{nk}^{\varphi}$ are compared with the exact solution $u^*(x)$,

$$\widetilde{d}_s = \left\|\widetilde{u}_n^{(N)} - P_n u^*\right\|_{\varphi,s} = \sqrt{\sum_{k=0}^{n-1}(k+1)^{2s}\left[\left\langle u_n^{(N)}, p_k^{\varphi}\right\rangle_{\varphi} - a_k^*\right]^2}\,.$$

n	N	$\widetilde{d}_{1/2}$	$\widetilde{d}_1$	$n^{2.0}\widetilde{d}_{1/2}$	$n^{1.5}\widetilde{d}_1$
31	74	1.059e-03	5.163e-03	1.017	0.891
63	144	2.716e-04	1.891e-03	1.078	0.946
127	268	6.887e-05	6.812e-04	1.111	0.975
255	487	1.734e-05	2.432e-04	1.128	0.990
511	867	4.352e-06	8.639e-05	1.136	0.998
1023	1511	1.090e-06	3.062e-05	1.141	1.002
2047	2573	2.727e-07	1.084e-05	1.143	1.004
4095	4242	6.819e-08	3.833e-06	1.143	1.004
8191	6677	1.702e-08	1.353e-06	1.142	1.003
16383	9749	4.205e-09	4.737e-07	1.129	0.993
32767	12399	9.948e-10	1.602e-07	1.068	0.950
65535	12773	2.312e-10	5.434e-08	0.993	0.912

Example 3,(b): $\varepsilon = 1.0$, $t = 1.0$, $toll = 10^{-12}$

n	N	$\widetilde{d}_{1/2}$	$\widetilde{d}_1$	$n^{2.0}\widetilde{d}_{1/2}$	$n^{1.5}\widetilde{d}_1$
31	93	1.003e-03	4.955e-03	0.964	0.855
63	178	2.581e-04	1.821e-03	1.025	0.910
127	332	6.555e-05	6.568e-04	1.057	0.940
255	604	1.652e-05	2.346e-04	1.074	0.955
511	1077	4.147e-06	8.338e-05	1.083	0.963
1023	1888	1.039e-06	2.956e-05	1.087	0.967
2047	3238	2.600e-07	1.046e-05	1.089	0.969
4095	5403	6.503e-08	3.702e-06	1.090	0.970
8191	8671	1.626e-08	1.309e-06	1.091	0.970
16383	13112	4.058e-09	4.622e-07	1.089	0.969
32767	17871	1.006e-09	1.622e-07	1.080	0.962
65535	19814	2.457e-10	5.646e-08	1.055	0.947

Example 3,(b): $\varepsilon = 0.5$, $t = 1.4$, $toll = 10^{-13}$

References

[1] D. Berthold, W. Hoppe, B. Silbermann, *A fast algorithm for solving the generalized airfoil equation*, J. Comp. Appl. Math., **43** (1992), 185–219.

[2] M.R. Capobianco, G. Criscuolo, P. Junghanns, *A fast algorithm for Prandtl's integro-differential equation*, J. Comp. Appl. Math., **77** (1997), 103–128.

[3] M.R. Capobianco, G. Criscuolo, P. Junghanns, U. Luther, *Uniform convergence of the collocation method for Prandtl's integro-differential equation*, ANZIAM J., **42** (2000), 151–168.

[4] M.R. Capobianco, G. Mastroianni, *Uniform boundedness of Lagrange operator in some weighted Sobolev-type space*, Math. Nachr., **187** (1997), 1–17.

[5] I. Gohberg, N. Krupnik, *One-Dimensional Linear Singular Integral Equations*, Volume I, Birkhäuser Verlag, 1992.

[6] N.I. Ioakimidis, *Application of finite-part integrals to the singular integral equations of crack problems in plane and three-dimensional elasticity*, Acta Mech., **45** (1982), 31–47.

[7] A.C. Kaya, F. Erdogan, *On the solution of integral equations with strong singularities*, Quart. Appl. Math., *45* (1987), 105–122.

[8] G. Mastroianni, M.G. Russo, *Lagrange interpolation in weighted Besov spaces*, Constr. Approx., **15** 2 (1999), 257–289.

[9] G. Mastroianni, M.G. Russo, *Weighted Marcinkiewicz inequalities and boundedness of the Lagrange operator*, Math. Anal. Appl., (2000), 149–182.

[10] S. Nemat-Nasser, M. Hori, *Toughening by partial or full bridging of cracks in ceramics and fiber reinforced composites*, Mech. Mat., **6** (1987), 245–269.

[11] S. Nemat-Nasser, M. Hori, *Asymptotic solution of a class of strongly singular integral equations*, SIAM J. Appl. Math., **50** 3 (1990), 716–725.

[12] P. Nevai, *Orthogonal Polynomials*, Mem. Amer. Math. Soc. 213, Providence, RI, 1979.

[13] D. Oestreich, *Approximated solution of a nonlinear singular equation of Prandtl's type*, Math. Nachr., **161** (1993), 95–105.

[14] M.A. Sheshko, G.A. Rasol'ko, V.S. Mastyanitsa, *On approximate solution of Prandtl's integro-differential equation*, Differential Equations, **29** (1993), 1345–1354.

[15] G. Steidl, *Fast radix-p discrete cosine transform*, AAECC, **3** (1992), 39–46.

[16] M. Tasche, *Fast algorithms for discrete Chebyshev-Vandermonde transforms and applications*, Numer. Algor., **5** (1993), 453–464.

[17] L. von Wolfersdorf, *Monotonicity methods for nonlinear singular integral and integro-differential equations*, ZAMM, **63** (1983), 249–259.

[18] E. Zeidler, *Nonlinear Functional Analysis and its Applications*, Part II, Springer Verlag, 1990.

M.R. Capobianco
C.N.R. – Istituto per le Applicazioni del
Calcolo "Mauro Picone", Sezione di Napoli
Via Pietro Castellino 111
I-80131 Napoli, Italy
e-mail: r.capobianco@na.iac.cnr.it

G. Criscuolo
Dipartimento di Matematica
Universitá degli Studi Napoli "Frederico II"
Edificio T Compless Monte Sant' Angelo
Via Cinthia
I-80126 Napoli, Italy
e-mail: giuliana.criscuolo@dma.unina.it

P. Junghanns
Fakultät für Mathematik
Technische Universität Chemnitz
D-09107 Chemnitz, Germany
e-mail: peter.junghanns@mathematik.tu-chemnitz.de

Operator Theory:
Advances and Applications, Vol. 160, 81–100

Fourier Integral Operators and Gelfand-Shilov Spaces

Marco Cappiello

Abstract. In this work, we study a class of Fourier integral operators of infinite order acting on the Gelfand-Shilov spaces of type S. We also define wave front sets in terms of Gelfand-Shilov classes and study the action of the previous Fourier integral operators on them.

Mathematics Subject Classification (2000). Primary 35S30; Secondary 35A18.

Keywords. Fourier integral operators, θ-wave front set, Gelfand-Shilov spaces.

1. Introduction

Fourier integral operators, as introduced by L. Hörmander [20], play a fundamental role in microlocal analysis and in the theory of the partial differential equations. In these fields, they find a natural application in the analysis of the Cauchy problem for some classes of hyperbolic equations. In particular, parametrices and solutions for these kinds of problems can be expressed in terms of Fourier integral operators. The propagation of singularities associated to the Cauchy problem can also be investigated by studying the action of Fourier integral operators on the wave front set of distributions. A large number of works concerning these operators and their applications in the study of the C^∞-well-posedness of hyperbolic problems appeared in the last thirty years, see [21], [23], [32] and the references there. Corresponding results have been obtained in the context of the Gevrey classes, see for example [19], [6], [30], [31], [17]. The Gevrey framework leads us to consider operators of infinite order, i.e., with symbols growing exponentially at infinity. In a different direction, S. Coriasco [8] has developed a global calculus for Fourier integral operators defined by symbols $a(x,\eta)$ satisfying estimates on $\mathbb{R}^{2n}_{x,\eta}$, called SG-symbols in the literature, see [25], [7], [28], [13], [29]. As an application, S. Coriasco [9], S. Coriasco and L. Rodino [12], S. Coriasco and P. Panarese [11] and S. Coriasco and L. Maniccia [10] have proved results on the well-posedness and propagation of singularities for some hyperbolic problems globally defined in the space variables, in the framework of the Schwartz spaces $\mathcal{S}(\mathbb{R}^n), \mathcal{S}'(\mathbb{R}^n)$. In some recent works, the author extends this global calculus in a Gevrey context first for symbols of finite order [2] and then for symbols of infinite order [3], [4]. In [3], [4], the functional framework is given by the Gelfand-Shilov space $S^{\nu}_{\mu}(\mathbb{R}^n), \mu > 0, \nu > 0, \mu + \nu \geq 1$, defined as the space of all functions $u \in C^\infty(\mathbb{R}^n)$ such that

$$\sup_{\alpha,\beta\in\mathbb{N}^n} \sup_{x\in\mathbb{R}^n} A^{-|\alpha|}B^{-|\beta|}(\alpha!)^{-\mu}(\beta!)^{-\nu} \left|x^\alpha \partial_x^\beta u(x)\right| < +\infty \tag{1.1}$$

for some positive constants A, B. More precisely, the results have been obtained in $S_\theta(\mathbb{R}^n) = S^\theta_\theta(\mathbb{R}^n)$, with $\theta > 1$ and in the dual space $S'_\theta(\mathbb{R}^n)$, corresponding to the case $\mu = \nu = \theta$ in (1.1) and representing a global version of the Gevrey classes $G^\theta(\mathbb{R}^n), \mathcal{D}'_\theta(\mathbb{R}^n)$.

In this work, we study SG-Fourier integral operators on Gelfand-Shilov spaces from a microlocal point of view. In Section 2, we recall the basic results concerning the SG-calculus on $S_\theta(\mathbb{R}^n)$. In Section 3, we define polyhomogeneous SG-symbols of finite order, which extend the standard notion of classical symbols in the SG-context. In Sections 4 and 5, we define the wave front sets for distributions $u \in S'_\theta(\mathbb{R}^n)$ and study the action of Fourier integral operators on them. These results are the starting point for the study of the propagation of singularities for the SG-hyperbolic Cauchy problem in the Gelfand-Shilov spaces, which we shall detail in a forthcoming paper (a construction of parametrices is given in [4]).

In the sequel we will use the following notation:

$\langle x\rangle = (1 + |x|^2)^{\frac{1}{2}}$ for $x \in \mathbb{R}^n$

$\nabla_x\varphi = \left(\frac{\partial\varphi}{\partial x_1}, \dots, \frac{\partial\varphi}{\partial x_n}\right)$

$D^\alpha_x = D^{\alpha_1}_{x_1} \dots D^{\alpha_n}_{x_n}$ for all $\alpha \in \mathbb{N}^n$, $x \in \mathbb{R}^n$, where $D_{x_h} = -i\partial_{x_h}$, $h = 1, \dots, n$

$e_1 = (1, 0), e_2 = (0, 1), e = (1, 1)$.

We will denote by $\mathbb{Z}_+$ the set of all positive integers and by $\mathbb{N}$ the set $\mathbb{Z}_+ \cup \{0\}$. We will also denote by $\mathcal{F}$ the Fourier transformation.

We start by giving the basic definitions and properties of the Gelfand-Shilov spaces $S_\theta(\mathbb{R}^n), \theta > 1$ and describing their relations with the Gevrey spaces. We will refer to [14], [15], [24] for proofs and details. Let $\theta > 1$, let A, B be positive integers and denote by $S_{\theta,A,B}(\mathbb{R}^n)$ the space of all functions u in $C^\infty(\mathbb{R}^n)$ such that

$$\sup_{\alpha,\beta\in\mathbb{N}^n} \sup_{x\in\mathbb{R}^n} A^{-|\alpha|}B^{-|\beta|}(\alpha!\beta!)^{-\theta} \left|x^\alpha \partial_x^\beta u(x)\right| < +\infty. \tag{1.2}$$

We have

$$S_\theta(\mathbb{R}^n) = \bigcup_{A,B\in\mathbb{Z}_+} S_{\theta,A,B}(\mathbb{R}^n).$$

For any $A, B \in \mathbb{Z}_+$, the space $S_{\theta,A,B}(\mathbb{R}^n)$ is a Banach space endowed with the norm given by the left-hand side of (1.2). Therefore, we can consider the space $S_\theta(\mathbb{R}^n)$ as an inductive limit of an increasing sequence of Banach spaces.

Let us give another characterization of the space $S_\theta(\mathbb{R}^n)$, providing another equivalent topology to $S_\theta(\mathbb{R}^n)$.

Proposition 1.1. *$S_\theta(\mathbb{R}^n)$ is the space of all functions $u \in C^\infty(\mathbb{R}^n)$ such that*

$$\sup_{\beta\in\mathbb{N}^n}\sup_{x\in\mathbb{R}^n} B^{-|\beta|}(\beta!)^{-\theta}e^{L|x|^{\frac{1}{\theta}}}|\partial_x^\beta u(x)| < +\infty$$

for some positive B, L.

Proposition 1.2. i) *$S_\theta(\mathbb{R}^n)$ is closed under differentiation;*

ii) *We have*

$$G_o^\theta(\mathbb{R}^n) \subset S_\theta(\mathbb{R}^n) \subset G^\theta(\mathbb{R}^n),$$

where $G^\theta(\mathbb{R}^n)$ is the Gevrey space of all functions $u \in C^\infty(\mathbb{R}^n)$ satisfying for every compact subset K of $\mathbb{R}^n$ estimates of the form:

$$\sup_{\beta\in\mathbb{N}^n} B^{-|\beta|}(\beta!)^{-\theta}\sup_{x\in K}|\partial_x^\beta u(x)| < +\infty$$

for some $B = B(K) > 0$, and $G_o^\theta(\mathbb{R}^n)$ is the space of all functions of $G^\theta(\mathbb{R}^n)$ with compact support.

We shall denote by $S'_\theta(\mathbb{R}^n)$ the space of all linear continuous forms on $S_\theta(\mathbb{R}^n)$, also known as temperate ultradistributions, cf. [26].

Remark 1.3. Given $u \in S'_\theta(\mathbb{R}^n)$, the restriction of u to $G_o^\theta(\mathbb{R}^n)$ is a Gevrey ultradistribution in $\mathcal{D}'_\theta(\mathbb{R}^n)$, topological dual of $G_o^\theta(\mathbb{R}^n)$. In this sense, we have $S'_\theta(\mathbb{R}^n) \subset \mathcal{D}'_\theta(\mathbb{R}^n)$. Similarly, the space of the ultradistributions with compact support $\mathcal{E}'_\theta(\mathbb{R}^n)$ can be regarded as subset of $S'_\theta(\mathbb{R}^n)$.

Theorem 1.4. *There exists an isomorphism between $\mathcal{L}(S_\theta(\mathbb{R}^n), S'_\theta(\mathbb{R}^n))$, the space of all linear continuous maps from $S_\theta(\mathbb{R}^n)$ to $S'_\theta(\mathbb{R}^n)$, and $S'_\theta(\mathbb{R}^{2n})$, which associates to every $T \in \mathcal{L}(S_\theta(\mathbb{R}^n), S'_\theta(\mathbb{R}^n))$ a distribution $K_T \in S'_\theta(\mathbb{R}^{2n})$ such that*

$$\langle Tu, v\rangle = \langle K_T, v \otimes u\rangle$$

for every $u, v \in S_\theta(\mathbb{R}^n)$. K_T is called the kernel of T.

Finally we give a result concerning the action of the Fourier transformation on $S_\theta(\mathbb{R}^n)$.

Proposition 1.5. *The Fourier transformation is an automorphism of $S_\theta(\mathbb{R}^n)$ and extends to an automorphism of $S'_\theta(\mathbb{R}^n)$.*

2. SG-calculus on Gelfand-Shilov spaces

In this section, we illustrate the main results concerning the action of Fourier integral operators of finite and infinite order on the spaces $S_\theta(\mathbb{R}^n), S'_\theta(\mathbb{R}^n)$. These results have been proved combining the standard arguments of the local theory of Fourier integral operators on the Gevrey classes, see [6], [18], [33], with the techniques coming from the SG-calculus on the Schwartz spaces $\mathcal{S}(\mathbb{R}^n), \mathcal{S}'(\mathbb{R}^n)$, see [7], [8]. For the sake of brevity, we omit or just sketch the proofs, since they are given in full detail in [3] and [4].

Let μ, ν, θ be real numbers such that $\mu > 1, \nu > 1$ and $\theta \geq \max\{\mu, \nu\}$.

Definition 2.1. For every $A > 0$ we denote by $\Gamma^{\infty}_{\mu,\nu,\theta}(\mathbb{R}^{2n}; A)$ the Fréchet space of all functions $a(x,\eta) \in C^{\infty}(\mathbb{R}^{2n})$ satisfying the following condition: for every $\varepsilon > 0$

$$\|a\|_{A,\varepsilon} = \sup_{\alpha,\beta\in\mathbb{N}^n} \sup_{(x,\eta)\in\mathbb{R}^{2n}} A^{-|\alpha|-|\beta|}(\alpha!)^{-\mu}(\beta!)^{-\nu}\langle\eta\rangle^{|\alpha|}\langle x\rangle^{|\beta|}\cdot$$

$$\cdot \exp\left[-\varepsilon(|x|^{\frac{1}{\theta}} + |\eta|^{\frac{1}{\theta}})\right] \left|D^{\alpha}_{\eta}D^{\beta}_{x}a(x,\eta)\right| < +\infty$$

endowed with the topology defined by the seminorms $\|\cdot\|_{A,\varepsilon}$, for $\varepsilon > 0$. We set

$$\Gamma^{\infty}_{\mu,\nu,\theta}(\mathbb{R}^{2n}) = \varinjlim_{A\to+\infty} \Gamma^{\infty}_{\mu,\nu,\theta}(\mathbb{R}^{2n}; A)$$

with the topology of inductive limit of Fréchet spaces.

An important subclass of $\Gamma^{\infty}_{\mu,\nu,\theta}(\mathbb{R}^{2n})$ is represented by SG-symbols of finite order which we will define as follows. Let μ, ν be real numbers such that $\mu > 1, \nu > 1$ and let $m = (m_1, m_2)$ be a vector of $\mathbb{R}^2$.

Definition 2.2. For every $B > 0$ we denote by $\Gamma^{m}_{\mu,\nu}(\mathbb{R}^{2n}; B)$ the Banach space of all functions $a(x,\eta) \in C^{\infty}(\mathbb{R}^{2n})$ such that

$$\|a\|_{B} = \sup_{\alpha,\beta\in\mathbb{N}^n} \sup_{(x,\eta)\in\mathbb{R}^{2n}} B^{-|\alpha|-|\beta|}(\alpha!)^{-\mu}(\beta!)^{-\nu}\cdot$$

$$\cdot\langle\eta\rangle^{-m_1+|\alpha|}\langle x\rangle^{-m_2+|\beta|} \cdot \left|D^{\alpha}_{\eta}D^{\beta}_{x}a(x,\eta)\right| < +\infty$$

endowed with the norm $\|\cdot\|_B$ and define

$$\Gamma^{m}_{\mu,\nu}(\mathbb{R}^{2n}) = \varinjlim_{B\to+\infty} \Gamma^{m}_{\mu,\nu}(\mathbb{R}^{2n}; B).$$

We observe that $\Gamma^{m}_{\mu,\nu}(\mathbb{R}^{2n}) \subset \Gamma^{\infty}_{\mu,\nu,\theta}(\mathbb{R}^{2n})$ for every $m \in \mathbb{R}^2$ and for all $\theta \geq \max\{\mu,\nu\}$.

Definition 2.3. A function $\varphi \in \Gamma^{e}_{\mu,\nu}(\mathbb{R}^{2n})$ will be called a phase function if it is real-valued and there exists a positive constant C_φ such that

$$C_{\varphi}^{-1}\langle x\rangle \leq \langle\nabla_\eta\varphi\rangle \leq C_\varphi\langle x\rangle \tag{2.1}$$

$$C_{\varphi}^{-1}\langle \eta\rangle \leq \langle\nabla_x\varphi\rangle \leq C_\varphi\langle \eta\rangle. \tag{2.2}$$

We shall denote by $\mathcal{P}_{\mu,\nu}$ the space of all phase functions from $\Gamma^{e}_{\mu,\nu}(\mathbb{R}^{2n})$.

Given $a \in \Gamma^{\infty}_{\mu,\nu,\theta}(\mathbb{R}^{2n})$ and $\varphi \in \mathcal{P}_{\mu,\nu}$, we can consider the Fourier integral operator

$$A_{a,\varphi}u(x) = \int_{\mathbb{R}^n} e^{i\varphi(x,\eta)}a(x,\eta)\hat{u}(\eta)đ\eta, \qquad u \in S_\theta(\mathbb{R}^n), \tag{2.3}$$

where we denote $đ\eta = (2\pi)^{-n}d\eta$. In particular, for $\varphi(x,\eta) = \langle x,\eta\rangle$, we obtain the pseudodifferential operator of symbol $a(x,\eta)$

$$Au(x) = \int_{\mathbb{R}^n} e^{i\langle x,\eta\rangle}a(x,\eta)\hat{u}(\eta)đ\eta. \tag{2.4}$$

In view of Proposition 1.5 and Definition 2.1, the integrals (2.3) and (2.4) are absolutely convergent. Given $m \in \mathbb{R}^2$, we shall denote by $OPS^m_{\mu,\nu}(\mathbb{R}^n)$ the space of all operators of the form (2.4) defined by a symbol $a \in \Gamma^m_{\mu,\nu}(\mathbb{R}^{2n})$. We set

$$OPS_{\mu,\nu}(\mathbb{R}^n) = \bigcup_{m\in\mathbb{R}^2} OPS^m_{\mu,\nu}(\mathbb{R}^n).$$

Lemma 2.4. *Let $\varphi \in \Gamma^e_{\mu,\nu}(\mathbb{R}^{2n})$. Then, for every α, β in $\mathbb{N}^n$, there exists a function $k_{\alpha,\beta}(x,\eta) \in C^\infty(\mathbb{R}^{2n})$ such that*

$$D^\alpha_\eta D^\beta_x e^{i\varphi(x,\eta)} = e^{i\varphi(x,\eta)} k_{\alpha,\beta}(x,\eta)$$

and

$$|k_{\alpha,\beta}(x,\eta)| \leq C^{|\alpha|+|\beta|}|\alpha|!^\mu|\beta|!^\nu \cdot \\ \cdot \sum_{h=0}^{\max\{0,|\alpha|-1\}} \frac{\langle x\rangle^{h+1-|\beta|}}{(h!)^\mu} \sum_{k=0}^{\max\{0,|\beta|-1\}} \frac{\langle \eta\rangle^{k+1-|\alpha|}}{(k!)^\nu} \tag{2.5}$$

for every $(x,\eta) \in \mathbb{R}^{2n}$ and for some $C > 0$ independent of α, β.

With the aid of Lemma 2.4, Propositions 1.1 and 1.5, we obtain the following result.

Theorem 2.5. *Let $\varphi \in \mathcal{P}_{\mu,\nu}$. Then, the map $(a,u) \to A_{a,\varphi}u$ is a bilinear and separately continuous map from $\Gamma^\infty_{\mu,\nu,\theta}(\mathbb{R}^{2n}) \times S_\theta(\mathbb{R}^n)$ to $S_\theta(\mathbb{R}^n)$ and it can be extended to a bilinear and separately continuous map from $\Gamma^\infty_{\mu,\nu,\theta}(\mathbb{R}^{2n}) \times S'_\theta(\mathbb{R}^n)$ to $S'_\theta(\mathbb{R}^n)$.*

Definition 2.6. An operator of the form (2.3) is said to be θ-regularizing if it can be extended to a linear continuous map from $S'_\theta(\mathbb{R}^n)$ to $S_\theta(\mathbb{R}^n)$.

Proposition 2.7. *Let $\varphi \in \mathcal{P}_{\mu,\nu}$ and let $a \in S_\theta(\mathbb{R}^{2n})$. Then, the operator $A_{a,\varphi}$ is θ-regularizing.*

We now give an asymptotic expansion of symbols from $\Gamma^\infty_{\mu,\nu,\theta}(\mathbb{R}^{2n})$. Let us denote, for $t > 0$

$$Q_t = \{(x,\eta) \in \mathbb{R}^{2n} : \langle\eta\rangle < t, \langle x\rangle < t\}$$

and

$$Q^e_t = \mathbb{R}^{2n} \setminus Q_t.$$

Definition 2.8. Let $B, C > 0$. We shall denote by $\mathcal{FS}^\infty_{\mu,\nu,\theta}(\mathbb{R}^{2n}; B, C)$ the space of all formal sums $\sum_{j\geq 0} a_j(x,\eta)$ such that $a_j(x,\eta) \in C^\infty(\mathbb{R}^{2n})$ for all $j \geq 0$ and for every $\varepsilon > 0$

$$\sup_{j\geq 0}\ \sup_{\alpha,\beta\in\mathbb{N}^n}\ \sup_{(x,\eta)\in Q^e_{Bj^{\mu+\nu-1}}} C^{-|\alpha|-|\beta|-2j}(\alpha!)^{-\mu}(\beta!)^{-\nu}(j!)^{-\mu-\nu+1}\langle\eta\rangle^{|\alpha|+j}\langle x\rangle^{|\beta|+j} \cdot \\ \cdot \exp\left[-\varepsilon(|x|^{\frac{1}{\theta}} + |\eta|^{\frac{1}{\theta}})\right] \left|D^\alpha_\eta D^\beta_x a_j(x,\eta)\right| < +\infty. \tag{2.6}$$

Consider the space $FS^\infty_{\mu,\nu,\theta}(\mathbb{R}^{2n};B,C)$ obtained from $\mathcal{FS}^\infty_{\mu,\nu,\theta}(\mathbb{R}^{2n};B,C)$ by taking its quotient by the subspace

$$E=\left\{\sum_{j\geq 0} a_j(x,\eta)\in \mathcal{FS}^\infty_{\mu,\nu,\theta}(\mathbb{R}^{2n};B,C): supp\,(a_j)\subset Q_{Bj^{\mu+\nu-1}}\quad \forall j\geq 0\right\}.$$

By abuse of notation, we shall denote the elements of $FS^\infty_{\mu,\nu,\theta}(\mathbb{R}^{2n};B,C)$ by formal sums of the form $\sum_{j\geq 0} a_j(x,\eta)$. The arguments in the following are independent of the choice of representative. We observe that $FS^\infty_{\mu,\nu,\theta}(\mathbb{R}^{2n};B,C)$ is a Fréchet space endowed with the seminorms given by the left-hand side of (2.6), for $\varepsilon>0$. We set

$$FS^\infty_{\mu,\nu,\theta}(\mathbb{R}^{2n})=\lim_{\substack{\longrightarrow\\ B,C\to+\infty}} FS^\infty_{\mu,\nu,\theta}(\mathbb{R}^{2n};B,C).$$

Each symbol $a\in\Gamma^\infty_{\mu,\nu,\theta}(\mathbb{R}^{2n})$ can be identified with an element $\sum_{j\geq 0} a_j$ of $FS^\infty_{\mu,\nu,\theta}(\mathbb{R}^{2n})$ by setting $a_0=a$ and $a_j=0$ for all $j\geq 1$.

Definition 2.9. We say that two sums $\sum_{j\geq 0} a_j$, $\sum_{j\geq 0} a'_j$ from $FS^\infty_{\mu,\nu,\theta}(\mathbb{R}^{2n})$ are equivalent $\left(\text{we write } \sum_{j\geq 0} a_j(x,\eta)\sim\sum_{j\geq 0} a'_j(x,\eta)\right)$ if there exist constants $B,C>0$ such that for all $\varepsilon>0$

$$\sup_{N\in\mathbb{Z}_+}\sup_{\alpha,\beta\in\mathbb{N}^n}\sup_{(x,\eta)\in Q^e_{BN^{\mu+\nu-1}}} C^{-|\alpha|-|\beta|-2N}(\alpha!)^{-\mu}(\beta!)^{-\nu}(N!)^{-\mu-\nu+1}\langle\eta\rangle^{|\alpha|+N}\langle x\rangle^{|\beta|+N}\cdot$$

$$\cdot\exp\left[-\varepsilon(|x|^{\frac{1}{\theta}}+|\eta|^{\frac{1}{\theta}})\right]\left|D^\alpha_\eta D^\beta_x\sum_{j<N}(a_j-a'_j)\right|<+\infty.$$

Theorem 2.10. *Given a sum* $\sum_{j\geq 0} a_j\in FS^\infty_{\mu,\nu,\theta}(\mathbb{R}^{2n})$, *we can find a symbol* a *in* $\Gamma^\infty_{\mu,\nu,\theta}(\mathbb{R}^{2n})$ *such that*

$$a\sim\sum_{j\geq 0} a_j \quad in \quad FS^\infty_{\mu,\nu,\theta}(\mathbb{R}^{2n}).$$

Proposition 2.11. *Let* $\varphi\in\mathcal{P}_{\mu,\nu}$ *and* $a\in\Gamma^\infty_{\mu,\nu,\theta}(\mathbb{R}^{2n})$ *such that* $a\sim 0$. *Then, the operator* $A_{a,\varphi}$ *is* θ*-regularizing.*

In the following statements, we will need stronger assumptions on μ,ν,θ. Namely, we will assume that

$$1<\mu\leq\nu,\qquad \theta\geq\mu+\nu-1. \tag{2.7}$$

These assumptions are crucial to define the product of a pseudodifferential and a Fourier integral operator in our classes. In particular, the condition $\theta\geq\mu+\nu-1$ is related to the loss of of Gevrey regularity occurring in the stationary phase method, cf. [5], [16], [17] and in the composition formula, cf. [1], [6], [19], [22], [33]. In some particular cases, the condition (2.7) can be relaxed, see Remark 2.13.

Theorem 2.12. *Let μ, ν, θ be real numbers satisfying (2.7) and let*

$$A_{a,\varphi}u(x) = \int_{\mathbb{R}^n} e^{i\varphi(x,\eta)} a(x,\eta)\hat{u}(\eta) đ\eta,$$

$$Pu(x) = \int_{\mathbb{R}^n} e^{i\langle x,\eta\rangle} p(x,\eta)\hat{u}(\eta) đ\eta,$$

where $\varphi \in \mathcal{P}_{\mu,\nu}, a \in \Gamma^{\infty}_{\mu,\nu,\theta}(\mathbb{R}^{2n}), p \in \Gamma^{m}_{\mu,\nu}(\mathbb{R}^{2n})$ for some $m = (m_1, m_2) \in \mathbb{R}^2$. Then, $PA_{a,\varphi}$ is, modulo θ-regularizing operators, a Fourier integral operator with phase function φ and symbol $q \in \Gamma^{\infty}_{\mu,\nu,\theta}(\mathbb{R}^{2n})$. Furthermore,

$$q(x,\eta) \sim \sum_{j\geq 0} \sum_{|\alpha|=j} (\alpha!)^{-1} D_z^{\alpha} \left((\partial_\eta^\alpha p)(x, \widetilde{\nabla}_x \varphi(x,z,\eta)) a(z,\eta) \right)_{|z=x} \tag{2.8}$$

in $FS^{\infty}_{\mu,\nu,\theta}(\mathbb{R}^{2n})$ with

$$\widetilde{\nabla}_x \varphi(x,z,\eta) = \int_0^1 (\nabla_x \varphi)(z + \tau(x-z), \eta) d\tau.$$

Remark 2.13. With the same notation as in Theorem 2.12, we see that if $a \sim \sum_{h\geq 0} a_h$ in $FS^{\infty}_{\mu,\nu,\theta}(\mathbb{R}^{2n})$, then

$$q(x,\eta) \sim \sum_{j\geq 0} \sum_{|\alpha|=j-h} (\alpha!)^{-1} D_z^{\alpha} \left((\partial_\eta^\alpha p)(x, \widetilde{\nabla}_x \varphi(x,z,\eta)) a_h(z,\eta) \right)_{|z=x}.$$

Moreover, when $\varphi(x,\eta) = \langle x, \eta\rangle$, we have from (2.8) the standard formula for the symbol of the product of two pseudodifferential operators. We emphasize that, as we are dealing only with pseudodifferential operators, we can simply assume $\mu > 1, \nu > 1$ in Theorem 2.12 instead of $1 < \mu \leq \nu$. Finally, we observe that if P is a differential operator, the sum in (2.8) is finite and Theorem 2.12 holds under the weaker assumptions $\mu > 1, \nu > 1, \theta \geq \max\{\mu, \nu\}$.

Remark 2.14. Given μ, ν, θ satisfying (2.7), if $\varphi \in \mathcal{P}_{\mu,\mu}, p \in \Gamma^m_{\mu,\mu}(\mathbb{R}^{2n})$ and $a \in \Gamma^{\infty}_{\nu,\mu,\theta}(\mathbb{R}^{2n})$, then the operator $PA_{a,\varphi}$ is a Fourier integral operator with phase φ and symbol $q \in \Gamma^{\infty}_{\nu,\mu,\theta}(\mathbb{R}^{2n})$ satisfying (2.8) in $FS^{\infty}_{\nu,\mu,\theta}(\mathbb{R}^{2n})$.

Under the same assumptions of Theorem 2.12, we are also interested in the study of the operator $A_{a,\varphi}P$, which will occur in the proofs of the statements of Section 5. Let us give a preliminary result.

Lemma 2.15. *Let $a \in \Gamma^{\infty}_{\mu,\nu,\theta}(\mathbb{R}^{2n})$ and $\varphi \in \mathcal{P}_{\mu,\nu}$ and consider the transpose ${}^tA_{a,\varphi}$ of the operator $A_{a,\varphi}$ defined by*

$$\langle {}^tA_{a,\varphi}u, v\rangle = \langle u, A_{a,\varphi}v\rangle, \qquad u \in S'_\theta(\mathbb{R}^n), v \in S_\theta(\mathbb{R}^n).$$

Then, we have

$${}^tA_{a,\varphi} = \mathcal{F} \circ A_{a^\sharp,\varphi^\sharp} \circ \mathcal{F}^{-1},$$

where we denote $a^\sharp(x,\eta) = a(\eta,x), \varphi^\sharp(x,\eta) = \varphi(\eta,x)$.

Theorem 2.16. *Let μ, ν, θ be real numbers satisfying (2.7) and let*

$$A_{a,\varphi}u(x) = \int_{\mathbb{R}^n} e^{i\varphi(x,\eta)} a(x,\eta)\hat{u}(\eta)đ\eta,$$

$$Pu(x) = \int_{\mathbb{R}^n} e^{i\langle x,\eta\rangle} p(x,\eta)\hat{u}(\eta)đ\eta,$$

where $\varphi \in \mathcal{P}_{\mu,\mu}, a \in \Gamma^{\infty}_{\mu,\nu,\theta}(\mathbb{R}^{2n}), p \in \Gamma^{m}_{\mu,\mu}(\mathbb{R}^{2n})$ for some $m = (m_1, m_2) \in \mathbb{R}^2$. Then, the operator $A_{a,\varphi}P$ is, modulo θ-regularizing operators, a Fourier integral operator with phase function φ and symbol $h \in \Gamma^{\infty}_{\mu,\nu,\theta}(\mathbb{R}^{2n})$ such that

$$h(x,\eta) \sim \sum_{j\geq 0}\sum_{|\alpha|=j} (\alpha!)^{-1} D^{\alpha}_{\zeta}\left((\partial^{\alpha}_{x}p)(\widetilde{\nabla}_{\eta}\varphi(x,\zeta,\eta),\eta)a(x,\zeta)\right)_{|\zeta=\eta} \tag{2.9}$$

in $FS^{\infty}_{\mu,\nu,\theta}(\mathbb{R}^{2n})$, where

$$\widetilde{\nabla}_{\eta}\varphi(x,\zeta,\eta) = \int_0^1 (\nabla_{\eta}\varphi)(x,\zeta+\tau(\eta-\zeta))d\tau.$$

Proof. By Lemma 2.15, Theorem 2.12 and Remark 2.14, denoting by $P^{\sharp}$ the operator with symbol $p^{\sharp}(x,\eta) = p(\eta,x)$, we can write

$$A_{a,\varphi}P = {}^t({}^tP\,{}^tA_{a,\varphi}) = {}^t[(\mathcal{F}(P^{\sharp}A_{a^{\sharp},\varphi^{\sharp}})\mathcal{F}^{-1})] = {}^t\,[\mathcal{F}A_{h^{\sharp},\varphi^{\sharp}}\mathcal{F}^{-1}]$$

with $h^{\sharp} \in \Gamma^{\infty}_{\nu,\mu,\theta}(\mathbb{R}^{2n})$ such that

$$h^{\sharp}(x,\eta) \sim \sum_{\alpha}(\alpha!)^{-1}D^{\alpha}_{z}\left((\partial^{\alpha}_{\eta}p)(\widetilde{\nabla}_{\eta}\varphi(\eta,z,x),x)a(\eta,z)\right)_{|z=x} \quad \textit{in} \quad FS^{\infty}_{\nu,\mu,\theta}(\mathbb{R}^{2n}).$$

Then, applying again Lemma 2.15, we deduce that $A_{a,\varphi}P = A_{h,\varphi}$ where h satisfies (2.9). □

Remark 2.17. The results of this section obviously hold also for symbols of finite order introduced in Definition 2.2. Nevertheless, in view of the applications in the next few sections, it is convenient to have for them more precise results, obtained by defining in a suitable way formal sums of finite order and a corresponding equivalence relation.

Let μ, ν be real numbers such that $\mu > 1, \nu > 1$, and let $m = (m_1, m_2) \in \mathbb{R}^2$.

Definition 2.18. Let $B, C > 0$. We shall denote by $FS^{m}_{\mu,\nu}(\mathbb{R}^{2n}; B, C)$ the space of all formal sums $\sum_{j\geq 0} p_j(x,\eta)$ such that $p_j(x,\eta) \in C^{\infty}(\mathbb{R}^{2n})$ for all $j \geq 0$ and

$$\sup_{j\geq 0}\sup_{\alpha,\beta\in\mathbb{N}^n}\sup_{(x,\eta)\in Q^e_{Bj^{\mu+\nu-1}}} C^{-|\alpha|-|\beta|-2j}(\alpha!)^{-\mu}(\beta!)^{-\nu}(j!)^{-\mu-\nu+1}\cdot$$

$$\cdot\langle\eta\rangle^{-m_1+|\alpha|+j}\langle x\rangle^{-m_2+|\beta|+j}\left|D^{\alpha}_{\eta}D^{\beta}_{x}p_j(x,\eta)\right| < +\infty, \tag{2.10}$$

where, as in Definition 2.8, we identify two sums $\sum_{j\geq 0} p_j$, $\sum_{j\geq 0} p'_j$ if $p_j - p'_j$ vanish in $Q_{Bj^{\mu+\nu-1}}$ for all $j \geq 0$. We set $FS^{m}_{\mu,\nu}(\mathbb{R}^{2n}) = \varinjlim_{B,C\to+\infty} FS^{m}_{\mu,\nu}(\mathbb{R}^{2n}; B, C)$.

Definition 2.19. We say that two sums $\sum_{j\geq 0} p_j$, $\sum_{j\geq 0} p'_j$ from $FS^m_{\mu,\nu}(\mathbb{R}^{2n})$ are equivalent $\left(\text{we write } \sum_{j\geq 0} p_j(x,\eta) \sim \sum_{j\geq 0} p'_j(x,\eta)\right)$ if there exist constants $B, C > 0$ such that

$$\sup_{N\in\mathbb{Z}_+} \sup_{\alpha,\beta\in\mathbb{N}^n} \sup_{(x,\eta)\in Q^e_{BN^{\mu+\nu-1}}} C^{-|\alpha|-|\beta|-2N}(\alpha!)^{-\mu}(\beta!)^{-\nu}(N!)^{-\mu-\nu+1}.$$

$$\cdot\langle\eta\rangle^{-m_1+|\alpha|+N}\langle x\rangle^{-m_2+|\beta|+N}\left|D^\alpha_\eta D^\beta_x \sum_{j<N}(p_j - p'_j)\right| < +\infty.$$

Theorem 2.10 and Proposition 2.11 can be formulated for symbols of finite order starting from Definitions 2.18 and 2.19. Moreover, with the same notation as in Theorems 2.12 and 2.16, if $a \in \Gamma^{m'}_{\mu,\nu}(\mathbb{R}^{2n})$ for some $m' \in \mathbb{R}^2$, then the operators $PA_{a,\varphi}$ and $A_{a,\varphi}P$ are Fourier integral operators with phase function φ and symbol q, respectively h, in $\Gamma^{m+m'}_{\mu,\nu}(\mathbb{R}^{2n})$ satisfying (2.8), respectively (2.9), in $FS^{m+m'}_{\mu,\nu}(\mathbb{R}^{2n})$.

3. Elliptic and polyhomogeneous symbols of finite order

In this section, we investigate two important typologies of SG-symbols of finite order and the relations between them. We will restrict our attention to pseudo-differential operators defined by such symbols. Let μ, ν be real numbers such that $\mu > 1, \nu > 1$.

Definition 3.1. A symbol $p \in \Gamma^m_{\mu,\nu}(\mathbb{R}^{2n})$ is said to be elliptic if there exist $B, C > 0$ such that

$$|p(x,\eta)| \geq C\langle\eta\rangle^{m_1}\langle x\rangle^{m_2} \qquad \forall(x,\eta) \in Q^e_B.$$

Theorem 3.2. *Given $p \in \Gamma^m_{\mu,\nu}(\mathbb{R}^{2n})$ elliptic, we can find $E, E' \in OPS^{-m}_{\mu,\nu}(\mathbb{R}^n)$ such that $EP = I + R, PE' = I + R'$, where I is the identity operator on $S'_\theta(\mathbb{R}^n)$ and R, R' are θ-regularizing operators.*

Proof. By Theorem 2.10, the symbols e and e' can be constructed starting from their asymptotic expansions $\sum_{j\geq 0} e_j$, $\sum_{j\geq 0} e'_j$. We can define $e_0 \in C^\infty(\mathbb{R}^{2n})$ such that

$$e_0(x,\eta) = p(x,\eta)^{-1} \quad \forall(x,\eta) \in Q^e_B$$

and by induction on $j \geq 1$

$$e_j(x,\eta) = -e_0(x,\eta) \sum_{0<|\alpha|\leq j} (\alpha!)^{-1}\partial^\alpha_\eta e_{j-|\alpha|}(x,\eta)D^\alpha_x p(x,\eta).$$

By induction on j, it is easy to prove that $\sum_{j\geq 0} e_j \in FS^{-m}_{\mu,\nu}(\mathbb{R}^{2n})$. Moreover, by Theorem 2.12 and Remark 2.13, the operator E is such that $EP = I + R$. The construction of E' is analogous. □

Corollary 3.3. *Let $p \in \Gamma^m_{\mu,\nu}(\mathbb{R}^{2n})$ be an elliptic symbol and let $f \in S_\theta(\mathbb{R}^n)$ for some $\theta \geq \mu+\nu-1$. Then, if $u \in S'_\theta(\mathbb{R}^n)$ is a solution of the equation $Pu = f$, then $u \in S_\theta(\mathbb{R}^n)$.*

We can also define the notion of ellipticity of a symbol with respect to another one.

Definition 3.4. Let $p \in \Gamma^m_{\mu,\nu}(\mathbb{R}^{2n}), q \in \Gamma^{m'}_{\mu,\nu}(\mathbb{R}^{2n})$ for some $m, m' \in \mathbb{R}^2$. We say that p is elliptic with respect to q if there exist $B, C > 0$ such that

$$|p(x,\eta)| \geq C\langle \eta\rangle^{m_1}\langle x\rangle^{m_2}$$

for all $(x,\eta) \in Q^e_B \cap supp(q)$.

In particular, the symbol p is elliptic if and only if p is elliptic with respect to $q \equiv 1$. Arguing as in the proof of Theorem 3.2, it is easy to prove the following result.

Proposition 3.5. *Given $p \in \Gamma^m_{\mu,\nu}(\mathbb{R}^{2n})$ elliptic with respect to $q \in \Gamma^{m'}_{\mu,\nu}(\mathbb{R}^{2n})$, we can find $E, E' \in OPS^{m'-m}_{\mu,\nu}(\mathbb{R}^n)$ such that $EP = Q + R, PE' = Q + R'$, where R, R' are θ-regularizing operators.*

We can now introduce polyhomogeneous SG-symbols. We follow the approach of Y. Egorov and B.-W. Schulze [13], [29], who have treated polyhomogeneous SG-symbols in the $\mathcal{S}$-$\mathcal{S}'$-framework. Namely, we will define three classes whose elements are polyhomogeneous in x, in η and in both x, η, respectively. Before giving precise definitions of these spaces, we need to introduce in our context a notion of asymptotic expansion with respect to x and η separately. Let μ, ν be real numbers such that $\mu > 1, \nu > 1$ and let $m = (m_1, m_2)$ be a vector of $\mathbb{R}^2$.

Definition 3.6. We denote by $FS^m_{\mu,\nu,\eta}(\mathbb{R}^{2n})$ the space of all formal sums $\sum_{j\geq 0} p_j(x,\eta)$ such that $p_j \in C^\infty(\mathbb{R}^{2n}) \forall j \geq 0$ and there exist $B, C > 0$ such that

$$\sup_{j\geq 0}\sup_{\alpha,\beta\in\mathbb{N}^n}\sup_{\substack{\langle\eta\rangle\geq Bj^{\mu+\nu-1}\\ x\in\mathbb{R}^n}} C^{-|\alpha|-|\beta|-j}(\alpha!)^{-\mu}(\beta!)^{-\nu}(j!)^{-\mu-\nu+1}\cdot$$

$$\cdot\langle\eta\rangle^{-m_1+|\alpha|+j}\langle x\rangle^{-m_2+|\beta|}\left|D^\alpha_\eta D^\beta_x p_j(x,\eta)\right| < +\infty. \tag{3.1}$$

As in Definition 2.19, we can define an equivalence relation among the elements of $FS^m_{\mu,\nu,\eta}(\mathbb{R}^{2n})$.

Definition 3.7. Two sums $\sum_{j\geq 0} p_j, \sum_{j\geq 0} p'_j \in FS^m_{\mu,\nu,\eta}(\mathbb{R}^{2n})$ are said to be equivalent $\left(\text{we write } \sum_{j\geq 0} p_j \sim_\eta \sum_{j\geq 0} p'_j\right)$ if there exist $B, C > 0$ such that

$$\sup_{N\in\mathbb{Z}_+}\sup_{\alpha,\beta\in\mathbb{N}^n}\sup_{\substack{\langle\eta\rangle\geq BN^{\mu+\nu-1}\\ x\in\mathbb{R}^n}} C^{-|\alpha|-|\beta|-N}(\alpha!)^{-\mu}(\beta!)^{-\nu}(N!)^{-\mu-\nu+1}\cdot$$

$$\cdot\langle\eta\rangle^{-m_1+|\alpha|+N}\langle x\rangle^{-m_2+|\beta|}\left|D^\alpha_\eta D^\beta_x \sum_{j<N}(p_j - p'_j)\right| < +\infty.$$

In an analogous way, we can define the space $FS^m_{\mu,\nu,x}(\mathbb{R}^{2n})$ and the corresponding relation $\sim_x$.

Remark 3.8. We observe that

$$FS^m_{\mu,\nu}(\mathbb{R}^{2n}) \subset FS^m_{\mu,\nu,\eta}(\mathbb{R}^{2n}) \cap FS^m_{\mu,\nu,x}(\mathbb{R}^{2n}).$$

Furthermore, if $\sum_{j\geq 0} p_j \sim \sum_{j\geq 0} p'_j$ in $FS^m_{\mu,\nu}(\mathbb{R}^{2n})$, then

$$\sum_{j\geq 0} p_j \sim_\eta \sum_{j\geq 0} p'_j \quad \textit{and} \quad \sum_{j\geq 0} p_j \sim_x \sum_{j\geq 0} p'_j.$$

Similarly to Theorem 2.10, we have the following result.

Proposition 3.9. *Given* $\sum_{j\geq 0} p_j \in FS^m_{\mu,\nu,\eta}(\mathbb{R}^{2n})$, $\sum_{j\geq 0} q_j \in FS^m_{\mu,\nu,x}(\mathbb{R}^{2n})$, *then there exist* $p, q \in \Gamma^m_{\mu,\nu}(\mathbb{R}^{2n})$ *such that*

$$p \sim_\eta \sum_{j\geq 0} p_j \quad \textit{in} \quad FS^m_{\mu,\nu,\eta}(\mathbb{R}^{2n}),$$

$$q \sim_x \sum_{j\geq 0} q_j \quad \textit{in} \quad FS^m_{\mu,\nu,x}(\mathbb{R}^{2n}).$$

We can define the following classes of homogeneous symbols.

Definition 3.10. We denote by $\Gamma^{[m_1],m_2}_{\mu,\nu}(\mathbb{R}^{2n})$ the space of all symbols $p \in \Gamma^m_{\mu,\nu}(\mathbb{R}^{2n})$ such that $p(x,\lambda\eta) = \lambda^{m_1} p(x,\eta) \quad \forall \lambda \geq 1, |\eta| \geq c > 0, x \in \mathbb{R}^n$. Analogously, we define the space $\Gamma^{m_1,[m_2]}_{\mu,\nu}(\mathbb{R}^{2n})$ by interchanging the roles of x and η. Finally, we set

$$\Gamma^{[m_1],[m_2]}_{\mu,\nu}(\mathbb{R}^{2n}) = \Gamma^{[m_1],m_2}_{\mu,\nu}(\mathbb{R}^{2n}) \cap \Gamma^{m_1,[m_2]}_{\mu,\nu}(\mathbb{R}^{2n}).$$

Using Definitions 3.6, 3.7, 3.10, we can now introduce polyhomogeneous symbols.

Definition 3.11. We denote by $\Gamma^{m_1,[m_2]}_{\mu,\nu,cl(\eta)}(\mathbb{R}^{2n})$ the space of all $p \in \Gamma^{m_1,[m_2]}_{\mu,\nu}(\mathbb{R}^{2n})$ satisfying the following condition: there exists a sum $\sum_{k\geq 0} p_k \in FS^m_{\mu,\nu,\eta}(\mathbb{R}^{2n})$ such that $p_k \in \Gamma^{[m_1-k],[m_2]}_{\mu,\nu}(\mathbb{R}^{2n}) \quad \forall k \geq 0$ and $p \sim_\eta \sum_{k\geq 0} p_k$ in $FS^m_{\mu,\nu,\eta}(\mathbb{R}^{2n})$.

Definition 3.12. We denote by $\Gamma^{m_1,m_2}_{\mu,\nu,cl(\eta)}(\mathbb{R}^{2n})$ the space of all symbols $p \in \Gamma^m_{\mu,\nu}(\mathbb{R}^{2n})$ satisfying the following condition: there exists a sum $\sum_{k\geq 0} p_k \in FS^m_{\mu,\nu,\eta}(\mathbb{R}^{2n})$ such that $p_k \in \Gamma^{[m_1-k],m_2}_{\mu,\nu}(\mathbb{R}^{2n}) \quad \forall k \geq 0$ and $p \sim_\eta \sum_{k\geq 0} p_k$ in $FS^m_{\mu,\nu,\eta}(\mathbb{R}^{2n})$.

Analogous definitions can be given for the spaces $\Gamma^{[m_1],m_2}_{\mu,\nu,cl(x)}(\mathbb{R}^{2n})$ and $\Gamma^{m_1,m_2}_{\mu,\nu,cl(x)}(\mathbb{R}^{2n})$, by interchanging the roles of x and η. Finally, we define a space of symbols which are polyhomogeneous with respect to both the variables.

Definition 3.13. We denote by $\Gamma^m_{\mu,\nu,cl}(\mathbb{R}^{2n})$ the space of all symbols $p \in \Gamma^m_{\mu,\nu}(\mathbb{R}^{2n})$ for which the following conditions hold:

i) there exists $\sum_{k\geq 0} p_k \in FS^m_{\mu,\nu,\eta}(\mathbb{R}^{2n})$ such that $p_k \in \Gamma^{[m_1-k],m_2}_{\mu,\nu,cl(x)}(\mathbb{R}^{2n}) \quad \forall k \in \mathbb{N}$,

$p \sim_\eta \sum_{k\geq 0} p_k \quad in \quad FS^m_{\mu,\nu,\eta}(\mathbb{R}^{2n})$ and $p - \sum_{k<N} p_k \in \Gamma^{m_1-N,m_2}_{\mu,\nu,cl(x)}(\mathbb{R}^{2n}) \quad \forall N \in \mathbb{Z}_+$;

ii) there exists $\sum_{h\geq 0} q_h \in FS^m_{\mu,\nu,x}(\mathbb{R}^{2n})$ such that $q_h \in \Gamma^{m_1,[m_2-h]}_{\mu,\nu,cl(\eta)}(\mathbb{R}^{2n}) \quad \forall h \in \mathbb{N}$,

$p \sim_x \sum_{h\geq 0} q_h \quad in \quad FS^m_{\mu,\nu,x}(\mathbb{R}^{2n})$ and $p - \sum_{h<N} q_h \in \Gamma^{m_1,m_2-N}_{\mu,\nu,cl(\eta)}(\mathbb{R}^{2n}) \quad \forall N \in \mathbb{Z}_+$.

The following inclusions hold:

$$\Gamma^{m_1,[m_2]}_{\mu,\nu,cl(\eta)}(\mathbb{R}^{2n}) \subset \Gamma^m_{\mu,\nu,cl}(\mathbb{R}^{2n}), \quad \Gamma^{[m_1],m_2}_{\mu,\nu,cl(x)}(\mathbb{R}^{2n}) \subset \Gamma^m_{\mu,\nu,cl}(\mathbb{R}^{2n}). \tag{3.2}$$

A simple homogeneity argument shows that for every $p \in \Gamma^m_{\mu,\nu,cl}(\mathbb{R}^{2n})$ and for every $k \in \mathbb{N}$, there exists a unique function $\sigma^{m_1-k}_\psi(p) \in C^\infty(\mathbb{R}^n \times (\mathbb{R}^n \setminus \{0\}))$ such that $\sigma^{m_1-k}_\psi(p)(x,\lambda\eta) = \lambda^{m_1-k}\sigma^{m_1-k}_\psi(p)(x,\eta)$ for all $\lambda > 0, x \in \mathbb{R}^n, \eta \neq 0$ and $\sigma^{m_1-k}_\psi(p)(x,\eta) = p_k(x,\eta)$ for $|\eta| \geq c > 0$. Analogously, in view of condition ii) of Definition 3.13, we can associate to every $p \in \Gamma^m_{\mu,\nu,cl}(\mathbb{R}^{2n})$ the functions $\sigma^{m_2-h}_e(p) \quad \forall h \in \mathbb{N}$ such that $\sigma^{m_2-h}_e(p) \in C^\infty((\mathbb{R}^n \setminus \{0\}) \times \mathbb{R}^n), \sigma^{m_2-h}_e(p)(x,\eta) = q_h(x,\eta)$ for $|x| \geq c > 0$ and $\sigma^{m_2-h}_e(p)(\lambda x,\eta) = \lambda^{m_2-h}\sigma^{m_2-h}_e(p)(x,\eta)$ for all $\lambda > 0, \eta \in \mathbb{R}^n, x \neq 0$. We also observe that if $\omega \in G^\mu(\mathbb{R}^n)$ is an excision function, i.e., $\omega = 0$ in a neighborhood of the origin and $\omega = 1$ in a neighborhood of ∞, then $\omega(\eta)\sigma^{m_1-k}_\psi(p)(x,\eta)$ is in $\Gamma^{[m_1-k],m_2}_{\mu,\nu,cl(x)}(\mathbb{R}^{2n})$. Similarly, if $\chi(x)$ is an excision function in $G^\nu(\mathbb{R}^n)$, then $\chi(x)\sigma^{m_2-h}_e(p)(x,\eta)$ is in $\Gamma^{m_1,[m_2-h]}_{\mu,\nu,cl(\eta)}(\mathbb{R}^{2n})$.

By these considerations and by the inclusions (3.2), we can also consider the functions $\sigma^{m_1-k}_\psi(\sigma^{m_2-h}_e(p))$ and $\sigma^{m_2-h}_e(\sigma^{m_1-k}_\psi(p))$. It is easy to show that

$$\sigma^{m_1-k}_\psi(\sigma^{m_2-h}_e(p)) = \sigma^{m_2-h}_e(\sigma^{m_1-k}_\psi(p)) \quad \text{for all } h,k \in \mathbb{N}.$$

In particular, given $p \in \Gamma^m_{\mu,\nu,cl}(\mathbb{R}^{2n})$, we can consider the triplet

$$\{\sigma^{m_1}_\psi(p), \sigma^{m_2}_e(p), \sigma^m_{\psi e}(p)\},$$

where we denote $\sigma^m_{\psi e}(p) = \sigma^{m_1}_\psi(\sigma^{m_2}_e(p))$.

The function $\sigma^{m_1}_\psi(p)$ is called the homogeneous principal interior symbol of p and the pair $\{\sigma^{m_2}_e(p), \sigma^m_{\psi e}(p)\}$ is the homogeneous principal exit symbol of p.

By the previous results, it turns out that, given two excision functions $\omega(\eta)$ in $G^\mu(\mathbb{R}^n)$ and $\chi(x) \in G^\nu(\mathbb{R}^n)$, we have also

$$p(x,\eta) - \omega(\eta)\sigma^{m_1}_\psi(p)(x,\eta) \in \Gamma^{(m_1-1,m_2)}_{\mu,\nu,cl}(\mathbb{R}^{2n}), \tag{3.3}$$

$$p(x,\eta) - \chi(x)\sigma^{m_2}_e(p)(x,\eta) \in \Gamma^{(m_1,m_2-1)}_{\mu,\nu,cl}(\mathbb{R}^{2n}), \tag{3.4}$$

$$\begin{aligned} p(x,\eta) - \omega(\eta)\sigma^{m_1}_\psi(p)(x,\eta) - \chi(x)(\sigma^{m_2}_e(p)(x,\eta) \\ - \omega(\eta)\sigma^m_{\psi e}(p)(x,\eta)) \in \Gamma^{m-e}_{\mu,\nu,cl}(\mathbb{R}^{2n}). \end{aligned} \tag{3.5}$$

We denote by $OPS^{m}_{\mu,\nu,cl}(\mathbb{R}^n)$ the set of all operators of the form (2.4) defined by a symbol $p \in \Gamma^{m}_{\mu,\nu,cl}(\mathbb{R}^{2n})$ and we set, for $\theta > 1$

$$OPS^{\theta}_{cl}(\mathbb{R}^n) = \bigcup_{\substack{m\in\mathbb{R}^2\\ \mu,\nu\in(1,+\infty)\\ \mu+\nu-1\leq\theta}} OPS^{m}_{\mu,\nu,cl}(\mathbb{R}^n).$$

Remark 3.14. Arguing as in the previous section and applying Remark 2.13, it is easy to prove that if $P \in OPS^{m}_{\mu,\nu,cl}(\mathbb{R}^n), Q \in OPS^{m'}_{\mu,\nu,cl}(\mathbb{R}^n)$, then the operator PQ is in $OPS^{m+m'}_{\mu,\nu,cl}(\mathbb{R}^n)$.

We recall (cf. Proposition 1.4.37 in [29]) that a symbol $p \in \Gamma^{m}_{\mu,\nu,cl}(\mathbb{R}^{2n})$ is elliptic if and only if the three following conditions hold:

$$\sigma^{m_1}_{\psi}(p)(x,\eta) \neq 0 \quad \forall (x,\eta) \in \mathbb{R}^n \times (\mathbb{R}^n \setminus \{0\}), \tag{3.6}$$

$$\sigma^{m_2}_{e}(p)(x,\eta) \neq 0 \quad \forall (x,\eta) \in (\mathbb{R}^n \setminus \{0\}) \times \mathbb{R}^n, \tag{3.7}$$

$$\sigma^{m}_{\psi e}(p)(x,\eta) \neq 0 \quad \forall (x,\eta) \in (\mathbb{R}^n \setminus \{0\}) \times (\mathbb{R}^n \setminus \{0\}). \tag{3.8}$$

Example. Consider a partial differential operator with polynomial coefficients

$$P = \sum_{\substack{|\alpha|\leq m_1\\ |\beta|\leq m_2}} c_{\alpha\beta} x^{\beta} D^{\alpha}.$$

The corresponding symbol belongs to $\Gamma^{m}_{\mu,\nu,cl}(\mathbb{R}^{2n})$ for every $\mu > 1, \nu > 1$ and $m = (m_1, m_2)$. The operator P is elliptic in the SG-sense if and only if

$$\sigma^{m_1}_{\psi} = \sum_{\substack{|\alpha|=m_1\\ |\beta|\leq m_2}} c_{\alpha\beta} x^{\beta} \eta^{\alpha} \neq 0 \text{ for } \eta \neq 0, \qquad \sigma^{m_2}_{e} = \sum_{\substack{|\alpha|\leq m_1\\ |\beta|= m_2}} c_{\alpha\beta} x^{\beta} \eta^{\alpha} \neq 0 \text{ for } x \neq 0$$

and

$$\sigma^{m}_{\psi e} = \sum_{\substack{|\alpha|=m_1\\ |\beta|= m_2}} c_{\alpha\beta} x^{\beta} \eta^{\alpha} \neq 0 \text{ for } x, \eta \neq 0.$$

4. θ-wave front set

In this section, we introduce an appropriate notion of wave front set for distributions $u \in S'_{\theta}(\mathbb{R}^n)$, and prove the standard properties of microellipticity with respect to the polyhomogeneous operators defined in Section 3. Similar results have been proved by S. Coriasco and L. Maniccia [10] for Schwartz tempered distributions. For every $\eta_o \in \mathbb{R}^n \setminus \{0\}$, we will denote by $\infty\eta_o$ the projection $\frac{\eta_o}{|\eta_o|}$ on the unit sphere S^{n-1}. In the following, an open set $V \subset \mathbb{R}^n$ is said to be a conic neighborhood of the direction $\infty\eta_o$ if it is the intersection of an open cone containing the direction $\infty\eta_o$ with the complementary set of a closed ball centered in the origin. The decomposition of the principal symbol into three components in the previous section suggests to define for the elements of $S'_{\theta}(\mathbb{R}^n)$ three sets which we will denote by $WF^\theta_\psi, WF^\theta_e, WF^\theta_{\psi e}, \theta > 1$. To give precise definitions, we need to introduce two types of cut-off functions.

Definition 4.1. Let $y_o \in \mathbb{R}^n$ and fix $\nu > 1$. We denote by $\mathcal{R}^\nu_{y_o}$ the set of all functions $\varphi \in G^\nu_o(\mathbb{R}^n)$ such that $0 \le \varphi \le 1$ and $\varphi \equiv 1$ in a neighborhood of y_o.

Definition 4.2. Let $\eta_o \in \mathbb{R}^n \setminus \{0\}$ and fix $\mu > 1$. We denote by $\mathcal{Z}^\mu_{\eta_o}$ the set of all functions $\psi \in C^\infty(\mathbb{R}^n)$ such that $\psi(\lambda\eta) = \psi(\eta) \forall \lambda \ge 1$ and $|\eta|$ large, $0 \le \psi \le 1, \psi \equiv 1$ in a conic neighborhood V of $\infty\eta_o, \psi \equiv 0$ outside a conic neighborhood V' of $\infty\eta_o, V \subset V'$ and

$$|D^\alpha_\eta \psi(\eta)| \le C^{|\alpha|+1}(\alpha!)^\mu \langle\eta\rangle^{-|\alpha|}, \quad \eta \in \mathbb{R}^n$$

for every $\alpha \in \mathbb{N}^n$ and for some $C > 0$.

Definition 4.3. Let θ be a positive real number such that $\theta > 1$ and let $u \in S'_\theta(\mathbb{R}^n)$.

- We say that $(x_o, \eta_o) \in \mathbb{R}^n \times (\mathbb{R}^n \setminus \{0\})$ is not in $WF^\theta_\psi u$ if there exist positive numbers $\mu, \nu \in (1, +\infty)$ such that $\theta \ge \max\{\mu, \nu\}$ and there exist cut-off functions φ_{x_o} in $\mathcal{R}^\nu_{x_o}, \psi_{\eta_o} \in \mathcal{Z}^\mu_{\eta_o}$ such that $\varphi_{x_o}(\psi_{\eta_o}(D)u) \in S_\theta(\mathbb{R}^n)$.
- We say that $(x_o, \eta_o) \in (\mathbb{R}^n \setminus \{0\}) \times \mathbb{R}^n$ is not in $WF^\theta_e u$ if there exist positive numbers $\mu, \nu \in (1, +\infty)$ such that $\theta \ge \max\{\mu, \nu\}$ and there exist cut-off functions φ_{η_o} in $\mathcal{R}^\mu_{\eta_o}, \psi_{x_o} \in \mathcal{Z}^\nu_{x_o}$ such that $\psi_{x_o}(\varphi_{\eta_o}(D)u) \in S_\theta(\mathbb{R}^n)$.
- We say that $(x_o, \eta_o) \in (\mathbb{R}^n \setminus \{0\}) \times (\mathbb{R}^n \setminus \{0\})$ is not in $WF^\theta_{\psi e} u$ if there exist positive numbers $\mu, \nu \in (1, +\infty)$ such that $\theta \ge \max\{\mu, \nu\}$ and there exist cut-off functions $\psi_{x_o} \in \mathcal{Z}^\nu_{x_o}, \psi_{\eta_o} \in \mathcal{Z}^\mu_{\eta_o}$ such that $\psi_{x_o}(\psi_{\eta_o}(D)u) \in S_\theta(\mathbb{R}^n)$.

Remark 4.4. It is easy to prove that Definition 4.3 is independent of the choice of μ and ν. In particular, if $(x_o, \infty\eta_o) \notin WF^\theta_\psi u$, then for any given $\mu > 1, \nu > 1$ with $\mu + \nu - 1 \le \theta$ we may actually find $\varphi_{x_o} \in \mathcal{R}^\nu_{x_o}, \psi_{\eta_o} \in \mathcal{Z}^\mu_{\psi_{\eta_o}}$ such that $\varphi_{x_o}(\psi_{\eta_o}(D)u) \in S_\theta(\mathbb{R}^n)$, and similarly for $WF^\theta_e u, WF^\theta_{\psi e} u$.

Remark 4.5. We can consider $WF^\theta_\psi u$ as a subset of $\mathbb{R}^n \times S^{n-1}$, being $WF^\theta_\psi u$ invariant with respect to the multiplication of the second variable η by positive scalars. Analogously, we can consider $WF^\theta_e u \subset S^{n-1} \times \mathbb{R}^n$ and $WF^\theta_{\psi e} u \subset S^{n-1} \times S^{n-1}$.

Remark 4.6. Every $u \in S'_\theta(\mathbb{R}^n)$ can be regarded as an element of $\mathcal{D}'_\theta(\mathbb{R}^n)$, according to Remark 1.3. It is easy to show that $WF^\theta_\psi u$ coincides with the standard Gevrey wave front set of u, cf. [27].

Let us characterize the sets defined before in terms of characteristic manifolds of polyhomogeneous operators. For $p \in \Gamma^m_{\mu,\nu,cl}(\mathbb{R}^{2n})$, we define

$$\mathrm{Char}_\psi(P) = \{(x, \eta) \in \mathbb{R}^n \times S^{n-1} : \sigma^{m_1}_\psi(p)(x, \eta) = 0\},$$

$$\mathrm{Char}_e(P) = \{(x, \eta) \in S^{n-1} \times \mathbb{R}^n : \sigma^{m_2}_e(p)(x, \eta) = 0\},$$

$$\mathrm{Char}_{\psi e}(P) = \{(x, \eta) \in S^{n-1} \times S^{n-1} : \sigma^m_{\psi e}(p)(x, \eta) = 0\}.$$

Proposition 4.7. *Let $u \in S'_\theta(\mathbb{R}^n)$. We have the following relations:*

$$WF^\theta_\psi u = \bigcap_{\substack{P \in OPS^\theta_{cl}(\mathbb{R}^n)\\ Pu \in S_\theta(\mathbb{R}^n)}} \mathrm{Char}_\psi(P), \quad WF^\theta_e u = \bigcap_{\substack{P \in OPS^\theta_{cl}(\mathbb{R}^n)\\ Pu \in S_\theta(\mathbb{R}^n)}} \mathrm{Char}_e(P),$$

$$WF^\theta_{\psi e} u = \bigcap_{\substack{P \in OPS^\theta_{cl}(\mathbb{R}^n)\\ Pu \in S_\theta(\mathbb{R}^n)}} \mathrm{Char}_{\psi e}(P)$$

Proof. Let $(x_o, \infty\eta_o) \notin WF^\theta_\psi u$. Then, by Remark 4.4, there exist $\mu, \nu \in (1, +\infty)$ such that $\theta \geq \mu+\nu-1$ and φ_{x_o} in $\mathcal{R}^\nu_{x_o}$, ψ_{η_o} in $\mathcal{Z}^\mu_{\eta_o}$ such that $Pu = \varphi_{x_o}(\psi_{\eta_o}(D)u) \in S_\theta(\mathbb{R}^n)$. Observe that P is a pseudodifferential operator with symbol $\varphi_{x_o}(x)\psi_{\eta_o}(\eta)$ in $\Gamma^{(0,0)}_{\mu,\nu,cl}(\mathbb{R}^{2n})$ and that $\varphi_{x_o}(x_o)\psi_{\eta_o}(\lambda\eta_o) = 1$ for $\lambda \in \mathbb{R}_+$ sufficiently large. Hence, $(x_o, \infty\eta_o) \notin \bigcap_{\substack{P \in OPS^\theta_{cl}(\mathbb{R}^n)\\ Pu \in S_\theta(\mathbb{R}^n)}} \mathrm{Char}_\psi(P)$. Conversely, let us assume that there exists $P = p(x, D)$ in $OPS^\theta_{cl}(\mathbb{R}^n)$ such that $Pu \in S_\theta(\mathbb{R}^n)$ and $\sigma_\psi(p)(x_o, \infty\eta_o) \neq 0$. Then, there exists a neighborhood U of x_o and a conic neighborhood V of $\infty\eta_o$ such that $\sigma_\psi(p)(x, \infty\eta) \neq 0 \quad \forall (x, \eta) \in U \times V$. Furthermore, by (3.3), it turns out that if $|\eta|$ is sufficiently large, we have

$$\frac{|p(x,\eta)|}{|\eta|^{m_1}\langle x\rangle^{m_2}} \geq \frac{|\sigma_\psi(p)(x,\eta)|}{|\eta|^{m_1}\langle x\rangle^{m_2}} - \frac{|p(x,\eta) - \sigma_\psi(p)(x,\eta)|}{|\eta|^{m_1}\langle x\rangle^{m_2}} \geq C > 0$$

for some $C > 0$. Hence we can construct two cut-off functions $\varphi_{x_o}, \psi_{\eta_o}$ supported in U and in V, respectively, such that p is elliptic with respect to $\varphi_{x_o}(x)\psi_{\eta_o}(\eta)$. By Proposition 3.5, there exists $E \in OPS^{-m}_{\mu,\nu}(\mathbb{R}^n)$ such that $EPu = \varphi_{x_o}(\psi_{\eta_o}(D)u) + Ru$, where R is θ-regularizing. Then, $\varphi_{x_o}(\psi_{\eta_o}(D)u) = Ru - EPu \in S_\theta(\mathbb{R}^n)$. This gives the statement for $WF^\theta_\psi u$. The corresponding relation for $WF^\theta_e u$ can be obtained with the same argument by simply interchanging the roles of x and η. As to the third relation, we obtain the inclusion $\bigcap_{\substack{P \in OPS^\theta_{cl}(\mathbb{R}^n)\\ Pu \in S_\theta(\mathbb{R}^n)}} \mathrm{Char}_{\psi e}(P) \subset WF^\theta_{\psi e} u$ directly again from Definition 4.3 and Remark 4.4. Assume now that there exists $P \in OPS^\theta_{cl}(\mathbb{R}^n)$ such that $Pu \in S_\theta(\mathbb{R}^n)$ and $\sigma_{\psi e}(p)(\infty x_0, \infty\eta_0) \neq 0$. Then, there exist two conic neighborhoods V_{x_0}, V_{η_0} such that $\sigma_{\psi e}(p)(x, \eta) \neq 0$ if (x, η) is in $V_{x_0} \times V_{\eta_0}$. Hence, by (3.5), we have

$$\frac{|p(x,\eta)|}{|\eta|^{m_1}|x|^{m_2}} \geq C > 0$$

if $|x|$ and $|\eta|$ are large enough. Then, we can conclude arguing as for $WF^\theta_\psi u$. □

Theorem 4.8. *Let $u \in S'_\theta(\mathbb{R}^n)$ and $p \in \Gamma^m_{\mu,\nu,cl}(\mathbb{R}^{2n})$, with $\mu + \nu - 1 \leq \theta$. Then, the following inclusions hold:*

$$WF^\theta_\psi(Pu) \subset WF^\theta_\psi u \subset WF^\theta_\psi(Pu) \cup \mathrm{Char}_\psi(P) \tag{4.1}$$

$$WF^\theta_e(Pu) \subset WF^\theta_e u \subset WF^\theta_e(Pu) \cup \mathrm{Char}_e(P) \tag{4.2}$$

$$WF^{\theta}_{\psi e}(Pu) \subset WF^{\theta}_{\psi e}u \subset WF_{\psi e}(Pu) \cup \mathrm{Char}_{\psi e}(P). \tag{4.3}$$

Proof. If $(x_o, \eta_o) \notin WF^{\theta}_{\psi}u$, then there exist cut-off functions $\varphi_{x_o} \in \mathcal{R}^{\nu}_{x_o}, \psi_{\eta_o} \in \mathcal{Z}^{\mu}_{\eta_o}$ such that $\varphi_{x_o}(\psi_{\eta_o}(D)u) \in S_\theta(\mathbb{R}^n)$, where, in view of Remark 4.4, we may take the same μ, ν as for the class $\Gamma^m_{\mu,\nu,cl}(\mathbb{R}^{2n})$. Shrinking the neighborhoods of $x_o, \infty\eta_o$, we can construct two cut-off functions $\widetilde{\varphi}_{x_o} \in \mathcal{R}^{\nu}_{x_o}, \widetilde{\psi}_{\eta_o} \in \mathcal{Z}^{\mu}_{\eta_o}$ such that $\widetilde{\varphi}_{x_o}\varphi_{x_o} = \widetilde{\varphi}_{x_o}$ and $\widetilde{\psi}_{\eta_o}\psi_{\eta_o} = \widetilde{\psi}_{\eta_o}$. Denote by Q the operator with symbol $\varphi_{x_o}\psi_{\eta_o}$ and by $\widetilde{Q}$ the operator with symbol $\widetilde{\varphi}_{x_o}\widetilde{\psi}_{\eta_o}$. By Theorem 2.12, we have

$$\widetilde{Q}QPu = \widetilde{Q}PQu + \widetilde{Q}[Q, P]u = \widetilde{Q}PQu + Ru,$$

where R is θ-regularizing. Observe that $\widetilde{Q}Q \in OPS^{\theta}_{cl}(\mathbb{R}^n)$ and $\sigma_\psi(\widetilde{Q}Q)(x_o, \infty\eta_o) = \sigma_\psi(\widetilde{Q})(x_o, \infty\eta_o)\sigma_\psi(Q)(x_o, \infty\eta_o) \neq 0$. Then, by Proposition 4.7, we conclude that $(x_o, \infty\eta_o) \notin WF^{\theta}_{\psi}(Pu)$. This proves the first inclusion in (4.1). Assume now that $(x_o, \infty\eta_o) \notin WF^{\theta}_{\psi}(Pu)$. By Proposition 4.7, there exists $Q = q(x, D) \in OPS^0_{\mu,\nu,cl}(\mathbb{R}^n)$ such that $QPu \in S_\theta(\mathbb{R}^n)$ and $\sigma_\psi(Q)(x_o, \infty\eta_o) \neq 0$. Furthermore, if $(x_o, \infty\eta_o) \notin \mathrm{Char}_\psi(P)$, then $\sigma_\psi(QP)(x_o, \infty\eta_o) = \sigma_\psi(Q)(x_0, \infty\eta_o)\sigma_\psi(P)(x_o, \infty\eta_o) \neq 0$. Moreover, $QP \in OPS^m_{\mu,\nu,cl}(\mathbb{R}^n)$. Hence, by Proposition 4.7, we conclude that $(x_o, \infty\eta_o) \notin WF^{\theta}_{\psi}u$. The proofs of (4.2) and (4.3) are analogous. □

Proposition 4.9. *Let $u \in S'_\theta(\mathbb{R}^n)$. Then, $u \in S_\theta(\mathbb{R}^n)$ if and only if $WF^{\theta}_{\psi}u = WF^{\theta}_{e}u = WF^{\theta}_{\psi e}u = \emptyset$.*

Proof. If $WF^{\theta}_{\psi e}u = \emptyset$, then, for every $(\infty x_o, \infty\eta_o) \in S^{n-1} \times S^{n-1}$, there exist $\psi_{x_o} \in \mathcal{Z}^{\nu}_{x_o}, \psi_{\eta_o} \in \mathcal{Z}^{\mu}_{\eta_o}$ such that $\psi_{x_o}(\psi_{\eta_o}(D)u) \in S_\theta(\mathbb{R}^n)$. In view of Remark 4.4, we may fix μ, ν independent of $(\infty x_o, \infty\eta_o)$. Let us observe that $\sigma^{(0,0)}_{\psi e}(\psi_{x_o}(x)\psi_{\eta_o}(\eta)) = 1$ in a conic set in $\mathbb{R}^{2n}$, obtained as a product of conic sets of $\mathbb{R}^n_x$ and $\mathbb{R}^n_\eta$, intersecting $S^{n-1} \times S^{n-1}$ in a neighborhood V_{x_o,η_o} of $(\infty x_o, \infty\eta_o)$. By the compactness of $S^{n-1} \times S^{n-1}$, we can find a finite family $(\infty x_j, \infty\eta_j), j = 1, \ldots, N$, such that $V_{x_j,\eta_j}, j = 1, \ldots, N$ cover $S^{n-1} \times S^{n-1}$. Define

$$q_o(x, \eta) = \sum_{j=1,\ldots,N} \psi_{x_j}(x)\psi_{\eta_j}(\eta).$$

If $|\eta| > R$ and $|x| > R$, with R sufficiently large, then $q_o(x, \eta) \geq C > 0$. Moreover, by construction, $q_o(x, D)u \in S_\theta(\mathbb{R}^n)$. Applying similar compactness arguments to $\{x \in \mathbb{R}^n : |x| \leq R\} \times S^{n-1}$ and to $S^{n-1} \times \{\eta \in \mathbb{R}^n : |\eta| \leq R\}$ and using the assumption $WF^{\theta}_{\psi}u = WF^{\theta}_{e}u = \emptyset$, we can construct $q_1(x, \eta), q_2(x, \eta)$ such that $q_1(x, D)u \in S_\theta(\mathbb{R}^n), q_2(x, D)u \in S_\theta(\mathbb{R}^n)$ and $q_1(x, \eta) \geq C_1 > 0$ if $|\eta| > R, |x| \leq R$ and $q_2(x, \eta) \geq C_2 > 0$ if $|x| > R, |\eta| \leq R$. Moreover, obviously $q_o, q_1, q_2 \in \Gamma^{(0,0)}_{\mu,\nu,cl}(\mathbb{R}^{2n})$. Then, the function $q(x, \eta) = q_o(x, \eta) + q_1(x, \eta) + q_2(x, \eta)$ is an elliptic symbol of order $(0, 0)$ and $q(x, D)u \in S_\theta(\mathbb{R}^n)$. Then, $u \in S_\theta(\mathbb{R}^n)$ in view of Corollary 3.3. The inverse implication is trivial. □

We conclude with a proposition which makes clear in what sense the exit components $WF^\theta_e, WF^\theta_{\psi e}$ determine the behavior of a distribution of $S'_\theta(\mathbb{R}^n)$ at infinity. The proof follows the same arguments of the proof of Proposition 4.9. We omit it for sake of brevity.

Proposition 4.10. *Let $u \in S'_\theta(\mathbb{R}^n)$ and denote by $\Pi_x : \mathbb{R}^{2n}_{x,\eta} \longrightarrow \mathbb{R}^n_x$ the standard projection on the variable x. If $x_0 \notin \Pi_x(WF^\theta_e u \cup WF^\theta_{\psi e} u)$, then there exists $\psi_{x_0} \in \mathcal{Z}^\theta_{x_0}$ such that $\psi_{x_0} u \in S_\theta(\mathbb{R}^n)$.*

5. Action of SG-Fourier integral operators on the θ-wave front set

In this section, we study the action of the Fourier integral operators of infinite order defined in Section 2 on the θ-wave front set of ultradistributions $u \in S'_\theta(\mathbb{R}^n)$. The results presented here have an analogous local version for the standard Gevrey ultradistributions from $\mathcal{D}'_\theta(\mathbb{R}^n)$, see [27] and the references there. The corresponding analysis of the action of SG-operators of finite order on the $\mathcal{S}$-wave front set is due to S. Coriasco and L. Maniccia [10]. We start introducing further assumptions on the phase functions. Given $\mu > 1$, we will denote by $\mathcal{P}_{\mu,\mu,cl}$ the space of all phase functions $\varphi \in \mathcal{P}_{\mu,\mu}$ such that φ is in $\Gamma^e_{\mu,\mu,cl}(\mathbb{R}^{2n})$. Given $\varphi \in \mathcal{P}_{\mu,\mu,cl}$, we can consider the following three maps:

$$\Phi_\psi : \left(x, \xi = \sigma^1_\psi(\nabla_x\varphi)(x,\eta)\right) \longrightarrow \left(y = \sigma^0_\psi(\nabla_\eta\varphi)(x,\eta), \eta\right), \tag{5.1}$$

$$\Phi_e : \left(x, \xi = \sigma^0_e(\nabla_x\varphi)(x,\eta)\right) \longrightarrow \left(y = \sigma^1_e(\nabla_\eta\varphi)(x,\eta), \eta\right), \tag{5.2}$$

$$\Phi_{\psi e} : \left(x, \xi = \sigma^{e_1}_{\psi e}(\nabla_x\varphi)(x,\eta)\right) \longrightarrow \left(y = \sigma^{e_2}_{\psi e}(\nabla_\eta\varphi)(x,\eta), \eta\right). \tag{5.3}$$

Definition 5.1. A phase function $\varphi \in \mathcal{P}_{\mu,\mu}$ is said to be regular if there exists $C > 0$ such that

$$\sup_{(x,\eta)\in\mathbb{R}^{2n}} \left| \det\left(\frac{\partial^2\varphi}{\partial x_j \partial\eta_k}\right)(x,\eta)\right| \geq C > 0 \tag{5.4}$$

for all $j, k = 1, \ldots, m$.

We can apply in particular to a regular phase function $\varphi \in \mathcal{P}_{\mu,\mu}$ the following results from [8], [10] in the C^∞-setting.

Proposition 5.2. *If $\varphi \in \mathcal{P}_{\mu,\mu,cl}$ is regular, then the maps $\Phi_\psi, \Phi_e, \Phi_{\psi e}$ are global diffeomorphisms acting on $\mathbb{R}^n \times S^{n-1}, S^{n-1} \times \mathbb{R}^n, S^{n-1} \times S^{n-1}$, respectively.*

Proof. See Proposition 12 in [8] for the proof. □

Theorem 5.3. *Let μ, ν, θ be real numbers satisfying (2.7) and let $a \in \Gamma^\infty_{\mu,\nu,\theta}(\mathbb{R}^{2n})$, $\varphi \in \mathcal{P}_{\mu,\mu,cl}$ regular. Then, for every $u \in S'_\theta(\mathbb{R}^n)$, we have the following inclusions:*

$$WF^\theta_\psi(A_{a,\varphi}u) \subset \Phi^{-1}_\psi(WF^\theta_\psi u) \tag{5.5}$$

$$WF^\theta_e(A_{a,\varphi}u) \subset \Phi^{-1}_e(WF^\theta_e u) \tag{5.6}$$

$$WF^\theta_{\psi e}(A_{a,\varphi}u) \subset \Phi^{-1}_{\psi e}(WF^\theta_{\psi e} u). \tag{5.7}$$

Proof. We will only prove (5.5) for sake of brevity. The proofs of (5.6) and (5.7) do not present further difficulties. Let $(y_o, \eta_o) \notin WF^\theta_\psi u$. Then, by Remark 4.4, there exist cut-off functions $\varphi_{y_o} \in \mathcal{R}^\mu_{y_o}, \psi_{\eta_o} \in \mathcal{Z}^\mu_{\eta_o}$ such that $Cu = \varphi_{y_o}(\psi_{\eta_o}(D)u) \in S_\theta(\mathbb{R}^n)$. We want to prove that there exist $\varphi_{x_o} \in \mathcal{R}^\nu_{x_o}, \psi_{\xi_o} \in \mathcal{Z}^\mu_{\xi_o}$ such that $\varphi_{x_o}(\psi_{\xi_o}(D)A_{a,\varphi}u) \in S_\theta(\mathbb{R}^n)$. We will fix the supports of φ_{x_o} and ψ_{ξ_o} later. Let us denote by T the operator with symbol $t(x,\eta) = \varphi_{x_o}(x)\psi_{\xi_o}(\eta)$. We can write

$$TA_{a,\varphi}u = TA_{a,\varphi}Cu + TA_{a,\varphi}Eu,$$

where $E = I - C$. We obviously have $TA_{a,\varphi}Cu \in S_\theta(\mathbb{R}^n)$. To prove (5.5), it is sufficient to show that also $TA_{a,\varphi}Eu \in S_\theta(\mathbb{R}^n)$. Indeed, we want to prove that, by suitably choosing the supports of φ_{x_o} and ψ_{ξ_o}, the operator $TA_{a,\varphi}E$ turns out to be θ-regularizing. Let us first consider the operator $B = TA_{a,\varphi}$. By Theorem 2.12, B is a Fourier integral operator with phase function φ and symbol $b \in \Gamma^\infty_{\mu,\nu,\theta}(\mathbb{R}^{2n})$ such that

$$b(x,\eta) \sim \sum_\alpha (\alpha!)^{-1} D^\alpha_z \left((\partial^\alpha_\eta t)(x, \widetilde{\nabla}_x\varphi(x,z,\eta))a(z,\eta) \right)_{|z=x} \tag{5.8}$$

in $FS^\infty_{\mu,\nu,\theta}(\mathbb{R}^{2n})$. We observe that all the terms in the sum (5.8) contain derivatives of t evaluated in $(x, \nabla_x\varphi(x,\eta))$. Moreover, by (3.3), we know that $\nabla_x\varphi(x,\eta) = \sigma^1_\psi(\nabla_x\varphi)(x,\eta) \bmod \Gamma^{(0,0)}_{\mu,\mu}$ for $|\eta|$ large. Then, by the properties of φ, we can assume that there exists a neighborhood U_{x_o} of x_o and a conic neighborhood V_{ξ_o} of $\infty\xi_o$ such that $b(x,\eta)$ vanishes when $(x,\xi) \in \mathbb{R}^{2n} \setminus (U_{x_o} \times V_{\xi_o})$. Furthermore, we can take U_{x_o} and V_{ξ_o} as small as we want by shrinking the supports of φ_{x_o} and ψ_{ξ_o}. Let us now consider BE. By Theorem 2.16, BE is a Fourier integral operator with phase function φ and symbol $h \in \Gamma^\infty_{\mu,\nu,\theta}(\mathbb{R}^{2n})$ such that

$$h(x,\eta) \sim \sum_\alpha (\alpha!)^{-1} D^\alpha_\zeta \left((\partial^\alpha_x e)(\widetilde{\nabla}_\eta\varphi(x,\zeta,\eta),\eta)b(x,\zeta) \right)_{|\zeta=\eta} \tag{5.9}$$

in $FS^\infty_{\mu,\nu,\theta}(\mathbb{R}^{2n})$. Observe that all the terms of the sum (5.9) contain some derivatives of e evaluated in $(\nabla_\eta\varphi(x,\eta),\eta)$. Arguing as before, we can conclude that there exists a neighborhood $\widetilde{U}_{y_o}$ of y_o and a conic neighborhood $\widetilde{V}_{\eta_o}$ of $\infty\eta_o$ depending only on the supports of φ_{y_o} and ψ_{η_o} such that $e(y,\eta) = 0$ for $(y,\eta) \in \widetilde{U}_{y_o} \times \widetilde{V}_{\eta_o}$. Furthermore, by the condition (5.4), it turns out that also the maps

$$M_1 : (x,\eta) \longrightarrow (y,\eta)$$

$$M_2 : (x,\eta) \longrightarrow (x,\xi)$$

defined in terms of (5.5) are global diffeomorphisms on $\mathbb{R}^n \times S^{n-1}$, cf. Proposition 12 in [8]. By homogeneity, we deduce that we can choose the supports of φ_{x_o} and ψ_{ξ_o} sufficiently small such that

$$M_1(M_2^{-1}(U_{x_o} \times V_{\xi_o})) \subset \widetilde{U}_{y_o} \times \widetilde{V}_{\eta_o}.$$

This gives $h \sim 0$ in $FS^\infty_{\mu,\nu,\theta}(\mathbb{R}^{2n})$. Then, (5.5) follows from Proposition 2.11. □

Acknowledgment

Thanks are due to Professor Luigi Rodino for helpful discussions and comments. The author also wishes to thank Professor Cornelis Van der Mee and the referees of the paper for several useful remarks which led to an improvement of the manuscript.

References

[1] L. Boutet de Monvel and P. Krée, *Pseudodifferential operators and Gevrey classes*, Ann. Inst. Fourier, Grenoble, **17** (1967), 295–323.

[2] M. Cappiello, *Pseudodifferential operators and spaces of type S*, in "Progress in Analysis" Proceedings 3rd Int. ISAAC Congress, Vol. I, Editors G.W. Begehr, R.B. Gilbert, M.W. Wong, World Scientific, Singapore (2003), 681–688.

[3] M. Cappiello, *Pseudodifferential parametrices of infinite order for SG-hyperbolic problems*, Rend. Sem. Mat. Univ. Pol. Torino, **61**, 4 (2003), 411–441.

[4] M. Cappiello, *Fourier integral operators of infinite order and applications to SG-hyperbolic equations*, Preprint 2003. To appear in Tsukuba J. Math.

[5] F. Cardin and A. Lovison, *Lack of critical phase points and exponentially faint illumination*, Preprint 2004.

[6] L. Cattabriga and L. Zanghirati, *Fourier integral operators of infinite order on Gevrey spaces. Application to the Cauchy problem for certain hyperbolic operators*, J. Math. Kyoto Univ., **30** (1990), 142–192.

[7] H.O. Cordes, *The technique of pseudodifferential operators*, Cambridge Univ. Press, 1995.

[8] S. Coriasco, *Fourier integral operators in SG classes.I. Composition theorems and action on SG-Sobolev spaces*, Rend. Sem. Mat. Univ. Pol. Torino, **57** n. 4 (1999), 249–302.

[9] S. Coriasco, *Fourier integral operators in SG classes.II. Application to SG hyperbolic Cauchy problems*, Ann. Univ. Ferrara Sez VII, **44** (1998), 81–122.

[10] S. Coriasco and L. Maniccia, *Wave front set at infinity and hyperbolic linear operators with multiple characteristics*, Ann. Global Anal. and Geom., **24** (2003), 375–400.

[11] S. Coriasco and P. Panarese, *Fourier integral operators defined by classical symbols with exit behaviour*, Math. Nachr., **242** (2002), 61–78.

[12] S. Coriasco and L. Rodino, *Cauchy problem for SG-hyperbolic equations with constant multiplicities*, Ricerche di Matematica, Suppl. Vol. XLVIII (1999), 25–43.

[13] Y. Egorov and B.-W. Schulze, *Pseudo-differential operators, Singularities, Applications*, Birkhäuser, 1997.

[14] I.M. Gelfand and G.E. Shilov, *Generalized functions, Vol. 2*, Academic Press, New York-London, 1968.

[15] I.M. Gelfand and N. Ya. Vilenkin, *Generalized functions, Vol. 4*, Academic Press, New York-London, 1964.

[16] T. Gramchev, *Stationary phase method in the Gevrey classes and the Gevrey wave front sets*, C. R. Acad. Bulgare Sci. **36** (1983), n. 12, 1487–1489.

[17] T. Gramchev, *The stationary phase method in Gevrey classes and Fourier integral operators on ultradistributions*, Banach Center Publ., PWN, Warsaw, **19** (1987), 101–111.

[18] T. Gramchev and P. Popivanov, *Partial differential equations: Approximate solutions in scales of functional spaces*, Math. Research, **108**, WILEY-VCH, Berlin, 2000.

[19] S. Hashimoto, T. Matsuzawa and Y. Morimoto, *Opérateurs pseudo-différentiels et classes de Gevrey*, Comm. Partial Differential Equations, **8** (1983), 1277–1289.

[20] L. Hörmander, *Fourier integral operators I*, Acta Math. **127** (1971), 79–183.

[21] H. Kumano-go, *Pseudodifferential operators*, MIT Press, 1981.

[22] R. Lascar, *Distributions intégrales de Fourier et classes de Denjoy-Carleman. Applications*, C. R. Acad. Sci. Paris Sér. A-B **284** (1977), no. 9, A485–A488.

[23] M. Mascarello and L. Rodino, *Partial differential equations with multiple characteristics*, Akademie Verlag, Berlin, 1997.

[24] B.S. Mitjagin, *Nuclearity and other properties of spaces of type S*, Amer. Math. Soc. Transl., Ser. 2 **93** (1970), 45–59.

[25] C. Parenti, *Operatori pseudodifferenziali in* $\mathbb{R}^n$ *e applicazioni*, Ann. Mat. Pura Appl. **93** (1972), 359–389.

[26] S. Pilipovic, *Tempered ultradistributions*, Boll. U.M.I. **7** 2-B (1988), 235–251.

[27] L. Rodino, *Linear Partial Differential Operators in Gevrey Spaces*, World Scientific Publishing Co., Singapore, 1993.

[28] E. Schrohe, *Spaces of weighted symbols and weighted Sobolev spaces on manifolds*, In H. O. Cordes, B. Gramsch and H. Widom editors, Proceedings, Oberwolfach, **1256** Springer LNM, New York (1986), 360–377.

[29] B.-W. Schulze, *Boundary value problems and singular pseudodifferential operators*, J. Wiley & sons, Chichester, 1998.

[30] K. Shinkai and K. Taniguchi, *On ultra wave front sets and Fourier integral operators of infinite order*, Osaka J. Math., **27** (1990), 709–720.

[31] K. Taniguchi, *Fourier integral operators in Gevrey class on* $\mathbb{R}^n$ *and the fundamental solution for a hyperbolic operator*, Publ. RIMS Kyoto Univ., **20** (1984), 491–542.

[32] K. Yagdjan, *The Cauchy problem for hyperbolic operators: multiple characteristics, microlocal approach*, Akademie Verlag, Berlin, 1997.

[33] L. Zanghirati, *Pseudodifferential operators of infinite order and Gevrey classes*, Ann. Univ Ferrara, Sez. VII, Sc. Mat., **31** (1985), 197–219.

Marco Cappiello
Dipartimento di Matematica
Università degli Studi di Torino
Via Carlo Alberto, 10
I-10123 Torino, Italy
e-mail: marco@dm.unito.it

Operator Theory:
Advances and Applications, Vol. 160, 101–160

Strongly Regular J-inner Matrix-valued Functions and Inverse Problems for Canonical Systems

Damir Z. Arov and Harry Dym

To Israel Gohberg, valued teacher, colleague and friend, on his 75th birthday.

Abstract. This paper provides an introduction to the role of strongly regular J-inner matrix-valued functions in the analysis of inverse problems for canonical integral and differential systems. A number of the main results that were developed in a series of papers by the authors are surveyed and examples and applications are presented, including an application to the matrix Schrödinger equation. The approach of M.G. Krein to inverse problems is discussed briefly.

Mathematics Subject Classification (2000). Primary 34A55, 45Q05, 47B32, 46E22 Secondary 34L40, 30E05.

Keywords. canonical systems, differential systems with potential, inverse problems, de Branges spaces, J-inner matrix-valued functions, interpolation, reproducing kernel Hilbert spaces, Dirac systems, Krein systems, Schrödinger systems.

1. Introduction

The purpose of this paper is to present a survey of the useful role played by the class of strongly regular J-inner mvf's (matrix-valued functions) in the theory of direct and inverse problems for canonical integral and differential systems. We shall not present proofs, unless they are short and instructive. A more complete analysis that includes all the missing details may be found in the cited references. A number of illustrative examples are included.

D.Z. Arov thanks the Weizmann Institute of Science for for hospitality and support, through The Minerva Foundation. H. Dym thanks Renee and Jay Weiss for endowing the Chair which supports his research and the Minerva Foundation.

We shall consider four systems of differential equations:

(a) canonical integral systems

$$y(t,\lambda) = y(0,\lambda) + i\lambda \int_0^t y(s,\lambda)dM(s)J\ , \quad 0 \le t < d\ .$$

(b) canonical differential systems

$$y'(t,\lambda) \ = \ i\lambda y(t,\lambda)H(t)J\,, \quad 0 \le t < d\,.$$

(c) differential systems with potential

$$y'(t,\lambda) = i\lambda y(t,\lambda)NJ + y(t,\lambda)\mathcal{V}(t)\ , \quad 0 \le t < d\ .$$

(d) matrix Schrödinger equations

$$-u''(t,\lambda) + u(t,\lambda)q(t) = \lambda u(t,\lambda)\,, \quad 0 \le t < d\,.$$

In these systems J is an $m \times m$ signature matrix with $\operatorname{rank}(I_m + J) = \operatorname{rank}(I_m - J) = p$, the mass function $M(t)$ in (a) is a continuous nondecreasing $m \times m$ mvf on the interval $[0,d)$ with $M(0) = 0$; the Hamiltonian $H(t)$ in (b) and the potential $\mathcal{V}(t)$ in (c) are locally summable $m \times m$ mvf's on the interval $[0,d)$ that are subject to the constraints

$$H(t) \ge 0 \quad \text{and} \quad \mathcal{V}(t)J + J\mathcal{V}(t)^* = 0 \quad \text{a.e. in} \quad [0,d)\,;$$

$N \in \mathbb{C}^{m \times m}$ is a positive semidefinite matrix and the potential $q(t)$ is a Hermitian locally summable $p \times p$ mvf on the interval $[0,d)$.

The imposed constraints insure that the matrizant $U_t(\lambda) = U(t,\lambda)$, $0 \le t < d$, for each of the systems (a)–(c) is an entire $m \times m$ mvf in the variable λ that belongs to the class $\mathcal{U}(J)$ of J-inner mvf's as a function of λ for each choice of $t \in [0,d)$, as does the fundamental matrix of the Schrödinger equation (d). This article focuses on the case where the matrizant belongs to the subclass $\mathcal{U}_{sR}(J)$ of strongly regular J-inner mvf's that is introduced in Section 4. This class includes the matrizants of systems of the form (b) when $H(t)$ is locally absolutely continuous on $[0,d)$, $H(t)JH(t) = J$ and $H(0) = I_m$. It also includes systems of the form (c) when $NJ = JN$, such as Dirac systems and Krein systems; see Sections 23 and 24. At first glance these restrictions may seem unduly restrictive. However, in Section 26 we shall explain how to exploit certain Dirac systems to study related systems of the form (c) with matrizants that do not belong to the class $\mathcal{U}_{sR}(J)$. This analysis will then be used to study a class of matrix Schrödinger equations in Section 26.

The paper is organized as follows: The next section lists the main notation. The subsequent ten sections present a brief survey of the preliminary material that is needed to formulate and explain the main developments in the article. These sections are short and are usually devoted to one topic: J-inner mvfs, reproducing kernel Hilbert spaces, linear fractional transformations, parametrization of $A \in \mathcal{U}_{sR}(J_p)$ and a description of the associated RKHS (reproducing kernel Hilbert space) $\mathcal{H}(A)$, chains of entire J-inner mvf's, canonical systems, chains of associated pairs, de Branges spaces associated with systems, generalized Carathéodory interpolation.

Sections 12–21 present a number of the authors' results on direct and inverse spectral problems for the canonical integral and differential systems described in items (a) and (b) of the previous list. (The two systems are really equivalent, but sometimes one form is more convenient than the other.) In particular we shall discuss the bitangential inverse spectral and input impedance problems (but not the inverse monodromy problem or the bitangential inverse input scattering problem that are considered in [ArD:00a], [ArD:00b], [ArD:02a]and [ArD:02b]). In our formulation of these problems, the given data includes a normalized monotonic continuous chain $\{b_3^t(\lambda), b_4^t(\lambda)\}$, $0 \leq t < d$, of entire inner $p \times p$ mvf's in addition to the spectral data (such as a Weyl-Titchmarsh function or a spectral function) that is usually specified. Thus, for example, the bitangential inverse spectral problem for the system (a) is: given $\{\sigma(\mu); b_3^t(\lambda), b_4^t(\lambda), 0 \leq t < d\}$, find a continuous nondecreasing $m \times m$ mvf $M(t)$ on the $[0, d)$ with $M(0) = 0$, such that

(1) $\sigma(\mu)$ is a spectral function of the corresponding system (a).
(2) The given pair $\{b_3^t(\lambda), b_4^t(\lambda)\}$ is associated with the matrizant $U_t(\lambda)$ of the system in a prescribed way, for each $t \in [0, d)$.
(3) $U_t \in \mathcal{U}_{sR}(J)$ for every $t \in [0, d)$.

The condition alluded to in (2) on the given chain of pairs serves to specify the class of systems in which a solution is sought. The third condition guarantees that there is at most one solution in the class specified by (2), up to a possible parameter $\alpha = \alpha^*$, $\alpha \in \mathbb{C}^{p \times p}$.

Sections 22–25 are devoted to differential systems with potential, i.e., systems of the type (c), whereas Section 26 considers matrix Schrödinger equations. Finally, Section 27 discusses the approach of M.G. Krein to inverse problems.

There is an extensive literature on the inverse spectral problem for assorted systems of differential and integral equations; see, e.g., [CG:02], [CG:01], [GKM:02], [Kr:55], [Kr:56], [LeMa:00], [LeSa:75], [MeA:67], [MeA:77], [MeA:99a], [MeA3:99b], [MeA:00], [Sak:96], [Sak:99], [Sak:00a], [Sak:00b], [Sak-A:92], [DK:78], [DI:84], [AlD:84], [AlD:85], [AG:95], [AG:01], [GKS:98], [GKS:02], and the references cited therein. The notion of associated pairs does not appear explicitly in any of these papers. Nevertheless, the restrictions imposed on the structure of the systems under study are such as to uniquely define a monotonic chain of normalized associated pairs, even if they are not mentioned explicitly. Moreover, in some of these problems, condition (3) is automatically in force; see, e.g., Theorem 22.3 and Corollary 22.4.

The bitangential inverse spectral problem with given data

$$\{\sigma(\mu); b_3^t(\lambda), b_4^t(\lambda), 0 \leq t < d\},$$

is solved by considering a bitangential inverse input impedance problem with given data

$$\{c^{(\alpha)}(\lambda); b_3^t(\lambda), b_4^t(\lambda), 0 \leq t < d\},$$

where the input impedance

$$c^{(\alpha)}(\lambda) = i\alpha + \frac{1}{\pi i}\int_{\infty}^{\infty}\left\{\frac{1}{\mu-\lambda} - \frac{\mu}{1+\mu^2}\right\} d\sigma(\mu)$$

is based on the given spectral function $\sigma(\mu)$, and the same chain $\{b_3^t(\lambda), b_4^t(\lambda)\}$, $0 \le t < d$, is considered in both problems; see Section 15. The matrizant of a system (a) that solves the inverse input impedance problem coincides with the resolvent matrix (in Krein's terminology) that describes the sets

$$\{c \in \mathcal{C}^{p\times p} : (b_3^t)^{-1}(c - c^{(\alpha)})(b_4^t)^{-1} \in \mathcal{N}_+^{p\times p}\}, \quad 0 \le t < d;$$

see Section 12 for additional details and references. This *chain of interpolation problems* (indexed by t) is equivalent to a chain of bitangential extension problems in the class of continuous $p \times p$ mvf's $g(s)$ on $\mathbb{R}$ for which $g(-s) = g(s)^*$ and

$$\int_0^{\infty} \varphi(s)^* \left\{\int_0^{\infty} \{g(s-u) - g(s) - g(-u) + g(0)\}\varphi(u)du\right\} ds \ge 0$$

for every $\varphi \in L_2^p([0,\infty))$ with compact support. These and other related bitangential extension problems that generalize a number of extension problems considered by Krein are studied in [ArD:98].

The link between these classes of problems rests on the formula

$$c(\lambda) = \lambda^2 \int_0^{\infty} e^{i\lambda s} g(s)ds\,, \quad \lambda \in \mathbb{C}_+\,,$$

which defines a one to one correspondence between mvf's in the Carathéodory class $\mathcal{C}^{p\times p}$ and the class of $p \times p$ mvf's $g(s)$ described just above that meet the extra condition $g(0) \le 0$. The extension problems that Krein considered in [Kr:44] correspond to the special choices $b_3^t(\lambda) = \exp\{i\lambda a_3 t\}$ and $b_4^t(\lambda) = \exp\{i\lambda a_4 t\}$, where $a_3 \ge 0$, $a_4 \ge 0$ and $a_3 + a_4 > 0$. In fact, if say $a_3 = a_4 = 1$ and $d < \infty$, then, as Krein pointed out in [Kr:55] and [Kr:56], only the values of $g(s)$ on the interval $[-2d, 2d]$ are relevant to the solution of an inverse spectral problem for a system of the form (c) or (d) on the interval $[0, d)$. This is the reason that extension problems are connected with inverse problems; additional discussion of this theme in the special case that $c(\lambda)$ is in the Wiener class may be found in [MeA:77], [DI:84], [KrL:85], [Dy:90], [ArD:??] and Section 27 below.

The interplay between a chain of extension problems (or, equivalently, a chain of interpolation problems) and inverse problems follows a general strategy envisioned by M.G. Krein some fifty years ago and is discussed in a little more detail in Section 27. Another generalization of Krein's extension problems and their application to inverse problems was developed by L.A. Sakhnovich [Sak:96], [Sak:99], [Sak:00a], [Sak:00b]. Some comparisons of his approach with ours are presented in [ArD:04b].

There seems to be a strong connection between the strategy proposed by Krein for solving inverse problems and the approach advocated in the recent papers [Sim:99], [GeSi:00], [RaSi:00] and [Rem:03]. In particular, the A-function introduced in [Sim:99] is remarkably close to Krein's transition function.

The approach to direct and inverse spectral problems that is described in this survey is, roughly speaking, a synthesis of the Krein strategy and RKHS methods that exploit the properties of two chains of RKHS's, $\mathcal{H}(U_t)$ and $\mathcal{B}(\mathfrak{E}_t)$, $0 \le t < d$, that are defined in terms of the fundamental solution $U_t(\lambda)$, $0 \le t < d$, of the system under study. These RKHS's were introduced and extensively studied by L. de Branges [dBr:63], [dBr:68a], [dBr:68b] and played a significant role in his celebrated proof that a suitably normalized real 2×2 Hamiltonian $H(t)$ of a system of the form (b) (with $m = 2$) is uniquely determined by a spectral function of the system. A number of other results on inverse problems that were announced without proof by Krein are established in the monograph [DMc:76] with the help of RKHS methods.

In general

$$\mathcal{H}(U_{t_1}) \subset \mathcal{H}(U_{t_2}) \quad \text{and} \quad \mathcal{B}(\mathfrak{E}_{t_1}) \subset \mathcal{B}(\mathfrak{E}_{t_2}) \quad \text{when} \quad 0 \le t_1 \le t_2 < d \tag{1.1}$$

and the inclusions are contractive. Moreover, there exists a partial isometry from $\mathcal{H}(U_t)$ onto $\mathcal{B}(\mathfrak{E}_t)$. However, if $U_t(\lambda)$ belongs to the class $\mathcal{U}_{sR}(J)$ of strongly regular J-inner mvf's for every $t \in [0, d)$, then:

(1) The inclusions in (1.1) are isometric.
(2) There exists a fixed $p \times m$ matrix T such that the map $f \longrightarrow Tf$ defines a unitary operator from $\mathcal{H}(U_t)$ onto $\mathcal{B}(\mathfrak{E}_t)$ for every $t \in [0, d)$.
(3) There is a useful parametrization of the space $\mathcal{H}(U_t)$ in terms of a unique pair $\{b_3^t(\lambda), b_4^t(\lambda)\}$ of normalized entire inner $p \times p$ mvf's that are uniquely defined by $U_t(\lambda)$.
(4) The chain of pairs $\{b_3^t(\lambda), b_4^t(\lambda)\}$, $0 \le t < d$, alluded to in (3) is continuous in t on the interval $[0, d)$ for each fixed choice of λ.
(5) The de Branges space $\mathcal{B}(\mathfrak{E}_t)$ and the RKHS $(H_2^p \ominus b_3^t H_2^p) \oplus (K_2^p \ominus (b_4^t)^{-1} K_2^p)$ based on the Hardy space H_2^p and its orthogonal complement $K_2^p = L_2^p \ominus H_2^p$ coincide as linear topological spaces, i.e., they contain the same elements and their norms are equivalent.

Item (5) in the last list is a generalization of the Paley-Wiener theorem. Thus, for example, if $b_3^t(\lambda) = e^{ia_3 t\lambda} I_p$ and $b_4^t(\lambda) = e^{ia_4 t\lambda} I_p$ for some choice of finite nonnegative numbers a_3 and a_4, then

$$\mathcal{B}(\mathfrak{E}_t) = \left\{ \int_{-a_4 t}^{a_3 t} e^{i\lambda s} g(s) ds \,:\, g \in L_2^p([-a_4, a_3]) \right\}$$

and the norms in the two spaces $\mathcal{B}(\mathfrak{E}_t)$ and $L_2^p([-a_4, a_3])$ are equivalent. Analogous conclusions prevail for the de Branges spaces associated with the matrix Schrödinger equation even though the fundamental matrix for this system of equations is not strongly regular. This result, which generalizes a theorem of Remling [Rem:02], [Rem:03] to the matrix case (though for what, at least as of this moment, appears to be a more restrictive class of potentials) is obtained in Section 26 as a byproduct of the methods of this paper, by exploiting the connections with an associated Dirac system.

The preceding discussion focuses on the direct problem, wherein one starts with a system. Sample inverse problems were discussed earlier. To supplement that discussion, it is worth noting that if $U_t(\lambda)$, $0 \le t < d$, is a given normalized chain of entire $m \times m$ mvf's in the class $\mathcal{U}_{sR}(J)$, and if the normalized pair $\{b_3^t(\lambda), b_4^t(\lambda)\}$, $0 \le t < d$, of entire inner $p \times p$ mvf's that are associated with $U_t(\lambda)$ is continuous as a function of t on $[0, d)$, then $U_t(\lambda)$ is automatically the matrizant of a system of the form (a) with

$$M(t) = i(\frac{\partial U_t}{\partial \lambda})(0)J \quad \text{for every} \quad t \in [0, d)\,.$$

This serves to connect the resolvent matrices of assorted classes of interpolation and extension problems with canonical systems.

2. Notation

In addition to the standard nomenclature such as $L_2^{m\times n}(\mathbb{R})$, $L_1^{m\times n}(\mathbb{R})$ and $L_\infty^{m\times n}(\mathbb{R})$ for the Lebesgue spaces of $m \times n$ mvf's on $\mathbb{R}$; $H_2^{m\times n}$ and $H_\infty^{m\times n}$ for the Hardy spaces of $m \times n$ mvf's in the open upper half plane $\mathbb{C}_+$, $\mathbb{C}_-$ for the open lower half plane, $J = J^* = J^{-1}$ for a general signature matrix, $(\Re f)(\lambda) = \{f(\lambda) + f(\lambda)^*\}/2$, $(\Im f)(\lambda) = \{f(\lambda) - f(\lambda)^*\}/2i$, $f^\#(\lambda) = f(\overline{\lambda})^*$, $f^\sim(\lambda) = f^\#(-\lambda)$ and the abbreviations $\mathcal{X}^m$ for $\mathcal{X}^{m\times 1}$ and $\mathcal{X}$ for $\mathcal{X}^1$, we shall make use of the following classes of functions:

$K_2^p = L_2^p \ominus H_2^p$ with respect to the standard inner product.

$\mathcal{E}^{p\times q} =$ the set of $p \times q$ mvf's with entire entries.

$\mathcal{C}^{p\times p} = \{p \times p$ mvf's $c(\lambda)$ which are holomorphic with $(\Re c)(\lambda) \ge 0$ in $\mathbb{C}_+\}$.

$\mathring{\mathcal{C}}^{p\times p} = \{c \in \mathcal{C}^{p\times p} :\ c \in H_\infty^{p\times p}$ and $(\Re c)^{-1} \in L_\infty^{p\times p}(\mathbb{R})\}$.

$\mathcal{S}^{p\times q} = \{p \times q$ mvf's $s(\lambda)$ which are holomorphic and contractive in $\mathbb{C}_+\}$.

$\mathcal{S}_{\text{in}}^{p\times p} = \{s \in \mathcal{S}^{p\times p} :\ s$ is an inner mvf$\}$.

$\mathcal{S}_{\text{out}}^{p\times p} = \{s \in \mathcal{S}^{p\times p} :\ s$ is an outer mvf$\}$.

$\mathring{\mathcal{S}}^{p\times q} = \{s \in \mathcal{S}^{p\times q} :\ \sup\{\|s(\lambda)\| : \lambda \in \mathbb{C}_+\} < 1\}$.

$\mathcal{H}(b) = H_2^r \ominus bH_2^r$ and $\mathcal{H}_*(b) = K_2^r \ominus b^{-1}K_2^r$ for $b \in \mathcal{S}_{\text{in}}^{r\times r}$.

$\mathcal{N}^{p\times q} = \{h^{-1}g : g \in \mathcal{S}^{p\times q}$ and $h \in \mathcal{S}\}$.

$\mathcal{N}_+^{p\times q} = \{h^{-1}g : g \in \mathcal{S}^{p\times q}$ and $h \in \mathcal{S}_{\text{out}}\}$.

$\mathcal{N}_{\text{out}}^{p\times p} = \{h^{-1}g : g \in \mathcal{S}_{\text{out}}^{p\times p}$ and $h \in \mathcal{S}_{\text{out}}\}$.

$\mathcal{U}(J) =$ the set of $m \times m$ mvf's that are J-inner with respect to $\mathbb{C}_+$.

$\mathcal{U}_{rR}(J) =$ the class of right regular J-inner mvf's.

$\mathcal{U}_{sR}(J) =$ the class of strongly regular J-inner mvf's.

$\mathcal{U}_S(J) =$ the class of singular J-inner mvf's.

$\mathcal{X}_{\text{const}}^{p\times q} =$ the set of constant functions in the set $\mathcal{X}^{p\times q}$.

$$\Pi^{p\times q} = \{f \in \mathcal{N}^{p\times q} : f \text{ admits a pseudocontinuation to } \mathbb{C}_- \text{ such that } f^\# \in \mathcal{N}^{q\times p}\}.$$

$$\mathcal{E}\cap\mathcal{X}^{p\times q} = \mathcal{E}^{p\times q}\cap\mathcal{X}^{p\times q}.$$

$$\mathcal{W}^{p\times q}(\gamma) = \{\gamma + \textstyle\int_{-\infty}^{\infty} e^{i\lambda t}h(t)dt \text{ with fixed } \gamma\in\mathbb{C}^{p\times q} \text{ and any } h\in L_1^{p\times q}(\mathbb{R})\}.$$

$$\mathcal{W}_+^{p\times q}(\gamma) = \{\gamma + \textstyle\int_{0}^{\infty} e^{i\lambda t}h(t)dt \text{ with fixed } \gamma\in\mathbb{C}^{p\times q} \text{ and any } h\in L_1^{p\times q}(0,\infty)\}.$$

$$\mathcal{W}_-^{p\times q}(\gamma) = \{\gamma + \textstyle\int_{-\infty}^{0} e^{i\lambda t}h(t)dt \text{ with fixed } \gamma\in\mathbb{C}^{p\times q} \text{ and any } h\in L_1^{p\times q}(-\infty,0)\}.$$

$$AC^{p\times q}([a,b)) = \{p\times q \text{ mvf's } g(t) : g(t) = \gamma + \textstyle\int_0^t h(s)ds \text{ where } \gamma\in\mathbb{C}^{p\times q} \text{ and } h\in L_1^{p\times q}([a,b))\}.$$

$$L_{1,\,\mathrm{loc}}^{p\times q}([a,b)) = \{p\times q \text{ mvf's } g(t) : g\in L_1^{p\times q}([a,c]) \text{ for every } c\in[a,b)\}.$$

$$AC_{\mathrm{loc}}^{p\times q}([a,b)) = \{p\times q \text{ mvf's } g(t) : g\in AC^{p\times q}([a,c)) \text{ for every } c\in[a,b)\}.$$

Here, $\mathcal{S}$ stands for the Schur class, $\mathcal{C}$ for the Carathéodory class, $\mathcal{N}$ for the Nevanlinna class, $\mathcal{N}_+$ for the Smirnov class, $\mathcal{N}_{\mathrm{out}}$ for outer functions from the Smirnov class and $\mathcal{W}$ stands for the Wiener class; $\mathcal{W}_\pm^{m\times m}(I_m)$ and $\mathcal{W}_\pm^{m\times m}(0)$ are closed under multiplication. Finally,

$e_a = e_a(\lambda) = e^{ia\lambda}$, $\rho_\omega(\lambda) = -2\pi i(\lambda - \bar{\omega})$,

$k_\omega^b(\lambda) = \rho_\omega(\lambda)^{-1}(I_p - b(\lambda)b(\omega)^*)$ and $\ell_\omega^b(\lambda) = \rho_\omega(\lambda)^{-1}(b(\lambda)^{-1}b(\omega)^{-*} - I_p)$ are the RK's for the RKHS's $\mathcal{H}(b)$ and $\mathcal{H}_*(b)$, respectively, when $b\in\mathcal{S}_{in}^{p\times p}$,

$\mathfrak{H}_f$ = the domain of holomorphy of the mvf $f(\lambda)$, $\mathfrak{H}_f^+ = \mathfrak{H}_f\cap\mathbb{C}_+$,

M_f denotes the operator of multiplication by the mvf $f(\lambda)$,

$\Pi_{\mathcal{M}}$ denotes the orthogonal projection onto the subspace $\mathcal{M}$,

$\Pi_+ = \Pi_{H_2^p}$, $\Pi_- = \Pi_{K_2^p}$,

$\widehat{f}(\lambda) = \int_{-\infty}^{\infty} e^{i\lambda x}f(x)dx$ denotes the Fourier transform of f,

$g^\vee(x) = \frac{1}{2\pi}\int_{-\infty}^{\infty} e^{-i\mu x}g(\mu)d\mu$ denotes the inverse Fourier transform of g,

$\langle f,g\rangle_{st} = \displaystyle\int_{-\infty}^{\infty} g(\mu)^* f(\mu)d\mu$ denotes the standard inner product

and

vvf stands for vector-valued function, while mvf stands for matrix-valued function.

$\mathcal{L}(X,Y)$ denotes the set of bounded linear operators from the Hilbert space X into the Hilbert space Y, and $\mathcal{L}(X)$ is short for $\mathcal{L}(X,X)$. All the Hilbert spaces considered in this paper are separable.

3. J-inner mvf's

An $m\times m$ constant matrix J is said to be a signature matrix, if $J = J^*$ and $JJ^* = I_m$, i.e., if it is both self-adjoint and unitary with respect to the standard inner product in $\mathbb{C}^m$. The main examples of signature matrices for this paper are

$$j_{pq} = \begin{bmatrix} I_p & 0 \\ 0 & -I_q \end{bmatrix}, \quad p+q=m,$$

and, if $2p = m$,

$$J_p = \begin{bmatrix} 0 & -I_p \\ -I_p & 0 \end{bmatrix}, \; j_p = \begin{bmatrix} I_p & 0 \\ 0 & -I_p \end{bmatrix} \text{ and } \mathcal{J}_p = \begin{bmatrix} 0 & -iI_p \\ iI_p & o \end{bmatrix}.$$

The signature matrices J_p and j_p are connected by the signature matrix

$$\mathfrak{V} = \frac{1}{\sqrt{2}} \begin{bmatrix} -I_p & I_p \\ I_p & I_p \end{bmatrix}, \quad \text{i.e.,} \quad \mathfrak{V} J_p \mathfrak{V} = j_p \quad \text{and} \quad \mathfrak{V} j_p \mathfrak{V} = J_p.$$

An $m \times m$ mvf $U(\lambda)$ is said to be J-inner with respect to the open upper half plane $\mathbb{C}_+$ if it is meromorphic in $\mathbb{C}_+$ and if

(1) $J - U(\lambda)^* J U(\lambda) \geq 0$ for every point $\lambda \in \mathfrak{H}_U^+$ and
(2) $J - U(\mu)^* J U(\mu) = 0$ a.e. on $\mathbb{R}$,

in which $\mathfrak{H}_U^+$ denotes the set of points in $\mathbb{C}_+$ at which U is holomorphic. This definition is meaningful because every mvf $U(\lambda)$ that is meromorphic in $\mathbb{C}_+$ and satisfies the first constraint automatically has nontangential boundary values. The second condition guarantees that $\det U(\lambda) \not\equiv 0$ in $\mathfrak{H}_U^+$ and hence permits us to define a pseudo-continuation of $U(\lambda)$ to the open lower half plane $\mathbb{C}_-$ by the symmetry principle

$$U(\lambda) = J\{U^\#(\lambda)\}^{-1} J \quad \text{for} \quad \lambda \in \mathbb{C}_-,$$

where $f^\#(\lambda) = f(\overline{\lambda})^*$. The symbol $\mathcal{U}(J)$ will denote the class of J-inner mvf's considered on the set $\mathfrak{H}_U$ of points of holomorphy of $U(\lambda)$ in the full complex plane $\mathbb{C}$.

4. Reproducing kernel Hilbert spaces

If $U \in \mathcal{U}(J)$ and

$$\rho_\omega(\lambda) = -2\pi i(\lambda - \overline{\omega}),$$

then the kernel

$$K_\omega^U(\lambda = \frac{J - U(\lambda) J U(\omega)^*}{\rho_\omega(\lambda)}$$

is positive on $\mathfrak{H}_U \times \mathfrak{H}_U$ in the sense that $\sum_{i,j=1}^n u_i^* K_{\omega_j}^U(\omega_i) u_j \geq 0$ for every set of vectors $u_1, \ldots, u_n \in \mathbb{C}^m$ and points $\omega_1, \ldots, \omega_n \in \mathfrak{H}_U$; see, e.g., [Dy:89]. Therefore, by the matrix version of a theorem of Aronszajn [Aron:50], there is an associated RKHS (reproducing kernel Hilbert space) $\mathcal{H}(U)$ with RK (reproducing kernel) $K_\omega^U(\lambda)$. This means that for every choice of $\omega \in \mathfrak{H}_U$, $u \in \mathbb{C}^m$ and $f \in \mathcal{H}(U)$,

(1) $K_\omega u \in \mathcal{H}(U)$ and
(2) $\langle f, K_\omega u \rangle_{\mathcal{H}(U)} = u^* f(\omega)$.

A vvf $f \in \mathcal{H}(U)$ is meromorphic in $\mathbb{C} \setminus \mathbb{R}$ and has nontangential limits $f(\mu)$ a.e. in $\mathbb{R}$ that may be used to identify $f(\lambda)$.

In particular,

$$f(\mu) = \lim_{\nu \downarrow 0} f(\mu \pm i\nu) \quad \text{a.e. in} \quad \mathbb{R}.$$

We shall say that a mvf $U \in \mathcal{U}(J)$ belongs to the class

(1) $\mathcal{U}_{sR}(J)$ of **strongly regular** J-inner mvf's if $\mathcal{H}(U) \subset L_2^m$.
(2) $\mathcal{U}_{rR}(J)$ of **right regular** J-inner mvf's if $\mathcal{H}(U) \cap L_2^m$ is dense in $\mathcal{H}(U)$.
(3) $\mathcal{U}_S(J)$ of **singular** J-inner mvf's if $\mathcal{H}(U) \cap L_2^m = \{0\}$.

Theorem 4.1. *Let $U \in \mathcal{U}(J)$. Then:*

(1) $\mathcal{U}(J) \cap L_\infty^{m\times m}(\mathbb{R}) \subset \mathcal{U}_{sR}(J)$.
(2) *The inclusion in* (1) *is proper.*
(3) *If $U \in \mathcal{U}_{sR}(J)$, then $(\mu + i)^{-1}U(\mu) \in L_2^{m\times m}$.*
(4) *If $U(\lambda)$ is an entire mvf, then $U \in \mathcal{U}_S(J)$ if and only if it is of minimal exponential type.*

Proof. (1) follows from the discussion of formula (3.25) in [ArD:97] and Remark 5.1 below; (2) follows from the example that starts on p. 293 in [ArD:01]; (3) is by definition and (4) follows from Lemma 3.8 and Theorem 3.8 in [ArD:97]. □

Item (1) of this theorem guarantees that J-inner mvf's in the Wiener class are automatically strongly regular:

$$\mathcal{U}(J) \cap \mathcal{W}^{m\times m} \subset \mathcal{U}_{sR}(J).$$

It also serves to guarantee that the matrizant $U_t(\lambda) = U(t,\lambda)$, $0 \le t < d$, of a number of classical first order differential systems with potential, such as Dirac systems and Krein systems, are automatically strongly regular; see Section 22 and the subsequent sections for additional details.

An entire $p \times m$ mvf

$$\mathfrak{E}(\lambda) = [E_-(\lambda) \qquad E_+(\lambda)]$$

with $p \times p$ components $E_+(\lambda)$ and $E_-(\lambda)$ is said to be a **de Branges function** if

(1) $\det E_+(\lambda) \not\equiv 0$ in $\mathbb{C}_+$ and
(2) the mvf $\chi(\lambda) = E_+(\lambda)^{-1}E_-(\lambda)$ is a $p \times p$ inner mvf.

The **de Branges space** $\mathcal{B}(\mathfrak{E})$ based on an entire de Branges function $\mathfrak{E}(\lambda)$ is equal to the set of entire $p \times 1$ vvf's $f(\lambda)$ such that $E_+^{-1}f \in H_2^p \ominus \chi H_2^p$. It is a RKHS with respect to the inner product

$$\langle f_1, f_2\rangle_{\mathcal{B}(\mathfrak{E})} = \langle E_+^{-1}f_1, E_+^{-1}f_2\rangle_{st}$$

and the RK is given by the formula

$$K_\omega^{\mathfrak{E}}(\lambda) = -\frac{\mathfrak{E}(\lambda)j_p\mathfrak{E}(\omega)^*}{\rho_\omega(\lambda)} = \frac{E_+(\lambda)E_+(\omega)^* - E_-(\lambda)E_-(\omega)^*}{\rho_\omega(\lambda)};$$

see, e.g., [dBr:68b] and [DI:84]. de Branges spaces can also be defined in the same way for a more general class of $p \times m$ mvf's that are meromorphic in $\mathbb{C}_+$, see, e.g., [ArD:04a], [ArD:04b] for additional information and references. However, the case of entire mvf's $\mathfrak{E}(\lambda)$ will suffice for the purposes of this article.

If $A \in \mathcal{E} \cap \mathcal{U}(J_p)$ and

$$B(\lambda) = A(\lambda)\mathfrak{V} = \begin{bmatrix} b_{11}(\lambda) & b_{12}(\lambda) \\ b_{21}(\lambda) & b_{22}(\lambda) \end{bmatrix},$$

with blocks $b_{ij}(\lambda)$ of size $p \times p$, then

$$\mathfrak{E}(\lambda) = \sqrt{2}[0 \quad I_p]\, B(\lambda),$$

is an entire de Branges function and the map

$$U_2 : f \in \mathcal{H}(A) \quad \overset{\text{onto}}{\longrightarrow} \quad \sqrt{2}[0 \quad I_p]f$$

is a coisometry from $\mathcal{H}(A)$ onto $\mathcal{B}(\mathfrak{E})$, i.e., it maps $\mathcal{H}(A) \ominus \ker U_2$ isometrically onto $\mathcal{B}(\mathfrak{E})$; see, e.g., Theorem 2.5 of [ArD:05] and Theorem 2.3 of [ArD:04b]. It will be an isometry, i.e.,

$$\langle f, f \rangle_{\mathcal{H}(A)} = \langle U_2 f, U_2 f \rangle_{\mathcal{B}(\mathfrak{E})} \quad \text{for every} \quad f \in \mathcal{H}(A)\,, \tag{4.1}$$

if and only if the mvf $c_0(\lambda) = b_{12}b_{22}^{-1}$ (which belongs to $\mathcal{C}^{p\times p}$) meets the condition

$$\lim_{\nu \uparrow \infty} \nu^{-1} \Re c_0(i\nu) = 0\,,$$

or equivalently, if and only if

$$\Re c_0(i) = \frac{1}{\pi} \int_{-\infty}^{\infty} \frac{1}{1+\mu^2} \Re\{c_0(\mu)\} d\mu\,.$$

In particular,

$$A \in \mathcal{U}_{sR}(J_p) \Longrightarrow \text{condition (4.1) prevails}.$$

5. Linear fractional transformations

The linear fractional transformation T_U based on the four block decomposition

$$U(\lambda) = \begin{bmatrix} u_{11}(\lambda) & u_{12}(\lambda) \\ u_{12}(\lambda) & u_{22}(\lambda) \end{bmatrix},$$

of an $m \times m$ mvf $U(\lambda)$ that is meromorphic in $\mathbb{C}_+$ with diagonal blocks $u_{11}(\lambda)$ of size $p \times p$ and $u_{22}(\lambda)$ of size $q \times q$ is defined on the set

$$\begin{aligned} \mathcal{D}(T_U) \quad = \quad & \{p \times q \text{ meromorphic mvf's } \varepsilon(\lambda) \text{ in } \mathbb{C}_+ \\ & \text{such that } \det\{u_{21}(\lambda)\varepsilon(\lambda) + u_{22}(\lambda)\} \not\equiv 0 \text{ in } \mathbb{C}_+\} \end{aligned}$$

by the formula

$$T_U[\varepsilon] = (u_{11}\varepsilon + u_{12})(u_{21}\varepsilon + u_{22})^{-1}.$$

If $U_1, U_2 \in \mathcal{U}(J)$ and if $\varepsilon \in \mathcal{D}(T_{U_2})$ and $T_{U_2}[\varepsilon] \in \mathcal{D}(T_{U_1})$ then

$$T_{U_1U_2}[\varepsilon] = T_{U_1}[T_{U_2}[\varepsilon]]\,.$$

The notation

$$T_U[E] = \{T_U[\varepsilon] : \varepsilon \in E\} \quad \text{for} \quad E \subset \mathcal{D}(T_U)$$

will be useful.

It is well known that if $W \in \mathcal{U}(j_{pq})$, then $\mathcal{S}^{p\times q} \subset \mathcal{D}(T_W)$ and $T_W[\mathcal{S}^{p\times q}] \subset \mathcal{S}^{p\times q}$. Moreover,

$$W \in \mathcal{U}_{sR}(j_{pq}) \iff T_W[\mathcal{S}^{p\times q}] \bigcap \mathring{\mathcal{S}}^{p\times q} \neq \emptyset .$$

Remark 5.1. The class $\mathcal{U}_{sR}(j_{pq})$ was introduced and defined by the condition on the right-hand side of this equivalence in [ArD:97] and then it was shown there that $W \in \mathcal{U}_{sR}(j_{pq}) \iff \mathcal{H}(W) \subset L_2^m$. Here, we have reversed the order and have defined the class $\mathcal{U}_{sR}(J)$ by the condition $\mathcal{H}(U) \subset L_2^m$ for arbitrary signature matrices J, including $J = \pm I_m$.

It is not hard to show that linear fractional transformations based on $A \in \mathcal{U}(J_p)$ map $\tau \in \mathcal{C}^{p\times p} \cap \mathcal{D}(T_A)$ into $\mathcal{C}^{p\times p}$. However, the set

$$\mathcal{C}(A) = T_B[\mathcal{S}^{p\times p} \cap \mathcal{D}(T_B)] .$$

based on the mvf $B(\lambda) = A(\lambda)\mathfrak{V}$ is more useful. In particular,

$$T_A[\mathcal{C}^{p\times p} \cap \mathcal{D}(T_A)] \subset T_B[\mathcal{S}^{p\times p} \cap \mathcal{D}(T_B)] \subset \mathcal{C}^{p\times p}$$

and

$$A \in \mathcal{U}_{sR}(J_p) \iff \mathcal{C}(A) \cap \mathring{\mathcal{C}}^{p\times p} \neq \emptyset .$$

We remark that

$$\mathcal{S}^{p\times p} \subset \mathcal{D}(T_B) \iff b_{22}(\omega)b_{22}(\omega)^* > b_{21}(\omega)b_{21}(\omega)^* \tag{5.1}$$

for some (and hence every) point $\omega \in \mathfrak{H}_A^+$; see Theorem 2.7 in [ArD:03b].

The set $\mathcal{C}(A)$ can also be described directly in terms of a linear fractional transformation $\widetilde{T}_A[\{a,b\}]$ based on the mvf $A(\lambda)$ acting on pairs $\{a(\lambda), b(\lambda)\}$ of $p \times p$ mvf's that are meromorphic in $\mathbb{C}_+$ by the formula

$$\widetilde{T}_A[\{a,b\}] = \{a_{11}(\lambda)a(\lambda) + a_{12}(\lambda)b(\lambda)\}\{a_{21}(\lambda)a(\lambda) + a_{22}(\lambda)b(\lambda)\}^{-1} . \tag{5.2}$$

The domain of definition of this transformation

$$\mathcal{D}(\widetilde{T}_A) = \left\{ \begin{array}{c} \{a,b\} : a \text{ and } b \text{ are meromorphic } p\times p \text{ mvf's in } \mathbb{C}_+ \\ \text{and } \det\{a_{21}(\lambda)a(\lambda) + a_{22}(\lambda)b(\lambda)\} \not\equiv 0 \text{ in } \mathbb{C}_+ \end{array} \right\} .$$

If J is a signature matrix that is unitarily equivalent to J_p, $\mathcal{F}(J)$ denotes the set of pairs $\{a(\lambda), b(\lambda)\}$ of $p \times p$ mvf's that are meromorphic in $\mathbb{C}_+$ and satisfy the following two conditions:

(1) $[a(\lambda)^* \quad b(\lambda)^*]J \begin{bmatrix} a(\lambda) \\ b(\lambda) \end{bmatrix} \leq 0$ for $\lambda \in \mathfrak{H}_a^+ \cap \mathfrak{H}_b^+$.

(2) $a(\lambda)^*a(\lambda) + b(\lambda)^*b(\lambda) > 0$ for at least one point $\lambda \in \mathfrak{H}_a^+ \cap \mathfrak{H}_b^+$.

It is not difficult to check that

$$\mathcal{C}(A) = \widetilde{T}_A[\mathcal{F}(J_p) \cap \mathcal{D}(\widetilde{T}_A)] ,$$

where

$$\widetilde{T}_A[\widetilde{E}] = \left\{ \widetilde{T}_A[\{a,b\}] : \{a,b\} \in \widetilde{E} \right\}$$

for every subset $\widetilde{E}$ of $\mathcal{D}(\widetilde{T}_A)$. Thus, if $A \in \mathcal{U}(J_p)$, then

$$A \in \mathcal{U}_{sR}(J_p) \Longleftrightarrow \widetilde{T}_A[\mathcal{F}(J_p) \cap \mathcal{D}(\widetilde{T}_A)] \cap \mathring{\mathcal{C}}^{p\times p} \neq \emptyset$$

and

$$\mathcal{S}^{p\times p} \subset \mathcal{D}(T_B) \Longleftrightarrow \mathcal{F}(J_p) \subset \mathcal{D}(\widetilde{T}_A)\,.$$

Other characterizations of the class $\mathcal{U}_{sR}(J)$ in terms of the Treil-Volberg matrix version of the Muckenhoupt $(A)_2$ condition are furnished in [ArD:01] and [ArD:03a].

6. Parameterization of $A \in \mathcal{U}_{sR}(J_p)$

If $A \in \mathcal{U}(J_p)$, then the formulas

$$A^{\#}(\lambda)J_pA(\lambda) = A(\lambda)J_pA^{\#}(\lambda) = J_p\,,$$

which are valid for every point $\lambda \in \mathfrak{H}_A \cap \mathfrak{H}_{A^{\#}}$, yield a number of relations between the blocks of $A(\lambda)$ and indicate that some of the blocks of $A(\lambda)$ may be computed in terms of the others. In fact, if $A \in \mathcal{U}_{sR}(J_p)$, then a pair of $p \times p$ inner mvf's $\{b_3(\lambda), b_4(\lambda)\}$ and a mvf $c \in \mathcal{C}^{p\times p}$ that are connected to $A(\lambda)$ by the rules given below serve to specify $A(\lambda)$ up to a constant J_p-unitary multiplier on the right. The first two rules are formulated in terms of the blocks of $B(\lambda) = A(\lambda)\mathfrak{V}$.

(1) $b_{21}^{\#}b_3 \in \mathcal{N}_{\text{out}}^{p\times p}$ and $b_3 \in \mathcal{S}_{\text{in}}^{p\times p}$.
(2) $b_4b_{22} \in \mathcal{N}_{\text{out}}^{p\times p}$ and $b_4 \in \mathcal{S}_{\text{in}}^{p\times p}$.
(3) $c \in \mathcal{C}(A)$.

The mvf's $b_3(\lambda)$ and $b_4(\lambda)$ are unique up to a constant $p\times p$ unitary multiplier (on the right for $b_3(\lambda)$ and on the left for $b_4(\lambda)$) and will be designated an **associated pair of the second kind** for $A(\lambda)$:

$$\{b_3(\lambda), b_4(\lambda)\} \in ap_{II}(A)\,.$$

(There is also a set of associated pairs $\{b_1(\lambda), b_2(\lambda)\}$ of the first kind that is more convenient to use in some other classes of problems that will not be discussed here.)

The main conclusions are summarized in the following theorems:

Theorem 6.1. *If $A \in \mathcal{U}(J_p)$ is an entire mvf and $\{b_3(\lambda), b_4(\lambda)\} \in ap_{II}(A)$, then $b_3(\lambda)$ and $b_4(\lambda)$ are also entire. Conversely, if $A \in \mathcal{U}_{rR}(J_p)$ (and hence a fortiori if $A \in \mathcal{U}_{sR}(J_p)$), $\{b_3, b_4\} \in ap_{II}(A)$ and $b_3(\lambda)$ and $b_4(\lambda)$ are entire inner mvf's, then $A(\lambda)$ is entire.*

Proof. See the discussion in Section 3.4 of [ArD:03b] □

Theorem 6.2. *If $A \in \mathcal{U}_{sR}(J_p)$, $\{b_3(\lambda), b_4(\lambda)\} \in ap_{II}(A)$ and $c \in \mathcal{C}(A) \cap H_\infty^{p\times p}$, then*

$$\mathcal{H}(A) = \left\{\begin{bmatrix} -\Pi_+c^*g + \Pi_-ch \\ g+h \end{bmatrix} : g \in \mathcal{H}(b_3) \quad and \quad h \in \mathcal{H}_*(b_4)\right\},$$

where Π_+ *denotes the orthogonal projection of* L_2^p *onto the Hardy space* H_2^p, $\Pi_- = I - \Pi_+$ *denotes the orthogonal projection of* L_2^p *onto* $K_2^p = L_2^p \ominus H_2^p$,

$$\mathcal{H}(b_3) = H_2^p \ominus b_3 H_2^p \quad \text{and} \quad \mathcal{H}_*(b_4) = K_2^p \ominus b_4^* K_2^p\,.$$

Moreover,

$$f = \begin{bmatrix} -\Pi_+ c^* g + \Pi_- ch \\ g + h \end{bmatrix} \Longrightarrow \langle f, f\rangle_{\mathcal{H}(A)} = \langle (c + c^*)(g+h), g+h\rangle_{st}\,,$$

where $g \in \mathcal{H}(b_3)$, $h \in \mathcal{H}_*(b_4)$ *and* $\langle\cdot,\cdot\rangle_{st}$ *denotes the standard inner product in* L_2^p.

Proof. See Theorem 3.8 in [ArD:05]. □

Theorem 6.3. *If* $A \in \mathcal{E} \cap \mathcal{U}_{sR}(J_p)$, $\{b_3, b_4\} \in ap_{II}(A)$ *and* $\mathfrak{E}(\lambda) = \sqrt{2}[0 \quad I_p]B(\lambda)$, *then*

$$\langle f, f\rangle_{\mathcal{H}(A)} = 2\|[0 \quad I_p]f\|^2_{\mathcal{B}(\mathfrak{E})}$$

for every $f \in \mathcal{H}(A)$ *and*

$$\mathcal{B}(\mathfrak{E}) = \mathcal{H}(b_3) \oplus \mathcal{H}_*(b_4) \quad \text{as Hilbert spaces with equivalent norms.}$$

Proof. See Theorem 3.8 in [ArD:05]. □

Remark 6.4. If $\mathcal{B}(\mathfrak{E}) = \mathcal{H}(b_3) \oplus \mathcal{H}_*(b_4)$ as linear spaces, then the two norms in these spaces are automatically equivalent, i.e., there exist a pair of positive constants γ_1, γ_2 such that

$$\gamma_1\|f\|_{st} \leq \|f\|_{\mathcal{B}(\mathfrak{E})} \leq \gamma_2\|f\|_{st}$$

for every $f \in \mathcal{B}(\mathfrak{E})$. This follows from the closed graph theorem and the fact that $\mathcal{B}(\mathfrak{E})$ and $\mathcal{H}(b_3) \oplus \mathcal{H}_*(b_4)$ are both RKHS's.

7. Chains of entire J-inner mvf's

A family $\{U_t(\lambda)\}$, $0 \leq t < d$, of entire J-inner mvf's is said to be

(1) **normalized**: if $U_t(0) = I_m$ for every $t \in [0, d)$ and $U_0(\lambda) = I_m$ for every $\lambda \in \mathbb{C}$,
(2_ℓ) **left monotonic**: if $(U_{t_1})^{-1}U_{t_2} \in \mathcal{U}(J)$ when $0 \leq t_1 \leq t_2 < d$,
(3) **continuous**: if $U_t(\lambda)$ is a continuous function of t on the interval $[0, d)$ for every fixed point $\lambda \in \mathbb{C}$.

Thus, a family $\{U_t(\lambda)\}$, $0 \leq t < d$, of entire J-inner mvf's is said to be a normalized left monotonic continuous chain of entire J-inner mvf's if the preceding three constraints are met. Similarly, a family $\{U_t(\lambda)\}$, $0 \leq t < d$, of entire J-inner mvf's is said to be a normalized right monotonic continuous chain of entire J-inner mvf's if the constraints (1), (3) and

(2_r) **right monotonic**: if $U_{t_2}(U_{t_1})^{-1} \in \mathcal{U}(J)$ when $0 \leq t_1 \leq t_2 < d$,

are met.

8. Canonical systems

The **matrizant** (fundamental solution) $U_t(\lambda) = U(t,\lambda)$, $0 \le t < d$, of the canonical integral system

$$y(t,\lambda) = y(0,\lambda) + i\lambda \int_0^t y(s,\lambda)dM(s)J\ , \quad 0 \le t < d\ , \tag{8.1}$$

based on a continuous nondecreasing $m \times m$ mvf $M(t)$ on the interval $[0,d)$ with $M(0) = 0$, is the unique continuous solution of the integral system

$$U(t,\lambda) = I_m + i\lambda \int_0^t U(s,\lambda)dM(s)J\ , \quad 0 \le t < d\ .$$

The mass function $M(t)$ of any such canonical integral system is uniquely determined by the matrizant via the formula

$$M(t) = -i\left(\frac{\partial U_t}{\partial \lambda}\right)(0)J\,, \quad 0 \le t < d\,.$$

Standard estimates lead to the conclusion that $U_t(\lambda)$ is an entire mvf of λ for each fixed $t \in [0,d)$. Moreover, as follows from the identity

$$J - U(t,\lambda)JU(t,\omega)^* = -i(\lambda - \overline{\omega}) \int_0^t U(s,\lambda)dM(s)U(s,\omega)^*, \tag{8.2}$$

$U_t \in \mathcal{U}(J)$ for every $t \in [0,d)$ and the corresponding RKHS's $\mathcal{H}(U_t)$ are ordered by inclusion as sets:

$$\mathcal{H}(U_{t_1}) \subset \mathcal{H}(U_{t_2}) \quad \text{if} \quad 0 \le t_1 \le t_2 < d \tag{8.3}$$

and

$$\|f\|_{\mathcal{H}(U_{t_2})} \le \|f\|_{\mathcal{H}(U_{t_1})} \quad \text{if} \quad f \in \mathcal{H}(U_{t_1})\,. \tag{8.4}$$

If $U_t \in \mathcal{U}_{sR}(J)$ for every $t \in [0,d)$, then the inclusion in relation (8.3) is isometric, i.e., equality prevails in the inequality (8.4); see, e.g., Theorem 2.5 of [ArD:05] and Theorem 2.3 of [ArD:04b].

The matrizant $U_t(\lambda)$, $0 \le t < d$, of every canonical integral system (8.1) with continuous nondecreasing mass function $M(t)$ is a normalized left monotonic continuous chain of entire J-inner mvf's. The converse of this statement is true under extra constraints:

Theorem 8.1. *Let $\{U_t(\lambda)\}$, $0 \le t < d$, be a normalized left monotonic continuous chain of entire J-inner mvf's such that*

$$U_t \in \mathcal{U}_{sR}(J) \quad \textit{for every} \quad t \in [0,d)\,. \tag{8.5}$$

Then there exists exactly one continuous nondecreasing mass function $M(t)$ on the interval $[0,d)$ such that $U_t(\lambda)$ is the matrizant of the corresponding canonical integral system (8.1).

Proof. This is a corollary of Theorem 4.6 in [ArD:97] and the discussion of formula (0.16) in [ArD:00a]. □

If $M \in AC_{\text{loc}}^{m\times m}([0,d))$, then

$$M(t) = \int_0^t H(s)ds$$

for a mvf

$$H \in L_{1,\text{loc}}^{m\times m}([0,d)) \quad \text{such that} \quad H(t) \geq 0 \quad \text{a.e. in} \quad [0,d)\,. \tag{8.6}$$

Thus, in this case, a continuous solution of (8.1) is automatically locally absolutely continuous on the interval $[0,d)$ and is a solution of the canonical differential system

$$y'(t,\lambda) \;=\; i\lambda y(t,\lambda)H(t)J \quad \text{for} \quad 0 \leq t < d\,. \tag{8.7}$$

We shall refer to a mvf $H(t)$ that meets the conditions in (8.6) as the **Hamiltonian** of the canonical differential system (8.7). The matrizant of the canonical integral system (8.1) coincides with the matrizant of this canonical differential system. Since $M(s)$ is absolutely continuous with respect to the strictly increasing function

$$\tau(s) = \operatorname{trace} M(s) + s \quad \text{for} \quad 0 \leq s < d\,,$$

it is always possible to reexpress the canonical integral system (8.1) as a canonical differential system with Hamiltonian $H = dM/d\tau$; see, e.g., Appendix I in [ArD:00a].

9. Chains of associated pairs

Let $A_t(\lambda)$ denote the matrizant of a canonical system of the form (8.1) with $J = J_p$:

$$A(t,\lambda) = I_m + i\lambda \int_0^t A(s,\lambda)dM(s)J_p\;,\; 0 \leq t < d\;, \tag{9.1}$$

where $M(t)$ is a continuous nondecreasing $m \times m$ mvf on the interval $[0,d)$ with $M(0) = 0$. Then the chain of associated pairs $\{b_3^t, b_4^t\} \in ap_{II}(A_t)$, $0 \leq t < d$, of entire inner $p \times p$ mvf's, is

(1) **monotonic** in the sense that $(b_3^{t_1})^{-1}b_3^{t_2} \in \mathcal{S}_{\text{in}}^{p\times p}$ and $b_4^{t_2}(b_4^{t_1})^{-1} \in \mathcal{S}_{\text{in}}^{p\times p}$ when $0 \leq t_1 \leq t_2 < d$.

Moreover,

(2) the chain may be **normalized** by the conditions $b_3^t(0) = b_4^t(0) = I_p$, for every $t \in [0,d)$ and $b_3^0(\lambda) = b_4^0(\lambda) = I_p$ for every $\lambda \in \mathbb{C}$. A normalized chain of associated pairs is uniquely determined by the matrizant $A_t(\lambda)$, $0 \leq t < d$.

If the matrizant $A_t(\lambda)$ also satisfies the condition

$$A_t \in \mathcal{U}_{sR}(J_p) \quad \text{for every} \quad t \in [0,d)\,, \tag{9.2}$$

then

(3) the normalized chain of associated pairs $\{b_3^t, b_4^t\} \in ap_{II}(A_t)$, $0 \leq t < d$, is **continuous** in the sense that both $b_3^t(\lambda)$ and $b_4^t(\lambda)$ are continuous mvf's of t on the interval $[0,d)$ for each fixed choice of $\lambda \in \mathbb{C}$.

Thus, if the matrizant $A_t(\lambda)$ of the canonical integral system (8.1) with $J = J_p$ satisfies condition (9.2), then the corresponding chain $\{b_3^t(\lambda), b_4^t(\lambda)\}$, $0 \le t < d$, of normalized associated pairs of the second kind for $A_t(\lambda)$ is uniquely defined by $A_t(\lambda)$ and is a **normalized monotonic continuous chain of pairs of entire inner** $p \times p$ **mvf's**; see Theorem 7.4 in [ArD:03b].

In future sections, we will discuss a number of inverse problems for canonical integral systems of the form (8.1) with $J = J_p$. In our formulation of these problems, the given data includes a normalized monotonic continuous chain of pairs $\{b_3^t(\lambda), b_4^t(\lambda)\}$, $0 \le t < d$, of entire inner $p \times p$ mvf's in addition to the spectral data that is furnished in traditional investigations. This chain helps to specify the class of admissible solutions by imposing the condition

$$\{b_3^t, b_4^t\} \in ap_{II}(A_t) \quad \text{for every} \quad t \in [0, d)$$

on the matrizant $A_t(\lambda)$ of the system (8.1) with $J = J_p$ and unknown mass function $M(t)$. Thus, it is of interest to describe the set of such chains. Moreover, since the RKHS's $\mathcal{H}(b_3^t) \subset L_2^p(\mathbb{R})$ and$\mathcal{H}_*(b_4^t) \subset L_2^p(\mathbb{R})$,

$$b_3^t \in \mathcal{U}_{sR}(I_p) \quad \text{and} \quad (b_4^t)^\tau \in \mathcal{U}_{sR}(I_p) \quad \text{for every} \quad t \in [0, d)$$

and hence, Theorem 8.1 guarantees that $b_3^t(\lambda)$ is the matrizant of a canonical integral system based on a continuous non decreasing $p \times p$ mvf $m_3(t)$ on the interval $[0, d)$. Similarly, $(b_4^t(\lambda))^\tau$ is the matrizant of a canonical integral system based on a continuous non decreasing $p \times p$ mvf $(m_4(t))^\tau$ on the interval $[0, d)$. Thus, we are led to the following conclusion:

Theorem 9.1. *There is a one to one correspondence between normalized monotonic continuous chains of pairs* $\{b_3^t(\lambda), b_4^t(\lambda)\}, 0 \le t < d$, *of entire inner* $p \times p$ *mvf's and the pairs* $\{m_3(t), m_4(t)\}$ *of continuous nondecreasing* $p \times p$ *mvf's on* $[0, d)$ *with* $m_3(0) = m_4(0) = 0$:
$\{b_3^t(\lambda), b_4^t(\lambda)\}, 0 \le t < d$, *are the unique continuous solutions of the integral equations*

$$b_3^t(\lambda) = I_p + i\lambda \int_0^t b_3^s(\lambda) dm_3(s), \quad b_4^t(\lambda) = I_p + i\lambda \int_0^t (dm_4(s)) b_4^s(\lambda), \; 0 \le t < d,$$

and

$$m_3(t) = -i\frac{\partial b_3^t}{\partial \lambda}(0) \quad \text{and} \quad m_4(t) = -i\frac{\partial b_4^t}{\partial \lambda}(0).$$

See Theorem 2.1 of [ArD:00a] for additional details.

10. de Branges spaces associated with systems

The matrizant $A_t(\lambda) = A(t, \lambda)$, $0 \le t < d$ of the canonical system (8.1) with $J = J_p$ satisfies the identity (8.2) with $J = J_p$ and $U(t, \lambda) = A(t, \lambda)$ for $0 \le t < d$.

Thus, for any matrix $L \in \mathbb{C}^{m\times p}$ such that $L^*J_pL = 0$, formula (8.2) implies that the kernel

$$\frac{-L^*A(t,\lambda)J_pA(t,\omega)^*L}{\rho_\omega(\lambda)} = \frac{1}{2\pi}\int_0^t L^*A(s,\lambda)dM(s)A(s,\omega)^*L$$

is positive on $\mathbb{C}\times\mathbb{C}$. Thus, if $L^*J_pL = 0$ and $\operatorname{rank} L = p$, then the mvf's $L^*A(t,\lambda)\mathfrak{V}$ are de Branges functions. The corresponding de Branges spaces play a significant role in the study of direct and inverse spectral problems. The matrix L may be chosen to conform to the initial conditions imposed on the solution of the canonical system that intervenes in the generalized Fourier transform that will be discussed in later sections. The particular choice $L^* = \sqrt{2}[0 \quad I_p]$, leads to the de Branges function

$$\mathfrak{E}_t(\lambda) = \sqrt{2}[0 \quad I_p]A_t(\lambda)\mathfrak{V}\,, \quad 0 \le t < d, \tag{10.1}$$

with $p \times p$ components

$$\mathfrak{E}_t(\lambda) = [E_-^t(\lambda) \quad E_+^t(\lambda)]\,.$$

Theorem 10.1. *Let $\mathfrak{E}_t(\lambda)$ denote the de Branges function associated with the matrizant $A_t(\lambda)$, $0 \le t < d$, of the canonical system (8.1) with $J = J_p$ by formula (10.1) and let $B_t(\lambda) = A_t(\lambda)\mathfrak{V}$ and $\{b_3^t, b_4^t\} \in ap_{II}(A_t)$ for every $t \in [0,d)$. Then:*

(1) *The spaces $\mathcal{B}(\mathfrak{E}_t)$ are ordered by contractive inclusion:*

$$\mathcal{B}(\mathfrak{E}_{t_1}) \subset \mathcal{B}(\mathfrak{E}_{t_2}) \quad \text{and} \quad \|f\|_{\mathcal{B}(\mathfrak{E}_{t_2})} \le \|f\|_{\mathcal{B}(\mathfrak{E}_{t_1})} \quad \text{for every} \quad f \in \mathcal{B}(\mathfrak{E}_{t_1})$$

when $0 \le t_1 \le t_2 < d$.

If $A_t \in \mathcal{U}_{sR}(J_p)$ for every $t \in [0,d)$, then more is true:

(2) *The inclusion in relation (8.3) for $U_t(\lambda) = A_t(\lambda)$ is isometric, i.e., equality prevails in the inequality (8.4).*
(3) *The map $f \in \mathcal{H}(A_t) \longrightarrow \sqrt{2}[0 \quad I_p]f \in \mathcal{B}(\mathfrak{E}_t)$ is unitary for every $t \in [0,d)$.*
(4) *The inclusions in (1) are isometric.*
(5) *$\mathcal{B}(\mathfrak{E}_t) = \mathcal{H}(b_3^t) \oplus \mathcal{H}_*(b_4^t)$ as Hilbert spaces with equivalent norms for every $t \in [0,d)$.*

Proof. Assertion (1) is due to de Branges; it also follows from Theorem 4.4 in [ArD:97] and Theorem 2.4 of [ArD:05]. The remaining assertions rest on Theorem 2.4 and Lemma 3.1 of [ArD:05] and the fact that if $A_t \in \mathcal{U}_{sR}(J_p)$, then the RKHS $\mathcal{H}_2(A_t)$ referred to in the cited theorem is equal to $\{0\}$. □

Expository introductions to the de Branges spaces associated with the solutions of generalized string equations may also be found in [DMc:76] and [Dy:70].

11. A generalized Carathéodory interpolation problem

Our next main objective is to discuss a bitangential inverse impedance problem. However, in keeping with a general strategy for studying such problems that was introduced by M.G. Krein, we first introduce a family of generalized Carathéodory interpolation problems GCIP$(b_3^t, b_4^t; c)$ based on a mvf $c \in \mathcal{C}^{p\times p}$ and a normalized monotonic continuous chain of pairs $\{b_3^t, b_4^t\}$, $0 \leq t < d$ of entire inner $p \times p$ mvf's.

In general, the GCIP$(b_3, b_4; c)$ based on a mvf $c \in \mathcal{C}^{p\times p}$ and a pair of mvf's $b_3, b_4 \in \mathcal{S}_{\text{in}}^{p\times p}$ is to describe the set

$$\mathcal{C}(b_3, b_4; c) = \{\widetilde{c} \in \mathcal{C}^{p\times p} : (b_3)^{-1}(\widetilde{c} - c)(b_4)^{-1} \in \mathcal{N}_+^{p\times p}\}.$$

There is a correspondence between problems of this sort that are subject to the extra condition

$$\mathcal{C}(b_3, b_4; c) \cap \mathring{\mathcal{C}}^{p\times p} \neq \emptyset \tag{11.1}$$

and the class $\mathcal{U}_{sR}(J_p)$:

Theorem 11.1. *Let $A \in \mathcal{U}_{sR}(J_p)$, $c \in \mathcal{C}(A)$ and let $\{b_3, b_4\} \in ap_{II}(A)$. Then the condition* (11.1) *is in force and*

$$\mathcal{C}(b_3, b_4; c) = \mathcal{C}(A). \tag{11.2}$$

If, in addition, $A(\lambda)$ is an entire mvf, then $b_3(\lambda)$ and $b_4(\lambda)$ are also entire mvf's.

Conversely, if the condition (11.1) *is in force for a given choice of $c \in \mathcal{C}^{p\times p}$ and $\{b_3, b_4\} \in \mathcal{S}_{\text{in}}^{p\times p}$, then there exists a mvf $A \in \mathcal{U}(J_p)$ such that* (11.2) *holds. Moreover, every mvf $A \in \mathcal{U}(J_p)$ for which* (11.2) *holds is automatically strongly regular and there is essentially only such mvf $A(\lambda)$ (up to a constant J_p-unitary factor on the right) such that $\{b_3, b_4\} \in ap_{II}(A)$.*

Proof. The stated results follow from the results established in [Ar:94]. □

In applications to inverse problems, the case in which $b_3(\lambda)$ and $b_4(\lambda)$ are entire inner mvf's is of particular interest. The following specialization of the preceding theorem is useful.

Theorem 11.2. *Let $b_3, b_4 \in \mathcal{E} \cap \mathcal{S}_{\text{in}}^{p\times p}$, let $c \in \mathcal{C}^{p\times p}$ and assume that condition* (11.1) *is in force. Then there exists exactly one mvf $A \in \mathcal{U}(J_p)$ such that*

(1) $\mathcal{C}(A) = \mathcal{C}(b_3, b_4;\, c)$.
(2) $\{b_3, b_4\} \in ap_{II}(A)$.
(3) $A(0) = I_m$.

Moreover, this mvf $A \in \mathcal{E} \cap \mathcal{U}_{sR}(J_p)$.

We remark that the bitangential version of the Krein helical extension problem is equivalent to a GCIP$(b_3, b_4; c)$ that is based on entire inner mvf's $b_3(\lambda)$ and $b_4(\lambda)$, whereas the classical Krein helical extension problem corresponds to the special case when $b_3^t(\lambda) = e_{\alpha t}(\lambda)$ and $b_4^t(\lambda) = e_{\beta t}(\lambda)$, where $\alpha \geq 0$, $\beta \geq 0$ and $\alpha + \beta > 0$. To be more precise, in the last setting it is only the sum $\alpha + \beta$ that is significant and not the specific choices of the nonnegative numbers α and β.

In future sections we shall have special interest in mvf's $c(\lambda)$ for which

$$\mathcal{C}(e_d I_p, I_p;\, c) \cap \mathcal{W}_+^{p\times p}(\gamma) \neq \emptyset \quad \text{for some} \quad \gamma \in \mathbb{C}^{p\times p}. \tag{11.3}$$

If the mvf

$$c^\circ(\lambda) = \gamma + 2\int_0^\infty e^{i\lambda t} h^\circ(t)dt \tag{11.4}$$

belongs to this intersection, then the constant matrix γ and the restriction of the $p \times p$ mvf $h^\circ(t)$ to the interval $[0, d]$ are uniquely determined by $c(\lambda)$. We shall refer to this restriction as the **accelerant** of $c(\lambda)$ on the interval $[0, d]$. If $c \in \mathcal{C}^{p\times p} \cap \mathcal{W}_+^{p\times p}(\gamma)$, then the mvf $h(t)$ in the representation

$$c(\lambda) = \gamma + 2\int_0^\infty e^{i\lambda t} h(t)dt \tag{11.5}$$

is called the accelerant of $c(\lambda)$ on the interval $[0, \infty)$. Additional information on the classical Krein extension problems and their bitangential generalizations is furnished in the last section and, in much more detail, in [ArD:98].

12. The bitangential inverse input impedance problem

The set $\mathcal{C}_{\text{imp}}^d(M)$ of input impedance matrices for a given canonical integral system (8.1) with $J = J_p$ is defined as the intersection of the sets $\mathcal{C}(A_t)$:

$$\mathcal{C}_{\text{imp}}^d(M) = \bigcap_{0\le t<d} \mathcal{C}(A_t)\,.$$

The mvf's $ic(\lambda)$, where $c \in \mathcal{C}_{\text{imp}}^d(M)$ are usually called **Weyl-Titchmarsh functions**. It may happen that $\mathcal{C}_{\text{imp}}^d(M) = \emptyset$. However, if

$$\mathcal{S}^{p\times p} \subset \mathcal{D}(T_{B_{t_0}}) \quad \text{for some} \quad t_0 \in (0, d), \tag{12.1}$$

then $\mathcal{C}_{\text{imp}}^d(M) \neq \emptyset$. In particular, the condition (12.1) will be in force if

$$e_{-a} b_3^{t_0} b_4^{t_0} \in \mathcal{S}_{\text{in}}^{p\times p} \quad \text{for some} \quad a > 0 \quad \text{and} \quad A_{t_0} \in \mathcal{U}_{sR}(J_p)\,;$$

see, e.g., Lemma 2.4 of [ArD:04b]. In our formulation of the **bitangential inverse input impedance problem**, the given data is a mvf $c \in \mathcal{C}^{p\times p}$, and a normalized monotonic continuous chain of pairs $\{b_3^t(\lambda), b_4^t(\lambda)\}$, $0 \le t < d$, of entire inner $p \times p$ mvf's. An $m \times m$ mvf $M(t)$ on the interval $[0, d)$ is said to be a **solution** of the bitangential inverse input impedance problem with data $\{c(\lambda); b_3^t(\lambda), b_4^t(\lambda),\ 0 \le t < d\}$ if $M(t)$ is a continuous nondecreasing $m \times m$ mvf on the interval $[0, d)$ with $M(0) = 0$ such that the matrizant $A_t(\lambda)$ of the corresponding canonical integral system (8.1) meets the following three conditions:

(1) $c \in \mathcal{C}_{\text{imp}}^d(M)$.
(2) $\{b_3^t, b_4^t\} \in ap_{II}(A_t)$ for every $t \in [0, d)$.
(3) $A_t \in \mathcal{U}_{sR}(J_p)$ for every $t \in [0, d)$.

Theorem 12.1. *Let $c \in \mathcal{C}^{p\times p}$, let $\{b_3^t(\lambda), b_4^t(\lambda)\}, 0 \le t < d$, be a normalized monotonic continuous chain of pairs of entire inner $p \times p$ mvf's. Then there exists at least one solution $M(t)$, $0 \le t < d$, of the bitangential inverse input impedance problem for the given set of data if and only if*

$$\mathcal{C}(b_3^t, b_4^t; c) \cap \mathring{\mathcal{C}}^{p\times p} \neq \emptyset \text{ for every } t \in [0, d)\,. \tag{12.2}$$

Moreover, if a solution exists then it is unique and the matrizant $A_t(\lambda)$ of this solution may be characterized as the unique mvf $A_t \in \mathcal{U}(J_p)$ such that the following three conditions are met for every $t \in [0, d)$:

(1) $\mathcal{C}(b_3^t, b_4^t; c) = \mathcal{C}(A_t)$.
(2) $\{b_3^t, b_4^t\} \in ap_{II}(A_t)$.
(3) $A_t(0) = I_m$.

Proof. See Theorem 7.9 in [ArD:03b]. □

In order to apply this theorem, we need to know when condition (12.2) is in force. In particular, the condition (12.2) is satisfied if $c \in \mathring{\mathcal{C}}^{p\times p}$. However, if the given matrix $c \in \mathcal{C}^{p\times p} \cap \mathcal{W}_+^{p\times p}(\gamma)$, then condition (12.2) will be in force if $\gamma + \gamma^* > 0$, even if $\det \Re c(\mu) = 0$ at some points $\mu \in \mathbb{R}$; see Theorem 5.2 in [ArD:05]. Moreover, if either

$$\lim_{\nu\uparrow\infty} b_3^{t_0}(i\nu) = 0 \quad \text{or} \quad \lim_{\nu\uparrow\infty} b_4^{t_0}(i\nu) = 0$$

for some point $t_0 \in [0, d)$, then the condition $\gamma + \gamma^* > 0$ is necessary for (12.2) to be in force and hence for the existence of a canonical system (8.1) with a matrizant $A_t(\lambda)$, $0 \le t < d$, that meets the conditions (1) (2) and (3); see Theorem 5.4 in [ArD:05].

Remark 12.2. The method of solution depends upon the interplay between the RKHS's that play a role in the parametrization formulas presented in Theorem 6.2 and their corresponding RK's. This method also yields the formulas for $M(t)$ and the corresponding matrizant $A_t(\lambda)$ that are discussed in the next section. It differs from the known methods of Gelfand-Levitan, Marchenko and Krein, which are not directly applicable to the bitangential problems under consideration.

13. A basic formula

In this section we shall assume that a mvf $c \in \mathcal{C}^{p\times p}$ and a normalized monotonic continuous chain of pairs $\{b_3^t(\lambda), b_4^t(\lambda)\}$, $0 \le t < d$, of entire inner $p \times p$ mvf's that meet the condition (12.2) have been specified. Then there exists a mvf

$$c^t \in \mathcal{C}(b_3^t, b_4^t; c) \cap H_\infty^{p\times p} \tag{13.1}$$

for every $t \in [0, d)$ and hence, the operators

$$\Phi_{11}^t = \Pi_{\mathcal{H}(b_3^t)} M_{c^t}\Big|_{H_2^p}, \quad \Phi_{22}^t = \Pi_- M_{c^t}\Big|_{\mathcal{H}_*(b_4^t)}, \quad \Phi_{12}^t = \Pi_{\mathcal{H}(b_3^t)} M_{c^t}\Big|_{\mathcal{H}_*(b_4^t)}, \tag{13.2}$$

$$Y_1^t = \Pi_{\mathcal{H}(b_3^t)}\left\{M_{c^t} + (M_{c^t})^*\right\}\Big|_{\mathcal{H}(b_3^t)} = 2\Re\left(\Phi_{11}^t\Big|_{\mathcal{H}(b_3^t)}\right) \tag{13.3}$$

and

$$Y_2^t = \Pi_{\mathcal{H}_*(b_4^t)}\left\{M_{c^t} + (M_{c^t})^*\right\}\Big|_{\mathcal{H}_*(b_4^t)} = 2\Re\left(\Pi_{\mathcal{H}_*(b_4^t)}\Phi_{22}^t\right) \tag{13.4}$$

are well defined. Moreover, they do not depend upon the specific choice of the mvf c^t in the set indicated in formula (13.1).

In order to keep the notation relatively simple, an operator T that acts in the space of $p \times 1$ vvf's will be applied to $p \times p$ mvf's with columns $f_1, \dots, f_p$ column by column: $T[f_1 \cdots f_p] = [Tf_1 \cdots Tf_p]$. We define three sets of $p \times p$ mvf's $\widehat{y}^t_{ij}(\lambda)$, $\widehat{u}^t_{ij}(\lambda)$ and $\widehat{x}^t_{ij}(\lambda)$ by the following system of equations, in which $\tau_3(t)$ and $\tau_4(t)$ denote the exponential types of $b^t_3(\lambda)$ and $b^t_4(\lambda)$, respectively and

$$(R_0 f)(\lambda) = \{f(\lambda) - f(0)\}/\lambda :$$

$$\begin{aligned} \widehat{y}^t_{11}(\lambda) &= i\left(\Phi^t_{11}(R_0 e_{\tau_3(t)} I_p)\right)(\lambda)\,, & \widehat{y}^t_{12}(\lambda) &= -i(R_0 b^t_3)(\lambda)\,, \\ \widehat{y}^t_{21}(\lambda) &= i\left((\Phi^t_{22})^*(R_0 e_{-\tau_4(t)} I_p)\right)(\lambda)\,, & \widehat{y}^t_{22}(\lambda) &= i(R_0 (b^t_4)^{-1})(\lambda)\,, \end{aligned} \tag{13.5}$$

$$\begin{aligned} Y^t_1 \widehat{u}^t_{1j} + \Phi^t_{12}\widehat{u}^t_{2j} &= \widehat{y}^t_{1j}(\lambda)\,, \\ (\Phi^t_{12})^* \widehat{u}^t_{1j} + Y^t_2 \widehat{u}^t_{2j} &= \widehat{y}^t_{2j}(\lambda)\,, \quad j = 1, 2\,, \end{aligned} \tag{13.6}$$

$$\widehat{x}^t_{1j}(\lambda) = -(\Phi^t_{11})^* \widehat{u}^t_{1j} \quad \text{and} \quad \widehat{x}^t_{2j}(\lambda) = \Phi^t_{22}\widehat{u}^t_{2j}\,, \quad j = 1, 2\,. \tag{13.7}$$

Theorem 13.1. *Let $\{c(\lambda); b^t_3(\lambda), b^t_4(\lambda),\ 0 \le t < d\}$ be given where $c \in \mathcal{C}^{p\times p}$, $\{b^t_3(\lambda), b^t_4(\lambda)\}$, $0 \le t < d$, is a normalized monotonic continuous chain of pairs of entire inner $p \times p$ mvf's and let assumption* (12.2) *be in force. Then the unique solution $M(t)$ of the inverse input impedance problem considered in Theorem* 12.1 *is given by the formula*

$$M(t) = \int_0^{\tau_3(t)} \begin{bmatrix} x^t_{11}(a) & x^t_{12}(a) \\ u^t_{11}(a) & u^t_{12}(a) \end{bmatrix} da + \int_{-\tau_4(t)}^{0} \begin{bmatrix} x^t_{21}(a) & x^t_{22}(a) \\ u^t_{21}(a) & u^t_{22}(a) \end{bmatrix} da\,, \tag{13.8}$$

where $x^t_{ij}(a)$ and $u^t_{ij}(a)$ designate the inverse Fourier transforms of the mvf's $\widehat{x}^t_{ij}(\lambda)$ and $\widehat{u}^t_{ij}(\lambda)$ defined earlier, respectively, and the corresponding matrizant

$$A_t(\lambda) = I_m + i\lambda \begin{bmatrix} \widehat{x}^t_{11}(\lambda) + \widehat{x}^t_{21}(\lambda) & \widehat{x}^t_{12}(\lambda) + \widehat{x}^t_{22}(\lambda) \\ \widehat{u}^t_{11}(\lambda) + \widehat{u}^t_{21}(\lambda) & \widehat{u}^t_{12}(\lambda) + \widehat{u}^t_{22}(\lambda) \end{bmatrix} J_p\,. \tag{13.9}$$

Proof. See Theorem 4.4 in [ArD:05]. □

14. Spectral functions

The term **spectral function** is defined in two different ways: The first definition is in terms of the generalized Fourier transform

$$(\mathcal{F}_2 f)(\lambda) = [0 \quad I_p]\frac{1}{\sqrt{\pi}}\int_0^d A(s,\lambda)dM(s)f(s) \tag{14.1}$$

based on the matrizant of the canonical system (8.1) with $J = J_p$ applied initially to the set of $f \in L^m_2(dM; [0,d))$ with compact support inside the interval $[0,d)$.

A nondecreasing $p \times p$ mvf $\sigma(\mu)$ on $\mathbb{R}$ is said to be a **spectral function** for the system (8.1) if the Parseval equality

$$\int_{-\infty}^{\infty} (\mathcal{F}_2 f)(\mu)^* d\sigma(\mu)(\mathcal{F}_2 f)(\mu), = \int_0^d f(t)^* dM(t) f(t) dt \tag{14.2}$$

holds for every $f \in L_2^m(dM; [0,d))$ with compact support. The notation $\Sigma_{sf}^d(M)$ will be used to denote the set of spectral functions of a canonical system of the form (8.1) with $J = J_p$.

Remark 14.1. The generalized Fourier transform introduced in formula (14.1) is a special case of of the transform

$$(\mathcal{F}^L f)(\lambda) = L^* \frac{1}{\sqrt{\pi}} \int_0^d A(s,\lambda) dM(s) f(s) \tag{14.3}$$

that is based on a fixed $m \times p$ matrix L that meets the conditions $L^* J_p L = 0$ and $L^* L = I_p$. The mvf $y(t,\lambda) = L^* A_t(\lambda)$ is the unique solution of the system (8.1) with $J = J_p$ that satisfies the initial condition $y(0,\lambda) = L^*$. Spectral functions may be defined relative to the transform $\mathcal{F}^L$ in just the same way that they were defined for the transform $\mathcal{F}_2$. Direct and inverse spectral problems for these spectral functions are easily reduced to the corresponding problems based on $\mathcal{F}_2$.; see Sections 4 and 5 of [ArD:04b] and Section 16 below for additional discussion.

The second definition of spectral function is based on the Riesz-Herglotz representation

$$c(\lambda) = i\alpha - i\beta\lambda + \frac{1}{\pi i} \int_{-\infty}^{\infty} \left\{ \frac{1}{\mu - \lambda} - \frac{1}{1+\mu^2} \right\} d\sigma(\mu), \quad \lambda \in \mathbb{C}_+, \tag{14.4}$$

which defines a correspondence between $p \times p$ mvf's $c \in \mathcal{C}^{p\times p}$ and a set $\{\alpha, \beta, \sigma\}$, in which $\sigma(\mu)$ is a nondecreasing $p \times p$ mvf on $\mathbb{R}$ that is normalized to be left continuous with $\sigma(0) = 0$ and is subject to the constraint

$$\int_{-\infty}^{\infty} \frac{d\text{trace}\sigma(\mu)}{1+\mu^2} < \infty, \tag{14.5}$$

and α and β are constant $p \times p$ matrices such that $\alpha = \alpha^*$ and $\beta \geq 0$.

The mvf $\sigma(\mu)$ in the representation (14.4) will be referred to as the **spectral function** of $c(\lambda)$. Correspondingly, if $\mathcal{F} \subset \mathcal{C}^{p\times p}$, then

$$(\mathcal{F})_{sf} = \{\sigma : \text{ such that } \sigma \text{ is the spectral function of some } c \in \mathcal{F}\}.$$

If $c(\lambda) = T_A[I_p]$ and $A \in \mathcal{U}_{sR}(J_p)$, then $\beta = 0$ and $\sigma(\mu)$ is absolutely continuous with $\sigma'(\mu) = \Re c(\mu)$ a.e. on $\mathbb{R}$; see Lemma 2.2 and the discussion following Lemma 2.3 in [ArD:05]. Moreover, if $A \in \mathcal{U}_{sR}(J_p)$ and $\mathcal{S}^{p\times p} \subset \mathcal{D}(T_{A\mathfrak{V}})$, then for each $\sigma \in (\mathcal{C}(A))_{sf}$, there exists at least one $p \times p$ Hermitian matrix α such that

$$c^{(\alpha)}(\lambda) = i\alpha + \frac{1}{\pi i} \int_{-\infty}^{\infty} \left\{ \frac{1}{\mu - \lambda} - \frac{\mu}{1+\mu^2} \right\} d\sigma(\mu) \tag{14.6}$$

belongs to $\mathcal{C}(A)$; see Theorem 2.14 in [ArD:04b].

We shall also make use of the following condition on the growth of the mvf $\chi_1^t(\lambda) = b_4^t(\lambda)b_3^t(\lambda)$:

$$\|\chi_1^a(re^{i\theta} + \omega)\| \le \gamma < 1 \quad \text{on the indicated ray in } \mathbb{C}_+ \,, \tag{14.7}$$

i.e., the inequality holds for some fixed choice of $\theta \in [0, \pi]$, $\omega \in \mathbb{C}_+$, $a \in (0, d)$ and all $r \ge 0$. It is readily checked that if this inequality is in force for some point $a \in (0, d)$, then it holds for all $t \in [a, d)$.

Remark 14.2. The condition (14.7) will be in force if

$$e_{-a}\chi_1^{t_0}(\lambda) \in \mathcal{S}_{\text{in}}^{p\times p}$$

for some choice of $a > 0$ and $t_0 \in (0, d)$.

These observations leads to the following conclusion:

Lemma 14.3. *If the matrizant $A_t(\lambda)$ of the canonical differential system* (8.1) *with $J = J_p$ satisfies the condition* (9.2) *and if condition* (14.7) *is in force for some $a \in [0, d)$ when $\{b_3^t, b_4^t\} \in ap_{II}(A_t)$ for $t \in [0, d)$ and if $c \in \mathcal{C}(A_a)$, then $\beta = 0$ in the representation* (14.4).

In view of the fact that

$$A \in \mathcal{U}_{sR}(J_p) \iff \mathcal{C}(A) \cap \mathring{\mathcal{C}}^{p\times p} \neq \emptyset \,, \tag{14.8}$$

the conditions (12.2) and (9.2) are equivalent if $\mathcal{C}(A_t) = \mathcal{C}(b_3^t, b_4^t; c)$ for every $0 \le t < d$. In particular, these conditions are satisfied if $\mathcal{C}_{\text{imp}}^d(M) \cap \mathring{\mathcal{C}}^{p\times p} \neq \emptyset$. They are also satisfied if $\mathcal{C}_{\text{imp}}^d(M) \cap \mathcal{W}^{m\times m}(\gamma) \neq \emptyset$ for some $\gamma \in \mathbb{C}^{m\times m}$ with $\gamma + \gamma^* > 0$, by Theorem 5.2 in [ArD:05].

The direct problem

The direct problem for a given canonical system with mass function $M(t)$ on the interval $[0, d)$ is to describe the set of input impedances $\mathcal{C}_{\text{imp}}^d(M)$ and the set $\Sigma_{sf}^d(M)$ of spectral functions of the system.

Theorem 14.4. *Let $A_t(\lambda)$ denote the matrizant of a canonical integral system of the form* (8.1) *with $J = J_p$ and suppose that the two conditions* (9.2) *and* (14.7) *are met. Then*

(1) $\Sigma_{sf}^d(M) = (\mathcal{C}_{\text{imp}}^d(M))_{sf}$.

(2) *To each $\sigma \in \Sigma_{sf}^d(M)$ there exists exactly one mvf $c(\lambda) \in \mathcal{C}_{\text{imp}}^d(M)$ with spectral function $\sigma(\mu)$. Moreover, this mvf $c(\lambda)$ is equal to one of the mvf's $c^{(\alpha)}(\lambda)$ defined by formula* (14.6) *for some Hermitian matrix α.*

(3) *If $d < \infty$ and* trace $M(t) < \delta < \infty$ *for every $t \in [0, d)$, then the equation* (8.1) *and the matrizant $A_t(\lambda)$ may be considered on the closed interval $[0, d]$ and $\mathcal{C}_{\text{imp}}^d(M) = \mathcal{C}(A_d)$.*

Proof. See Theorem 2.21 in [ArD:04b]. □

A spectral function $\sigma \in \Sigma_{sf}^d(M)$ of the canonical integral system (8.1) with $J = J_p$ is said to be **orthogonal** if the isometric operator that extends the generalized Fourier transform $\mathcal{F}_2$ defined by formula (14.1) maps $L_2^m(dM;\,[0,d))$ onto $L_2^p(d\sigma;\,\mathbb{R})$.

Theorem 14.5. *Let the canonical integral system* (8.1) *with* $J = J_p$, *mass function* $M(t)$ *and matrizant* $A_t(\lambda)$ *be considered on a finite closed interval* $[0,d]$ *(so that* $\operatorname{trace} M(d) < \infty$*) and let* $A(\lambda) = A_d(\lambda)$, $B(\lambda) = A(\lambda)\mathfrak{V}$ *and* $\mathfrak{E}(\lambda) = \sqrt{2}[0 \quad I_p]B(\lambda)$. *Suppose further that*

(a) $(\mathcal{C}(A))_{sf} = \Sigma_{sf}^d(M)$ *and*

(b) $K_\omega^{\mathfrak{E}}(\omega) > 0$ *for at least one (and hence every) point* $\omega \in \mathbb{C}_+$.

Then:

(1) $\mathcal{S}^{p\times p} \subset \mathcal{D}(T_A)$.

(2) *The spectral function* $\sigma(\mu)$ *of the mvf* $c(\lambda) = T_B[\varepsilon]$ *is an orthogonal spectral function of the given canonical system if* ε *is a constant* $p\times p$ *unitary matrix.*

Proof. The first assertion is equivalent to condition (b); see (5.1). The proof of assertion (2) will be given elsewhere. □

Remark 14.6. The conclusions of the last theorem can be reformulated in terms of the linear fractional transformations of pairs that were discussed briefly in Section 5: The inclusion (1) in the last theorem is equivalent to the inclusion $\mathcal{F}(J_p) \subset \mathcal{D}(\widetilde{T}_A)$. Moreover, $c(\lambda) = T_B[\varepsilon]$ for some constant unitary $p \times p$ matrix ε if and only if $c(\lambda) = \widetilde{T}_A[\{a,b\}]$ for some pair of constant $p\times p$ matrices $\{a,b\}$ that meet the conditions $a^*b + b^*a = 0$ and $a^*a + b^*b > 0$. Assertion (2) is obtained under somewhat stronger assumptions in Theorem 2.5 of Chapter 4 of [Sak:99].

15. The bitangential inverse spectral problem

In our formulation of the **bitangential inverse spectral problem** the given data $\{\sigma; b_3^t, b_4^t,\ 0 \le t < d\}$ is a $p \times p$ nondecreasing mvf $\sigma(\mu)$ on $\mathbb{R}$ that meets the constraint (14.5) and a normalized monotonic continuous chain $\{b_3^t, b_4^t\}$, $0 \le t < d$, of pairs of entire inner $p \times p$ mvf's. An $m \times m$ mvf $M(t)$ on the interval $[0,d)$ is said to be a solution of the bitangential inverse spectral problem with data $\{\sigma(\mu); b_3^t(\lambda), b_4^t(\lambda),\ 0 \le t < d\}$ if $M(t)$ is a continuous nondecreasing $m \times m$ mvf on the interval $[0,d)$ with $M(0) = 0$ such that the matrizant $A_t(\lambda)$ of the corresponding canonical integral system (8.1) with $J = J_p$ meets the following three conditions:

(i) $\sigma(\mu)$ is a spectral function for this system, i.e., $\sigma \in \Sigma_{sf}^d(M)$.

(ii) $\{b_3^t, b_4^t\} \in ap_{II}(A_t)$ for every $t \in [0,d)$.

(iii) $A_t \in \mathcal{U}_{sR}(J_p)$ for every $t \in [0,d)$.

The constraint (ii) defines the class of canonical integral systems in which we look for a solution of the inverse problem for the given spectral function $\sigma(\mu)$. Subsequently, the condition (iii) guarantees that in this class there is at most one solution.

The solution of this problem rests on the preceding analysis of the bitangential inverse input impedance problem with data $\{c^{(\alpha)};\, b_3^t, b_4^t,\ 0 \le t < d\}$, where $c^{(\alpha)}(\lambda)$ is given by formula (14.6).

Theorem 15.1. *If the data $\{\sigma; b_3^t, b_4^t,\ 0 \le t < d\}$ for a bitangential inverse spectral problem meets the conditions (14.5) and (14.7) and the mvf $c(\lambda) = c^{(0)}(\lambda)$ satisfies the constraint (12.2), then the following conclusions hold:*

(1) *For each Hermitian matrix $\alpha \in \mathbb{C}^{p\times p}$, there exists exactly one solution $M^{(\alpha)}(t)$ of the bitangential inverse input spectral problem such that $c^{(\alpha)}(\lambda)$ is an input impedance for the corresponding canonical integral system (8.1) with $J = J_p$ based on the mass function $M^{(\alpha)}(t)$.*

(2) *The solutions $M^{(\alpha)}(t)$ are related to $M^{(0)}(t)$ by the formula*

$$M^{(\alpha)}(t) = \begin{bmatrix} I_p & i\alpha \\ 0 & I_p \end{bmatrix} M^{(0)}(t) \begin{bmatrix} I_p & 0 \\ -i\alpha & I_p \end{bmatrix} . \tag{15.1}$$

The corresponding matrizants are related by the formula

$$A_t^{(\alpha)}(\lambda) = \begin{bmatrix} I_p & i\alpha \\ 0 & I_p \end{bmatrix} A_t^{(0)}(\lambda) \begin{bmatrix} I_p & -i\alpha \\ 0 & I_p \end{bmatrix} . \tag{15.2}$$

(3) *If $M(t)$ is a solution of the bitangential inverse spectral problem, then $M(t) = M^{(\alpha)}(t)$ for exactly one Hermitian matrix $\alpha \in \mathbb{C}^{p\times p}$.*

(4) *The solution $M^{(0)}(t)$ and matrizant $A_t^{(0)}(\lambda)$ may be obtained from the formulas for the solution of the bitangential inverse input impedance problem with data $\{c^{(0)}; b_3^t, b_4^t, 0 \le t < d\}$ that are given in Theorem 13.1.*

Proof. See Theorem 2.20 in [ArD:04b]. □

The condition (12.2) is clearly satisfied if $c^{(0)} \in \mathring{\mathcal{C}}^{p\times p}$. However, this condition is far from necessary. If, for example, $c^{(0)} \in \mathcal{C}^{p\times p} \cap \mathcal{W}^{p\times p}(\gamma)$, then, as noted earlier, condition (12.2) holds if $\gamma + \gamma^* > 0$, even if $\det\{\Re c(\mu)\} = 0$ on some set of points $\mu \in \mathbb{R}$.

16. Spectral problems for systems with $J \ne J_p$

The preceding results for systems of the form (8.1) with $J = J_p$ may be adapted to systems of the form (8.1) with any signature matrix J that is unitarily equivalent to J_p. If

$$J = V^* J_p V \quad \text{for some unitary matrix} \quad V\,, \tag{16.1}$$

then $\widetilde{y}(t,\lambda) = y(t,\lambda)V^*$ is a solution of the system

$$\widetilde{y}(t,\lambda) = \widetilde{y}(t,0) + i\lambda \int_0^t \widetilde{y}(s,\lambda)d\widetilde{M}(s)J_p \tag{16.2}$$

with mass function $\widetilde{M}(t) = VM(t)V^*$ and matrizant $\widetilde{U}_t(\lambda) = VU_t(\lambda)V^*$ that belongs to class $\mathcal{U}(J_p)$. Correspondingly, the set of input impedance matrices $\mathcal{C}^d_{\rm imp}(M)$ for the system (8.1) with J unitarily equivalent to J_p is defined as the intersection of the sets $\mathcal{C}(VU_tV^*)$, $0 \leq t < d$. This set coincides with $\mathcal{C}^d_{\rm imp}(\widetilde{M})$.

A nondecreasing $p \times p$ mvf $\sigma(\mu)$ on $\mathbb{R}$ is said to be a spectral function for the system (8.1) if the Parseval equality holds for the transform

$$(\mathcal{F}_2 f)(\lambda) = [0 \quad I_p]\, V \frac{1}{\sqrt{\pi}} \int_0^d U(s,\lambda)dM(s)f(s) \tag{16.3}$$

for every $f \in L_2^m(dM;[0,d))$ with compact support. Such a spectral function is said to be orthogonal if the isometry that extends this transform to a mapping from $L_2^m(dM;[0,d))$ into $L_2^p(d\sigma;\mathbb{R})$ is onto.

The data for the inverse spectral problem for a canonical integral system of the form (8.1) with J unitarily equivalent to J_p is: a fixed constant unitary matrix V such that $J = V^*J_pV$, a $p \times p$ nondecreasing mvf $\sigma(\mu)$ on $\mathbb{R}$ that meets the constraint (14.5) and a normalized monotonic continuous chain $\{b_3^t, b_4^t\}$, $0 \leq t < d$, of pairs of entire inner $p \times p$ mvf's. A continuous nondecreasing $m \times m$ mvf $M(t)$ on the interval $[0,d)$ with $M(0) = 0$ is said to be a solution of the inverse spectral problem for this given set of data if the matrizant $U_t(\lambda)$ of the canonical integral system (8.1) with mass function $M(t)$ meets the following three conditions:

(i) $\sigma(\mu)$ is a spectral function for this system, i.e., the Parseval formula based on the transform (16.3) holds.

(ii) $\{b_3^t, b_4^t\} \in ap_{II}(VU_tV^*)$ for every $t \in [0,d)$.

(iii) $U_t \in \mathcal{U}_{sR}(J_p)$ for every $t \in [0,d)$.

The results discussed earlier for the case $J = J_p$, as well as those that will be discussed in the next section, are easily adapted to the case $J = V^*J_pV$ considered in this section.

The reduction considered above is also useful in the case $J = J_p$: It may be used to reduce direct and inverse spectral problems based on the generalized Fourier transform $\mathcal{F}^L$ defined in (14.3) to the corresponding problems based on the transform $\mathcal{F}_2$ defined in (14.1) with the help of the unitary matrix $V^* = [-J_pL \quad L]$, since $L^*V^* = [0 \quad I_p]$:

$$\begin{aligned} \mathcal{F}^L(f) &= L^* \frac{1}{\sqrt{\pi}} \int_0^d A(s,\lambda)dM(s)f(s) \\ &= L^*V^* \frac{1}{\sqrt{\pi}} \int_0^d \widetilde{A}(s,\lambda)d\widetilde{M}(s)Vf(s) = \mathcal{F}_2(Vf)\,, \end{aligned}$$

where $\widetilde{A}(s,\lambda) = VA(s,\lambda)V^*$ and $\widetilde{M}(s) = VM(s)V^*$.

17. Weyl limit balls

Let $A_t(\lambda) = A(t, \lambda)$, $0 \le t < d$, denote the matrizant of a canonical integral system of the form (8.1) with $J = J_p$, let $B_t(\lambda) = A_t(\lambda)\mathfrak{V}$ for every $t \in [0, d)$ and suppose that $\mathcal{S}^{p\times p} \subset \mathcal{D}(T_{B_{t_0}})$ for some $t_0 \in [0, d)$. Then

$$\mathcal{S}^{p\times p} \subset \mathcal{D}(T_{B_t}) \quad \text{for every} \quad t \in [t_0, d) \tag{17.1}$$

and hence,

$$\mathcal{C}^d_{\text{imp}}(M) = \bigcap_{t_0 \le t < d} T_{B_t}[\mathcal{S}^{p\times p}] \quad \text{and} \quad \mathcal{C}^d_{\text{imp}}(M) \neq \emptyset\,.$$

In view of (5.1), the condition (17.1) is in force if and only if

$$K^{\mathfrak{E}_{t_0}}_\omega(\omega) > 0 \quad \text{for some (and hence every) point} \quad \omega \in \mathbb{C}_+\,. \tag{17.2}$$

If condition (17.2) (or, equivalently, (17.1)) is in force, then the set

$$\mathcal{B}_*(\omega) = \{c(\omega) : \ c \in \mathcal{C}^d_{\text{imp}}(M)\}$$

is a matrix ball with center $\gamma(\omega)$ and left and right semiradii $R_\ell(\omega) \ge 0$ and $R_r(\omega) \ge 0$ for each point $\omega \in \mathbb{C}_+$:

$$\mathcal{B}_*(\omega) = \{\delta(\omega) + R_\ell(\omega)\varepsilon R_r(\omega) : \varepsilon \in \mathcal{S}^{p\times p}_{\text{const}}\}\,.$$

The center $\gamma(\omega)$ is unique and the semiradii $R_\ell(\omega)$ and $R_r(\omega)$ are defined up to a positive scalar multiplier δ: $R_\ell(\omega) \longrightarrow \delta R_\ell(\omega)$ and $R_r(\omega) \longrightarrow \delta^{-1}R_r(\omega)$. The set $\mathcal{B}_*(\omega)$ is called the Weyl limit ball. By a theorem of Orlov [Or:76], the ranks of the semiradii $\mathfrak{n}_\ell = \operatorname{rank} R_\ell(\omega)$ and $\mathfrak{n}_r = \operatorname{rank} R_r(\omega)$ are independent of the choice of the point $\omega \in \mathbb{C}_+$. If $A_{t_0} \in \mathcal{U}_{sR}(J_p)$, then the three conditions (17.1), (17.2) and

$$\chi^{t_0}_1(\omega)\chi^{t_0}_1(\omega)^* < I_p \quad \text{for at least one (and hence every) point} \quad \omega \in \mathbb{C}_+ \tag{17.3}$$

are equivalent; see Theorem 3.3 in [ArD:05].

Theorem 17.1. *Let $A_t(\lambda) = A(t, \lambda)$, $0 \le t < d$, be the matrizant for the system (8.1) with $J = J_p$, let $\{b^t_3, b^t_4\} \in ap_{II}(A_t)$ for every $t \in [0, d)$ and suppose that the mvf $\chi^t_1(\lambda) = b^t_4(\lambda)b^t_3(\lambda)$ meets the condition (14.7) and that $c \in \mathcal{C}^d_{\text{imp}} \cap \mathring{\mathcal{C}}^{p\times p}$. Then the ranks $\mathfrak{n}_\ell$ and $\mathfrak{n}_r$ of the left and right semiradii of the limit ball $\mathcal{B}_*(\omega)$ (that are independent of $\omega \in \mathbb{C}_+$) may be computed by the formulas*

$$\mathfrak{n}_\ell = \operatorname{rank} \lim_{t\uparrow d} b^t_3(\omega)b^t_3(\omega)^* \text{ and } \mathfrak{n}_r = \operatorname{rank} \lim_{t\uparrow d} b^t_4(\omega)^*b^t_4(\omega)\,. \tag{17.4}$$

Proof. See Theorem 3.16 in [ArD:05]. □

The two extreme cases:

(a) the limit point case, in which at least one of the two indices $\mathfrak{n}_\ell$, $\mathfrak{n}_r$ is equal to zero, and
(b) the full rank case, in which $\mathfrak{n}_\ell = \mathfrak{n}_r = p$,

are of particular interest.

In the setting of the previous theorem, the following conclusions prevail:

(1) The limit point case is in force $\Longleftrightarrow$ $\mathcal{C}^d_{\rm imp}(M)$ contains exactly one mvf $c(\lambda)$.

(2) If $\mathfrak{n}_\ell = 0$, then for any two vectors $\xi, \eta \in \mathbb{C}^p$ and any point $\lambda \in \mathbb{C}_+$,

$$[\xi^* \quad \eta^*]A(s,\overline{\lambda}) \in L_2^{1\times m}(dM;[0,d)) \Longleftrightarrow \eta = c(\lambda)\xi\,,$$

which serves to characterize the single mvf $c(\lambda)$ in $\mathcal{C}^d_{\rm imp}(M)$. This is really the Weyl-Titchmarsh characterization of input impedances for this setting.

(3) The full rank case is in force $\Longleftrightarrow$ $\lim_{t\uparrow d} \mathrm{trace} M(t) < \infty$.

The last two conclusions follow from Theorems 6.4 and 7.5 of [ArD:03b], respectively.

18. The bitangential inverse spectral problem for $\sigma(\mu) = \mu I_p$

Theorem 18.1. *Let $\sigma(\mu) = \mu I_p$ and let $\{b_3^t, b_4^t\}$ $0 \le t < d$, be a normalized monotonic continuous chain of pairs of entire inner $p \times p$ mvf's that satisfies the condition* (14.7). *Then every solution $M(t)$ of the bitangential inverse spectral problem with given data $\{\mu I_p; b_3^t, b_4^t, 0 \le t < d\}$ is of the form*

$$M^{(\alpha)}(t) = \begin{bmatrix} I_p & i\alpha \\ 0 & I_p \end{bmatrix} \mathfrak{V} \begin{bmatrix} m_3(t) & 0 \\ 0 & m_4(t) \end{bmatrix} \mathfrak{V} \begin{bmatrix} I_p & 0 \\ -i\alpha & I_p \end{bmatrix},$$

where $\alpha = \alpha^*$ is a $p \times p$ Hermitian matrix and

$$m_j(t) = -i\frac{\partial b_j^t}{\partial \lambda}(0) = 2\pi k_0^{b_j^t}(0), \quad j = 3, 4\,. \tag{18.1}$$

Moreover,

$$\mathcal{B}(\mathfrak{E}_t) = \mathcal{H}(b_3^t) \oplus \mathcal{H}_*(b_4^t)$$

as Hilbert spaces and

$$\mathcal{C}(A_t) = \{(I_p - b_3^t \varepsilon b_4^t)(I_p + b_3^t \varepsilon b_4^t)^{-1} : \varepsilon \in \mathcal{S}^{p\times p}\}\,.$$

Proof. Since $\sigma(\mu) = \mu I_p$ is the spectral function of the mvf $c^{(0)}(\lambda) = I_p$, it is readily checked that Theorem 15.1 is applicable and guarantees the existence of a solution $M^{(0)}(t)$ to the bitangential inverse spectral problem for the given data $\{\mu I_p; b_3^t, b_4^t, 0 \le t < d\}$: $M^{(0)}(t)$ is the solution of the bitangential inverse input impedance problem with data $\{I_p; b_3^t, b_4^t, 0 \le t < d\}$ and may be obtained by invoking the formulas in Section 13. In particular, it follows readily from the formulas in Section 13 that the matrizant $A_t(\lambda)$ of the system (8.1) with mass function $M^{(0)}(t)$ is given by the formula

$$A_t(\lambda) = I_m + i\lambda\sqrt{2}\mathfrak{V} \begin{bmatrix} \widehat{u}_{11}^t(\lambda) & \widehat{u}_{12}^t(\lambda) \\ \widehat{u}_{21}^t(\lambda) & \widehat{u}_{22}^t(\lambda) \end{bmatrix} J_p\,,$$

where

$$\widehat{u}_{11}^t(\lambda) = -\frac{b_3^t(\lambda) - I_p}{2i\lambda} = -\widehat{u}_{12}^t(\lambda), \quad \widehat{u}_{21}^t(\lambda) = -\frac{b_4^t(\lambda)^{-1} - I_p}{2i\lambda} = \widehat{u}_{22}^t(\lambda)$$

(and the mvf's $\widehat{x}_{ij}^t(\lambda)$ that appear in the cited formulas are: $\widehat{x}_{1j}^t(\lambda) = -\widehat{u}_{1j}^t(\lambda)$ and $\widehat{x}_{2j}^t(\lambda) = \widehat{u}_{2j}^t(\lambda)$ for $j = 1, 2$). Therefore,

$$M^{(0)}(t) = \sqrt{2}\mathfrak{V} \begin{bmatrix} \widehat{u}_{11}^t(0) & \widehat{u}_{12}^t(0) \\ \widehat{u}_{21}^t(0) & \widehat{u}_{22}^t(0) \end{bmatrix},$$

which can be reexpressed in terms of the mvf's (18.1) as

$$M^{(0)}(t) = \mathfrak{V} \begin{bmatrix} m_3(t) & 0 \\ 0 & m_4(t) \end{bmatrix} \mathfrak{V}.$$

Moreover, since

$$B_t(\lambda) = A_t(\lambda)\mathfrak{V} = \frac{1}{\sqrt{2}} \begin{bmatrix} -b_3^t(\lambda) & b_4^t(\lambda)^{-1} \\ b_3^t(\lambda) & b_4^t(\lambda)^{-1} \end{bmatrix},$$

$$\mathfrak{E}_t(\lambda) = \sqrt{2}N_2^* B_t(\lambda) = [b_3^t(\lambda) \quad b_4^t(\lambda)^{-1}].$$

Thus, in this case $\|f\|_{\mathcal{B}(\mathfrak{E}_t)}^2 = \|f\|_{st}^2$ for $f \in \mathcal{B}(\mathfrak{E}_t)$ and

$$\mathcal{B}(\mathfrak{E}_t) = \mathcal{H}(b_3^t) \oplus \mathcal{H}_*(b_4^t)$$

as Hilbert spaces. Moreover, Theorem 14.4 is applicable,

$$\mathcal{C}(A_t) = \{(I_p - b_3^t \varepsilon b_4^t)(I_p + b_3^t \varepsilon b_4^t)^{-1} : \varepsilon \in \mathcal{S}^{p\times p}\},$$

$\beta_c = 0$ for every mvf $c \in \mathcal{C}(A_t)$ and

$$\Sigma_{sf}^d(M) = \cap_{0\le t<d}(\mathcal{C}(A_t))_{sf}. \qquad \square$$

As a concrete application of this theorem, let $b_3^t(\lambda) = \exp\{i\lambda t D_3\}$ and $b_4^t(\lambda) = \exp\{i\lambda t D_4\}$, where D_3 and D_4 are positive semidefinite $p \times p$ matrices, then the preceding formulas imply that

$$M^{(0)}(t) = \mathfrak{V} \begin{bmatrix} tD_3 & 0 \\ 0 & tD_4 \end{bmatrix} \mathfrak{V}$$

and

$$A_t(\lambda) = \exp\left\{ i\lambda t \mathfrak{V} \begin{bmatrix} D_3 & 0 \\ 0 & D_4 \end{bmatrix} \mathfrak{V} J_p \right\} = \mathfrak{V} \begin{bmatrix} e^{i\lambda D_3 t} & 0 \\ 0 & e^{-i\lambda D_4 t} \end{bmatrix} \mathfrak{V},$$

for $0 \le t < d$.

If $d = \infty$, then $\mathfrak{n}_\ell = p - \operatorname{rank} D_3$ and $\mathfrak{n}_r = p - \operatorname{rank} D_4$.

If $d < \infty$, then $\mathfrak{n}_\ell = \mathfrak{n}_r = p$.

19. Spectral functions for the space $\mathcal{H}(b_3) \oplus \mathcal{H}_*(b_4)$

A non decreasing $p \times p$ mvf $\sigma(\mu)$ on $\mathbb{R}$ is said to be a **spectral function** for the de Branges space $\mathcal{B}(\mathfrak{E})$ based on an entire $p \times m$ de Branges function $\mathfrak{E}(\lambda)$ if

$$\int_{-\infty}^{\infty} f(\mu)^* d\sigma(\mu) f(\mu) = \|f\|_{\mathcal{B}(\mathfrak{E})}$$

for every $f \in \mathcal{B}(\mathfrak{E})$. A spectral function of a de Branges space $\mathcal{B}(\mathfrak{E})$ is said to be an **orthogonal spectral function** for $\mathcal{B}(\mathfrak{E})$ if $L_2^p(d\sigma) = \mathcal{B}(\mathfrak{E})$. In this section we shall describe the set of all spectral functions for the de Branges spaces that arose in the preceding section.

Let $b_3(\lambda)$ and $b_4(\lambda)$ be a normalized pair of $p \times p$ entire inner mvf's. Then the mvf

$$A(\lambda) = \frac{1}{\sqrt{2}} \begin{bmatrix} -b_3(\lambda) & b_4(\lambda)^{-1} \\ b_3(\lambda) & b_4(\lambda)^{-1} \end{bmatrix} \mathfrak{V} \tag{19.1}$$

is a normalized entire J_p-inner mvf and

$$\mathfrak{E}(\lambda) = \sqrt{2}[I_p \quad 0]B(\lambda) = \begin{bmatrix} b_3(\lambda) & (b_4(\lambda))^{-1} \end{bmatrix}.$$

The corresponding de Branges space

$$\mathcal{B}(\mathfrak{E}) = \mathcal{H}(b_3) \oplus \mathcal{H}_*(b_4).$$

The analysis in the preceding example lends itself to consideration of the following problem:
Find all nondecreasing $p \times p$ mvf's $\sigma(\mu)$ on $\mathbb{R}$ such that

$$\int_{\infty}^{\infty} f(\mu)^* f(\mu) d\mu = \int_{\infty}^{\infty} f(\mu)^* d\sigma(\mu) f(\mu)$$

for every $f \in \mathcal{B}(\mathfrak{E})$.

If the mvf $\chi_1(\lambda) = b_4(\lambda) b_3(\lambda)$ is uniformly contractive on a ray in $\mathbb{C}_+$, then, by Lemma 2.4 and Theorem 2.14 of [ArD:04b],

$$\mathcal{C}(A) = \left\{ (I_p - b_3 \varepsilon b_4)(I_p + b_3 \varepsilon b_4)^{-1} : \varepsilon \in \mathcal{S}^{p \times p} \right\}$$

and $\sigma(\mu)$ is a solution of the problem stated just above if and only if

$$\sigma \in (\mathcal{C}(A))_{sf}.$$

If ε is a constant unitary matrix and $b(\lambda) = b_3(\lambda) \varepsilon b_4(\lambda)$, then

$$c(\lambda) = (I_p - b(\lambda))(I_p + b(\lambda))^{-1}$$

is a meromorphic $p \times p$ mvf in $\mathbb{C}$ with poles in $\mathbb{R}$ at the points $\mu \in \mathbb{R}$ at which $\det(I_p + b(\mu)) = 0$. Let $\mu_0 = 0$ if $\det(I_p + b(0)) = 0$ and let $\mu_1, \mu_2, \ldots,$ denote the remaining poles of $c(\lambda)$. Then, since $\Re c(\mu) = 0$ for all points $\mu \in \mathbb{R}$ that are not poles of $c(\lambda)$, the spectral function $\sigma_c(\mu)$ of $c(\lambda)$ is a step function with jumps m_j at the points μ_j. In this setting, the Riesz-Herglotz formula for $c(\lambda)$ reduces to

$$c(\lambda) = i\gamma + \frac{m_0}{-\pi i \lambda} + \frac{\lambda}{\pi i} \sum_{j \geq 1} \frac{m_j}{\mu_j(\mu_j - \lambda)},$$

where $\gamma = \gamma^*$ is a constant $p \times p$ Hermitian matrix and the condition (14.5) is equivalent to the constraint

$$\sum_{j\geq 1} \frac{\operatorname{trace} m_j}{\mu_j^2} < \infty\,.$$

By Theorem 14.5, the spectral function

$$\sigma(\mu) = \sum_{\{j:\mu_j<\mu\}} m_j \tag{19.2}$$

of this mvf $c(\lambda)$ is orthogonal for the space $\mathcal{B}(\mathfrak{E})$, i.e., $L_2^p(d\sigma;\mathbb{R}) = \mathcal{B}(\mathfrak{E})$. The corresponding Parseval formula states that

$$\int_{\infty}^{\infty} f(\mu)^* f(\mu) d\mu = \sum_{j\geq 0} f(\mu_j)^* m_j f(\mu_j)$$

for every $f \in \mathcal{H}(b_3) \oplus \mathcal{H}_*(b_4)$. Moreover, if $\xi_0, \xi_1, \ldots$ is any sequence of vectors in $\mathbb{C}^p$ such that

$$\xi_j \in \operatorname{range} m_j \quad \text{and} \quad \sum_{j\geq 0} \xi_j^* m_j \xi_j < \infty\,,$$

then there exists a unique $f \in \mathcal{H}(b_3) \oplus \mathcal{H}_*(b_4)$ such that $f(\mu_j) = \xi_j$.

The matrix-valued jumps m_j of $\sigma(\mu)$ at the points μ_j can be recovered from the formula for $c(\lambda)$:

$$m_j = \pi i \lim_{\lambda\to\mu_j} (\mu_j - \lambda) c(\lambda) = 2\pi i \lim_{\lambda\to\mu_j} (\mu_j - \lambda)(I_p + b(\lambda))^{-1}\,.$$

Moreover, upon using $d\sigma(\mu)$ to calculate the inner product in $\mathcal{B}(\mathfrak{E})$ instead of $d\mu$, the formula

$$\xi^* f(\omega) = \langle f, K_\omega^{\mathfrak{E}} \xi \rangle\,,$$

which is valid for every $f \in \mathcal{B}(\mathfrak{E})$ and every $\xi \in \mathbb{C}^p$, yields the sampling formula

$$\xi^* f(\omega) = \sum_{j\geq 0} \xi^* K_{\mu_j}^{\mathfrak{E}}(\omega) m_j f(\mu_j)\,.$$

In the special case that $m = 1$ and $b(\lambda) = e^{ia\lambda}$ for some $a > 0$ this reduces to the well-known Shannon-Kotelnikov sampling formula.

We remark that for each given normalized pair $\{b_3, b_4\}$ of entire inner $p \times p$ mvf's, there exists at least one normalized monotonic continuous chain $\{b_3^t, b_4^t\}$, $0 \leq t < d$, of entire inner $p \times p$ mvf's such that $b_3(\lambda) = b_3^d(\lambda)$ and $b_4(\lambda) = b_4^d(\lambda)$; see, e.g., Theorem 2.4 in [ArD:00a]. Consequently, the set $(\mathcal{C}(A))_{sf}$ for the mvf $A(\lambda)$ defined by formula (19.1) is equal to the set of all spectral functions for the systems considered in Example 1 with this chain $\{b_3^t, b_4^t\}$, $0 \leq t < d$. In particular, the spectral function defined by formula (19.2) is an orthogonal spectral function for each such system. The same analysis is applicable to the orthogonal spectral functions of general canonical integral systems of the form (8.1) with $J = J_p$ that are considered in the last assertion of Theorem 14.5 and will be discussed in more detail elsewhere.

20. A left tangential accelerant inverse problem

Let the data $\{c(\lambda); b_3^t(\lambda), b_4^t(\lambda),\ 0 \le t < \infty\}$ for the inverse impedance problem for the system (8.1) with $J = J_p$ be specified by the formulas:

$$c(\lambda) = I_p - 2\int_0^\infty e^{i\lambda a} h(a) da, \tag{20.1}$$

where

$$h \in L_1^{p\times p}([0,\infty)) \quad \text{and} \quad h(a) = h(-a)^* \text{ for a.e. point } a \in (-\infty, 0)\ , \tag{20.2}$$

i.e., $c \in \mathcal{C}^{p\times p} \cap \mathcal{W}_+^{p\times p}(I_p)$ and $-h(t)$ is the accelerant of $c(\lambda)$ on the interval $[0,\infty)$, and

$$b_3^t(\lambda) = e^{i\lambda t D} \quad \text{and} \quad b_4^t(\lambda) = I_p \quad \text{for} \quad t \ge 0 \tag{20.3}$$

where $D = \operatorname{diag}\{\alpha_1, \ldots, \alpha_p\}$ is a positive definite diagonal matrix. Let

$$f_D(a) = \sum_{j=1}^p \mathfrak{e}_j \mathfrak{e}_j^* f(\alpha_j a) \quad \text{and} \quad h_D(a,b) = \sum_{j,k=1}^p \mathfrak{e}_j \mathfrak{e}_j^* h(\alpha_j a - \alpha_k b) \mathfrak{e}_k \mathfrak{e}_k^*\ ,$$

where $\mathfrak{e}_j$, $j = 1, \ldots, p$, denotes the standard basis for $\mathbb{C}^p$ and let $\gamma_D^t(a,b)$ denote the resolvent kernel for the Fredholm integral operator with kernel $D^{1/2} h_D(a,b) D^{1/2}$ on the square $[0,t] \times [0,t]$:

$$-D^{1/2} h_D(a,b) D^{1/2} + \gamma_D^t(a,b) - \int_0^t D^{1/2} h_D(a,c) D^{1/2} \gamma_D^t(c,b) dc = 0$$

and

$$-D^{1/2} h_D(a,b) D^{1/2} + \gamma_D^t(a,b) - \int_0^t \gamma_D^t(a,c) D^{1/2} h_D(c,b) D^{1/2} dc = 0\ .$$

Theorem 20.1. *Let* (20.1)–(20.3) *be in force, let*

$$v_1(a) = -I_p + 2\int_0^a h(b) db \text{ and } v_2(a) = I_p \text{ for } a \ge 0\ , \tag{20.4}$$

and let

$$\varphi_D(a,0) = [(v_1)_D(a) \quad (v_2)_D(a)]\ .$$

Then the inverse input impedance problem has a unique solution

$$M(t) = \frac{1}{2}\int_0^t \varphi_D(a,0)^* D^{1/2} \left\{ D^{1/2} \varphi_D(a,0) + \int_0^t \gamma_D^t(a,b) D^{1/2} \varphi_D(b,0) db \right\} da\ . \tag{20.5}$$

If it is also assumed that $h(t)$ is continuous, then $M(t)$ is differentiable, $M'(t)$ is locally absolutely continuous on $[0,\infty)$ and

$$M'(t) = Y_1(t) Y_1(t)^*\ , \tag{20.6}$$

where

$$Y_1(t)^* = \frac{1}{\sqrt{2}}\left(D^{1/2} \varphi_D(t,0) + \int_0^t \gamma_D^t(t,b) D^{1/2} \varphi_D(b,0) db \right)\ . \tag{20.7}$$

Proof. See Theorems 5.11 and 5.12 in [ArD:05]. □

Remark 20.2. The conclusions of the last theorem are in force under the less restrictive assumption that $c(\lambda)$ has an accelerant $h^\circ(t)$ on every finite interval $[0, d]$ and that that $\gamma = I_p$ in formula (11.4).

If $D = I_p$, then we shall write $\gamma^t(a, b)$ instead of $\gamma^t_D(a, b)$ and the formulas in the preceding theorem simplify:

Theorem 20.3. *Let $c \in \mathcal{C}^{p\times p} \cap \mathcal{W}_+(I_p)$ and suppose that in the representation (20.1) the mvf $h(t)$ is continuous on $[0, d)$ and let $\gamma^t(a, b)$ denote the solution of the resolvent equation*

$$\gamma^t(a,b) - \int_0^t h(a-c)\gamma^t(c,b)dc = h(a-b)\ ,\ 0 \le a, b \le t < d\ . \tag{20.8}$$

Then:

(1) *There exists exactly one solution $M(t)$ of the inverse impedance problem $(c(\lambda); e^{i\lambda t}I_p, I_p, 0 \le t < d)$ for the system (8.1) with $J = J_p$. It has a continuous second order derivative $M''(t)$ on $[0, d)$ and*

$$M'(t) = Y(t)\begin{bmatrix} I_p & 0 \\ 0 & 0 \end{bmatrix} Y(t)^*\ , \tag{20.9}$$

where

$$Y(t) = \begin{bmatrix} Y_1(t)\ Y_2(t) \end{bmatrix} \tag{20.10}$$

is the solution of the Cauchy problem

$$Y'(t) = Y(t)\begin{bmatrix} 0 & \gamma^t(t,0) \\ \gamma^t(0,t) & 0 \end{bmatrix}\ ,\quad 0 \le t < d\ , \tag{20.11}$$

$$Y(0) = \mathfrak{V}\,,$$

$$Y_1(t)^* = v(t) + \int_0^t \gamma^t(t,b)v(b)db\ , \tag{20.12}$$

$$Y_2(t)^* = \frac{1}{\sqrt{2}}[I_p\ I_p] + \int_0^t \gamma^t(0,b)v(b)db\ ,\ 0 \le t < d\ . \tag{20.13}$$

$$v(b) = \frac{1}{\sqrt{2}}\left[-I_p + 2\int_0^b h(a)da \quad I_p \right] \tag{20.14}$$

and

(2)

$$Y(t)j_pY(t)^* = J_p \quad \textit{for every} \quad t \in [0, d)\ . \tag{20.15}$$

Proof. See Theorem 5.13 in [ArD:05]. □

21. The rational homogeneous case

In this section we shall check that if the given input impedance matrix $c \in \mathcal{C}^{p\times p}$ is a rational $p \times p$ mvf that has no poles on $\mathbb{R}$ such that $c(\infty) + c(\infty)^* > 0$, then the formulas in the previous subsection lead to the same realization formulas for $M(t)$ that were presented in Section 5 of [ArD:02b]. Let

$$c(\lambda) = I_p - 2\int_0^\infty e^{i\lambda t}h(t)dt = I_p - 2C(\lambda I_n - A)^{-1}B \ , \ \text{for } \lambda \in \mathbb{C}_+,$$

where the exhibited realization is minimal. By Theorem 20.3, $M'(t) = Y_1(t)Y_1(t)^*$, where

$$Y_1(t) \ = \ \mathfrak{V}\begin{bmatrix} I_p + iC\int_0^t R_{12}(b)R_{22}(t)^{-1}C^* \\ iB^*\int_0^t R_{22}(b)dbR_{22}(t)^{-1}C^* \end{bmatrix},$$

and

$$\begin{bmatrix} R_{11}(t) & R_{12}(t) \\ R_{21}(t) & R_{22}(t) \end{bmatrix} = \exp\left\{-it\begin{bmatrix} A+BC & BB^* \\ C^*C & A^*+C^*B^* \end{bmatrix}\right\}.$$

which is consistent with formula (5.31) of [ArD:02b]. For additional information and formulas on inverse problems with rational data, see, e.g., [AG:95], [AG:01], [GKS:98], [GKS:02] and the references cited therein. We wish to emphasize that we do not assume that

$$\Re c(\mu) \ = \ I_p - \int_{-\infty}^\infty e^{i\mu t}h(t)dt > 0$$

for every point $\mu \in \mathbb{R}$. Thus, $\det\{\Re c(\mu)\}$ may be equal to zero on some subset of $\mathbb{R}$.

22. Differential systems with potential

In this section we consider differential systems of the form

$$u'(t,\lambda) = i\lambda u(t,\lambda)NJ + u(t,\lambda)\mathcal{V}(t) \ , \quad 0 \le t < d \ , \tag{22.1}$$

with an $m \times m$ signature matrix J, a constant $m \times m$ matrix N such that

$$N \ge 0 \tag{22.2}$$

and an $m \times m$ matrix-valued potential $\mathcal{V}(t)$ such that

$$\mathcal{V} \in L_{1,\,\mathrm{loc}}^{m\times m}([0,d)) \quad \text{and } \mathcal{V}(t)J + J\mathcal{V}(t)^* = 0 \quad \text{a.e.}. \tag{22.3}$$

It is readily checked that the matrizant $U_t(\lambda) = U(t,\lambda)$, $0 \le t < d$, of this system satisfies the identity

$$\{U_t(\lambda)JU_t(\omega)^*\}' \ = \ i(\lambda - \overline{\omega})U_t(\lambda)NU_t(\omega)^* \quad \text{for} \quad 0 \le t < d, \tag{22.4}$$

and hence that

$$\frac{J - U_t(\lambda)JU_t(\omega)^*}{\rho_\omega(\lambda)} \ = \ \frac{1}{2\pi}\int_0^t U_s(\lambda)NU_s(\omega)^*ds \quad \text{for} \quad 0 \le t < d. \tag{22.5}$$

This in turn leads easily to the conclusion that

$$U_t \in \mathcal{U}(J) \quad \text{for every} \quad t \in [0, d)\,. \tag{22.6}$$

In particular, $U_t(0)$ is J-unitary and so invertible. Moreover, the mvf

$$Y_t(\lambda) \;=\; U_t(\lambda)U_t(0)^{-1} \quad \text{for} \quad 0 \le t < d,$$

is the matrizant of the canonical differential system

$$y'(t,\lambda) \;=\; i\lambda y(t,\lambda)H(t)J, \quad 0 \le t < d, \tag{22.7}$$

with Hamiltonian

$$H(t) \;=\; U_t(0)NU_t(0)^*, \quad 0 \le t < d. \tag{22.8}$$

If

$$J \;=\; V^*J_pV \quad \text{for some constant unitary matrix} \quad V\,, \tag{22.9}$$

then

$$VU_tV^* \in \mathcal{U}(J_p) \quad \text{for every} \quad t \in [0, d)$$

and, in keeping with the conventions initiated in Section 16, we say that

$$\{b_3^t, b_4^t\} \in ap_{II}(U_t) \quad \text{if} \quad \{b_3^t, b_4^t\} \in ap_{II}(VU_tV^*)\,.$$

Moreover, when (22.9) is in force, we shall introduce the following V dependent definitions for differential systems of the form (22.1) with potential $\mathcal{V}(t)$ and matrizant $U_t(\lambda)$:

(1) The set $\mathcal{C}^d_{\rm imp}(\mathcal{V})$ of **input impedances** is defined by the formula

$$\mathcal{C}^d_{\rm imp}(\mathcal{V}) \;=\; \bigcap_{0\le t<d} \mathcal{C}(VU_tV^*)\,. \tag{22.10}$$

(2) The **generalized Fourier transform**

$$g^{\triangle}(\lambda) \;=\; [0 \quad I_p]\,V\frac{1}{\sqrt{\pi}}\int_0^d U(s,\lambda)Ng(s)ds \tag{22.11}$$

for every $g \in L_2^m(Nds; [0, d))$ with compact support in $[0, d)$.

(3) The set $\Sigma^d_{sf}(\mathcal{V})$ of **spectral functions** for the system (22.1) is the set of non-decreasing $p \times p$ mvf's $\sigma(\mu)$ on $\mathbb{R}$ for which the Parseval equality

$$\int_{\infty}^{\infty} g^{\triangle}(\mu)^*d\sigma(\mu)g^{\triangle}(\mu) = \int_0^d g(s)^*Ng(s)ds \tag{22.12}$$

holds for every $g \in L_2^m(Nds; [0, d))$ with compact support in $[0, d)$.

Remark 22.1. The set $\mathcal{C}^d_{\rm imp}(\mathcal{V})$ coincides with the set $\mathcal{C}^d_{\rm imp}(H)$ of input impedance matrices for the canonical differential system (22.7) with Hamiltonian $H(t)$ given by formula (22.8). The generalized Fourier transform (16.3) for the canonical system with $dM(t) = H(t)dt$ and $H(t) = U(t)NU(t)^*$ for $0 \le t < d$, is related to the transform (22.11):

$$(\mathcal{F}_2 f)(\lambda) = [0 \quad I_p]\, V \frac{1}{\sqrt{\pi}} \int_0^d U(s,\lambda)U(s,0)^{-1}H(s)f(s)ds \tag{22.13}$$

$$= [0 \quad I_p]\, V \frac{1}{\sqrt{\pi}} \int_0^d U(s,\lambda)NU(s,0)^* f(s)ds \tag{22.14}$$

for $f \in L_2^m(H(s)ds; [0,d))$ with compact support in $[0,d)$.

Theorem 22.2. *If*

$$NJ = JN\,, \tag{22.15}$$

then the matrizants $U_t(\lambda)$ and $Y_t(\lambda)$ of the systems (22.1) and (22.7) are both strongly regular:

$$U_t \in \mathcal{U}_{sR}(J) \quad \text{and} \quad Y_t \in \mathcal{U}_{sR}(J) \quad \text{for every} \quad t \in [0,d)\,. \tag{22.16}$$

Proof. Since N and J are both Hermitian matrices,

$$NJ = JN \iff NJ = (NJ)^*$$

and hence the assumption $NJ = JN$ guarantees that the mvf $\exp\{i\mu tNJ\}$ is unitary for $\mu t \in \mathbb{R}$. Consequently, standard estimates show that

$$U_t \in L_\infty^{m\times m} \quad \text{and} \quad Y_t \in L_\infty^{m\times m} \quad \text{for every} \quad t \in [0,d)\,.$$

Therefore, by Theorem 4.1,

$$Y_t \in \mathcal{U}_{sR}(J) \quad \text{and} \quad U_t \in \mathcal{U}_{sR}(J) \quad \text{for every} \quad t \in [0,d)\,. \qquad \square$$

Systems of the form (22.1) with $NJ - JN \neq 0$ and matrizants $U_t \notin \mathcal{U}_{sR}(J)$ will be considered in Section 25. However, for the moment we focus on systems for which the condition $NJ = JN$ is met.

In particular, the condition $NJ = JN$ is met if N is a convex combination of the orthogonal projections

$$P_J = \frac{I_m + J}{2} \quad \text{and} \quad Q_J = \frac{I_m - J}{2}\,,$$

or, even more generally, if

$$N = \delta_1 P_J + \delta_2 Q_J\,, \quad \text{where} \quad \delta_1 \ge 0\,,\ \delta_2 \ge 0 \quad \text{and} \quad \delta_1 + \delta_2 > 0\,. \tag{22.17}$$

The direct problem

The following results on the direct problem are established in [ArD:??]:

Theorem 22.3. *Let $U_t(\lambda) = U(t,\lambda)$, $0 \le t < d$, be the matrizant of the system (22.1). Assume that (22.2), (22.3), (22.9) and (22.17) are in force, that $A_t(\lambda) = VU_t(\lambda)V^*$ for every $t \in [0,d)$ and that the potential $\mathcal{V}(t) = \mathcal{V}(t)^*$ a.e. on the interval $[0,d)$. Then:*

1. *$A_t \in \mathcal{U}_{sR}(J_p)$ for every $t \in [0,d)$.*
2. *$\{e_{\delta_1 t}I_p, e_{\delta_2 t}I_p\} \in ap_{II}(A_t)$ for every $t \in [0,d)$.*

(3) *The de Branges spaces* $\mathcal{B}(\mathfrak{E}_t)$ *based on* $\mathfrak{E}_t(\lambda) = \sqrt{2}A_t(\lambda)\mathfrak{V}$ *are independent of the potential* $\mathcal{V}(t)$ *as linear topological spaces, i.e.,*

$$\mathcal{B}(\mathfrak{E}_t) = \left\{ \int_{-\delta_2 t}^{\delta_1 t} e^{i\lambda s} h(s)ds \,:\, h \in L_2^p([-\delta_2 t, \delta_1 t]) \right\}$$

as linear spaces and for each $t \in [0, d)$, *there exist a pair of positive constants* $\gamma_1 = \gamma_1(t)$ *and* $\gamma_2 = \gamma_2(t)$ *such that*

$$\gamma_1 \|f\|_2 \leq \|f\|_{\mathcal{B}(\mathfrak{E}_t)} \leq \gamma_2 \|f\|_2$$

for every $f \in \mathcal{B}(\mathfrak{E}_t)$.

(4) $\mathcal{S}^{p\times p} \subset \mathcal{D}(T_{A_t\mathfrak{V}})$ *for every* $t \in (0, d)$.

(5) $\mathcal{C}(A_t) = T_{A_t\mathfrak{V}}[\mathcal{S}^{p\times p}]$ *for every* $t \in (0, d)$ *and* $\beta = 0$ *in the integral representation* (14.4) *of every mvf* $c \in \mathcal{C}(A_t)$, $0 < t < d$.

(6) $\mathcal{C}^d_{\text{imp}}(\mathcal{V}) = \bigcap_{0\leq t<d} \mathcal{C}(A_t) \neq \emptyset$.

(7) $\Sigma^d_{sf}(\mathcal{V}) = (\mathcal{C}^d_{\text{imp}}(\mathcal{V}))_{sf}$ *and the integral representation* (14.4) *defines a one to one correspondence between these two sets.*

(8) *If* $d < \infty$ *and* $\mathcal{V} \in L_1([0, d])$, *then* $\mathcal{C}^d_{\text{imp}}(\mathcal{V}) = \mathcal{C}(A_d)$.

Proof. A proof will be supplied in [ArD:??]. □

Corollary 22.4. *If* $A_t(\lambda) = A(t, \lambda)$, $0 \leq t < d$, *is the matrizant of the system* (22.1) *with* $J = J_p$ *and if the conditions* (22.2) *and* (22.3) *are in force, then:*

(a) $N = \kappa I_m \implies \{e^{i\kappa t\lambda} I_p, e^{i\kappa t\lambda} I_p\} \in ap_{II}(A_t)$.

(b) $N = \kappa P_J \implies \{e^{i\kappa t\lambda} I_p, I_p\} \in ap_{II}(A_t)$.

(c) $N = \kappa Q_J \implies \{I_p, e^{i\kappa t\lambda} I_p\} \in ap_{II}(A_t)$.

Remark 22.5. In the preceding theorem, the sets $\mathcal{C}(A_t)$ depend only upon the potential $\mathcal{V}(t)$ and the positive number $\kappa = \delta_1 + \delta_2$ and not on the particular choices $\delta_1 \geq 0$ and $\delta_2 \geq 0$. This follows from the fact that the mvf $e^{i\delta\lambda t} U_t(\lambda)$ is the matrizant of the system (22.1) with potential $\mathcal{V}(t)$ that is independent of δ and with $N = (\delta_1 + \delta)P_J + (\delta_2 - \delta)Q_J$ for every number δ in the interval $-\delta_1 \leq \delta \leq \delta_2$. Consequently, the sets $\mathcal{C}^d_{\text{imp}}(\mathcal{V})$ and $\Sigma^d_{sf}(\mathcal{V})$ depend only upon the potential $\mathcal{V}(t)$ and the number κ.

The inverse input impedance problem

The data for the inverse input impedance problem for differential systems of the form (22.1) on an interval $[0, d)$ is a mvf $c \in \mathcal{C}^{p\times p}$ and the right-hand endpoint d, $0 < d \leq \infty$, of the interval and the problem is to find a locally summable potential $\mathcal{V}(t)$ of the prescribed form on $[0, d)$ such that $c \in \mathcal{C}^d_{\text{imp}}(\mathcal{V})$. In the setting of Theorem 22.3, it is not necessary to specify a chain $\{b_3^t, b_4^t\}$, $0 \leq t < d$, to solve this inverse problem.

Theorem 22.6. *Let* $c \in \mathcal{C}^{p\times p}$ *and* $0 < d \leq \infty$ *be given and let* N *of the form* (22.17) *and* V *as in* (22.9) *be fixed. Then:*

(1) *There exists at most one differential system of the form* (22.1) *with the given* N *and* J *and potential* $\mathcal{V}(t) = \mathcal{V}(t)^*$ *a.e. on* $[0, d)$ *that meets the condition* (22.3) *with an input impedance equal to* $c(\lambda)$.

(2) *If* $c \in \mathcal{C}^{p\times p}$ *has a continuous accelerant* $h^\circ(t)$ *on the interval* $[0, d)$ *and if the real part of the matrix* γ *in definition* (11.4) *is positive definite and* $\kappa > 0$, *then there exists exactly one locally summable potential* $\mathcal{V}(t)$, $0 \le t < d$, *such that*

$$\mathcal{V}(t) = \mathcal{V}(t)^* \quad \text{and} \quad \mathcal{V}(t)J + J\mathcal{V}(t)^* = 0 \quad \text{a.e. on the interval} \quad [0, d) \qquad (22.18)$$

and $c \in \mathcal{C}^d_{\rm imp}(\mathcal{V})$. *Moreover, this potential* $\mathcal{V}(t)$ *is continuous on the interval* $[0, d)$ *and is of the form*

$$\mathcal{V}(t) = V^*\mathfrak{V}\begin{bmatrix} 0 & a(t) \\ a(t)^* & 0 \end{bmatrix}\mathfrak{V}V \quad \text{for} \quad 0 \le t < d. \qquad (22.19)$$

If $\kappa = 1$ *and* $c(\lambda)$ *is given by formula* (20.1), *then* $a(t) = \gamma^t(t, 0)$, *where* $\gamma^t(a, b)$ *is the unique solution of the integral equation* (20.8).

Proof. Assertion (1) follows from (1) and (2) of Theorem 22.3, Theorem 12.1 and the fact that the set $\mathcal{C}^d_{\rm imp}(\mathcal{V})$ for the system (22.1) coincides with the set $\mathcal{C}^d_{\rm imp}(H)$ for the corresponding canonical differential system (22.7) with Hamiltonian (22.8). Assertion (2) follows from Theorem 20.3, Remark 22.5, the connection between the systems (22.1) and (22.7) and the reduction of the matrix γ in formula (11.4) with $\Re\gamma > 0$ to the case $\gamma = I_p$ considered in Section 5.2 in [ArD:05]. □

If $c \in H^{p\times p}_\infty \cap \mathcal{C}^d_{\rm imp}(\mathcal{V})$ and $d = \infty$, then $\mathcal{C}^d_{\rm imp}(\mathcal{V}) = \{c\}$ for every N of the form (22.17) with $\delta_1 + \delta_2 > 0$, i.e., the limit point case prevails for all such $\kappa = \delta_1 + \delta_2$. This follows from the upper bounds on the left and right semiradii of the Weyl balls that are given in formulas (3.36) and (3.37) of [ArD:05]. Consequently, $\mathcal{V} \notin L^{m\times m}_1([0, \infty))$ in this case. Moreover, if $\delta_1 > 0$, then the values of the input impedance $c(\lambda)$ may be characterized by the Weyl-Titchmarsh property that is discussed in Section 17:

$$[\xi^* \quad \eta^*]VU_t(\overline{\lambda})V^* \in L^{m\times m}_2 \Longleftrightarrow \eta = c(\lambda)\xi$$

for every point $\lambda \in \mathbb{C}_+$.

The inverse spectral problem

The data for the inverse spectral problem is a $p \times p$ nondecreasing mvf $\sigma(\mu)$ on $\mathbb{R}$ that meets the condition (14.5). The special form of N in (22.17) automatically insures that the matrizant will be strongly regular and prescribes the associated pair of the matrizant in accordance with (1) and (2) of Theorem 22.3. Moreover, for a fixed pair of nonnegative numbers $\delta_1 \ge 0, \delta_2 \ge 0$ with $\kappa = \delta_1 + \delta_2 > 0$, there is at most one mvf $c \in \mathcal{C}^d_{\rm imp}(\mathcal{V})$ with the spectral function in its Riesz-Herglotz representation (14.4) equal to the given spectral function $\sigma(\mu)$. This will be established in Theorem 23.4 for the case $\delta_1 = \delta_2$. The case $\delta_1 \ne \delta_2$ may be reduced to the case $\delta_1 = \delta_2$ by invoking Remark 22.5. Consequently, Theorem 22.6 yields exactly one solution for the inverse spectral problem.

23. Dirac systems

Differential systems of the form (22.1) with $\operatorname{rank} P_J = \operatorname{rank} Q_J = p$ and $N = \kappa I_m$ for some $\kappa > 0$ and potentials $\mathcal{V}(t) \in L_{1,\mathrm{loc}}^{m\times m}$ that meet the two conditions

(1) $\mathcal{V}(t)J + J\mathcal{V}(t)^* = 0$ for a.e $t \in [0,d)$
(2) $\mathcal{V}(t) = \mathcal{V}(t)^*$ for a.e $t \in [0,d)$,

are called Dirac systems. Without loss of generality we may assume that $\kappa = 1$ and $J = J_p$. Then the potential must of the form

$$\mathcal{V}(t) = \begin{bmatrix} v_1(t) & -iv_2(t) \\ iv_2(t)^* & -v_1(t) \end{bmatrix} , \quad 0 \le t < d , \tag{23.1}$$

where $v_1(t) = v_1(t)^*$ and $v_2(t) = v_2(t)^*$ a.e. and the matrizant $U_t(\lambda) = U(t,\lambda)$ is a solution of the system

$$u_t'(\lambda) = i\lambda u_t(\lambda)J_p + u_t(\lambda)\mathcal{V}(t) , \quad \text{for} \quad 0 \le t < d , \tag{23.2}$$

with potential $\mathcal{V}(t)$ of the form (23.1).

The generalized Fourier transform for this system is given by the formula

$$g^{\triangle}(\lambda) = \frac{1}{\sqrt{\pi}} \int_0^d [u_{21}(s,\lambda) \quad u_{22}(s,\lambda)]g(s)ds , \tag{23.3}$$

where $g \in L_2^m([0,d))$ and has compact support in $[0,d)$. Consequently, a nondecreasing $p\times p$ mvf $\sigma(\mu)$ on $\mathbb{R}$ is said to be a spectral function for the Dirac system with $N = I_m$ and $J = J_p$ if the Parseval identity

$$\int_{-\infty}^{\infty} g^{\triangle}(\mu)^* d\sigma(\mu) g^{\triangle}(\mu) = \int_0^d g(s)^* g(s) ds$$

holds for every $g \in L_2^m([0,d))$ with compact support in $[0,d)$.

The direct problem

Theorem 23.1. *Let $A_t(\lambda) = A(t,\lambda)$, $0 \le t < d$, be the matrizant of the system (23.2) with a locally summable potential $\mathcal{V}(t)$ of the form (23.1). Then:*

(1) *$A_t \in \mathcal{U}_{sR}(J_p)$ for every $t \in [0,d)$.*
(2) *$\{e_tI_p, e_tI_p\} \in ap_{II}(A_t)$ for every $t \in [0,d)$.*
(3) *The de Branges spaces $\mathcal{B}(\mathfrak{E}_t)$ based on $\mathfrak{E}_t(\lambda) = \sqrt{2}A_t(\lambda)\mathfrak{V}$ are independent of the potential $\mathcal{V}(t)$ as linear topological spaces, i.e.,*

$$\mathcal{B}(\mathfrak{E}_t) = \left\{ \int_{-t}^{t} e^{i\lambda s} h(s) ds : h \in L_2^p([-t,t]) \right\}$$

as linear spaces and for each $t \in [0,d)$, there exist a pair of positive constants $\gamma_1 = \gamma_1(t)$ and $\gamma_2 = \gamma_2(t)$ such that

$$\gamma_1 \|f\|_2 \le \|f\|_{\mathcal{B}(\mathfrak{E}_t)} \le \gamma_2 \|f\|_2$$

for every $f \in \mathcal{B}(\mathfrak{E}_t)$.

(4) *Assertions (4)–(8) of Theorem 22.3 are in force.*

Proof. This theorem is a special case of Theorem 22.3. □

The inverse input impedance problem

The inverse input impedance problem for differential systems of the form (23.2) on an interval $[0, d)$ with a locally summable potential $\mathcal{V}(t)$ of the form (23.1) is to find $\mathcal{V}(t)$, given a mvf $c \in \mathcal{C}^{p\times p}$ and the right-hand endpoint d, $0 < d \leq \infty$, of the interval.

Theorem 23.2. *Let $c \in \mathcal{C}^{p\times p}$ and $0 < d \leq \infty$ be given. Then:*

(1) *There exists at most one differential system of the form* (23.2) *with a locally summable potential $\mathcal{V}(t)$ of the form* (23.1) *and an input impedance equal to $c(\lambda)$.*
(2) *There exists exactly one such differential system if $c \in \mathcal{C}^{p\times p}$ has a continuous accelerant $h^\circ(t)$ on the interval $[0, 2d)$ and $\gamma = I_p$ in the representation* (22.19). *Moreover, in this case, the potential $\mathcal{V}(t)$ is given by formula* (23.1) *with $v_1(t) = -\Re a(2t)$, $v_2(t) = \Im a(2t)$, where $a(t) = \gamma^t(t, 0)$ and $\gamma^t(a, b)$ is the unique solution of the integral equation* (20.8) *with d replaced by $2d$.*

Proof. This theorem is a special case of Theorem 22.6. □

Remark 23.3. The condition in (2) is automatically met if $c \in \mathcal{C}^{p\times p} \cap \mathcal{W}_+(I_p)$ and the mvf $h(t)$ in the representation (20.1) is continuous on $[0, \infty)$. The solution of the inverse input impedance problem for a Dirac system on a finite interval $[0, d)$ depends only on the accelerant $h^\circ(t)$ of the mvf $c(\lambda)$ on the interval $[0, 2d)$. In other words, the solution will be the same for all input impedances with the same accelerant on the interval $[0, 2d)$.

The inverse spectral problem

Theorem 23.4. *Let $\sigma(\mu)$ be a nondecreasing $p\times p$ mvf on $\mathbb{R}$ that meets the constraint* (14.5) *and let $0 < d \leq \infty$. Then:*

(1) *There exists at most one differential system of the form* (23.2) *with a locally summable potential $\mathcal{V}(t)$ of the form* (23.1) *with spectral function $\sigma(\mu)$.*
(2) *If the given nondecreasing $p\times p$ mvf $\sigma(\mu)$ is differentiable at every point $\mu \in \mathbb{R}$ and*
$$\sigma'(\mu) = I_p - \int_{-\infty}^{\infty} e^{i\mu t} h(t)dt\,,$$
where $h(t)$ is a continuous summable $p \times p$ mvf on $\mathbb{R}$ such that $h(t) = h(-t)^*$ for every point $t \in \mathbb{R}$, then there exists exactly one such differential system on the interval $[0, \infty)$. Moreover, in this case, the potential $\mathcal{V}(t)$ is given by formula (23.1) *with $v_1(t) = -\Re a(2t)$ and $v_2(t) = \Im a(2t)$, where $a(t) = \gamma^t(t, 0)$ and $\gamma^t(a, b)$ is the unique solution of the integral equation* (20.8) *with d replaced by $2d$.*

Proof. The asserted conclusions follow from the preceding theorem and Theorem 15.1 applied to the canonical system (8.1) that corresponds to the differential system (23.2) with potential. The latter theorem implies that there is a family of possible Hermitians $H^{(\alpha)}(t)$ that are parametrized by a Hermitian $p\times p$ matrix α and are connected by the formula

$$H^{(\alpha)}(t) = \begin{bmatrix} I_p & i\alpha \\ 0 & I_p \end{bmatrix} H^{(0)}(t) \begin{bmatrix} I_p & 0 \\ -i\alpha & I_p \end{bmatrix} .$$

However, since any Hermitian that corresponds to a Dirac system must also satisfy the condition

$$H^{(\alpha)}(0) \;=\; I_m\,,$$

it follows that there is only one choice of α that yields a solution. □

24. Krein systems

Differential systems of the form (22.1) with $\operatorname{rank} P_J = \operatorname{rank} Q_J = p$ and $N = \kappa P_J$ or $N = \kappa Q_J$ for some $\kappa > 0$ and potentials $\mathcal{V}(t) \in L_{1,\mathrm{loc}}^{m\times m}$ that meet the two conditions

(1) $\mathcal{V}(t)J + J\mathcal{V}(t)^* = 0$ for a.e $t\in[0,d)$
(2) $\mathcal{V}(t) = \mathcal{V}(t)^*$ for a.e $t\in[0,d)$,

are called Krein systems.

Without loss of generality, we may assume that $\kappa = 1$ and $J = j_p$. Then the matrizant $U_t(\lambda) = U(t,\lambda)$ is a solution of the system

$$u_t'(\lambda) = i\lambda u_t(\lambda) \begin{bmatrix} I_p & 0 \\ 0 & 0 \end{bmatrix} j_p + u_t(\lambda)\mathcal{V}(t) \quad \text{for} \quad 0\le t<d \tag{24.1}$$

with potential $\mathcal{V}(t)$ of the form

$$\mathcal{V}(t) = \begin{bmatrix} 0 & a(t) \\ a(t)^* & 0 \end{bmatrix}\,,\ 0\le t<d\,. \tag{24.2}$$

Therefore, since $\mathfrak{V}j_p\mathfrak{V}^* = J_p$ and $\mathfrak{V} = \mathfrak{V}^*$, the general formulas (22.10)–(22.12) imply that

$$\mathcal{C}_{\mathrm{imp}}^d(\mathcal{V}) \;=\; \bigcap_{0\le t<d} \mathcal{C}(\mathfrak{V}U_t\mathfrak{V})$$

and, as

$$\sqrt{2}[0 \quad I_p]\mathfrak{V}U_s(\lambda) = [I_p \quad I_p]U_s(\lambda)\,,$$

the generalized Fourier transform for the system (24.1) may be taken equal to

$$g^{\triangle}(\lambda) \;=\; \frac{1}{\sqrt{2\pi}}\int_0^d (u_{11}(s,\lambda) + u_{21}(s,\lambda))g(s)ds \tag{24.3}$$

for every function $g\in L_2^p([0,d))$ with compact support in $[0,d)$.

Correspondingly, a nondecreasing $p \times p$ mvf $\sigma(\mu)$ on $\mathbb{R}$ is a spectral function for the system (24.1) if

$$\int_{-\infty}^{\infty} g^{\triangle}(\mu)^* d\sigma(\mu) g^{\triangle}(\mu) = \int_0^d g(s)^* g(s) ds$$

for every function $g \in L_2^p([0,d))$ with compact support in $[0,d)$.

The direct problem

Theorem 24.1. *Let $U_t(\lambda) = U(t,\lambda)$, $0 \le t < d$, be the matrizant of the system (24.1) with a locally summable potential $\mathcal{V}(t)$ of the form (24.2) and let $A_t(\lambda) = \mathfrak{V} U_t(\lambda) \mathfrak{V}$ for every $t \in [0,d)$. Then:*

1. *$A_t \in \mathcal{U}_{sR}(J_p)$ for every $t \in [0,d)$.*
2. *$\{e_t I_p, I_p\} \in ap_{II}(A_t)$ for every $t \in [0,d)$.*
3. *The de Branges spaces $\mathcal{B}(\mathfrak{E}_t)$ based on $\mathfrak{E}_t(\lambda) = \sqrt{2} A_t(\lambda) \mathfrak{V}$ are independent of the potential $\mathcal{V}(t)$ as linear topological spaces, i.e.,*

$$\mathcal{B}(\mathfrak{E}_t) = \left\{ \int_0^t e^{i\lambda s} h(s) ds : h \in L_2^p([0,t]) \right\}$$

as linear spaces and for each $t \in [0,d)$, there exist a pair of positive constants $\gamma_1 = \gamma_1(t)$ and $\gamma_2 = \gamma_2(t)$ such that

$$\gamma_1 \|f\|_2 \le \|f\|_{\mathcal{B}(\mathfrak{E}_t)} \le \gamma_2 \|f\|_2$$

for every $f \in \mathcal{B}(\mathfrak{E}_t)$.

4. *Assertions (4)–(8) of Theorem 22.3 are in force.*

Proof. This theorem is a special case of Theorem 22.3. □

The inverse input impedance problem

The inverse input impedance problem for differential systems of the form (24.1) on an interval $[0,d)$ with a locally summable potential $\mathcal{V}(t)$ of the form (24.2) is to find $\mathcal{V}(t)$, given a mvf $c \in \mathcal{C}^{p\times p}$ and the right-hand endpoint d, $0 < d \le \infty$, of the interval.

Theorem 24.2. *Let $c \in \mathcal{C}^{p\times p}$ and $0 < d \le \infty$ be given. Then:*

1. *There exists at most one differential system of the form (24.1) with a locally summable potential $\mathcal{V}(t)$ of the form (24.2) with input impedance equal to $c(\lambda)$.*
2. *There exists exactly one such differential system if $c \in \mathcal{C}^{p\times p}$ has a continuous accelerant $h^\circ(t)$ on the interval $[0,d)$ and $\gamma = I_p$ in the representation formula (11.4). Moreover, in this case, the potential $\mathcal{V}(t)$ may be obtained from formula (24.2), where $a(t) = \gamma^t(t,0)$ and $\gamma^t(a,b)$ is the unique solution of the integral equation (20.8).*

Proof. This theorem is a special case of Theorem 22.6. □

Remark 24.3. The condition in (2) is automatically met if $c \in \mathcal{C}^{p\times p} \cap \mathcal{W}_+(I_p)$ and the mvf $h(t)$ in the representation (20.1) is continuous on $[0,\infty)$. Moreover, the solution of the inverse impedance problem for a Krein system on a finite interval $[0,d)$ depends only on the accelerant $h^\circ(t)$ on the interval $[0,d)$. In other words, the solution will be the same for all input impedances with the same accelerant on the interval $[0,d)$.

The inverse spectral problem

Theorem 24.4. *Let $\sigma(\mu)$ be a nondecreasing $p\times p$ mvf on $\mathbb{R}$ that meets the constraint* (14.5) *and let $0 < d \le \infty$. Then:*

(1) *There exists at most one differential system of the form* (24.1) *with a locally summable potential $\mathcal{V}(t)$ of the form* (24.2) *with spectral function $\sigma(\mu)$.*
(2) *If the given nondecreasing $p\times p$ mvf $\sigma(\mu)$ is differentiable at every point $\mu \in \mathbb{R}$ and*

$$\sigma'(\mu) = I_p - \int_{-\infty}^{\infty} e^{i\mu t} h(t)dt\,,$$

where $h(t)$ is a continuous summable $p\times p$ mvf on $\mathbb{R}$ such that $h(t) = h(-t)^*$ for every point $t \in \mathbb{R}$, then there exists exactly one such differential system on the interval $[0,\infty)$. Moreover, in this case it may be obtained from formula (24.2)*, where $a(t) = \gamma^t(t,0)$ and $\gamma^t(a,b)$ is the unique solution of the integral equation* (20.8).

Proof. The asserted conclusions follow from Theorem 23.4 and Remark 22.5. □

Theorem 20.3, and the solution of the inverse spectral problem for the differential systems with potential that are now called Dirac and Krein systems were first announced by M. G. Krein [Kr:55], [Kr:56], given a continuous accelerant on an appropriate interval (see also [Sak:00a]). Krein also formulated a converse statement: if the potential $\mathcal{V}(t)$ of such a system is continuous, then $c \in \mathcal{C}^d_{\mathrm{imp}}(\mathcal{V})$ has a continuous accelerant on an appropriate interval. Proofs of a number of related statements may be found in [KrL:85]. For additional discussion, see also [MeA:67] and [KrMA:86].

25. A differential system with potential with $NJ \neq JN$

In this section we shall consider differential systems of the form (22.1) with

$$N = \begin{bmatrix} I_p & 0 \\ 0 & 0 \end{bmatrix}, \quad J = J_p \quad \text{and potentials} \quad \mathcal{V} \in L_{1,\mathrm{loc}}^{m\times m}([0,d)) \tag{25.1}$$

that are subject to the two additional constraints

$$\mathcal{V}(t)J_p + J_p\mathcal{V}(t)^* = 0 \quad \text{a.e. in} \quad [0,d) \tag{25.2}$$

and

$$\det\left\{[0 \quad I_p]\mathcal{V}(t)\begin{bmatrix} I_p \\ 0 \end{bmatrix}\right\} \neq 0 \quad \text{a.e. in} \quad [0,d)\,. \tag{25.3}$$

The two specific choices of potential

$$\mathcal{V}(t) = \begin{bmatrix} v(t) & 0 \\ -iI_p & -v(t) \end{bmatrix}, \quad 0 \le t < d, \quad \text{with } v(t) = v(t)^* \text{ a.e. in } [0,d) \tag{25.4}$$

and

$$\mathcal{V}(t) = \begin{bmatrix} 0 & iq(t) \\ -iI_p & 0 \end{bmatrix}, \quad 0 \le t < d, \quad \text{with } q(t) = q(t)^* \text{ a.e. in } [0,d), \tag{25.5}$$

which both fit into this setting, will play a useful role in our analysis of the Schrödinger equation in the next section. This stems from the fact that for the given choice of N and these two potentials

$$NJ_pN = 0 \quad \text{and} \quad NJ_p\mathcal{V}(t) + \mathcal{V}(t)NJ_p = iI_m \quad \text{a.e. in } [0,d).$$

Consequently, if $\mathcal{V} \in AC_{\text{loc}}^{m\times m}([0,d))$ and

$$Y_t'(\lambda) = i\lambda Y_t(\lambda)NJ_p + Y_t(\lambda)\mathcal{V}(t) \quad \text{for} \quad 0 \le t < d,$$

then $Y_t' \in AC_{\text{loc}}^{m\times m}([0,d))$ and

$$\begin{aligned} Y_t''(\lambda) &= Y_t'(\lambda)(i\lambda NJ_p + \mathcal{V}(t)) + Y_t(\lambda)\mathcal{V}'(t) \\ &= Y_t(\lambda)(i\lambda NJ_p + \mathcal{V}(t))\,(i\lambda NJ_p + \mathcal{V}(t) + Y_t(\lambda))\mathcal{V}'(t) \\ &= -\lambda Y_t(\lambda) + Y_t(\lambda)(\mathcal{V}(t)^2 + \mathcal{V}'(t)), \end{aligned}$$

i.e., $Y_t(\lambda)$ is automatically the solution of a Schrödinger equation with potential $\mathcal{V}(t)^2 + \mathcal{V}'(t)$. Notice that

$$NJ_p - J_pN = -i\mathcal{J}_p$$

for this choice of N. Therefore, we cannot invoke Theorem 22.2 to conclude that the matrizant $Y_t(\lambda)$ of the corresponding system is strongly regular. In fact, $Y_t(\lambda)$ is an entire mvf of minimal exponential type and hence of class $\mathcal{U}_S(J_p)$. Nevertheless, it will turn out to be possible to use the interplay between these systems and some related systems with strongly regular matrizants to obtain useful conclusions for a class of Schrödinger equations.

Theorem 25.1. *Let $Y_t(\lambda) = Y(t,\lambda)$, $0 \le t < d$, denote the matrizant of the differential system* (22.1) *with N, J and $\mathcal{V}(t)$ specified by* (25.1)–(25.3) *and let $B_t(\lambda) = Y_t(\lambda)\mathfrak{V}$ for $0 \le t < d$. Then the generalized Fourier transform is given by the formula*

$$g^{\triangle}(\lambda) = \frac{1}{\sqrt{\pi}} \int_0^d y_{21}(s,\lambda)g(s)ds \tag{25.6}$$

for every $g \in L_2^p([0,d))$ with compact support in $[0,d)$ and:

1. (1) *$Y_t \in \mathcal{U}(J_p)$ for every $t \in [0,d)$.*
2. (2) *$\mathcal{S}^{p\times p} \subset \mathcal{D}(T_{Y_t\mathfrak{V}})$ for every $t \in (0,d)$.*
3. (3) *$\mathcal{C}(Y_t) = T_{Y_t\mathfrak{V}}[\mathcal{S}^{p\times p}]$ for every $t \in (0,d)$.*
4. (4) *$\mathcal{C}_{\text{imp}}^d(\mathcal{V}) = \bigcap_{0\le t<d} \mathcal{C}(Y_t) \ne \emptyset$.*
5. (5) *$\Sigma_{sf}^d(\mathcal{V}) = (\mathcal{C}_{\text{imp}}^d(\mathcal{V}))_{sf}$.*
6. (6) *If $d < \infty$ and $\mathcal{V} \in L_1([0,d])$, then $\mathcal{C}_{\text{imp}}^d(\mathcal{V}) = \mathcal{C}(Y_d)$.*

Proof. In view of (25.2), the matrizant $Y_t \in \mathcal{U}(J_p)$. Therefore, the $m \times m$ mvf

$$Y(t) = Y(t,0)$$

is J_p unitary and the mvf

$$A_t(\lambda) = Y(t,\lambda)Y(t)^{-1}$$

is the matrizant of the canonical differential system (22.7) with continuous Hamiltonian

$$H(t) = Y(t)NY(t)^* = Y_1(t)Y_1(t)^* \quad \text{for} \quad t \in [0,d),$$

where the $m \times p$ mvf $Y_1(t)$ is the first block in the block column decomposition

$$Y(t) = [Y_1(t) \quad Y_2(t)].$$

Moreover, if $d < \infty$ and $\mathcal{V} \in L_1^{m\times m}([0,d])$, then the mvf $A(\lambda) = A_d(\lambda)$ is the characteristic function $A(\lambda) = I_m + i\lambda F(I - \lambda A)^{-1}F^*J_p$ of the Livsic-Brodskii J_p-node $\mathfrak{N} = (K, F; X, \mathbb{C}^m)$, based on the bounded linear operators $K \in \mathcal{L}(X)$ and $F \in \mathcal{L}(X, \mathbb{C}^p)$ that are defined by the formulas

$$(Kg)(t) = iY_1(t)^*J_p \int_t^d Y_1(s)g(s)ds \quad \text{and} \quad Fg = \int_0^d Y_1(s)g(s)ds,$$

where $X = L_2^p([0,d])$ and $g \in X$ (and $K - K^* = iFJ_pF^*$).

Let

$$L = \begin{bmatrix} 0 \\ I_p \end{bmatrix} \quad \text{and} \quad F_2 = L^*F.$$

Then, by Corollary 5.9 in [ArD:04b],

$$\bigvee_{n\geq 0} K^nF_2^*\mathbb{C}^p = L_2^p([0,d)) \Longleftrightarrow \ker K \cap \ker F = \{0\}.$$

Moreover, if this last condition is in force, then, by Theorems 5.10 and 2.14 in [ArD:04b],

(a) $\mathcal{S}^{p\times p} \subset \mathcal{D}(T_{A\mathfrak{B}}) \Longleftrightarrow \ker F_2^* = \{0\}$.
(b) $\ker K^* = \{0\}$ and $\operatorname{range} K^* \cap \operatorname{range} F_2^* = \{0\} \Longrightarrow \Sigma_{sf}^d(H) = (\mathcal{C}(A))_{sf}$.

Therefore, to complete the proof of the theorem, it suffices to check that

(a) $\ker K = \{0\}$.
(b) $\ker K^* = \{0\}$.
(c) $\ker F_2^* = \{0\}$.
(d) $\operatorname{range} K^* \cap \operatorname{range} F_2^* = \{0\}$.

This can be verified by straightforward calculations that exploit the identity

$$\begin{bmatrix} Y_1(t)^*J_pY_1(t) & Y_1(t)^*J_pY_2(t) \\ Y_2(t)^*J_pY_1(t) & Y_2(t)^*J_pY_2(t) \end{bmatrix} = \begin{bmatrix} 0 & -I_p \\ -I_p & 0 \end{bmatrix},$$

which is valid for every $t \in [0,d)$, the equation $Y'(t) = Y(t)\mathcal{V}(t)$ and the fact that the 21 block $v_{21}(t)$ of the potential $\mathcal{V}(t)$ is invertible a.e. on the interval $[0,d)$. Details will be presented elsewhere. □

26. Spectral problems for the Schrödinger equation

In this section we shall indicate how to extract information on direct and inverse spectral problems for the Schrödinger equation

$$-u''(x,\lambda)+u(x,\lambda)q(x)=\lambda u(x,\lambda)\,,\quad 0\le x<d\,, \tag{26.1}$$

with a $p\times p$ matrix-valued potential $q(x)$ from the corresponding results that were discussed earlier for differential systems of the form (22.1). The direct spectral problem will be considered in the class of potentials $q(t)$ that satisfy the conditions

(A1) $q(t)=q(t)^*$ a.e. on $[0,d)$ and $q\in L_{1,\,\mathrm{loc}}^{m\times m}([0,d))$.

The inverse spectral problem will be discussed under a more stringent condition on $q(t)$:

(A2) There exists a solution $v(t)$ of the Riccati equation

$$v'(t)+v(t)^2\;=\;q(t)\,,\quad \text{for every}\quad t\in[0,d)\,, \tag{26.2}$$

in the class of $p\times p$ mvf's $v(t)$ such that

$$v\in AC_{\mathrm{loc}}^{p\times p}([0,d))\quad\text{and}\quad v(t)=v(t)^*\quad\text{for every}\quad t\in[0,d)\,. \tag{26.3}$$

Let $\psi(t,\lambda)$ and $\varphi(t,\lambda)$ be the unique solutions of equation (26.1) that meet the initial conditions

$$\psi(0,\lambda)=I_p\,,\quad \psi'(0,\lambda)=0\,,\quad \varphi(0,\lambda)=0\quad\text{and}\quad \varphi'(0,\lambda)=I_p\,,$$

respectively, and let

$$U(t,\lambda)\;=\;\begin{bmatrix}\psi(t,\lambda) & \psi'(t,\lambda)\\ \varphi(t,\lambda) & \varphi'(t,\lambda)\end{bmatrix} \tag{26.4}$$

be the fundamental matrix of equation (26.1).

In the present discussion, we focus on spectral problems for the Schrödinger equation (26.1) that are related to the generalized Fourier transform

$$g^{\triangle}(\lambda)\;=\;\frac{1}{\sqrt{\pi}}\int_0^d \varphi(s,\lambda)g(s)ds \tag{26.5}$$

of vvf's $g\in L_2^p([0,d))$ with compact support in $[0,d)$.

A nondecreasing $p\times p$ mvf $\sigma(\mu)$ on $\mathbb{R}$ is said to be a spectral function of (26.1) with respect to this transform if the Parseval equality

$$\int_{\infty}^{\infty} g^{\triangle}(\mu)^*d\sigma(\mu)g^{\triangle}(\mu)=\int_0^d g(s)^*g(s)ds \tag{26.6}$$

holds for every $g\in L_2^p([0,d))$ with compact support in $[0,d)$. The symbol $\Sigma_{sf}^d(q)$ will be used to denote the set of all spectral functions of (26.1) with respect to this transform.

The direct spectral problem when (A1) is in force

The direct spectral problem is to describe the set $\Sigma_{sf}^d(q)$ for a given potential $q(t)$ on $[0, d)$. The solution of this problem will be given under assumption (A1).

Theorem 26.1. *Let $U(t,\lambda)$ denote the fundamental matrix of equation* (26.1), *let*

$$X(t,\lambda) = \begin{bmatrix} iI_p & 0 \\ 0 & I_p \end{bmatrix} U(t,\lambda) \begin{bmatrix} -iI_p & 0 \\ 0 & I_p \end{bmatrix} \quad \text{for} \quad 0 \le t < d \tag{26.7}$$

and assume that the potential $q(t)$ satisfies the condition (A1). Then

(1) *$U_t(\lambda) = U(t,\lambda)$ is the matrizant of the differential system* (22.1) *with*

$$N = \begin{bmatrix} I_p & 0 \\ 0 & 0 \end{bmatrix}, \quad J = -\mathcal{J}_p \quad \text{and potential} \quad \mathcal{V}(t) = \begin{bmatrix} 0 & q(t) \\ I_p & 0 \end{bmatrix},$$

i.e.,

$$U'(t,\lambda) = i\lambda U(t,\lambda) \begin{bmatrix} I_p & 0 \\ 0 & 0 \end{bmatrix} (-\mathcal{J}_p) + U(t,\lambda) \begin{bmatrix} 0 & q(t) \\ I_p & 0 \end{bmatrix} \tag{26.8}$$

for $0 \le t < d$ and $U(0,\lambda) = I_m$.

(2) *$X_t(\lambda) = X(t,\lambda)$ is the matrizant of the differential system* (22.1) *with*

$$N = \begin{bmatrix} I_p & 0 \\ 0 & 0 \end{bmatrix}, \quad J = J_p \quad \text{and potential} \quad \mathcal{V}(t) = \begin{bmatrix} 0 & iq(t) \\ -iI_p & 0 \end{bmatrix},$$

i.e.,

$$X'(t,\lambda) = i\lambda X(t,\lambda) \begin{bmatrix} I_p & 0 \\ 0 & 0 \end{bmatrix} J_p + X(t,\lambda) \begin{bmatrix} 0 & iq(t) \\ -iI_p & 0 \end{bmatrix} \tag{26.9}$$

for $0 \le t < d$, and $X(0,\lambda) = I_m$.

(3) *$U_t \in \mathcal{U}(-\mathcal{J}_p)$ and $X_t \in \mathcal{U}(J_p)$ for every $t \in [0, d)$.*

(4) *$\mathcal{S}^{p\times p} \subset \mathcal{D}(T_{X_t\mathfrak{V}})$ for every $t \in (0, d)$.*

(5) *$\mathcal{C}(X_t) = T_{X_t\mathfrak{V}}[\mathcal{S}^{p\times p}]$ for every $t \in (0, d)$.*

(6) *$\mathcal{C}_{\rm imp}^t(q) = \mathcal{C}(X_t)$ for every $t \in [0, d)$.*

(7) *$\mathcal{C}_{\rm imp}^d(q) = \bigcap_{0\le t<d} \mathcal{C}(X_t) \ne \emptyset$.*

(8) *$\Sigma_{sf}^d(q) = (\mathcal{C}_{\rm imp}^d(q))_{sf} = \Sigma_{sf}^d(\mathcal{V})$.*

(9) *If $d < \infty$ and $q \in L_1^{p\times p}([0, d])$, then $\mathcal{C}_{\rm imp}^d(q) = \mathcal{C}(X_d)$.*

Proof. Let $u(t,\lambda)$ be a solution of equation (26.1) and let

$$y(t,\lambda) = [u(t,\lambda) \quad u'(t,\lambda)].$$

Then

$$\begin{aligned} y'(t,\lambda) &= [u'(t,\lambda) \quad u''(t,\lambda)] \\ &= [u'(t,\lambda) \quad u(t,\lambda)((q(t) - \lambda I_p)] \\ &= \lambda y(t,\lambda) \begin{bmatrix} 0 & -I_p \\ 0 & 0 \end{bmatrix} + y(t,\lambda) \begin{bmatrix} 0 & q(t) \\ I_p & 0 \end{bmatrix} \\ &= i\lambda y(t,\lambda) \begin{bmatrix} I_p & 0 \\ 0 & 0 \end{bmatrix} (-\mathcal{J}_p) + y(t,\lambda) \begin{bmatrix} 0 & q(t) \\ I_p & 0 \end{bmatrix} \end{aligned}$$

This justifies the first assertion. The remaining assertions follow from Theorem 25.1 and the identity

$$-\mathcal{J}_p = \begin{bmatrix} -iI_p & 0 \\ 0 & I_p \end{bmatrix} J_p \begin{bmatrix} iI_p & 0 \\ 0 & I_p \end{bmatrix}. \qquad \square$$

Remark 26.2. Assertions (2) and (3) of the theorem can be formulated directly in terms of the fundamental matrix $U(t,\lambda)$:

(2′) $\mathcal{F}(-\mathcal{J}_p) \subset \mathcal{D}(\widetilde{T}_{U_t})$ for every $t \in [0,d)$.

(3′) $\mathcal{C}(A_t) = i\widetilde{T}_{U_t}[\mathcal{F}(-\mathcal{J}_p)]$ for every $t \in [0,d)$.

The direct spectral problem when (A2) is in force

If assumption (A2) is in force and $q(t) = v(t)^2 + v'(t)$, then the mvf's

$$Y(t,\lambda) = \begin{bmatrix} I_p & v(0) \\ 0 & -iI_p \end{bmatrix} U(t,\lambda) \begin{bmatrix} I_p & -iv(t) \\ 0 & iI_p \end{bmatrix} \quad \text{for} \quad 0 \le t < d \tag{26.10}$$

and

$$A_t(\lambda) = L_\lambda Y_t(\lambda^2) L_\lambda^{-1} \quad \text{for} \quad 0 \le t < d\,, \quad \text{where} \quad L_\lambda = \begin{bmatrix} I_p & 0 \\ 0 & \lambda I_p \end{bmatrix} \tag{26.11}$$

are matrizants of differential systems of the form (22.1). These systems play a useful role in the study of spectral problems for the Schrödinger equation with potential $q(t) = v(t)^2 + v'(t)$, because the generalized Fourier transforms of all three systems of equations are simply related. The following table, summarizes the main facts concerning the four matrizants that have been introduced in this section and the corresponding transforms for the convenience of the reader.

Matr.	N	J	$\mathcal{V}(t)$	$g^{\triangle}(\lambda)$
$U_t(\lambda)$	$\begin{bmatrix} I_p & 0 \\ 0 & 0 \end{bmatrix}$	$-\mathcal{J}_p$	$\begin{bmatrix} 0 & q(t) \\ I_p & 0 \end{bmatrix}$	$-i\int_0^d u_{21}(s,\lambda)g(s)ds\,,$ $g \in L_2^p$
$X_t(\lambda)$	$\begin{bmatrix} I_p & 0 \\ 0 & 0 \end{bmatrix}$	J_p	$\begin{bmatrix} 0 & iq(t) \\ -iI_p & 0 \end{bmatrix}$	$-i\int_0^d x_{21}(s,\lambda)g(s)ds\,,$ $g \in L_2^p$
$Y_t(\lambda)$	$\begin{bmatrix} I_p & 0 \\ 0 & 0 \end{bmatrix}$	J_p	$\begin{bmatrix} v(t) & 0 \\ -iI_p & -v(t) \end{bmatrix}$	$-i\int_0^d y_{21}(s,\lambda)g(s)ds\,,$ $g \in L_2^p$
$A_t(\lambda)$	I_m	J_p	$\begin{bmatrix} v(t) & 0 \\ 0 & -v(t) \end{bmatrix}$	$-i\int_0^d a_{21}(s,\lambda)g(s)ds$ $-i\int_0^d a_{22}(s)h(s)ds\,,$ $g, h \in L_2^p$

Moreover,

$$x_{21}(s,\lambda) = y_{21}(s,\lambda) = -iu_{21}(s,\lambda) = \frac{a_{21}(s,\sqrt{\lambda})}{\sqrt{\lambda}} = -i\varphi_{21}(s,\lambda)$$

for $s \in [0,d)$. This connection permits one to reduce the spectral problem for the Schrödinger equation (26.1) to a spectral problem for Dirac systems. This strategy was initiated by M.G. Krein in [Kr:55].

Theorem 26.3. *Let $U(t,\lambda)$ be the fundamental matrix for the Schrödinger equation (26.1) with a potential $q(t)$ that satisfies condition (A2), let $v(t)$ be the $p\times p$ mvf that appears in this condition and let the mvf $Y(t,\lambda)$ be defined by (26.10). Then:*

(1) *$Y_t(\lambda)=Y(t,\lambda)$ is the matrizant of the differential system (22.1) with*

$$N=\begin{bmatrix} I_p & 0 \\ 0 & 0 \end{bmatrix}, \quad J=J_p \quad \text{and potential} \quad \mathcal{V}(t)=\begin{bmatrix} v(t) & 0 \\ -iI_p & -v(t) \end{bmatrix}, \tag{26.12}$$

i.e.,

$$Y'(t,\lambda)=i\lambda Y(t,\lambda)\begin{bmatrix} I_p & 0 \\ 0 & 0 \end{bmatrix}J_p+Y(t,\lambda)\begin{bmatrix} v(t) & 0 \\ -iI_p & -v(t) \end{bmatrix} \tag{26.13}$$

for $0\le t<d$, and $Y(0,\lambda)=I_m$.

(2) *$Y_t\in\mathcal{U}(J_p)$ for every $t\in[0,d)$.*

(3) *$\mathcal{S}^{p\times p}\subset\mathcal{D}(T_{Y_t\mathfrak{V}})$ for every $t\in(0,d)$.*

(4) *$\mathcal{C}(Y_t)=T_{Y_t\mathfrak{V}}[\mathcal{S}^{p\times p}]$ for every $t\in(0,d)$.*

(5) *$\mathcal{C}^d_{\rm imp}(q)=-iv(0)+\mathcal{C}(Y_t)$ for every $t\in[0,d)$.*

(6) *$\mathcal{C}^t_{\rm imp}(q)=-iv(0)+\mathcal{C}^d_{\rm imp}(\mathcal{V})$.*

(7) *$\Sigma^d_{sf}(q)=\Sigma^d_{sf}(\mathcal{V})=(\mathcal{C}^d_{\rm imp}(\mathcal{V}))_{sf}$.*

(8) *If $d<\infty$ and $q\in L_1^{p\times p}([0,d])$, then $\mathcal{C}^d_{\rm imp}(q)=-iv(0)+\mathcal{C}(Y_d)$ and $\Sigma^d_{sf}(q)=(\mathcal{C}(Y_d))_{sf}$.*

Proof. Let

$$T(t)=\begin{bmatrix} I_p & -iv(t) \\ 0 & iI_p \end{bmatrix} \quad \text{for} \quad 0\le t<d.$$

Then

$$Y(t,\lambda)=T(0)^{-1}U(t,\lambda)T(t) \quad \text{for} \quad 0\le t<d,$$

and, in view of formula (26.8),

$$\begin{aligned} Y'(t,\lambda) &= T(0)^{-1}U'(t,\lambda)T(t)+T(0)^{-1}U(t,\lambda)T'(t) \\ &= T(0)^{-1}U(t,\lambda)\left\{i\lambda N(-\mathcal{J}_p)+\begin{bmatrix} 0 & q(t) \\ I_p & 0 \end{bmatrix}\right\}T(t) \\ &\quad +T(0)^{-1}U(t,\lambda)T'(t). \end{aligned}$$

Formula (26.13) now follows easily from the fact that for the given choice of N

$$N(-\mathcal{J}_p)T(t)=T(t)NJ_p \quad \text{and} \quad \begin{bmatrix} 0 & q(t) \\ I_p & 0 \end{bmatrix}T(t)+T'(t)=T(t)\begin{bmatrix} v(t) & 0 \\ -iI_p & -v(t) \end{bmatrix}$$

when

$$q(t)=v(t)^2+v'(t).$$

Assertion (2) follows from the fact that $\mathcal{V}(t)J_p+J_p\mathcal{V}(t)^*=0$ a.e. on $[0,d)$.

Next, since

$$[0 \quad I_p]Y(t,\lambda)\begin{bmatrix} I_p \\ 0 \end{bmatrix}=-i[0 \quad I_p]U(t,\lambda)\begin{bmatrix} I_p \\ 0 \end{bmatrix}=-i\varphi(t,\lambda) \quad \text{for} \quad 0\le t<d,$$

the generalized Fourier transform (25.6)) based on the matrizant $Y(t,\lambda)$ coincides with the generalized Fourier transform (26.5) up to a constant factor of modulus one. Therefore,

$$\Sigma^t_{sf}(q) = \Sigma^t_{sf}(\mathcal{V}) \quad \text{for} \quad 0 \le t < d\,.$$

Consequently, assertions (3)–(8) follow from Theorem 26.1 and formula (26.10). □

Theorem 26.4. *Let assumption (A2) be in force for the potential $q(t) = v(t)^2 + v'(t)$ of the Schrödinger equation, let $Y_t(\lambda)$ denote the matrizant considered in Theorem 26.3. Then the mvf $A_t(\lambda) = L_\lambda Y_t(\lambda^2)L_\lambda^{-1}$ is the matrizant of the Dirac system (23.2) and*

$$\mathcal{V}(t) = \begin{bmatrix} v(t) & 0 \\ 0 & -v(t) \end{bmatrix}, \quad 0 \le t < d\,,$$

i.e., $A_t(\lambda)$ is a solution of the Cauchy problem

$$\begin{aligned} A_t'(\lambda) &= i\lambda A_t(\lambda)J_p + A_t(\lambda)\mathcal{V}(t)\,, \quad 0 \le t < d\,, \\ A_0(\lambda) &= I_m\,. \end{aligned}$$

Proof. Clearly

$$A_t'(\lambda) = L_\lambda Y_t'(\lambda^2)L_\lambda^{-1} = L_\lambda\left\{i\lambda^2 Y(t,\lambda^2)NJ_p + Y(t,\lambda^2)\mathcal{V}(t)\right\}\,,$$

with N and $\mathcal{V}(t)$ as in (26.12). Therefore, since

$$i\lambda^2 L_\lambda NJ_pL_\lambda^{-1} = i\lambda NJ_p$$

and

$$L_\lambda\mathcal{V}(t)L_\lambda^{-1} = \mathcal{V}(t) + i\lambda\begin{bmatrix} 0 & 0 \\ 0 & I_p \end{bmatrix} J_p\,,$$

for this choice of N and $\mathcal{V}(t)$, the proof is easily completed. □

Theorem 26.5. *The fundamental matrix $U_t(\lambda) = U(t,\lambda)$ for the Schrödinger equation (26.1) with potential $q(t) = v(t)^2 + v'(t)$ that satisfies assumption (A2) enjoys the following properties:*

(1) $U_t \in \mathcal{U}_S(-\mathcal{J}_p)$ *for every* $t \in [0,d)$.

(2) $\limsup_{r\uparrow\infty} \dfrac{\ln\max\{\|U_t(\lambda)\| : |z| \le r\}}{r^{1/2}} = \limsup_{\mu\downarrow-\infty} \dfrac{\ln\|U_t(\mu)\|}{|\mu|^{1/2}} = t\,.$

Proof. By Theorem 23.1 $\{e_tI_p, e_tI_p\} \in ap_{II}(A_t)$ and

$$\limsup_{r\uparrow\infty} \frac{\ln\max\{\|A_t(\lambda)\| : |\lambda| \le r\}}{r} = \limsup_{\nu\uparrow\infty} \frac{\ln\|A_t(\pm i\nu)\|}{\nu} = t\,.$$

But this serves to establish the second assertion, since

$$\|A_t(\lambda)\| \le |\lambda|\|Y_t(\lambda^2)\| \quad \text{and} \quad \|Y_t(\lambda^2)\| \le |\lambda|\|A_t(\lambda)\| \quad \text{when} \quad |\lambda| \ge 1$$

and the matrizants $Y_t(\lambda)$ and $U_t(\lambda)$ are related by formula (26.11). Thus, as $U_t(\lambda)$ is an entire $m \times m$ mvf of minimal exponential type, assertion (1) is in force. □

Remark 26.6. Analogous conclusions hold for the matrizants $X_t(\lambda)$ and $Y_t(\lambda)$.

de Branges spaces

Let

$$\mathfrak{E}_t^X(\lambda) = \sqrt{2}[0 \quad I_p]X_t(\lambda)\mathfrak{V}$$

denote the de Branges function based on the matrizant $X_t(\lambda)$ that was introduced in Theorem 26.1. Then, in view of formulas (26.4) and (26.7), it is readily checked that

$$\mathfrak{E}_t^X(\lambda) = [\varphi'(t,\lambda) + i\varphi(t,\lambda) \quad \varphi'(t,\lambda) - i\varphi(t,\lambda)]$$

and, since $x_{21}(s,\lambda) = -iu_{21}(s,\lambda) = -i\varphi(s,\lambda)$, that the corresponding de Branges space

$$\mathcal{B}(\mathfrak{E}_t^X) = \left\{\frac{1}{\sqrt{\pi}}\int_0^t \varphi(s,\lambda)g(s)ds : g \in L_2^p([0,t]) \quad \text{for every} \quad t \in [0,d)\right\} \tag{26.14}$$

with norm

$$\langle g^\triangle, g^\triangle\rangle_{\mathcal{B}(\mathfrak{E}_t^X)} = \int_{-\infty}^{\infty} g^\triangle(\mu)^*\Delta_t^X(\mu)g^\triangle(\mu)d\mu\,,$$

where

$$g^\triangle(\mu) = \frac{1}{\sqrt{\pi}}\int_0^t \varphi(s,\lambda)g(s)ds$$

and

$$\begin{aligned}\Delta_t^X(\mu)^{-1} &= (\varphi'(t,\mu) - i\varphi(t,\mu))(\varphi'(t,\mu) - i\varphi(t,\mu))^* \\ &= \varphi'(t,\mu)\varphi'(t,\mu)^* + \varphi(t,\mu)\varphi(t,\mu)^*\,,\end{aligned}$$

since the fundamental matrix $U_t \in \mathcal{U}(-\mathcal{J}_p)$.

An analogous set of calculations for the de Branges function

$$\mathfrak{E}_t^A(\lambda) = \sqrt{2}[0 \quad I_p]A_t(\lambda)\mathfrak{V}$$

based on the matrizant $A_t(\lambda)$ that was introduced in Theorem 26.4 leads to the conclusion that

$$\mathfrak{E}_t^A(\lambda) = [a_{22}(t,\lambda) - a_{21}(t,\lambda) \quad a_{22}(t,\lambda) + a_{21}(t,\lambda)]$$

and that the corresponding de Branges space

$$\mathcal{B}(\mathfrak{E}_t^A) = \left\{\frac{1}{\sqrt{\pi}}\int_0^t [a_{21}(s,\lambda) \quad a_{22}(s,\lambda)]f(s)ds : f \in L_2^m([0,t]) \text{ for every } t \in [0,d)\right\}$$

with norm

$$\langle f^\triangle, f^\triangle\rangle_{\mathcal{B}(\mathfrak{E}_t^X)} = \int_{-\infty}^{\infty} f^\triangle(\mu)^*\Delta_t^A(\mu)f^\triangle(\mu)d\mu\,,$$

where, upon writing $f = \mathrm{col}[g\, h]$, with components $g, h \in L_2^p([0,t])$,

$$\begin{aligned}f^\triangle(\mu) &= \frac{1}{\sqrt{\pi}}\int_0^t [a_{21}(s,\lambda) \quad a_{22}(s,\lambda)f(s)ds \\ &= \frac{1}{\sqrt{\pi}}\int_0^t a_{21}(s,\lambda)g(s) + a_{22}(s,\lambda)h(s)ds\end{aligned}$$

and

$$\begin{aligned}\Delta_t^A(\mu)^{-1} &= (a_{22}(t,\mu)+a_{21}(t,\mu))(a_{22}(t,\mu)+a_{21}(t,\mu))^* \\ &= a_{22}(t,\mu)a_{22}(t,\mu)^* + a_{21}(t,\mu)a_{21}(t,\mu)^*\,,\end{aligned}$$

since $A_t \in \mathcal{U}(J_p)$. Moreover, formula (26.10) implies that $a_{21}(t,\lambda)$ is an odd function of λ, whereas $a_{22}(t,\lambda)$ is an even function of λ. Thus,

$$\mathcal{B}(\mathfrak{E}_t^A) = \mathcal{B}(\mathfrak{E}_t^A)_{odd} \oplus \mathcal{B}(\mathfrak{E}_t^A)_{ev}\,,$$

where

$$\mathcal{B}(\mathfrak{E}_t^A)_{odd} = \left\{\int_0^t a_{21}(s,\lambda)g(s)ds\,:\, g\in L_2^p([0,t])\right\}$$

and

$$\mathcal{B}(\mathfrak{E}_t^A)_{ev} = \left\{\int_0^t a_{22}(s,\lambda)g(s)ds\,:\, g\in L_2^p([0,t])\right\}.$$

At the same time, Theorem 23.1 implies that

$$\mathcal{B}(\mathfrak{E}_t^A) = \left\{\int_{-t}^t e^{i\lambda s}g(s)ds\,:\, g\in L_2^p([-t,t])\right\}$$

and hence that

$$\mathcal{B}(\mathfrak{E}_t^A)_{odd} = \left\{\int_0^t \sin(s\lambda)g(s)ds\,:\, g\in L_2^p([0,t])\right\}$$

and

$$\mathcal{B}(\mathfrak{E}_t^A)_{ev} = \left\{\int_0^t \cos(s\lambda)g(s)ds\,:\, g\in L_2^p([0,t])\right\}.$$

Thus, as

$$y_{21}(s,\lambda) = \frac{a_{21}(s,\sqrt{\lambda})}{\sqrt{\lambda}} = -i\varphi(s,\lambda)\,,$$

we obtain the following conclusion:

Theorem 26.7. *If the potential $q(t)$ of the Schrödinger equation (26.1) satisfies assumption (A2), then the de Branges space*

$$\mathcal{B}(\mathfrak{E}_t^X) = \left\{\frac{1}{\sqrt{\pi}}\int_0^t \frac{\sin\sqrt{\lambda}s}{\sqrt{\lambda}}g(s)ds\,:\, g\in L_2^p([0,t]) \quad \text{for every} \quad t\in[0,d)\right\}, \tag{26.15}$$

as linear spaces and hence these spaces do not depend upon the potential.

In view of the indicated connection between Dirac systems and Schrödinger equations, Theorems 23.2 and 23.4 may be applied to yield existence and uniqueness theorems for the inverse input impedance problem and the inverse spectral problem for the latter when assumption (A2) is in force, as well as recipes for the solution. A detailed analysis will be presented elsewhere.

Remark 26.8. The identification (26.15) for the scalar case $p = 1$ is obtained in [Rem:03] under assumption (A1) on the potential $q(t)$ of the Schrödinger equation, which is less restrictive than the assumption (A2) that is imposed here.

27. Epilogue

In the early fifties M.G. Krein published a series of notes on inverse problems for second order differential equations and canonical differential systems. Most of these notes were short and did not contain detailed proofs. In fact, the task of filling in the details is far from trivial and is not complete even to this day. Nevertheless, Krein did try to convey a picture of the ideas that guided him. The strategy that he followed was, to paraphrase his own words [Kr:47], based

> *on the following idea. Just as every Jacobi matrix may be uniquely defined by the solution of a power moment problem, every second order differential operator (of appropriate form) with boundary conditions at one end is uniquely determined by the solution of a generalized moment problem. For differential operators of sufficiently regular type, this generalized moment problem is the extension problem for Hermitian positive functions, that was investigated by the author...*

Krein's point of view is described in more detail in his 1964 lecture at the Jubilee session of the Moscow Mathematical Society. A translation of this lecture is reprinted in [GG:97].

Every mvf $c \in \mathcal{C}^{p\times p}$ can be represented as the Fourier transform of a positive definite $p \times p$ matrix-valued distribution of order at most two. This is equivalent to the fact that the formula

$$c(\lambda) = \lambda^2 \int_0^\infty e^{i\lambda t} g(t)dt \quad \text{for} \quad \lambda \in \mathbb{C}_+ , \tag{27.1}$$

defines a one to one correspondence between the class of mvf's $c \in \mathcal{C}^{p\times p}$ and the class of mvf's $g \in \mathcal{G}_\infty^{p\times p}(0)$ that is defined by the following three conditions:

(1) $g(t)$ is a continuous $p \times p$ mvf on $\mathbb{R}$ with $g(-t) = g(t)^*$ for every point $t \in \mathbb{R}$.
(2) The kernel

$$k(t,s) = g(t-s) - g(t) - g(-s) + g(0)$$

is positive on $[0,\infty) \times [0,\infty)$.
(3) $g(0) \le 0$.

If $c \in \mathcal{C}^{p\times p}$, we shall let $g_c(t)$ denote the unique mvf in $\mathcal{G}_\infty^{p\times p}(0)$ that corresponds to $c(\lambda)$ by formula (27.1); conversely, if $g \in \mathcal{G}_\infty^{p\times p}(0)$, then we shall let $c_g(\lambda)$ denote the unique mvf in $\mathcal{C}^{p\times p}$ that is defined by formula (27.1). Thus, the interplay between the Riesz-Herglotz integral representation (14.4) for mvf's in $\mathcal{C}^{p\times p}$ and formula (27.1) leads to the following complementary pair of integral representation formulas for $g \in \mathcal{G}_\infty^{p\times p}(0)$:

$$c(\lambda) = \quad i\alpha - i\beta\lambda + \frac{1}{\pi i}\int_{-\infty}^{\infty}\left\{\frac{1}{\mu-\lambda} - \frac{\mu}{1+\mu^2}\right\}d\sigma(\mu) \quad = c_g(\lambda) \quad (27.2)$$

$$g_c(t) = \quad -\beta + i\alpha t + \frac{1}{\pi}\int_{-\infty}^{\infty}\left(e^{i\mu t} - 1 - \frac{i\mu t}{1+\mu^2}\right)\frac{d\sigma(\mu)}{\mu^2} \quad = g(t), \quad (27.3)$$

in which the parameters α, β and the the spectral function $\sigma(\mu)$ are the same in both.

Given $c \in \mathcal{C}^{p\times p}$, we shall refer to the restriction of the mvf $g_c(t)$ to the the interval $[-a, a]$ as the **helical trace of $c(\lambda)$ on the interval** $[-a, a]$. One of Krein's fundamental observations was that

> *if $c \in \mathcal{C}_{\rm imp}^d(q)$, then the restriction of the potential $q(s)$ to the interval $[0, t]$, depend only upon the helical trace of $c(\lambda)$ on the interval $[-2t, 2t]$.*

Krein also characterized the set of helical traces $\{g_c(t)|_{[-a,a]} : c \in \mathcal{C}^{p\times p}\}$ on a fixed interval $[-a, a]$:

Theorem 27.1. *Let $\mathcal{G}_a^{p\times p}$ denote the set of continuous $p \times p$ mvf's on $[-a, a]$ such that the kernel $k(t, s)$ defined above is positive on $[0, a] \times [0, a]$ and let*

$$\mathcal{G}_a^{p\times p}(0) = \{g \in \mathcal{G}_a^{p\times p} : g(0) \le 0\}.$$

(1) *$g \in \mathcal{G}_a^{p\times p}(0) \iff$ there exists a mvf $c \in \mathcal{C}^{p\times p}$ such that $g_c(t) = g(t)$ for $|t| \le a$.*

(2) *If $g^\circ \in \mathcal{G}_a^{p\times p}(0)$ and $g^\circ(t)$ coincides with the helical trace on the interval $[-a, a]$ of a mvf $c^\circ \in \mathcal{C}^{p\times p}$, then*

$$\begin{aligned} \{g \quad &\in \quad \mathcal{G}_\infty^{p\times p}(0) : g(t) = g^\circ(t) \quad \textit{for} \quad |t| \le a\} \\ &= \quad \{g_c : c \in \mathcal{C}^{p\times p} \quad \textit{and} \quad e_a^{-1}(c - c^\circ) \in \mathcal{N}_+^{p\times p}\}. \end{aligned} \quad (27.4)$$

Proof. See Theorem 3.12 of [GG:97] for a proof of assertion (1) and [Ar:94] or Theorem 5.1 of [ArD:98] for a proof of (2). □

This establishes a connection between the set of helical traces of a given mvf $c^\circ \in \mathcal{C}^{p\times p}$ on the interval $[-a, a]$ and the solutions of the generalized Carathéodory interpolation problem that was considered in Section 11, i.e., the set in item (2) is equal to

$$\mathcal{C}(e_a I_p, I_p; c^\circ) = \mathcal{C}(e_{a/2} I_p, e_{a/2} I_p; c^\circ).$$

We remark that if $g \in \mathcal{G}_\infty^{p\times p}(0)$, then

$$g(t) = g(t)^* \quad \text{for every} \quad t \in \mathbb{R} \iff c_g(\lambda) = (c_g)^\sim(\lambda) \quad \text{for every} \quad \lambda \in \mathbb{C}_+. \quad (27.5)$$

Moreover, if (27.5) is in force, then $\alpha = 0$ and $\sigma(-\mu + 0) = \sigma(\mu)$ for $\mu > 0$ in the integral representation (27.3). Thus, if $\beta = 0$, then

$$g(t) = \frac{2}{\pi}\int_{0+}^{\infty}\frac{\cos\mu t - 1}{\mu^2}d\sigma(\mu) - \frac{1}{2\pi}\sigma(0+)t^2,$$

which can be reexpressed as

$$g(t) = \frac{2}{\pi}\int_{0-}^{\infty} \frac{\cos\sqrt{\mu}t - 1}{\mu} d\tau(\mu)\,,$$

where

$$\tau(0+) - \tau(0) = \frac{1}{2}\sigma(0+) - \sigma(0) = \frac{1}{2}\sigma(0+) \quad \text{and} \quad \tau(\mu) = \sigma(\sqrt{\mu}) \quad \text{on} \quad [0,\infty)$$

is a nondecreasing $p \times p$ mvf on $[0, \infty)$ that meets the constraint

$$\int_0^\infty \frac{d\text{trace}\tau(\mu)}{1+\mu} < \infty\,.$$

Remark 27.2. The class of nondecreasing scalar functions $\tau(\mu)$ on the interval $[0, \infty)$ that satisfy this last constraint coincides with the class of spectral functions with nonnegative support for a class of strings with arbitrary mass distribution on $[0, \infty)$. In this setting, Krein called the Hermitian function $g(t)$ the transition function of the string. Additional information on direct and inverse spectral problems for the string may be found in [KaKr:68] and [DMc:76] and the references cited therein, particularly a number of Doklady notes by Krein.

Theorem 27.3. *Let $g^\circ \in \mathcal{G}_{2d}^{p\times p}(0)$ be given, let the helical trace of the mvf $c^\circ \in \mathcal{C}^{p\times p}$ coincide with $g^\circ(t)$ on the interval $(-2d, 2d)$ and suppose that*

$$\mathcal{C}(e_tI_p, e_tI_p; c^\circ) \cap \mathring{\mathcal{C}}^{p\times p} \neq \emptyset \quad \textit{for every} \quad t \in [0,d)\,. \tag{27.6}$$

Then there exists a unique normalized left monotonic continuous chain of entire J_p-inner mvf's $\{A_t(\lambda)\}$, $0 \le t < d$, such that:

(1) $\{e_tI_p, e_tI_p\} \in ap_{II}(A_t)$ *for every* $t \in [0,d)$.
(2) $\mathcal{C}(e_tI_p, e_tI_p; c^\circ)\} = \mathcal{C}(A_t)$ *for every* $t \in [0,d)$.
(3) $A_t \in \mathcal{U}_{sR}(J_p)$ *for every* $t \in [0,d)$.

Moreover, the mvf

$$M(t) = -i\frac{\partial A_t}{\partial \lambda}(0)J_p\,, \quad 0 \le t < d,$$

is the unique solution of the inverse input impedance problem with data

$$\{c^\circ; e_tI_p, e_tI_p, 0 \le t < d\}\,.$$

This solution depends only upon $g^\circ(t)$ and not upon the choice of the mvf $c^\circ(\lambda)$.

Proof. The last theorem may be viewed as a corollary of Theorem 12.1 applied to the chain $\{b_3^t, b_4^t\} = \{e_tI_p, e_tI_p\}$, $0 \le t < d$. □

Necessary and sufficient conditions on $g^\circ \in \mathcal{G}_{2d}^{p\times p}(0)$ to insure that condition (27.6) is in force are given in [ArD:98].

If the mvf $g^\circ \in \mathcal{G}_{2d}^{p\times p}(0)$ that is given in the preceding theorem is Hermitian, i.e., if $g(t) = g(t)^*$ for every $t \in [0,d)$, then the mvf $c^\circ \in \mathcal{C}^{p\times p}$ may be chosen to meet the extra symmetry condition $c^\circ(\lambda) = (c^\circ)^\sim(\lambda)$ and then, if the condition (27.6) is met, the mvf's $\{A_t(\lambda)\}$ meet the extra condition

$$(A_t)^\sim(\lambda)\mathcal{J}_pA_t(\lambda) = \mathcal{J}_p \quad \text{for every} \quad t \in [0,d)\,.$$

This property is exploited in the spectral theory of strings and the Schrödinger equation when $p = 1$; see, e.g., [KrL:85].

There are two particular subclasses of $\mathcal{G}_a^{p\times p}(0)$ that are of particular interest and have useful descriptions:

Theorem 27.4. *Let $g^\circ \in \mathcal{G}_a^{p\times p}(0)$ for some $a \in [0, \infty)$. Then:*

(1) $g \in C^2([-a, a])$ *and* $g'(0) = 0 \iff g(t) = \int_0^t (t-s)f(s)ds$ *for some* $p \times p$ *mvf $f(s)$ that is continuous on the interval $[-a, a]$ and meets the positivity condition*

$$\int_0^a \varphi(t)^* \{ \int_0^a f(t-s)\varphi(s)ds \} dt \geq 0 \quad \textit{for every} \quad \varphi \in L_2^p([0,a]]) .$$

(2) $g' \in AC([-a, 0)) \cap AC((0, a])$ *and* $\Re g'(0^+) \geq 0 \iff$

$$g(t) = \begin{cases} -\gamma t + \int_0^t (t-s)h(s)ds & \textit{for} \quad t \in [0, a] \\ -g(-t)^* & \textit{for} \quad t \in [-a, 0] , \end{cases}$$

where $\gamma \in \mathbb{C}^{p\times p}$ *and* $h \in L_1^p([0, a])$ *meets the positivity condition*

$$\int_0^a \varphi(t)^* \{\gamma\varphi(s) + \int_0^a h(t-s)\varphi(s)ds\} dt \geq 0 \quad \textit{for every} \quad \varphi \in L_2^p([0,a]]) .$$

The mvf $h(t)$ considered in case (2) is also referred to as the accelerant of $g(t)$ on the interval $[0, a]$. Additional details on extension problems that are formulated in $g^\circ \in \mathcal{G}_a^{p\times p}(0)$ and the indicated subclasses may be found in [ArD:98]. Connections of extension problems in the Wiener class with inverse problems are discussed in [MeA:67], [DI:84], [KrL:85], [KrMA:86] and [Dy:90].

References

[AlD:84] D. Alpay and H. Dym, Hilbert spaces of analytic functions, inverse scattering and operator models, I, Integral Equations Operator Theory 7 (1984) 589–641.

[AlD:85] D. Alpay and H. Dym, Hilbert spaces of analytic functions, inverse scattering and operator models, II, Integral Equations Operator Theory 8 (1985) 145–180.

[AG:95] D. Alpay and I. Gohberg, Inverse spectral problems for differential operators with rational scattering matrix functions, J. Differential Equations 118 (1995) 1–19.

[AG:01] D. Alpay and I. Gohberg, Inverse problems associated to a canonical differential system, in: Recent Advances in Operator Theory and Related Topics (L. Kérchy, C. Foias, I. Gohberg and H. Langer, eds.), Oper. Theor. Adv. Appl. 127, Birkhäuser , Basel, 2001, pp. 1–27.

[Ar:94] D.Z. Arov, The generalized bitangent Carathéodory-Nevanlinna-Pick problem and (j, J_0)-inner matrix-valued functions, *Russian Acad. Sci. Izvestija* **42** (1994), 1–26.

[Ar:97] D.Z. Arov, On monotone families of J-contractive matrix functions, Algebra i Analiz 9 (1997), No. 6, 3–37; English transl. St. Petersburg Math. J. 9 (1998), No. 6, 1025–1051.

[ArD:97] D.Z. Arov and H. Dym, J-inner matrix functions, interpolation and inverse problems for canonical systems, I: Foundations, Integral Equations Operator Theory 29 (1997), No. 4, 373–454.

[Aron:50] N. Aronszajn, Theory of reproducing kernels, Trans. Amer. Math. Soc., **68**(1950), 337–404.

[ArD:98] D.Z. Arov and H. Dym, On three Krein extension problems and some generalizations, Integral Equations Operator Theory 31 (1998) 1–91.

[ArD:00a] D.Z. Arov and H. Dym, J-inner matrix functions, interpolation and inverse problems for canonical systems, II: The inverse monodromy problem, Integral Equations Operator Theory 36 (2000), No. 1, 11–70.

[ArD:00b] D.Z. Arov and H. Dym, J-inner matrix functions, interpolation and inverse problems for canonical systems, III: More on the inverse monodromy problem, Integral Equations Operator Theory 36 (2000), No. 2, 127–181.

[ArD:01] [ArD9] D.Z. Arov and H. Dym, Matricial Nehari problems, J-inner matrix functions and the Muckenhoupt condition, J. Funct. Anal. 181 (2001) 227–299.

[ArD:02a] D.Z. Arov and H. Dym, J-inner matrix functions, interpolation and inverse problems for canonical systems, IV: Direct and inverse bitangential input scattering problems, Integral Equations Operator Theory 43 (2002), No. 1, 1–67.

[ArD:02b] D.Z. Arov and H. Dym, J-inner matrix functions, interpolation and inverse problems for canonical systems, V: The inverse input scattering problem for Wiener class and rational $p \times q$ input scattering matrices, Integral Equations Operator Theory 43 (2002), No. 1, 68–129.

[ArD:03a] D.Z. Arov and H. Dym, Criteria for the strong regularity of J-inner functions and γ-generating matrices, J. Math. Anal. Appl. 280 (2003) 387–399.

[ArD:03b] D.Z. Arov and H. Dym, The bitangential inverse input impedance problem for canonical systems, I.: Weyl-Titchmarsh classification, existence and uniqueness, Integral Equations Operator Theory 47 (2003) 3–49.

[ArD:04a] D.Z. Arov and H. Dym, Strongly regular J-inner matrix functions and related problems, in: Current Trends in Operator Theory and its Applications (J.A. Ball, J.W. Helton, M. Klaus and L. Rodman, eds.), Oper. Theor. Adv. Appl., **149**, Birkhäuser, Basel, 2004, pp. 79–106.

[ArD:04b] D.Z. Arov and H. Dym, The bitangential inverse spectral problem for canonical systems, J. Funct. Anal., 214 (2004), 312–385.

[ArD:05] D.Z. Arov and H. Dym, The bitangential inverse input impedance problem for canonical systems, II.: Formulas and examples, Integral Equations and Operator Theory 51(2), 155–213 (2005).

[ArD:??] D.Z. Arov and H. Dym, Direct and inverse problems for differential systems connected with Dirac systems and related factorization problems, in preparation.

[dBr:63] L. de Branges, Some Hilbert spaces of analytic functions I, Trans. Amer. Math. Soc. 106 (1963) 445–668.

[dBr:68a] L. de Branges, Hilbert Spaces of Entire Functions, Prentice-Hall, Englewood Cliffs, 1968.

[dBr:68b] L. de Branges, The expansion theorem for Hilbert spaces of entire functions, in: Entire Functions and Related Parts of Analysis, Amer. Math. Soc., Providence, 1968, pp. 79–148.

[Bro:71] M.S. Brodskii, Triangular and Jordan Representations of Linear Operators, Transl. of Math. Monographs, 32, Amer. Math. Soc., Providence, 1971.

[BrL:60] M.S. Brodskii and M.S. Livsic, Spectral analysis of non-selfadjoint operators and intermmediate systems, Amer. Math. Soc. Transl. (2) 13 (1960) 265–346.

[CG:02] S. Clark and F. Gesztesy, Weyl-Titchmarsh M-function asymptotics, local uniqueness results, trace formulas, and Borg-type theorems for Dirac operators, Trans. Amer. Math. Soc. 354 (2002), 3475–3534.

[CG:01] S. Clark and F. Gesztesy, Weyl-Titchmarsh M-function asymptotics for matrix-valued Schrödinger operators, Proc. London Math. Soc. **82** (2001), 701–724.

[Dy:70] H. Dym, An introduction to de Branges spaces of entire functions with applications to differential equations of Sturm-Liouville type, Advances in Math., **5** (1970), 395–471.

[Dy:89] H. Dym, J Contractive Matrix Functions, Reproducing Kernel Hilbert Spaces and Interpolation, CBMS Regional Conference Series, number 71, Amer. Math. Soc., Providence, R.I., 1989.

[Dy:90] H. Dym, On reproducing kernels and the covariance extension problem, in: Analysis and Partial Differential Equations (C. Sadosky, ed.), Marcel Dekker, New York, 1990, pp. 427–482.

[DI:84] H. Dym and A. Iacob, Positive definite extensions, canonical equations and inverse problems, in: Topics in Operator Theory, Systems and Networks (H. Dym and I. Gohberg, eds.), Oper. Theory Adv. Appl. 12, Birkhäuser, Basel, 1984, pp. 141–240.

[DK:78] H. Dym and N. Kravitsky, On recovering the mass distribution function of a string from its spectral function, in: Topics in Functional Analysis (I. Gohberg and M. Kac, eds.), Academic Press, New York, 1978, pp. 45–90.

[DMc:76] H. Dym and H.P. McKean, Gaussian Processes, Function Theory, and the Inverse Spectral Problem, Academic Press, New York, 1976.

[GKM:02] F. Gesztesy, A. Kiselev and K.A. Makarov, Uniqueness results for matrix-valued Schrödinger, Jacobi and Dirac-type operators, Math. Nachr. 239/240 (2002) 103–145.

[GeSi:00] F. Gesztesy and B. Simon, A new approach to inverse spectral theory, II. General real potentials and the connection to the spectral measure, Ann. Math., **152** (2000), 593–643.

[GKS:98] I. Gohberg, M.A. Kaashoek and A.L. Sakhnovich, Canonical systems with rational spectral densities: Explicit formulas and applications, Math. Nachr. 149 (1998) 93–125.

[GKS:02] I. Gohberg, M.A. Kaashoek and A.L. Sakhnovich, Scattering problems with a pseudo-exponential potential, Asympt. Anal. 29(2002), no. 1, 1–38.

[GKr:70] I. Gohberg and M.G. Krein, Theory and applications of Volterra operators in Hilbert space, Trans. Math. Monographs, 24, Amer. Math. Soc., Providence, R.I., 1970.

[GG:97] M.L. Gorbachuk and V.I. Gorbachuk, M.G. Krein's Lectures on Entire Operators, Operator Theory: Advances and Applications, **97**, Birkhäuser, Basel, 1997.

[Ia:86] A. Iacob, On the Spectral Theory of a Class of Canonical Systems of Differential Equations, PhD Thesis, The Weizmann Institute of Science, Rehovot, Israel, 1986.

[Ka:03] I.S. Kats, "Linear relations generated by the canonical differential equation of phase dimension 2, and eigenfunction expansions, St. Petersburg Math. J. 14 (2003), no.–3, 429–452.

[KaKr:68] I.S. Kac and M.G. Krein, On the spectral functions of the string, Transl. (2) Amer. Math. Soc., **103**(1974), 19–102.

[Kr:44] M.G. Krein, On the logarithm of an infinitely decomposable Hermite-positive function, Dokl. Akad. Nauk SSSR **45** (1944), no. 3, 91–94.

[Kr:47] M.G. Krein, A contribution to the theory of entire functions of exponential type, Izv. Akad. Nauk SSSR 11 (1947) 309–326.

[Kr:51] M.G. Krein, On the theory of entire matrix functions of exponential type, Ukrain. Mat. Zh. 3 (1951), no. 2, 154–173.

[Kr:55] M.G. Krein, Continuous analogs of theorems on polynomials orthogonal on the unit circle, Dokl. Akad. Nauk SSSR 105 (1955) 433–436.

[Kr:56] M.G. Krein, On the theory of accelerants and S–matrices of canonical differential systems, Dokl. Akad. Nauk 111 (1956), no. 6, 1167–1170.

[KrL:85] M.G. Krein and H. Langer, On some continuation problems which are closely related to the theory of operators in spaces Π_κ. IV: Continuous analogues of orthogonal polynomials on the unit circle with respect to an indefinite weight and related continuation problems for some classes of functions, J. Oper. Theory, **13** (1985), 299–417.

[KrMA:86] M.G. Krein and F.E. Melik-Adamyan, The matrix continual analogues of Schur and Carathéodory-Toeplitz problems, Izv. Akad. Nauk Armyan SSR, Ser. Mat. **21** (1986), no. 2, 107–141.

[LeMa:00] M. Lesch and M.M. Malamud, The inverse spectral problem for first order systems on the half line, in: Differential operators and related topics, Vol. I (Odessa, 1997), Oper. Theory Adv. Appl. 117 (2000), Birkhäuser, Basel, pp. 199–238.

[LeSa:75] B.M. Levitan and I.S. Sargsjan, Introduction to Spectral Theory, Transl. Math. Mon. 39, Amer. Math. Soc., Providence, 1975.

[Li:73] M.S. Livsic, Operators, Oscillations, Waves, Open Systems, Trans. Math. Monographs 34 Amer. Math. Soc., Providence, R.I., 1973.

[Ma:99] M.M. Malamud, Uniqueness questions in inverse problems for systems of ordinary differential equations on a finite interval, Trans. Moscow Math. Soc. 60 (1999) 173–124.

[MeA:67] F.E. Melik-Adamyan, On the theory of matrix accelerants and spectral matrix functions of canonical differential systems, Dokl. Akad. Nauk Armyan SSR, **45** (1967), 145–151.

[MeA:77] F.E. Melik-Adamyan, On canonical differential operators in Hilbert space, Izv. Akad. Nauk Armyan SSR, Ser. Mat. 12 (1977) 10–31.

[MeA:99a] F.E. Melik-Adamyan, Description of spectral functions for a class of differential operators, J. Contemp. Math. Anal. 34 (1999), no. 2, 54–70 (2000).

[MeA3:99b] F.E. Melik-Adamyan, Description of spectral functions for a class of differential operators with decaying boundary conditions, J. Contemp. Math. Anal. 34 (1999), no. 3, 64–74 (2000).

[MeA:00] F.E. Melik-Adamyan, Spectral functions of canonical differential equations, J. Contemp. Math. Anal. 35 (2000), no. 2, 42–60 (2001).

[Or:76] S.A. Orlov, Nested matrix discs that depend analytically on a parameter and theorems on the invariance of the ranks of the radii of the limit matrix discs, Izv. Akad. Nauk. SSSR Ser. Mat. **40** (1976), No. 3, 593–644, 710.

[P:55] V.P. Potapov, The multiplicative structure of J-contractive matrix functions, Trudy Mosk. Mat. Obshch. 4 (1955) 125–236, English: Amer. Math. Soc. Transl. (2) 15 (1960) 131–243.

[RaSi:00] A. Ramm and B. Simon, A new approach to inverse spectral theory, III. Short range potentials, J. d'Analyse Math., **80** (2000), 319–334.

[Rem:02] C. Remling, Schrödinger operators and de Branges spaces, J. Funct. Anal., **196** (2002), 323–394.

[Rem:03] C. Remling, Inverse spectral theory for one dimensional Schrödinger operators: The A function, Math. Z., **245** (2003), 597–617.

[RR:77] M. Rosenblum and J. Rovnyak, Hardy Classes and Operator Theory, Dover Reprint, New York, 1977.

[Sak-A:92] A.L. Sakhnovich, Spectral functions of a canonical system of order $2n$, Math. USSR Sbornik, 71 (1992), No. 2, 355–369.

[Sak:96] L.A. Sakhnovich, Spectral problems on half-axis. Methods Funct. Anal. Topology 2 (1996), no. 3-4, 128–140.

[Sak:99] L.A. Sakhnovich, Spectral Theory of Canonical Differential Systems. Method of Operator Identities, Birkhäuser, Basel, 1999.

[Sak:00a] L.A. Sakhnovich, Works by M.G. Krein on inverse problems, Differential operators and related topics, Vol. I (Odessa, 1997), 59–69, Oper. Theory Adv. Appl., 117, Birkhäuser, Basel, 2000.

[Sak:00b] L.A. Sakhnovich, Spectral theory of a class of canonical differential systems, (Russian) Funktsional. Anal. i Prilozhen. 34 (2000), no. 2, 50–62, 96; translation in Funct. Anal. Appl. 34 (2000), no. 2, 119–128.

[Sim:99] B. Simon, A new approach to inverse spectral theory, I. Fundamental formalism, Ann. Math., **150** (1999), 1029–1057.

[Sm:90] Ju.L. Smul'yan, Operator Balls, translation in: Integral Equations Operator Theory, **13** (1990), No. 6, 864–882.

[W:00] H. Winkler, Small perturbations of canonical systems, Integral Equations Operator Theory, 38 (2000) 222–250.

Damir Z. Arov
Department of Mathematics
South-Ukranian Pedagogical University
65020 Odessa, Ukraine
e-mail: `aspect@farlep.net`

Harry Dym
Department of Mathematics
The Weizmann Institute of Science
Rehovot 76100, Israel
e-mail: `dym@wisdom.weizmann.ac.il`

Operator Theory:
Advances and Applications, Vol. 160, 161–178

Regularization Processes for Real Functions and Ill-posed Toeplitz Problems

Claudio Estatico

This work is dedicated to Prof. Israel Gohberg, on the occasion of his 75th birthday.

Abstract. Most preconditioners for Toeplitz systems $A_n(f)$ arising in the discretization of ill-posed problems give rise to instability and noise amplification. Indeed, since these preconditioners are constructed from linear approximation processes of the generating function f, they inherit the ill-posedness of the problem.

Here we first identify a novel set of approximation processes which regularizes the inversion of real functions. Then, such processes are used as a basic tool for the computation of preconditioners endowed with regularizing properties. We show that these preconditioners provide fast convergence and noise control of iterative methods for discrete ill-posed Toeplitz systems.

Mathematics Subject Classification (2000). 47A52,65F22,65F10,15A29.

Keywords. preconditioning, regularization, linear approximation operators, matrix algebras, Toeplitz matrices.

1. Introduction

Preconditioning techniques for Toeplitz systems are widely used in order to speed up the convergence of iterative methods [10]. In this paper we consider $n \times n$ Hermitian Toeplitz matrices $A_n(f)$ generated by a 2π-periodic Lebesgue integrable real function f, that is, the entries along the kth diagonal of $A_n(f)$ are equal to the kth Fourier coefficient of f [21, 20]. Since the spectral distribution of $A_n(f)$ is asymptotically equivalent to the distribution of the generating function f, most preconditioners are constructed by means of approximations of f [28]. Main examples are the linear approximation processes such as the Fourier partial sum $F_n(f) = \sum_{j=0}^{n} a_j e^{\mathbf{i}jx}$ and the Césaro sum $C_n(f) = \frac{1}{n+1}\sum_{j=0}^{n}\sum_{k=-j}^{j} a_k e^{\mathbf{i}kx}$,

This work was partially supported by MIUR, grant numbers 2002014121 and 2004015437.

where $a_k = \frac{1}{2\pi}\int_0^{2\pi} f(x)e^{-\mathbf{i}kx}dx$ are the values on the diagonals of $A_n(f)$. These linear approximation processes lead to the G. Strang natural and the T. Chan optimal preconditioners respectively [29, 12, 32, 11, 28]. Generally, if the generating function f is sufficiently smooth, the linear approximation processes converge to f uniformly. In that case, the corresponding preconditioner is a close approximation of the system matrix $A_n(f)$. If $A_n(f)$ is derived from the discretization of an ill-posed problem, these preconditioners can yield numerical instability and amplification of the errors due to noise on input data [23, 25]. The rank of these preconditioners is asymptotically ill determined as well as the rank of $A_n(f)$, that is, these preconditioners inherit the ill-posedness of the problem.

In this paper, we first characterize a class of approximation processes which allow preconditioners for ill-posed Toeplitz systems to be constructed effectively. In particular, linear approximation processes are filtered by continuous regularization algorithms [15]. These procedures lead to approximation operators which are called regularization processes. Basically, if a continuous real function f has a root, a regularization process gives rise to a family of bounded functions which approximate the unbounded function $1/f$.

The properties of preconditioners constructed from regularization processes are then analyzed. We show that these preconditioners have bounded inverses and the spectrum of the preconditioned matrix is clustered at unity.

With reference to the effectiveness for ill-posed linear systems, we prove that the proposed preconditioners belong to the class of regularizing preconditioners [18]. If a linear system comes from the discretization of an ill-posed problem, regularization preconditioners can improve the convergence of appropriate iterative methods without amplifying the components related to the noise in the data. This is a very favorable property, which is absent in other preconditioning strategies. For instance, preconditioners constructed from linear approximation processes behave differently and often do not yield good results, since they give rise to fast reconstruction on components corrupted by noise.

The paper is organized as follows. In Section 2 we introduce notations and basic results about approximation techniques for Toeplitz preconditioning in trigonometric matrix algebras. In Section 3 we define the class of regularization processes for bounded approximations of the unbounded inverse of 2π-periodic real functions. In Section 4 we study properties of matrices constructed from regularization processes of the previous section. We prove that the inverse of any $n \times n$ matrix associated with a regularization process of a real function f converges, with respect to n, to the Toeplitz matrix $A_n(f)$ and that the spectrum of the preconditioned matrix has a cluster at unity. In Section 5 we study such matrices in the context of Toeplitz preconditioning. We show that these matrices belong to the class of regularization preconditioners [18], and therefore are suitable for preconditioning of linear systems arising in the discretization of ill-posed problems. Section 6 collects some final remarks and future goals.

2. Toeplitz preconditioning and linear approximation processes

An $n \times n$ matrix $A_n = (a_{i,j})_{i,j=1}^n \in \mathbb{C}^{n\times n}$ is said to be a *Toeplitz matrix* if $a_{i,j} = a_{r,s}$ for $i - j = r - s$, that is, A_n is constant along any diagonal. Toeplitz matrices arise in a wide range of applications, such as the resolution of Fredholm integral operators with space-invariant integral kernels [5, 20, 10].

The family of Hermitian Toeplitz matrices $\{A_n = A_n(f)\}_{n=1}^{+\infty}$ is said to be generated by a Lebesgue-integrable scalar function $f : I \longrightarrow \mathbb{R},\ I = [-\pi, \pi]$, if the entries along the kth diagonal are equal to the kth Fourier Transform coefficient a_k of f, that is,

$$[A_n(f)]_{r,s} = a_{r-s}\,, \qquad a_k = \frac{1}{2\pi}\int_I f(x)e^{-\mathbf{i}kx}dx \qquad (\mathbf{i}^2 = -1,\ k \in \mathbf{Z})\,.$$

The Szegő-Tyrtyshnikov results state that the distribution of eigenvalues of $A_n(f)$ asymptotically converge to $f \in L^1(I)$ [21, 30, 31].

Since there is a connection between the Toeplitz matrix $A_n(f)$ and the trigonometric Fourier Transform of the generating function f, preconditioners for $A_n(f)$ usually belong to trigonometric matrix algebras. If U_n denotes an $n \times n$ complex unitary matrix of eigenvectors, then the matrix algebra $M_n = M(U_n)$ is a matrix space defined as follows

$$M_n = M(U_n) = \{X = U_n \Delta_n U_n^* \in \mathbb{C}^{n\times n}\}\,, \tag{2.1}$$

where $\Delta_n = \mathrm{diag}(d_0, d_1, \ldots, d_{n-1})$ is the complex diagonal matrix of eigenvalues of X. If the columns of U_n are trigonometric vectors, then the matrix algebra is said to be trigonometric. In particular, let $\{v_n\}_{n\in\mathbb{N}}$ denote a sequence of trigonometric functions on I and let $\{W_n\}_{n\in\mathbb{N}}$ denote a sequence of grids of n points on I, that is, $W_n = \{x_s^{(n)}\}_{s=0}^{n-1} \subset I$. If the $n \times n$ Vandermonde matrix $\tilde{V}_n = \left(v_r(x_s^{(n)})\right)_{r,s=0}^{n-1}$, is unitary, the corresponding $n \times n$ matrix space $M_n = M(\tilde{V}_n^*)$ is a trigonometric matrix algebra [14].

Thus, for any continuous function g defined in $[0, 2\pi]$, $M_n(g)$ denotes the matrix

$$M_n(g) = U_n G_n U_n^* \in M(U_n) \tag{2.2}$$

such that the diagonal matrix G_n is $(G_n)_{s,s} = g(x_s^{(n)})$, for $s = 0, \ldots, n-1$.

Widely used trigonometric matrix algebras M_n for Toeplitz preconditioners are the circulant, $\{\omega\}$-circulant, Tau and Hartley matrix spaces [13, 4, 7, 8]. These algebras are related to the Fast Fourier Transform, the Fast Sine Transform and Fast Hartley Transform; all of them allow fast, i.e., $O(n \log n)$, matrix-vector multiplication and diagonalization.

The convergence speed of iterative system solvers depends on the distribution of singular values of its system matrix: basically the speed is high when the spectrum is "close" to the unity [1].

In order to speed up the convergence, the linear system $A_n x = b$ is replaced by the algebraic equivalent one $P_n^{-1} A_n x = P_n^{-1} b$, where P_n is an $n \times n$ preconditioner. Since the rate of convergence can be improved if the spectrum of the preconditioned matrix $P_n^{-1} A_n$ is clustered at unity, often we have that $P_n^{-1} A_n \approx I$, that is, $P_n \approx A_n$.

The approximation of a Toeplitz matrix $A_n(f)$ by a preconditioner derives from the approximation of the generating function f. The linear approximation processes in trigonometric functional spaces [32] are widely used approximation schemes for Toeplitz preconditioning [28]. Let $\{V_n\}_{n \in \mathbb{N}}$ be the sequence of spaces of all the trigonometric polynomials of degree n. Note that $V_n \subset V_{n+1}$ and $\cup_{n \in \mathbb{N}} V_n$ is dense in the space $(C_{2\pi}, \| \bullet \|_\infty)$ of all the globally continuous and 2π-periodic functions endowed with the supremum norm. A linear approximation process for a (generating) function $f \in C_{2\pi}$ is a sequence of linear approximation operators $\{S_n\}_{n \in \mathbb{N}}$, with $S_n : C_{2\pi} \longrightarrow V_n$, whose images uniformly converge to f, that is,

$$\lim_{n \longrightarrow +\infty} \|S_n(f) - f\|_\infty = 0 . \tag{2.3}$$

On the grounds of (2.3) and (2.2), if $f \in C_{2\pi}$, a trigonometric preconditioner $P_n \in M_n = M(U_n)$ of a Toeplitz matrix $A_n(f)$ can be defined as follows

$$P_n = M_n(S_n(f)) = U_n D_n U_n^* , \tag{2.4}$$

where D_n is the $n \times n$ diagonal matrix such that $(D_n)_{s,s} = [S_n(f)](x_s^{(n)})$ for $s = 0, \ldots, n-1$. Notice that the eigenvalues of $A_n(f)$ are distributed as the generating function f, while the eigenvalues of its preconditioner P_n are distributed as $S_n(f)$. This yields that the preconditioner $M_n(S_n(f))$ is an accurate approximation in the algebra M_n of the Toeplitz matrix $A_n(f)$. Many preconditioners can be represented as (2.4), such as, for instance, the Strang natural and the T.Chan optimal ones [29, 12]. We recall that if A_n is Hermitian, the T.Chan optimal preconditioner $P_{opt}(A_n)$ solves the minimization problem $P_{opt}(A_n) = \arg\min_{X \in M_n} \|A_n - X\|_F$, where $\|\cdot\|_F$ is the Frobenius norm. According to notation (2.2), the optimal preconditioner can be written as $P_{opt}(A_n) = M_n(C_n(f))$, where $C_n(f) = \frac{1}{n+1} \sum_{j=0}^{n} \sum_{k=-j}^{j} a_k e^{ikx}$ is the linear approximation process of the Césaro sums [11].

As already mentioned, if a continuous generating function f has a root, as in ill-posed problems, the rank of the matrices $A_n(f)$ is asymptotically ill-determined, since the smallest non-null eigenvalues tend to zero. Any preconditioner (2.4) "inherits" the spectral distribution of the system matrix and give numerical instability. For instance, it is known that the T. Chan optimal approximation leads to bad numerical results due to amplification of the noise of the data [23]. In these cases, all preconditioners are usually modified by means of spectral filtering procedures [23, 9, 27, 24, 25, 3, 2, 6, 17]. These procedures give rise to preconditioners which approximate the system matrix only in the space less sensitive to noise. In that way, they are suitable for ill-posed linear systems, as explained in Section 5.

3. Regularization processes for real functions

The regularization theory gives mathematical tools for obtaining low noise-sensitive solutions of ill-posed problems [15, 5].

Let X, Y denote two (infinite dimensional) Hilbert spaces. Given a datum $y \in Y$ and a bounded linear operator $T : X \longrightarrow Y$, we consider the linear equation $Tx = y$, where $x \in X$ is the output solution. We look for the minimum norm solution $x^\dagger$ of the Gaussian normal equation $T^*Tx = T^*y$, that is, we solve the equation in the generalized sense.

If the generalized solution $x^\dagger$ exists, it is given by $x^\dagger = \int_0^{\|T\|^2} 1/\lambda \, d\,E_\lambda T^* y$, where $\{E_\lambda\}$ is the spectral family of the self-adjoint operator T^*T [22]. According to Hadamard, a problem is said to be ill-posed if its solution may not exist, may not be unique, or it does not depend continuously on the data. The regularization theory for solving ill-posed problems states that the problem of computing the generalized solution $x^\dagger$ is ill-posed if and only if the range $R(T)$ is non-closed [15]. Indeed, if $y \notin R(T) \oplus R(T)^\perp \subset Y$ then the integrand $1/\lambda$ has a non-integrable pole in zero with respect to the "data-depending" measure $d\,E_\lambda T^* y$. On these bases, the computation of $x^\dagger$ needs procedures which filter the pole in zero of the integrand function $1/\lambda$, whenever $R(T)$ is a non-closed set. These procedures are called regularization algorithms. Basically, if $T^\dagger$ denotes the (Moore-Penrose) generalized inverse $T^\dagger : R(T) \oplus R(T)^\perp \longrightarrow X$ such that $T^\dagger y = x^\dagger$ for $y \in R(T) \oplus R(T)^\perp$, a regularization algorithm is a family of continuous, i.e., stable, operators which approximate the unbounded, i.e., unstable, operator $T^\dagger$.

Simple regularization algorithms come from the approximation of the integrand $1/\lambda$ by a family of neighboring functions, which are piecewise continuous over the closure $[0, \|T\|^2]$ of the spectrum of T^*T (see [15], except from the monotonic characterization added here).

Definition 3.1. Let $\bar{I}$ denotes the closure of $I = [0, \beta)$, with $\beta \in \mathbb{R}^+ \cup \{+\infty\}$ and let $\alpha_0 > 0$. A family $\{\tilde{R}_\alpha\}_{\alpha \in (0,\alpha_0)}$ of real functions $\tilde{R}_\alpha : \bar{I} \subseteq \mathbb{R} \longrightarrow \mathbb{R}$, is called regularization inverse on I if the following three conditions hold:

(i) $\forall \alpha \in (0, \alpha_0)$, $\tilde{R}_\alpha$ is piecewise continuous and globally continuous from the right;

(ii) $\forall \alpha \in (0, \alpha_0)$, the function $x\tilde{R}_\alpha(x)$ is uniformly bounded, i.e., there exists a constant $C > 0$ such that $|x\tilde{R}_\alpha(x)| \leq C, \quad \forall x \in \bar{I}$,

(iii) $\tilde{R}_\alpha(x)$ approximates $1/x$ as $\alpha \longrightarrow 0^+$, that is,

$$\lim_{\alpha \longrightarrow 0^+} \tilde{R}_\alpha(x) = \frac{1}{x} \qquad \forall x \in \bar{I} \setminus \{0\}.$$

Moreover, the regularization inverse is called monotone if, for $x \in (0, \eta)$ with $\eta > 0$, the following inequality holds

$$\tilde{R}_{\alpha_1}(x) \geq \tilde{R}_{\alpha_2}(x) \qquad \forall\, 0 < \alpha_1 < \alpha_2 < \alpha_0 . \tag{3.1}$$

With the help of the latter definition, we have the continuous regularization algorithms.

Lemma 3.2. [15] *Let* $\{\tilde{R}_\alpha\}_{\alpha>0}$ *be a regularization inverse on* $[0, \|T\|^2]$. *The family of operators* $\{R_\alpha\}_{\alpha>0}$, $R_\alpha : Y \longrightarrow X$, *such that*

$$R_\alpha y = \tilde{R}_\alpha(T^*T)T^*y := \int_0^{\|T\|^2} \tilde{R}_\alpha(\lambda)\, d\, E_\lambda T^* y\,, \tag{3.2}$$

is a regularization linear algorithm for $T^\dagger$, *called continuous regularization algorithm.*

For instance, the family of operators $\{R_\alpha\}_{\alpha>0}$ of Tikhonov regularization [5, 15] defined as $R_\alpha = (T^*T + \alpha I)^{-1}T^*$, can be represented according to (3.2), with $\tilde{R}_\alpha(\lambda) = (\lambda+\alpha)^{-1}$.

We remark that inequality (3.1) guarantees that the larger the value of the parameter α is, the stronger the filtering capabilities are. If the result of the regularization is unstable, we can adopt a larger regularization parameter in order to improve the noise filtering.

Since it is very simple to design regularization inverses, formula (3.2) is a constructive method for having regularization linear algorithms. Indeed, regularization inverses can be constructed by means of filters for spectral control. Below, we collect and propose some useful filters, including the Tikhonov one.

(I) *Tikhonov Filter* [5, 15]

$$\tilde{R}_\alpha(x) = \frac{1}{x+\alpha} \qquad x \geq 0$$

(II) *Low Pass Filter*

$$\tilde{R}_\alpha(x) = \begin{cases} 0 & 0 \leq x < \alpha \\ x^{-1} & x \geq \alpha \end{cases}$$

(III) *M. Hanke, J.G. Nagy and R.J. Plemmons' Filter* [23]

$$\tilde{R}_\alpha(x) = \begin{cases} 1 & 0 \leq x < \alpha \\ x^{-1} & x \geq \alpha \end{cases}$$

(IV) *p-Polynomial Low Pass Filter* $(p \geq 0)$

$$\tilde{R}_\alpha(x) = \begin{cases} \alpha^{-(p+1)}x^p & 0 \leq x < \alpha \\ x^{-1} & x \geq \alpha \end{cases}$$

(V) $(1/\alpha)$*-Polynomial Low Pass Filter*

$$\tilde{R}_\alpha(x) = \begin{cases} \alpha^{-\frac{\alpha+1}{\alpha}} x^{\frac{1}{\alpha}} & 0 \leq x < \alpha \\ x^{-1} & x \geq \alpha \end{cases}$$

(VI) *Exponential Low Pass Filter*

$$\tilde{R}_\alpha(x) = \begin{cases} 0 & x = 0 \\ \alpha^{-1} e^{\frac{x-\alpha}{\alpha x}} & 0 < x < \alpha \\ x^{-1} & x \geq \alpha \end{cases}$$

(VII) *Showalter's Filter for asymptotic regularization* [15]

$$\tilde{R}_\alpha(x) = \frac{1}{x}\left(1 - e^{-\frac{x}{\alpha}}\right) = \int_0^{1/\alpha} e^{-xs} ds \,, \qquad x \geq 0 \,.$$

It is simple to verify that all these regularizing inverses, except (V), are monotone. The list can be extended by considering their linear combinations, with suitable normalizing factors.

On the basis of regularization inverses and linear approximation processes (2.3), now we characterize a class of approximation schemes which simultaneously approximate and regularize periodic real functions.

Definition 3.3. Let $V_{2\pi}$ be the space of the 2π-periodic real functions, and let $C_{2\pi} \subset V_{2\pi}$ denote the subset of the continuous ones.

A family $\{R_{n;\alpha}\}_{n\in\mathbb{N};\alpha\in(0,\alpha_0)}$ of operators $R_{n;\alpha} : C_{2\pi} \longrightarrow V_{2\pi}$, is said to be a regularization process if there exists a linear approximation process $\{S_n\}_{n\in\mathbb{N}}$ in the sense of (2.3), such that, for any function $f \in C_{2\pi}$ and $n \in \mathbb{N}$, the following three conditions hold:

(i) $\forall \alpha \in (0, \alpha_0)$, the function $R_{n,\alpha}(f)$ is piecewise continuous in $[0, 2\pi]$ and continuous from the right at each point $x \in [0, 2\pi]$ such that $[S_n(f)](x) = 0$;

(ii) $\forall \alpha \in (0, \alpha_0)$, the product function $S_n(f)R_{n,\alpha}(f)$ is uniformly bounded, i.e., there exists a constant $C > 0$ such that $|\,[S_n(f)](x)\,[R_{n,\alpha}(f)](x)\,| \leq C$ for any $x \in [0, 2\pi]$;

(iii) $R_{n,\alpha}(f)$ "approximates the inverse" of $S_n(f)$, that is,

$$\lim_{\alpha \longrightarrow 0^+} [R_{n,\alpha}(f)](x) = \Big([S_n(f)](x)\Big)^{-1} \quad \forall x \in \{x \in [0, 2\pi] : [S_n(f)](x) \neq 0\} \,.$$

Summarizing, a regularization process is able to regularize the approximation of $1/f$ in all the points $x \in [0, 2\pi]$ such that $f(x) = 0$. On the other hand, if $f(x) \neq 0$, the regularization process must guarantee the convergence to $f(x)^{-1}$, resulting in the following lemma.

Theorem 3.4. *Let* $\{R_{n;\alpha}\}_{n\in\mathbb{N};\alpha\in(0,\alpha_0)}$ *be a regularization process and* $f \in C_{2\pi}$. *Let* $x \in [0, 2\pi]$ *with* $f(x) \neq 0$.

Then, for any $\epsilon > 0$, *there exist* $n_{\epsilon,x} \in \mathbb{N}$ *and* $\alpha_{\epsilon,x} > 0$ *such that*

$$|[R_{n,\alpha}(f)](x) - f(x)^{-1}| < \epsilon \,,$$

if $n > n_{\epsilon,x}$ *and* $0 < \alpha < \alpha_{\epsilon,x}$.

Proof. Let $\{S_n\}$ be the linear approximation process associated to the regularization process $\{R_{n;\alpha}\}$ in the sense of Definition 3.3, and let s_n denote the function such that $s_n(y) = [S_n(f)](y)$, $\forall y \in [0, 2\pi]$.

Let us suppose $f(x) > 0$ (the case $f(x) < 0$ is analogous). Since $f(x) > 0$, there exists an $n_x \in \mathbb{N}$ such that $s_n(x) \neq 0$ for $n > n_x$. This can be easily shown by recalling that the sequence of spaces $\{V_n\}_{n\in\mathbb{N}}$ is dense in $(C_{2\pi}, \| \bullet \|_\infty)$ and the limit (2.3) holds. Hence, if $n > n_x$, we can write

$$|[R_{n,\alpha}(f)](x) - f(x)^{-1}| \leq |s_n(x)^{-1} - f(x)^{-1}| + |[R_{n,\alpha}(f)](x) - s_n(x)^{-1}|. \quad (3.3)$$

By virtue of the uniform approximation of f by s_n, if $\delta \in (0, f(x)/2)$ there exists an $n_\delta \in \mathbb{N}$ such that $|s_n(x) - f(x)| < \delta$ for $n > n_\delta$. By the first addendum of (3.3), if $n > \tilde{n} = \max\{n_x, n_\delta\}$, we have that

$$\begin{aligned} |s_n(x) - f(x)| < \delta \quad &\Longleftrightarrow \quad |s_n(x)^{-1} - f(x)^{-1}| < \delta/(s_n(x)f(x)) \\ &\Longrightarrow \quad |s_n(x)^{-1} - f(x)^{-1}| < 2\delta/f(x)^2 , \end{aligned}$$

since $s_n(x) > f(x)/2 \neq 0$. If we substitute $\delta = (f(x)^2\epsilon)/4$ in the latter inequality, then $|s_n(x)^{-1} - f(x)^{-1}|$ is bounded by $\epsilon/2$ for any $n > \tilde{n} =: n_{\epsilon,x}$.

Finally, by virtue of part (iii) of Definition 3.3, we note that the second addendum of (3.3) is also bounded by $\epsilon/2$ for any $0 < \alpha < \alpha_{\epsilon,x}$, with $\alpha_{\epsilon,x}$ sufficiently small, which concludes the proof. □

If $f(x) > C > 0 \ \forall x \in [0, 2\pi]$, the family of functions $R_{n,\alpha}$ defined as $[R_{n,\alpha}(f)](x) \equiv \Big([S_n(f)](x)\Big)^{-1}$ is a regularization process for f. Nevertheless, it is evident that such a regularization process is useless for preconditioning of ill-posed Toeplitz systems, since generating functions $f(x) > C > 0$ are associated to well-posed problems [18].

The following lemma states that the application of a regularization inverse to a linear approximation process gives rise to a regularization process. Since Definition 3.3 has been introduced according to the three conditions of Definition 3.1, the lemma is proved straightforwardly.

Lemma 3.5. *Let $f \in C_{2\pi}$ and let $\{S_n(f)\}_{n\in\mathbb{N}}$ be a linear approximation process such as in* (2.3). *Moreover, let $\{\tilde{R}_\alpha\}_{\alpha\in(0,\alpha_0)}$, $\alpha_0 > 0$, denote a regularization inverse on $(0, +\infty)$ of Definition* 3.1.

Then the family of operators $\{R_{n;\alpha}\}_{n\in\mathbb{N};\alpha\in(0,\alpha_0)}$ such that

$$[R_{n;\alpha}(f)](x) = \tilde{R}_\alpha\Big([S_n(f)](x)\Big), \qquad \forall x \in [0, 2\pi], \quad (3.4)$$

is a regularization process in the sense of Definition 3.3.

In the following sections, we consider regularization processes applied to generating functions of Toeplitz matrices, and Lemma 3.5 will be useful there.

4. Regularization processes for Toeplitz matrices

Given a Toeplitz matrix $A_n(f)$, with $f \in C_{2\pi}$, and a sequence of matrix algebras M_n, let $\{Q_{n;\alpha}\}_{n\in\mathbb{N};\alpha\in(0,\alpha_0)}$ be the family of preconditioners $Q_{n;\alpha} \in M_n$ such that

$$Q_{n;\alpha} = M_n(R_{n;\alpha}(f)), \quad (4.1)$$

where the notation was introduced in (2.2). Notice that $\{R_{n;\alpha}\}$ in (4.1) works in place of $\{S_n\}$ in (2.4). We will show that, under suitable hypotheses, such preconditioners $Q_{n;\alpha}$ converge to the inverse of $A_n(f)$, and the eigenvalues of the preconditioned matrix $Q_{n;\alpha}A_n(f)$ are well clustered at unity.

Before continuing, we need to characterize what are the matrix algebras which allow good approximations of Toeplitz matrices.

Definition 4.1. [28] A sequence $\{M_n\}_{n\in\mathbb{N}}$ of matrix algebras of order n, is called a good sequence of algebras if and only if, for any $\epsilon > 0$, there exists an integer $n_\epsilon \in \mathbb{N}$ such that, for any trigonometric polynomial p of fixed degree and $n > n_\epsilon$, the eigenvalues of the matrix $A_n(p) - M_n(p)$ are contained in $(-\epsilon, \epsilon)$ except for a constant number H_ϵ of outliers.

We remark that all the trigonometric matrix algebras of Section 2 are good sequence of algebras. Now, we may give the first theorem which connects regularization processes to generating functions $f \in C_{2\pi}$ of Toeplitz matrices.

Theorem 4.2. *Let $A_n(f)$ be a sequence of Toeplitz matrices generated by a real function $f \in C_{2\pi}$, and let $\{M_n\}$ be a good sequence of algebras with grid points W_n. According to Definition* 3.3, *let $\{R_{n;\alpha}\}_{\alpha>0}$ denote a regularization process.*

Let us suppose that, for any $\epsilon > 0$, there exist $n_\epsilon \in \mathbb{N}$ and $\alpha_\epsilon > 0$ such that, for $n > n_\epsilon$ and $0 < \alpha < \alpha_\epsilon$,

$$[R_{n;\alpha}(f)](x_i^{(n)}) \neq 0\,, \qquad x_i^{(n)} \in W_n\,, \qquad \forall\ i \in \{1,\dots,n\} \tag{4.2}$$

and

$$|[R_{n;\alpha}(f)](x_i^{(n)})^{-1} - f(x_i^{(n)})| < \epsilon \qquad x_i^{(n)} \in W_n \quad \forall\ i \in \{1,\dots,n\} \setminus J_\epsilon\,, \tag{4.3}$$

where the number of elements of J_ϵ is $o(n)$.

Let us consider the family of sequences of matrices $\{B_{n;\alpha}\}_{n\in\mathbb{N};\alpha\in(0,\alpha_0)}$, with $B_{n;\alpha} \in M_n$ defined as follows

$$B_{n;\alpha} = \Big(M_n(R_{n;\alpha}(f))\Big)^{-1}. \tag{4.4}$$

Then, for any $\epsilon > 0$ there exist $n_\epsilon \in \mathbb{N}$ and $\alpha_\epsilon \in (0, \alpha_0)$ such that, for $n > n_\epsilon$ and $0 < \alpha < \alpha_\epsilon$, the eigenvalues of the matrix $A_n(f) - B_{n;\alpha}$ are contained in $(-\epsilon, \epsilon)$ except for a number $H_\epsilon = o(n)$ of outliers. We shall say that $B_{n;\alpha}$ weakly converges to $A_n(f)$.

Proof. Let $\{p_k\}_{k\in\mathbb{N}}$ denote a family of trigonometric polynomials of degree k which uniformly approximates the function f. Then, fixed $\epsilon > 0$, there exists an $n'_\epsilon \in \mathbb{N}$ such that

$$\|A_n(f) - A_n(p_{n'_\epsilon})\|_2 < \|f - p_{n'_\epsilon}\|_\infty \leq \epsilon/3$$

for $n > n'_\epsilon$.

Let $Z_{n,\alpha}(f)$ denote a family of real functions defined on W_n such that

$$[Z_{n,\alpha}(f)](x_i^{(n)}) = \begin{cases} [R_{n;\alpha}(f)](x_i^{(n)})^{-1} & \text{if} \quad i \notin J_\epsilon\,, \\ f(x_i^{(n)}) & \text{if} \quad i \in J_\epsilon\,. \end{cases}$$

By virtue of (4.3), there exist $n''_\epsilon > n'_\epsilon \in \mathbb{N}$ and $\alpha_\epsilon > 0$ such that

$$\|M_n(Z_{n;\alpha}(f)) - M_n(p_{n'_\epsilon})\|_2 = \|M_n(Z_{n;\alpha}(f) - p_{n'_\epsilon})\|_2 < \epsilon/3\,,$$

for $n > n''_\epsilon$ and $0 < \alpha < \alpha_\epsilon$.

Due to (4.2), the matrix $M_n(R_{n;\alpha}(f))$ is invertible. Thus we can write

$$\begin{aligned} A_n(f) - B_{n;\alpha} &= A_n(f) - M_n(R_{n;\alpha}(f))^{-1} = A_n(f) - M_n(R_{n;\alpha}(f)^{-1}) \\ &= \Big(A_n(f) - A_n(p_{n'_\epsilon})\Big) + \Big(A_n(p_{n'_\epsilon}) - M_n(p_{n'_\epsilon})\Big) \\ &+ \Big(M_n(p_{n'_\epsilon}) - M_n(Z_{n;\alpha}(f))\Big) + \Big(M_n(Z_{n;\alpha}(f)) - M_n(R_{n;\alpha}(f)^{-1})\Big)\,. \end{aligned}$$

The first and the third addendum have the 2-norm bounded by $\epsilon/3$, for $n > n''_\epsilon$ and $\alpha < \alpha_\epsilon$, as shown before.

Since $\{M_n\}$ is a good sequence of algebras, the second addendum can be split to two parts, the former with norm bounded by $\epsilon/3$ and the latter with constant rank, for n sufficiently larger than a suitable n'''_ϵ.

The fourth addendum $M_n(Z_{n;\alpha}(f)) - M_n(R_{n;\alpha}(f)^{-1})$ is the difference of matrices of the same algebra, with the same generating function, except at most $\#J_\epsilon$ points of the grid W_n of the algebra M_n. It follows that the rank of the last addendum is at most equal to $\#J_\epsilon = o(n)$, for all $\alpha > 0$.

Therefore we have shown that the matrix $A_n(f) - B_{n;\alpha}$ is the sum of two parts, one with 2-norm bounded by ϵ and the other with rank equal to $o(n)$, if $n > n_\epsilon := \max\{n''_\epsilon, n'''_\epsilon\}$ and $0 < \alpha < \alpha_\epsilon$. Hence, we have only to invoke the Cauchy interlace Theorem, and the result is proved. □

The latter theorem leads to the following lemma which can be used to analyze the clustering at unity of the preconditioned matrices $M_n(R_{n;\alpha}(f))\,A_n(f)$. We recall that eigenvalues' clustering at unity can give a fast convergence of iterative methods such as the conjugate gradient and the Landweber ones [5, 10].

Lemma 4.3. *Under the assumptions of Theorem 4.2 and with f positive, then, for any $\epsilon > 0$, there exist $n_\epsilon \in \mathbb{N}$ and $\alpha_\epsilon > 0$ such that, for $n > n_\epsilon$ and $0 < \alpha < \alpha_\epsilon$, the preconditioned matrix*

$$M_n(R_{n;\alpha}(f))\,A_n(f)$$

has eigenvalues contained in $(1-\epsilon, 1+\epsilon)$ except at most $o(n)$ outliers.

Proof. From Theorem 4.2, for $\epsilon' > 0$ there exist $n_{\epsilon'} \in \mathbb{N}$ and $\alpha_{\epsilon'} > 0$ such that the eigenvalues of the matrix $A_n(f) - M_n(R_{n;\alpha}(f))^{-1}$ are contained in $(-\epsilon', \epsilon')$ except for a number $o(n)$ of outliers.

Since f is continuous and positive in the closed set $[0,2\pi]$, the functions $R_{n;\alpha}(f)$ are uniformly bounded; this implies that all the operators $M_n(R_{n;\alpha}(f))$ are uniformly bounded. With the help of the identity

$$M_n(R_{n;\alpha}(f))A_n(f) = M_n(R_{n;\alpha}(f))\left(A_n(f) - M_n(R_{n;\alpha}(f))^{-1}\right) + I\,,$$

the claimed result is proved, by invoking Theorem 4.2 with $\epsilon' = \epsilon/K$, where $K \in \mathbb{R}$ is

$$K = \sup_{\alpha\in(0,\alpha_0);n\in\mathbb{N}} \|M_n(R_{n;\alpha}(f))\|_2\,. \qquad \square$$

5. Regularization preconditioners from regularization processes

If the continuous generating function has a root, most preconditioners for Toeplitz systems are asymptotically ill-conditioned and give rise to high numerical instability. In order to improve the stability of the preconditioned system by filtering the components related to the noise, in [18] we introduced the class of regularization preconditioners.

Definition 5.1. [18] Let $\{A_n\}_{n\in\mathbb{N}}$ be a sequence of $n\times n$ matrices and let $\{M_n\}_{n\in\mathbb{N}}$ be a sequence of matrix algebras.

A family of matrices $\{Q_{n;\alpha}\}_{n\in\mathbb{N};\alpha\in(0,\alpha_0)}$, $Q_{n;\alpha}\in M_n$, with $\alpha_0>0$, is a family of regularization preconditioners for $\{A_n\}$ if and only if there exists a sequence of preconditioners $\{P_n\}_{n\in\mathbb{N}}$, $P_n\in M_n$, such that:

(1) $\{P_n\}$ weakly converges to $\{A_n\}$, that is, for any $\epsilon>0$, there exists an $n_\epsilon\in\mathbb{N}$ such that, for $n>n_\epsilon$, the singular values of the matrix P_n-A_n are contained in the disc $B(0,\epsilon)$ except for a number $o(n)$ of outliers.

(2) For any $n\in\mathbb{N}$, we have that

$$\lim_{\alpha\longrightarrow 0^+} \sup_{y_n\in\mathbb{C}^n} \|Q_{n;\alpha}y_n - P_n^\dagger y_n\| = 0\,, \tag{5.1}$$

where the matrix $P_n^\dagger$ is the Moore-Penrose inverse of P_n.

(3) For any $n\in\mathbb{N}$, let $|l_{\min}^{(n)}| = |l_1^{(n)}| \le |l_2^{(n)}| \le \cdots \le |l_n^{(n)}| = |l_{\max}^{(n)}|$ be the singular values of P_n associated with a basis B of singular vectors of the algebra M_n, and let $l_{1;\alpha}^{(n)}, l_{2;\alpha}^{(n)}, \ldots, l_{n;\alpha}^{(n)}$ denote the singular values of the matrix $Q_{n;\alpha}$ associated with the same basis B.

If $|l_{\min}^{(n)}| \longrightarrow 0$ $(n\longrightarrow+\infty)$ then, for $\alpha\in(0,\alpha_0')$ with $0<\alpha_0'<\alpha_0$, there exist an index function $j_\alpha(n):\mathbb{N}\longrightarrow\mathbb{N}$, which satisfies $j_\alpha(n)\le n$, $j_\alpha(n)\longrightarrow+\infty$ $(n\longrightarrow+\infty)$, and a constant $0<C_\alpha<1$, such that

$$0 \le |l_{i;\alpha}^{(n)}|\,|l_i^{(n)}| \le C_\alpha < 1\,, \qquad \text{if } i\le j_\alpha(n)\,. \tag{5.2}$$

In addition, if

$$|l_{i;\alpha_1}^{(n)}| \ge |l_{i;\alpha_2}^{(n)}| \qquad \forall\, i\le j'(n) \tag{5.3}$$

for $0<\alpha_1<\alpha_2\le\alpha_0$, where $j':\mathbb{N}\longrightarrow\mathbb{N}$ is an index function such that $j'(n)\longrightarrow+\infty$, then the family $\{Q_{n;\alpha}\}$ is said to be monotone.

In summary, the "unstable" inversion of the smallest singular values of a (non-regularizing) preconditioner P_n, is controlled by using the regularization preconditioners $Q_{n;\alpha}$, for $\alpha \in (0, \alpha_0)$.

Now, using regularization processes of type (3.4), we build families of regularization preconditioners in the sense of the latter definition.

Let $\{R_\alpha\}_{\alpha\in(0,\alpha_0)}$ be the family of operators $R_\alpha : C_{2\pi} \longrightarrow V_{2\pi}$ such that

$$[R_\alpha(g)](x) = \begin{cases} 0 & \text{if } \; g(x) = 0 \\ \tilde{R}_\alpha(g(x)) & \text{otherwise} \end{cases} \tag{5.4}$$

for any $g \in C_{2\pi}$, where $\{\tilde{R}_\alpha\}_{\alpha\in(0,\alpha_0)}$ is a regularization inverse on $[0, +\infty)$ in the sense of Definition 3.1. Under the hypotheses of Lemma 3.5 and according to (4.1), we consider the family of Hermitian preconditioners $\{Q_{n;\alpha}\}_{n\in\mathbb{N};\alpha\in(0,\alpha_0)}$, with $Q_{n;\alpha} \in M_n$, defined as follows

$$Q_{n;\alpha} = M_n(R_{n;\alpha}(f)) = M_n(R_\alpha(S_n(f)))\,. \tag{5.5}$$

The following theorem shows that such a family of preconditioners belongs to the class of Definition 5.1.

Theorem 5.2. *Let $\{A_n(f)\}_{n\in\mathbb{N}}$ be a sequence of Toeplitz matrices generated by a real function $f \in C_{2\pi}$. Let $\bar{x} \in [0, 2\pi]$ be a point such that $f(\bar{x}) = 0$ and let $\{M_n\}_{n\in\mathbb{N}}$ be a good sequence of algebras.*

If S_n denotes a linear approximation process such that (2.3) *holds, and the family of operators* (5.4) *is denoted by $\{R_\alpha\}_{\alpha\in(0,\alpha_0)}$, then the family of matrices $Q_{n;\alpha}$ defined by* (5.5) *is a family of regularization preconditioners for $\{A_n(f)\}$ according to Definition* 5.1.

Furthermore, if the regularization inverse $\{\tilde{R}_\alpha\}$ of (5.4) *is monotone, then the family $\{Q_{n;\alpha}\}$ is monotone.*

Proof. Let P_n be the $n \times n$ Hermitian matrix defined as $P_n = M_n(S_n(f))$ and let $\{p_k\}_{k\in\mathbb{N}}$ denote a family of real trigonometric polynomials of degree k which uniformly approximates the function f. With the same notations of Theorem 4.2, we have

$$A_n(f) - P_n = \Big(A_n(f) - A_n(p_{n'_\epsilon})\Big) + \Big(A_n(p_{n'_\epsilon}) - M_n(p_{n'_\epsilon})\Big) + \Big(M_n(p_{n'_\epsilon}) - M_n(S_n(f))\Big).$$

The norms of the first and third addendum are small, for a sufficiently large n, since $p_{n'_\epsilon}$ and $S_n(f)$ converge to f uniformly. Since M_n is a good sequence of algebras, the second addendum can be split to two parts, the former with small norm and the latter with constant rank, for a sufficiently large n. As a result of the Cauchy interlace theorem, the family $\{P_n\}$ satisfies condition (1) of Definition 5.1.

Let us consider the eigenvalues of $Q_{n;\alpha} = M_n(R_{n;\alpha}(f))$, that is, the values of $[R_\alpha(g)](x_s^{(n)})$, where $\{x_s^{(n)}\}_{s=0}^{n-1}$ is the grid W_n of the algebra M_n.

Due to property (iii) of Definition 3.1 for the regularization inverse $\{\tilde{R}_\alpha\}_{\alpha\in(0,\alpha_0)}$, we have that $[R_\alpha(S_n(f))](x_s^{(n)}) = 0$, if $[S_n(f)](x_s^{(n)}) = 0$, and

$$\lim_{\alpha\longrightarrow 0^+}[R_\alpha(S_n(f))](x_s^{(n)}) = \lim_{\alpha\longrightarrow 0^+}\tilde{R}_\alpha([S_n(f)](x_s^{(n)})) = ([S_n(f)](x_s^{(n)}))^{-1},$$

if $[S_n(f)](x_s^{(n)}) \neq 0$.

Let $K(P_n)$ denote the kernel of the preconditioner P_n and let $y_n = u_n + u_n^\perp$ be the unique decomposition of $y_n \in \mathbb{C}^n$ such that $u_n \in K(P_n)$ and $u_n^\perp \in K(P_n)^\perp$.

Recalling that the eigenvalues of P_n are $[S_n(f)](x_s^{(n)})$, $s = 0,\ldots,n-1$, then we have that $Q_{n;\alpha}u_n = 0$, since $[R_\alpha(S_n(f))](x_s^{(n)}) = 0$.

Furthermore, since $P_n^\dagger u_n = 0$, we obtain that

$$\lim_{\alpha\longrightarrow 0^+} Q_{n;\alpha}y_n = \lim_{\alpha\longrightarrow 0^+} Q_{n;\alpha}u_n^\perp = P_n^\dagger u_n^\perp = P_n^\dagger y_n$$

for all $y_n \in \mathbb{C}^n$, which states that condition (2) of Definition 5.1 holds for the family (5.5).

Now we discuss condition (3) of Definition 5.1.

Since $f \in C_{2\pi}$ and $f(\bar{x}) = 0$, we have that $\lim_{n\longrightarrow+\infty}|l_{\min}^{(n)}| = 0$ in light of the Szegő Theorem ([21] Section 5.2). From Definition 3.1, any regularization inverse $\{\tilde{R}_\alpha(x)\}$ is bounded in a right neighborhood of $x = 0$, since it is continuous from the right on $[0,+\infty)$. Thus, for any $\alpha \in (0,\alpha_0)$ there exists $\xi_\alpha > 0$ such that

$$\tilde{R}_\alpha(x) < C_\alpha x^{-1}$$

for $x \in (0,\xi_\alpha)$, where C_α is a constant value $0 < C_\alpha < 1$. This implies that for $\alpha \in (0,\alpha_0)$

$$|[R_\alpha(f)](x)\,f(x)\,| \le C < 1, \quad \forall x \in L_\alpha = \Big\{x \in [0,2\pi] \,:\, |f(x)| < \xi_\alpha\Big\}. \tag{5.6}$$

Since $f \in C_{2\pi}$, the Lebesgue measure of L_α is positive. The number of points of the grid W_n of M_n whose images under f are contained in the set $(-\xi_\alpha,\xi_\alpha)$ tends to infinity, as n increases. More precisely, if we define the set

$$K_{n;\alpha} = \Big\{s \in \mathbb{N} \,:\, x_s^{(n)} \in W_n\,,\;\; |f(x_s^{(n)})| < \xi_\alpha\Big\}$$

of indices related to the smallest eigenvalues of P_n, we have that

$$\lim_{n\longrightarrow+\infty} \#K_{n;\alpha} = +\infty\,.$$

If we consider the function $j_\alpha(n)$ of Definition 5.1 and, for $\alpha \in (0,\alpha_0)$, we set $j_\alpha(n) \equiv \#K_{n;\alpha}$, then the second condition for the regularization preconditioners holds. In fact, for $i \in K_{n;\alpha}$, we have that

$$\lim_{n\longrightarrow+\infty} l_i^{(n)} = f(\tilde{x}) \qquad \text{and} \qquad \lim_{n\longrightarrow+\infty} l_{i;\alpha}^{(n)} = [R_\alpha(f)](\tilde{x})$$

for a suitable $\tilde{x} \in L_\alpha$, so that inequality (5.2) is a direct consequence of (5.6).

Finally, we show that, if the regularization inverse $\{\tilde{R}_\alpha\}_{\alpha\in(0,\alpha_0)}$ is monotone, then the regularization family of preconditioners is monotone too.

Let η be the value introduced in Definition 3.1. From (3.1), for $0 < \alpha_1 < \alpha_2 < \alpha_0$ we can write

$$\begin{aligned}
|l_{i;\alpha_1}^{(n)}| &= |l_i(R_{n;\alpha_1})| = |l_i(M_n(R_{\alpha_1}(S_n(f))))| = |[R_{\alpha_1}(S_n(f))](x_i^{(n)})| \\
&= |\tilde{R}_{\alpha_1}([S_n(f)](x_i^{(n)}))| \geq |\tilde{R}_{\alpha_2}([S_n(f)](x_i^{(n)}))| \\
&= |[R_{\alpha_2}(S_n(f))](x_i^{(n)})| = |l_i(M_n(R_{\alpha_2}(S_n(f))))| = |l_i(R_{n;\alpha_2})| = |l_{i;\alpha_2}^{(n)}|
\end{aligned}$$

for all $i \in H_n = \left\{s \in \mathbb{N} : x_s^{(n)} \in W_n\,,\ \ |(S_n(f))(x_s^{(n)})| \in [0,\eta)\right\}$.
If the function j' of (5.3) is defined as $j'(n) \equiv \#H_n$, we have that $\{Q_{n;\alpha}\}$ is monotone, since

$$\lim_{n\longrightarrow+\infty} \#H_n = +\infty\,. \qquad \square$$

The class of regularization preconditioners of Definition 5.1 includes many preconditioners of the literature for Toeplitz systems derived from the discretization of ill-posed problems [23, 26, 25, 10]. Here we show that the basic preconditioner developed in 1993 by M. Hanke, J.G. Nagy and R.J. Plemmons [23], is a preconditioner obeying the hypotheses of Theorem 5.2. This preconditioner will be denoted by HNP.

According to Section 2, let F_n be the unitary Fourier matrix which diagonalizes the matrix algebra $M_n = M_n(F_n)$ of the circulant matrices and let $P_{opt}(A_n) \in M_n$ denote the optimal circulant preconditioner for the Toeplitz matrix $A_n = A_n(f)$, with $f \in C_{2\pi}$.

If $\lambda_1(B), \lambda_2(B), \ldots, \lambda_n(B)$ denote the eigenvalues of the circulant matrix B with respect to the set of eigenvectors of F_n, the HNP preconditioner $P_{opt,\tau}(A_n)$ with truncation parameter $\tau > 0$, is the circulant matrix such that

$$\lambda_i(P_{opt,\tau}(A_n)) = \begin{cases} 1 & \text{if } \ |\lambda_i(P_{opt}(A_n))| < \tau \\ \lambda_i(P_{opt}(A_n)) & \text{otherwise} \end{cases}\,. \tag{5.7}$$

Lemma 5.3. *Let $\{A_n(f)\}_{n\in\mathbb{N}}$ be a sequence of Toeplitz matrices generated by a real function $f \in C_{2\pi}$ and let $\bar{x} \in [0,2\pi]$ be a point such that $f(\bar{x}) = 0$.*

Let $K(P_{opt}(A_n))$ denote the kernel of the optimal circulant preconditioner $P_{opt}(A_n) \in M_n(F_n)$ and let $y_n = u_n + u_n^\perp$ be the unique decomposition of $y_n \in \mathbb{C}^n$ such that $u_n \in K(P_{opt}(A_n))$ and $u_n^\perp \in K(P_{opt}(A_n))^\perp$.

Then the family of circulant preconditioners $\{Q_{n;\alpha}\}_{n\in\mathbb{N};\alpha>0}$, $Q_{n;\alpha} \in M_n(F_n)$, such that

$$Q_{n;\alpha} y_n = P_{opt,\alpha}(A_n)^\dagger u_n^\perp$$

is a family of monotone regularization preconditioners for $\{A_n(f)\}$ in the sense of Definition 5.1.

Proof. The preconditioners $Q_{n;\alpha}$ belong to the class (5.5) since:

(i) the sequence $\{M(F_n)\}$ of trigonometric algebras of circulant matrices is a good sequence of matrix algebras;

(ii) R_α is the operator (5.4) based on the monotone regularization inverse $\tilde{R}_\alpha$ of the Hanke, Nagy and Plemmons' filter (III) described in Section 3;

(iii) the linear approximation process S_n is the Césaro sum C_n described in Section 2.

The thesis holds by invoking Theorem 5.2. □

We argue that the latter lemma could be applied to other preconditioners based on similar filtering procedures [25, 26], with few generalizations of the arguments.

Numerical 1-D applications of regularization preconditioners (5.5) can be found in [18], for the solution of a Fredholm equation of the first kind. In [6, 17] several regularization preconditioners for the Landweber and the conjugate gradient methods have been tested for 2-D deconvolution problems in image restoration. An application to an interferometric multi-image problem related to the astronomical image restoration of the Large Binocular Telescope can be found in [19].

6. Concluding remarks

In the case of Hermitian Toeplitz matrices $A_n(f)$, the eigenvalues of most trigonometric preconditioners are the values of linear approximation processes of the generating function f on a uniform grid of $[0, 2\pi]$ [10, 11, 28]. These preconditioners approximate the Toeplitz matrix in the noise space, that is, in the subspace related to components of the data mainly corrupted by noise. If the Toeplitz system comes from the discretization of an ill-posed problem, such preconditioners must be endowed with regularization features. Otherwise, preconditioned iterative methods may provide inaccurate results, due to a fast reconstruction from components with the highest noise [23, 26, 16].

Here we have introduced a different kind of approximation operators for real functions, which has been called regularization process. If a regularization process is applied to the generating function of a discrete ill-posed Toeplitz matrix, we can design and compute efficient preconditioners for the related linear systems. In this paper, regularization processes have been constructed by applying well-known continuous regularization algorithms for inverse problems on linear approximation processes [15].

Preconditioned iterative system solvers with such regularization preconditioners give rise to fast convergence in the components of the signal space only. On the other hand, in the noise space the convergence is slow, providing a resolution less sensitive to data errors.

Some widely used preconditioners for ill-conditioned linear systems belong to the general family of regularizing preconditioners from regularization processes introduced here. On these grounds, the arguments of the paper can be considered as an extension. Some families based on the above techniques have been proposed and tested in [18, 6, 17, 19]. Those numerical tests showed that regularization preconditioners give a fast and "clean" reconstruction of the solution. It is important to notice that regularization preconditioners lead to solutions which can be better than in the non-preconditioned case. This unusual feature is due to the filtering capabilities of the regularization preconditioners, which provide fast and accurate reconstruction simultaneously.

Regularization preconditioners depend on a real value, say α, which plays the role of regularization parameter, that is, it allows both convergence speed and noise filtering to be controlled. Such a real value α is related to a regularization parameter of the regularization process which approximates the generating function of the Toeplitz matrix. The choice of this parameter α is crucial for the effectiveness of the regularization preconditioning procedure. This aspect deserves more attention and will be considered in a future work.

Acknowledgment

The author is grateful to Prof. F. Di Benedetto for useful suggestions and discussions. In addition, the author wishes to thank the anonymous referee for his many advices to improve the paper.

References

[1] O. Axelsson and G. Lindskög, The rate of convergence of the preconditioned conjugate gradient method, *Numer. Math.*, **52**, 1986, pp. 499–523.

[2] D. Bertaccini, Reliable preconditioned iterative linear solvers for some integrators, *Numerical Linear Algebra Appl.*, **8**, 2001, pp. 111–125.

[3] D. Bertaccini, The spectrum of circulant-like preconditioners for some general linear multistep formulas for linear boundary value problems, *SIAM J. Numer. Anal.*, **40**, 2002, pp. 1798–1822.

[4] D. Bertaccini and M.K. Ng, Block $\{\omega\}$-circulant preconditioners for the systems of differential equations, *Calcolo*, **40**, 2003, pp. 71–90.

[5] M. Bertero and P. Boccacci, *Introduction to Inverse Problem in Imaging*, Institute of Physics Publishing, London, 1998.

[6] C. Biamino, *Applicazioni del metodo di Landweber per la ricostruzione di immagini*, Graduate Thesis in Mathematics, Dipartimento di Matematica, Università di Genova, 2003.

[7] D.A. Bini and F. Di Benedetto, A new preconditioner for the parallel solution of positive definite Toeplitz systems, in *Proc. 2nd SPAA*, Crete, Greece, July 1990, ACM Press, New York, pp. 220–223.

[8] D.A. Bini and P. Favati, On a matrix algebra related to the discrete Hartley transform, *SIAM J. Matrix Anal. Appl.*, **14**, 1993, pp. 500–507.

[9] D.A. Bini, P. Favati and O. Menchi, A family of modified regularizing circulant preconditioners for image reconstruction problems, *Computers & Mathematics with Applications*, **48**, 2004, pp. 755–768.

[10] R.H. Chan and M.K. Ng, Conjugate gradient methods for Toeplitz systems, *SIAM Rev.*, **38**, 1996, pp. 427–482.

[11] R.H. Chan and M. Yeung, Circulant preconditioners constructed from kernels, *SIAM J. Numer. Anal.*, **30**, 1993, pp. 1193–1207.

[12] T. Chan, An optimal circulant preconditioner for Toeplitz systems, *SIAM J. Sci. Stat. Comp.*, **9**, 1988, pp. 766–771.

[13] P. Davis, *Circulant matrices*, John Wiley & Sons, 1979.

[14] F. Di Benedetto and S. Serra Capizzano, A unifying approach to abstract matrix algebra preconditioning, *Numer. Math.*, **82-1**, 1999, pp. 57–90.

[15] H.W. Engl, M. Hanke and A. Neubauer, *Regularization of Inverse Problems*, Kluwer, Dordrecht, 1996.

[16] C. Estatico, A class of filtering superoptimal preconditioners for highly ill-conditioned linear systems, *BIT*, **42**, 2002, pp. 753–778.

[17] C. Estatico, Classes of regularization preconditioners for image processing, *Proc. SPIE, 2003 – Advanced Signal Processing: Algorithms, Architectures, and Implementations XIII*, Ed. F.T. Luk, Vol. 5205, pagg. 336–347, 2003, Bellingham, WA, USA.

[18] C. Estatico, "A classification scheme for regularizing preconditioners, with application to Toeplitz systems", *Linear Algebra Appl.*, **397**, 2005, pp. 107–131.

[19] C. Estatico, Regularized fast deblurring for the Large Binocular Telescope, *Tech. Rep. N. 490, Dipartimento di Matematica, Università di Genova*, 2003

[20] R. Gray, *Toeplitz and Circulant matrices: a review*, http: www-isl.stanford.edu/~gray/toeplitz.pdf, 2000.

[21] U. Grenander and G. Szegő, *Toeplitz Forms and Their Applications*, Second edition, Chelsea, New York, 1984.

[22] C.W. Groetsch, *Generalized inverses of linear operators: representation and approximation*, Pure and Applied Mathematics, **37**, Marcel Dekker, New York, 1977.

[23] M. Hanke, J.G. Nagy and R. Plemmons, Preconditioned iterative regularization for ill-posed problems, *Numerical Linear Algebra and Scientific Computing*, L. Reichel, A. Ruttan and R.S. Varga, eds., Berlin, de Gruyter, 1993, pp. 141–163.

[24] J. Kamm and J.G. Nagy, Kronecker product and SVD approximations in image restoration, *Linear Algebra Appl.*, **284**, 1998, pp. 177–192.

[25] M. Kilmer, Cauchy-like Preconditioners for Two-Dimensional Ill-Posed Problems, *SIAM J. Matrix Anal. Appl.*, **20**, No. 3, 1999, pp. 777–799.

[26] M. Kilmer and D. O'Learly, Pivoted Cauchy-like preconditioners for regularized solution of ill-posed problems, *SIAM J. Sci. Comp.*, **21**, 1999, pp. 88–110.

[27] J.G. Nagy, M.K. Ng and L. Perrone, Kronecker product approximation for image restoration with reflexive boundary conditions, *SIAM J. Matrix Anal. Appl.*, **25**, 2004, pp. 829–841.

[28] S. Serra Capizzano, Toeplitz preconditioners constructed from linear approximation processes, *SIAM J. Matrix Anal. Appl.*, **20-2**, 1998, pp. 446–465.

[29] G. Strang, A proposal for Toeplitz matrix calculations, *Stud. Appl. Math.*, **74**, 1986, pp. 171–176.

[30] E. Tyrtyshnikov, A unifying approach to some old and new theorems in distribution and clustering, *Linear Algebra Appl.*, **232**, 1996, pp. 1–43.

[31] E. Tyrtyshnikov and N. Zamarashkin, Spectra of multilevel Toeplitz matrices: advanced theory via simple matrix relationship, *Linear Algebra Appl.*, **270**, 1997, pp. 15–27.

[32] N. Trefethen, *Approximation theory and numerical linear algebra*, J. Mason and M. Cox, eds., Chapman and Hall, London, 1990, pp. 336–360.

Claudio Estatico
Dipartimento di Matematica, Università di Genova
Via Dodecaneso 35
I-16146 Genova, Italy
e-mail: estatico@dima.unige.it

Operator Theory:
Advances and Applications, Vol. 160, 179–194

Minimal State-space Realization for a Class of nD Systems

K. Galkowski

Abstract. Minimal realizations play a key role in system analysis and synthesis. Among a variety of realizations they are characterised by a minimal dimension, which guarantees that no pole zero cancellations occur, a very important feature for the stability analysis, and are also of importance from the numerical point of view. This paper provides a simple method for minimal realization construction for multi-linear odd rational functions an important class from the practical point of view due to strong links to so-called reactance functions.

Keywords. Multidimensional systems, multi-linear, odd rational functions.

1. Introduction

The past two to three decades, in particular, have seen a continually growing interest in multidimensional (nD) systems which is clearly linked to the wide variety of applications arising in both the theory and practical applications domains. The key unique feature of an nD system is that the plant or process dynamics (inputs, states and outputs) are functions of more than one independent variable as the result of the fact that information is propagated in independent directions. This is an essential difference with the classical, or 1D, case where the process dynamics (inputs, states and outputs) are functions of only one variable. In both cases, i.e., 1D and nD, the process can be single-input single-output (SISO) or multiple-input multiple-output (MIMO). Hence, for example, a SISO nD linear system can be represented by a transfer function, which is a rational function in n indeterminates.

Many physical systems, data analysis procedures, computational algorithms and (more recently) learning algorithms have a natural (and underexploited) two-dimensional (2D) structure due to the presence of more than one spatial variable, the combined effect of space and time or the combined effect of a spatial/time variable and an integer index representing iteration, pass or trial number. Physical examples of such systems include bench mining systems, metal rolling, automatic ploughing aids and vehicle convoy coordination on motorways whilst algorithmic examples include image processing, discrete models of spatial behavior, point mapping algorithms and recursive learning schemes as illustrated by trajectory learning in iterative learning control.

Focusing on discrete nD linear systems, two basic state space models have been developed. The first is due to Roesser ([19]) and clearly has a first order structure. Among the key features of this model is that the state vector is partitioned into sub-vectors – one for each of the two directions of information propagation (usually termed horizontal and vertical respectively). One main alternative to the Roesser model is the Fornasini-Marchesini model class [7]. Note, however, that Roesser and Fornasini-Marchesini models are not fully independent and it is possible to transform one into the other.

A key task in 2D/nD systems theory and applications is the construction of state-space realizations of the [19] or [7] types from input-output data, often in the form of a 2D/nD transfer function matrix. This problem is well studied (see, for example, [3], [4], [6], [8], [11], [14], [20], [21] and references therein). To date, however, the key systems theoretic and applications relevant question of how to construct a so-called minimal realization has not been solved in the general case. For further details, see, for example, [15], [16], or [5]. It is also an interesting mathematical problem on its own and it plays an important role in multivariable interpolation (see, e.g., [1], and [2]).

This paper aims to extend the class of multidimensional linear systems for which the solution of this key problem is known. In particular, the existence and construction of a minimal realization for the class of single-input single-output nD linear systems characterized by transfer functions with multi-linear (i.e., of first degree in each indeterminate) numerator and denominator has been developed in [8]. In turn, an assumption for a transfer function to be odd, which is the topic of the paper, also makes the analysis significantly easier and provides interesting results. Such systems are of some practical interest as this subclass consists of so-called reactance functions, a subclass of positive real functions, frequently used in circuits theory.

2. Background

A multidimensional (nD), linear system can be described in the state-space form by the well-known Roesser model

$$\begin{bmatrix} x^1(i_1+1,\ldots,i_n) \\ \vdots \\ x^n(i_1,\ldots,i_n+1) \end{bmatrix} = \begin{bmatrix} A_{11} & \cdots & A_{1n} \\ \vdots & \ddots & \vdots \\ A_{n1} & \cdots & A_{nn} \end{bmatrix} \begin{bmatrix} x^1(i_1,\ldots,i_n) \\ \vdots \\ x^n(i_1,\ldots,i_n) \end{bmatrix} + \begin{bmatrix} B_1 \\ \vdots \\ B_n \end{bmatrix} u(i_1,\ldots,i_d),$$

$$y(i_1,\ldots,i_n) = \begin{bmatrix} C_1 & \cdots & C_n \end{bmatrix} \begin{bmatrix} x^1(i_1,\ldots,i_n) \\ \vdots \\ x^n(i_1,\ldots,i_n) \end{bmatrix} + Du(i_1,\ldots,i_n), \quad (1)$$

where

$x^i(i_1,\ldots,i_n) \in \mathbb{R}^{p_i}$ is an ith *local state sub-vector* $(i = 1,\ldots,n)$, $u(i_1,\ldots,i_n) \in \mathbb{R}^r$ is an *input (control) vector* , $y(i_1,\ldots,i_n) \in \mathbb{R}^m$ is an *output vector*, A_{ij}, B_i, C_i, D are the *real matrices* of appropriate dimensions $(i,j = 1,\ldots,n)$, $i_1,\ldots,i_n \in \mathbb{Z}^+ \cup \{0\}$ are the *discrete independent variables.* In block form

$$\begin{aligned} A &= \begin{bmatrix} A_{11} & \cdots & A_{1n} \\ \vdots & \ddots & \vdots \\ A_{n1} & \cdots & A_{nn} \end{bmatrix}, B = \begin{bmatrix} B_1 \\ \vdots \\ B_n \end{bmatrix}, \\ C &= \begin{bmatrix} C_1 & \cdots & C_n \end{bmatrix}. \end{aligned} \quad (2)$$

In this paper, SISO systems are investigated, i.e., the input u and the output y are scalars. A linear, multidimensional (nD) system of the form (1) can also be described in the generalised frequency domain by an n-variable, rational function matrix. In the SISO case a transfer function matrix becomes a single, rational transfer function

$$f(s_1,\ldots,s_n) = \frac{a(s_1,\ldots,s_n)}{b(s_1,\ldots,s_n)}, \quad (3)$$

where $a(s_1,\ldots,s_n)$ and $b(s_1,\ldots,s_n)$ are real, n-variable polynomials. It is well known that the transfer function description (3) is linked to the Roesser model (1) by

$$f(s_1,\ldots,s_n) = D + C\left(s_1 I_{t_1} \oplus \cdots \oplus s_n I_{t_n} - A\right)^{-1} B \quad (4)$$

where $\oplus$ denotes the direct sum, provided that the rational function (4) is proper.

A very important problem in nD systems theory and practice (for the SISO case) is: *given a (scalar) rational matrix function (in n complex variables) $f(s_1,\ldots,s_n)$ as in* (3)*, construct matrices A,B,C,D as in* (2) *so that f is realized in the form* (4) *as the transfer function of the associated linear system* (1). Note that a necessary condition for the problem to have a solution is that f be proper in each variable. Conversely, under the assumption that this necessary condition is satisfied, it is known that the realization problem always has a solution (see, e.g., [7]).

For the 1D case it is well understood how to construct *state-space minimal* or *least-order* realizations, i.e., realizations for which the dimension of the state space is as small as possible among all possible realizations. Moreover, in the 1D case, it is well known that the minimal state space dimension is equal to the degree of the rational transfer function denominator b, provided that there are no pole-zero cancellations between the denominator b and numerator a, i.e., provided that the pair of polynomials $\{a,b\}$ is coprime. This suggests our definition of denominator-degree minimal realization for the nD case.

Definition 1. *Suppose that $f(s_1, \ldots, s_n)$ is a rational function of n complex variables as in* (3). *Without loss of generality we may assume that the numerator polynomial $a(s_1, \ldots, s_n)$ and the denominator polynomial $b(s_1, \ldots, s_n)$ are* coprime, *i.e., have no nontrivial common polynomial factors. Then the realization* (4) *for f is said to be* denominator-degree minimal *provided that the dimension $p_1 + \cdots + p_n$ of the state space of the realization is equal to the total degree of the denominator b, i.e., the sum of the degrees of b in each variable.*

The realization (3) *of f is said to be* least-order *minimal if the dimension $p_1 + \cdots + p_n$ of the state space is as small as possible among all possible realizations* (3) *of f.*

In the 1D case we have that least-order minimal and denominator-degree minimal are equivalent, and that it is always possible to obtain a minimal (in either equivalent sense) realization of a given proper rational function. Also in the 1D case, the properties of controllability and observability of the state space realization are equivalent to minimality. For the nD case, on the other hand, the equivalence between minimality and simultaneous controllability and observability fails. Moreover, the concepts of denominator-degree minimal and state-space least-order minimal are not equivalent in general. It is even possible that a given rational function of n variables (proper in each variable separately) may not have a denominator-degree minimal realization. This phenomenon is related to the complication in the nD case that there are at least three distinct notions of coprimeness, termed factor, minor and zero coprimeness, for polynomials in n variables. By definition, however, given that the realization problem always has a solution as observed above, it follows that the state-space minimal (i.e., least-order minimal) realization problem also always has a solution: the issue is how to compute such a realization, where now the size of a least-order minimal realization may be greater than the total denominator degree in a factor-coprime fractional representation of the rational function f. The Elementary Operation Algorithm developed by Gałkowski ([8]) is an attempt to produce efficient solutions to this problem and is based on symbolic calculations.

Until now the problem of the existence and construction of denominator-degree minimal realizations for nD systems has been solved only in some particular cases, as, e.g., when the transfer function has a separable denominator or numerator. Due to its great practical importance (digital filter applications) there exists a very rich literature devoted to this class, see for example [18]. However, the problem solution then is much easier than in the general case and in fact it is based on 1D techniques.

3. Problem formulation

Note first that the state-space realization of a SISO system can be derived in the following way, see, e.g., [8]

1. Define the $n+1$-variable polynomial

$$a_f(s_1,\ldots,s_{n+1}) = s_{n+1}b(s_1,\ldots,s_n) - a(s_1,\ldots,s_n) \tag{5}$$

2. Find the companion matrix H for a polynomial, i.e., the matrix H, which satisfies

$$a_f(s_1,\ldots,s_{n+1}) = \det\left[\oplus_{i=1}^{n} s_i I_{t_i} \oplus s_{n+1} - H\right] \tag{6}$$

where t_i denotes both a polynomial degree in the ith variable and a dimension of ith state sub-vector in (1) in an obvious manner, and the polynomial a_f can be written as

$$a_f(s_1,\ldots,s_{n+1}) = \sum_{j_1=0}^{t_1}\cdots\sum_{j_n=0}^{t_n}\sum_{j_{n+1}=0}^{t_{n+1}(=1)} a_{j_{n+1}j_n\cdots j_1}\prod_{k=1}^{n+1} s_k^{t_k-j_k} \tag{7}$$

where, due to (6), the polynomial $a_f(s_1,\ldots,s_{n+1})$ has to be monic, i.e.,

$$a_{0\ldots0} = 1. \tag{8}$$

3. Write the matrix H in the block form $H = \begin{bmatrix} H_{11} & H_{12} \\ H_{21} & H_{22} \end{bmatrix}$, where H_{22} is a scalar and refers to s_{n+1}. Then, $H_{11} = A, H_{12} = B, H_{21} = C, H_{22} = D$.

Thus the problem of finding a state-space realization for a rational, n-variate transfer function has been replaced by the problem of finding the companion matrix for the $n+1$-variate polynomial a_f. Hence, given a multi-linear polynomial a_f it is aimed to find the matrix that satisfies a polynomial equation of the form (6) where $t_i = 1$, $i = 1,2,\ldots,n$, which requires solving this polynomial equation. The problem is solved in [9]. This approach generalizes also to the general MIMO case ([8]).

In this paper, we present a simple efficient construction algorithm for the solution of the problem for the particular case of an odd multi-linear rational function. The solution algorithm which we obtain here is much simpler than that obtained for the general case in [9]. In the multi-linear case, the polynomial a_f of (7) can be written with $t_1 = 1$, $i = 1,2,\ldots,n$, and then the only coefficients $a_{j_1\cdots j_{n+1}}$ appearing are those with indices $j_1\ldots j_{n+1}$ equal to 0 or 1. For simplicity, the coefficients of the polynomial (7) are denoted as

$$a_{0\ldots\underbrace{010\ldots\ldots\ldots\underbrace{010\ldots\underbrace{010\ldots0}_{i_1}}_{i_2}}_{i_k}} := a_{i_1 i_2\cdots i_k} \tag{9}$$

We say that the rational function $f(s_1,\ldots,s_n)$ is *odd* if $f(s_1,\ldots,s_n) = -f(-s_1,\ldots,-s_n)$. Thus, if f given by (3) is odd, then either

1. a is odd and b is even, or
2. a is even and b is odd.

In the first case the associated polynomial a_f (5) is odd while in the second case a_f is even. For the case where f is also multi-linear, a_f being odd in terms of the coefficients (9) means that

$$\begin{cases} a_{i_1 i_2 \cdots i_{2\ell+1}} = 0 \text{ for all } \ell = 0, 1, \dots & \text{in case } n \text{ is odd,} \\ a_{i_1 i_2 \cdots i_{2\ell}} = 0 \text{ for all } \ell = 1, \dots & \text{in case } n \text{ is even.} \end{cases}$$

while a_f being even means that

$$\begin{cases} a_{i_1 i_2 \cdots i_{2\ell+1}} = 0 \text{ for all } \ell = 0, 1, \dots & \text{in case } n \text{ is even,} \\ a_{i_1 i_2 \cdots i_{2\ell}} = 0 \text{ for all } \ell = 1, 2, \dots & \text{in case } n \text{ is odd.} \end{cases}$$

For example, consider the two odd rational functions

$$\frac{-s_1 - s_2}{s_1 s_2 + 2}$$

and

$$\frac{-s_1 s_2 - s_1 s_3 - s_2 s_3 - 2}{s_1 s_2 s_3 + s + s_2 + s_3}.$$

In the first case the polynomial a_f is

$$\begin{aligned} a_f &= s_1 s_2 s_3 + 2 s_3 + s_2 + s_1 \\ &= a_{000} s_1 s_2 s_3 + a_{011} s_3 + a_{101} s_2 + a_{110} s_1 \\ &= a_{000} s_1 s_2 s_3 + a_{12} s_3 + a_{13} s_2 + a_{23} s_1 \end{aligned}$$

while in the second case a_f is given by

$$\begin{aligned} a_f &= s_1 s_2 s_3 s_4 + s_1 s_4 + s_2 s_4 + s_3 s_4 + s_1 s_2 + s_1 s_3 + s_2 s_3 + 2 \\ &= a_{0000} s_1 s_2 s_3 s_4 + a_{0110} s_1 s_4 + a_{0101} s_2 s_4 + a_{0011} s_3 s_4 \\ &\quad + a_{1100} s_1 s_2 + a_{1010} s_1 s_3 + a_{1001} s_2 s_3 + a_{1111} \\ &= a_{0000} s_1 s_2 s_3 s_4 + a_{23} s_1 s_4 + a_{13} s_2 s_4 + a_{12} s_3 s_4 \\ &\quad + a_{34} s_1 s_2 + a_{24} s_1 s_3 + a_{14} s_2 s_3 + a_{1234}. \end{aligned}$$

Now the problem to be solved can be stated as follows: *given a multi-linear polynomial a_f of the form (7) with $t_1 = 1$ for $i = 1, 2, \dots, n+1$, find a matrix H which solves equation (6).*

4. The denominator-degree minimal realization and existence conditions

The general case has been considered in [9] and basing ourselves on this we encounter the particular case of odd transfer functions. First, recall that the polynomial equation (6) for the denominator-degree minimal realization can be rewritten as a system of polynomial equations with the entries h_{ij} of the $(n+1) \times (n+1)$ matrix H being taken as the unknowns, i.e.,

$$(-1)^k M_{i_1 i_2 \dots i_k} = a_{i_1 i_2 \dots i_k} \tag{10}$$

where, for each $k \in \{1, 2, \ldots, n+1\}$, the set $\{i_1, i_2, \ldots, i_k\}$ is an element of the set $C_k\{1, 2, \ldots, n+1\}$ of k-element subsets of the set $\{1, 2, \ldots, n+1\}$, and where $M_{i_1 i_2 \ldots i_k} = M_{i_1 i_2 \ldots i_k}(h_{ij})$ denotes the principal minor of the matrix H corresponding to row and column indices $i_1, i_2, \ldots, i_k$. To avoid misunderstandings, $\mathbb{M}_{i_1 i_2 \cdots i_k}$ denotes here the respective sub-matrix and $M_{i_1 i_2 \cdots i_k}$ the value of the minor, i.e., $\det \mathbb{M}_{i_1 i_2 \cdots i_k} := M_{i_1 i_2 \cdots i_k}$. Obviously, the minor $M_{i_1 i_2 \cdots i_k}$ depends on the entries of the matrix H and hence (10) constitutes the set of equations whose solution is a required denominator-degree minimal realization. Note that this equation set can be significantly simplified, which is possible due to the fact that the equations (10) for a given k contain some equation parts for values smaller than k.

We define a *transversale* of a minor $M_{i_1 i_2 \ldots i_k}$ to be any terms of the form $A := h_{i_1 \alpha} h_{i_2 \beta} \cdots h_{i_k \psi}$ where $\{\alpha, \beta, \ldots, \psi\}$ is any permutation of the set $\{i_1, i_2, \ldots, i_k\}$. Then the value of a minor $M_{i_1 i_2 \ldots i_k}$ can be expressed as

$$M_{i_1 i_2 \ldots i_k} \sum \pm A$$

where $k \in \{1, 2, \ldots, n+1\}$, $\{i_1, i_2, \ldots, i_k\} \in C_k\{1, 2, \ldots, n+1\}$ and where A sweeps over all transversales of $M_{i_1 i_2 \ldots i_k}$, and the signs are given by an appropriate expansion of the determinant along rows or columns respectively. Note that transversales A can be elements of products of two or more lower order minors $\mathbb{M}_{i_1 i_2 \cdots i_\alpha}$, which constitute a mutually exclusive, exhaustive partition of $\{i_1, i_2, \ldots, i_k\}$ or cannot be presented in that way. In the first case, the transversale can be calculated in terms of the coefficients of a polynomial a_f for smaller values of k while transversales of the second type remain unchanged in the reduced system of equations.

To introduce a formal way to achieve the 'minimal' equation set equivalent to (10) the following notations are necessary. Denote, hence, by $r = \{j_1, j_2, \ldots, j_k\}$ such a permutation of the set $\{i_1, i_2, \ldots, i_k\}$ that

$$j_1 = i_1; j_2 \neq i_k; j_k > j_2 \tag{11}$$

and call $\mathbb{R}\{i_1, i_2, \ldots, i_k\}$ the set of all such permutations. Next introduce for any $r \in \mathbb{R}\{i_1, i_2, \ldots, i_k\}$

$$\begin{aligned} A_r &= h_{i_1 j_2} h_{j_2 j_3} \cdots h_{j_{k-1} j_k} h_{j_k i_1}, \\ A_{r^*} &= h_{i_1 j_k} h_{j_k j_{k-1}} \cdots h_{j_3 j_2} h_{j_2 i_1}, \end{aligned} \tag{12}$$

which represent the 'non-partitioned' transversales and enables rewriting (10) in the form of

$$\sum_{r \in \mathbb{R}\{i_1, i_2, \ldots, i_k\}} (A_r + A_{r^*}) = -\widetilde{a}_{i_1 i_2 \cdots i_k}, \tag{13}$$

where

$$\widetilde{a}_{i_1 i_2 \cdots i_k} = \sum_{z \in \mathbb{Z}\{i_1 i_2 \cdots i_k\}} (-1)^{u-1} (u-1)! \prod_{v=1}^{u} a_{z_v}. \tag{14}$$

Here $\mathbb{Z}\{i_1, i_2, \ldots, i_k\}$ is the set of all mutually exclusive, exhaustive partitions of the index set $\{i_1, i_2, \ldots, i_k\}$, i.e.,

$$\begin{aligned} z &:= \{z_1, \ldots, z_v, \ldots, z_w, \ldots, z_u\} \in \mathbb{Z}\{i_1 i_2 \cdots i_k\} : u = 1, 2, \ldots k; \\ z_v, z_w &\subseteq \{i_1 i_2 \cdots i_k\}; z_v \cap z_w = \emptyset; \cup_{v=1}^{u} z_v = \{i_1 i_2 \cdots i_k\}. \end{aligned}$$

The analogous equation set has been derived and exploited in [9] for a general multi-linear case.

In the following, assume that all $a_{i_1 i_2} \neq 0, \{i_1, i_2\} \in C_2\{1, 2, \ldots, n+1\}$, which makes simpler the solution and , moreover, is valid for the reactance functions case. The case when this assumption is not valid requires more complicated methodology and was solved for the general multi-linear case in [9].

At this stage it is instructive to recall the form of equations (13) for $k = 1, 2, 3, 4$ with the oddness assumption introduced previously.

$$h_{ii} = -\widetilde{a}_i = -a_i = 0, \tag{15}$$

$i = 1, 2, \ldots, n+1,$

$$h_{i_1 i_2} h_{i_2 i_1} = -\widetilde{a}_{i_1 i_2} := -a_{i_1 i_2} + a_{i_1} a_{i_2} = -a_{i_1 i_2}, \tag{16}$$

$\{i_1, i_2\} \in C_2\{1, 2, \ldots, n+1\}$

$$\begin{aligned} h_{i_1 i_2} h_{i_2 i_3} h_{i_3 i_1} + h_{i_1 i_3} h_{i_3 i_2} h_{i_2 i_1} &= -\widetilde{a}_{i_1 i_2 i_3} \\ := -a_{i_1 i_2 i_3} + a_{i_1} a_{i_2 i_3} + a_{i_2} a_{i_1 i_3} + a_{i_3} a_{i_1 i_2} &= 0 \end{aligned} \tag{17}$$

$\{i_1, i_2, i_3\} \in C_3\{1, 2, \ldots, n+1\}$

$$\begin{aligned} & h_{i_1 i_2} h_{i_2 i_3} h_{i_3 i_4} h_{i_4 i_1} + h_{i_1 i_3} h_{i_3 i_4} h_{i_4 i_2} h_{i_2 i_1} \\ & + h_{i_1 i_2} h_{i_2 i_4} h_{i_4 i_3} h_{i_3 i_1} + h_{i_1 i_4} h_{i_4 i_2} h_{i_2 i_3} h_{i_3 i_1} \\ & + h_{i_1 i_3} h_{i_3 i_2} h_{i_2 i_4} h_{i_4 i_1} + h_{i_1 i_4} h_{i_4 i_3} h_{i_3 i_2} h_{i_2 i_1} \\ = \; & -\widetilde{a}_{i_1 i_2 i_3 i_4} := -a_{i_1 i_2 i_3 i_4} + a_{i_1 i_4} a_{i_2 i_3} + a_{i_2 i_4} a_{i_1 i_3} + a_{i_3 i_4} a_{i_1 i_2}, \end{aligned} \tag{18}$$

$\{i_1, i_2, i_3, i_4\} \in C_4\{1, 2, \ldots, n+1\}$.

It is immediate to see from (15) that all diagonal elements of the matrix H must be zero. Moreover, it is straightforward from (16) and (17) that

$$h_{i_1 i_2}^2 h_{i_2 i_3}^2 h_{i_3 i_1}^2 = a_{i_1 i_3} a_{i_2 i_3} a_{i_1 i_2} \tag{19}$$

$\forall \{i_1, i_2, i_3\} \in C_3\{1, 2, \ldots, n+1\}$

Also, it is straightforward to see from (16)–(18) that for any $k > 3$ any transversale A_r (see (12)) can be presented in the form of

$$A_r = \frac{A_{i_1 j_2 j_3} A_{i_1 j_3 j_4} \cdots A_{i_1 j_{k-1} j_k}}{(-a_{i_1 j_3})(-a_{i_1 j_4}) \cdots (-a_{i_1 j_{k-1}})}. \tag{20}$$

For example,

$$h_{i_1 i_2} h_{i_2 i_3} h_{i_3 i_4} h_{i_4 i_1} = \frac{(h_{i_1 i_2} h_{i_2 i_3} h_{i_3 i_1})(h_{i_1 i_3} h_{i_3 i_4} h_{i_4 i_1})}{-a_{i_1 i_3}}.$$

The following two lemmas can be proved directly in a straightforward way.

Lemma 1. *For each* $\{i_1, i_2, \ldots, i_k\} \in C_k \{1, 2, \ldots, n+1\}$, $k \geq 3$ *and permutation* $r = \{i_1, j_2, \ldots, j_k\} \in \mathbb{R}\{i_1, i_2, \ldots, i_k\}$

$$A_r = \sqrt{a_{i_1 j_2} a_{j_2 j_3} \cdots a_{j_{k-1} j_k} a_{j_k i_1}}. \tag{21}$$

Lemma 2. *For each* $\{i_1, i_2, \ldots, i_k\} \in C_k \{1, 2, \ldots, n+1\}$, $k \geq 3$ *and permutation* $r = \{i_1, j_2, \ldots, j_k\} \in \mathbb{R}\{i_1, i_2, \ldots, i_k\}$

$$A_{r^*} = \begin{cases} A_r & k = 2l \\ -A_r & k = 2l+1. \end{cases} \tag{22}$$

Now, we are in a position to characterize if the denominator-degree minimal realization is real or complex (provides it exists).

Theorem 1. *The denominator-degree minimal realization H is real (provided that such an H exists) if and only if* $\forall \{i_1, i_2, i_3\} \in C_3 \{1, 2, \ldots, n+1\}$

$$a_{i_1 i_2} a_{i_2 i_3} a_{i_1 i_3} > 0. \tag{23}$$

Proof. Sufficiency. If (23) holds then by (19), (21) and (20) all transversales A_r for $3 < k \leq n+1$ are real, and hence from (12) all h_{ij} can be real too.

Necessity. If (23) does not hold then by (21) there has to exist at least one complex h_{ij}. □

The condition of Theorem 1 for $n = 3$ holds obviously if all polynomial coefficients a_{ij} are positive but also for example, when $a_{12}, a_{23}, a_{14}, a_{34} < 0$ and $a_{13}, a_{24} > 0$, for the case of $(n+1 = 4)$.

In what follows, we can calculate possible values of the matrix H elements by solving (15–17). This however does not prejudge that such a matrix is definitely a denominator-degree minimal realization for a given multi-variate, multi-linear polynomial. In what follows a set of necessary and sufficient conditions will be presented.

Theorem 2. *The elements of the matrix H (provided that such an H exists), can be calculated as*

$$h_{ii} = 0 \tag{24}$$

$i = 1, 2, \ldots, n+1$ *and*

1. *If for some* $\{i_1, i_2, i_3\} \in C_3 \{1, 2, \ldots, n+1\}$ (23) *holds then* $\forall \{i, j\} \subset \{i_1, i_2, i_3\}$
   (a) *for* $a_{ij} > 0$
   $$h_{ij} = -h_{ji} = \pm\sqrt{a_{ij}} \tag{25}$$
   (b) *for* $a_{ij} < 0$
   $$h_{ij} = -h_{ji} = \pm\sqrt{|a_{ij}|} \tag{26}$$
2. *If for some* $\{i_1, i_2, i_3\} \in C_3 \{1, 2, \ldots, n+1\}$ (23) *does not hold then for such* $\{i, j\} \subset \{i_1, i_2, i_3\}$ *that* $a_{ij} < 0$
   $$h_{ij} = -h_{ji} = \pm \jmath \sqrt{|a_{ij}|}, \jmath^2 = -1 \tag{27}$$
   and as in (25) *when* $a_{ij} > 0$.

Proof. It is straightforward by (15–17), Lemmas 1, 2 and the above discussion. □

Theorem 2 shows immediately that the polynomial coefficients $a_{i_1 i_2}$, when nonzero, can be assumed to be the basic ones (they determine the elements of the matrix H), and the remaining coefficients required for a denominator-degree minimal realization are to be recovered as functions of these basic ones. When some of the coefficients $a_{i_1 i_2}$ are zero then the general procedure of [9] has to be applied.

Note also that the solution of Theorem 2 is not unique and every similar matrix, with diagonal similarity matrix, is also a solution.

The next necessary stage is obviously to obtain conditions for the existence of a denominator-degree minimal realization, which can be achieved by substitution the values of the elements of the matrix H already calculated into the remaining equations of (13–14) (with exception of (15–17)).

Lemma 3. *Given an $n+1$ variate polynomial $a_f(s_1,\ldots,s_{n+1})$ of (5) with nonzero coefficients a_{ij}. Then the necessary condition for solvability of the equation set of (10) with a companion matrix H entries h_{ij} as indeterminates, i.e., the necessary and sufficient conditions for the existence of a denominator-degree minimal realization is that all the polynomial $a_f(s_1,\ldots,s_{n+1})$ coefficients satisfy*

$$a_{i_1 i_2 \cdots i_{2l+1}} = 0,\ \forall \{i_1, i_2, \ldots, i_{2l+1}\} \in C_{2l+1}\{1,2,\ldots,n+1\}$$

$$\begin{aligned} a_{i_1 i_2 \cdots i_{2l}} &= \sum_{z\in\mathbb{Z}^2\{i_1 i_2\cdots i_k\}} \prod_{z_v\in z} a_{z_v} \\ &+ \sum_{\substack{z',z"\in\mathbb{Z}^2\{i_1 i_2\cdots i_k\}\\ z'\neq z"}} \pm 2 \prod_{z_v\in z'\cup z"} \sqrt{a_{z_v}} \end{aligned} \tag{28}$$

$$\forall\ \{i_1, i_2, \ldots, i_{2l}\} \in C_{2l}\{1,2,\ldots,n+1\}$$

where $\mathbb{Z}^2\{i_1,i_2,\ldots,i_k\}$ is the set of all mutually exclusive, exhaustive two element partitions of the index set $\{i_1,i_2,\ldots,i_k\}$, i.e.,

$$\begin{aligned} z \ &: = \{z_1,\ldots,z_v,\ldots,z_l\} \in \mathbb{Z}^2\{i_1 i_2\cdots i_{2l}\} \\ z_v, z_w &\subseteq \{i_1 i_2\cdots i_k\}; z_v\cap z_w = \emptyset; \cup_{v=1}^{u} z_v = \{i_1 i_2\cdots i_{2l}\}. \end{aligned}$$

Proof. It is straightforward when substituting the results of Theorem 2 in the equations (13–14) (with exception of (15-17)) and making use of Lemmas 1, 2. □

For example for $k=4$ we have

$$\begin{aligned} a_{i_1 i_2 i_3 i_4} = a_{i_1 i_2}a_{i_3 i_4} + a_{i_1 i_3}a_{i_2 i_4} + a_{i_1 i_4}a_{i_2 i_3} \pm 2\sqrt{a_{i_1 i_2}a_{i_3 i_4}a_{i_1 i_3}a_{i_2 i_4}} \\ \pm 2\sqrt{a_{i_1 i_2}a_{i_3 i_4}a_{i_1 i_4}a_{i_2 i_3}} \pm 2\sqrt{a_{i_1 i_3}a_{i_2 i_4}a_{i_1 i_4}a_{i_2 i_3}}. \end{aligned}$$

However, Lemma 3 does not constitute the sufficient conditions as some sign combinations in (28) are not allowed. In what follows, to achieve the necessary and sufficient existence conditions we solve the problem of appropriate signs in (28) and their relationships to the signs of the elements h_{ij} signs in (25)–(27). First, however introduce the following notations

$$s(i,j) := \begin{cases} 1 & a_{ij} > 0 \\ -1 & a_{ij} < 0 \end{cases} \tag{29}$$

$$s_{kl} = \begin{cases} 1 & h_{kl} > 0 \\ -1 & h_{kl} < 0 \end{cases} \tag{30}$$

$\aleph(\{\cdots\})$, which denotes the cardinality (the number of elements) of the set $\{\cdots\}$, and $S(i_1, j_2, j_3, j_4)$ denotes the sign of the transversale $A_r := A_{i_1 j_2 j_3 j_4}$. Basing ourselves on these notations, it is possible to obtain the following "sign" relationships, first for $k = 4$.

Lemma 4. *While* (25)–(26) *are valid then*

$$S(i_1, j_2, j_3, j_4) = s_{i_1 j_2} s_{j_2 j_3} s_{j_3 j_4} s_{j_4 j_1} \tag{31}$$

and while (25), (27) *are valid then*

$$S(i_1, j_2, j_3, j_4) = (-1)^{\frac{1}{2} w(i_1, j_2, j_3, j_4)} s_{i_1 j_2} s_{j_2 j_3} s_{j_3 j_4} s_{j_4 j_1} \tag{32}$$

where $w(i_1, j_2, j_3, j_4) := \aleph \{\{k, l\}$
$\in \{\{i_1, j_2\}, \{j_2, j_3\}, \{j_3, j_4\}, \{j_4, j_1\}\} : s(k,l) = -1\}$.

Proof. The part "1" is straightforward when the part "2" is to guarantee reality of the respective A_r when possible complex matrix elements. Note also that due to this $w(i_1, j_2, j_3, j_4)$ can be equal 0, 2 or 4. □

Lemma 5. $\forall \{i_1, i_2, i_3, i_4\} \in C_4 \{1, 2, \ldots, n+1\}$

$$S(i_1, i_3, i_2, i_4) = (-1)^{\omega(i_1, i_2, i_3, i_4)} S(i_1, i_2, i_3, i_4) S(i_1, i_2, i_4, i_3) \tag{33}$$

where

1. $\omega(i_1, i_2, i_3, i_4) := \aleph \{\{k, l\} \in \{\{i_1, i_3\}, \{i_2, i_3\}, \{i_3, i_4\}\} : s(k,l) = -1\} + 1$ (34)

if (25)–(26) *are valid, and*

2. $\omega(i_1, i_2, i_3, i_4) := \dfrac{1}{2} [w(i_1, i_2, i_3, i_4) + w(i_1, i_2, i_4, i_3) + w(i_1, i_3, i_2, i_4)] + 1$ (35)

if (25), (27) *are valid.*

Proof.

1. It is straightforward by noting that

$$\begin{aligned} S(i_1, i_2, i_3, i_4) &= (s_{i_1 i_2} s_{i_3 i_4})(s_{i_2 i_3} s_{i_4 i_1}) := \sigma_{i_3 i_4} \sigma_{i_2 i_3} \\ S(i_1, i_2, i_4, i_3) &= (s_{i_1 i_2} s_{i_4 i_3})(s_{i_2 i_4} s_{i_3 i_1}) := \sigma'_{i_3 i_4} \sigma_{i_3 i_1} \\ S(i_1, i_3, i_2, i_4) &= (s_{i_1 i_3} s_{i_2 i_4})(s_{i_3 i_2} s_{i_4 i_1}) := \sigma'_{i_3 i_1} \sigma'_{i_2 i_3} \end{aligned}$$

and

$$\sigma'_{ij} = \begin{cases} \sigma_{ij} & s(i,j) = -1 \\ -\sigma_{ij} & s(i,j) = 1. \end{cases}$$

2. Note that due to

$$\begin{aligned} (s_{i_1 i_2})^2 &= 1, s_{i_3 i_4} s_{i_4 i_3} = -1 \\ s_{i_1 i_3} s_{i_3 i_2} &= s_{i_3 i_1} s_{i_3 i_2} \end{aligned}$$

we have

$$\begin{aligned} S(i_1, i_2, i_3, i_4)\, S(i_1, i_2, i_4, i_3) &= (-1)^{\frac{1}{2}[w(i_1,i_2,i_3,i_4)+w(i_1,i_2,i_4,i_3)]+1} \\ &\quad \times s_{i_1 i_3} s_{i_2 i_4} s_{i_3 i_2} s_{i_4 i_1} \end{aligned}$$

which together with (32) completes the proof. □

Note that from the above analysis it is straightforward to see that not all signs

$$(j_1, j_2, j_3, j_4), \{j_1, j_2, j_3, j_4\} \in \mathbb{R}\{i_1, i_2, i_3, i_4\}$$

where $\{i_1, i_2, i_3, i_4\} \in C_4\{1, 2, \ldots, n+1\}$ can be arbitrarily chosen. However, the next "sign" Lemma will play a crucial role in development of the so-called sign basis, i.e., these $\{j_1, j_2, j_3, j_4\}$ $\in \mathbb{R}\{i_1, i_2, i_3, i_4\}$, $\{i_1, i_2, i_3, i_4\} \in C_4\{1, 2, \ldots, n+1\}$ for which signs can be arbitrarily chosen.

Lemma 6. $\forall\ \{j_1, j_2, j_3, j_4\} \in \mathbb{R}\{i_1, i_2, i_3, i_4\}$, $\{i_1, i_2, i_3, i_4\} \in C_4\{1, 2, \ldots, n+1\}$, $\forall k \in \{1, 2, \ldots, n+1\} \setminus \{j_1, j_2, j_3, j_4\}$ *the sign* $S(j_1, j_2, j_3, j_4)$ *can be determined as*

$$S(i_1, j_2, j_3, j_4) = s(j'_1, k) s(k, j'_3) S(j'_1, j'_2, j'_3, k)\, S(j'_1, k, j'_3, j'_4) \tag{36}$$

where $\{j'_1, j'_2, j'_3, j'_4\}$ *is any cyclic permutation of* $\{j_1, j_2, j_3, j_4\}$.

Proof. It is a clear consequence of

$$h_{j_1 j_2} h_{j_2 j_3} h_{j_3 j_4} h_{j_4 j_1} = \frac{\left(h_{j'_1 j'_2} h_{j'_2 j'_3} h_{j'_3 k} h_{k j'_1}\right)\left(h_{j'_1 k} h_{k j'_3} h_{j'_3 j'_4} h_{j'_4 j'_1}\right)}{a_{j'_1 k} a_{k j'_3}}.$$ □

Due to these results, we can choose arbitrarily $S(1, 2, 3, l)$ and $S(1, 2, l, 3)$, which yields

$$S(1, 3, 2, l) = (-1)^{\omega(1,2,3,l)}\, S(1, 2, 3, l) S(1, 2, l, 3) \tag{37}$$

$l = 4, 5, \ldots, n+1$. Moreover, the following signs are determined by this choice

$$S(1, k, 2, l) = s(1, 3) s(2, 3) S(1, 3, 2, k) S(1, 3, 2, l) \tag{38}$$

$k = 4, 5, \ldots, n+1$, $\quad l = k+1, k+2, \ldots, n+1$. Also, the signs $S(1, 2, k, l)$ and $S(1, 2, l, k)$ can be partially arbitrarily chosen, i.e., under the condition

$$S(1, 2, k, l) S(1, 2, l, k) = (-1)^{\omega(1,2,k,l)}\, s(1, 3) s(2, 3) S(1, k, 2, 3) S(1, 3, 2, l) \tag{39}$$

Note also that the signs of A_r for $k > 4$ are not independent but are related to the aforementioned arbitrarily chosen signs. For example,

$$S(i_1, j_2, j_3, j_4, j_5, j_6) = -s(i_1, j_4) S(i_1, j_2, j_3, j_4)\, S(i_1, j_4, j_5, j_6). \tag{40}$$

Now, applying aforementioned sign relationships allows us to present the necessary and sufficient conditions for the existence of a denominator-degree minimal realization.

Theorem 3. *Given an $n+1$ variate polynomial $a_f(s_1,\ldots,s_{n+1})$ of (5) with nonzero coefficients a_{ij}. Then the necessary and sufficient conditions for the existence of a denominator-degree minimal realization are that all the polynomial $a_f(s_1,\ldots,s_{n+1})$ coefficients have to satisfy*

$$a_{i_1 i_2 \cdots i_{2l+1}} = 0, \forall \{i_1, i_2, \ldots, i_{2l+1}\} \in C_{2l+1}\{1,2,\ldots,n+1\}$$

$$\begin{aligned} a_{i_1 i_2 \cdots i_{2l}} &= \left(\sum_{z \in \mathbb{Z}^2\{i_1 i_2 \cdots i_k\}} \prod_{z_v \in z} \pm\sqrt{a_{z_v}} \right)^2 \\ \forall \{i_1, i_2, \ldots, i_{2l}\} &\in C_{2l}\{1,2,\ldots,n+1\} \end{aligned} \tag{41}$$

or

$$a_{i_1 i_2 \cdots i_{2l}} = \left(\sum_{z \in \mathbb{Z}^2\{i_1 i_2 \cdots i_k\}} \prod_{z_v \in z} \pm i_{z_v}\sqrt{a_{z_v}} \right)^2 \tag{42}$$

where $i_{z_v} = 1$ *or* $, i_{z'_v} \neq i_{z"_v}, i^2_{z'_v} = i^2_{z"_v} = 1, i_{z'_v} i_{z"_v} = -1.$

Proof. It is a straightforward consequence of the analysis above. □

Note that in the first case all coefficients must be positive.

The last problem to solve is here to find appropriate signs of elements h_{ij} with respect to signs in the conditions (41) or (42), which is accomplished by the following sign algorithm based on previous considerations. Start hence from arbitrarily chosen basic signs $S(1,3,2,k)$, $S(1,3,2,l), k,l = 4,5,\ldots,n+1$, and from $S(1,2,k,l), k = 4,5,\ldots,n$, $l = k+1, k+2, \ldots, n+1$, chosen to satisfy (39). Note that they can be rewritten as

$$\begin{aligned} S(1,2,k,l) &= h_{k-2} g^k_{l-k}, \\ S(1,2,l,k) &= h_{l-2} d^k_{l-k} \end{aligned} \tag{43}$$

$k = 3,\ldots,n, l = k+1,\ldots,n+1$, where

$$\begin{aligned} h_i &:= s_{12} s_{2,i+2}, i = 1,\ldots,n-1, \\ g^j_i &:= s_{i,i+j} s_{i+j,1}, \\ d^j_i &:= s_{j1} s_{i+j,i}, \\ j &= 3,\ldots,n, i = 1,\ldots,n+1-j. \end{aligned} \tag{44}$$

Now, assuming the sign base $S(1,3,2,k)$, $S(1,3,2,l), k,l = 4,5,\ldots,n+1$, and $S(1,2,k,l), k = 4,5,\ldots,n$, $l = k+1, k+2, \ldots, n+1$, which satisfy (39) are known and then we can calculate

$$\begin{aligned} g^k_{l-k} &= S(1,2,k,l)\, h_{k-2} \\ d^k_{l-k} &= S(1,2,l,k)\, h_{k-2} \end{aligned} \tag{45}$$

if (25)–(26) are valid for h_{ij}, and

$$\begin{aligned} g^k_{l-k} &= (-1)^{w(1,2,k,l)/2}\, S\,(1,2,k,l)\, h_{k-2} \\ d^k_{l-k} &= (-1)^{w(1,2,l,k)/2}\, S\,(1,2,l,k)\, h_{k-2} \end{aligned} \tag{46}$$

if (25), (27) are valid. In both cases the signs h_{k-2} are arbitrary.

At the final stage we can determine the appropriate signs of matrix H elements h_{ij} using the following algorithm.

1. Put the signs $s_{12}, s_{23}, s_{24}, \ldots, s_{2,n+1}, s_{31}$ arbitrary,
2. Choose $s_{l3}, l = 4, \ldots, n+1$ according to

$$s_{l3} = d^3_{l-3} s_{31} \tag{47}$$

3. and $s_{l1}, l = 4, \ldots, n+1$ according to

$$s_{l1} = g^3_{l-3} s_{31} = \begin{cases} g^3_{l-3} s_{13} & \text{(25)–(26) are valid and } s(3,l) = -1 \\ -g^3_{l-3} s_{13} & \text{in the rest of cases} \end{cases} \tag{48}$$

4. and finally for $k = 4, 5, \ldots, n$, $l = k+1, k+2, \ldots, n+1$,

$$s_{lk} = s_{k1} d_{l-k,k} \tag{49}$$

which finishes the algorithm.

5. Conclusions

The problem of how to construct the state-space realizations for a given 2D MIMO system, written, for example, in the 2D transfer function matrix form, is central to various applications of 2D systems theory, such as, multidimensional filters analysis and synthesis, and has received a considerable attention in the literature. There is no general solution yet to the problem of obtaining a minimal realization (both the denominator-degree and the least-order realization) in the general nD systems case but there are several approaches, which aim at developing a general solution of the problem of determining the minimal possible dimension. These include the work of Guiver, Bose (1982) [10], [11], [21], [5] and the EOA algorithm due to the author. In this paper we have developed a method of examining if there exists a denominator-degree minimal realization and constructing it for the particular class of SISO nD systems characterized by multi-linear transfer function polynomials, but moreover odd. This subclass of systems play a significant role in practical implementations of system and circuit theory as it has strong links to important classes of so-called reactance functions [13] and so-called positive systems [12].

Acknowledgments

This work is partially supported by Ministry of Scientific Research and Information Technology under the project 3 T11A 008 26.

The author feels indebted to express his gratitude to the reviewers for their valuable comments.

Krzysztof Galkowski is with The University of Zielona Gora but also is a visiting Professor of The University of Southampton and during the academic year 2004–2005 is on sabbatical leave to The University of Wuppertal under the Gerhard Mercator Guest Professorship founded by DFG. In 2004 he has received the Siemens Poland scientific award for the research in the nD systems area.

References

[1] Agler J. and McCarthy J.E. (2002) Pick interpolation and Hilbert function spaces, Amer. Math. Soc., Providence, RI.

[2] Ball J.A. and Trent T.T. (1998) Unitary colligations, reproducing kernel Hilbert spaces, and Nevanlinna-Pick interpolation in several variables, J. Functional Analysis, vol. 157, 1–61.

[3] Bose N.K. (1976) New techniques and results in multidimensional problems, Journal of the Franklin Institute, Special issue on recent trends in systems theory.

[4] Bose N.K. (1982) Applied Multidimensional Systems Theory, New York, Van Nostrand Reinhold.

[5] Cockburn J.C., Morton B.G. (1997) Linear fractional representations of uncertain systems, Automatica, vol. 33, no.7, 1263–1271.

[6] Eising R. (1978) Realization and stabilization of 2-D systems IEEE Trans. on Automatic Control, AC-23, 793–799.

[7] Fornasini E., Marchesini G. (1978) Doubly-indexed dynamical systems, Math. Syst. Theory, vol. 12, 59–72.

[8] K. Galkowski (2002), State-space Realizations of Linear 2-D Systems with Extensions to the General nD ($n > 2$) Case, Lecture Notes in Control and Information Sciences, vol. 263, Springer, London.

[9] Galkowski K. (2001), Minimal state-space realization of the particular case of SISO nD discrete linear systems, International Journal of Control, Vol. 74, No. 13, 1279–1294.

[10] Guiver J.P., Bose N.K. (1982), Polynomial matrix primitive factorization over arbitrary coefficient field and related results, IEEE Trans. on Circuits and Systems, Vol. CAS-29, No. 10, 649–657.

[11] Kaczorek T. (1985) Two Dimensional Linear Systems, Lecture Notes in Control and Information Sciences, No. 68, Berlin: Springer-Verlag.

[12] Kaczorek T. (2002) Positive 1D and 2D Systems, Berlin: Springer-Verlag.

[13] Koga T. (1968) Synthesis of finite passive N-ports with prescribed positive real matrices of several variables, IEEE Trans. on Circuit Theory, Vol. CT-15, No. 1, 2–23.

[14] Kung S.Y. et al. (1977) New results in 2-D systems theory, part II, Proc. IEEE, 945–961.

[15] Mentzelopoulos S.H., Theodorou N.J. (1991) N-dimensional minimal state-space realization, IEEE Trans. on Circuits and Systems, vol. 38, no. 3, 340–343.

[16] Premaratne K., Jury E.I., Mansour M. (1997) Multivariable canonical forms for model reduction of a 2-D discrete time systems, IEEE Trans. on Circuits and Systems, vol. 37, no. 4, 488–501.

[17] Pugh A.C., McInerney S.J., Boudellioua M.S., Hayton G.E. (1998) Matrix pencil of a general 2-D polynomial matrix, Int. J. Control, vol. 71, no. 6, 1027–1050.

[18] Raghuramireddy D., Unbehauen R., 1991, Realization of 2-D denominator-separable digital filter transfer functions using complex arithmetic, Multidimensional Systems and Signal Processing, vol. 2, 319–336.

[19] R. Roesser, A discrete state space model for linear image processing, IEEE Trans. Automatic Control, vol. 20, pp. 1–10, 1975.

[20] Sontag E. (1978) On first-order equations for multidimensional filters, IEEE Trans. Acoust. Speech Signal Processing, Vol. 26, No. 5, 480–482.

[21] Żak S. H., Lee E. B., Lu W. S. (1986) Realizations of 2-D filters and time delay systems, IEEE Trans on Circuits and Systems, Vol. CAS-33, No. 12, 1241–1244.

K. Galkowski
University of Zielona Gora
Institute of Control and Computational Engineering
Podgorna Str. 50
65-246 Zielona Gora, Poland
e-mail: `k.galkowski@issi.uz.zgora.pl`

Operator Theory:
Advances and Applications, Vol. 160, 195–216

Continuity in Weighted Besov Spaces for Pseudodifferential Operators with Non-regular Symbols

Gianluca Garello and Alessandro Morando

Abstract. The authors state and prove a result of continuity in weighted Besov spaces for a class of pseudodifferential operators whose symbol $a(x,\xi)$ admits a finite number of bounded derivatives with respect to ξ and is of weighted Besov type in the x variable.

Mathematics Subject Classification (2000). 35S05; 35A17.

Keywords. Pseudodifferential operators, Besov spaces.

1. Introduction

Let $a(x,\xi)$ be in the Schwartz class of tempered distributions $\mathcal{S}'(\mathbb{R}^n_x \times \mathbb{R}^n_\xi)$. We consider the pseudodifferential operator defined in a formal way, for any rapidly decreasing function $u(x) \in \mathcal{S}(\mathbb{R}^n)$, by:

$$a(x,D)u = (2\pi)^{-n} \int e^{ix\cdot\xi} a(x,\xi)\hat{u}(\xi)\, d\xi, \tag{1.1}$$

where $\hat{u}(\xi)$ is the Fourier transform of u and the other notations are standard, in particular $x\cdot\xi = \sum_{j=1}^n x_j\xi_j$. We are primarily interested in studying the conditions on the symbol $a(x,\xi)$ which allow $a(x,D)$ to belong to the space $\mathcal{L}(L^p)$ of bounded linear operators on L^p, $1 < p < \infty$.

Aiming at non-specialists we begin by giving a short review of known results. In applications to linear partial differential equations with C^∞ coefficients one deals, as a rule, with symbols $a(x,\xi)$ which are smooth functions both in x and ξ. We first recall some basic result in this case. Namely, let us refer to the Hörmander

The authors are supported by F.I.R.B. grant of Italian Government.

symbol classes $S^m_{\rho,\delta}$, $m \in \mathbb{R}$, $0 \le \delta \le \rho \le 1$, given by the sets of those $a(x,\xi)$ satisfying

$$|\partial_\xi^\alpha \partial_x^\beta a(x,\xi)| \le c_{\alpha,\beta}(1+|\xi|)^{m-\rho|\alpha|+\delta|\beta|}, \quad x,\xi \in \mathbb{R}^n, \quad \alpha,\beta \in \mathbb{Z}^n_+. \tag{1.2}$$

As a natural extension of the Mikhlin-Hörmander Lemma on Fourier multipliers we have $a(x,D) \in \mathcal{L}(L^p)$ provided that $a(x,\xi) \in S^0_{1,0}$, i.e., $a(x,D)$ is a classical pseudodifferential operator. More generally $a(x,D) \in \mathcal{L}(L^p)$ if $a(x,\xi) \in S^0_{1,\delta}$, for some $\delta < 1$, for the proof see for example Taylor [[18], Ch. XI, §1, §2, §3]. On the other hand it is well known since a counter-example of Hirschmann and Wainger, well summarized in [5], that in general $a(x,D) \notin \mathcal{L}(L^p)$ when $a(x,\xi)$ belongs to $S^0_{\rho,\delta}$ with $\rho < 1$. Namely Hörmander [8], Fefferman [5] proved that the class Op $S^{-m}_{\rho,\delta}$ of pseudodifferential operators with symbols in $S^{-m}_{\rho,\delta}$, $0 \le \delta \le \rho < 1$, is contained in $\mathcal{L}(L^p)$, for $1 < p < \infty$ when $m \ge m_p$ where the critical order is given by $m_p = n(1-\rho)\left|\frac{1}{2} - \frac{1}{p}\right|$.

Let us now deal with pseudodifferential operators with non-regular symbols. Their importance in the literature is now increasing, because of the applications to linear partial differential equations with non-smooth coefficients and non-linear equations, as well as to applicative problems of a different nature (signal theory, quantization etc ...). Willing to review some results in this connection, we may quote as a basic example pseudodifferential operators with symbols which are differentiable with respect to the ξ variable a finite number of times and which belong to generalized Hölder classes with respect to x. As to L^p continuity of such a kind of operators, a very important part is covered by the works of M. Nagase [13], [14], [15], [16]. As a significant example we recall the main result of [15]. For $1 < p < \infty$, $0 \le \delta < \rho \le 1$ Nagase assumes that for some suitable positive constants $k = k(n)$, $\mu = \mu(n,\rho,\delta)$ the symbol $a(x,\xi)$ satisfies for any $\xi \in \mathbb{R}^n$:

$$\begin{aligned} &\sup_x |\partial_\xi^\alpha \partial_x^\beta a(x,\xi)| \le c_{\alpha,\beta}(1+|\xi|)^{-m_p-\rho|\alpha|+\delta|\beta|} \text{ for } |\alpha| \le k, \ |\beta| < \mu; \\ &\|\partial_\xi^\alpha a(x,\xi)\|_{C^\mu} \le c_\alpha(1+|\xi|)^{-m_p-\rho|\alpha|+\delta\mu}, \quad |\alpha| \le k, \end{aligned} \tag{1.3}$$

where $\|\cdot\|_{C^\mu}$ is the standard Hölder norm of order μ. Then, provided that as before $m_p = n(1-\rho)\left|\frac{1}{p} - \frac{1}{2}\right|$, it follows that $a(x,D) \in \mathcal{L}(L^p)$. More precisely $k(n) = \left[\frac{n}{2}\right]$ if $2 \le p < \infty$ and $k(n) = n+1$ if $1 < p < 2$.

Among the results that are worthy to be noticed we mention those of J. Marschall [10], [11] obtained using techniques of the paradifferential calculus of Bony and Meyer (see [1], [12]) which in some way generalize the results of Nagase.

Let us quote also the Sugimoto paper [17], where L^p continuity is studied for pseudodifferential operators with symbols $a(x,\xi)$ in weighted Besov spaces with respect to both variables, with a loss of Besov regularity estimated by means of $n\left(\frac{1}{2} - \frac{1}{p}\right)$, for $2 \le p < \infty$. In fact, the guiding thread in most of the literature on L^p continuity of pseudodifferential operators is characterized by the critical order

$m_p \sim n \left| \frac{1}{p} - \frac{1}{2} \right|$ and much effort is made in estimating it as well as possible, under minimal assumptions on the symbol.

Returning to smooth symbols, a different point of view was offered by Taylor: keeping as a model the proof of the L^p continuity of the operators in Op $S^0_{1,0}$, he found that it may be adapted to proving the boundedness of a suitable subclass of Op $S^0_{\rho,0}$, $0 < \rho < 1$, by replacing the Mikhlin-Hörmander Lemma on Fourier multipliers [[18], Ch. XI] with the analogous one due to Marcinkiewicz-Lizorkin, [9], see the next Lemma 2.4. Namely Taylor proved that Op $M^0_\rho \subset \mathcal{L}(L^p)$, $0 < \rho < 1$, $1 < p < \infty$, where the smooth symbol classes M^m_ρ, $m \in \mathbb{R}$, are given by the functions $a(x,\xi) \in S^m_{\rho,0}$ such that $\xi^\gamma \partial^\gamma_\xi a(x,\xi) \in S^m_{\rho,0}$, for any multi-index γ with components 0 or 1. In some sense we have in this case the critical order $m_p = 0$.

The present paper, which follows Taylor's basic layout, is an attempt to give a general result of L^p continuity for pseudodifferential operators with similar non-regular symbols. More precisely pseudodifferential operators are considered, corresponding to symbols $a(x,\xi)$ of Taylor's type, but with a finite number of derivatives with respect to ξ and of weighted Besov type $B^{s,\Lambda}_{p,q}$ with respect to x; here $s > 0$, $1 < p < \infty$, $1 < q < \infty$ and Λ is a suitable weight function.

The paper runs as follows. In Section 2 we first introduce the weight functions $\Lambda(\xi)$. The weighted Besov spaces $B^{s,\Lambda}_{p,q}$ are then characterized by means of a suitable partition of unity due to Triebel, see [20], [21]. Corresponding properties are given to be used in the following part of the paper.

In §3 we introduce the non-regular symbols $a(x,\xi)$, first defined as finitely differentiable with respect to ξ and in a general Banach space with respect to x. It is shown in this section that such non-regular symbols can be decomposed in an expansion of elementary symbols, see Definition 3.2, following a technique of Coifman-Meyer [4].

The remaining part of the paper is devoted to the proof of the principal result: Theorem 3.4 about the boundedness of the pseudodifferential operators $a(x,D)$ between two weighted Besov spaces under suitable conditions. Let us notice that our result appears to be new also in the classical case when $\Lambda(\xi) = \sqrt{1+|\xi|^2}$. Namely in this case we get an extension of Proposition 4.5 of Taylor [[18], Ch. XI] for smooth symbols. Let us emphasize that, with respect to Nagase and Marschall, our assumptions are stronger, but we get the effective L^p continuity of zero order pseudodifferential operators, without any loss of regularity, i.e., $m_p = 0$.

In fact, sharp L^p estimates are essential for the applications to non-linear equations, by following the line of Bony-Meyer [1], [12], Beals and Reed [2], Taylor [[19], Ch. 2–3]. After further development of symbolic calculus, applications of our result are expected in the study of the regularity of solutions to some kind of non-linear partial differential equations, generalizing the multi-quasi-elliptic equations considered in Garello, Morando [6], [7]. This will be detailed in future papers.

2. Weighted Besov spaces

In the whole paper $\Lambda(\xi)$ will be a *weight function* satisfying the following definition.

Definition 2.1. $\Lambda(\xi) \in C^\infty(\mathbb{R}^n)$ is a weight function provided that the following assumptions are satisfied with some positive constants $C > 0$, $\mu_0 \geq 1$.

1. $\Lambda(\xi) \geq \frac{1}{C}(1+|\xi|)^{\mu_0}, \qquad \xi \in \mathbb{R}^n$;
2. for every $\gamma \in \mathbb{Z}^n_+$ there exists $C_\gamma > 0$ such that
$$\prod_{j=1}^n (1+\xi_j^2)^{\frac{\gamma_j}{2}} |\partial^\gamma \Lambda(\xi)| \leq C_\gamma \Lambda(\xi), \qquad \xi \in \mathbb{R}^n;$$
3. $\Lambda(t\xi) \leq C\Lambda(\xi), \qquad t,\xi \in \mathbb{R}^n, \quad \max_{1\leq j\leq n} |t_j| \leq 1, \quad t\xi := (t_1\xi_1, \dots, t_n\xi_n)$;
4. *(δ-condition)* for some $0 < \delta < 1$ there holds
$$\Lambda(\xi) \leq C\left(\Lambda(\eta) + \Lambda(\xi-\eta) + \Lambda(\eta)^\delta \Lambda(\xi-\eta)^\delta\right), \quad \xi, \eta \in \mathbb{R}^n. \tag{2.1}$$

As it has been shown in Triebel [[20], Lemma 2.1/2] we can always find $\mu_1 \geq \mu_0$ such that $\Lambda(\xi) < C(1+|\xi|)^{\mu_1}$.

Example. The basic examples of weight functions are the *elliptic weights* $\Lambda_m(\xi) := \sqrt{1+\sum_{j=1}^n \xi_j^{2m}}$. Of greater interest are the *multi-quasi-elliptic* weights defined as $\Lambda_{\mathcal{P}}(\xi) := \sqrt{\sum_{\alpha\in\mathcal{V}(\mathcal{P})} \xi^{2\alpha}}$ where $\mathcal{V}(\mathcal{P})$ are the vertices of a *complete Newton polyhedron*, see for instance [7]. Other examples, such as $\langle\xi\rangle^s[\log(2+\langle\xi\rangle)]^t$, where $s,t > 1$ and $\langle\xi\rangle = \sqrt{1+|\xi|^2}$, are given in Triebel [20].

For a fixed $H > 1$ let us consider the decomposition of $\mathbb{R}^n$, given by the sequence of n-intervals $P^{(H)}_{h,\lambda}$ defined, for $h = (h_1, \dots, h_n) \in \mathbb{Z}^n_+$, $\lambda = (\lambda_1, \dots, \lambda_n) \in \mathbb{E}^n = \{-1,1\}^n$, by
$$P^{(H)}_{h,\lambda} := \left\{\xi \in \mathbb{R}^n \quad ; \quad \frac{1}{H} 2^{h_j}\eta_{h_j} \leq \lambda_j\xi_j \leq H2^{h_j+1} \quad ; \quad j = 1, \dots, n\right\}, \tag{2.2}$$
where $\eta_h = -1$ if $h = 0$ and $\eta_h = 1$ if $h > 0$.

Remark 2.2. It is easy to prove the existence of a positive number $N_0 = N_0(H)$ such that, for any $\lambda, \epsilon \in \mathbb{E}^n$, $P^{(H)}_{h,\lambda} \cap P^{(H)}_{k,\varepsilon} = \emptyset$ when $|h_j - k_j| > N_0$ for some $j = 1, \dots n$; see [7] and [20].

Following again Triebel [20], we introduce also the next

Definition 2.3 (Partition of unity). For a fixed $H > 1$, $\phi^{(H)}$ is the set of all sequences $\{\varphi_{h,\lambda}\}_{\substack{h\in\mathbb{Z}^n_+\\ \lambda\in\mathbb{E}^n}} \subset C_0^\infty$ such that, for every $h \in \mathbb{Z}^n_+$, $\lambda \in \mathbb{E}^n$, supp $\varphi_{h,\lambda} \subset P^{(H)}_{h,\lambda}$, $\sum_{h\in\mathbb{Z}^n_+,\lambda\in\mathbb{E}^n} \varphi_{h,\lambda}(\xi) = 1$ and for any $\alpha \in \mathbb{Z}^n_+$ there exists a positive constant C_α such that
$$|\partial^\alpha_\xi \varphi_{h,\lambda}(\xi)| \leq C_\alpha 2^{-h\cdot\alpha}, \qquad \xi \in \mathbb{R}^n, \quad h \in \mathbb{Z}^n_+, \quad \lambda \in \mathbb{E}^n. \tag{2.3}$$

In the remaining part of the paper, the subscripts h and λ will be always understood to run through the sets $\mathbb{Z}_+^n$ and $\mathbb{E}^n$ respectively.

For $1 \le p \le \infty$, $1 \le q \le \infty$ and any sequence $\{u_{h,\lambda}\} \subset L^p(\mathbb{R}^n)$ we define $\|\{u_{h,\lambda}\}\|_{\ell^q(L^p)} := (\sum_{h,\lambda} \|u_{h,\lambda}\|_p^q)^{\frac{1}{q}}$, with obvious modification for $q = \infty$.

Moreover we will write $\mathcal{F}_{x\to\xi}u(\xi) = \hat{u}(\xi)$ for the Fourier transform of a distribution $u \in \mathcal{S}'(\mathbb{R}^n)$ and $\mathcal{F}^{-1}_{\xi\to x}$ for the inverse Fourier transformation.

Let us consider a function $m(\xi)$ on $\mathbb{R}^n$; we set $m(D)u(x) = \mathcal{F}^{-1}_{\xi\to x}(m(\xi)\hat{u}(\xi))$ for any $u \in \mathcal{S}'(\mathbb{R}^n)$, provided that the expressions involved make sense; as usual $m(D)$ is called Fourier multiplier. The next is a classical result in the theory of Fourier multipliers (cf. [9], [18]).

Lemma 2.4 (of Lizorkin-Marcinckiewicz, [18], **Ch.XI, Prop. 4.5).** *Let $m(\xi)$ be a continuous function together with its derivatives $\partial^\gamma m(\xi)$ for any $\gamma \in \mathbb{K}^n := \{0,1\}^n$. If there exists a constant $B > 0$ such that*

$$|\xi^\gamma \partial^\gamma m(\xi)| \le B, \qquad \xi \in \mathbb{R}^n,\ \gamma \in \mathbb{K}^n, \tag{2.4}$$

then for every $1 < p < \infty$ we can find a constant $A_p > 0$, only depending on p, B and the dimension n, such that:

$$\|m(D)u\|_p \le A_p\|u\|_p, \qquad u \in \mathcal{S}(\mathbb{R}^n). \tag{2.5}$$

Remark 2.5. The Fourier multipliers $\varphi_{h,\lambda}(D)$ are L^p continuous, for every $1 < p < \infty$; indeed by using the inclusions $\mathrm{supp}\varphi_{h,\lambda} \subset P^{(H)}_{h,\lambda}$ and inequalities (2.3), the functions $\varphi_{h,\lambda}$ are shown to satisfy the estimates (2.4) with a positive constant B independent of h and λ.

For $\{\varphi_{h,\lambda}\} \in \phi^{(H)}$, $1 < p < \infty$, $1 \le q \le \infty$, $s \in \mathbb{R}$ we can introduce, with obvious modification for $q = \infty$, the following norms:

$$\|u\|_{B^{s,\Lambda}_{p,q}} := \|\{\Lambda(c^{(H)}_{h,\lambda})^s u_{h,\lambda}\}\|_{\ell^q(L^p)} := \left(\sum_{h,\lambda} \|\Lambda(c^{(H)}_{h,\lambda})^s u_{h,\lambda}\|_p^q\right)^{\frac{1}{q}}. \tag{2.6}$$

Here and later on, for any $u \in \mathcal{S}'(\mathbb{R}^n)$, we set $u_{h,\lambda} = \varphi_{h,\lambda}(D)u$ where $c^{(H)}_{h,\lambda}$ is the center of the n-interval $P^{(H)}_{h,\lambda}$.

We denote by $B^{s,\Lambda}_{p,q}$ the Banach space of tempered distributions $u \in \mathcal{S}'(\mathbb{R}^n)$ whose norm in (2.6) is finite.

Remark 2.6. It may be shown that for different choices of systems $\{\varphi_{h,\lambda}(\xi)\} \in \Phi^{(H)}$ the norms in (2.6) are equivalent. For H, K greater than 1, there holds $\frac{1}{C} < \frac{\Lambda(c^{(H)}_{h,\lambda})}{\Lambda(c^{(K)}_{k,\epsilon})} < C$ when $|h-k| \le A$, for some $A > 0$ and $C > 1$ independent of k, h.

The next propositions 2.7, 2.8 are proved by adapting to our context the arguments used by Triebel [22] in the framework of classical Besov spaces $B^s_{p,q}$ (cf. also [21]).

Proposition 2.7 (Nikol'skij representation, [21] Theorem 2.1/2). *Let us consider* $u = \sum_{h,\lambda} u_{h,\lambda}$, *with convergence in* $\mathcal{S}'(\mathbb{R}^n)$ *and assume that* supp $\hat{u}_{h,\lambda} \subset P^{(H)}_{h,\lambda}$. *Then for any* $s \in \mathbb{R}$, $1 < p < \infty$ *and* $1 \leq q \leq \infty$ *there exists a constant* $C = C_{s,p,q} > 0$ *such that:*

$$\|u\|_{B^{s,\Lambda}_{p,q}} \leq C \left(\sum_{h,\lambda} \Lambda(c^{(H)}_{h,\lambda})^{sq} \|u_{h,\lambda}\|_p^q \right)^{\frac{1}{q}}. \tag{2.7}$$

Since $\mathbb{Z}^n_+$ does not have a natural order, here and in the following the convergence in $\mathcal{S}'(\mathbb{R}^n)$ of the series $\sum_{h,\lambda} u_{h,\lambda}$ must be assumed independent of the particular order of the terms.

Proposition 2.8 ([7], Proposition 5.2). *For every* $s \in \mathbb{R}$ *and* $1 < p < \infty$ *the following continuous embeddings hold true:*

$$\mathcal{S}(\mathbb{R}^n) \subset B^{s,\Lambda}_{p,q_1} \subset B^{s,\Lambda}_{p,q_2} \subset \mathcal{S}'(\mathbb{R}^n), \quad \text{if} \quad 1 \leq q_1 < q_2 \leq \infty; \tag{2.8}$$

$$B^{s+\varepsilon,\Lambda}_{p,q_1} \subset B^{s,\Lambda}_{p,q_2}, \quad \text{if} \quad 1 \leq q_1, q_2 \leq \infty, \quad \varepsilon > 0. \tag{2.9}$$

Moreover $\mathcal{S}(\mathbb{R}^n)$ *is dense in* $B^{s,\Lambda}_{p,q}$ *for any* $1 \leq q < \infty$.

In view of the Nikol'skij inequality, see Triebel [22], we can even prove the following

Lemma 2.9 ([7], Lemma 5.1). *For any* $\alpha \in \mathbb{Z}^n_+$ *and* $1 \leq p_1 \leq p_2 \leq \infty$ *there exists a positive constant* $C = C(\alpha, p_1, p_2)$ *such that*

$$\|\partial^\alpha v\|_{p_2} \leq C 2^{h\cdot\alpha + \left(\frac{1}{p_1} - \frac{1}{p_2}\right)|h|} \|v\|_{p_1}, \tag{2.10}$$

for any $v \in \mathcal{S}'(\mathbb{R}^n)$ *such that* supp $\hat{v} \subset P^{(H)}_{h,\lambda}$.

Proposition 2.10 ([7], Proposition 5.3). *For any* $s \in \mathbb{R}$, $1 < p_1 < p_2 < \infty$ *and* $1 \leq q \leq \infty$, *the following continuous embedding holds true:*

$$B^{s+\frac{n}{\mu_0}\left(\frac{1}{p_1} - \frac{1}{p_2}\right),\Lambda}_{p_1,q} \subset B^{s,\Lambda}_{p_2,q}. \tag{2.11}$$

Proposition 2.11. *For any* $s \in \mathbb{R}$, $1 < p < \infty$, $1 \leq q \leq \infty$ *and* $H > 1$, *we can find a positive constant* $M = M_{s,p,q,H}$ *such that for every* $u \in \mathcal{S}'(\mathbb{R}^n)$:

$$\|u_{h,\lambda}\|_\infty \leq M \|u\|_{B^{s,\Lambda}_{p,q}} \Lambda(c^{(H)}_{h,\lambda})^{-s+\frac{n}{\mu_0 p}}. \tag{2.12}$$

Proof. Since the continuous embedding $B^{s,\Lambda}_{p,q} \subset B^{s,\Lambda}_{p,\infty}$ holds true for any $1 \leq q < \infty$, we may restrict ourselves to proving (2.12) only for $q = \infty$ without loss of generality. Let u belong to $B^{s,\Lambda}_{p,\infty}$. By definition (cf. (2.6)) there holds

$$\|u_{h,\lambda}\|_p \leq \|u\|_{B^{s,\Lambda}_{p,\infty}} \Lambda(c^{(H)}_{h,\lambda})^{-s}, \quad h, \lambda, \tag{2.13}$$

for a given $H > 1$ ((2.13) would be trivial if $u \notin B^{s,\Lambda}_{p,\infty}$, since $\|u\|_{B^{s,\Lambda}_{p,\infty}} = \infty$). Applying to $u_{h,\lambda}$ the Nikol'skij type inequality (2.10), with $\alpha = 0$, $p_1 = p$ and $p_2 = \infty$, gives

$$\|u_{h,\lambda}\|_\infty \le C_0 2^{\frac{|h|}{p}} \|u_{h,\lambda}\|_p, \quad h,\, \lambda, \tag{2.14}$$

with a positive C_0 independent of h, λ. Then we find estimate (2.12), gathering (2.13), (2.14) and using also $2^{|h|} \le C_1 \Lambda(c^{(H)}_{h,\lambda})^{\frac{n}{\mu_0}}$, h, λ, where C_1 depends only on H and the dimension n; the latter inequalities are an easy consequence of assumption 1 of Definition 2.1. □

For $r \in \mathbb{Z}_+$ and $\kappa \in \{-1, 1\}$ we now set:

$$L^{(H)}_{r,\kappa} := \left\{ t \in \mathbb{R};\quad \frac{1}{H} 2^r \eta_r \le \kappa t \le H 2^{r+1} \right\}, \tag{2.15}$$

with $\eta_r = 1$ if $r > 0$, $\eta_0 = -1$.

Lemma 2.12. *Let us consider $u = \sum_{h,\lambda} u_{h,\lambda}$, with convergence in $\mathcal{S}'(\mathbb{R}^n)$ such that*

$$\operatorname{supp} \hat{u}_{h,\lambda} \subset J^{(H)}_{h_1,\lambda_1} \times \cdots \times J^{(H)}_{h_n,\lambda_n}, \tag{2.16}$$

where $J^{(H)}_{h_j,\lambda_j}$ are either $L^{(H)}_{h_j,\lambda_j}$ defined in (2.15) or $\left[-H2^{h_j+1}, H2^{h_j+1}\right]$. Then for every $s \ge 0$, $\gamma > 0$, $1 < p < \infty$ and $1 \le q \le \infty$ there exists a positive constant $C = C_{s,\gamma,p,q}$ such that

$$\|u\|_{B^{s,\Lambda}_{p,q}} \le C \left(\sum_{h,\lambda} \Lambda(c^{(H)}_{h,\lambda})^{qs} 2^{q\gamma\sigma(h)\cdot h} \|u_h, \lambda\|_p^q \right)^{\frac{1}{q}}, \tag{2.17}$$

where $\sigma(h) = (\chi(h_1), \dots, \chi(h_n))$ and

$$\chi(h_j) := \begin{cases} 1, & \textit{if } J^{(H)} = [-H2^{h_j+1}, H2^{h_j+1}] \\ 0, & \textit{otherwise.} \end{cases} \tag{2.18}$$

Proof. Let us consider a partition of unity $\{\varphi_{k,\epsilon}\} \in \phi^{(K)}$, where $k \in \mathbb{Z}^n_+$, $\epsilon \in \mathbb{E}^n$, $K > 1$. Then following the arguments in [[7], Lemma 7.3], we obtain:

$$\varphi_{k,\epsilon}(D)u = \sum_{\substack{h \in E^{(N_0),n_1}_k \\ \lambda \in \mathbb{E}^n}} \varphi_{k,\epsilon}(D) u_{h,\lambda}, \tag{2.19}$$

where for any fixed $N_0 > \log_2(2HK)$ we set

$$E^{(N_0),n_1}_k := \left\{ h \in \mathbb{Z}^n_+ \colon \begin{array}{ll} h_j \ge k_j - N_0, & j = 1, \dots, n_1 \\ k_j - N_0 \le h_j \le k_j + N_0, & j = n_1 + 1, \dots, n \end{array} \right\}; \tag{2.20}$$

here, without loss of generality, we have assumed $J^{(H)}_{h_j,\lambda_j} = \left[-H2^{h+1}, h2^{h_j+1}\right]$, for $1 \le j \le n_1$ and $J^{(H)}_{h_j,\lambda_j} = L^{(H)}_{h_j,\lambda_j}$ when $n_1 + 1 \le j \le n$, for a given $1 \le n_1 \le n$. For the case $q \neq \infty$ we obtain

$$\begin{aligned}\|u\|_{B^{s,\Lambda}_{p,q}} &= \left(\sum_{k,\epsilon} \Lambda(c^{(K)}_{k,\epsilon})^{qs} \| \sum_{\substack{h\in E_k^{(N_0),n_1}\\ \lambda\in\mathbb{E}^n}} \varphi_{k,\epsilon}(D) u_{h,\lambda} \|_p^q \right)^{\frac{1}{q}} \\ &= \left(\sum_{k,\epsilon} \Lambda(c^{(K)}_{k,\epsilon})^{qs} \| \sum_{\substack{t\in E^{(N0),n_1}\\ \lambda\in\mathbb{E}^n}} \varphi_{k,\epsilon}(D) u_{k+t,\lambda} \|_p^q \right)^{\frac{1}{q}},\end{aligned} \tag{2.21}$$

where $t = h - k$ and

$$E^{(N_0),n_1} := \left\{ t \in \mathbb{Z}^n : \begin{array}{ll} t_j \geq -N_0, & j = 1, \dots, n_1 \\ -N_0 \leq t_j \leq N_0, & j = n_1 + 1, \dots, n \end{array} \right\}, \tag{2.22}$$

agreeing that $u_{k+t,\lambda} \equiv 0$ when $k_j + t_j < 0$ for some $1 \leq j \leq n$.

By the triangle inequality applied to the norm $\| \cdot \|_{\ell^q(L^p)}$ we obtain

$$\|u\|_{B^{s,\Lambda}_{p,q}} \leq \sum_{t\in E^{(N_0),n_1}} \left\| \left\{ \varphi_{k,\epsilon}(D) \left(\sum_{\lambda\in\mathbb{E}^n} \Lambda(c^{(K)}_{k,\epsilon})^s u_{k+t,\lambda} \right) \right\}_{k,\epsilon} \right\|_{\ell^q(L^p)}. \tag{2.23}$$

Thanks to Remark 2.5 we obtain

$$\left\| \varphi_{k,\epsilon}(D) \sum_{\lambda\in\mathbb{E}^n} \Lambda(c^{(K)}_{k,\epsilon})^s u_{k+t,\lambda} \right\|_p < C \left\| \sum_{\lambda\in\mathbb{E}^n} \Lambda(c^{(K)}_{k,\epsilon})^s u_{k+t,\lambda} \right\|_p ; \tag{2.24}$$

moreover, there exists $C = C_{s,n,N_0} > 0$ such that $\Lambda(c^{(K)}_{k,\epsilon}) \leq C\Lambda(c^{(H)}_{k+t,\lambda})$, for $t \in E^{(N_0),n_1}$. It then follows

$$\|u\|_{B^{s,\Lambda}_{p,q}} \leq C \sum_{t\in E^{(N_0),n_1}} \left(\sum_{k,\lambda} \Lambda(c^{(H)}_{k+t,\lambda})^{qs} \|u_{k+t,\lambda}\|_p^q \right)^{\frac{1}{q}}. \tag{2.25}$$

For an arbitrary $\gamma > 0$, let us multiply any term depending on t in the right-hand side of (2.25) by $2^{\gamma t_j}$ and its own inverse, as $j = 1, \dots, n_1$; then by setting $\sum_{t\in E^{(N_0),n_1}} 2^{-\gamma|t|} = C_{N_0,\gamma,n_1}$ we obtain

$$\|u\|_{B^{s,\Lambda}_{p,q}} \leq CC_{N_0,\gamma,n_1} \left(\sum_{h,\lambda} \Lambda(c^{(H)}_{h,\lambda})^{qs} 2^{\gamma h_1} \dots 2^{\gamma h_{n_1}} \|u_{h,\lambda}\|_p^q \right)^{\frac{1}{q}}, \tag{2.26}$$

which ends the proof. All the arguments may be repeated with few changes for the case $q = \infty$. □

At the end of this section let us consider two interpolation results which directly follow from Calderón [3] and Triebel [20], respectively. In the notation of these authors $[\cdot, \cdot]_\Theta$, $0 < \Theta < 1$ is the complex interpolation functor.

Proposition 2.13. *Let (B^0, B^1) and (C^0, C^1) be two interpolation couples. Let L be a linear mapping from $B^0 + B^1$ to $C^0 + C^1$ such that $x \in B^i$ implies $L(x) \in C^i$ and*

$$\|L(x)\|_{C^i} \leq M_i \|x\|_{B^i}, \qquad i = 0, 1. \tag{2.27}$$

Then $x \in B_\Theta := [B^0, B^1]_\Theta$ implies $L(x) \in C_\Theta := [C^0, C^1]_\Theta$ and

$$\|L(x)\|_{C_\Theta} \leq M_0^{1-\Theta} M_1^{\Theta} \|x\|_{B_\Theta}. \tag{2.28}$$

Proposition 2.14 ([20]. **Theorem 4.2/2).** *For any weight function $\Lambda(\xi)$ and $1 < p < \infty$, $1 \leq q < \infty$:*

$$\left[B^{0,\Lambda}_{p,q}, B^{r,\Lambda}_{p,q}\right]_\Theta = B^{r\Theta,\Lambda}_{p,q}, \qquad 0 < \Theta < 1. \tag{2.29}$$

3. Pseudodifferential operators with non-regular symbols

In the remainder of the paper X will be a generic Banach space with norm $\|\cdot\|$.

Definition 3.1. For any non-negative integer N and $m \in \mathbb{R}$, we define $XM^m_\Lambda(N)$ as the class of all measurable functions $a(x, \xi)$ on $\mathbb{R}^{2n}$ such that

$$\begin{aligned} &\textstyle\prod_{j=1}^n (1 + \xi_j^2)^{\frac{\gamma_j}{2}} |\partial_\xi^\gamma a(x, \xi)| \leq C_N \Lambda(\xi)^m, && |\gamma| \leq N, \quad x, \xi \in \mathbb{R}^n; \\ &\textstyle\prod_{j=1}^n (1 + \xi_j^2)^{\frac{\gamma_j}{2}} \|\partial_\xi^\gamma a(\cdot, \xi)\| \leq C_N \Lambda(\xi)^m, && |\gamma| \leq N, \quad \xi \in \mathbb{R}^n. \end{aligned} \tag{3.1}$$

Definition 3.2. For any integer $N \geq 0$, we define $XM_E(N)$ as the class of all expansions

$$\sigma(x, \xi) := \sum_{h,\lambda} d_{h,\lambda}(x) \psi_{h,\lambda}(\xi) \tag{3.2}$$

whose terms $d_{h,\lambda} \in L^\infty(\mathbb{R}^n) \cap X$ and $\psi_{h,\lambda} \in C_0^\infty(\mathbb{R}^n)$ satisfy for some $M > 0$, $H > 1$ and $C > 0$

$$\|d_{h,\lambda}\|_\infty < M; \qquad \|d_{h,\lambda}\| < M; \qquad \operatorname{supp} \psi_{h,\lambda} \subset P^{(H)}_{h,\lambda}; \tag{3.3}$$

$$|\partial^\alpha \psi_{h,\lambda}(\xi)| < C 2^{-h \cdot \alpha}, \quad \text{for any } \xi \in \mathbb{R}^n \quad \text{and} \quad |\alpha| \leq N. \tag{3.4}$$

We say that $\sigma(x, \xi)$ is an *elementary symbol.*

The expansion in (3.2) is trivially convergent, since, in view of Remark 2.2, for any fixed $\xi \in \mathbb{R}^n$ all but a finite number of its terms vanish. Moreover $XM_E(N) \subset XM^0_\Lambda(N)$, for any weight function $\Lambda(\xi)$ and integer $N \geq 0$.

Proposition 3.3. *Provided that $N \geq n + 1$, any symbol $a(x, \xi) \in XM^0_\Lambda(N)$ may be written as an expansion of elementary symbols $a_m(x, \xi) \in XM_E(N - n - 1)$, $m \in \mathbb{Z}^n$, in the following way:*

$$a(x, \xi) = \sum_{m \in \mathbb{Z}^n} \frac{1}{(1 + |m|)^{n+1}} a_m(x, \xi), \tag{3.5}$$

where $|m| = |m_1| + \cdots + |m_n|$ *and the expansion is absolutely convergent in* $L^\infty(\mathbb{R}^n_x \times \mathbb{R}^n_\xi)$. *Moreover the elementary symbols* $a_m(x,\xi)$ *may be constructed in such a way that the assumptions* (3.3), (3.4) *are satisfied with constants* M *and* C *independent of* m.

Proposition 3.3 is proved by slightly modifying the arguments used to show Proposition 6.1 in [7].

Theorem 3.4. *For any weight function* $\Lambda(\xi)$, *let* $a(x,\xi)$ *be a symbol in* $B^{r,\Lambda}_{p,q}M^m_\Lambda(N)$ *with* $r > \frac{n}{(1-\delta)\mu_0 p}$, $N \geq 2n+1$, $1 < p < \infty$, $1 \leq q < \infty$ *and* $m \in \mathbb{R}$. *Then:*

$$a(x,D) : B^{s+m,\Lambda}_{p,q} \longmapsto B^{s,\Lambda}_{p,q}, \quad \textit{continuously for any } 0 \leq s \leq r. \tag{3.6}$$

The following remarks will be useful in proving the previous statement.

1) Thanks to Remark 2.1/2 in [20], there exists a constant $C > 1$ such that $1/C < \Lambda(\xi)/\Lambda(c^{(H)}_{h,\lambda}) < C$ for every $\xi \in P^{(H)}_{h,\lambda}$ and C does not depend on h, λ.

Let us assume that $\{\chi_{h,\lambda}\}$ belongs to $\phi^{(K)}$, with $K > H$, and satisfies $\chi_{h,\lambda} = 1$ in supp $\varphi_{h,\lambda}$; it is quite easy to prove that $\chi_{h,\lambda}(D)\Lambda^m(D)\Lambda(c^{(H)}_{h,\lambda})^{-m}$ is a continuous Fourier multiplier on $L^p(\mathbb{R}^n)$, $1 < p < \infty$, thanks to Lemma 2.4. We can then show, for any $u \in \mathcal{S}(\mathbb{R}^n)$:

$$\begin{aligned}\|\Lambda^m(D)u\|_{B^{s,\Lambda}_{p,q}} &= \Big(\sum_{h,\lambda} \Lambda(c^{(H)}_{h,\lambda})^{(s+m)q} \|\Lambda(c^{(H)}_{h,\lambda})^{-m}\chi_{h,\lambda}(D)\Lambda^m(D)\varphi_{h,\lambda}(D)u\|^q_p\Big)^{\frac{1}{q}} \\ &\leq C\Big(\sum\nolimits_{h,\lambda} \Lambda(c^{(H)}_{h,\lambda})^{(s+m)q} \|\varphi_{h,\lambda}(D)u\|^q_p\Big)^{\frac{1}{q}} = C\|u\|_{B^{s+m,\Lambda}_{p,q}}.\end{aligned}$$

Thus we conclude that for any $m, s \in \mathbb{R}$, $1 < p < \infty$, $1 \leq q < \infty$, $\Lambda^m(D)$ is bounded from $B^{s+m,\Lambda}_{p,q}$ into $B^{s,\Lambda}_{p,q}$, because of the density of $\mathcal{S}(\mathbb{R}^n)$ in $B^{s,\Lambda}_{p,q}$ for $q \neq \infty$. Since $a(x,D)\Lambda^{-m}(D)$ has symbol $a(x,\xi)\Lambda^{-m}(\xi) \in B^{r,\Lambda}_{p,q}M^0_\Lambda(N)$, assuming that Theorem 3.4 holds in the case $m = 0$, we obtain that the operator $a(x,D) = (a(x,D)\Lambda^{-m}(D))\Lambda^m(D)$ is bounded from $B^{s+m,\Lambda}_{p,q}$ into $B^{s,\Lambda}_{p,q}$, when $0 \leq s \leq r$.

2) Proposition 3.3 and the Dominated Convergence Theorem assure that for any $a(x,\xi) \in B^{r,\Lambda}_{p,q}M^0_\Lambda(N)$, we can write for $u \in \mathcal{S}(\mathbb{R}^n)$:

$$a(x,D)u(x) = \sum_{m\in\mathbb{Z}^n} \frac{1}{(1+|m|)^{n+1}} a_m(x,D)u(x), \quad |m| = \sum_{j=1}^n |m_j|, \tag{3.7}$$

where $a_m(x,\xi) \in B^{r,\Lambda}_{p,q}M_E(N-n-1)$ and under the assumption of Theorem 3.4 we have $N - n - 1 \geq n$. As a first step, let us then prove Theorem 3.4 for $\sigma(x,\xi) \in B^{r,\Lambda}_{p,q}M^0_E(N)$, with $N \geq n$.

3) For any $\sigma(x,\xi) \in B^{s,\Lambda}_{p,q}M^0_E(N)$ having the form (3.2) we can write:

$$\sigma(x,D)u(x) = \sum_{h,\lambda} d_{h,\lambda}(x)u_{h,\lambda}(x), \qquad u \in \mathcal{S}(\mathbb{R}^n), \tag{3.8}$$

where $u_{h,\lambda}(x) = \psi_{h,\lambda}(D)u(x)$ and $d_{h,\lambda}(x), \psi_{h,\lambda}(\xi)$ satisfy (3.3), (3.4). The expansion in (3.8) is absolutely convergent in $L^\infty(\mathbb{R}^n)$. In fact in view of (3.4), for any

$T > 0$ and some suitable positive constant C we have:

$$\|u_{h,\lambda}\|_\infty \le C \frac{C_T}{\left(1 + \frac{1}{H^2}\sum_{j=1}^n \chi_{h_j} 2^{2h_j}\right)^T} \le C C_T a_{h_1} \dots a_{h_n}, \tag{3.9}$$

where $C_T := \int ((1+|\xi|^2)^T |\hat{u}(\xi)|\, d\xi$ is finite, $\chi_h = 0$ if $h = 0$, $\chi_h = 1$, if $h > 0$ and $a_h = 1$ for $h = 0$, $a_h = \frac{1}{\frac{1}{H^2} 2^{\frac{2hM}{n}}}$ for $h > 0$.

Using also (3.3), the estimate

$$\sum_{h,\lambda} \|d_{h,\lambda}\|_\infty \|u_{h,\lambda}\|_\infty \le M 2^n C_T \sum_{h_1=0}^\infty a_{h_1} \cdots \sum_{h_n=0}^\infty a_{h_n} < \infty \tag{3.10}$$

shows that (3.8) is absolutely convergent in $L^\infty(\mathbb{R}^n)$.

Let us also remark that for every $v \in \mathcal{S}'(\mathbb{R}^n)$ and $\{\varphi_{h,\lambda}\} \in \phi^{(H)}$ there holds

$$v = \sum_{h,\lambda} \varphi_{h,\lambda}(D) v = \sum_{h,\lambda} v_{h,\lambda}, \text{ with convergence in } \mathcal{S}'(\mathbb{R}^n). \tag{3.11}$$

4) For $\{\psi_{k,\epsilon}(\xi)\} \in \Phi^{(K)}$, $K > 1$, let us set $d_{h,\lambda}^{k,\epsilon}(x) := \psi_{k,\epsilon}(D) d_{h,\lambda}(x)$ and consider

$$a(x,D)u(x) = \sum_{h,\lambda} \sum_{k,\epsilon} d_{h,\lambda}^{k,\epsilon}(x) u_{h,\lambda}(x). \tag{3.12}$$

It follows from Proposition 2.11 and (3.3) that $\|d_{h,\lambda}^{k,\epsilon}\|_\infty < M \Lambda(c_{k,\epsilon}^{(K)})^{-\left(r - \frac{n}{\mu_0 p}\right)}$, for some $M > 0$. Then using (3.9) and provided that $r > \frac{n}{\mu_0 p}$ we can conclude that the expansion in (3.12) is absolutely convergent in $L^\infty(\mathbb{R}_x^n)$.

5) Thanks to the absolute convergence we can change the order of the terms in the expansion in (3.12) and choose a useful order. Let us first introduce some notations. Namely for a fixed $N_0 \in \mathbb{N}$ and any $j \in \mathbb{Z}_+$ we set:

$$E_{1,j}^{(N_0)} := \begin{cases} \emptyset\,, & j \le N_0; \\ \mathbb{Z}_+ \cap [0, j - N_0[\,, & j > N_0; \end{cases} \tag{3.13}$$

$$E_{2,j}^{(N_0)} := \mathbb{Z}_+ \cap [j - N_0, j + N_0[; \tag{3.14}$$

$$E_{3,j}^{(N_0)} := \mathbb{Z}_+ \cap [j + N_0, \infty[. \tag{3.15}$$

For $A := \{1, 2, \dots, n\}$ and $B := \{1, 2, 3\}$, let B^A be the set of all the functions $\omega : A \mapsto B$. For any $h \in \mathbb{Z}_+^n$ and $\omega \in B^A$ we set $E_{\omega,h}^{(N_0)} := \prod_{i=1}^n E_{\omega(i),h_i}^{(N_0)}$. Agreeing with the previous notation we can write:

$$\sigma(x,D)u(x) = \sum_{\omega \in B^A} \sum_{h,\lambda} \sum_{k \in E_{\omega,h}^{(N_0)},\ \epsilon \in \mathbb{E}^n} d_{h,\lambda}^{k,\epsilon}(x) u_{h,\lambda}(x). \tag{3.16}$$

6) For every $h,k \in \mathbb{Z}_+^n$ and $\lambda, \epsilon \in \mathbb{E}^n$: $\mathrm{supp}(\widehat{d_{h,\lambda}^{k,\epsilon} u_{h,\lambda}}) \subset P_{k,\epsilon}^{(K)} + P_{h,\lambda}^{(H)}$. The n-intervals $P_{h,\lambda}^{(H)}$ and $P_{k,\epsilon}^{(K)}$ are obtained as a superposition of n real intervals of the type $L_{r,\kappa}^{(H)}$ and $L_{s,\delta}^{(K)}$ introduced in (2.15). Therefore we have reduced the study of the n-dimensional sum $P_{h,\lambda}^{(H)} + P_{k,\epsilon}^{(K)}$ to an argument involving the one-dimensional sums $L_{r,\kappa}^{(H)} + L_{s,\delta}^{(K)}$. We now need the following technical lemma, for the proof of which we refer to [7].

Lemma 3.5 ([7], **Lemma 7.1**). *Let us consider $r, s \in \mathbb{Z}_+$, $\kappa, \delta \in \{-1, 1\}$, H, K greater than 1. For any N_0 positive integer such that $N_0 > \log_2(2HK)$, we can always find two positive constants T, M such that $T > H+K$, $\frac{1}{T} < \min\{\frac{1}{K} - \frac{2H}{2^{N_0}}, \frac{1}{H} - \frac{2K}{2^{N_0}}\}$ and $M > 2^{N_0+1}K + 2H$, which fulfill the following statements, with $\eta_j = 1$ if $j > 0$, and $\eta_0 = -1$:*

(a) *if $s \in E_{1,r}^{(N_0)}$ and $r > N_0$ then*

$$L_{r,\kappa}^{(H)} + L_{s,\delta}^{(K)} \subset \left\{ \theta \in \mathbb{R} : \frac{2^r \eta_r}{T} \leq \kappa\theta \leq T2^{r+1} \right\} =: L_{r,\kappa}^{(T)}; \tag{3.17}$$

(b) *if $s \in E_{2,r}^{(N_0)}$ then*

$$L_{r,\kappa}^{(H)} + L_{s,\delta}^{(K)} \subset \{\theta \in \mathbb{R} : \ |\theta| \leq M2^r\} =: [-M2^r, M2^r]; \tag{3.18}$$

(c) *if $s \in E_{3,r}^{(N_0)}$ then*

$$L_{r,\kappa}^{(H)} + L_{s,\delta}^{(K)} \subset \left\{ \theta \in \mathbb{R} : \frac{2^s \eta_s}{T} \leq \delta\theta \leq T2^{s+1} \right\} =: L_{s,\delta}^{(T)}. \tag{3.19}$$

It then follows that $\widehat{d_{h,\lambda}^{k,\epsilon} u_{h,\lambda}}$ is supported in the product of n real intervals of the type (3.17)–(3.19). This suggests to split B^A in the following way:

$$\begin{aligned}
&C_1 := \{\omega \in B^A : \omega(A) = \{1\}\}; && C_2 := \{\omega \in B^A : \omega(A) = \{2\}\};\\
&C_3 := \{\omega \in B^A : \omega(A) = \{3\}\}; && C_4 := \{\omega \in B^A : \omega(A) = \{1,2\}\};\\
&C_5 := \{\omega \in B^A : \omega(A) = \{1,3\}\}; && C_6 := \{\omega \in B^A : \omega(A) = \{2,3\}\};\\
&C_7 := \{\omega \in B^A : \ \omega(A) = \{1,2,3\}\}.
\end{aligned} \tag{3.20}$$

The sets C_1, C_2 and C_3 reduce to a singleton set $\{\omega\}$, while C_4-C_7 contain several functions, for any dimension $n \geq 2$.

For any $\sigma(x, D) \in H_\Lambda^{r,p} M_E(N)$ we can write

$$\sigma(x, D)u(x) = \sum_{j=1}^{7} T_j u(x), \quad u \in \mathcal{S}(\mathbb{R}^n), \tag{3.21}$$

where for $j = 1, \dots 7$:

$$T_j u(x) := \sum_{\omega \in C_j} \sum_{h,\lambda} \sum_{k \in E^{(N_0)}_{\omega,h},\ \epsilon \in \mathbb{E}^n} d^{k,\epsilon}_{h,\lambda}(x) u_{h,\lambda}(x), \qquad u \in \mathcal{S}(\mathbb{R}^n). \tag{3.22}$$

In the following we will work under the conditions obtained step by step in the remarks 1)–6). In particular, in the statements of the following propositions 3.6–3.9 we will always assume $\sigma(x,\xi) \in B^{r,\Lambda}_{p,q} M^0_E(N)$, with $1 < p < \infty$, $1 \leq q < \infty$, $\Lambda(\xi)$ weight function, $N \geq n$ and $r > \frac{n}{\mu_0 p}$.

Proposition 3.6.

$$T_1 : B^{s,\Lambda}_{p,t} \longmapsto B^{s,\Lambda}_{p,t} \quad \textit{continuously, for any} \quad s \in \mathbb{R}, \quad 1 \leq t < \infty. \tag{3.23}$$

Proof. Assuming $N_0 > \log_2(2HK)$, from Lemma 3.5 we find a constant $T > 1$ such that $\operatorname{supp} \widehat{d^{k,\epsilon}_{h,\lambda} u_{h,\lambda}} \subset P^{(T)}_{h,\lambda}$, for any $h, k \in \mathbb{Z}^n_+$, with $k_j < h_j - N_0$ $(j = 1, \dots, n)$, and any $\lambda, \epsilon \in \mathbb{E}^n$.

In view of Proposition 2.7, for every $s \in \mathbb{R}$ and $1 < p < \infty$ we get:

$$\|T_1 u\|_{B^{s,\Lambda}_{p,t}} \leq C\Big(\sum_{h,\lambda} \Lambda(c^{(T)}_{h,\lambda})^{ts} \|u_{h,\lambda}\|^t_p \Big(\sum_{k \in E^{(N_0)}_{1,h}, \epsilon \in \mathbb{E}^n} \|d^{k,\epsilon}_{h,\lambda}\|_\infty\Big)^t\Big)^{\frac{1}{t}}, \tag{3.24}$$

where $E^{(N_0)}_{1,h} := \prod_{j=1}^n E^{(N_0)}_{1,h_j}$ and C is independent of u. Since the sequence $\{d_{h,\lambda}\}$ is bounded in $B^{r,\Lambda}_{p,q}$ and $r > \frac{n}{\mu_0 p}$, from Proposition 2.11 we have:

$$\sum_{k \in E^{(N_0)}_{1,h}, \epsilon \in \mathbb{E}^n} \|d^{k,\epsilon}_{h,\lambda}(x)\|_\infty \leq M\Big(\sum_{k,\epsilon} \Lambda(c^{(K)}_{k,\epsilon})^{-\left(r - \frac{n}{\mu_0 p}\right)}\Big) \sup_{h,\lambda} \|d_{h,\lambda}\|_{B^{r,\Lambda}_{p,q}}. \tag{3.25}$$

Using now (3.25) jointly with Remark 2.2, $s \in \mathbb{R}$ and $1 \leq t < \infty$, we get for any $u \in \mathcal{S}(\mathbb{R}^n)$:

$$\begin{aligned} \|T_1 u\| &\leq CM \sup_{h,\lambda} \|d_{h,\lambda}\|_{B^{r,\Lambda}_{p,q}} \Big(\sum_{h,\lambda} \Lambda(c^{(H)}_{h,\lambda})^{ts} \|u_{h,\lambda}\|^t_p \Big)^{\frac{1}{t}} \\ &\leq CM \sup_{h,\lambda} \|d_{h,\lambda}\|_{B^{r,\Lambda}_{p,q}} \|u\|_{B^{s,\Lambda}_{p,t}}. \end{aligned} \tag{3.26}$$

In order to get the last inequality in (3.26), we need to observe that the functions $\psi_{h,\lambda}(\xi)$ involved in the expression (3.2) give L^p continuous Fourier multipliers $\psi_{h,\lambda}(D)$ for every $1 < p < \infty$; indeed it is enough to apply the Lizorkin-Marcinckiewicz Lemma in view of (3.3), (3.4) and follow the arguments used in Remark 2.5. Let us also point out that the estimate (3.4), with $N \geq n$, is essential in order to apply Lemma 2.4 to $\psi_{h,\lambda}(\xi)$. Since $\mathcal{S}(\mathbb{R}^n)$ is dense in $B^{s,\Lambda}_{p,t}$ (cf. Proposition 2.8) the proof is concluded. □

Proposition 3.7.

$$T_2 : B^{s,\Lambda}_{p,t} \longmapsto B^{s+r-\frac{n}{\mu_0 p}-\theta,\Lambda}_{p,t}, \qquad 1 \leq t < \infty \tag{3.27}$$

continuously for any $s > -r + \frac{n}{\mu_0 p}$, $0 < \theta < s + r - \frac{n}{\mu_0 p}$.

Proof. Let us set $U_{h,\lambda}(x) := \sum_{\substack{k\in E_{2,h}^{(N_0)}\\ \epsilon\in\mathbb{E}^n}} d_{h,\lambda}^{k,\epsilon}(x)u_{h,\lambda}(x)$, where $E_{2,h}^{(N_0)} := \prod_{j=1}^{n} E_{2,h_j}^{(N_0)}$. From Lemma 3.5 it follows that $T_2u(x) = \sum_{h,\lambda} U_{h,\lambda}(x)$ fulfills the assumptions of Lemma 2.12. Since $s+r-\frac{n}{\mu^0 p}-\theta>0$, we may estimate the $B_{p,t}^{s+r-\frac{n}{\mu_0 p}-\theta,\Lambda}$ norm of $T_2u(x)$ by means of (2.17), with $\gamma=\frac{\mu_0\theta}{n}$, remembering also that $2^{|h|}\le C\Lambda(c_{h,\lambda}^{(H)})^{\frac{n}{\mu_0}}$. Then for a positive constant C depending only on r,s,p,t,μ_0,n and $\theta>0$:

$$\|T_2u\|_{B_{p,t}^{s+r-\frac{n}{\mu_0 p}-\theta,\Lambda}} \le C\left(\sum_{h,\lambda}\Lambda(c_{h,\lambda}^{(T)})^{t(s+r-\frac{n}{\mu_0 p})}\|U_{h,\lambda}\|_p^t\right)^{\frac{1}{t}}. \tag{3.28}$$

Using now Proposition 2.11, Remark 2.6 and observing moreover that $E_{2,h}^{(N_0)}$ is a finite set, we have:

$$\begin{aligned}\|U_{h,\lambda}\|_p &\le M\|u_{h,\lambda}\|_p \sup_{h,\lambda}\|d_{h,\lambda}\|_{B_{p,q}^{r,\Lambda}} \sum_{k\in E_{2,h}^{(N_0)},\epsilon\in\mathbb{E}^n}\Lambda(c_{k,\epsilon}^{(K)})^{-\left(r-\frac{n}{\mu_0 p}\right)}\\ &\le MC\|u_{h,\lambda}\|_p\Lambda(c_{h,\lambda}^{(H)})^{-\left(r-\frac{n}{\mu_0 p}\right)}\sup_{h,\lambda}\|d_{h,\lambda}\|_{B_{p,q}^{r,\Lambda}}.\end{aligned} \tag{3.29}$$

Then we conclude that

$$\begin{aligned}\|T_2u\|_{B_{p,t}^{s+r-\frac{n}{\mu_0 p}-\theta,\Lambda}} &\le CM\sup_{h,\lambda}\|d_{h,\lambda}\|_{B_{p,q}^{r,\Lambda}}(\textstyle\sum_{h,\lambda}\Lambda(c_{h,\lambda}^{(H)})^{st}\|u_{h,\lambda}\|_p^t)^{\frac{1}{t}}\\ &\le CM\sup_{h,\lambda}\|d_{h,\lambda}\|_{B_{p,q}^{r,\Lambda}}\|u\|_{B_{p,t}^{s,\Lambda}},\end{aligned} \tag{3.30}$$

with positive constants C, M independent of $u\in\mathcal{S}(\mathbb{R}^n)$. □

Proposition 3.8.

$$T_3 : B_{p,t}^{s-r+\theta+\frac{n}{\mu_0 p},\Lambda} \longmapsto B_{p,t}^{s,\Lambda} \tag{3.31}$$

continuously, for any $s<r$, $1\le t<\infty$, $\theta>0$.

Proof. Setting $V_{k,\epsilon}(x) := \sum_{\substack{h\in E_{1,k}^{(N_0-1)}\\ \lambda\in\mathbb{E}^n}} d_{h,\lambda}^{k,\epsilon}(x)u_{h,\lambda}(x)$ and exploiting the absolute convergence of the expansion in (3.12) we may write $T_3u(x)=\sum_{k,\epsilon}V_{k,\epsilon}(x)$, for any $u\in\mathcal{S}(\mathbb{R}^n)$. It follows from Lemma 3.5 that $\mathrm{supp}\widehat{V_{k,\epsilon}}\subset P_{k,\epsilon}^{(K)}$. Then using Proposition 2.7 and the embedding (2.11) we obtain for any $1<p_1<p<\infty$ and some positive constant $C=C_{t,s,p,p_1}$:

$$\|T_3u\|_{B_{p,t}^{s,\Lambda}} \le C\left(\sum_{k,\epsilon}\Lambda(c_{k,\epsilon}^{(K)})^{\left(s+\frac{n}{\mu_0}\left(\frac{1}{p_1}-\frac{1}{p}\right)\right)^t}\|V_{k,\epsilon}\|_{p_1}^t\right)^{\frac{1}{t}}. \tag{3.32}$$

Setting now $\eta=\frac{1}{p_1}-\frac{1}{p}$ and applying the Hölder inequality it follows:

$$\begin{aligned}\|V_{k,\epsilon}\|_{p_1} &\le \sum_{h\in E_{1,k}^{(N_0-1)},\lambda\in\mathbb{E}^n}\|d_{h,\lambda}^{k,\epsilon}\|_p\|u_{h,\lambda}\|_{\frac{1}{\eta}}\\ &\le \sup_{h,\lambda}\|d_{h,\lambda}\|_{B_{p,\infty}^{r,\Lambda}}\Lambda(c_{k,\epsilon}^{(K)})^{-r}\sum_{h\in E_{1,k}^{(N_0-1)},\lambda\in\mathbb{E}^n}\|u_{h,\lambda}\|_{\frac{1}{\eta}}.\end{aligned} \tag{3.33}$$

Then, writing for the sake of brevity $\Lambda_k := \Lambda(c_{k,\epsilon}^{(K)})$, we get

$$\begin{aligned}\|T_3 u\|_{B^{s,\Lambda}_{p,t}} &\leq C \sup_{h,\lambda} \|d_{h,\lambda}\|_{B^{r,\Lambda}_{p,q}} \left(\sum_{k,\epsilon} (\Lambda_k^{s-r+\frac{n}{\mu_0}\eta} \sum_{\substack{h\in E^{(N_0-1)}_{1,k}\\ \lambda\in\mathbb{E}^n}} \|u_{h,\lambda}\|_{\frac{1}{\eta}})^t \right)^{\frac{1}{t}} \\ &\leq C \sup_{h,\lambda} \|d_{h,\lambda}\|_{B^{r,\Lambda}_{p,q}} \sum_{k,\epsilon} \Lambda_k^{s-r+\frac{n}{\mu_0}\eta+\eta'} \Lambda_k^{-\eta'} \sum_{\substack{h\in E^{(N_0-1)}_{1,k}\\ \lambda\in\mathbb{E}^n}} \|u_{h,\lambda}\|_{\frac{1}{\eta}}.\end{aligned} \tag{3.34}$$

Assuming now without any restriction that $s - r + \frac{n}{\mu_0}\eta + \eta' < 0$, thanks to the assumption 3 in Definition 2.1 and Remark 2.6 we have: $\Lambda(c_{k,\epsilon}^{(K)})^{s-r+\frac{n}{\mu_0}\eta+\eta'} \leq T\Lambda(c_{h,\lambda}^{(H)})^{s-r+\frac{n}{\mu_0}\eta+\eta'}$, with $T > 0$ independent of k, h, ϵ, λ. Then the $B^{s,\Lambda}_{p,t}$ norm of $T_3 u(x)$ can be bounded from above by

$$\begin{aligned}&CT \sup_{h,\lambda} \|d_{h,\lambda}\|_{B^{r,\Lambda}_{p,q}} \textstyle\sum_{k,\epsilon} \Lambda(c_{k,\epsilon}^{(K)})^{-\eta'} \sum_{h,\lambda} \Lambda(c_{k,\lambda}^{(H)})^{s-r+\frac{n}{\mu_0}\eta+\eta'} \|u_{h,\lambda}\|_{\frac{1}{\eta}} \\ &\leq CT \sup_{h,\lambda} \|d_{h,\lambda}\|_{B^{r,\Lambda}_{p,q}} \|u\|_{B^{s-r+\frac{n}{\mu_0}\eta+\eta',\Lambda}_{\frac{1}{\eta},1}}.\end{aligned} \tag{3.35}$$

To get the bound (3.35) it is essential to exploit that the operators $\psi_{h,\lambda}(D)$, from the expression in (3.2), are continuous Fourier multipliers in $L^{\frac{1}{\eta}}(\mathbb{R}^n)$ when $N \geq n$. We have thus proved the continuity of T_3 from $B^{s-r+\frac{n}{\mu_0}\eta+\eta',\Lambda}_{\frac{1}{\eta},1}$ into $B^{s,\Lambda}_{p,t}$. As a result of Proposition 2.10 we obtain

$$B^{s-r+\frac{n}{\mu_0 p}+\eta',\Lambda}_{p,1} \subset B^{s-r+\frac{n}{\mu_0 p}+\eta'-\frac{n}{\mu_0}(\frac{1}{p}-\eta),\Lambda}_{\frac{1}{\eta},1} = B^{s-r+\frac{n}{\mu_0}\eta+\eta',\Lambda}_{\frac{1}{\eta},1}, \tag{3.36}$$

with continuous embedding. Furthermore, using (2.9), we prove the following continuous inclusion

$$B^{s-r+\frac{n}{\mu_0 p}+\theta,\Lambda}_{p,t} \subset B^{s-r+\frac{n}{\mu_0 p}+\eta',\Lambda}_{p,1}, \tag{3.37}$$

for an arbitrary η' such that $0 < \eta' < \theta$. Gathering estimates (3.34)–(3.37) and using the density of $\mathcal{S}(\mathbb{R}^n)$ in $B^{s-r+\frac{n}{\mu_0 p}+\theta,\Lambda}_{p,t}$ we obtain (3.31). □

Proposition 3.9.

$$T_3 : B^{\theta+\frac{n}{\mu_0 p},\Lambda}_{p,q} \longmapsto B^{r,\Lambda}_{p,q} \quad \textit{continuously for any } \theta > 0. \tag{3.38}$$

Proof. We can write $\|T_3 u\|_{B^{r,\Lambda}_{p,q}} \leq C\|\{\Lambda(c_{k,\epsilon}^{(K)})^r V_{k,\epsilon}\}\|_{\ell^q(L^p)}$, in the notation of the previous proof. Applying the Hölder-Schwarz inequality we have for the conjugate order q' such that $\frac{1}{q} + \frac{1}{q'} = 1$ and any $\tau > 0$:

$$\|V_{k,\epsilon}\|_p \leq (\sum_{\substack{h\in E^{(N_0-1)}_{1,k}\\ \lambda\in\mathbb{E}^n}} \Lambda(c_{h,\lambda}^{(H)})^{-q\tau} \|d^{k,\epsilon}_{h,\lambda}\|_p^q)^{\frac{1}{q}} (\sum_{\substack{h\in E^{(N_0-1)}_{1,k}\\ \lambda\in\mathbb{E}^n}} \Lambda(c_{h,\lambda}^{(H)})^{q'\tau} \|u_{h,\lambda}\|_\infty^{q'})^{\frac{1}{q'}}. \tag{3.39}$$

We obtain for the $B^{r,\Lambda}_{p,q}$ norm of $T_3 u$ the following bound

$$
\begin{aligned}
&(\sum_{h,\lambda} \Lambda(c^{(H)}_{h,\lambda})^{q'\tau}\|u_{h,\lambda}\|^{q'}_{\infty})^{\frac{1}{q'}} \sum_{h,\lambda} \Lambda(c^{(H)}_{h,\lambda})^{-\tau}(\sum_{k,\epsilon} \Lambda(c^{(K)}_{k,\epsilon})^{qr}\|d^{k,\epsilon}_{h,\lambda}\|^{q}_{p})^{\frac{1}{q}} \\
&\leq C \sup_{h,\lambda} \|d_{h,\lambda}\|_{B^{r,\Lambda}_{p,q}} \|\{\Lambda(c^{(H)}_{h,\lambda})^{\tau} u_{h,\lambda}\|_{\ell^{q'}(L^\infty)},
\end{aligned}
\tag{3.40}
$$

where $\sum_{h,\lambda} \Lambda(c^{(H)}_{h,\lambda})^{-\tau}$ is finite. From Lemma 2.9, with $p_2 = \infty$, and the L^p continuity of the Fourier multiplier $\psi_{h,\lambda}(D)$, it follows, for suitable $C > 0$,

$$
\begin{aligned}
&\|\{\Lambda(c^{(H)}_{h,\lambda})^{\tau} u_{h,\lambda}\}\|_{\ell^{q'}(L^\infty)} \leq \textstyle\sum_{h,\lambda} \Lambda(c^{(H)}_{h,\lambda})^{\tau}\|u_{h,\lambda}\|_{\infty} \\
&\leq C \textstyle\sum_{h,\lambda} \Lambda(c^{(H)}_{h,\lambda})^{\tau+\frac{n}{\mu_0 p}}\|u_{h,\lambda}\|_{p} \leq C\|u\|_{B^{\tau+\frac{n}{\mu_0 p},\Lambda}_{p,1}}.
\end{aligned}
\tag{3.41}
$$

Using again the embedding (2.9) we now obtain for any $\tau > 0$, $\varepsilon > 0$, $1 \leq t < \infty$ and suitable $C > 0$:

$$
\|T_3 u\|_{B^{r,\Lambda}_{p,q}} \leq C \sup_{h,\lambda} \|d_{h,\lambda}\|_{B^{r,\Lambda}_{p,q}} \|u\|_{B^{\tau+\frac{n}{\mu_0 p},\Lambda}_{p,1}} \leq C\|d_{h,\lambda}\|_{B^{r,\Lambda}_{p,q}} \|u\|_{B^{\tau+\varepsilon+\frac{n}{\mu_0 p},\Lambda}_{p,t}}. \tag{3.42}
$$

Since $\theta = \tau + \varepsilon$ ranges over all of $(0, \infty)$, using usual density arguments, we conclude the proof. We have really proved that $T_3 : B^{\theta+\frac{n}{\mu_0 p},\Lambda}_{p,t} \mapsto B^{r,\Lambda}_{p,q}$ as a bounded linear operator for any $1 \leq t < \infty$. □

Let us remark that any operator T_j, $j = 4, \ldots, 7$, may be written as a finite sum of operators with the following form

$$
Ru(x) = \sum_{h,\lambda} \sum_{\substack{k \in E^{(N_0),n_1,n_2,\pi}_h, \\ \epsilon \in \mathbb{E}^n}} d^{k,\epsilon}_{h,\lambda}(x) u_{h,\lambda}(x), \qquad u \in \mathcal{S}(\mathbb{R}^n). \tag{3.43}
$$

Here n_1, n_2 are integers such that $0 \leq n_1 \leq n_2 \leq n$ and at least two of these inequalities must be strict; π is any permutation of the set $\{1, 2, \ldots, n\}$, $E^{(N_0)}_{1,h_{\pi(j)}}$, $E^{(N_0)}_{2,h_{\pi(j)}}$, $E^{(N_0)}_{3,h_{\pi(j)}}$ are defined by (3.13)–(3.15) and

$$
E^{(N_0),n_1,n_2,\pi}_h := \prod_{j=1}^{n_1} E^{(N_0)}_{1,h_{\pi(j)}} \times \prod_{j=n_1+1}^{n_2} E^{(N_0)}_{2,h_{\pi(j)}} \times \prod_{j=n_2+1}^{n} E^{(N_0)}_{3,h_{\pi(j)}}. \tag{3.44}
$$

Therefore, one only needs to study the $B^{s,\Lambda}_{p,q}$-continuity of an operator having the form (3.43).

In order to simplify the notation we assume from this moment, without loss of generality, that the permutation π in (3.43) is the identity of $\{1, 2, \ldots, n\}$ and restrict ourselves to the case $n_1 = 1$, $n_2 = 2$ and $n = 3$, that is $Ru(x) := \sum_{h,\lambda} \sum_{\substack{k_j \in E^{(N_0)}_{j,h_j},\ j=1,2,3 \\ \epsilon \in \mathbb{E}^n}} d^{k,\epsilon}_{h,\lambda}(x) u_{h,\lambda}(x)$, for any $u \in \mathcal{S}(\mathbb{R}^3)$. Because of the absolute convergence of the expansion in the L^∞ norm, we can write

$$Ru(x) := \sum_{\substack{h_1,h_2,k_3\\ \lambda_1,\lambda_2,\epsilon_3}} \sum_{\substack{k_j\in E^{(N_0)}_{j,h_j},\ j=1,2,\\ h_3\in E^{(N_0-1)}_{1,k_3},\\ \epsilon_1,\epsilon_2,\lambda_3}} d^{k,\epsilon}_{h,\lambda}(x)u_{h,\lambda}(x). \tag{3.45}$$

From Lemma 3.5, we find $T>1$ such that supp $\widehat{d^{k,\epsilon}_{h,\lambda}u_{h,\lambda}} \subset L^{(T)}_{h_1,\lambda_1}\times[-T2^{h_2},T2^{h_2}]\times L^{(T)}_{k_3,\epsilon_3}$, for any (k_1,k_2,h_3) satisfying $k_1<h_1-N_0$, $h_2-N_0\le k_2<h_2+N_0$ and $h_3\le k_3-N_0$, and all $\epsilon_1,\epsilon_2,\lambda_3$.

For shortness, we set $t:=(h_1,h_2,k_3)$, $\sigma:=(\lambda_1,\lambda_2,\epsilon_3)$ and $E^{(N_0)}_t := E^{(N_0)}_{1,h_1}\times E^{(N_0)}_{2,h_2}\times E^{(N_0-1)}_{1,k_3}$; moreover $e^{(K)}_{1,r,\epsilon} := (c_{1,K}\epsilon 2^r,0,0)$, $e^{(K)}_{2,r,\epsilon} := (0,c_{2,K}\epsilon 2^r,0)$, $e^{(K)}_{3,r,\epsilon} := (0,0,c_{3,K}\epsilon 2^r)$, for any integer r, $K>1$, $\epsilon\in\{-1,1\}$ and $c_{j,K}:=K\pm\frac{1}{2K}$, $j=1,2,3$. Using now Lemma 2.12 we find that for every $s\ge 0$, $1<p<\infty$, $1\le q\le\infty$ and $\gamma>0$ there exists $C=C_{s,p,q,\gamma}>0$ such that

$$\|Ru\|_{B^{s,\Lambda}_{p,q}} \le C\left(\sum_{t,\sigma}\Lambda(c^{(T)}_{t,\sigma})^{qs}2^{q\gamma h_2}\|U_{t,\sigma}\|^p_q\right)^{\frac{1}{p}}, \tag{3.46}$$

where $c_{T,j} := T\pm\frac{1}{2T}$, $j=1,2,3$, $c^{(T)}_{t,\sigma} = \left(c_{T,1}\lambda_1 2^{h_1}, c_{T,2}\lambda_2 2^{h_2}, c_{T,3}\epsilon_3 2^{k_3}\right)$ and $U_{t,\sigma}(x) := \sum_{\substack{(k_1,k_2,h_3)\in E^{(N_0)}_t\\ \epsilon_1,\epsilon_2,\lambda_3}} d^{k,\epsilon}_{h,\lambda}(x)u_{h,\lambda}(x)$. Let us notice that $c^{(T)}_{t,\sigma} = c^{(T)}_{h,\sigma} + \tau_3 e^{(T)}_{3,k_3,\epsilon_3}$, where $\tau_3 := 1-2^{h_3-k_3}$ satisfies $0<\tau_3<1$ as $h_3\le k_3-N_0$; by using the assumptions 3 and 4 (δ-condition) of Definition 2.1, we get a positive constant C such that

$$\Lambda(c^{(T)}_{t,\sigma}) \le C\big(\Lambda(c^{(T)}_{h,\sigma}) + \Lambda(e^{(T)}_{3,k_3,\epsilon_3}) + \Lambda(c^{(T)}_{h,\sigma})^\delta\Lambda(e^{(T)}_{3,k_3,\epsilon_3})^\delta\big), \tag{3.47}$$

for any t, σ, $k_3\ge h_3+N_0$. It then follows:

$$\|Ru\|_{B^{s,\Lambda}_{p,q}} \le C(I_1+I_2+I_3), \tag{3.48}$$

where

$$I_1 := \Big(\sum_{t,\sigma}2^{q\gamma h_2}\|\sum_{\substack{(k_1,k_2,h_3)\in E^{(N_0)}_t\\ \epsilon_1,\epsilon_2,\lambda_3}}\Lambda(c^{(T)}_{h,\sigma})^s d^{k,\epsilon}_{h,\lambda}u_{h,\lambda}\|^q_p\Big)^{\frac{1}{q}}, \tag{3.49}$$

$$I_2 := \Big(\sum_{t,\sigma}2^{q\gamma h_2}\Lambda(e^{(T)}_{3,k_3,\epsilon_3})^{qs}\|U_{t,\sigma}\|^q_p\Big)^{\frac{1}{q}}, \tag{3.50}$$

$$I_3 := \Big(\sum_{t,\sigma}2^{q\gamma h_2}\Lambda(e^{(T)}_{3,k_3,\epsilon_3})^{qs\delta}\|\sum_{\substack{(k_1,k_2,h_3)\in E^{(N_0)}_t\\ \epsilon_1,\epsilon_2,\lambda_3}}\Lambda(c^{(T)}_{h,\sigma})^{\delta s}d^{k,\epsilon}_{h,\lambda}u_{h,\lambda}\|^q_p\Big)^{\frac{1}{q}}. \tag{3.51}$$

In order to estimate separately I_1, I_2, I_3, let us compute in the **case s=r**.

1) *Estimate of* I_1. Thanks to Proposition 2.11:

$$\begin{aligned}\Big\|\sum_{\substack{(k_1,k_2,h_3)\in E_t^{(N_0)}\\ \epsilon_1,\epsilon_2,\lambda_3}} \Lambda(c^{(T)}_{h,\sigma})^r d^{k,\epsilon}_{h,\lambda} u_{h,\lambda}\Big\|_p &\le \sum_{\substack{(k_1,k_2,h_3)\in E_t^{(N_0)}\\ \epsilon_1,\epsilon_2,\lambda_3}} \Lambda(c^{(T)}_{h,\sigma})^r \|d^{k,\epsilon}_{h,\lambda}\|_\infty \|u_{h,\lambda}\|_p \\ &\le M \sup_{h,\lambda}\|d_{h,\lambda}\|_{B^{r,\Lambda}_{p,q}} \sum_{\substack{h_3\in E^{(N_0-1)}_{1,k_3}\\ \lambda_3\in\mathbb{E}^n}} \Lambda(c^{(T)}_{h,\sigma})^r\|u_{h,\lambda}\|_p \sum_{\substack{k_j\in E^{(N_0)}_{j,h_j}\\ \epsilon_j\in\mathbb{E}^n, j=1,2}} \Lambda(c^{(K)}_{k,\epsilon})^{-r+\frac{n}{\mu_0 p}}.\end{aligned} \tag{3.52}$$

Since $r > \frac{n}{\mu_0 p}$ we can find $\theta > 0$ in such a way that $\Lambda(c^{(K)}_{k,\epsilon})^{-\left(r-\frac{n}{\mu_0 p}\right)}$ is bounded above by

$$\Lambda(e^{(K)}_{1,k_1,\epsilon_1})^{-\left(r-\frac{n}{\mu_0 p}-\theta\right)} \Lambda(e^{(K)}_{3,k_3,\epsilon_3})^{-\frac{\theta}{3}} \Lambda(e^{(K)}_{2,h_2,\epsilon_2})^{-\frac{\theta}{3}} \Lambda(e^{(K)}_{3,h_3,\epsilon_3})^{-\frac{\theta}{3}}, \tag{3.53}$$

for all $k \in \mathbb{Z}^3_+$, $k_2 - N_0 < h_2 \le k_2 + N_0$, $h_3 \le k_3 - N_0$ and $\epsilon \in \mathbb{E}^n$.

From (3.53) it follows, for k_j, $j = 1, 2$ running as above,

$$\sum_{\substack{k_j\in E^{(N_0)}_{j,h_j}\\ \epsilon_j\in\mathbb{E}^n, j=1,2}} \Lambda(c^{(K)}_{k,\epsilon})^{-\left(r-\frac{n}{\mu_0 p}\right)} \le C_2 \Lambda(e^{(K)}_{3,k_3,\epsilon_3})^{-\frac{\theta}{3}} \Lambda(e^{(K)}_{2,h_2,\epsilon_2})^{-\frac{\theta}{3}} \Lambda(e^{(K)}_{3,h_3,\epsilon_3})^{-\frac{\theta}{3}}. \tag{3.54}$$

Now from (3.52), together with (3.54) and the Hölder inequality, we obtain the following bound for $\|\sum_{\substack{(k_1,k_2,h_3)\in E_t^{(N_0)}\\ \epsilon_1,\epsilon_2,\lambda_3}} \Lambda(c^{(T)}_{h,\sigma})^r d^{k,\epsilon}_{h,\lambda} u_{h,\lambda}\|_p$:

$$MC \sup_{h,\lambda}\|d_{h,\lambda}\|_{B^{r,\Lambda}_{p,q}} \Lambda_2^{-\frac{\theta}{3}} \Lambda_3^{-\frac{\theta}{3}} \Big(\sum_{h_3,\lambda_3} \Lambda(c^{(T)}_{h,\sigma})^{qr} \|u_{h,\lambda}\|^p_q\Big)^{\frac{1}{q}}, \tag{3.55}$$

where $\Lambda_2 := \Lambda(e^{(K)}_{2,h_2,\epsilon_2})$, $\Lambda_3 := \Lambda(e^{(K)}_{3,k_3,\epsilon_3})$ and C is a suitable positive constant dependent only on q and θ. If we now choose $\gamma > 0$ suitably small in such a way that $2^{\gamma h_2}\Lambda_2^{-\frac{\theta}{3}}$ is uniformly bounded with respect to h_2 and using $\Lambda(c^{(T)}_{h,\sigma}) \le C\Lambda(c^{(H)}_{h,\lambda})$, we can conclude

$$\begin{aligned} I_1 &\le CM \sup_{h,\lambda}\|d_{h,\lambda}\|_{B^{s\Lambda}_{p,q}} \Big(\sum_{\substack{h_1,h_2,k_3\\ \lambda_1,\lambda_2,\epsilon_3}} 2^{qh_2\gamma} \Lambda_3^{-q\frac{\theta}{3}} \max_{\epsilon_2=\pm 1} \Lambda^{-q\frac{\theta}{3}} \sum_{h_3,\lambda_3} \Lambda(c^{(T)}_{h,\sigma})^{qr}\|u_{h,\lambda}\|^q_p\Big)^{\frac{1}{q}} \\ &\le CM \sup_{h,\lambda}\|d_{h,\lambda}\|_{B^{r,\Lambda}_{p,q}} \Big(\sum_{h,\lambda} \Lambda(c^{(H)}_{h,\lambda})^{qr}\|u_{h,\lambda}\|^q_p\Big)^{\frac{1}{q}} \le CM \sup_{h,\lambda}\|d_{h,\lambda}\|_{B^{r,\Lambda}_{p,q}} \|u\|_{B^{r,\Lambda}_{p,q}}. \end{aligned}$$

2) *Estimate of* I_2. Let us set $\Lambda_1 := \Lambda(e^{(K)}_{1,k_1,\epsilon_1})$ and $\Lambda_3 := \Lambda(e^{(K)}_{3,h_3,\lambda_3})$; then for generic positive ρ, ϱ let us multiply $U_{t,\sigma}$ by Λ_1^ρ, Λ_3^ϱ and their own inverses. Since $k_1 < h_1 - N_0$, in view of assumption 3 in Definition 2.1 we have $\Lambda_1 \le C\Lambda(e^{(K)}_{1,h_1,\lambda_1})$. Then using also the Cauchy-Schwarz inequality we obtain

$$\begin{aligned} \|U_{t,\sigma}\|_p &\le C_{\rho,\varrho,q'} \Lambda(e^{(K)}_{1,h_1,\lambda_1})^\rho \Big(\sum_{\substack{(k_1,k_2,h_3)\in E_t^{(N_0)}\\ \epsilon_1,\epsilon_2,\lambda_3}} \Lambda_3^{q\varrho} \|d^{k,\epsilon}_{h,\lambda}\|^q_p \|u_{h,\lambda}\|^q_\infty\Big)^{\frac{1}{q}} \\ &\le C_{\rho,\varrho,q'} \Lambda(e^{(K)}_{1,h_1,\lambda 1})^\rho \Big(\textstyle\sum_{h_3,\lambda_3} \Lambda_3^{q\varrho}\|u_{h,\lambda}\|^q_\infty \sum_{\substack{k_1,k_2\\ \epsilon_1,\epsilon_2}} \|d^{k,\epsilon}_{h,\lambda}\|^q_p\Big)^{\frac{1}{q}}, \end{aligned} \tag{3.56}$$

where $C^{q'}_{\rho,\varrho,q'} = \sum_{\substack{k_1,k_2,h_3\in E_t^{(N_0)}\\ \epsilon_1,\epsilon_2,\lambda_3}} \Lambda_1^{-q'\rho}\Lambda_3^{-q'\varrho}$ is bounded and $\frac{1}{q}+\frac{1}{q'}=1$. Then for a suitable $C=C_{\rho,\varrho,q}>0$, I_2 is bounded by

$$\begin{aligned}
&C(\sum_{\substack{h_1,h_2\\ \lambda_1,\lambda_2}} 2^{q\gamma h_2}\Lambda(e^{(K)}_{1,h_1,\lambda_1})^{q\rho}\sum_{k_3,\epsilon_3}\Lambda(e^{(K)}_{3,k_3,\epsilon_3})^{qr}\sum_{h_3,\lambda_3}\Lambda_3^{q\varrho}\|u_{h,\lambda}\|^q_\infty\sum_{\substack{k_1,k_2\\ \epsilon_1,\epsilon_2}}\|d^{k,\epsilon}_{h,\lambda}\|^q_p)^{\frac{1}{q}}\\
&\le C\sum\nolimits_{h,\lambda}2^{\gamma h_2}\Lambda(e^{(K)}_{1,h_1,\lambda_1})^{\rho}\Lambda_3^{\varrho}\|u_{h,\lambda}\|_\infty(\sum\nolimits_{k,\epsilon}\Lambda(c^{(K)}_{k,\epsilon})^{qr}\|d^{k,\epsilon}_{h,\lambda}\|^q_p)^{\frac{1}{q}}\\
&\le C\sup\nolimits_{h,\lambda}\|d_{h,\lambda}\|_{B^{r,\Lambda}_{p,q}}\sum\nolimits_{h,\lambda}\Lambda(c^{(H)}_{h,\lambda})^{\rho+\varrho+\frac{\gamma}{\mu_0}}\|u_{h,\lambda}\|_\infty,
\end{aligned} \tag{3.57}$$

where the inequality $2^{\gamma h_2}\Lambda(e^{(K)}_{1,h_1,\lambda_1})^{\rho}\Lambda(e^{(K)}_{3,h_3,\lambda_3})^{\varrho}\le C\Lambda(c^{H)}_{h,\lambda})^{\rho+\varrho+\frac{\gamma}{\mu_0}}$ is used for proving the last estimate. Let us remark that the statement of Proposition 2.11 holds true even if a smooth partition of unity in $\phi^{(H)}$ is replaced by any system $\{\psi_{h,\lambda}\}$ satisfying (3.3) and (3.4) up to a finite order $N\ge n$; indeed, provided that $N\ge n$, it amounts to $\psi_{h,\lambda}(D)$ being L^p continuous Fourier multipliers. Then by using Proposition 2.11, we compute

$$\Lambda(c^{(H)}_{h,\lambda})^{\rho+\varrho+\frac{\gamma}{\mu_0}}\|u_{h,\lambda}\|_\infty\le M\Lambda(c^{(H)}_{h,\lambda})^{\rho+\varrho+\frac{\gamma}{\mu_0}-r+\frac{n}{\mu_0 p}}\|u\|_{B^{r,\Lambda}_{p,q}}. \tag{3.58}$$

Since $r>\frac{n}{\mu_0 p}$, for suitable ρ, ϱ, γ we can set $-\theta=\rho+\varrho+\frac{\gamma}{\mu_0}-r+\frac{n}{\mu_0 p}$ in such a way that $\sum_{h,\lambda}\Lambda(c^{(H)}_{h,\lambda})^{-\theta}$ is bounded. We then conclude that $I_2\le C\sup_{h,\lambda}\|d_{h,\lambda}\|_{B^{r,\Lambda}_{p,q}}\|u\|_{B^{r,\Lambda}_{p,q}}$.

3) *Estimate for I_3*. Let us assume in this part that $r>\frac{n}{\mu_0(1-\delta)p}$ with $0<\delta<1$ introduced in (2.1). Using $(1-\delta)r$ and $\Lambda(c^{(T)}_{h,\sigma})^{\delta r}u_{h,\lambda}$ instead of ϱ and $u_{h,\lambda}$ respectively, now we can estimate $\|\sum_{\substack{(k_1,k_2,h_3)\in E_t^{(N_0)}\\ \epsilon_1,\epsilon_2,\lambda_3\in\mathbb{E}^n}}\Lambda(c^{(T)}_{h,\sigma})^{\delta r}d^{k,\epsilon}_{h,\lambda}u_{h,\lambda}\|_p$ as we did for $\|U_{t,\sigma}\|_p$ in (3.56); we then obtain the following estimate:

$$\begin{aligned}
&\|\sum\nolimits_{\substack{(k_1,k_2,h_3)\in E_t^{(N_0)}\\ \epsilon_1,\epsilon_2,\lambda_3\in\mathbb{E}^n}}\Lambda(c^{(T)}_{h,\sigma})^{\delta r}d^{k,\epsilon}_{h,\lambda}u_{h,\lambda}\|_p\le\sum_{\substack{k_1,k_2,h_3\\ \epsilon_1,\epsilon_2,\lambda_3}}\Lambda(c^{(T)}_{h,\sigma})^{\delta r}\|d^{k,\epsilon}_{h,\lambda}\|_p\|u_{h,\lambda}\|_\infty\\
&\le C\Lambda(e^{(K)}_{1,h_1,\lambda_1})^{\rho}(\sum_{h_3,\lambda_3}\Lambda(e^{(K)}_{3,h_3,\lambda_3})^{(1-\delta)rq}\Lambda(c^{(K)}_{h,\sigma})^{\delta rq}\|u_{h,\lambda}\|^q_\infty\sum_{\substack{k_1,k_2\\ \epsilon_1,\epsilon_2}}\|d^{k,\epsilon}_{h,\lambda}\|^q_p)^{\frac{1}{q}}.
\end{aligned}$$

Pointing out now that

$$\max\{\Lambda(e^{(T)}_{3,k_3,\epsilon_3}),\Lambda(e^{(K)}_{3,h_3,\lambda_3})\}\le C\Lambda(e^{(K)}_{3,k_3,\epsilon_3})=:C\Lambda_3,\quad \Lambda(c^{(K)}_{h,\sigma})\le C\Lambda(c^{K)}_{h,\lambda})$$

and setting moreover $\Lambda_1:=\Lambda(e^{(K)}_{1,h_1,\lambda_1})$, we obtain the following bound for I_3, with a suitable $C>0$ independent of u,γ,ρ:

$$\begin{aligned}
&C(\sum_{\substack{h_1,h_2,k_3\\ \lambda_1,\lambda_2,\epsilon_3}}2^{q\gamma h_2}\Lambda_3^{qr\delta}\Lambda_1^{q\rho}\sum_{h_3,\lambda_3}\Lambda(c^{(K)}_{h,\sigma})^{\delta rq}\|u_{h,\lambda}\|_\infty\sum_{\substack{k_1,k_2\\ \epsilon_1,\epsilon_2}}\Lambda_3^{(1-\delta)rq}\|d^{k,\epsilon}_{h,\lambda}\|^q_p)^{\frac{1}{q}}\\
&\le C\sum\nolimits_{h,\lambda}2^{\gamma h_2}\Lambda_1^{\rho}\Lambda(c^{(K)}_{h,\lambda})^{\delta r}\|u_{h,\lambda}\|_\infty(\sum\nolimits_{k,\epsilon}\Lambda(c^{(K)}_{k,\epsilon})^{rq}\|d^{k,\epsilon}_{h,\lambda}\|^q_p)^{\frac{1}{q}}\\
&\le C\sup\nolimits_{h,\lambda}\|d_{h,\lambda}\|_{B^{r,\Lambda}_{p,q}}(\sum\nolimits_{h,\lambda}\Lambda(c^{(H)}_{h,\lambda})^{\rho+\delta r+\frac{\gamma}{\mu_0}-r+\frac{n}{\mu_0 p}})\|u\|_{B^{r,\Lambda}_{p,q}}.
\end{aligned} \tag{3.59}$$

Arguing as in 2), since $(1-\delta)r > \frac{n}{\mu_0 p}$, we can now choose positive numbers θ, ρ, γ such that $\rho + \delta r + \frac{\gamma}{\mu_0} = r - \frac{n}{\mu_0 p} - \theta$; we have thus proved that $I_3 \le C \sup_{h,\lambda} \|d_{h,\lambda}\|_{B^{r,\Lambda}_{p,q}} \|u\|_{B^{r,\Lambda}_{p,q}}$.

Concerning the **case s=0**, we can just repeat the arguments used to estimate I_1, starting from (3.46) with $s = 0$ and $\gamma > 0$ sufficiently small.

Summing up the above computations leads to proving the following

Proposition 3.10. *For* $r > \frac{n}{\mu_0(1-\delta)p}$, $1 < p < \infty$, $1 \le q < \infty$:

$$R : B^{r,\Lambda}_{p,q} \longmapsto B^{r,\Lambda}_{p,q} \qquad R : B^{0,\Lambda}_{p,q} \longmapsto B^{0,\Lambda}_{p,q}, \quad \textit{continuously.} \tag{3.60}$$

With the help of the interpolation results in the propositions 2.13, 2.14, we are able to show the following continuity result about the operators T_j, $j = 4, \dots, 7$.

Proposition 3.11. *For* $r > \frac{n}{\mu_0(1-\delta)p}$, $1 < p < \infty$, $1 \le q < \infty$:

$$T_j : B^{s,\Lambda}_{p,q} \longmapsto B^{s,\Lambda}_{p,q}, \quad \textit{continuously for} \quad 0 \le s \le r, \quad j = 4, \dots, 7. \tag{3.61}$$

We end this section with the proof of Theorem 3.4.

Proof. of Theorem 3.4. By using the Propositions 3.6–3.11 we immediately get the statement for an elementary symbol $\sigma(x,\xi) \in B^{r,\Lambda}_{p,q} M_E(N)$ with $N \ge n$. More precisely for any $0 \le s \le r$, $1 < p < \infty$ and $1 \le q < \infty$

$$\|\sigma(x,D)u\|_{B^{s,\Lambda}_{p,q}} \le C \sup_{h,\lambda} \|d_{h,\lambda}\|_{B^{r,\Lambda}_{p,q}} \|u\|_{B^{s,\Lambda}_{p,q}}, \qquad u \in \mathcal{S}(\mathbb{R}^n), \tag{3.62}$$

where the constant $C > 0$ depends only on r, s, p, q and n.

Let us now take an arbitrary symbol $a(x,\xi)$ in $B^{r,\Lambda}_{p,q} M^0_\Lambda(N)$ for $N \ge 2n+1$; in view of (3.7), where the elementary symbols $a_m(x,\xi)$ are in $B^{r,\Lambda}_{p,q} M_E(N-n-1)$ with $N - n - 1 \ge n$, for every $0 \le s \le r$, $1 < p < \infty$, $1 \le q < \infty$ and $u \in \mathcal{S}(\mathbb{R}^n)$ we obtain

$$\|a(x,D)u\|_{B^{s,\Lambda}_{p,q}} \le C \|u\|_{B^{s,\Lambda}_{p,q}} \sum_{m \in \mathbb{Z}^n} \frac{1}{(1+|m|)^{n+1}} \sup_{h,\lambda} \|d^m_{h,\lambda}\|_{B^{r,\Lambda}_{p,q}}, \tag{3.63}$$

with $C > 0$ depending only on r, s, p, q and the dimension n.

Since the sequences $\{d^m_{h,\lambda}\}_{h,\lambda}$ are bounded in $B^{r,\Lambda}_{p,q}$ uniformly in $m \in \mathbb{Z}^n$, the series $\sum_{m \in \mathbb{Z}^n} \frac{1}{(1+|m|)^{n+1}}$ converges and $\mathcal{S}(\mathbb{R}^n)$ is dense in $B^{s,\Lambda}_{p,q}$, (3.63) implies that $a(x,D)$ is $B^{s,\Lambda}_{p,q}$ bounded.

When the symbol $a(x,\xi) \in B^{r,\Lambda}_{p,q} M^m_\Lambda$ has an arbitrary order m, we easily reduce to the case of order zero as it was already noticed in this section. □

Acknowledgements

We thank Prof. C. Van der Meee, Prof. L. Rodino and the two unknown referees for the suggestion they offer us, above all in order to let the paper be more clear and more understandable to non-specialist readers.

References

[1] J.M. Bony, *Calcul symbolique et propagation des singularités pour les équations aux dérivées partielles non linéaires*, Ann. Sc. Ec. Norm. Sup. **14** (1981), 161–205.

[2] M. Beals, M.C. Reeds, *Microlocal regularity theorems for non smooth pseudodifferential operators and applications to non-linear problems*, Trans. Am. Math. Soc. **285** (1984), 159–184.

[3] A.P. Calderón, *Intermediate spaces and interpolation, the complex method*, Studia Math. **24** (1964), 113–190.

[4] R. Coifman, Y. Meyer, *Au delà des opérateurs pseudo-differentiels*; Astérisque 57, Soc. Math. France, 1978.

[5] C. Fefferman, L^p bounds for pseudodifferential operators, Israel J. Math. **14** (1973), 413–417.

[6] G. Garello, A. Morando, *L^p-bounded pseudodifferential operators and regularity for multi-quasi-elliptic equations*, Quad. Dip. Mat. Univ. Torino **46/2001**, Integral Equations and Operator Theory 51(4) (2005), 501–517.

[7] G. Garello, A. Morando, *L^p boundedness for pseudodifferential operators with non-smooth symbols and applications*, Quad. Dip. Mat. Univ. Torino **44/2002**, to appear on Boll. Un. Mat. It.

[8] L. Hörmander, *Pseudodifferential operators and hypoelliptic equations*, Proc. Symp. Singular Integral AMS **10** (1967), 138–183.

[9] P.I. Lizorkin, (L_p, L_q)-multipliers of Fourier integrals, Dokl. Akad. Nauk SSSR **152**(1963), 808–811. (Engl. transl. Sov. Math. Dokl. **4** (1963), 1420–1424)

[10] J. Marschall, *Pseudo-differential operators with non regular symbols of the class $S^m_{\rho,\delta}$*, Comm. in Part. Diff. Eq. **12(8)** (1987), 921–965. corr. Comm. in Part. Diff. Eq. **13(1)** (1988), 129–130.

[11] J. Marschall, *Weighted L^p estimates for pseudo-differential operators with non regular symbols*, Z. Anal. Anwendungen **10(4)**(1991), 493–501.

[12] Y. Meyer, *Remarques sur un théorème de J.-M. Bony*, Proceedings of the Seminar on Harmonic Analysis (Pisa, 1980). Rend. Circ. Mat. Palermo **(2)suppl. 1** (1981), 1–20.

[13] M. Nagase, *On a class of L^p-bounded pseudodifferential operators*, Sci. Rep. College Gen. Ed. Osaka Univ. **33(4)**(1985), 1–7.

[14] M. Nagase, *On some classes of L^p-bounded pseudodifferential operarators*, **23(2)** (1986), 425–440.

[15] M. Nagase, *On sufficient conditions for pseudodifferential operators to be L^p-bounded. Pseudodifferential operators* (Oberwolfach,1986), Lecture Notes in Math. **1256**, Springer, Berlin 1987.

[16] M. Nagase, *On L^p boundedness of a class of pseudodifferential operators. Harmonic analysis and nonlinear partial differential equations*, (Japanese) (Kyoto, 2001).

[17] M. Sugimoto, *L^p-boundedness of pseudo-differential operators satisfying Besov estimates II*, J. Fac. Sci. Univ. Tokyo Sect. IA, Math. **35** (1988), 149–162.

[18] M.E. Taylor, "Pseudodifferential Operators", Princeton, Univ. Press 1981.

[19] M.E. Taylor, "Pseudodifferential operators and nonlinear PDE", Birkhäuser, Basel-Boston-Berlin, 1991.

[20] H. Triebel, *General Function Spaces, III. Spaces $B^{g(x)}_{p,q}$ and $F^{g(x)}_{p,q}$, $1 < p < \infty$: basic properties*, Anal. Math. **3(3)** (1977), 221–249.

[21] H. Triebel, *General Function Spaces, IV. Spaces $B^{g(x)}_{p,q}$ and $F^{g(x)}_{p,q}$, $1 < p < \infty$: special properties*, Anal. Math. **3(4)** (1977), 299–315.

[22] H. Triebel, "Theory of Function Spaces", Birkhäuser Verlag, Basel, Boston, Stuttgart, 1983.

Gianluca Garello
Dipartimento di Matematica
Università di Torino
Via Carlo Alberto 10,
I-10123 Torino, Italy
e-mail: `gianluca.garello@unito.it`

Alessandro Morando
Dipartimento di Matematica
Facoltà di Ingegneria
Università di Brescia
Via Valotti 9,
I-25133 Brescia, Italy
e-mail: `morando@ing.unibs.it`

Operator Theory:
Advances and Applications, Vol. 160, 217–232

A New Proof of an Ellis-Gohberg Theorem on Orthogonal Matrix Functions Related to the Nehari Problem

G.J. Groenewald and M.A. Kaashoek

To Israel Gohberg on the occasion of his 75th birthday, with gratitude and admiration.

Abstract. The state space method for rational matrix functions and a classical inertia theorem are used to give a new proof of the main step in a recent theorem of R.L. Ellis and I. Gohberg on orthogonal matrix functions related to the Nehari problem. Also we comment on a connection with the Nehari–Takagi interpolation problem.

Mathematics Subject Classification (2000). Primary 33C47, 42C05, 47B35; Secondary 47A57.

Keywords. Orthogonal matrix function, inertia theorem, Nehari-Takagi problem.

0. Introduction

This paper concerns the following theorem which was stated and proved in Ellis-Gohberg [4]:

Theorem 0.1. *Let $k \in L_1^{m\times m}(a,\infty)$ with $a \geq 0$. Assume that there exist solutions g_a in $L_1^{m\times m}(a,\infty)$ and h_a in $L_1^{m\times m}(-\infty,-a)$ of the equations*

$$g_a(t) + \int_a^\infty k(t+s-a)h_a(-s)\,ds = 0, \qquad (t \geq a), \tag{1}$$

and

$$\int_a^\infty k(t+s-a)^* g_a(s)\,ds + h_a(-t) = -k(t)^*, \qquad (t \geq a). \tag{2}$$

The research of the first author is supported by the National Research Foundation, South Africa, under Grant number 2053733.

Assume also that there exist γ_a in $L_1^{m\times m}(a,\infty)$ and χ_a in $L_1^{m\times m}(-\infty,-a)$ satisfying the equations

$$\gamma_a(t) + \int_a^\infty k(t+s-a)\chi_a(-s)\,ds = -k(t), \qquad (t \geq a), \tag{3}$$

and

$$\int_a^\infty k(t+s-a)^*\gamma_a(s)\,ds + \chi_a(-t) = 0, \qquad (t \geq a). \tag{4}$$

Define

$$\Phi_a(\lambda) = e^{i\lambda a} I + \int_a^\infty e^{i\lambda t} g_a(t)\,dt, \qquad (\Im\lambda \geq 0), \tag{5}$$

and

$$\Theta_a(\lambda) = e^{-i\lambda a} I + \int_a^\infty e^{-i\lambda t}\chi_a(-t)\,dt, \qquad (\Im\lambda \leq 0). \tag{6}$$

Then $\Phi_a(\lambda)$ and $\Theta_a(\lambda)$ are invertible for all real λ, and counting multiplicities, the number of zeros of $\det\Phi_a$ (respectively, $\det\Theta_a$) in the upper (respectively, lower) half-plane is finite and equals the number of negative eigenvalues of the operator

$$T = \begin{pmatrix} I & K_a \\ K_a^* & I \end{pmatrix}, \tag{7}$$

on $L_1^m(a,\infty) \times L_1^m(-\infty,-a)$. Here

$$(K_a\psi)(t) = \int_a^\infty k(t+s-a)\psi(-s)\,ds, \qquad (t \geq a), \tag{8}$$

and

$$(K_a^*\phi)(-t) = \int_a^\infty k(t+s-a)^*\phi(s)\,ds, \qquad (t \geq a), \tag{9}$$

for $\phi \in L_1^m(a,\infty)$ and $\psi \in L_1^m(-\infty,-a)$.[1]

The proof of this theorem in [4] (see also Chapter 12 of [5]) is given in three steps. In the first step it is shown that the operator T in (7) is invertible whenever the equations (1) and (2) have solutions g_a in $L_1^{m\times m}(a,\infty)$ and h_a in $L_1^{m\times m}(-\infty,-a)$ and the equations (3) and (4) have solutions γ_a in $L_1^{m\times m}(a,\infty)$ and χ_a in $L_1^{m\times m}(-\infty,-a)$. In the second step T is assumed to be invertible, k is the limit in $L_1^{m\times m}(a,\infty)$ of a sequence $\{k_n\}$ consisting of continuous functions of compact support, and the authors show that it suffices to prove the theorem when k is replaced by k_n for n sufficiently large. Since the continuous $m\times m$ matrix functions with compact support in (a,∞) are dense in $L_1^{m\times m}(a,\infty)$, the second step shows that it is enough to prove the theorem for such a function. The latter is done in Step 3 by converting the operator T into an operator $I - B$, where B is a self adjoint convolution integral operator on a finite interval with a kernel function depending on the difference of arguments. By applying to B the main theorem of [6] the proof is completed.

[1] In [4] and also in [5] the operator T is considered on $L_1^{m\times m}(a,\infty)\times L_1^{m\times m}(-\infty,-a)$ but from the context it is clear that the space $L_1^m(a,\infty)\times L_1^m(-\infty,-a)$ is meant.

In this paper we replace the role of continuous functions k with compact support by functions of the form

$$k(t) = Ce^{tA}B, \tag{10}$$

where A, B and C are matrices of sizes $n \times n$, $n \times m$ and $m \times n$, respectively, A is assumed to be stable, i.e., all eigenvalues of A are in the open left half-plane Π^-, and the triple (A, B, C) is minimal, i.e.,

$$\bigcap_{j=0}^{n} \ker CA^j = \{0\}, \qquad \operatorname{span}\{A^j B\mathbb{C}^m : j = 0, \ldots, n\} = \mathbb{C}^n. \tag{11}$$

We shall refer to a function k with such a representation as a *stable kernel function of exponential type* on (a, ∞) and we call (10) a *stable exponential representation* of k. Since rational functions with poles off $\mathbb{R} \cup \infty$ are dense in the Wiener algebra on the real line (see [3, page 63]), stable kernel functions of exponential type on (a, ∞) are dense in $L_1^{m\times m}(a, \infty)$, and hence by repeating Step 2 in [4] with such a k in place of a continuous function with compact support, it is clear that it suffices to prove the above theorem for functions k of the form (10). We do this by reformulating the underlying problem as a linear algebra problem which is solved by using classical inertia theorems ([10, page 448]). The choice of the representation (10) is inspired by [8].

The paper consists of four sections (not counting this introduction). In the first section we associate with k in (10) the $n \times n$ matrix

$$M_a = I_n - P_a e^{-aA^*} Q_a e^{-aA}, \tag{12}$$

where P_a and Q_a are given by

$$P_a = \int_a^\infty e^{sA} BB^* e^{sA^*}\, ds, \quad Q_a = \int_a^\infty e^{sA^*} C^* C e^{sA}\, ds. \tag{13}$$

Since A is assumed to be stable, P_a and Q_a are well-defined $n \times n$ matrices, and (11) is equivalent to P_a and Q_a being positive definite. We refer to M_a as the *indicator* of the operator T in (7) associated to the stable exponential representation of k in (10). We show (Theorem 2.1 below) that T in (7) is invertible if and only if M_a is invertible, and the number of negative eigenvalues of T is equal to the number of negative eigenvalues of M_a, multiplicities taken into account. We also rewrite the equations (1)–(4) in terms of the indicator M_a.

In the second section we show that for M_a invertible, the inertia of M_a is equal to the inertia of the matrix $C^*CP_a(M_a^{-1})^* - A^*$. In Section 3 we use this result and those of Section 1 to prove Theorem 0.1 for the case when k is a stable kernel function of exponential type. In the final section we comment on the connection with the Nehari–Takagi interpolation problem.

In conclusion we mention that our approach has its roots in the theory of input output systems. In fact (see, e.g., [2], page 6), a function k of the form (10) is the impulse response of the system

$$\Sigma \begin{cases} x'(t) = Ax(t) + Bu(t), & t \geq 0, \\ y(t) \;\; = Cx(t), \end{cases}$$

at time t to a unit impulse at time 0. Furthermore, the conditions in (11) are equivalent to the requirement that the system Σ is minimal, that is, the order of A is minimal among all systems with the same impulse response as Σ. When A is stable, then for $a = 0$ the matrix P_a in (13) is the controllability gramian and Q_a is the observability gramian of Σ ([2], page 62), and in that case the operator K_a is the Hankel operator which can be written as the product $\Lambda_a \Gamma_a$, where Γ_a is the controllability operator which maps the past input ($t < 0$) to the present state (t=0), and Λ_a is the observability operator mapping the present state to future outputs. This representation of K_a and the corresponding representation of K_a^* play an essential role in our analysis (see the proof of Theorem 1.1 below).

Finally, for the connection with the theory of orthogonal polynomials we refer the reader to [5], where Theorem 0.1 is presented as a continuous infinite analogue of Kreĭn's theorem [9]. The latter theorem is a generalization of the classical Szegö theorem to the case in which the weight function is not necessarily positive. In Theorem 0.1 the operator T plays the same role as the Toeplitz matrix in Kreĭn's theorem. See the first chapter of [5], where these results are described in detail and additional references can be found.

1. The operator T and its indicator

In the sequel we assume throughout that k is a stable kernel function of exponential type given by the stable exponential representation (10). In particular, A is stable and (11) holds. Furthermore, P_a and Q_a are the $n \times n$ matrices defined by (13). Since A and hence also A^* is stable, it is well–known that P_a and Q_a are positive definite and satisfy the following Lyapunov equations:

$$AP_a + P_a A^* = -e^{aA} BB^* e^{aA^*}, \tag{14}$$

and

$$A^* Q_a + Q_a A = -e^{aA^*} C^* C e^{aA}. \tag{15}$$

Recall that the indicator M_a associated with (10) is given by (12).

Theorem 1.1. *Assume k has the stable exponential representation* (10). *Then the operator T in* (7) *is invertible if and only if the corresponding indicator M_a is invertible and the number of negative eigenvalues of T is equal to the number of negative eigenvalues of M_a, multiplicities taken into account.*

Proof. Since K_a is of finite rank and T is self adjoint, $\sigma(T) \setminus \{1\}$ consists of a finite number of eigenvalues of finite multiplicity. Introduce the following auxiliary operators:

$$\Lambda_a : \mathbb{C}^n \mapsto L_1^m(a,\infty), \qquad (\Lambda_a x)(t) = Ce^{(t-a)A}x, \quad (t \geq a)$$

$$\Lambda_a^\# : \mathbb{C}^n \mapsto L_1^m(-\infty,-a), \qquad (\Lambda_a^\# x)(-t) = B^* e^{(t-a)A^*}x, \quad (t \geq a)$$

$$\Gamma_a : L_1^m(-\infty,-a) \mapsto \mathbb{C}^n, \qquad \Gamma_a f = \int_a^\infty e^{sA}Bf(-s)\,ds,$$

$$\Gamma_a^\# : L_1^m(a,\infty) \mapsto \mathbb{C}^n, \qquad \Gamma_a^\# f = \int_a^\infty e^{sA^*}C^* f(s)\,ds.$$

We have

$$K_a = \Lambda_a\Gamma_a, \quad K_a^* = \Lambda_a^\#\Gamma_a^\#, \text{ and } \Gamma_a\Lambda_a^\# = P_a e^{-aA^*}, \quad \Gamma_a^\#\Lambda_a = Q_a e^{-aA}.$$

It follows that

$$T = \begin{pmatrix} I & 0 \\ 0 & I \end{pmatrix} + \begin{pmatrix} \Lambda_a & 0 \\ 0 & \Lambda_a^\# \end{pmatrix}\begin{pmatrix} 0 & \Gamma_a \\ \Gamma_a^\# & 0 \end{pmatrix}, \tag{16}$$

$$M_a = I_n - \Gamma_a\Lambda_a^\#\Gamma_a^\#\Lambda_a. \tag{17}$$

Put

$$\widehat{T} = \begin{pmatrix} I & 0 \\ 0 & I \end{pmatrix} + \begin{pmatrix} 0 & \Gamma_a \\ \Gamma_a^\# & 0 \end{pmatrix}\begin{pmatrix} \Lambda_a & 0 \\ 0 & \Lambda_a^\# \end{pmatrix} = \begin{pmatrix} I & \Gamma_a\Lambda_a^\# \\ \Gamma_a^\#\Lambda_a & I \end{pmatrix}. \tag{18}$$

From (16) it follows that on the domain $\mathbb{C}\setminus\{1\}$ the operator function $\lambda I - T$ is globally equivalent (see Section III.2 in [7] for this terminology) to the matrix–valued function $\lambda I - \widehat{T}$. In particular,

(a) T is invertible if and only if $\widehat{T}$ is invertible,
(b) the number of negative eigenvalues of T is equal to the number of negative eigenvalues of $\widehat{T}$.

Notice that

$$\begin{aligned}\widehat{T} &= \begin{pmatrix} I & P_a e^{-aA^*} \\ Q_a e^{-aA} & I \end{pmatrix} \\ &= \begin{pmatrix} P_a^{\frac12} & 0 \\ 0 & Q_a^{\frac12} \end{pmatrix}\begin{pmatrix} I & P_a^{\frac12} e^{-aA^*} Q_a^{\frac12} \\ Q_a^{\frac12} e^{-aA} P_a^{\frac12} & I \end{pmatrix}\begin{pmatrix} P_a^{-\frac12} & 0 \\ 0 & Q_a^{-\frac12} \end{pmatrix}.\end{aligned}$$

The previous identity is a similarity relation. It follows that (a) and (b) remain true if $\widehat{T}$ is replaced by $\widehat{L}$, where

$$\widehat{L} = \begin{pmatrix} I & L_a \\ L_a^* & I \end{pmatrix}, \qquad L_a = P_a^{\frac12} e^{-aA^*} Q_a^{\frac12}.$$

Now

$$\widehat{L} = \begin{pmatrix} I & L_a \\ 0 & I \end{pmatrix}\begin{pmatrix} I - L_aL_a^* & 0 \\ 0 & I \end{pmatrix}\begin{pmatrix} I & 0 \\ L_a^* & I \end{pmatrix}, \tag{19}$$

and

$$I - L_aL_a^* = P_a^{-\frac12} M_a P_a^{\frac12}. \tag{20}$$

The relation (19) is a congruence relation, and (20) is a similarity relation. It follows that

(c) $\widehat{L}$ is invertible if and only if M_a is invertible,

(d) the number of negative eigenvalues of $\widehat{L}$ is equal to the number of negative eigenvalues of M_a multiplicities taken into account.

Since (a) and (b) remain true with $\widehat{L}$ in place of $\widehat{T}$, the theorem is proved. □

Proposition 1.2. *Assume k has the stable exponential representation* (10). *Then there exists $g_a \in L_1^{m\times m}(a,\infty)$ and $h_a \in L_1^{m\times m}(-\infty,-a)$ such that* (1) *and* (2) *hold if and only if the matrix equation*

$$M_a X = -P_a C^* \tag{21}$$

is solvable. In this case, if X is a solution of (21), *then the $m\times m$ matrix functions*

$$g_a(t) = -Ce^{(t-a)A}X, \tag{22}$$

$$h_a(-t) = B^*e^{(t-a)A^*}(Q_a e^{-aA}X - e^{aA^*}C^*), \tag{23}$$

satisfy equations (1) *and* (2), *respectively. Furthermore, for this choice of g_a the function Φ_a in* (5) *is given by*

$$\Phi_a(\lambda) = e^{i\lambda a}[I - iC(\lambda - iA)^{-1}X], \qquad \Im\lambda \geq 0. \tag{24}$$

Proof. Suppose that g_a and h_a satisfy (1) and (2). Define X by

$$X = \int_a^\infty e^{sA}Bh_a(-s)\,ds. \tag{25}$$

Then clearly,

$$g_a(t) = -Ce^{(t-a)A}(\int_a^\infty e^{sA}Bh_a(-s)\,ds),$$

so g_a has the representation (22). Next, note that

$$k(t)^* = B^*e^{tA^*}C^*, \qquad t \geq a. \tag{26}$$

Substituting (26) for $k(t)^*$ in (2) yields a formula for $h_a(-t)$, namely,

$$\begin{aligned} h_a(-t) &= B^*e^{(t-a)A^*}\{(\int_a^\infty e^{sA^*}C^*Ce^{sA}\,ds)e^{-aA}X - e^{aA^*}C^*\} \\ &= B^*e^{(t-a)A^*}(Q_a e^{-aA}X - e^{aA^*}C^*). \end{aligned}$$

Thus h_a is of the form (23). Next we show that X in (25) is a solution of (21). Indeed, from (25), (23), the first equality in (13) and (12) we get that

$$\begin{aligned} X &= \int_a^\infty e^{sA}Bh_a(-s)\,ds = (\int_a^\infty e^{sA}BB^*e^{(s-a)A^*}\,ds)(Q_a e^{-aA}X - e^{aA^*}C^*) \\ &= (\int_a^\infty e^{sA}BB^*e^{sA^*}\,ds)(e^{-aA^*}Q_a e^{-aA}X - C^*) \\ &= P_a e^{-aA^*}Q_a e^{-aA}X - P_a C^*, \end{aligned}$$

which yields (21).

Conversely, suppose that X is a solution of the matrix equation (21), and that g_a and h_a are defined by (22) and (23), respectively. Then it follows that

$$\begin{aligned} g_a(t) &+ \int_a^\infty k(t+s-a)h_a(-s)\,ds \\ &= -Ce^{(t-a)A}X + \int_a^\infty Ce^{(t+s-a)A}BB^*e^{(s-a)A^*}(Q_a e^{-aA}X - e^{aA^*}C^*)\,ds \\ &= -Ce^{(t-a)A}X + Ce^{(t-a)A}(\int_a^\infty e^{sA}BB^*e^{sA^*}\,ds)(e^{-aA^*}Q_a e^{-aA}X - C^*) \\ &= -Ce^{(t-a)A}X + Ce^{(t-a)A}(P_a e^{-aA^*}Q_a e^{-aA}X - P_a C^*) \\ &= -Ce^{(t-a)A}X + Ce^{(t-a)A}X = 0. \end{aligned}$$

So (1) is satisfied. Furthermore, it follows from (13), (22) and (23) that

$$\begin{aligned} &\int_a^\infty k(t+s-a)^* g_a(s)\,ds + h_a(-t) \\ &= -\int_a^\infty B^*e^{(t+s-a)A^*}C^*Ce^{(s-a)A}X\,ds + B^*e^{(t-a)A^*}(Q_a e^{-aA}X - e^{aA^*}C^*) \\ &= -B^*e^{(t-a)A^*}(\int_a^\infty e^{sA^*}C^*Ce^{sA}\,ds)e^{-aA}X + B^*e^{(t-a)A^*}\times \\ &\qquad\qquad \times(Q_a e^{-aA}X - e^{aA^*}C^*) \\ &= -B^*e^{(t-a)A^*}Q_a e^{-aA}X + B^*e^{(t-a)A^*}Q_a e^{-aA}X - B^*e^{(t-a)A^*}e^{aA^*}C^* \\ &= -B^*e^{tA^*}C^* = -k(t)^*. \end{aligned}$$

So (2) is satisfied.
Finally, let g_a be given by (22), and take $\Im\lambda \geq 0$. Using the Fundamental Theorem of Calculus and the stability of A we get from (5) that

$$\begin{aligned} \Phi_a(\lambda) &= e^{i\lambda a}I - \int_a^\infty e^{i\lambda t}Ce^{(t-a)A}X\,dt \\ &= e^{i\lambda a}I - C(\int_a^\infty e^{i(\lambda - iA)t}\,dt)e^{-aA}X \\ &= e^{i\lambda a}I - C[-i(\lambda - iA)^{-1}e^{i(\lambda-iA)t}]_a^\infty e^{-aA}X \\ &= e^{i\lambda a}I - iC(\lambda - iA)^{-1}e^{ia(\lambda-iA)}e^{-aA}X = e^{i\lambda a}[I - iC(\lambda - iA)^{-1}X]. \end{aligned}$$

This yields (24) and completes the proof. □

Put $\tilde{k}(t) = k(t)^* = B^*e^{tA^*}C^*$, $t \geq 0$, and apply the previous proposition with k replaced by $\tilde{k}$. In this way (we omit the details) it is straightforward to derive the following result.

Proposition 1.3. *Assume k has the stable exponential representation* (10). *Then there exist $\gamma_a \in L_1^{m\times m}(a,\infty)$ and $\chi_a \in L_1^{m\times m}(-\infty,-a)$ such that* (3) *and* (4) *hold if and only if the matrix equation*

$$M_a P_a e^{-aA^*}U = -P_a e^{-aA^*}Q_a B \tag{27}$$

is solvable. In this case, if U is a solution of (27), then the $m\times m$ matrix functions

$$\chi_a(-t) = -B^* e^{(t-a)A^*}U, \tag{28}$$

$$\gamma_a(t) = Ce^{(t-a)A}(P_a e^{-aA^*}U - e^{aA}B), \tag{29}$$

satisfy equations (3) and (4), respectively. Furthermore, for this choice of χ_a the function Θ_a in (6) is given by

$$\Theta_a(\lambda) = e^{-i\lambda a}[I + iB^*(\lambda + iA^*)^{-1}U], \qquad \Im\lambda \le 0. \tag{30}$$

2. The inertia of M_a

We continue with the assumptions that (A, B, C) is a minimal triple and that A and thus A^* are both stable. Let M_a be defined by (12), and let the positive definite operators P_a and Q_a satisfy the Lyapunov equations (14) and (15).

Lemma 2.1. *Let M_a be given by (12). Then*

$$M_aA - AM_a = P_aC^*C - e^{aA}BB^*Q_ae^{-aA}. \tag{31}$$

Proof. Using (12) and the Lyapunov equations (14) and (15) we readily obtain (31). Indeed,

$$\begin{aligned}
M_aA &= A - P_ae^{-aA^*}(Q_aA)e^{-aA}\\
&= A - P_ae^{-aA^*}(-e^{aA^*}C^*Ce^{aA} - A^*Q_a)e^{-aA}\\
&= A + P_aC^*C + (P_aA^*)e^{-aA^*}Q_ae^{-aA}\\
&= A + P_aC^*C + (-e^{aA}BB^*e^{aA^*} - AP_a)e^{-aA^*}Q_ae^{-aA}\\
&= A + P_aC^*C - e^{aA}BB^*Q_ae^{-aA} - AP_ae^{-aA^*}Q_ae^{-aA}\\
&= AM_a + P_aC^*C - e^{aA}BB^*Q_ae^{-aA},
\end{aligned}$$

and so (31) is proved. □

The next proposition gives necessary and sufficient conditions for the matrix M_a to be invertible.

Proposition 2.2. *The indicator M_a is invertible if and only if the following matrix equations are solvable*

$$M_aX = -P_aC^*, \quad M_aP_ae^{-aA^*}U = -P_ae^{-aA^*}Q_aB. \tag{32}$$

Proof. Clearly, if M_a is invertible, then the matrix equations in (32) are solvable. Conversely, suppose that the matrix equations (32) are solvable and that M_a is not invertible. Then there exists a nonzero x in $\mathbb{C}^n$ such that $x^*M_a = 0$. Note that this implies from (12) that

$$x^*e^{aA} = x^*P_ae^{-aA^*}Q_a. \tag{33}$$

Since $x^*M_a = 0$ and the matrix equations in (32) are solvable, we have

$$x^*P_aC^* = -x^*M_aX = 0, \tag{34}$$

$$x^*P_ae^{-aA^*}Q_aB = -x^*M_aP_ae^{-aA^*}U = 0. \tag{35}$$

Using (33) it follows from (35) that

$$x^* e^{aA} B = 0. \tag{36}$$

Employing (31), (34) and (36) we conclude that

$$\begin{aligned} -x^* A M_a &= x^*(M_a A - A M_a) \\ &= x^* P_a C^* C - x^* e^{aA} B B^* Q_a e^{-aA} = 0. \end{aligned}$$

Thus $x^* A M_a = 0$. Repeating this argument with x^* replaced by $x^* A$ we obtain that $x^* A^n M_a = 0$ for $n = 0, 1, 2, \ldots$. According to (36) with x^* replaced by $x^* A^n$ we have

$$x^* e^{aA} A^n B = 0, \qquad n = 0,\ 1,\ 2, \ldots. \tag{37}$$

But the pair (A, B) satisfies the second identity in (11). Hence (37) implies that $x^* e^{aA} = 0$ and thus $x^* = 0$. So M_a is invertible. □

An important ingredient in the sequel is the following inertia theorem. We use the symbol In M to denote the inertia of the square matrix M.

Theorem 2.3. *Assume M_a is invertible, and put $A^\times = A - M_a^{-1} P_a C^* C$. Then*

$$\text{In } M_a = \text{In } (-A^\times)^*. \tag{38}$$

Proof. The proof is divided into four steps.

Step 1. First we show that without loss of generality we may take $P_a = I$. Indeed, replace the triple (A, B, C) by $(\widehat{A}, \widehat{B}, \widehat{C})$, where

$$\widehat{A} = S^{-1} A S, \qquad \widehat{B} = S^{-1} B, \qquad \widehat{C} = C S, \tag{39}$$

for some invertible matrix S. Let $\widehat{P_a}$, $\widehat{Q_a}$ and $\widehat{M_a}$ be defined by (13) and (12) with $\widehat{A}$, $\widehat{B}$, $\widehat{C}$ in place of A, B, C. From (13) it follows that

$$\begin{aligned} \widehat{P_a} &= \int_a^\infty e^{s\widehat{A}} \widehat{B} \widehat{B}^* e^{s\widehat{A}^*} \, ds = \int_a^\infty S^{-1} e^{sA} S S^{-1} B B^* S^{*-1} S^* e^{sA^*} S^{*-1} \, ds \\ &= S^{-1} (\int_a^\infty e^{sA} B B^* e^{sA^*} \, ds) S^{*-1} \end{aligned}$$

thus

$$\widehat{P_a} = S^{-1} P_a S^{*-1}. \tag{40}$$

Likewise, it follows from (13) that

$$\widehat{Q_a} = S^* Q_a S. \tag{41}$$

From (12) it follows that $\widehat{M_a}$ is given by

$$\begin{aligned} \widehat{M_a} &= I_n - \widehat{P_a} e^{-a\widehat{A}^*} \widehat{Q_a} e^{-a\widehat{A}} \\ &= I_n - (S^{-1} P_a S^{*-1}) S^* e^{-aA^*} S^{*-1} (S^* Q_a S)(S^{-1} e^{-aA} S) \\ &= I_n - S^{-1} P_a e^{-aA^*} Q_a e^{-aA} S, \end{aligned}$$

thus

$$\widehat{M_a} = S^{-1} M_a S. \tag{42}$$

Since M_a is invertible, the same holds true for $\widehat{M_a}$, and we can define $(\widehat{A})^\times = \widehat{A} - \widehat{M_a}^{-1}\widehat{P_a}\widehat{C}^*\widehat{C}$. Note that

$$(\widehat{A})^\times = S^{-1}A^\times S. \tag{43}$$

In fact, it follows from (39), (40) and (42) that

$$\begin{aligned} S^{-1}A^\times &= S^{-1}A - S^{-1}M_a^{-1}P_aC^*C = \widehat{A}S^{-1} - \widehat{M_a}^{-1}S^{-1}P_aC^*C \\ &= \widehat{A}S^{-1} - \widehat{M_a}^{-1}\widehat{P_a}S^*C^*C = \widehat{A}S^{-1} - \widehat{M_a}^{-1}\widehat{P_a}\widehat{C}^*\widehat{C}S^{-1} \\ &= (\widehat{A} - \widehat{M_a}^{-1}\widehat{P_a}\widehat{C}^*\widehat{C})S^{-1} = (\widehat{A})^\times S^{-1}, \end{aligned}$$

and hence (43) is proved. It follows from (42) and (43) that

$$\text{In } \widehat{M_a} = \text{In } M_a, \quad \text{In } ((-\widehat{A})^\times)^* = \text{In } (-A^\times)^*. \tag{44}$$

Hence it suffices to prove (38) with $\widehat{M_a}$ in place of M_a and $(\widehat{A})^\times$ in place of $A^\times$. In particular, by choosing $S = P_a^{\frac{1}{2}}$ we see from (40) that it suffices to prove the theorem for $P_a = I$.

Step 2. In the following we assume that $P_a = I$. Note that this implies that M_a is Hermitian. We shall show that the following identity holds:

$$M_a(-A^\times) + (-A^\times)^*M_a = e^{aA}BB^*e^{aA^*} + C^*C, \tag{45}$$

where now $M_a = I_n - e^{-aA^*}Q_ae^{-aA}$ and $A^\times = A - M_a^{-1}C^*C$. Note that, $M_aA^\times = M_aA - C^*C$. On the other hand, by using (14), (15) and (12) we obtain:

$$\begin{aligned} (A^\times)^*M_a &= (M_aA^\times)^* = A^*M_a - C^*C \\ &= A^* - e^{-aA^*}A^*Q_ae^{-aA} - C^*C \\ &= A^* - e^{-aA^*}(-Q_aA - e^{aA^*}C^*Ce^{aA})e^{-aA} - C^*C \\ &= A^* + e^{-aA^*}Q_ae^{-aA}A + C^*C - C^*C \\ &= A^* + (I_n - M_a)A = A^* + A - M_aA. \end{aligned}$$

Then adding we get (using (14) with $P_a = I$) that

$$M_aA^\times + (A^\times)^*M_a = A^* + A - C^*C = -e^{aA}BB^*e^{aA^*} - C^*C.$$

Hence (45) is proved.

Step 3. We show that $A^\times$ has no eigenvalue on $i\mathbb{R}$. Suppose that $A^\times$ has an imaginary eigenvalue, i.e., there exists a nonzero vector $x \in \mathbb{C}^n$ such that $A^\times x = i\alpha x$, $\alpha \in \mathbb{R}$. Then using (45) with $W = e^{aA}BB^*e^{aA^*} + C^*C$ and premultiplying by x^* and postmultiplying by x we obtain:

$$x^*Wx = -x^*(A^\times)^*M_ax - x^*M_aA^\times x = i\alpha x^*M_ax - i\alpha x^*M_ax = 0.$$

But then $x^*e^{aA}BB^*e^{aA^*}x + x^*C^*Cx = 0$, i.e., $\|B^*e^{aA^*}x\|^2 + \|Cx\|^2 = 0$. Thus $Cx = 0$ and $B^*e^{aA^*}x = 0$. Moreover, $A^\times x = i\alpha x$ and $Cx = 0$, together imply that

$$Ax = (A - M_a^{-1}C^*C)x = A^\times x = i\alpha x, \quad \alpha \in \mathbb{R}.$$

Hence $\sigma(A) \cap i\mathbb{R} \neq \emptyset$. This contradicts the stability of A. Therefore, $A^\times$ has no eigenvalue on $i\mathbb{R}$.

Step 4. To finish the proof we use a classical inertia theorem due to D. Carlson and H. Schneider, which can be found in [10, page 448]. We apply this inertia theorem with

$$A = (-A^\times)^*, \quad H = M_a, \quad \text{and} \quad W = e^{aA}BB^*e^{aA^*} + C^*C.$$

From Steps 2–3 we know that $A = (-A^\times)^*$ has no eigenvalue on the imaginary axis, and that the Hermitian nonsingular matrix $H = M_a$ satisfies $AH + HA^* = W$, where $W = e^{aA}BB^*e^{aA^*} + C^*C \geq 0$. Thus the Carlson and Schneider inertia theorem yields $\text{In}\,(-A^\times)^* = \text{In}\,M_a$, and we are done. □

3. Proof of Theorem 0.1 for kernel functions of stable exponential type

Throughout this section k is given by the stable exponential representation (10) with (A, B, C) being a minimal triple. In particular, A and hence A^* are stable. Let M_a be given by (12) and let the positive definite operators P_a and Q_a satisfy the Lyapunov equations (14) and (15).

Proof of Theorem 0.1 *for k as in (*10*).* We divide the proof into six steps.

Step 1. Let $\Im\lambda \geq 0$. From (21) and (24) we get that

$$\begin{aligned}\det\Phi_a(\lambda) &= \det e^{i\lambda a}I \det[I + iC(\lambda - iA)^{-1}M_a^{-1}P_aC^*] \\ &= \det e^{i\lambda a}I \det[I + i(\lambda - iA)^{-1}M_a^{-1}P_aC^*C] \\ &= \det e^{i\lambda a}I \det(\lambda - iA)^{-1}\det[\lambda - i(A - M_a^{-1}P_aC^*C)].\end{aligned}$$

Therefore

$$\det\Phi_a(\lambda) = \det e^{i\lambda a}I \det(\lambda - iA)^{-1}\det(\lambda - iA^\times), \tag{46}$$

where $A^\times = A - M_a^{-1}P_aC^*C$. We know that $|\det e^{i\lambda a}I| \neq 0$ and that $\det(\lambda - iA)^{-1} \neq 0$ since A is stable. Hence from (46), we see that $\det\Phi_a(\lambda) \neq 0$ for $\Im\lambda \geq 0$ if and only if $\det(\lambda - iA^\times) \neq 0$ for $\Im\lambda \geq 0$.

Step 2. From Step 3 of Theorem 2.3 we know that $\sigma(A^\times) \cap i\mathbb{R} = \emptyset$, equivalently that $\sigma(iA^\times) \cap \mathbb{R} = \emptyset$, hence $\det(\lambda - iA^\times) \neq 0$ for each $\lambda \in \mathbb{R}$. It follows immediately from (46) that $\det\Phi_a(\lambda) \neq 0$ for each $\lambda \in \mathbb{R}$. So $\Phi_a(\lambda)$ is invertible for each $\lambda \in \mathbb{R}$.

Step 3. It follows from (46) and the Inertia Theorem 2.3 that

$$\begin{aligned}&\#\text{ zeros of }\det\Phi_a\text{ in the upper half-plane } = \\ &\quad = \#\text{ eigenvalues of } iA^\times \text{ in the upper half-plane} \\ &\quad = \#\text{ eigenvalues of } (-A^\times)^* \text{ in the left half-plane} \\ &\quad = \#\text{ negative eigenvalues of } M_a\,.\end{aligned}$$

Step 4. Suppose that equations (1)–(4) are satisfied. Then the matrix equations (21) and (27) are both solvable (see Propositions 1.2 and 1.3). Therefore by Proposition 2.2, M_a is invertible. Hence T is invertible by Theorem 1.1.

Step 5. Next, it follows from Theorem 1.1 and the result of Step 3 above that

$$\begin{aligned}\#\text{ negative eigenvalues of } T &= \#\text{ negative eigenvalues of } M_a\\ &= \#\text{ zeros of } \det\Phi_a(\lambda) \text{ in the upper half-plane.}\end{aligned}$$

Step 6. Finally, we prove the statement about the number of zeros of $\det\Theta_a$. In fact, we first replace the system triple (A,B,C) by (A^*,C^*,B^*). Define $P_a^\#$, $Q_a^\#$ and $M_a^\#$ by (13) and (12) with (A^*,C^*,B^*) in place of (A,B,C). Then, clearly $P_a^\# = Q_a$, $Q_a^\# = P_a$ and $M_a^\#$ is defined by

$$M_a^\# = I_n - Q_a e^{-aA} P_a e^{-aA^*}. \tag{47}$$

Observe that

$$(M_a^\#)^{-1} = Q_a e^{-aA} M_a^{-1} e^{aA} Q_a^{-1}. \tag{48}$$

Next, it follows from (24) that the transformed function $\Phi_a^\#(\lambda)$ with (A^*,C^*,B^*) and $(P_a^\#,Q_a^\#,M_a^\#)$ instead of (A,B,C) and (P_a,Q_a,M_a), respectively, is defined by

$$\Phi_a^\#(\lambda) = e^{i\lambda a}[I + iB^*(\lambda - iA^*)^{-1}(M_a^\#)^{-1}P_a^\# B], \quad \Im\lambda \geq 0. \tag{49}$$

Using (48) and $P_a^\# = Q_a$ we can recast $\Phi_a^\#(\lambda)$ as:

$$\Phi_a^\#(\lambda) = e^{i\lambda a}[I + iB^*(\lambda - iA^*)^{-1}Q_a e^{-aA} M_a^{-1} e^{aA} B], \quad \Im\lambda \geq 0.$$

Then by comparing the realization formulas for $\Phi_a^\#$ above and Θ_a in (30) we see that

$$Q_a e^{-aA} M_a^{-1} e^{aA} B = e^{aA^*} P_a^{-1} M_a^{-1} P_a e^{-aA^*} Q_a B. \tag{50}$$

Indeed, first note that

$$M_a^\# = e^{aA^*} P_a^{-1} M_a P_a e^{-aA^*}. \tag{51}$$

Taking inverses of both sides of equation (51) yields:

$$(M_a^\#)^{-1} = e^{aA^*} P_a^{-1} M_a^{-1} P_a e^{-aA^*}. \tag{52}$$

So, using (48) and (52) proves (50). Clearly, from (50) we see that

$$\Phi_a^\#(-\lambda) = \Theta_a(\lambda). \tag{53}$$

Formula (53) and Step 5 together imply that:

$$\begin{aligned}&\#\text{ zeros of } \det\Theta_a \text{ in the lower half-plane}\\ &\quad = \#\text{ zeros of } \det\Phi_a^\# \text{ in the upper half-plane}\\ &\quad = \#\text{ eigenvalues of } (-iA^\times)^* \text{ in the upper half-plane}\\ &\quad = \#\text{ eigenvalues of } iA^\times \text{ in the upper half-plane}\\ &\quad = \#\text{ zeros of } \det\Phi_a \text{ in the upper half-plane}\\ &\quad = \#\text{ negative eigenvalues of } T\,.\end{aligned}$$

□

4. A connection with the Nehari–Takagi interpolation problem

We conclude with some comments on the connection with the Nehari–Takagi interpolation problem. We consider the rational matrix case, that is, we use the formulation of the Nehari–Takagi problem given in [1, page 452]). Thus K is a rational $m \times m$ matrix[2] function given by a minimal realization of the form

$$K(z) = C(zI - A)^{-1}B, \tag{54}$$

where $\sigma(A) \subset \Pi^-$. Here Π^- denotes the open left half-plane. We want to obtain all rational matrix functions F of the form $F = K + R$ such that R is an $m \times m$ rational matrix function with at most κ poles in Π^- and

$$\|F\|_\infty = \sup\{\|F(z)\| : z \in i\mathbb{R}\} \leq 1.$$

If $\kappa = 0$, we obtain the Nehari problem, (see [1, page 443]).

The solution of the above Nehari–Takagi problem is given by the following theorem (see [1, page 452]).

Theorem 4.1. *Let* (54) *be a minimal realization for a rational $m \times m$ matrix function K with $\sigma(A) \subset \Pi^-$. Let P and Q be the controllability and observability gramians corresponding to* (54), *that is*

$$P = \int_0^\infty e^{sA}BB^*e^{sA^*}\,ds, \quad Q = \int_0^\infty e^{sA^*}C^*Ce^{sA}\,ds.$$

Assume that 1 *is not an eigenvalue of PQ. Then there is a rational matrix function R with at most κ poles (counted with multiplicities) in Π^- such that*

$$\|K + R\|_\infty \leq 1 \tag{55}$$

if and only if the matrix PQ has at most κ eigenvalues (counted with multiplicities) bigger than 1. *Moreover, if κ_0 is the number of eigenvalues of PQ bigger than* 1, *then the rational matrix functions $F = K + R$ satisfying* (55) *and such that R has precisely κ_0 poles in Π^-, are given by the linear fractional formula*

$$F = (\Theta_{11}G + \Theta_{12})(\Theta_{21}G + \Theta_{22})^{-1}, \tag{56}$$

where G is an arbitrary rational $m \times m$ matrix function satisfying

$$\sup_{z\in\Pi^-} \|G(z)\| \leq 1. \tag{57}$$

Here

$$\begin{aligned}\Theta(z) = \quad & \begin{pmatrix} I_M & 0 \\ 0 & I_N \end{pmatrix} + \begin{pmatrix} C & 0 \\ 0 & B^* \end{pmatrix} \begin{pmatrix} (zI-A)^{-1} & 0 \\ 0 & (zI+A^*)^{-1} \end{pmatrix} \\ & \times \begin{pmatrix} -ZP & Z \\ Z^* & -QZ \end{pmatrix} \begin{pmatrix} -C^* & 0 \\ 0 & B \end{pmatrix}\end{aligned}$$

where $Z = (I - PQ)^{-1}$.

[2] In [1] the matrix functions are allowed to be non-square but in this section we restrict ourselves to the square case.

Let us associate with the minimal realization (54) the kernel function $k(t) = Ce^{tA}B, t \geq 0$. Then for this k the entries $\Theta_{ij}, 1 \leq i, j \leq 2$, in the 2×2 block coefficient matrix

$$\Theta(z) = \begin{pmatrix} \Theta_{11}(z) & \Theta_{12}(z) \\ \Theta_{21}(z) & \Theta_{22}(z) \end{pmatrix}$$

in Theorem 4.1 are closely related to functions $g_a, h_a, \gamma_a, \chi_a$ (with $a = 0$) appearing in Theorem 0.1. In fact we have the following proposition.

Proposition 4.2. *Given the stable minimal realization* (54), *put* $k(t) = Ce^{tA}B, t \geq 0$. *Then the entries* $\Theta_{ij}, 1 \leq i, j \leq 2$, *in the* 2×2 *block coefficient matrix* Θ *are given by*

$$\begin{aligned} \Theta_{11}(z) &= I + \int_0^\infty e^{-zt} g(t)dt, \quad \Re z \geq 0, \\ \Theta_{21}(z) &= -\int_0^\infty e^{zt} h(-t)dt, \quad \Re z \leq 0, \\ \Theta_{12}(z) &= -\int_0^\infty e^{-zt} \gamma(t)dt, \quad \Re z \geq 0, \\ \Theta_{22}(z) &= I + \int_0^\infty e^{zt} \chi(-t)dt, \quad \Re z \leq 0. \end{aligned}$$

Here the functions g, h, γ, χ *are, respectively, equal to the functions* $g_a, h_a, \gamma_a, \chi_a$, *with* $a = 0$, *appearing in Theorem* 0.1 *with* $k(t) = Ce^{tA}B$.

Proof. First notice that $P = P_0$ and $Q = Q_0$. Since $M_0 = I - P_0 Q_0$, we see that $Z = M_0^{-1}$. From $Z = (I - PQ)^{-1}$, it also follows that

$$ZP = PZ^*, \quad QZ = Z^*Q. \tag{58}$$

Now take $g = g_0$. Then we can use (24), (21) and (5) to show that

$$I + \int_0^\infty e^{-zt} g(t)dt = I - iC(izI - iA)^{-1}(-ZPC^*) = \Theta_{11}(z), \quad \Re z \geq 0.$$

In the same way, using the identities in (58), we see that with $\chi = \chi_0$ the formulas (30), (27) and (6) yield

$$I + \int_0^\infty e^{zt} \chi(-t)dt = I + iB^*(izI + iA^*)^{-1}(-Z^*QB) = \Theta_{22}(z), \quad \Re z \leq 0.$$

To get the two remaining formulas we first use formulas (21) and (27) to show that for $a = 0$ we have

$$Q_0 X - C^* = -Z^* C^*, \quad P_0 U - B = -ZB.$$

But then we can use (23) and (29) to prove that

$$h(-t) = h_0(-t) = -B^* e^{tA^*} Z^* C^*, \quad \gamma(t) = \gamma_0(t) = -Ce^{tA} ZB \quad (t \geq 0).$$

Using these identities together with the stability of A and A^* we obtain

$$\begin{aligned} -\int_0^\infty e^{zt}h(-t)dt &= -B^*(z+A^*)^{-1}Z^*C^* = \Theta_{21}(z), \quad \Re z \leq 0, \\ -\int_0^\infty e^{-zt}\gamma(-t)dt &= C(z-A)^{-1}ZB = \Theta_{12}(z), \quad \Re z \geq 0, \end{aligned}$$

which completes the proof. □

The preceding proposition, together with the approximation argument described in Section 12.3 of [5], can be used to obtain the solution of the Nehari-Takagi problem in a Wiener algebra setting.

References

[1] J.A. Ball, I. Gohberg and L. Rodman, *Interpolation of rational matrix functions*, OT **45**, Birkhäuser Verlag, Basel, 1990.

[2] M.J. Corless, A.E. Frazho: *Linear systems and control*, Marcel Dekker, Inc., New York, NY, 2003.

[3] K. Clancey and I. Gohberg, *Factorization of matrix functions and singular integral operators*, OT **3**, Birkhäuser Verlag, Basel, 1981.

[4] R.L. Ellis and I. Gohberg, Distribution of zeros of orthogonal functions related to the Nehari problem, in: *Singular integral operators and related topics. Joint German-Israeli Workshop*, OT **90**, Birkhäuser Verlag, Basel, 1996, pp. 244–263.

[5] R.L. Ellis and I. Gohberg, *Orthogonal Systems and Convolution Operators*, OT **140**, Birkhäuser Verlag, Basel, 2003.

[6] R.L. Ellis, I. Gohberg and D.C. Lay, Distribution of zeros of matrix-valued continuous analogues of orthogonal polynomials, in: *Continuous and discrete Fourier transforms, extension problems and Wiener-Hopf equations*, OT **58**, Birkhäuser Verlag, Basel, 1992, pp. 26–70.

[7] I. Gohberg, S. Goldberg and M.A. Kaashoek, *Classes of Linear operators I*, OT **49**, Birkhäuser Verlag, Basel, 1990.

[8] I. Gohberg, M.A. Kaashoek and F. van Schagen, On inversion of convolution integral operators on a finite interval, in: *Operator Theoretical Methods and Applications to Mathematical Physics. The Erhard Meister Memorial Volume*, OT **147**, Birkhäuser Verlag, Basel, 2004, pp. 277–285.

[9] M.G. Kreĭn, On the location of roots of polynomials which are orthogonal on the unit circle with respect to an indefinite weight, *Teor. Funkcii, Funkcional. Anal. i Prilozen* **2** (1966), 131-137 (Russian).

[10] P. Lancaster and M. Tismenetsky, *The theory of matrices with applications*, Second Edition, Academic Press, 1985.

G.J. Groenewald
Department of Mathematics
North-West University
Private Bag X6001
Potchefstroom 2520, South Africa
e-mail: wskgjg@puknet.puk.ac.za

M.A. Kaashoek
Afdeling Wiskunde, Faculteit der Exacte Wetenschappen
Vrije Universiteit
De Boelelaan 1081a
1081 HV Amsterdam, The Netherlands
e-mail: ma.kaashoek@few.vu.nl

Operator Theory:
Advances and Applications, Vol. 160, 233–252

Schur-type Algorithms for the Solution of Hermitian Toeplitz Systems via Factorization

Georg Heinig and Karla Rost

Dedicated to our teacher and friend Israel Gohberg

Abstract. In this paper fast algorithms for the solution of systems $T\mathbf{u} = \mathbf{b}$ with a strongly nonsingular hermitian Toeplitz coefficient matrix T via different kinds of factorizations of the matrix T are discussed. The first aim is to show that ZW-factorization of T is more efficient than the corresponding LU-factorization. The second aim is to design and compare different Schur-type algorithms for LU- and ZW-factorization of T. This concerns the classical Schur-Bareiss algorithm, 3-term one-step and double-step algorithms, and the Schur-type analogue of a Levinson-type algorithm of B. Krishna and H. Krishna. The latter one reduces the number of the multiplications by almost 50% compared with the classical Schur-Bareiss algorithm.

Mathematics Subject Classification (2000). Primary 15A23; Secondary 65F05.

Keywords. Toeplitz matrix, Schur algorithm, LU-factorization, ZW-factorization.

1. Introduction

This paper is dedicated to fast algorithms for the solution of linear systems of equations $T\mathbf{u} = \mathbf{b}$ with a nonsingular hermitian Toeplitz matrix $T = [\, a_{i-j}\,]_{i,j=1}^{n}$, $a_{-j} = \overline{a}_j$. We assume that T is strongly nonsingular, which means that all leading principal submatrices $T_k = [\, a_{i-j}\,]_{i,j=1}^{k}$ $(k = 1, \ldots, n)$ are nonsingular. This condition is in particular fulfilled if T is positive definite.

There are mainly two types of direct algorithms to solve a system $T\mathbf{u} = \mathbf{b}$ with computational complexity $O(n^2)$: Levinson-type and Schur-type. The original Levinson algorithm is closely related to Szegö's recursion formulas for orthogonal polynomials on the unit circle and to factorizations of the inverse matrix T^{-1}. A Levinson-type algorithm can be combined with an inversion formula, like the Gohberg-Semençul formula. Schur-type algorithms are related to factorizations of the matrix itself. Using the factorization of the matrix a linear system of equations can be solved by back substitution.

Practical experience (see [26]) and theoretical results (see [4], [9], [5]) indicate that Schur-type algorithms have, in general, a better stability behavior than Levinson-type algorithms. For this reason we restrict ourselves in this paper to Schur-type algorithms, i.e., algorithms for fast factorization of T, despite they have, in general, a higher computational complexity. However, we develop Schur-type algorithms on the basis of their corresponding Levinson-type counterpart, so that in this paper also Levinson-type algorithms can be found despite it is not always mentioned explicitly.

The classical Schur algorithm in its original form is an algorithm in complex function theory (see [19] and references therein) to decide whether an analytic function maps the unit disk into itself. But it can also be applied to compute the LU-factorization of a Toeplitz matrix. An algorithm for solving Toeplitz systems via factorization was first proposed by Bareiss in [1] (not mentioning Schur).

The property of a matrix to be hermitian is reflected in the LU-factorization in such a way that the U-factor is the adjoint of the L-factor. But hermitian Toeplitz matrices have an additional symmetry property. They are *centro-hermitian.* This means that $J_n T J_n = \overline{T}$. Here J_n denotes the $n \times n$ matrix of counteridentity with ones on the antidiagonal and zeros elsewhere. The bar denotes the matrix with conjugate complex entries. This property is not reflected in the LU-factorization in an obvious way.

For this reason we consider another type of factorization: the ZW-factorization. This is a representation of T in the form $T = ZXZ^*$, in which Z is a Z- and X an X-matrix (for the definition of these concepts see Section 2). The factors Z and X in this factorization will also be centro-hermitian. The ZW-factorization is closely related to the "quadrant interlocking" or WZ-factorization, which was originally introduced and studied by D. J. Evans and his coworkers for the parallel solution of tridiagonal systems (see [24], [8] and references therein). While the LU-factorization of a matrix $A = [\,a_{ij}\,]_{i,j=1}^{n}$ relies on the leading principal submatrices $A_k = [\,a_{ij}\,]_{i,j=1}^{k}$, $k = 1, \dots, n$, the ZW-factorization relies on the central submatrices $[\,a_{ij}\,]_{i,j=l}^{n+1-l}$ $(l = 1 \dots, [(n+1)/2])$.

The ZW-factorization for real symmetric Toeplitz matrices was first mentioned by C. J. Demeure in [7]. In our papers [15] (see also [18]), [16] and [17] ZW-factorizations for skewsymmetric Toeplitz, centrosymmetric and centro-skewsymmetric Toeplitz–plus–Hankel, and general Toeplitz–plus–Hankel matrices were described, respectively. Note that for skewsymmetric Toeplitz matrices (and with it for purely imaginary hermitian Toeplitz matrices) the factors of the ZW-factorization have some surprising additional symmetry properties which are not shared by the symmetric case.

The first aim of the present paper is to show that for hermitian Toeplitz matrices the ZW-factorization leads to more efficient algorithms for the solution of linear systems than LU-factorization. In Section 2 we consider three types of factorizations for matrices which are hermitian and centro-hermitian. We show that the ZW-factorization reflects, in contrast to the LU-factorization, both symmetry properties. This leads to a computational gain in solving linear systems. The number of additions can still be slightly reduced if instead of the standard ZW-factorization a modification, which we call "column conjugate-symmetric ZW-factorization", is considered. In Section 3 the factors of the factorizations are described in terms of the residuals of solutions of special systems.

The second aim of the paper is to present and to compare different algorithms for LU- and ZW-factorization of hermitian Toeplitz matrices. We present first the classical Schur algorithm in Section 4, then a one-step Schur algorithm based on 3-term recursions in Section 5, and a double-step version of it in Section 6. These algorithms are somehow related to the split Schur algorithm for real symmetric Toeplitz matrices of Delsarte and Genin presented in [6].

The split Schur algorithm for real symmetric Toeplitz matrices requires, compared with the classical Schur algorithm, only half of the number of multiplications while keeping the number of additions. This gain is not achieved for the algorithms in Sections 5 and 6. Actually, the split Schur algorithm cannot be directly generalized from the real-symmetric to the hermitian case. However, algorithms with a saving of about 50% of the multiplications do exist in the literature (see [22], [21], [3]).

In Section 7 we recall a Levinson-type algorithm presented in [21] and derive a Schur version of it, which will be called *Krishna-Schur algorithm* and which is new. We show how ZW-factorizations of T are obtained with the help of the data computed in this algorithm. In Section 8 we compare the complexity of all algorithms for the solution of hermitian Toeplitz systems. It turns out that the Krishna-Schur algorithm gives the lowest computational amount.

For sake of simplicity of notation we assume that n is even, $n = 2m$, throughout the paper. The case of odd n can be considered in an analogous way.

We use the following notations:

- $\mathbf{0}_k$ will be a zero vector of length k.
- $\mathbf{1}_k$ will be a vector of length k with all components equal to 1.
- $\mathbf{e}_k$ will be the kth vector in the standard basis of $\mathbb{C}^n$.

2. LU- versus ZW-factorization

Throughout this section, let $A = [\, a_{ij} \,]_{i,j=1}^{n}$ be a nonsingular hermitian matrix that is also centro-hermitian where n is even and $m = n/2$. We compare the efficiency of three kinds of factorizations of A for solving a system $A\mathbf{u} = \mathbf{b}$, namely the classical LU-factorization and two types of ZW-factorizations with different symmetry properties. By "efficiency" we mean here in the first place the computational complexity of an algorithm for solving a linear system with coefficient matrix A. This complexity will be $O(n^2)$. Therefore, we care only for the n^2-term and neglect lower order terms. Furthermore we will compare the storage requirement (number of real parameters) for the factorizations which is also $O(n^2)$.

CA and CM will stand for complex additions and multiplications, respectively, and RA and RM for their real counterparts. We count 1 CA as 2 RA and 1 CM as 4 RM plus 2 RA.

2.1. LU-factorization

If A is a strongly nonsingular matrix, then it admits an LU-factorization $A = LDL^*$, in which D is real diagonal and L is lower triangular. Among the LU-factorizations there is a unique one in which the matrix L has ones on the main diagonal. This will be referred to as the *unit LU-factorization.* If $A = L_0 D_0 L_0^*$ is the unit LU-factorization of A, then the factorization $A = LDL^*$ with $L = L_0 D_0$ and $D = D_0^{-1}$ will be called *standard LU-factorization.* The reason for introducing this concept is that it is often more efficient to produce the standard than the unit LU-factorization.

If an LU-factorization of A is known, then a system $A\mathbf{u} = \mathbf{b}$ can be solved via the solution of two triangular systems and a diagonal system. The diagonal system can be solved with $O(n)$ complexity, so it can be neglected in the complexity estimation. The solution of a complex triangular system requires $0.5\,n^2$ CM plus $0.5\,n^2$ CA, which is equivalent to $2\,n^2$ RA and $2\,n^2$ RM.

Proposition 2.1. If an LU-factorization of A is known, then the solution of $A\mathbf{u} = \mathbf{b}$ requires $4\,n^2$ RM and $4\,n^2$ RA.

The property of the matrix A to be centro-hermitian must be somehow hidden in the structure of the factor L. In the general case the characterization of the factor L for a centro-hermitian A is, to the best of our knowledge, unknown. But in the case of a hermitian Toeplitz matrix some relations between the entries of L follow from the theory of orthogonal polynomials on the unit circle. (see [23], relation (1.3) on p.113). However, it is not clear how to get any computational advantage from it. Thus the number of real parameters in an LU-factorization, which is the storage requirement, will be about n^2.

2.2. Centro-hermitian ZW-factorization

Since ZW-factorizations are not commonly used, we recall the basic concepts (compare [24], [8] and references therein). A matrix $A = [\,a_{ij}\,]_{i,j=1}^n$ is called a *W-matrix* (or a bow tie matrix) if $a_{ij} = 0$ for all (i,j) for which $i > j$ and $i + j > n$ or $i < j$ and $i + j \leq n$. The matrix A will be called a *unit W-matrix* if in addition $a_{ii} = 1$ and $a_{i,n+1-i} = 0$ for $i = 1, \dots, n$. The transpose of a W-matrix is called a *Z-matrix* (or hourglass matrix). A matrix which is both a Z- and a W-matrix will be called an *X-matrix.* A matrix which is either a Z-matrix or a W-matrix will be called *butterfly matrix.*

These names are suggested by the shapes of the set of all possible positions for nonzero entries, which are as follows:

$$W = \begin{bmatrix} \bullet & & & & & \bullet \\ \bullet & \circ & & & \circ & \bullet \\ \bullet & \circ & \circ & \circ & \circ & \bullet \\ \bullet & \circ & \bullet & \bullet & \circ & \bullet \\ \bullet & \bullet & & & \bullet & \bullet \\ \bullet & & & & & \bullet \end{bmatrix}, \quad Z = \begin{bmatrix} \bullet & \bullet & \bullet & \bullet & \bullet & \bullet \\ & \circ & \circ & \circ & \bullet & \\ & & \circ & \bullet & & \\ & & \bullet & \circ & & \\ & \bullet & \circ & \circ & \circ & \\ \bullet & \bullet & \bullet & \bullet & \bullet & \bullet \end{bmatrix},$$

$$X = \begin{bmatrix} \bullet & & & & & \bullet \\ & \bullet & & & \bullet & \\ & & \bullet & \bullet & & \\ & & \bullet & \bullet & & \\ & \bullet & & & \bullet & \\ \bullet & & & & & \bullet \end{bmatrix}.$$

A representation of a nonsingular matrix A in the form $A = ZXW$ in which Z is a Z-matrix, W a W-matrix and X an X-matrix is called a *ZW-factorization of A*. It is called *unit ZW-factorization* if Z is a unit Z-matrix and W is a unit W-matrix. Clearly, if A admits a ZW-factorization, then it admits a unique unit ZW-factorization.

A necessary and sufficient condition for a matrix $A = [\, a_{jk} \,]_{j,k=1}^{n}$ to admit a ZW-factorization is that the central submatrices $A^c_{n+2-2l} = [\, a_{jk} \,]_{j,k=l}^{n+1-l}$ are nonsingular for $l = 1, \dots, m = \frac{n}{2}$. A matrix with this property will be called *centro-nonsingular*.

Since for a Toeplitz matrix the central submatrices coincide with principal submatrices, a strongly nonsingular Toeplitz matrix is also centro-nonsingular.

It follows from the uniqueness of the unit ZW-factorization that if a centro-nonsingular matrix A is hermitian and centro-hermitian, then its unit ZW-factorization is of the form $A = Z_0 X_0 Z_0^*$, in which Z_0 is centro-hermitian and X_0 is hermitian and centro-hermitian. That means that in the unit ZW-factorization both the hermitian and the centro-hermitian structure are reflected. Thus the number of real parameters that characterize a centro-hermitian butterfly matrix is only about half the number of parameters describing a triangular matrix.

Any ZW-factorization in which Z is centro-hermitian will be called *centro-hermitian ZW-factorization.* If $A = Z_0 X_0 Z_0^*$ is the unit ZW-factorization of A, then $A = ZXZ^*$ with $Z = Z_0 X_0$ and $X = X_0^{-1}$ will be called *standard ZW-factorization.* Clearly, it is also centro-hermitian.

If a centro-hermitian ZW-factorization of A is known, then a system $A\mathbf{u} = \mathbf{b}$ can be solved via the solution of two centro-hermitian butterfly systems and an X-system. The X-system can be solved with $O(n)$ complexity, so it can be neglected in the complexity estimation. We show how a centro-hermitian Z-system $Z\mathbf{u} = \mathbf{b}$ can be solved and estimate the complexity.

For $\mathbf{u} \in \mathbb{C}^n$, we denote by $\mathbf{u}^\#$ the vector $\mathbf{u}^\# = J_n\overline{\mathbf{u}}$. A vector $\mathbf{u} \in \mathbb{C}^n$ is called *conjugate-symmetric* if $\mathbf{u} = \mathbf{u}^\#$.

First we observe that a centro-hermitian matrix transforms conjugate-symmetric vectors into conjugate-symmetric ones, so that the solution $\mathbf{u}$ of $Z\mathbf{u} = \mathbf{b}$ with a conjugate-symmetric $\mathbf{b}$ is conjugate-symmetric again.

The solution of a linear system $Z\mathbf{u} = \mathbf{b}$ with general $\mathbf{b}$ can be reduced to the solution of two systems with conjugate-symmetric right-hand sides. For this we represent $\mathbf{b} = \mathbf{b}_+ + \mathrm{i}\mathbf{b}_-$, where $\mathbf{b}_+ = \frac{1}{2}(\mathbf{b} + \mathbf{b}^\#)$ and $\mathbf{b}_- = \frac{1}{2\mathrm{i}}(\mathbf{b} - \mathbf{b}^\#)$. Then $\mathbf{b}_\pm$ are conjugate-symmetric, and the solution $\mathbf{u}$ is obtained from the solutions of $Z\mathbf{u}_\pm = \mathbf{b}_\pm$ via $\mathbf{u} = \mathbf{u}_+ + \mathrm{i}\mathbf{u}_-$.

We consider now a system $Z\mathbf{u}_+ = \mathbf{b}_+$ with a conjugate-symmetric right-hand side $\mathbf{b}_+ = \begin{bmatrix} \mathbf{c}^\# \\ \mathbf{c} \end{bmatrix}$, $\mathbf{c} \in \mathbb{C}^m$. Suppose that the solution is $\mathbf{u}_+ = \begin{bmatrix} \mathbf{v}^\# \\ \mathbf{v} \end{bmatrix}$ for some $\mathbf{v} \in \mathbb{C}^m$.

A centro-hermitian Z-matrix is of the form

$$Z = \begin{bmatrix} J_m\overline{L}_0 J_m & J_m\overline{L}_1 \\ L_1 J_m & L_0 \end{bmatrix}, \tag{2.1}$$

where L_0 and L_1 are lower triangular. Hence $Z\mathbf{u}_+ = \mathbf{b}_+$ is equivalent to

$$L_1\overline{\mathbf{v}} + L_0\mathbf{v} = \mathbf{c}\,. \tag{2.2}$$

Let the subscript r designate the real and i the imaginary part of a vector or of a matrix. Then (2.2) is equivalent to

$$\begin{aligned} (L_{0,r} + L_{1,r})\mathbf{v}_r + (-L_{0,i} + L_{1,i})\mathbf{v}_i &= \mathbf{c}_r\,, \\ (L_{0,i} + L_{1,i})\mathbf{v}_r + (L_{0,r} - L_{1,r})\mathbf{v}_i &= \mathbf{c}_i\,. \end{aligned}$$

This can be written as a real Z-system

$$Z' \begin{bmatrix} J_m\mathbf{v}_i \\ \mathbf{v}_r \end{bmatrix} = \begin{bmatrix} J_m\mathbf{c}_i \\ \mathbf{c}_r \end{bmatrix}, \tag{2.3}$$

where

$$Z' = \begin{bmatrix} J_m(L_{0,r} - L_{1,r})J_m & J_m(L_{0,i} + L_{1,i}) \\ (-L_{0,i} + L_{1,i})J_m & L_{0,r} + L_{1,r} \end{bmatrix}.$$

Here all matrices L are lower triangular.

In this way the solution of the complex system $Z\mathbf{u} = \mathbf{b}$ is reduced to the solution of two systems (2.3) with a real Z-coefficient matrix. Since a Z-system is equivalent to a block triangular system with 2×2 blocks the solution of such a system requires $0.5\,n^2$ RM and $0.5\,n^2$ RA. Thus for 2 systems n^2 RM plus n^2 RA are sufficient.

Similar arguments can be used for W-systems with the coefficient matrix Z^*.

For building the matrix Z' we need 4 additions of $m \times m$ real, lower triangular matrices, which results in $2\,m^2 = 0.5\,n^2$ real additions. Let us summarize.

Proposition 2.2. If a centro-hermitian ZW-factorization of A is known, then the solution of $A\mathbf{u} = \mathbf{b}$ requires $2\,n^2$ RM and $2.5\,n^2$ RA.

2.3. Column conjugate-symmetric ZW-factorization

We show now that the number of additions can be still reduced if another kind of ZW-factorization of A is given.

We introduce the $n \times n$ X-matrix

$$\Sigma = \begin{bmatrix} -\mathrm{i} & & & & & 1 \\ & \ddots & & & \iddots & \\ & & -\mathrm{i} & 1 & & \\ & & \mathrm{i} & 1 & & \\ & \iddots & & & \ddots & \\ \mathrm{i} & & & & & 1 \end{bmatrix}.$$

Obviously, $\Sigma^{-1} = \frac{1}{2}\Sigma^*$.

If Z is an $n \times n$ centro-hermitian Z-matrix, then the matrix $Z_h = Z\Sigma$ has the property $J_n Z_h = \overline{Z\Sigma} = \overline{Z_h}$. That means that Z_h has conjugate-symmetric columns. Let us call a matrix with this property *column conjugate-symmetric.* If moreover the X-matrix built from the diagonal and antidiagonal of Z_h is equal to Σ, then Z_h will be referred to as *unit.*

A centro-hermitian ZW-factorization $A = ZXZ^*$ can be transformed into a ZW-factorization $A = Z_h X_h Z_h^*$ in which Z_h is unit column conjugate-symmetric. We will call this factorization *unit column conjugate-symmetric ZW-factorization.* The *standard column conjugate-symmetric ZW-factorization* $A = ZXZ^*$ is given by $Z = Z_h X_h$ and $X = X_h^{-1}$. Concerning the factor X_h we obtain $X_h = \frac{1}{4}\Sigma^* X \Sigma$. For X is hermitian, X_h is hermitian. Moreover, X_h is real. In fact, we have

$$\overline{X}_h = \frac{1}{4}\overline{\Sigma}^*\overline{X\Sigma} = \frac{1}{4}\overline{\Sigma}^* J_n X J_n \overline{\Sigma} = \frac{1}{4}\Sigma^* X \Sigma = X_h\,.$$

Obviously, a column conjugate-symmetric Z-matrix Z_h has the form

$$Z_h = \begin{bmatrix} J_m \overline{L}_1 J_m & J_m \overline{L}_0 \\ L_1 J_m & L_0 \end{bmatrix}, \tag{2.4}$$

where $L_0 = L_{0,r} + \mathrm{i}L_{0,i}$, $L_1 = L_{1,r} + L_{1,i}$ are lower triangular matrices, $L_{0,r}$ and $L_{1,i}$ are unit, and $L_{0,i}$ and $L_{1,r}$ have zeros on their main diagonal.

From this representation it can be seen that Z_h transforms real vectors to conjugate-symmetric vectors, so that the solution of $Z_h\mathbf{u} = \mathbf{b}_+$ with conjugate-symmetric $\mathbf{b}_+$ is real.

Suppose that $\mathbf{b}_+ = \begin{bmatrix} \mathbf{c}^\# \\ \mathbf{c} \end{bmatrix}$, $\mathbf{c} = \mathbf{c}_r + \mathrm{i}\mathbf{c}_i$, and let $\mathbf{u} = \begin{bmatrix} \mathbf{v} \\ \mathbf{w} \end{bmatrix}$ with $\mathbf{v}, \mathbf{w} \in \mathbb{R}^m$ be the solution of $Z_h\mathbf{u} = \mathbf{b}$. Then we have

$$Z_h \begin{bmatrix} \mathbf{v} \\ \mathbf{w} \end{bmatrix} = \begin{bmatrix} J_m\mathbf{c}_r \\ \mathbf{c}_r \end{bmatrix} + \mathrm{i}\begin{bmatrix} -J_m\mathbf{c}_i \\ \mathbf{c}_i \end{bmatrix},$$

which is equivalent to

$$\begin{aligned} L_{1,i}J_m\mathbf{v} + L_{0,i}\mathbf{w} &= \mathbf{c}_i\,, \\ L_{1,r}J_m\mathbf{v} + L_{0,r}\mathbf{w} &= \mathbf{c}_r\,. \end{aligned}$$

This system can be written as a real unit Z-system

$$Z_h' \begin{bmatrix} \mathbf{v} \\ \mathbf{w} \end{bmatrix} = \begin{bmatrix} J_m \mathbf{c}_i \\ \mathbf{c}_r \end{bmatrix},$$

where

$$Z_h' = \begin{bmatrix} J_m L_{1,i} J_m & J_m L_{0,i} \\ L_{1,r} J_m & L_{0,r} \end{bmatrix}.$$

In this way the solution of a system $Z_h \mathbf{u} = \mathbf{b}$ is reduced, like in 2.2, to two real unit Z-systems. In contrast to 2.2 no additional amount is necessary to built the matrix Z_h'. The same can be done for a system with the coefficient matrix Z_h^*. Let us summarize.

Proposition 2.3. If a column conjugate-symmetric ZW-factorization of the matrix A is known, then the solution of $A\mathbf{u} = \mathbf{b}$ requires $2\,n^2$ RM and $2\,n^2$ RA.

3. Description of the factors

We show how the factors in the three types of factorizations can be characterized via the solutions of some equations. As in the previous section, let $A = [\,a_{ij}\,]_{i,j=1}^n$ be nonsingular hermitian and centro-hermitian, n even and $m = n/2$. Besides A we consider the leading principal submatrices $A_k = [\,a_{ij}\,]_{i,j=1}^k$ for $k = 1, \ldots, n$ and the central submatrices $A_{2k}^c = [\,a_{ij}\,]_{i,j=l}^{n+1-l}$, $k + l = m + 1$ for $k = 1, \ldots, m$. All general observations we specify for the case of a strongly nonsingular hermitian Toeplitz matrix $T = [\,a_{i-j}\,]_{i,j=1}^n$.

3.1. LU-factorization

For strongly nonsingular A, we consider equations

$$A_k \mathbf{u}_k = \rho_k\, \mathbf{e}_k \quad (k = 1, \ldots, n)\,, \tag{3.1}$$

where ρ_k are nonzero real numbers. Then the factors of the LU-factorization $A = LDL^*$ can be characterized as follows. The kth column of L is given by $L\mathbf{e}_k = A \begin{bmatrix} \mathbf{u}_k \\ \mathbf{0}_{n-k} \end{bmatrix}$ and

$$D = \operatorname{diag}\,(\xi_k^{-1} \rho_k^{-1})_{k=1}^n\,,$$

where ξ_k is the last component of $\mathbf{u}_k$. If we choose $\rho_k = 1$ for all k, then we obtain the unit LU-factorization. If we demand that $\xi_k = 1$ for all k we obtain the standard LU-factorization. In this specific case we write $\mathbf{x}_k$ instead of $\mathbf{u}_k$.

Consider the case of a Toeplitz matrix T. We introduce the residuals

$$r_{jk}^+ = \begin{bmatrix} a_{k+j-1} & \ldots & a_j \end{bmatrix} \mathbf{x}_k \tag{3.2}$$

for $j = 0, \ldots, n-k$. By definition, $r_{0k}^+ = \rho_k$, and ρ_k is real. The kth column of L consists of $k-1$ zeros and the numbers r_{jk}^+ for $j = 0, \ldots, n-k$.

As a conclusion we can state: *The standard LU-factorization of T is given if the residuals r_{jk}^+ for $j = 0, \ldots, n-k$ and $k = 1, \ldots, n$ are known .*

3.2. Centro-hermitian ZW-factorization

For centro-nonsingular A, we consider equations of the form

$$A_{2k}^c \mathbf{w}_k = \rho_k^- \mathbf{e}_1 + \rho_k^+ \mathbf{e}_{2k} \quad (k = 1, \ldots, m)$$

where $\rho_k^{\pm}$ are numbers satisfying $|\rho_k^+| \neq |\rho_k^-|$. This condition guarantees that $\mathbf{w}_k$ and $\mathbf{w}_k^{\#}$ are linearly independent. Then

$$Z\mathbf{e}_{m+k} = A \begin{bmatrix} \mathbf{0}_{m-k} \\ \mathbf{w}_k \\ \mathbf{0}_{m-k} \end{bmatrix}$$

is the $(m+k)$th column of Z for $k = 1, \ldots, m$. The remaining columns are obtained using the property of Z to be centro-hermitian by

$$Z\mathbf{e}_{m+1-k} = (Z\mathbf{e}_{m+k})^{\#} .$$

In order to describe the X-factor we introduce a notation for X-matrices that is analogous to the "diag" notation for diagonal matrices. If $M_k = \begin{bmatrix} \alpha_k & \beta_k \\ \gamma_k & \delta_k \end{bmatrix}$ $(k = 1, \ldots, m)$, then we set

$$\operatorname{xma}(M_k)_{k=1}^m = \begin{bmatrix} \alpha_m & & & & & \beta_m \\ & \ddots & & & \iddots & \\ & & \alpha_1 & \beta_1 & & \\ & & \gamma_1 & \delta_1 & & \\ & \iddots & & & \ddots & \\ \gamma_m & & & & & \delta_m \end{bmatrix} .$$

Clearly, $\operatorname{xma}(M_k)_{k=1}^m$ is nonsingular if and only if all M_k are nonsingular and

$$(\operatorname{xma}(M_k)_{k=1}^m)^{-1} = \operatorname{xma}(M_k^{-1})_{k=1}^m .$$

Now the X-factor is given by

$$X = \operatorname{xma}\left(\begin{bmatrix} \overline{\xi}_k^+ & \xi_k^- \\ \overline{\xi}_k^- & \xi_k^+ \end{bmatrix}^{-1} \begin{bmatrix} \overline{\rho}_k^+ & \rho_k^- \\ \overline{\rho}_k^- & \rho_k^+ \end{bmatrix}^{-1} \right)_{k=1}^m , \tag{3.3}$$

where ξ_k^+ is the last and ξ_k^- is the first component of $\mathbf{w}_k$.

We obtain the factors of the unit ZW-factorization for $\rho_k^+ = 1$, $\rho_k^- = 0$ and the standard ZW-factorization for the choice $\xi_k^+ = 1$, $\xi_k^- = 0$.

A crucial observation is that for a Toeplitz matrix T we have $T_{2k}^c = T_{2k}$. Let $\mathbf{x}_k$ denote, as in the previous subsection, the solution of $T_k \mathbf{x}_k = \rho_k \mathbf{e}_k$ with last component equal to 1, and let r_{jk}^+ be defined by (3.2). Besides the numbers r_{jk}^+ we consider the residuals

$$r_{jk}^- = \begin{bmatrix} a_{k+j-1} & \ldots & a_j \end{bmatrix} \mathbf{x}_k^{\#} = \begin{bmatrix} \overline{a}_j & \ldots & \overline{a}_{k+j-1} \end{bmatrix} \overline{\mathbf{x}}_k \tag{3.4}$$

for $j = 0, \dots, n-k$. Then $r_{0k}^- = 0$ and

$$T_{2k}\begin{bmatrix} 0 \\ \mathbf{x}_{2k-1} \end{bmatrix} = \overline{r}_{1,2k-1}^- \mathbf{e}_1 + r_{0,2k-1}^+ \mathbf{e}_{2k}.$$

Recall that $r_{0,2k-1}^+ = \rho_{2k-1}$ and that ρ_{2k-1} is real.

That means that we have $\mathbf{w}_k = \begin{bmatrix} 0 \\ \mathbf{x}_{2k-1} \end{bmatrix}$ for the standard centro-hermitian ZW-factorization. Thus, the $(m+k)$th column of the Z-factor of the standard ZW-factorization of T is given by

$$Z\mathbf{e}_{m+k} = \begin{bmatrix} (\overline{r}_{j,2k-1}^-)_{j=m-k+1}^1 \\ \mathbf{0}_{2k-2} \\ (r_{j,2k-1}^+)_{j=0}^{m-k} \end{bmatrix}$$

and the X-factor by

$$X = \operatorname{xma}\left(\begin{bmatrix} r_{0,2k-1}^+ & \overline{r}_{1,2k-1}^- \\ r_{1,2k-1}^- & r_{0,2k-1}^+ \end{bmatrix}^{-1} \right)_{k=1}^m ,$$

As a conclusion we can state: *The standard centro-hermitian ZW-factorization of T is given if the residuals $r_{j,2k-1}^\pm$ for $j = 0, \dots, m-k$ and $k = 1, \dots, m$ are known.*

In Section 7 we will construct a centro-hermitian ZW-factorization which is not standard. In this case besides the residuals the first and last components of $\mathbf{w}_k$ are needed to apply formula (3.3).

3.3. Column conjugate-symmetric ZW-factorization

For the construction of a column conjugate-symmetric ZW-factorization $A = Z_h X_h Z_h^*$ we consider two families of equations

$$A_{2k}^c \mathbf{w}_k^\pm = \overline{\rho}_k^\pm \, \mathbf{e}_1 + \rho_k^\pm \, \mathbf{e}_{2k} \quad (k = 1, \dots, m)$$

with $\operatorname{Im} \rho_k^- \overline{\rho}_k^+ \neq 0$. This condition guarantees the linear independence of the solutions $\mathbf{w}_k^+$ and $\mathbf{w}_k^-$. Note that the vectors $\mathbf{w}_k^\pm$ are conjugate-symmetric, since the right-hand sides are conjugate-symmetric.

Now

$$Z_h \mathbf{e}_{m+k} = A \begin{bmatrix} \mathbf{0}_{m-k} \\ \mathbf{w}_k^+ \\ \mathbf{0}_{m-k} \end{bmatrix} \quad \text{and} \quad Z_h \mathbf{e}_{m-k+1} = A \begin{bmatrix} \mathbf{0}_{m-k} \\ \mathbf{w}_k^- \\ \mathbf{0}_{m-k} \end{bmatrix}$$

are the $(m+k)$th and $(m-k+1)$th columns of Z_h, respectively, for $k = 1, \dots, m$.

It remains to describe the middle factor X_h. Let $\xi_k^\pm$, denote the last component of $\mathbf{w}_k^\pm$.

We take advantage of the relation

$$\begin{bmatrix} \overline{z}_1 & \overline{z}_2 \\ z_1 & z_2 \end{bmatrix} = \begin{bmatrix} -\mathrm{i} & 1 \\ \mathrm{i} & 1 \end{bmatrix} \begin{bmatrix} \operatorname{Im} z_1 & \operatorname{Im} z_2 \\ \operatorname{Re} z_1 & \operatorname{Re} z_2 \end{bmatrix} \tag{3.5}$$

for complex numbers z_1, z_2, to observe that $Z_h X_Z^{-1}$ is unit for

$$X_Z = \operatorname{xma}\left(\begin{bmatrix} \operatorname{Im} \rho_k^- & \operatorname{Im} \rho_k^+ \\ \operatorname{Re} \rho_k^- & \operatorname{Re} \rho_k^+ \end{bmatrix} \right)_{k=1}^m .$$

Similarly, $W_h X_W^{-1}$ is unit for

$$W_h = \begin{bmatrix} & 0 & & \mathbf{0} & \mathbf{0} & & 0 & \\ \mathbf{w}_m^- & \mathbf{w}_{m-1}^- & \cdots & \mathbf{w}_1^- & \mathbf{w}_1^+ & \cdots & \mathbf{w}_{m-1}^+ & \mathbf{w}_m^+ \\ & 0 & & \mathbf{0} & \mathbf{0} & & 0 & \end{bmatrix}$$

and

$$X_W = \operatorname{xma}\left(\begin{bmatrix} \operatorname{Im} \xi_k^- & \operatorname{Im} \xi_k^+ \\ \operatorname{Re} \xi_k^- & \operatorname{Re} \xi_k^+ \end{bmatrix} \right)_{k=1}^m .$$

From the uniqueness of the unit column conjugate-symmetric ZW-factorization we conclude now

$$X_h = \frac{1}{2} \operatorname{xma}\left(\begin{bmatrix} \operatorname{Im} \xi_k^- & \operatorname{Im} \xi_k^+ \\ \operatorname{Re} \xi_k^- & \operatorname{Re} \xi_k^+ \end{bmatrix}^{-1} \begin{bmatrix} \operatorname{Im} \rho_k^- & \operatorname{Im} \rho_k^+ \\ \operatorname{Re} \rho_k^- & \operatorname{Re} \rho_k^+ \end{bmatrix}^{-1} \right)_{k=1}^m . \tag{3.6}$$

The factor $\frac{1}{2}$ appears in view of this factor in $\Sigma^{-1} = \frac{1}{2} \Sigma^*$.

For a Toeplitz matrix T we have $T_{2k}^c = T_{2k}$. We consider the residuals

$$r_{jk}^{\pm} = \begin{bmatrix} a_{k+j-1} & \cdots & a_j \end{bmatrix} \mathbf{w}_k^{\pm},$$

for $j = 0, \ldots, n-k$. The $(m-k+1)$th and $(m+k)$th columns of Z is given by

$$Z\mathbf{e}_{m-k+1} = \begin{bmatrix} (\overline{r}_{j,2k}^-)_{j=m-k}^0 \\ \mathbf{0}_{2k-2} \\ (r_{j,2k}^-)_{j=0}^{m-k} \end{bmatrix}, \quad Z\mathbf{e}_{m+k} = \begin{bmatrix} (\overline{r}_{j,2k}^+)_{j=m-k}^0 \\ \mathbf{0}_{2k-2} \\ (r_{j,2k}^+)_{j=0}^{m-k} \end{bmatrix}.$$

As a conclusion we can state: *A column conjugate-symmetric ZW-factorization of T is given if the residuals $r_{j,2k}^{\pm}$ for $j = 0, \ldots, m-k$ and the last components of $\mathbf{w}_k^{\pm}$ for $k = 1, \ldots, m$ are known.*

4. Classical Schur algorithm for LU- and ZW-factorizations

Throughout this section, let T be a strongly nonsingular hermitian Toeplitz matrix. We use the notations from the previous section.

The classical Schur algorithm is an algorithm that computes in a fast way the residuals $r_{jk}^{\pm}$ defined by (3.2) and (3.4), so it can be used to construct both the LU- and the ZW-factorizations of T considered in the previous section. For convenience of the reader we present a derivation of the algorithm.

We collect the residuals $r_{jk}^{\pm}$ defined by (3.2) and (3.4) to vectors

$$\mathbf{r}_k^+ = (r_{jk}^+)_{j=0}^{n-k}, \mathbf{r}_k^- = (r_{jk}^-)_{j=0}^{n-k}.$$

Recall that $r_{0k}^- = 0$ and $r_{0k}^+ = \rho_k$. If T_k' denotes the $(n-k)\times n$ matrix

$$T_k' = \begin{bmatrix} a_k & \dots & a_1 \\ \vdots & & \vdots \\ a_{n-1} & \dots & a_{n-k} \end{bmatrix},$$

then $T_k'\mathbf{x}_k^\pm = (r_{jk}^\pm)_{j=1}^{n-k}$.

Let us first introduce a notation. For a vector $\mathbf{v} = (v_j)_{j=1}^m \in \mathbb{C}^m$, we denote by $[\mathbf{v}]_+$, $[\mathbf{v}]_-$, and by $[\mathbf{v}]_-^+$ the vectors

$$[\mathbf{v}]_+ = (v_j)_{j=2}^m, \quad [\mathbf{v}]_- = (v_j)_{j=1}^{m-1}, \quad [\mathbf{v}]_-^+ = (v_j)_{j=2}^{m-1}. \tag{4.1}$$

It is easily checked that

$$\begin{bmatrix} T_{k+1} \\ T_{k+1}' \end{bmatrix} \begin{bmatrix} \mathbf{x}_k^\# & 0 \\ 0 & \mathbf{x}_k \end{bmatrix} = \begin{bmatrix} r_{0k}^+ & \overline{r_{1k}^-} \\ \mathbf{0}_{k-1} & \mathbf{0}_{k-1} \\ [\mathbf{r}_k^-]_+ & [\mathbf{r}_k^+]_- \end{bmatrix}.$$

From this relation we obtain

$$\begin{bmatrix} \mathbf{x}_{k+1}^\# & \mathbf{x}_{k+1} \end{bmatrix} = \begin{bmatrix} \mathbf{x}_k^\# & 0 \\ 0 & \mathbf{x}_k \end{bmatrix} \Theta_k, \tag{4.2}$$

where

$$\Theta_k = \begin{bmatrix} 1 & \overline{\gamma_k} \\ \gamma_k & 1 \end{bmatrix}, \quad \gamma_k = -\frac{r_{1k}^-}{r_{0k}^+}.$$

This leads to the following.

Proposition 4.1. The residual vectors $\mathbf{r}_k^\pm$ satisfy the recursion

$$\begin{bmatrix} \mathbf{r}_{k+1}^- & \mathbf{r}_{k+1}^+ \end{bmatrix} = \begin{bmatrix} [\mathbf{r}_k^-]_+ & [\mathbf{r}_k^+]_- \end{bmatrix} \Theta_k. \tag{4.3}$$

The recursion starts with $\mathbf{r}_1^- = \mathbf{r}_1^+ = (a_j)_{j=0}^{n-1}$.

Recall that for the L-factor of the standard LU-factorization we need the numbers r_{jk}^+ for $j = 0, \dots, n-k$ and for the diagonal factor the numbers r_{0k}^+ $(k = 1 \dots, n)$. For the Z-factor of the standard centro-hermitian ZW-factorization we need the numbers $r_{j,2k-1}^\pm$ for $j = 1, \dots, m-k$ and $k = 1, \dots, m$ and for the X-factor the numbers $r_{0,2k-1}^+$ and $r_{1,2k-1}^-$ for these k.

Remark 4.1. At the first glance one might think that for the computation of the parameters of the ZW-factorization only a part of the numbers $r_{jk}^\pm$ have to be computed. A closer look, however, reveals that this is not the case.

Let us estimate the complexity of the algorithm emerging from Proposition 4.1. The step $k \to k+1$ consists in 2 complex vector additions, 2 multiplications of a complex vector by a complex number. The lengths of the vectors are about $n-k$. This results in $4\,n^2$ RM and $4\,n^2$ RA.

Remark 4.2. There is a recursion similar (4.3) for the columns of *unit* LU- and ZW-factorizations. The difference to (4.3) is that the matrix Θ_k has not ones on the main diagonal but some real number. This increases the complexity by n^2 RM. That means this algorithm is less efficient than that generated by (4.3).

5. ZW-Factorization via one-step three-term Schur algorithm

To find the standard column conjugate-symmetric ZW-factorization we consider the equations

$$T_k \mathbf{x}_k^\pm = \overline{\rho_k^\pm}\mathbf{e}_1 + \rho_k^\pm \mathbf{e}_k$$

for complex numbers $\rho_k^\pm$ under the assumption that

$$\begin{bmatrix} \mathbf{e}_1^T \\ \mathbf{e}_k^T \end{bmatrix} \begin{bmatrix} \mathbf{x}_k^- & \mathbf{x}_k^+ \end{bmatrix} = \begin{bmatrix} -\mathrm{i} & 1 \\ \mathrm{i} & 1 \end{bmatrix}.$$

This guarantees the linear independence of $\mathbf{x}_k^-$ and $\mathbf{x}_k^+$. Besides the $\mathbf{x}_k^\pm$ we consider the residual vectors $\mathbf{s}_k^\pm = (s_{jk}^\pm)_{j=0}^{n-k}$ where $s_{0k}^\pm = \rho_k^\pm$ and $(s_{jk}^\pm)_{j=1}^{n-k} = T_k'\mathbf{x}_k^\pm$.

We have

$$T_{k+1}\left(\begin{bmatrix} 0 \\ \mathbf{x}_k^\pm \end{bmatrix} + \begin{bmatrix} \mathbf{x}_k^\pm \\ 0 \end{bmatrix}\right) = \begin{bmatrix} \overline{s}_{1k}^\pm + \overline{s}_{0k} \\ \overline{s}_{0k}^\pm \\ \mathbf{0}_{k-3} \\ s_{0k}^\pm \\ s_{1k}^\pm + s_{0k}^\pm \end{bmatrix} \qquad T_{k+1}\begin{bmatrix} 0 \\ \mathbf{x}_{k-1}^\pm \\ 0 \end{bmatrix} = \begin{bmatrix} \overline{s}_{1,k-1}^\pm \\ \overline{s}_{0,k-1}^\pm \\ \mathbf{0}_{k-3} \\ s_{0,k-1}^\pm \\ s_{1,k-1}^\pm \end{bmatrix}.$$

We are looking for real numbers $\alpha_k^\pm$ and $\beta_k^\pm$ such that

$$\mathbf{x}_{k+1}^\pm = \begin{bmatrix} 0 \\ \mathbf{x}_k^\pm \end{bmatrix} + \begin{bmatrix} \mathbf{x}_k^\pm \\ 0 \end{bmatrix} - \alpha_k^\pm \begin{bmatrix} 0 \\ \mathbf{x}_{k-1}^- \\ 0 \end{bmatrix} - \beta_k^\pm \begin{bmatrix} 0 \\ \mathbf{x}_{k-1}^+ \\ 0 \end{bmatrix}. \tag{5.1}$$

We introduce the matrices

$$\Gamma_k = \begin{bmatrix} \operatorname{Re} s_{0k}^- & \operatorname{Re} s_{0k}^+ \\ \operatorname{Im} s_{0k}^- & \operatorname{Im} s_{0k}^+ \end{bmatrix}. \tag{5.2}$$

It follows from the linear independence of $\mathbf{x}_k^\pm$ that the matrices Γ_k are nonsingular. A comparison of coefficients reveals that (5.1) holds if we choose

$$\begin{bmatrix} \alpha_k^- & \alpha_k^+ \\ \beta_k^- & \beta_k^+ \end{bmatrix} = \Gamma_{k-1}^{-1}\Gamma_k. \tag{5.3}$$

The recursion for the vectors $\mathbf{x}_k^\pm$ transfers to a recursion for the residual vectors.

Proposition 5.1. The vectors $\mathbf{s}_k^\pm$ satisfy the recursion

$$\mathbf{s}_{k+1}^\pm = [\mathbf{s}_k^\pm]_- + [\mathbf{s}_k^\pm]_+ - \alpha_k^\pm[\mathbf{s}_{k-1}^-]_\pm - \beta_k^\pm[\mathbf{s}_{k-1}^+]_\pm,$$

where $\alpha_k^\pm$ and $\beta_k^\pm$ are given by (5.3).

The initialization is given by

$$\begin{bmatrix} \mathbf{s}_2^- & \mathbf{s}_2^+ \end{bmatrix} = \begin{bmatrix} \mathbf{e}_2^T T_2 \\ T_2' \end{bmatrix} \begin{bmatrix} -\mathrm{i} & 1 \\ \mathrm{i} & 1 \end{bmatrix},$$

$$\begin{bmatrix} \mathbf{s}_3^- & \mathbf{s}_3^+ \end{bmatrix} = \begin{bmatrix} \mathbf{e}_3^T T_3 \\ T_3' \end{bmatrix} \begin{bmatrix} 1 & -\mathrm{i} \\ b\,\mathrm{Re}\,a_1 & b\,\mathrm{Im}\,a_1 \\ 1 & \mathrm{i} \end{bmatrix},$$

where $b = -\dfrac{2}{a_0}$.

Recall that the Z-factor in the standard column conjugate-symmetric ZW-factorization is given by the numbers $s_{j,2k}^{\pm}$ for $j = 0, \dots, m-k$ and $k = 1, \dots, m$, and the X-factor is given by the numbers $s_{0,2k}^{\pm}$.

Let us estimate the complexity. In the step $k \to k+1$ we have 4 multiplications of a complex vector by a real number and 6 additions of complex vectors. The lengths of the vectors are about $n-k$. This results in $4\,n^2$ RM and $6\,n^2$ RA.

Remark 5.1. There is a recursion similar to that in Proposition 5.1 for computing the columns of the *unit* column conjugate-symmetric ZW-factorization. In contrast to the classical Schur algorithm, the complexity of this algorithm is the same as for the *standard* column conjugate-symmetric ZW-factorization.

6. ZW-factorization via double-step three-term Schur algorithm

Since for the standard column conjugate-symmetric ZW-factorization we need only the numbers $s_{j,2k}^{\pm}$, i.e., only every second residual vector, it is reasonable to think about a double-step algorithm.

We are looking for a recursion of the form

$$\mathbf{x}_{2k+2}^{\pm} = \begin{bmatrix} 0 \\ 0 \\ \mathbf{x}_{2k}^{\pm} \end{bmatrix} + \begin{bmatrix} \mathbf{x}_{2k}^{\pm} \\ 0 \\ 0 \end{bmatrix} - \begin{bmatrix} 0 & 0 \\ \mathbf{x}_{2k}^{-} & \mathbf{x}_{2k}^{+} \\ 0 & 0 \end{bmatrix} \begin{bmatrix} \gamma_k^{\pm} \\ \delta_k^{\pm} \end{bmatrix} - \begin{bmatrix} \mathbf{0}_2 & \mathbf{0}_2 \\ \mathbf{x}_{2k-2}^{-} & \mathbf{x}_{2k-2}^{+} \\ \mathbf{0}_2 & \mathbf{0}_2 \end{bmatrix} \begin{bmatrix} \alpha_k^{\pm} \\ \beta_k^{\pm} \end{bmatrix}.$$

If we multiply a vector of this form by T_{2k+2}, then only the first and last 3 components of the resulting (conjugate-symmetric) vector are nonzero. First we find numbers $\alpha_k^{\pm}$, $\beta_k^{\pm}$ such that the third last component vanishes, which means

$$s_{0,2k}^{\pm} = \alpha_k^{\pm} s_{0,2k-2}^{-} - \beta_k^{\pm} s_{0,2k-2}^{+}.$$

This is equivalent to

$$\begin{bmatrix} \alpha_k^- & \alpha_k^+ \\ \beta_k^- & \beta_k^+ \end{bmatrix} = \Gamma_{2k-2}^{-1} \Gamma_{2k} \tag{6.1}$$

where Γ_{2k} is defined by (5.2).

Next we find $\gamma_k^{\pm}$ and $\delta_k^{\pm}$ such that the last but one component vanishes. This is equivalent to

$$s_{1,2k}^{\pm} - \alpha_k^{\pm} s_{1,2k-2}^{-} - \beta_k^{\pm} s_{1,2k-2}^{+} = \gamma_k^{\pm} s_{0,2k}^{-} + \delta_k^{\pm} s_{0,2k}^{+}.$$

To write this in matrix form we introduce

$$\widetilde{\Gamma}_k = \begin{bmatrix} \operatorname{Re} s_{1k}^- & \operatorname{Re} s_{1k}^+ \\ \operatorname{Im} s_{1k}^- & \operatorname{Im} s_{1k}^+ \end{bmatrix}. \tag{6.2}$$

After some calculation we find that

$$\begin{bmatrix} \gamma_k^- & \gamma_k^+ \\ \delta_k^- & \delta_k^+ \end{bmatrix} = \widetilde{\Gamma}_{2k}\Gamma_{2k}^{-1} - \widetilde{\Gamma}_{2k-2}\Gamma_{2k-2}^{-1}. \tag{6.3}$$

The recursion of the vectors $\mathbf{x}_{2k}^{\pm}$ transfers to the recursion of the residuals. In order to present this recursion for the residual vectors $\mathbf{s}_{2k}^{\pm}$ we extend the notation (4.1) as follows

$$[\,\mathbf{s}\,]_{++} = [\,[\,\mathbf{s}\,]_+]_+,\ [\,\mathbf{s}\,]_{--} = [\,[\,\mathbf{s}\,]_-]_-,\ [\,\mathbf{s}\,]_{--}^{++} = [\,[\,\mathbf{s}\,]_-^+]_-^+.$$

If S is a matrix then $[S]_-^+$ means that the $[.]_-^+$ operator is applied to each column of S.

Proposition 6.1. The vectors $\mathbf{s}_{2k}^{\pm}$ satisfy the recursions

$$\mathbf{s}_{2k+2}^+ = [\mathbf{s}_{2k}^+]_{--} + [\mathbf{s}_{2k}^+]_{++} - [\ \mathbf{s}_{2k}^- \quad \mathbf{s}_{2k}^+\ ]_-^+ \begin{bmatrix} \gamma_k^+ \\ \delta_k^+ \end{bmatrix} - [\ \mathbf{s}_{2k-2}^- \quad \mathbf{s}_{2k-2}^+\ ]_{--}^{++} \begin{bmatrix} \alpha_k^+ \\ \beta_k^+ \end{bmatrix},$$

$$\mathbf{s}_{2k+2}^- = [\mathbf{s}_{2k}^-]_{--} + [\mathbf{s}_{2k}^-]_{++} - [\ \mathbf{s}_{2k}^- \quad \mathbf{s}_{2k}^+\ ]_-^+ \begin{bmatrix} \gamma_k^- \\ \delta_k^- \end{bmatrix} - [\ \mathbf{s}_{2k-2}^- \quad \mathbf{s}_{2k-2}^+\ ]_{--}^{++} \begin{bmatrix} \alpha_k^- \\ \beta_k^- \end{bmatrix},$$

where coefficients are given by (6.3) and (5.3).

We start this recursion with $k = 1$, where we put $\widetilde{\Gamma}_0 = 0$ and $\mathbf{s}_0^- = \mathbf{s}_0^+ = 0$. The vectors $\mathbf{s}_2^{\pm}$ are given in the previous section.

We estimate the complexity. In the step $k \to k+1$ we have 8 multiplications of a complex vector by a real number and 10 additions of complex vectors. The lengths of the vectors are about $n - 2k$. But the number of steps is only m. This results in $4\,n^2$ RM and $5\,n^2$ RA. That means the number of multiplications is the same as for the one-step algorithm of Section 5, but we save n^2 RA.

7. ZW-factorizations via the Schur version of Krishna's algorithm

The algorithms presented in the previous sections do not lead to a further reduction in the complexity compared with the classical Schur algorithm for ZW-factorizations. The reason is that two families of vectors are computed. It is desirable to have all information contained in only one family. In [21] a Levinson algorithm of this kind was presented that leads to about 50% reduction of the number of multiplications. A similar algorithm was proposed in [22]. For other algorithms (based on different ideas) we refer to [3].

Here we present a Schur version of the algorithm in [21] and show how it can be used to find a centro-hermitian as well as a column conjugate-symmetric ZW-factorization of T.

Let $\mathbf{q}_k$ be the solution of an equation

$$T_k \mathbf{q}_k = \theta_k \, \mathbf{1}_k \,,$$

where θ_k is a nonzero real number. We allow the freedom in admitting a factor θ_k and not demanding something about $\mathbf{q}_k$ in order to save operations. Clearly, $\mathbf{q}_k$ is conjugate-symmetric. We set $\mathbf{s}_k = (s_{jk})_{j=0}^{n-k}$ with $s_{0k} = \theta_k$ and $(s_{jk})_{j=1}^{n-k} = T'_k \mathbf{q}_k$.

We have

$$\begin{bmatrix} T_{k+1} \\ T'_{k+1} \end{bmatrix} \begin{bmatrix} 0 & \mathbf{q}_k \\ \mathbf{q}_k & 0 \end{bmatrix} = \begin{bmatrix} \overline{s}_{1k} & \theta_k \\ \theta_k \mathbf{1}_{k-1} & \theta_k \mathbf{1}_{k-1} \\ [\mathbf{s}_k]_- & [\mathbf{s}_k]_+ \end{bmatrix},$$

$$\begin{bmatrix} T_{k+1} \\ T'_{k+1} \end{bmatrix} \begin{bmatrix} 0 \\ \mathbf{q}_{k-1} \\ 0 \end{bmatrix} = \begin{bmatrix} \overline{s}_{1,k-1} \\ \theta_{k-1} \mathbf{1}_{k-1} \\ [\mathbf{s}_{k-1}]_-^+ \end{bmatrix}.$$

The number $s_{1k} - \theta_k$ is nonzero, since otherwise the nonzero vector $\begin{bmatrix} 0 \\ \mathbf{q}_k \end{bmatrix} - \begin{bmatrix} \mathbf{q}_k \\ 0 \end{bmatrix}$ would belong to the kernel of T_{k+1}, which contradicts the nonsingularity of T_{k+1}.

We are looking for $\mathbf{q}_{k+1}$ to be of the form

$$\mathbf{q}_{k+1} = \begin{bmatrix} 0 & \mathbf{q}_k \\ \mathbf{q}_k & 0 \end{bmatrix} \begin{bmatrix} \overline{\alpha}_k \\ \alpha_k \end{bmatrix} - \begin{bmatrix} 0 \\ \mathbf{q}_{k-1} \\ 0 \end{bmatrix}.$$

If we choose

$$\alpha_k = \frac{s_{1,k-1} - \theta_{k-1}}{s_{1,k} - \theta_k} \,,$$

then we really have

$$T_{k+1} \mathbf{q}_{k+1} = \theta_{k+1} \mathbf{1}_{k+1},$$

with $\theta_{k+1} = 2\,\theta_k \operatorname{Re} \alpha_k - \theta_{k-1} \neq 0$. The recursion for the solutions leads to a recursion of the residual vectors as follows.

Proposition 7.1. The vectors $\mathbf{s}_k$ satisfy a recursion

$$\mathbf{s}_{k+1} = \alpha_k [\mathbf{s}_k]_+ + \overline{\alpha}_k [\mathbf{s}_k]_- - [\mathbf{s}_{k-1}]_-^+ \qquad \text{where} \qquad \alpha_k = \frac{s_{1,k-1} - s_{0,k-1}}{s_{1,k} - s_{0,k}} \,.$$

To start the recursion we observe that we can choose $\mathbf{q}_1 = 1$ and $\mathbf{q}_2 = \begin{bmatrix} a_0 - \overline{a}_1 \\ a_0 - a_1 \end{bmatrix}$. Hence

$$\mathbf{s}_1 = (a_j)_{j=0}^{n-1}, \quad \mathbf{s}_2 = ((a_0 - \overline{a}_1) a_{j+1} + (a_0 - a_1) a_j)_{j=0}^{n-2} \,.$$

Besides the vectors $\mathbf{s}_k$ we have to compute the last component ν_k of $\mathbf{q}_k$. The recursion for these numbers follows from the recursion of $\mathbf{q}_k$ as

$$\nu_{k+1} = \overline{\alpha}_k \nu_k \,.$$

In each step of the Schur-type algorithm for computing the vectors $\mathbf{s}_k$ we have 1 multiplication of a complex vector by a complex number and by its conjugate complex. This is equivalent to 4 multiplications of a real vector by a real number and 4 real vector additions. Besides this we have 2 complex vector additions. Since the length of the vectors is $n-k$ we end up with $2\,n^2$ RM plus $4\,n^2$ RA.

We show how we can obtain the data for a centro-hermitian ZW-factorization. For this we observe that the vectors

$$\mathbf{w}_k = \begin{bmatrix} \mathbf{q}_{2k-1} \\ 0 \end{bmatrix} - \frac{s_{0,2k-1}}{s_{0,2k}} \mathbf{q}_{2k}$$

satisfy $T_{2k}\mathbf{w}_k = (s_{1,2k-1} - s_{0,2k-1})\,\mathbf{e}_{2k}$. Since $\mathbf{q}_k$ are conjugate-symmetric and $s_{0,k}$ are real, we have

$$\mathbf{w}_k^{\#} = \begin{bmatrix} 0 \\ \mathbf{q}_{2k-1} \end{bmatrix} - \frac{s_{0,2k-1}}{s_{0,2k}} \mathbf{q}_{2k}.$$

The first and last components ξ_k^- and ξ_k^+ of $\mathbf{w}_k$ are given by

$$\xi_k^- = -\frac{s_{0,2k-1}}{s_{0,2k}} \nu_{2k}^-, \quad \xi_k^+ = \nu_{2k-1}^+ - \frac{s_{0,2k-1}}{s_{0,2k}} \nu_{2k}^+. \tag{7.1}$$

We introduce the vectors

$$\mathbf{z}_k^+ = \left(s_{j+1,2k-1} - \frac{s_{0,2k-1}}{s_{0,2k}} s_{j,2k} \right)_{j=0}^{m-k}, \quad \mathbf{z}_k^- = \left(s_{j,2k-1} - \frac{s_{0,2k-1}}{s_{0,2k}} s_{j,2k} \right)_{j=0}^{m-k}$$

and set

$$\mathbf{z}_k = \begin{bmatrix} (\mathbf{z}_k^-)^{\#} \\ \mathbf{0}_{2k-2} \\ \mathbf{z}_k^+ \end{bmatrix}.$$

We obtain the following.

Proposition 7.2. The matrix

$$Z = \begin{bmatrix} \mathbf{z}_m^{\#} & \dots & \mathbf{z}_1^{\#} & \mathbf{z}_1 & \dots & \mathbf{z}_m \end{bmatrix}$$

is the Z-factor of a centro-hermitian ZW-factorization of T, and the corresponding X-factor is given by

$$X = \operatorname{xma}\left(\begin{bmatrix} \overline{\xi}_k^+ & \xi_k^- \\ \overline{\xi}_k^- & \xi_k^+ \end{bmatrix}^{-1} \begin{bmatrix} \overline{\rho}_k^{-1} & 0 \\ 0 & \rho_k^{-1} \end{bmatrix} \right)_{k=1}^{m},$$

where $\xi_k^{\pm}$ are given by (7.1) and $\rho_k = s_{1,2k-1} - s_{0,2k-1}$.

For computing these data each step involves 1 multiplication of a complex vector by a real number and 2 additions of complex vectors. The lengths of the vectors are $m-k$ and the number of steps is m. This results in $m^2 = 0.25\,n^2$ RM and $2\,m^2 = 0.5\,n^2$ RA. Hence the total amount for computing a centro-hermitian ZW-factorization of T will be $2.25\,n^2$ RM and $4.5\,n^2$ RA.

We show how to compute the data for a column conjugate-symmetric ZW-factorization. Introduce the conjugate-symmetric vectors

$$\mathbf{w}_k^- = \begin{bmatrix} 0 \\ \mathbf{q}_{2k-2} \\ 0 \end{bmatrix} - \frac{s_{0,2k-2}}{s_{0,2k}} \mathbf{q}_{2k}, \quad \mathbf{w}_k^+ = \mathrm{i}\left(\begin{bmatrix} \mathbf{q}_{2k-1} \\ 0 \end{bmatrix} - \begin{bmatrix} 0 \\ \mathbf{q}_{2k-1} \end{bmatrix} \right),$$

which are obviously linearly independent. Then

$$T_{2k}\left[\begin{array}{cc}\mathbf{w}_k^- & \mathbf{w}_k^+\end{array}\right]=\left[\begin{array}{cc}\overline{\rho_k^-} & \overline{\rho_k^+}\\ \mathbf{0}_{2k-2} & \mathbf{0}_{2k-2}\\ \rho_k^- & \rho_k^+\end{array}\right],$$

where

$$\rho_k^- = s_{1,2k-2} - \frac{s_{0,2k-2}}{s_{0,2k}}\, s_{1,2k}, \quad \rho_k^+ = \mathrm{i}\,(s_{1,2k-1} - s_{0,2k-1})\,. \tag{7.2}$$

The last components $\xi_k^\pm$ of $\mathbf{w}_k^\pm$ are given by

$$\xi_k^- = -\frac{s_{0,2k-2}}{s_{0,2k}}\,\nu_{2k}, \quad \xi_k^+ = -\mathrm{i}\,\nu_{2k-1}\,. \tag{7.3}$$

We introduce vectors $\mathbf{t}_k^- = \left(s_{j+1,2k-2} - \frac{s_{0,2k-1}}{s_{0,2k}}\, s_{j,2k}\right)_{j=0}^{m-k}$,

$$\mathbf{t}_k^+ = \mathrm{i}\,\left(s_{j+1,2k-1} - s_{j,2k-1}\right)_{j=0}^{m-k} \quad \text{and} \quad \mathbf{z}_k^\pm = \left[\begin{array}{c}(\mathbf{t}_k^\pm)^\#\\ \mathbf{0}_{2k-2}\\ \mathbf{t}_k^\pm\end{array}\right].$$

Then we have the following.

Proposition 7.3. The matrix

$$Z_h = \left[\begin{array}{cccccc}\mathbf{z}_m^- & \dots & \mathbf{z}_1^- & \mathbf{z}_1^+ & \dots & \mathbf{z}_m^+\end{array}\right]$$

is the Z-factor of a column conjugate-complex ZW-factorization of T. The corresponding X-factor is given by (3.6) with $\rho_k^\pm$, and $\xi_k^\pm$ defined by (7.2) and (7.3)..

The amount for computing the data of this column conjugate-complex ZW-factorization is the same as for the centro-hermitian ZW-factorization above.

8. Summary

In the following table we compare the computational complexities for the algorithms discussed in this paper.

Method	n^2 RM	n^2 RA
LU and classical Schur-Bareiss	8	8
ch ZW and classical Schur	6	6.5
ccs ZW and one-step 3-term Schur	6	8
ccs ZW and double-step 3-term Schur	6	7
ch ZW and Krishna-Schur	4.25	7
ccs ZW and Krishna-Schur	4.25	6.5

Here "ch" stands for "centro-hermitian" and "ccs" stands for "column conjugate-symmetric".

References

[1] E.H. Bareiss, *Numerical solution of linear equations with Toeplitz and vector Toeplitz matrices*, Numer. Math., **13** (1969), 404–424.

[2] D. Bini, V. Pan, *Polynomial and Matrix Computations*, Birkhäuser Verlag, Basel, Boston, Berlin, 1994.

[3] Y. Bistritz, H. Lev-Ari, T. Kailath, *Immitance-type three-term Schur and Levinson recursions for quasi-Toeplitz complex Hermitian matrices*, SIAM J. Matrix. Analysis Appl., **12**, 3 (1991), 497–520.

[4] A. Bojanczyk, R. Brent, F. de Hoog, D. Sweet, *On the stability of Bareiss and related Toeplitz factorization algorithms*, SIAM J. Matrix. Analysis Appl., **16**, 1 (1995), 40–57.

[5] R.P. Brent, *Stability of fast algorithms for structured linear systems*, In: T. Kailath, A.H. Sayed (Eds.), *Fast Reliable Algorithms for Matrices with Structure*, SIAM, Philadelphia, 1999.

[6] P. Delsarte, Y. Genin, *On the splitting of classical algorithms in linear prediction theory*, IEEE Transactions on Acoustics Speech and Signal Processing, ASSP-**35** (1987), 645–653.

[7] C.J. Demeure, *Bowtie factors of Toeplitz matrices by means of split algorithms*, IEEE Transactions on Acoustics Speech and Signal Processing, ASSP-**37**, 10 (1989), 1601–1603.

[8] D.J. Evans, M. Hatzopoulos, *A parallel linear systems solver*, Internat. J. Comput. Math., **7**, 3 (1979), 227–238.

[9] I. Gohberg, I. Koltracht, T. Xiao, *Solution of the Yule-Walker equations*, Advanced Signal Processing Algorithms, Architectures, and Implementation II, Proceedings of SPIE, **1566** (1991).

[10] I. Gohberg, A. A. Semençul, *On the inversion of finite Toeplitz matrices and their continuous analogs* (in Russian), Matemat. Issledovanya, **7**, 2 (1972), 201–223.

[11] G. Golub, C. Van Loan, *Matrix Computations*, John Hopkins University Press, Baltimore, 1996.

[12] G. Heinig, *Chebyshev-Hankel matrices and the splitting approach for centrosymmetric Toeplitz-plus-Hankel matrices*, Linear Algebra Appl., **327**, 1-3 (2001), 181–196.

[13] G. Heinig, K. Rost, *Algebraic Methods for Toeplitz-like Matrices and Operators*, Birkhäuser Verlag, Basel, Boston, Stuttgart, 1984.

[14] G. Heinig, K. Rost, *DFT representations of Toeplitz-plus-Hankel Bezoutians with application to fast matrix-vector multiplication*, Linear Algebra Appl., **284** (1998), 157–175.

[15] G. Heinig, K. Rost, *Fast algorithms for skewsymmetric Toeplitz matrices*, Operator Theory: Advances and Applications, Birkhäuser Verlag, Basel, Boston, Berlin, **135** (2002), 193–208.

[16] G. Heinig, K. Rost, *Fast algorithms for centro-symmetric and centro-skewsymmetric Toeplitz-plus-Hankel matrices*, Numerical Algorithms, **33** (2003), 305–317.

[17] G. Heinig, K. Rost, *New fast algorithms for Toeplitz-plus-Hankel matrices*, SIAM Journal Matrix Anal. Appl. 25(3), 842–857 (2004).

[18] G. Heinig, K. Rost, *Split algorithms for skewsymmetric Toeplitz matrices with arbitrary rank profile*, Theoretical Computer Science 315 (2–3), 453–468 (2004).

[19] T. Kailath, *A theorem of I. Schur and its impact on modern signal processing*, Operator Theory: Advances and Applications, Birkhäuser Verlag, Basel, Boston, Stuttgart, **18** (1986), 9–30.

[20] T. Kailath, A.H. Sayed, *Fast Reliable Algorithms for Matrices with Structure*, SIAM, Philadelphia, 1999.

[21] B. Krishna, H. Krishna, *Computationally efficient reduced polynomial based algorithms for hermitian Toeplitz matrices*, SIAM J. Appl. Math., **49**, 4 (1989), 1275–1282.

[22] H. Krishna, S.D. Morgera, *The Levinson recurrence and fast algorithms for solving Toeplitz systems of linear equations*, IEEE Transactions on Acoustics Speech and Signal Processing, ASSP-**35** (1987), 839–848.

[23] E.M. Nikishin, V.N. Sorokin, *Rational approximation and orthogonality* (in Russian), Nauka, Moscow 1988; English: Transl. of Mathematical Monographs 92, Providence, AMS 1991.

[24] S. Chandra Sekhara Rao, *Existence and uniqueness of WZ factorization*, Parallel Comp., **23**, 8 (1997), 1129–1139.

[25] W.F. Trench, *An algorithm for the inversion of finite Toeplitz matrices*, J. Soc. Indust. Appl. Math., **12** (1964), 515–522.

[26] J.M. Varah, *The prolate matrix*, Linear Algebra Appl., **187** (1993), 269–278.

Georg Heinig
Department of Mathematics and Computer Sciences
P.O.Box 5969
Safat 1306, Kuwait
e-mail: `georg@mcs.sci.kuniv.edu.kw`

Karla Rost
Department of Mathematics
Chemnitz University of Technology
D-09107 Chemnitz, Germany
e-mail: `karla.rost@mathematik.tu-chemnitz.de`

Operator Theory:
Advances and Applications, Vol. 160, 253–271

Almost Pontryagin Spaces

M. Kaltenbäck, H. Winkler and H. Woracek

Abstract. The purpose of this note is to provide an axiomatic treatment of a generalization of the Pontryagin space concept to the case of degenerated inner product spaces.

Mathematics Subject Classification (2000). Primary 46C20; Secondary 46C05.

Keywords. Pontryagin space, indefinite scalar product.

1. Introduction

In this note we provide an axiomatic treatment of a generalization of the Pontryagin space concept to the case of degenerated inner product spaces. Pontryagin spaces are inner product spaces which can be written as the direct and orthogonal sum of a Hilbert space and a finite-dimensional anti Hilbert space. The subject of our paper are spaces which can be written as the direct and orthogonal sum of a Hilbert space, a finite-dimensional anti Hilbert space and a finite-dimensional neutral space.

The necessity of a systematic approach to such "almost" Pontryagin spaces became clear in the study of various topics: For example in the investigation of indefinite versions of various classical interpolation problems (e.g., [7]). Related to these questions is the generalization of Krein's formula for generalized resolvents of a symmetric operator (e.g., [8]). Another topic where the occurrence of degeneracy plays a crucial role is the theory of Pontryagin spaces of entire functions which generalizes the theory of Louis de Branges on Hilbert spaces of entire functions.

In Section 2 we generalize the concept of Pontryagin spaces by giving the definition of almost Pontryagin spaces and investigating the basic notion of Gram operator and fundamental decomposition. Moreover, the role played by the topology of an almost Pontryagin space is made clear. In the subsequent Section 3 we investigate some elementary constructions which can be made with almost Pontryagin spaces. We deal with subspaces, product spaces and factor spaces. Related to the last one of these constructions is the notion of morphism between almost Pontryagin spaces. Section 4 deals with the concept of completion. This topic is much more involved than the previous constructions. However, it is clearly of particular importance to be able to construct almost Pontryagin spaces from given linear spaces carrying an inner product. In Section 5 we turn our attention to a particular class of almost Pontryagin spaces, so-called almost reproducing kernel Pontryagin spaces. The intention there is to prove the existence of the correct analogue of a reproducing kernel of a reproducing kernel Pontryagin space. Finally, in Section 6, we explain some circumstances where almost Pontryagin spaces actually occur.

2. Almost Pontryagin spaces

Before we give the definition of almost Pontryagin spaces, recall the definition of Pontryagin spaces. A pair $(\mathfrak{P}, [.,.])$ where $\mathfrak{P}$ is a complex vector space and $[.,.]$ is a hermitian inner product on $\mathfrak{P}$ is called a Pontryagin space if one can decompose $\mathfrak{P}$ as

$$\mathfrak{P} = \mathfrak{P}_-[\dot{+}]\mathfrak{P}_+, \tag{2.1}$$

where $(\mathfrak{P}_-, [.,.])$ is a finite-dimensional anti Hilbert space, $(\mathfrak{P}_+, [.,.])$ is a Hilbert space, and $[\dot{+}]$ denotes the direct and $[.,.]$-orthogonal sum. Such decompositions of $\mathfrak{P}$ are called fundamental decompositions. It is worthwhile to note (see [2] or see below) that every Pontryagin space carries a unique Hilbert space topology $\mathcal{O}$ (there exists an inner product $(.,.)$ such that $(\mathfrak{P}, (.,.))$ is a Hilbert space and such that $(.,.)$ induces the topology $\mathcal{O}$, i.e., $\mathcal{O} = \mathcal{O}_{(.,.)}$) such that the inner product $[.,.]$ is continuous with respect to $\mathcal{O}$. This topology is also called the Pontryagin space topology on $\mathfrak{P}$.

With respect to this topology the subspace $\mathfrak{P}_+$ is closed for any fundamental decomposition (2.1). Conversely, the product topology induced on $\mathfrak{P}$ by any fundamental decomposition (2.1) coincides with the unique Hilbert space topology.

It will turn out that for almost Pontryagin spaces the uniqueness assertion about the topology is no longer true. Thus we will include the topology into the definition.

Definition 2.1. Let $\mathfrak{L}$ be a linear space, $[.,.]$ an inner product on $\mathfrak{L}$ and $\mathcal{O}$ a Hilbert space topology on $\mathfrak{L}$. The triplet $(\mathfrak{L}, [.,.], \mathcal{O})$ is called an almost Pontryagin space, if

(aPS1) $[.,.]$ is $\mathcal{O}$-continuous.

(aPS2) There exists a $\mathcal{O}$-closed linear subspace $\mathfrak{M}$ of $\mathfrak{L}$ with finite codimension such that $(\mathfrak{M}, [.,.])$ is a Hilbert space.

Let $(\mathfrak{R}, [.,.])$ be any linear space equipped with a inner product $[.,.]$ and assume that

$$\sup\left\{ \dim \mathfrak{U} : \mathfrak{U} \text{ negative definite subspace of } \mathfrak{R} \right\} < \infty.$$

Then (see for example [2, Corollary I.3.4]) the dimensions of all maximal negative definite subspaces of $\mathfrak{R}$ are equal. We denote this number by $\kappa_-(\mathfrak{R}, [.,.])$ and refer to it as the negative index (or the degree of negativity) of $(\mathfrak{R}, [.,.])$. If the above supremum is not finite, we set $\kappa_-(\mathfrak{R}, [.,.]) = \infty$.

The isotropic part of an inner product space $(\mathfrak{R}, [.,.])$ is defined as

$$\mathfrak{R}^{[\circ]} = \{x \in \mathfrak{R} : [x,y] = 0, y \in \mathfrak{R}\}.$$

We will denote its dimension by $\Delta(\mathfrak{R}, [.,.]) \in \mathbb{N} \cup \{0, \infty\}$ and call this number the degree of degeneracy of $(\mathfrak{R}, [.,.])$.

Remark 2.2. It immediately follows from the definition that if $(\mathfrak{L}, [.,.], \mathcal{O})$ is an almost Pontryagin space, then $\kappa_-(\mathfrak{L}, [.,.])$ and $\Delta(\mathfrak{L}, [.,.])$ are both finite.

The fact that a given triplet $(\mathfrak{L}, [.,.], \mathcal{O})$ is an almost Pontryagin space can be characterized in several ways. First let us give one characterization via a spectral property of a Gram operator.

Proposition 2.3. *Let $\mathfrak{L}$ be a linear space, $[.,.]$ an inner product on $\mathfrak{L}$ and $\mathcal{O}$ a Hilbert space topology on $\mathfrak{L}$.*

(i) *Assume that $(\mathfrak{L}, [.,.], \mathcal{O})$ is an almost Pontryagin space and let $(.,.)$ be any Hilbert space inner product which induces the topology $\mathcal{O}$. Then there exists a unique $(.,.)$-selfadjoint bounded operator $G_{(.,.)}$ with*

$$[x,y] = (G_{(.,.)}x, y), \quad x, y \in \mathfrak{L}.$$

There exists $\epsilon > 0$ such that $\sigma(G_{(.,.)}) \cap (-\infty, \epsilon)$ consists of finitely many eigenvalues of finite multiplicity. If we denote by $E(M)$ the spectral measure of $G_{(.,.)}$, this just means that

$$\dim \operatorname{ran} E(-\infty, \epsilon) < \infty. \tag{2.2}$$

Moreover,

$$\Delta(\mathfrak{L}, [.,.]) = \dim \ker G_{(.,.)}, \quad \kappa_-(\mathfrak{L}, [.,.]) = \dim \operatorname{ran} E(-\infty, 0).$$

We will refer to $G_{(.,.)}$ as the Gram operator corresponding to $(.,.)$.

(ii) *Let $(\mathfrak{L}, (.,.))$ be a Hilbert space, and let G be a bounded selfadjoint operator on $(\mathfrak{L}, (.,.))$ which satisfies (2.2) where $E(M)$ denotes the spectral measure of G. Moreover, let $\mathcal{O}$ be the topology induced by $(.,.)$ and define $[.,.] = (G.,.)$. Then $(\mathfrak{L}, [.,.], \mathcal{O})$ is an almost Pontryagin space.*

Proof. ad (i): Since $[.,.]$ is continuous with respect to the topology $\mathcal{O}$, the Lax-Milgram theorem ensures the existence and uniqueness of $G_{(.,.)}$. Moreover, since $[.,.]$ is an inner product, $G_{(.,.)}$ is selfadjoint.

Let $\mathfrak{M}$ be a $\mathcal{O}$-closed linear subspace of $\mathfrak{L}$ with finite codimension such that $(\mathfrak{M}, [.,.])$ is a Hilbert space. By the open mapping theorem $[.,.]|_{\mathfrak{M}}$ and $(.,.)|_{\mathfrak{M}}$ are equivalent. Hence $P_{\mathfrak{M}} G_{(.,.)}|_{\mathfrak{M}}$, where $P_{\mathfrak{M}}$ denotes the $(.,.)$-orthogonal projection onto $\mathfrak{M}$, is strictly positive. Choose $\epsilon > 0$ such that $\epsilon I_{\mathfrak{M}} < P_{\mathfrak{M}} G_{(.,.)}|_{\mathfrak{M}}$. Assume that $\dim \operatorname{ran} E(-\infty, \epsilon) > \operatorname{codim}_{\mathfrak{L}} \mathfrak{M}$, then $\operatorname{ran} E(-\infty, \epsilon) \cap \mathfrak{M} \neq \{0\}$. For any $x \in \operatorname{ran} E(-\infty, \epsilon) \cap \mathfrak{M}$, $(x,x) = 1$, we have

$$\epsilon < (P_{\mathfrak{M}} G_{(.,.)}|_{\mathfrak{M}} x, x) = (G_{(.,.)}x, x) \leq \epsilon,$$

a contradiction.

ad (ii): Choose $\mathfrak{M} = \operatorname{ran} E[\epsilon, \infty)$.

Then $\mathfrak{M}$ is $(.,.)$-closed, $\operatorname{codim}_{\mathfrak{L}} \mathfrak{M} = \dim \operatorname{ran} E(-\infty, \epsilon) < \infty$ and $[.,.] = (G.,.)$ is a Hilbert space inner product on $\mathfrak{M}$ since $G|_{\mathfrak{M}}$ is strictly positive. □

Corollary 2.4. *Let an almost Pontryagin space $(\mathfrak{L}, [.,.], \mathcal{O})$ be given. If $(.,.)$ and $\langle .,. \rangle$ are two Hilbert space inner products on $\mathfrak{L}$ which both induce the topology $\mathcal{O}$ and T is the $(.,.)$-strictly positive bounded operator on $\mathfrak{L}$ with $\langle .,. \rangle = (T.,.)$, then the Gram operators $G_{(.,.)}$ and $G_{\langle .,. \rangle}$ are connected by*

$$G_{(.,.)} = TG_{\langle .,. \rangle}.$$

There exists a Hilbert space inner product $(.,.)$ on $\mathfrak{L}$ which induces $\mathcal{O}$ such that its Gram operator $G_{(.,.)}$ is a finite-dimensional perturbation of the identity.

Proof. The first assertion is clear from

$$(G_{(.,.)}x, y) = [x, y] = \langle G_{\langle .,. \rangle}x, y \rangle = (TG_{\langle .,. \rangle}x, y), \ x, y \in \mathfrak{L}.$$

For the second assertion choose a Hilbert space inner product $\langle .,. \rangle$ on $\mathfrak{L}$ which induces $\mathcal{O}$. Let $G_{\langle .,. \rangle}$ be the corresponding Gram operator, $E(M)$ its spectral measure, and choose $\epsilon > 0$ as in Proposition 2.3, (i). Define

$$(x, y) = \langle (E[\epsilon, \infty)G_{\langle .,. \rangle} + E(-\infty, \epsilon))x, y \rangle, \ x, y \in \mathfrak{L}.$$

Then

$$G_{(.,.)} = (E[\epsilon, \infty)G_{\langle .,. \rangle} + E(-\infty, \epsilon))^{-1}G_{\langle .,. \rangle} = E[\epsilon, \infty) + E(-\infty, \epsilon)G_{\langle .,. \rangle}.$$ □

In the study of Pontryagin spaces so-called fundamental decompositions play an important role. The following is the correct analogue for almost Pontryagin spaces. In particular, it gives us another characterization of this notion.

Proposition 2.5. *The following assertions hold true:*

(i) *Let $(\mathfrak{L}, [.,.], \mathcal{O})$ be an almost Pontryagin space. Then there exists a direct and $[.,.]$-orthogonal decomposition*

$$\mathfrak{L} = \mathfrak{L}_+ [\dot{+}] \mathfrak{L}_- [\dot{+}] \mathfrak{L}^{[\circ]}, \tag{2.3}$$

where $\mathfrak{L}_+$ is $\mathcal{O}$-closed, $(\mathfrak{L}_+, [.,.])$ is a Hilbert space and $\mathfrak{L}_-$ is negative definite, $\dim \mathfrak{L}_- = \kappa_-(\mathfrak{L}, [.,.])$.

(ii) *Let $(\mathfrak{L}_+, (.,.)_+)$ be a Hilbert space, $(\mathfrak{L}_-, (.,.)_-)$ be a finite-dimensional Hilbert space, and let $\mathfrak{L}_0$ be a finite-dimensional linear space. Define a linear space*

$$\mathfrak{L} = \mathfrak{L}_+ \dot{+} \mathfrak{L}_- \dot{+} \mathfrak{L}_0,$$

and inner products

$$[x_+ + x_- + x_0, y_+ + y_- + y_0] = (x_+, y_+) - (x_-, y_-),$$

$$(x_+ + x_- + x_0, y_+ + y_- + y_0) = (x_+, y_+) + (x_-, y_-) + (x_0, y_0)_0,$$

where $(.,.)_0$ is any Hilbert space inner product on $\mathfrak{L}_0$. Moreover, let $\mathcal{O}$ be the topology on $\mathfrak{L}$ induced by the Hilbert space inner product $(.,.)$. Then $(\mathfrak{L}, [.,.], \mathcal{O})$ is an almost Pontryagin space. Thereby $\kappa_-(\mathfrak{L}, [.,.]) = \dim \mathfrak{L}_-$ and $\mathfrak{L}^{[\circ]} = \mathfrak{L}_0$.

Proof. Let $(\mathfrak{L}, [.,.], \mathcal{O})$ be an almost Pontryagin space. Choose a Hilbert space inner product $(.,.)$ which induces $\mathcal{O}$, let $G_{(.,.)}$ be the corresponding Gram operator, and denote by $E(M)$ the spectral measure of $G_{(.,.)}$. Define

$$\mathfrak{L}_+ = \operatorname{ran} E(0,\infty),\ \ \mathfrak{L}_- = \operatorname{ran} E(-\infty,0).$$

Then $\mathfrak{L}_+$ is $\mathcal{O}$-closed. The inner products $[.,.]$ and $(.,.)$ are equivalent on $\mathfrak{L}_+$ since $G|_{\mathfrak{L}_+}$ is strictly positive. Hence $(\mathfrak{L}_+, [.,.])$ is a Hilbert space. Clearly $(\mathfrak{L}_-, [.,.])$ is negative definite and $\dim \mathfrak{L}_- = \kappa_-(\mathfrak{L}, [.,.])$. Since $E(-\infty,0)+E\{0\}+E(0,\infty) = I$, the space $\mathfrak{L}$ is decomposed as in (2.3).

Conversely, let $(\mathfrak{L}_+, (.,.)_+)$, $(\mathfrak{L}_-, (.,.)_-)$ and $\mathfrak{L}_0$ be given. The Gram operator of $[.,.]$ with respect to $(.,.)$ is equal to

$$G = \begin{pmatrix} I & 0 & 0 \\ 0 & -I & 0 \\ 0 & 0 & 0 \end{pmatrix}.$$

Obviously, $\operatorname{ran} E(-\infty, \frac{1}{2}) = \mathfrak{L}_- + \mathfrak{L}_0$, $\ker G = \mathfrak{L}_0$ and $\operatorname{ran} E(-\infty, 0) = \mathfrak{L}_-$. □

Corollary 2.6. *We have*

(i) *Let* $(\mathfrak{L}_+, (.,.)_+)$, $(\mathfrak{L}_-, (.,.)_-)$ *and* $\mathfrak{L}_0$ *be as in (ii) of Proposition* 2.5, *and let* $(\mathfrak{L}, [.,.], \mathcal{O})$ *be the almost Pontryagin space constructed there. Then* $\mathfrak{L} = \mathfrak{L}_+[\dot{+}]\mathfrak{L}_-[\dot{+}]\mathfrak{L}_0$ *is a decomposition of the same kind as in* (2.3).

(ii) *Let* $(\mathfrak{L}, [.,.], \mathcal{O})$ *be an almost Pontryagin space, and assume that* $\mathfrak{L}$ *is decomposed as* $\mathfrak{L} = \mathfrak{L}_+[\dot{+}]\mathfrak{L}_-[\dot{+}]\mathfrak{L}^{[\circ]}$ *where* $(\mathfrak{L}_+, [.,.])$ *is a Hilbert space and* $(\mathfrak{L}_-, [.,.])$ *is negative definite.*
Let $(\mathfrak{L}_1, [.,.]_1, \mathcal{O}_1)$ *be the almost Pontryagin space constructed by means of Proposition* 2.5, *(ii), from* $(\mathfrak{L}_+, [.,.])$, $(\mathfrak{L}_-, -[.,.])$, $\mathfrak{L}_0 = \mathfrak{L}^{[\circ]}$. *Then* $\mathfrak{L}_1 = \mathfrak{L}$ *and* $[.,.]_1 = [.,.]$. *We have* $\mathcal{O}_1 = \mathcal{O}$ *if and only if* $\mathfrak{L}_+$ *is* $\mathcal{O}$*-closed.*

Proof. The assertion (i) follows immediately since $\mathfrak{L}_+$ is $(.,.)$-closed. We come to the proof of (ii). The facts that $\mathfrak{L}_1 = \mathfrak{L}$ and $[.,.]_1 = [.,.]$ are obvious.

Assume that $\mathfrak{L}_+$ is $\mathcal{O}$-closed. Note that by the assumption on their dimensions the subspaces $\mathfrak{L}_-$ and $\mathfrak{L}_0$ are closed, too. By the Open Mapping Theorem the linear bijection

$$(x_+; x_-; x_0) \mapsto x_+ + x_- + x_0,$$

is bicontinuous from $\mathfrak{L}_+ \times \mathfrak{L}_- \times \mathfrak{L}_0$ provided with the product topology onto $\mathfrak{L}$ provided with $\mathcal{O}$. On the other hand by the definition of $\mathcal{O}_1$ this mapping is also bicontinuous if we provide $\mathfrak{L}$ with $\mathcal{O}_1$. Thus $\mathcal{O}_1 = \mathcal{O}$.

Finally, assume that $\mathcal{O}_1 = \mathcal{O}$. By the construction of $(.,.)_1$ the space $\mathfrak{L}_+$ is $\mathcal{O}_1$-closed and, therefore, also $\mathcal{O}$-closed. □

From the above results we obtain a statement which shows from another point of view that almost Pontryagin spaces can be viewed as a generalization of Pontryagin spaces.

Corollary 2.7. *Let $(\mathfrak{P}, [.,.])$ be a Pontryagin space, and let $\mathcal{O}$ be the unique topology on $\mathfrak{P}$ such that $[.,.]$ is continuous (see [2], compare also Corollary 2.10). Then $(\mathfrak{P}, [.,.], \mathcal{O})$ is an almost Pontryagin space. Moreover, $\Delta(\mathfrak{P}, [.,.]) = 0$.*

Conversely, if $(\mathfrak{P}, [.,.], \mathcal{O})$ is an almost Pontryagin space with $\Delta(\mathfrak{P}, [.,.]) = 0$, then $(\mathfrak{P}, [.,.])$ is a Pontryagin space.

Proof. Let $(\mathfrak{P}, [.,.])$ be a Pontryagin space. Choose a fundamental decomposition $\mathfrak{P} = \mathfrak{P}_+[\dot{+}]\mathfrak{P}_-$. Then $\mathfrak{P}_+$ is $\mathcal{O}$-closed, $(\mathfrak{P}_+, [.,.])$ is a Hilbert space and $\operatorname{codim}_{\mathfrak{P}} \mathfrak{P}_+ = \dim \mathfrak{P}_- < \infty$.

Let $(\mathfrak{P}, [.,.], \mathcal{O})$ be an almost Pontryagin space with $\Delta(\mathfrak{P}, [.,.]) = 0$. Choose a decomposition $\mathfrak{P} = \mathfrak{P}_+[\dot{+}]\mathfrak{P}_-$ according to (2.3). By Corollary 2.6, (ii), the topology $\mathcal{O}$ coincides with the topology of the Pontryagin space $(\mathfrak{P}_+[\dot{+}]\mathfrak{P}_-, [.,.])$. □

It is a noteworthy fact that in certain cases the topology of an almost Pontryagin space $(\mathfrak{L}, [.,.], \mathcal{O})$ is uniquely determined by the inner product, see the Proposition 2.9 below. However, in general this is not true. This fact goes back to [4], [5].

Lemma 2.8. *For any infinite-dimensional almost Pontryagin space $(\mathfrak{L}, [.,.], \mathcal{O})$ with $\Delta(\mathfrak{L}, [.,.]) > 0$ there exists a topology $\mathcal{T}$ different from $\mathcal{O}$ such that also $(\mathfrak{L}, [.,.], \mathcal{T})$ is an almost Pontryagin space.*

Proof. Choose a Hilbert space inner product $(.,.)$ on $\mathfrak{L}$ inducing $\mathcal{O}$. Let $h \in \mathfrak{L}^{[\circ]}$ and $\mathfrak{K} = h^{(\perp)}$, and let f be a non-continuous linear functional on $\mathfrak{K}$. We define the linear mapping U from $\mathfrak{L} = \mathfrak{K}(\dot{+}) \operatorname{span}\{h\}$ onto itself by

$$U(x + \xi h) = x + (\xi + f(x))h, \; x \in \mathfrak{K}, \; \xi \in \mathbb{C}.$$

The mapping U is bijective and non-continuous. In fact, if it were continuous, then also f would be continuous. Nevertheless, U is isometric:

$$[U(x+\xi h), U(y+\eta h)] = [x+(\xi+f(x))h, y+(\eta+f(y))h] = [x, y] = [x+\xi h, y+\eta h].$$

Therefore, with $(\mathfrak{L}, [.,.], \mathcal{O})$ also its isometric copy $(\mathfrak{L}, [.,.], U^{-1}(\mathcal{O}))$ is an almost Pontryagin space. As U is not continuous we have $\mathcal{T} = U^{-1}(\mathcal{O}) \neq \mathcal{O}$. □

The existence of a sufficiently large family of functionals which are required to be continuous guarantees the uniqueness of the topology. Such a family of functionals will show up, in particular, when we deal with spaces consisting of functions such that the point evaluation functionals are continuous.

A family $(f_i)_{i \in I}$ of linear functionals on a linear space $\mathfrak{L}$ is said to be point separating if for each two $x, y \in \mathfrak{L}$, $x \neq y$, there exists $i \in I$ such that $f_i(x) \neq f_i(y)$.

Proposition 2.9. *Let $(\mathfrak{L}, [.,.], \mathcal{O})$ be an almost Pontryagin space and assume that there exists a point separating family of continuous linear functionals $(f_i)_{i \in I}$. Then $\mathcal{O}$ is the unique Banach space topology on $\mathfrak{L}$ such that all functionals f_i, $i \in I$, are continuous.*

Proof. Let $\mathcal{T}$ be a Banach space topology on $\mathfrak{L}$ such that every f_i is continuous. The identity mapping $\mathrm{id} : (\mathfrak{L}, \mathcal{O}) \to (\mathfrak{L}, \mathcal{T})$ has a closed graph. In fact, if $x_n \to x$ with respect to $\mathcal{O}$ and $x_n \to y$ with respect to $\mathcal{T}$, then by assumption

$$f_i(x) = \lim_{n\to\infty} f_i(x_n) = f_i(y), \text{ for all } i \in I,$$

and hence $x = y$. By the Closed Graph Theorem the identity map is bicontinuous, and therefore $\mathcal{T} = \mathcal{O}$. □

As a corollary we obtain the well-known result that a Pontryagin space carries a unique Hilbert space topology with respect to which $[.,.]$ is continuous.

Corollary 2.10. *If an almost Pontryagin space $(\mathfrak{P}, [.,.], \mathcal{O})$ is a Pontryagin space, i.e., $\Delta(\mathfrak{P}, [.,.]) = 0$, then $\mathcal{O}$ is the unique Banach space topology $\mathcal{T}$ on $\mathfrak{P}$ such that $[.,.]$ is continuous with respect to $\mathcal{T}$. In particular, it is the unique Hilbert space topology $\mathcal{T}$ on $\mathfrak{P}$ such that $(\mathfrak{P}, [.,.], \mathcal{T})$ is an almost Pontryagin space.*

Proof. The assumption $\Delta(\mathfrak{P}, [.,.]) = 0$ is equivalent to the fact that the family of functionals $f_x = [.,x]$, $x \in \mathfrak{P}$, is point separating. Hence we can apply Lemma 2.9. □

3. Subspaces, products, factors

The next result shows that the class of almost Pontryagin spaces is closed under the formation of subspaces and finite direct products. Note that the first half of this statement is not true for Pontryagin spaces.

Proposition 3.1. *Let $(\mathfrak{L}, [.,.], \mathcal{O})$ be an almost Pontryagin space, $\mathfrak{K}$ a closed linear subspace of $\mathfrak{L}$, and denote by $\mathcal{O} \cap \mathfrak{K}$ the subspace topology induced by $\mathcal{O}$ on $\mathfrak{K}$. Then $(\mathfrak{K}, [.,.], \mathcal{O} \cap \mathfrak{K})$ is an almost Pontryagin space. We have $\kappa_-(\mathfrak{K}, [.,.]) \leq \kappa_-(\mathfrak{L}, [.,.])$.*

Let $(\mathfrak{L}_1, [.,.]_1, \mathcal{O}_1)$ and $(\mathfrak{L}_2, [.,.]_2, \mathcal{O}_2)$ be two almost Pontryagin spaces, and denote by $\mathcal{O}_1 \times \mathcal{O}_2$ the product topology on $\mathfrak{L}_1 \times \mathfrak{L}_2$. Define the inner product

$$[(u; v), (x; y)] = [u, x]_1 + [v, y]_2, \ (u; v), (x; y) \in \mathfrak{L}_1 \times \mathfrak{L}_2.$$

Then $(\mathfrak{L}_1 \times \mathfrak{L}_2, [.,.], \mathcal{O}_1 \times \mathcal{O}_2)$ is an almost Pontryagin space. We have

$$\kappa_-(\mathfrak{L}_1 \times \mathfrak{L}_2, [.,.]) = \kappa_-(\mathfrak{L}_1, [.,.]_1) + \kappa_-(\mathfrak{L}_2, [.,.]_2),$$
$$\Delta(\mathfrak{L}_1 \times \mathfrak{L}_2, [.,.]) = \Delta(\mathfrak{L}_1, [.,.]_1) + \Delta(\mathfrak{L}_2, [.,.]_2).$$

Proof. To establish the first part of the assertion choose an $\mathcal{O}$-closed linear subspace $\mathfrak{M}$ of $\mathfrak{L}$ with finite codimension such that $(\mathfrak{M}, [.,.])$ is a Hilbert space. We already saw that by the closed graph theorem $[.,.]$ induces the topology $\mathcal{O} \cap \mathfrak{M}$ on $\mathfrak{M}$. Thus $\mathfrak{K} \cap \mathfrak{M}$ is at the same time $\mathcal{O} \cap \mathfrak{K}$-closed linear subspace of $\mathfrak{K}$ with finite codimension in $\mathfrak{K}$, and a $\mathcal{O} \cap \mathfrak{M}$-closed (i.e., $[.,.]$-closed) subspace of $\mathfrak{M}$. Hence $(\mathfrak{K} \cap \mathfrak{M}, [.,.])$ is a Hilbert space. Thus $(\mathfrak{K}, [.,.], \mathcal{O} \cap \mathfrak{K})$ is an almost Pontryagin space. The relation between the negative indices is clear.

To prove the second assertion take for $j = 1, 2$ a $\mathcal{O}_j$-closed subspace $\mathfrak{M}_j$ with finite codimension in $\mathfrak{L}_j$ such that $(\mathfrak{M}_j, [.,.]_j)$ is a Hilbert space. Then $\mathfrak{M}_1 \times \mathfrak{M}_2$ is a

$\mathcal{O}_1 \times \mathcal{O}_2$-closed subspace of $\mathfrak{L}_1 \times \mathfrak{L}_2$ of finite codimension such that $(\mathfrak{M}_1 \times \mathfrak{M}_2, [.,.])$ is a Hilbert space. □

We conclude from Corollary 2.7 together with Proposition 3.1 that every closed subspace of a Pontryagin space is an almost Pontryagin space. Also the converse holds true:

Proposition 3.2. *Let $(\mathfrak{L}, [.,.], \mathcal{O})$ be an almost Pontryagin space. Then there exists a Pontryagin space $(\mathfrak{P}, [.,.])$ such that $\mathfrak{L}$ is a closed subspace of $\mathfrak{P}$ with codimension $\Delta(\mathfrak{L}, [.,.])$ and $\mathcal{O}$ is the subspace topology on $\mathfrak{L}$ induced by the Pontryagin space topology on $\mathfrak{P}$. Moreover, $\kappa_-(\mathfrak{P}, [.,.]) = \kappa_-(\mathfrak{L}, [.,.]) + \Delta(\mathfrak{L}, [.,.])$. Any two Pontryagin spaces with the listed properties are isometrically isomorphic.*

Conversely, let $(\mathfrak{P}, [.,.])$ be a Pontryagin space. If $\mathfrak{L}$ is a closed subspace of $\mathfrak{P}$, so that $\mathfrak{L}$ with the inner product and topology inherited from $(\mathfrak{P}, [.,.])$ is an almost Pontryagin space, then $\operatorname{codim}_{\mathfrak{P}} \mathfrak{L} \geq \Delta(\mathfrak{L}, [.,.])$.

Proof. Fix a decomposition $\mathfrak{L} = \mathfrak{L}_+[\dot{+}]\mathfrak{L}_-[\dot{+}]\mathfrak{L}^{[\circ]}$ according to (2.3). Let $\mathfrak{L}'$ be a linear space of dimension $\Delta(\mathfrak{L}, [.,.])$ and define

$$\mathfrak{P} = \mathfrak{L}_+ \dot{+} \mathfrak{L}_- \dot{+} \mathfrak{L}^{[\circ]} \dot{+} \mathfrak{L}'.$$

We declare an inner product $[.,.]_1$ on $\mathfrak{P}$ by

$$[x,y]_1 = [x,y], \;\; x,y \in \mathfrak{L}_+ + \mathfrak{L}_-, \;\; (\mathfrak{L}_+ + \mathfrak{L}_-)[\perp]_1(\mathfrak{L}^{[\circ]} + \mathfrak{L}'),$$

and the requirement that $\mathfrak{L}^{[\circ]}$ and $\mathfrak{L}'$ are skewly linked neutral subspaces, i.e., for every non-zero $x \in \mathfrak{L}^{[\circ]}$ there exists a $y \in \mathfrak{L}'$ such that $[x,y] \neq 0$ and, conversely, for every non-zero $y \in \mathfrak{L}'$ there exists an $x \in \mathfrak{L}^{[\circ]}$ such that $[x,y] \neq 0$.

Then $(\mathfrak{P}, [.,.]_1)$ is a Pontryagin space because it can be seen as the product of the Pontryagin spaces $(\mathfrak{L}_+ \dot{+} \mathfrak{L}_-, [.,.])$ and $(\mathfrak{L}^{[\circ]} \dot{+} \mathfrak{L}', [.,.]_1)$.

Clearly, $\operatorname{codim}_{\mathfrak{P}} \mathfrak{L} = \Delta(\mathfrak{L}, [.,.])$ and $[.,.]_1|_{\mathfrak{L}} = [.,.]$. Since $\mathfrak{L} = (\mathfrak{L}^{[\circ]})^{[\perp]_1}$, the space $\mathfrak{L}$ is a closed subspace of $\mathfrak{P}$. Let $\mathcal{T}$ be the subspace topology on $\mathfrak{L}$ induced by the Pontryagin space topology on $\mathfrak{P}$. Then $\mathcal{T}$ coincides with the topology on $\mathfrak{L}$ obtained from the construction of Proposition 2.5, (ii), applied with $(\mathfrak{L}_+, [.,.])$, $(\mathfrak{L}_-, -[.,.])$ and $\mathfrak{L}^{[\circ]}$. Since $\mathfrak{L}_+$ is $\mathcal{O}$-closed, Corollary 2.6, (ii), yields $\mathcal{T} = \mathcal{O}$.

Let $(\mathfrak{P}_2, [.,.]_2)$ be another Pontryagin space which contains $\mathfrak{L}$ with codimension $\Delta(\mathfrak{L}, [.,.])$. Then $\mathfrak{P}_2$ can be decomposed as

$$\mathfrak{P}_2 = \mathfrak{L}_+[\dot{+}]\mathfrak{L}_-[\dot{+}](\mathfrak{L}^{[\circ]} \dot{+} \mathfrak{L}''),$$

where $\mathfrak{L}''$ is a neutral subspace skewly linked to $\mathfrak{L}^{[\circ]}$. It is now obvious that there exists an isometric isomorphism of $\mathfrak{P}_2$ onto the above constructed space $\mathfrak{P}$.

The second part of the assertion follows from [2, Theorem I.10.9]: Consider the $\Delta(\mathfrak{L}, [.,.])$-dimensional subspace $\mathfrak{L}^{[\circ]}$ of $\mathfrak{P}$. Then certainly $\mathfrak{L} \subseteq (\mathfrak{L}^{[\circ]})^{[\perp]}$ and thus

$$\operatorname{codim}_{\mathfrak{P}} \mathfrak{L} \geq \operatorname{codim}_{\mathfrak{P}} (\mathfrak{L}^{[\circ]})^{[\perp]} = \dim \mathfrak{L}^{[\circ]} = \Delta(\mathfrak{L}, [.,.]).$$ □

Let us introduce the correct notion of morphism between almost Pontryagin spaces.

Definition 3.3. Let $(\mathfrak{L}_1, [.,.]_1, \mathcal{O}_1)$ and $(\mathfrak{L}_2, [.,.]_2, \mathcal{O}_2)$ be almost Pontryagin spaces. A map $\phi : \mathfrak{L}_1 \to \mathfrak{L}_2$ is called a morphism between $(\mathfrak{L}_1, [.,.]_1, \mathcal{O}_1)$ and $(\mathfrak{L}_2, [.,.]_2, \mathcal{O}_2)$ if ϕ is linear, isometric, continuous and maps $\mathcal{O}_1$-closed subspaces of $\mathfrak{L}_1$ onto $\mathfrak{O}_2$-closed subspaces of $\mathfrak{L}_2$.

A linear mapping ϕ from an almost Pontryagin space $(\mathfrak{L}_1, [.,.]_1, \mathcal{O}_1)$ onto an almost Pontryagin space $(\mathfrak{L}_2, [.,.]_2, \mathcal{O}_2)$ is called an isomorphism if ϕ is bijective, bicontinuous and isometric with respect to $[.,.]_1$ and $[.,.]_2$.

Let us collect a couple of elementary facts.

Lemma 3.4. *The identity map of an almost Pontryagin space onto itself is an isomorphism. Every isomorphism is a morphism. The composition of two (iso)morphisms is a(n) (iso)morphism.*

Let $\phi : (\mathfrak{L}_1, [.,.]_1, \mathcal{O}_1) \to (\mathfrak{L}_2, [.,.]_2, \mathcal{O}_2)$ be a morphism. Then

(i) $\ker \phi \subseteq \mathfrak{L}^{[\circ]_1}$
(ii) $(\operatorname{ran} \phi, [.,.]_2, \mathcal{O}_2 \cap \operatorname{ran} \phi)$ *is an almost Pontryagin space.*
(iii) *If ϕ is surjective, then ϕ is open.*
(iv) *If ϕ is bijective, then ϕ is an isomorphism.*

If $\mathfrak{K}$ is a closed subspace of an almost Pontryagin space $(\mathfrak{L}, [.,.], \mathcal{O})$, then the inclusion map $\iota : (\mathfrak{K}, [.,.], \mathcal{O} \cap \mathfrak{K}) \to (\mathfrak{L}, [.,.], \mathcal{O})$ is a morphism.

Proof. The first statement of the lemma is obvious.

ad (i): Since ϕ is isometric an element $x \in \ker \phi$ must satisfy

$$[x, y]_1 = [\phi x, \phi y]_2 = 0, \ \ y \in \mathfrak{L},$$

and hence $x \in \mathfrak{L}^{[\circ]_1}$.

ad (ii): Since $\operatorname{ran} \phi$ is $\mathcal{O}_2$-closed, we may refer to Proposition 3.1.

ad (iii): Apply the Open Mapping Theorem.

ad (iv): This is an immediate consequence of the previous assertion.

The last statement follows since $\mathfrak{K}$ is a closed subspace of $\mathfrak{L}$. □

Morphisms can be constructed in a canonical way from subspaces of $\mathfrak{L}^{[\circ]}$.

Proposition 3.5. *Let $(\mathfrak{L}, [.,.], \mathcal{O})$ be an almost Pontryagin space and let $\mathfrak{R}$ be a subspace of $\mathfrak{L}^{[\circ]}$. We consider the factor space $\mathfrak{L}/\mathfrak{R}$ endowed with an inner product $[.,.]_1$ defined by*

$$[x + \mathfrak{R}, y + \mathfrak{R}]_1 = [x, y], \tag{3.1}$$

and with the quotient topology $\mathcal{O}/\mathfrak{R}$. Then $(\mathfrak{L}/\mathfrak{R}, [.,.]_1, \mathcal{O}/\mathfrak{R})$ is an almost Pontryagin space. We have

$$\kappa_-(\mathfrak{L}/\mathfrak{R}, [.,.]_1) = \kappa_-(\mathfrak{L}, [.,.]),$$
$$\Delta(\mathfrak{L}/\mathfrak{R}, [.,.]_1) = \Delta(\mathfrak{L}, [.,.]) - \dim \mathfrak{R}.$$

The quotient map $\pi : \mathfrak{L} \to \mathfrak{L}/\mathfrak{R}$ is a morphism of

$$(\mathfrak{L}, [.,.], \mathcal{O}) \quad \textit{onto} \quad (\mathfrak{L}/\mathfrak{R}, [.,.]_1, \mathcal{O}/\mathfrak{R}).$$

Proof. The inner product on $\mathfrak{L}/\mathfrak{R}$ is well defined by (3.1) because of $\mathfrak{R} \subseteq \mathfrak{L}^{[\circ]}$. Since $\mathcal{O}$ is a Hilbert space topology and $\mathfrak{R}$ is a finite-dimensional and, hence, closed subspace of $\mathfrak{L}$, the topology $\mathcal{O}/\mathfrak{R}$ is also a Hilbert space topology.

Denote by $\pi : \mathfrak{L} \to \mathfrak{L}/\mathfrak{R}$ the canonical projection. Since the inner product on $\mathfrak{L}/\mathfrak{R}$ is defined according to

$$\begin{array}{ccc} (\mathfrak{L}/\mathfrak{R})^2 & \overset{[.,.]_1}{\dashrightarrow} & \mathbb{C}\,, \\ {\scriptstyle \pi\times\pi}\Big\uparrow & \nearrow{\scriptstyle [.,.]} & \\ \mathfrak{L}^2 & & \end{array}$$

and the quotient topology is the final topology with respect to π, we obtain that $[.,.]_1$ is $\mathcal{O}/\mathfrak{R}$-continuous.

Choose an $\mathcal{O}$-closed subspace $\mathfrak{M}$ of $\mathfrak{L}$ such that $\operatorname{codim}_{\mathfrak{L}} \mathfrak{M} < \infty$ and such that $(\mathfrak{M}, [.,.])$ is a Hilbert space. Since $\mathfrak{R}$ is finite-dimensional $\mathfrak{M} + \mathfrak{R}$ is $\mathcal{O}$-closed. Thus $\pi(\mathfrak{M}) = (\mathfrak{M} + \mathfrak{R})/\mathfrak{R}$ satisfies the requirements of axiom (aPS2).

The formulas for the negative index and the degree of degeneracy are obvious.

The quotient map π is clearly linear, isometric and continuous. If $\mathfrak{U}$ is any $\mathcal{O}$-closed subspace of $\mathfrak{L}$, then also $\mathfrak{U} + \mathfrak{R}$ is $\mathcal{O}$-closed and therefore $\pi(\mathfrak{U}) = (\mathfrak{U} + \mathfrak{R})/\mathfrak{R}$ is $\mathcal{O}/\mathfrak{R}$-closed. This shows that π is a morphism. □

We conclude this section with the 1st homomorphy theorem.

Lemma 3.6. *Let $\phi : (\mathfrak{L}_1, [.,.]_1, \mathcal{O}_1) \to (\mathfrak{L}_2, [.,.]_2, \mathcal{O}_2)$ be a morphism. Then ϕ induces an isomorphism $\hat{\phi}$ between $(\mathfrak{L}_1/\ker\phi, [.,.]_1, \mathcal{O}_1/\ker\phi)$ and $(\operatorname{ran}\phi, [.,.]_2, \mathcal{O}_2 \cap \operatorname{ran}\phi)$ with*

$$\begin{array}{ccc} (\mathfrak{L}_1, [.,.]_1, \mathcal{O}_1) & \xrightarrow{\quad\phi\quad} & (\mathfrak{L}_2, [.,.]_2, \mathcal{O}_2) \\ {\scriptstyle \pi}\Big\downarrow & & \Big\uparrow{\scriptstyle \iota} \\ (\mathfrak{L}_1/\ker\phi, [.,.]_1, \mathcal{O}_1/\ker\phi) & \underset{\hat{\phi}}{\dashrightarrow} & (\operatorname{ran}\phi, [.,.]_2, \mathcal{O}_2 \cap \operatorname{ran}\phi) \end{array}$$

Proof. The induced mapping $\hat{\phi}$ is bijective, isometric and continuous. By the Open Mapping Theorem it is also open. Thus it is an isomorphism. □

4. Completions

The generalization of the concept of completion to the almost Pontryagin space setting is a much more delicate topic.

Remark 4.1. Let an inner product space $(\mathfrak{A}, [.,.])$ with $\kappa_-(\mathfrak{A}, [.,.]) = \kappa < \infty$ be given. Then there always exists a Pontryagin space which contains $\mathfrak{A}/\mathfrak{A}^{[\circ]}$ as a dense subspace. We are going to sketch the construction of this so-called completion of $(\mathfrak{A}, [.,.])$ (see, e.g., [4]).

Take any subspace $\mathfrak{M}$ of $\mathfrak{A}$ which is maximal with respect to the property that $(\mathfrak{M}, [.,.])$ is an anti Hilbert space. If $e_1, \dots, e_\kappa$ is an orthonormal basis of $(\mathfrak{M}, -[.,.])$, then

$$P_{\mathfrak{M}} = -[.,e_1]e_1 \cdots - [.,e_\kappa]e_\kappa,$$

is the orthogonal projection of $\mathfrak{A}$ onto $\mathfrak{M}$. By the maximality property of $\mathfrak{M}$ the orthogonal complement $((I-P_{\mathfrak{M}})\mathfrak{A}, [.,.])$ is positive semidefinite. Therefore, setting $J_{\mathfrak{M}} = I - 2P_{\mathfrak{M}}$ we see that $[J_{\mathfrak{M}}.,.] = (.,.)_{\mathfrak{M}}$ is a positive semidefinite product on $\mathfrak{A}$. We then have $(J_{\mathfrak{M}}.,.) = [.,.]_{\mathfrak{M}}$, and $J_{\mathfrak{M}}$ and $[.,.]$ are continuous with respect to the topology induced by $(.,.)_{\mathfrak{M}}$.

Note that if $\mathfrak{M}'$ is another subspace of $\mathfrak{L}$ which is maximal with respect to the property that $(\mathfrak{M}', [.,.])$ is an anti Hilbert space, then $P_{\mathfrak{M}'}$, and hence $J_{\mathfrak{M}'}$ and $(.,.)_{\mathfrak{M}'}$ are continuous with respect to $(.,.)_{\mathfrak{M}}$. By symmetry we obtain that $(.,.)_{\mathfrak{M}}$ and $(.,.)_{\mathfrak{M}'}$ are equivalent scalar products, i.e., there exist $\alpha, \beta > 0$ such that

$$\alpha(x,x)_{\mathfrak{M}} \leq (x,x)_{\mathfrak{M}'} \leq \beta(x,x)_{\mathfrak{M}}, \ x \in \mathfrak{A}. \tag{4.1}$$

This in turn means that these two scalar products induce the same topology $\mathcal{T}$ on $\mathfrak{A}$. In particular, $\mathcal{T}$ is determined by $[.,.]$ and not by a particularly chosen $\mathfrak{M}$.

A completion of $(\mathfrak{A}, [.,.])$ is given by $(\mathfrak{P}, [.,.])$, where $\mathfrak{P}$ is the completion of $\mathfrak{A}/\mathfrak{A}^{[\circ]}$ with respect to $(.,.)_{\mathfrak{M}}$. Note that

$$\mathfrak{A}^{(\circ)_{\mathfrak{M}}} = \{x \in \mathfrak{A} : (x,x)_{\mathfrak{M}} = 0\} = \{x \in \mathfrak{A} : [x,y] = 0 \text{ for all } y \in \mathfrak{A}\} = \mathfrak{A}^{[\circ]},$$

and that $\mathfrak{A}^{[\circ]} \cap \mathfrak{M} = \{0\}$.

After factoring out $\mathfrak{A}^{[\circ]}$ by continuity we can extend $P_{\mathfrak{M}}, J_{\mathfrak{M}}, [.,.]$ to $\mathfrak{P}$. Then we have $J_{\mathfrak{M}} = I - 2P_{\mathfrak{M}}$ and $[.,.] = (J_{\mathfrak{M}}.,.)_{\mathfrak{M}}$ also on $\mathfrak{P}$. The extension $P_{\mathfrak{M}}$ is the orthogonal projection of $\mathfrak{P}$ onto $\mathfrak{M}/\mathfrak{A}^{[\circ]}$. It is straightforward to check that

$$\mathfrak{P} = P_{\mathfrak{M}}\mathfrak{P}[\dot{+}]((I - P_{\mathfrak{M}})\mathfrak{P}) \tag{4.2}$$

is a fundamental decomposition of $(\mathfrak{P}, [.,.])$. Therefore, $(\mathfrak{P}, [.,.])$ is a Pontryagin space and by (4.1) its construction does not depend on the chosen space $\mathfrak{M}$. Moreover, it is the unique Pontryagin space (up to isomorphisms) which contains $\mathfrak{A}/\mathfrak{A}^{[\circ]}$ such that $[.,.]$ on $\mathfrak{P}$ is a continuation of $[.,.]$ on $\mathfrak{A}/\mathfrak{A}^{[\circ]}$.

To see this let $(\mathfrak{P}', [.,.])$ be another such Pontryagin space, and let $\mathfrak{P}' = \mathfrak{P}_-[\dot{+}]\mathfrak{P}_+$ be a fundamental decomposition of $\mathfrak{P}'$. By a density argument we find a subspace $\mathfrak{M}$ of $\mathfrak{A}$ with the same dimension as $\mathfrak{P}_-$ such that $\mathfrak{M}/\mathfrak{A}^{[\circ]}$ is sufficiently close to $\mathfrak{P}_-$ in order that $(\mathfrak{M}, [.,.])$ is an anti Hilbert space. It follows that $\mathfrak{M}$ is maximal with respect to this property. Let P be the orthogonal projection of $\mathfrak{P}'$ onto $\mathfrak{M}/\mathfrak{A}^{[\circ]}$, and let $(.,.)$ be the Hilbert space inner product $[(I-2P).,.]$ on $\mathfrak{P}$.

If $(\mathfrak{P}, [.,.])$ is the completion as constructed above, then the identity ϕ on $\mathfrak{A}/\mathfrak{A}^{[\circ]}$ is a $[.,.]$-isometric linear mapping from a dense subspace of $\mathfrak{P}$ onto a dense subspace of $\mathfrak{P}'$. By construction $\phi P_{\mathfrak{M}} = P\phi$. Hence ϕ is isometric with respect to $(.,.)_{\mathfrak{M}}$ and $(.,.)$. As both induce the topology on the respective spaces $\mathfrak{P}$ and $\mathfrak{P}'$ we see that ϕ can be extended to an isomorphism from $\mathfrak{P}$ onto $\mathfrak{P}'$.

Definition 4.2. Let $(\mathfrak{A}, [.,.])$ be an inner product space such that $\kappa_-(\mathfrak{A}, [.,.]) = \kappa < \infty$. An almost Pontryagin space with a linear mapping $((\mathfrak{L}, [.,.], \mathcal{O}), \iota)$ is called a completion of $\mathfrak{A}$, if ι is an isometric mapping (with respect to $[.,.]$) from $\mathfrak{A}$ onto a dense subspace $\iota(\mathfrak{A})$ of $\mathfrak{L}$.

Two completions $((\mathfrak{L}_1, [.,.]_1, \mathcal{O}_1), \iota_1)$ and $((\mathfrak{L}_2, [.,.]_2, \mathcal{O}_2), \iota_2)$ are called isomorphic if there exists an isomorphism ϕ from $(\mathfrak{L}_1, [.,.]_1, \mathcal{O}_1)$ onto $(\mathfrak{L}_2, [.,.]_2, \mathcal{O}_2)$ such that $\phi \circ \iota_1 = \iota_2$.

Remark 4.3. We saw above that, up to isomorphism, there always exists a unique Pontryagin space which is a completion of $(\mathfrak{A}, [.,.])$.

If we allow the almost Pontryagin space of a completion $((\mathfrak{L}, [.,.], \mathcal{O}), \iota)$ to be degenerated, i.e., $\Delta(\mathfrak{L}, [.,.]) > 0$, then $(\mathfrak{L}, [.,.], \mathcal{O})$ is not uniquely determined if we assume $\dim \mathfrak{A}/\mathfrak{A}^{[\circ]} = \infty$. This can be derived immediately from Proposition 2.8.

For $\dim \mathfrak{A}/\mathfrak{A}^{[\circ]} = \infty$ it follows from the subsequent result that for any $\Delta \geq 0$ there exists a completion $((\mathfrak{L}, [.,.], \mathcal{O}), \iota)$ of $(\mathfrak{A}, [.,.])$ such that $\Delta(\mathfrak{L}, [.,.]) = \Delta$. Also for fixed Δ Proposition 2.8 shows that $(\mathfrak{L}, [.,.], \mathcal{O})$ is not uniquely determined.

Proposition 4.4. *Let $(\mathfrak{A}, [.,.])$ be an inner product space with $\kappa_-(\mathfrak{A}, [.,.]) = \kappa < \infty$, and let $\mathcal{T}$ be the topology determined by $[.,.]$ on $\mathfrak{A}$ (see Remark 4.1).*

If $f_1, \dots, f_\Delta$ are complex linear functionals on $\mathfrak{A}$ such that no linear combination of them is continuous with respect to $\mathcal{T}$, then there exists an (up to isomorphic copies) unique completion $((\mathfrak{L}, [.,.], \mathcal{O}), \iota)$ with $\Delta(\mathfrak{L}, [.,.]) = \Delta$ such that $f_1, \dots, f_\Delta$ are continuous with respect to $\iota^{-1}(\mathcal{O})$.

Proof. The construction made in this proof stems from [6].

Let $(\mathfrak{P}, [.,.])$ be the unique Pontryagin space completion of $(\mathfrak{A}, [.,.])$, i.e., the completion with respect to $\mathcal{T}$. Let $(.,.)_\mathfrak{M}$ be the Hilbert space inner product on $\mathfrak{P}$ from Remark 4.1 constructed with the help of a subspace $\mathfrak{M}$ of $\mathfrak{A}$ being maximal with respect to the property that $(\mathfrak{M}, [.,.])$ is an anti Hilbert space. We define

$$\mathfrak{L} = \mathfrak{P} \times \mathbb{C}^\Delta,$$

and provide $\mathfrak{L}$ with the inner product $(.,.)$ such that $(.,.)$ coincides with $(.,.)_\mathfrak{M}$ on $\mathfrak{P}$ and with the Euclidean product on $\mathbb{C}^\Delta$, and such that $\mathfrak{L} = \mathfrak{P}(\dot{+})\mathbb{C}^\Delta$. Let $[.,.]$ be defined on $\mathfrak{L}$ by

$$[(x;\xi), (y;\eta)] = [x, y].$$

By definition $(\mathfrak{L}, [.,.], \mathcal{O}_{(.,.)})$ is an almost Pontryagin space. Hereby $\mathcal{O}_{(.,.)}$ is the topology induced by $(.,.)$ on $\mathfrak{L}$.

Now we embed $\mathfrak{A}$ in $\mathfrak{L}$ via the mapping ι

$$\iota(x) = (x + \mathfrak{A}^{[\circ]}; (f_1(x), \dots, f_\Delta(x))).$$

Then $\iota(\mathfrak{A})$ is dense in $\mathfrak{L}$. In fact, if not, then we could find $(y;\eta) \in \mathfrak{L}$ such that $(y;\eta)(\perp)\iota(\mathfrak{A})$. It would follow that

$$(x + \mathfrak{A}^{[\circ]}, -y) = \sum_{j=1}^{\Delta} \bar{\eta}_j f_j(x), \;\; x \in \mathfrak{A},$$

and, therefore, the right-hand side would be continuous with respect to $\mathcal{T}$. By assumption $\eta = 0$ and further $y(\perp)\mathfrak{A}$ in $\mathfrak{P}$. This is not possible as $\mathfrak{A}$ is dense in $\mathfrak{P}$.

The mapping ι is isometric with respect to $[.,.]$. Thus by defining $\mathcal{O} = \mathcal{O}_{(.,.)}$, $((\mathfrak{L}, [.,.], \mathcal{O}), \iota)$ is a completion of $(\mathfrak{A}, [.,.])$. By the definition of ι the functionals $f_1, \dots, f_\Delta$ are continuous with respect to $\iota^{-1}(\mathcal{O})$.

Assume now that $((\mathfrak{L}', [.,.]', \mathcal{O}'), \iota')$ is another completion of $(\mathfrak{A}, [.,.])$ such that $\Delta(\mathfrak{L}', [.,.]') = \Delta$ and such that $f_1, \dots, f_\Delta$ are continuous with respect to $\iota'^{-1}(\mathcal{O}')$. Let $(.,.)'$ be a Hilbert space scalar product on $\mathfrak{L}'$ which induces $\mathcal{O}'$. By elementary considerations from the theory of locally convex vector spaces we can factor $f_1, \dots, f_\Delta$ through the isotropic part $\mathfrak{A}^{(\circ)'}$ of $\mathfrak{A}$ with respect to $(\iota(.), \iota(.))'$. Note that $\mathfrak{A}^{(\circ)'}$ is also the set of all points in $\mathfrak{A}$ which have exactly the same neighborhoods as 0 with respect to the topology $\iota'^{-1}(\mathcal{O}')$.

Clearly, $(\mathfrak{L}', (.,.)')$ is isomorphic to the completion of $\mathfrak{A}/\mathfrak{A}^{(\circ)'}$ with respect to $(\iota(.), \iota(.))'$. Hence by continuation to the completion we obtain continuous linear functionals $g_1, \dots, g_\Delta$ on $(\mathfrak{L}', [.,.]', \mathcal{O}')$ such that $f_1 = g_1 \circ \iota', \dots, f_\Delta = g_\Delta \circ \iota'$.

By Proposition 3.5 $(\mathfrak{L}'/\mathfrak{L}'^{[\circ]'}, [.,.]')$ is a Pontryagin space. We denote by π the factorization mapping. As $(\mathfrak{A}/\mathfrak{A}^{[\circ]}, [.,.])$ is isometrically embedded by $\pi \circ \iota'$ in this Pontryagin space Remark 4.1 shows that $(\mathfrak{L}'/\mathfrak{L}'^{[\circ]'}, [.,.]')$ is an isomorphic copy of $(\mathfrak{P}, [.,.])$. Let $\phi : \mathfrak{L}'/\mathfrak{L}'^{[\circ]'} \to \mathfrak{P}$ be this isomorphism, which satisfies $\phi \circ \pi \circ \iota' = \mathrm{id}_{\mathfrak{A}}$.

Let $0 \neq x \in \mathfrak{L}'^{[\circ]'}$ be such that $g_1(x) = \dots = g_\Delta(x) = 0$. By elementary linear algebra we find a non-trivial linear combination g of the functionals $g_1, \dots, g_\Delta$ which vanishes on $\mathfrak{L}'^{[\circ]'}$. Hence, we find a functional f on $\mathfrak{P}$ such that $g = f \circ \phi \circ \pi$. But then $a \mapsto f(a + \mathfrak{A}^{[\circ]})$ is a non-trivial linear combination of $f_1, \dots, f_\Delta$ which is continuous with respect to $\mathcal{T}$. By assumption this is ruled out. Thus the intersection of the kernels of g_j, $j = 1, \dots, \Delta$, has no point in common with $\mathfrak{L}'^{[\circ]'}$ except of 0. Since the intersection of Δ hyperplanes has codimension at most Δ, we have

$$\mathfrak{L}'^{[\circ]'} \dot{+} (\ker(g_1) \cap \dots \cap \ker(g_\Delta)) = \mathfrak{L}', \tag{4.3}$$

and see that the mapping

$$\varphi : \mathfrak{L}' \to \mathfrak{L}, \ x \mapsto (\phi \circ \pi(x); (g_1(x), \dots, g_\Delta(x))),$$

is bijective. Moreover, φ is isometrically with respect to $[.,.]$ and satisfies $\varphi \circ \iota' = \iota$. Since in the decomposition (4.3) all subspaces are closed, the Open Mapping Theorem implies that φ is bicontinuous with respect to $\mathcal{O}'$ and $\mathcal{O}$. □

Remark 4.5. With the notation from Proposition 4.4

$$\langle .,. \rangle = (.,.)_{\mathfrak{M}} + \sum_{j=1}^{\Delta} f_j(.)\bar{f}_j(.), \tag{4.4}$$

is a non-negative inner product on $\mathfrak{A}$. It is easy to see that ι induces an isomorphism from the completion of $(\mathfrak{A}/\mathfrak{A}^{\langle\circ\rangle}, \langle .,. \rangle)$ onto $(\mathfrak{L}, (.,.))$. In particular, $\mathcal{O}_{\langle .,. \rangle} = \iota^{-1}(\mathcal{O})$.

The completion constructed in Proposition 4.4 appeared in implicit forms already in various papers. See for example [7].

Definition 4.6. We call the completion of $(\mathfrak{A}, [.,.])$ constructed in Proposition 4.4 the completion of $(\mathfrak{A}, [.,.], (f_i)_{i=1,\dots,\Delta})$.

Corollary 4.7. *Let $(\mathfrak{A}, [.,.])$ be an inner product space with $\kappa_-(\mathfrak{A}, [.,.]) = \kappa < \infty$, and let $\mathcal{T}$ be the topology determined by $[.,.]$ on $\mathfrak{A}$ (see Remark 4.1).*

Let $(f_i)_{i=1,\dots,\Delta}$ and $(f_i')_{i=1,\dots,\Delta}$ be two sets of complex linear functionals on $\mathfrak{A}$ such that no linear combination of $(f_i)_{i=1,\dots,\Delta}$ and no linear combination of $(f_i')_{i=1,\dots,\Delta}$ is continuous with respect to $\mathcal{T}$.

The completion of $(\mathfrak{A}, [.,.], (f_i)_{i=1,\dots,\Delta})$ is isomorphic to the completion of $(\mathfrak{A}, [.,.], (f_i')_{i=1,\dots,\Delta})$ if and only if the functionals $f_1', \dots, f_\Delta'$ are continuous with respect to the topology induced by $\langle .,. \rangle$ defined in (4.4) on $\mathfrak{A}$.

Proof. We denote by $((\mathfrak{L}, [.,.], \mathcal{O}), \iota)$ and $((\mathfrak{L}', [.,.]', \mathcal{O}'), \iota')$ the completions of the triplets $(\mathfrak{A}, [.,.], (f_i)_{i=1,\dots,\Delta})$ and $(\mathfrak{A}, [.,.], (f_i')_{i=1,\dots,\Delta})$, respectively. Moreover, let $(g_i)_{i=1,\dots,\Delta}$ and $(g_i')_{i=1,\dots,\Delta}$ be the continuous linear functionals on $\mathfrak{L}$ and $\mathfrak{L}'$, respectively, such that $f_i = g_i \circ \iota$ and $f_i' = g_i' \circ \iota'$, respectively.

If the two completions are isomorphic by the isomorphism $\phi : (\mathfrak{L}, [.,.], \mathcal{O}) \to (\mathfrak{L}', [.,.]', \mathcal{O}')$, then $g_i' \circ \phi$, $i = 1, \dots, \Delta$, are continuous functionals on $(\mathfrak{L}, [.,.], \mathcal{O})$. By Remark 4.5 $f_i' = g_i' \circ \iota' = g_i' \circ \phi \circ \iota$ is continuous on $\mathfrak{A}$ with respect to the topology induced by $\langle .,. \rangle$.

Conversely, if $f_1', \dots, f_\Delta'$ are continuous with respect to the topology induced by $\langle .,. \rangle$, then by continuation to the completion we obtain continuous linear functionals $(h_i')_{i=1,\dots,\Delta}$ on $(\mathfrak{L}, [.,.], \mathcal{O})$ such that $f_i' = h_i' \circ \iota$. By the uniqueness assertion in Proposition 4.4 $((\mathfrak{L}, [.,.], \mathcal{O}), \iota)$ is also a completion of $(\mathfrak{A}, [.,.], (f_i')_{i=1,\dots,\Delta})$. □

5. Almost reproducing kernel Pontryagin spaces

Objects of intensive studies are the so-called reproducing kernel Pontryagin spaces. These are Pontryagin spaces $(\mathfrak{P}, [.,.])$ which consist of functions F mapping some set M into $\mathbb{C}$ such that there exist $K(.,t) \in \mathfrak{P}$, $t \in M$, with

$$F(t) = [F, K(.,t)], \quad F \in \mathfrak{P}, t \in M. \tag{5.1}$$

An equivalent definition of reproducing kernel Pontryagin spaces is the assumption that $(\mathfrak{P}, [.,.])$ consists of complex valued functions on a set M such that the point evaluations are continuous at all points of M.

The first approach to reproducing kernel Pontryagin spaces does not have an immediate generalization to almost Pontryagin spaces but the second does.

Definition 5.1. Let $(\mathfrak{L}, [.,.], \mathcal{O})$ be an almost Pontryagin space, and assume that the elements of $\mathfrak{L}$ are complex valued functions on a set M. This space is a called an almost reproducing kernel Pontryagin spaces on M, if for any $t \in M$ the linear functional

$$f_t : F \mapsto F(t), \ F \in \mathfrak{L},$$

is continuous on $\mathfrak{L}$ with respect to $\mathcal{O}$.

Remark 5.2. As the elements of $\mathfrak{L}$ are functions we see that the family $(f_t)_{t\in M}$ of point evaluation functionals is point separating. Hence Proposition 2.9 yields the uniqueness of the topology $\mathcal{O}$ for which the functionals f_t, $t \in M$, are continuous. Consequently, we are going to skip the topology and write almost reproducing kernel Pontryagin spaces as pairs $(\mathfrak{L}, [.,.])$.

A major setback to the study of almost reproducing kernel Pontryagin spaces is the fact that in the case $\Delta(\mathfrak{L}, [.,.]) > 0$ we do not find a reproducing kernel $K(s,t)$ which satisfies (5.1). However, we do have the following

Proposition 5.3. *Let $(\mathfrak{L}, [.,.])$ be an almost reproducing kernel Pontryagin space on a set M and put $\Delta = \Delta(\mathfrak{L}, [.,.])$. Moreover, let N be a separating subset of M, i.e., assume that the family $(f_t)_{t\in N}$ is point separating. Then there exist $t_1, \dots, t_\Delta \in N$, $c \in \mathbb{R}$, and $R(.,t) \in \mathfrak{L}$ such that*

$$F(t) = [F, R(.,t)] + c\big(F(t_1)R(t,t_1) + \cdots + F(t_\Delta)R(t,t_\Delta)\big), \ F \in \mathfrak{L}, \ t \in M.$$

Proof. The number Δ is by definition the dimension of $\mathfrak{L}^{[\circ]} = \ker G$, where $G = G_{(.,.)}$ is the Gram operator with respect to a Hilbert space product $(.,.)$ inducing the topology of $(\mathfrak{L}, [.,.])$. By Corollary 2.4 we may choose $(.,.)$ such that $G = I + L$, where L is a selfadjoint finite rank operator.

Because of the assumption on N by induction one can easily show the existence of points $t_1, \dots, t_\Delta \in N$, such that $h \in \mathfrak{L}^{[\circ]}$ and $h(t_j) = 0$, $j = 1, \dots, \Delta$, implies $h = 0$.

Because of the continuity of point evaluations we find elements $K(.,t) \in \mathfrak{L}$ with

$$F(t) = (F, K(.,t)), \ F \in \mathfrak{L}.$$

We define the following selfadjoint operator H of finite rank on $\mathfrak{L}$

$$H(F) = \sum_{j=1}^{\Delta} F(t_j)K(.,t_j).$$

Let $\mathfrak{K} = \ker(H) \cap \ker(L)$, then $\mathfrak{K}^{(\perp)}$ is finite-dimensional since the selfadjoint operators H and L are of finite rank, and $\mathfrak{K}^{(\perp)}$ contains the range of L and H. For $z \in \mathbb{C}$ it follows that the restriction of the operator $I + L + zH$ onto $\mathfrak{K}$ is equal to the identity. Hence $I + L + zH$ is invertible on $\mathfrak{L}$ if and only if it is invertible on $\mathfrak{K}^{(\perp)}$. To show that $I + L + zH$ is invertible for $z = i$, let $(I + L + iH)F = 0$. Then $(HF, F) = 0 = ((I+L)F, F)$, and the form of H implies that $F(t_j) = 0$, $j = 1, \dots, \Delta$. It follows that $H(F) = 0$, and hence $(I + L)F = 0$, or $F \in \mathfrak{L}^{[\circ]}$ as $I + L$ is the Gram operator. The definition of the points $t_1, \dots, t_\Delta$ implies that $F = 0$, that is, the operator $(I + L + iH)$ is invertible.

Thus $\det((I+L+zH)|_{\mathfrak{K}[\perp]})$ is not identically zero, and therefore has only a discrete zero set. In particular, we find $c \in \mathbb{R}$ such that $(I+L+cH)$ is invertible. Now set

$$R(.,t) = (I+L+cH)^{-1}K(.,t).$$

Note that because of the selfadjointness of $I+L+cH$,

$$R(s,t) = (R(.,t), K(.,s)) = (K(.,t), R(.,s)) = \overline{R(t,s)}.$$

For $F \in \mathfrak{L}$ and $t \in M$ we have

$$[F, R(.,t)] = ((I+L+cH)F, R(.,t)) - c(HF, R(.,t))$$
$$= F(t) - c\sum_{j=1}^{\Delta} F(t_j)R(t,t_j).$$

□

6. Examples of almost Pontryagin spaces

As the first topic of this section we are going to sketch the continuation problem for hermitian functions with finitely many negative squares on intervals $[-2a, 2a]$ to the whole real axis. We will meet inner product spaces and completions in the sense of Section 4. Taking into account also a possible degeneracy of this completion one obtains a refinement of classical results on the number of all possible extensions of the given hermitian functions with finitely many negative squares to $\mathbb{R}$. For a complete treatment of this topic see [7].

Definition 6.1. Let $a > 0$ be a real number, and assume that $f : [-2a, 2a] \to \mathbb{C}$ is a continuous function. We say that f is hermitian if it satisfies $f(-t) = \overline{f(t)}$, $t \in [-2a, 2a]$, and f is said to be hermitian with $\kappa(\in \mathbb{N} \cup \{0\})$ many negative squares if the kernel $f(t-s)$, $s,t \in (-a,a)$, has κ negative squares. The set of all such functions we denote by $\mathcal{P}_{\kappa,a}$.

By $\mathcal{P}_\kappa$ we denote the set of all continuous hermitian functions with κ negative squares on $\mathbb{R}$, i.e., $f(t-s)$, $s,t \in \mathbb{R}$, has κ negative squares.

For $\kappa = 0$ the function f is called positive definite.

The continuation problem is to find for given $f \in \mathcal{P}_{\kappa,a}$ and $\tilde{\kappa} \in \mathbb{N} \cup \{0\}$ all possible extensions $\tilde{f}$ of f to the whole real axis such that $\tilde{f} \in \mathcal{P}_{\tilde{\kappa}}$. Trivially, by the definition of the respective classes a necessary condition for the existence of such extensions is $\kappa \leq \tilde{\kappa}$. The following classical result can be found for example in [3].

Theorem 6.2. *Let $f \in \mathcal{P}_{\kappa,a}$. Then either f has exactly one extension belonging to $\mathcal{P}_\kappa$, or it has infinitely many extensions in $\mathcal{P}_\kappa$. In the latter case f also has infinitely many extensions in $\mathcal{P}_{\tilde{\kappa}}$ for every $\tilde{\kappa} \geq \kappa$.*

This result originates from the following operator theoretic considerations. First let $(\mathfrak{P}(f), [.,.])$ be the reproducing kernel Pontryagin space on $(-a,a)$ having $k(s,t) = f(s-t)$ as its reproducing kernel. As we assume $f \in \mathcal{P}_{\kappa,a}$ the degree of negativity of $(\mathfrak{P}(f), [.,.])$ is κ. Clearly, $(\mathfrak{P}(f), [.,.])$ is the completion of $(\mathfrak{A}(f), [.,.])$ where $\mathfrak{A}(f)$ is the linear hull of $\{k(.,t) : t \in (-a,a)\}$.

Moreover, a certain differential operator $S(f)$ is constructed on $(\mathfrak{P}(f), [.,.])$. This operator is symmetric and densely defined. Its defect elements are given by

$$\ker (S(f)^* - z) = e^{izs}, \; z \in \mathbb{C} \setminus \mathbb{R},$$

as a function of $s \in (-a, a)$ if they belong to $\mathfrak{P}(f)$. Thus $S(f)$ has either defect indices $(1, 1)$ or $(0, 0)$ depending on whether e^{izs} belongs to this space or not.

A crucial fact in verifying Theorem 6.2 is that all extensions of f belonging to $\mathcal{P}_\kappa$ correspond bijectively to all $\mathfrak{P}(f)$-minimal selfadjoint extensions A of $S(f)$ in a possibly larger Pontryagin space $\tilde{\mathfrak{P}} \supseteq \mathfrak{P}(f)$ with $\kappa_-(\tilde{\mathfrak{P}}, [.,.]) = \kappa$. Hereby $\mathfrak{P}(f)$-minimal means

$$\operatorname{cls}(\mathfrak{P}(f) \cup \{(A - z)^{-1}x : x \in \mathfrak{P}(f), \; z \in \rho(A)\}) = \tilde{\mathfrak{P}}.$$

Hence, in the case that $S(f)$ has defect index $(0, 0)$ or, equivalently, that $S(f)$ is selfadjoint there are no $\mathfrak{P}(f)$-minimal selfadjoint extensions of $S(f)$ other than $S(f)$ itself. Therefore, f has exactly one extension in $\mathcal{P}_\kappa$.

If $S(f)$ has defect index $(1, 1)$, then there are infinitely many $\mathfrak{P}(f)$-minimal selfadjoint extensions A of $S(f)$ and, hence, infinitely many extensions in $\mathcal{P}_\kappa$. Moreover, in this case the extensions of f in $\mathcal{P}_{\tilde{\kappa}}$ for $\tilde{\kappa} \geq \kappa$ correspond bijectively to all $\mathfrak{P}(f)$-minimal selfadjoint extensions A of $S(f)$ in a Pontryagin space $\tilde{\mathfrak{P}} \supseteq \mathfrak{P}(f)$ with $\kappa_-(\tilde{\mathfrak{P}}, [.,.]) = \tilde{\kappa}$, and there are also infinitely many of them for an arbitrary $\tilde{\kappa} \geq \kappa$.

Theorem 6.2 seems to give a sufficiently satisfactory answer to the continuation problem. But as some examples show it can happen that f has exactly one extension in $\mathcal{P}_\kappa$ but infinitely many extensions in $\mathcal{P}_{\tilde{\kappa}}$ for some $\tilde{\kappa} > \kappa$. How does this fit in with the operator theoretic approach mentioned above?

Here almost Pontryagin spaces come into play. In the case that $S(f)$ has defect index $(1, 1)$ the fact that e^{izs}, $z \in \mathbb{C} \setminus \mathbb{R}$ belongs to $\mathfrak{P}(f)$ can be reformulated by saying that

$$F_z : \sum_j \alpha_j k(., t_j) \mapsto \sum_j \alpha_j e^{izt_j}$$

are continuous linear functionals on $(\mathfrak{A}, [.,.])$ for all $z \in \mathbb{C} \setminus \mathbb{R}$.

If f has a unique extension $f_0 \in \mathcal{P}_\kappa$, i.e., $S(f)$ has defect $(0, 0)$, then these functionals are not continuous. But it can happen that by refining the topology on $(\mathfrak{A}, [.,.])$ by finitely many functionals $F_{z_1}, \dots, F_{z_\Delta}$, $z_j \in \mathbb{C} \setminus \mathbb{R}$ as in Remark 4.5 we obtain a topology $\mathcal{O}$ on $(\mathfrak{A}, [.,.])$ such that all functionals F_z, $z \in \mathbb{C} \setminus \mathbb{R}$, are continuous. Hereby let $\Delta \in \mathbb{N}$ always be chosen such that $F_{z_1}, \dots, F_{z_\Delta}$ is a minimal set of functionals such that all the functionals F_z, $z \in \mathbb{C} \setminus \mathbb{R}$, are continuous with respect to $\mathcal{O}$. Then no linear combination of $F_{z_1}, \dots, F_{z_\Delta}$ is continuous with respect to the topology induced by $[.,.]$ on $\mathfrak{A}$ as in Remark 4.1.

Now let $((\mathfrak{Q}(f), [.,.], \mathcal{O}(f)), \iota)$ be the completion of $(\mathfrak{A}, [.,.], (F_{z_j})_{j=1,\dots,\Delta})$. On $(\mathfrak{Q}(f), [.,.], \mathcal{O}(f))$ one can find a symmetric operator $T(f)$ with defect index $(1, 1)$. For the concept of symmetric operators on almost Pontryagin spaces see [8]. In that paper almost Pontryagin spaces were always considered as degenerate subspaces of Pontryagin spaces, and they were not yet called almost Pontryagin spaces. Similarly as for $S(f)$ the extensions $\tilde{f} \in \mathcal{P}_{\tilde{\kappa}}$ of f which differ from f_0 correspond bijectively to all $\mathfrak{Q}(f)$-minimal selfadjoint extensions A of $T(f)$ in a Pontryagin space $\tilde{\mathfrak{P}} \supseteq \mathfrak{Q}(f)$ with $\kappa_-(\tilde{\mathfrak{P}}, [.,.]) = \tilde{\kappa}$.

Since every Pontryagin space $\tilde{\mathfrak{P}}$ which contains $\mathfrak{Q}$ must satisfy $\kappa_-(\tilde{\mathfrak{P}}, [.,.]) \geq \Delta(\mathfrak{Q}, [.,.]) + \kappa_-(\tilde{\mathfrak{Q}}, [.,.])$, there exist extensions $\tilde{f} \in \mathcal{P}_{\tilde{\kappa}}$, $\tilde{f} \neq f_0$ of f if only if $\tilde{\kappa} \geq \Delta + \kappa$. In fact, for these $\tilde{\kappa}$ there always exist infinitely many extensions in $\mathcal{P}_{\tilde{\kappa}}$. These considerations yield the following refinement of Theorem 6.2.

Theorem 6.3. *Let $f \in \mathcal{P}_{\kappa,a}$. Then there exists $\Delta \in \{0\} \cup \mathbb{N} \cup \{\infty\}$ such that*

- *If $\Delta > 0$, then f has a unique extension in $\mathcal{P}_\kappa$.*
- *f has no extensions in $\mathcal{P}_{\tilde{\kappa}}$ for $\kappa < \tilde{\kappa} < \Delta + \kappa$.*
- *f has infinitely many extensions in $\mathcal{P}_{\tilde{\kappa}}$ for $\tilde{\kappa} \geq \Delta + \kappa$.*

As a second topic in the present section we give an example of an interesting class of almost reproducing kernel Pontryagin spaces. In fact, we are going to consider the indefinite generalization of Hilbert space of entire functions introduced by Louis de Branges (see [1], [9], [10], [11]).

Definition 6.4. An inner product space $(\mathfrak{L}, [.,.])$ is called a de Branges space (dB-space) if the following three axioms hold true:

(dB1) $(\mathfrak{L}, [.,.])$ is an almost reproducing kernel Pontryagin space on $\mathbb{C}$ consisting of entire functions.

(dB2) If $F \in \mathfrak{L}$ then $F^\# \in \mathfrak{L}$, where $F^\#(z) = \overline{F(\bar{z})}$. Moreover,

$$[F^\#, G^\#] = [G, F].$$

(dB3) If $F \in \mathfrak{L}$ and $z_0 \in \mathbb{C} \setminus \mathbb{R}$ with $F(z_0) = 0$, then

$$\frac{z - \bar{z}_0}{z - z_0} F(z) \in \mathfrak{L},$$

as a function of z. Moreover, if also $G \in \mathfrak{L}$ with $G(z_0) = 0$, then

$$\Big[\frac{z - \bar{z}_0}{z - z_0} F(z), \frac{z - \bar{z}_0}{z - z_0} G(z)\Big] = [F, G].$$

In many cases one can assume that a dB-space also satisfies

$$\text{For all } t \in \mathbb{R} \text{ there exists } F \in \mathfrak{L} \text{ such that } F(t) \neq 0. \tag{6.1}$$

One of the main results about dB-spaces is that the set of all admissible dB-subspaces of a given dB-space is totally ordered. To explain this in more detail, let us start with a dB-space satisfying (6.1). We call a subspace $\mathfrak{K}$ of $\mathfrak{L}$ a dB-subspace of $(\mathfrak{L}, [.,.])$ if $(\mathfrak{K}, [.,.])$ itself is a dB-space. It is called an admissible dB-subspace if $(\mathfrak{K}, [.,.])$ also satisfies (6.1). The following result originates from [1] and was generalized to the indefinite situation in [9].

Theorem 6.5. *Let $(\mathfrak{L}, [.,.])$ be a dB-space satisfying* (6.1). *Then the set of all admissible dB-subspaces is totally ordered with respect to inclusion, i.e., if $\mathfrak{P}$ and $\mathfrak{Q}$ are two admissible dB-subspaces of $(\mathfrak{L}, [.,.])$, then $\mathfrak{P} \subseteq \mathfrak{Q}$ or $\mathfrak{Q} \subseteq \mathfrak{P}$.*

One may think of the degenerate members of the chain of admissible dB-subspaces of a given dB-space as singularities. Thus it is desirable not to have too many of this kind. In [9] the following result was obtained.

Theorem 6.6. *With the same assumptions as in Theorem* 6.5 *the number of admissible dB-subspaces* $\mathfrak{K}$ *of* $(\mathfrak{L}, [.,.])$ *with* $\Delta(\mathfrak{K}, [.,.]) > 0$ *is finite.*

The presence of singularities is exactly what distinguishes the classical – positive definite – case from the indefinite situation. Thus, to obtain a thorough understanding of the structure of an indefinite dB-space, it is inevitable to deal with degenerated spaces.

References

[1] L. de Branges, *Hilbert spaces of entire functions*, Prentice-Hall, London 1968

[2] J. Bognár, *Indefinite inner product spaces*, Springer Verlag, Berlin 1974

[3] M. Grossmann, H. Langer, *Über indexerhaltende Erweiterungen eines hermiteschen Operators im Pontrjaginraum*, Math. Nachr. **64**(1974), 289–317

[4] I.S. Iohvidov, M.G. Kreĭn, *Spectral theory of operators in spaces with indefinite metric I.*, Trudy Moskov. Mat. Obšč. **5** (1956), 367–432

[5] I.S. Iohvidov, M. G. Kreĭn, *Spectral theory of operators in spaces with indefinite metric II.*, Trudy Moskov. Mat. Obšč. **8** (1959), 413–496

[6] P. Jonas, H. Langer, B. Textorius, *Models and unitary equivalence of cyclic selfadjoint operators in Pontryagin spaces*, Oper. Theory Adv. Appl. **59**(1992), 252–284

[7] M. Kaltenbäck, H. Woracek, *On extensions of Hermitian functions with a finite number of negative squares*, J. Operator Theory **40** (1998), no. 1, 147–183

[8] M. Kaltenbäck, H. Woracek, *The Krein formula for generalized resolvents in degenerated inner product spaces*, Monatsh. Math. **127** (1999), no. 2, 119–140

[9] M. Kaltenbäck, H. Woracek, *Pontryagin spaces of entire functions I*, Integral Equations Operator Theory **33**(1999), 34–97

[10] M. Kaltenbäck, H. Woracek, *Pontryagin spaces of entire functions II*, Integral Equations Operator Theory **33**(1999), 305–380

[11] M. Kaltenbäck, H. Woracek, *Pontryagin spaces of entire functions III*, Acta Sci. Math. (Szeged) **69** (2003), 241–310

M. Kaltenbäck and H. Woracek
Institut für Analysis und Scientific Computing
Technische Universität Wien
Wiedner Hauptstraße 8–10
A-1040 Wien, Austria
e-mail: michael.kaltenbaeck@tuwien.ac.at
e-mail: harald.woracek@tuwien.ac.at

H. Winkler
Faculteit der Wiskunde en Natuurwetenschappen
Rijksuniversiteit Groningen
NL-9700 AV Groningen, The Netherlands
e-mail: winkler@math.rug.nl

Operator Theory:
Advances and Applications, Vol. 160, 273–298

Multivariable ρ-contractions

Dmitry S. Kalyuzhnyĭ-Verbovetzkiĭ

Dedicated to Israel Gohberg on his 75th birthday

Abstract. We suggest a new version of the notion of ρ-dilation ($\rho > 0$) of an N-tuple $\mathbf{A} = (A_1, \dots, A_N)$ of bounded linear operators on a common Hilbert space. We say that $\mathbf{A}$ belongs to the class $C_{\rho,N}$ if $\mathbf{A}$ admits a ρ-dilation $\widetilde{\mathbf{A}} = (\widetilde{A}_1, \dots, \widetilde{A}_N)$ for which $\zeta\widetilde{\mathbf{A}} := \zeta_1\widetilde{A}_1 + \cdots + \zeta_N\widetilde{A}_N$ is a unitary operator for each $\zeta := (\zeta_1, \dots, \zeta_N)$ in the unit torus $\mathbb{T}^N$. For $N = 1$ this class coincides with the class C_ρ of B. Sz.-Nagy and C. Foiaş. We generalize the known descriptions of $C_{\rho,1} = C_\rho$ to the case of $C_{\rho,N}$, $N > 1$, using so-called Agler kernels. Also, the notion of operator radii $w_\rho, \rho > 0$, is generalized to the case of N-tuples of operators, and to the case of bounded (in a certain strong sense) holomorphic operator-valued functions in the open unit polydisk $\mathbb{D}^N$, with preservation of all the most important their properties. Finally, we show that for each $\rho > 1$ and $N > 1$ there exists an $\mathbf{A} = (A_1, \dots, A_N) \in C_{\rho,N}$ which is not simultaneously similar to any $\mathbf{T} = (T_1, \dots, T_N) \in C_{1,N}$, however if $\mathbf{A} \in C_{\rho,N}$ admits a uniform unitary ρ-dilation then $\mathbf{A}$ is simultaneously similar to some $\mathbf{T} \in C_{1,N}$.

Mathematics Subject Classification (2000). Primary 47A13; Secondary 47A20, 47A56.

Keywords. Multivariable, ρ-dilations, linear pencils of operators, operator radii, Agler kernels, similarity to a 1-contraction.

1. Introduction

Linear pencils of operators $L_{\mathbf{A}}(z) := A_0 + z_1A_1 + \cdots + z_NA_N$ on a Hilbert space which take contractive (resp., unitary or J-unitary for some signature operator $J = J^* = J^{-1}$) values for all $z = (z_1, \dots, z_N)$ in the *unit torus* $\mathbb{T}^N := \{\zeta \in \mathbb{C}^N : |\zeta_k| = 1,\ k = 1, \dots, N\}$ serve as one of possible generalizations of a single contractive (resp., unitary, J-unitary) operator on a Hilbert space. They appear in constructions of Agler's unitary colligation and corresponding conservative (unitary) scattering N-dimensional discrete-time linear system of Roesser type [1, 8], and also of Fornasini–Marchesini type [7], and dissipative (contractive), conservative (unitary) or J-conservative (J-unitary) scattering N-dimensional linear systems of one more form introduced in our paper [17] and studied in [17, 18, 19, 23, 20, 21, 7]. These constructions, in particular, provide the transfer function realization formulae for certain classes of holomorphic functions [1, 8, 17, 19, 21, 7], the solutions to the Nevanlinna–Pick interpolation problem [2, 8], the Toeplitz corona problem [2, 8], and the commutant lifting problem [6] in several variables.

In [18] we developed the dilation theory for multidimensional linear systems, and in particular gave a necessary and sufficient condition for such a system to have a conservative dilation. As a special case, this gave a criterion for the existence of a unitary dilation of a contractive (on $\mathbb{T}^N$) linear pencil of operators on a Hilbert space. Linear pencils of operators satisfying this criterion inherit the most important properties of single contraction operators on a Hilbert space (note that, due to [22], not all linear pencils which take contractive operator values on $\mathbb{T}^N$ satisfy this criterion).

The purpose of the present paper is to develop the theory of ρ-contractions in several variables in the framework of "linear pencils approach". We introduce the notion of *ρ-dilation* of an N-tuple $\mathbf{A} = (A_1, \dots, A_N)$ of bounded linear operators on a common Hilbert space by means of a simultaneous ρ-dilation, in the sense of B. Sz.-Nagy and C. Foiaş [32, 34], of the values of a homogeneous linear pencil of operators $z\mathbf{A} := \sum_{k=1}^N z_k A_k$. The class $C_{\rho,N}$ consists of those N-tuples of operators $\mathbf{A} = (A_1, \dots, A_N)$ (*ρ-contractions*) for which there exists a ρ-dilation $\widetilde{\mathbf{A}} = (\widetilde{A}_1, \dots, \widetilde{A}_N)$ such that the operators $\zeta\widetilde{\mathbf{A}} = \sum_{k=1}^N \zeta_k \widetilde{A}_k$ are unitary for all $\zeta = (\zeta_1, \dots, \zeta_N) \in \mathbb{T}^N$. On the one hand, this class generalizes the class $C_{\rho,1} = C_\rho$ of Sz.-Nagy and Foiaş [32, 34] consisting of operators which admit a unitary ρ-dilation to the case $N > 1$. On the other hand, this class generalizes the class of N-tuples of operators $\mathbf{A}$ for which the associated linear pencil of operators $z\mathbf{A}$ admits a unitary dilation in the sense of [18] (this corresponds to $\rho = 1$) to the case of N-tuples of operators $\mathbf{A}$ which have a unitary ρ-dilation for $\rho \neq 1$.

The paper is organized as follows. Section 2 gives preliminaries on ρ-contractions for the case $N = 1$. Namely, we recall the relevant definitions, the known criteria for an operator to be a ρ-contraction, i.e., to belong to the class C_ρ of Sz.-Nagy and Foiaş, the notion of operator radii w_ρ and their properties, and the theorem on similarity of ρ-contractions to contractions. In Section 3 we give the definitions of a ρ-dilation of an N-tuple of operators, and of the class $C_{\rho,N}$ of ρ-contractions for the case $N > 1$, and prove a theorem which generalizes the criteria of ρ-contractiveness to this case, as well as to the case $0 < \rho \neq 1$. Some properties of classes $C_{\rho,N}$ are discussed. Then it is shown that the notions of a ρ-contraction and of the corresponding class $C_{\rho,N}$, as well as the theorem just mentioned, can be extended to holomorphic functions on the *open unit polydisk* $\mathbb{D}^N := \{z \in \mathbb{C}^N : |z_k| < 1,\ k = 1, \dots, N\}$ that are bounded in a certain strong sense, though the notion of unitary ρ-dilation is not relevant any more in this case. In Section 4 we define operator radii $w_{\rho,N}$ of N-tuples of operators, and operator-function radii $w_{\rho,N}^{(\infty)}$ of bounded holomorphic functions on $\mathbb{D}^N$, $\rho > 0$. These radii generalize w_ρ's and inherit all the most important properties of them. In Section 5 we prove that for each $\rho > 1$ and $N > 1$ there exists an $\mathbf{A} = (A_1, \dots, A_N) \in C_{\rho,N}$ which is not simultaneously similar to any $\mathbf{T} = (T_1, \dots, T_N) \in C_{1,N}$. Then we introduce the classes $C^u_{\rho,N}$, $\rho > 0$, of N-variable ρ-contractions $\mathbf{A} = (A_1, \dots, A_N)$ which admit a uniform unitary ρ-dilation. We prove that if $\mathbf{A} \in C^u_{\rho,N}$ for some $\rho > 1$ then $\mathbf{A}$ is simultaneously similar to some $\mathbf{T} \in C^u_{1,N}$. Note, that since the class $C^u_{\rho,N}$ (as well as $C_{\rho,N}$) increases as a function of ρ, for any $\rho \leq 1$ an $\mathbf{A} \in C^u_{\rho,N}$ (resp., $\mathbf{A} \in C_{\rho,N}$) belongs to $C^u_{1,N}$ (resp., $C_{1,N}$) itself. We show the relation of our results to ones of G. Popescu [30] where a different notion of multivariable ρ-contractions has been introduced, and the relevant theory has been developed. The classes $C^u_{\rho,N}$, $\rho > 0$, which appear in Section 5 in connection with the similarity problem discussed there, certainly deserve a further investigation.

2. Preliminaries

Let $L(\mathcal{X}, \mathcal{Y})$ denote the Banach space of bounded linear operators mapping a Hilbert space $\mathcal{X}$ into a Hilbert space $\mathcal{Y}$, and $L(\mathcal{X}) := L(\mathcal{X}, \mathcal{X})$. For $\rho > 0$, an operator $\widetilde{A} \in L(\widetilde{\mathcal{X}})$ is said to be a *ρ-dilation of an operator* $A \in L(\mathcal{X})$ if $\widetilde{\mathcal{X}} \supset \mathcal{X}$ and

$$A^n = \rho P_{\mathcal{X}} \widetilde{A}^n | \mathcal{X}, \quad n \in \mathbb{N}, \tag{2.1}$$

where $P_{\mathcal{X}}$ denotes the orthogonal projection onto the subspace $\mathcal{X}$ in $\widetilde{\mathcal{X}}$. If, moreover, $\widetilde{A}$ is a unitary operator then $\widetilde{A}$ is called a *unitary ρ-dilation of* A. In [32] (see also [34]) B. Sz.-Nagy and C. Foiaş introduced the classes C_ρ, $\rho > 0$, consisting of operators which admit a unitary ρ-dilation. Due to B. Sz.-Nagy [31], the class C_1 is precisely the class of all contractions, i.e., operators A such that $\|A\| \leq 1$. C.A. Berger [9] showed that the class C_2 is precisely the class of all operators $A \in L(\mathcal{X})$, for some Hilbert space $\mathcal{X}$, which have the *numerical radius*

$$w(A) = \sup\{|\langle Ax, x\rangle| : \ x \in \mathcal{X}, \ \|x\| = 1\}$$

equal to at most one. Thus, the classes C_ρ, $\rho > 0$, provide a framework for simultaneous investigation of these two important classes of operators.

Recall that the *Herglotz* (or *Carathéodory*) *class* $\mathcal{H}(\mathcal{X})$ (respectively, the *Schur class* $\mathcal{S}(\mathcal{X})$) consists of holomorphic functions f on the open unit disk $\mathbb{D}$ which take values in $L(\mathcal{X})$ and satisfy $\operatorname{Re} f(z) = f(z) + f(z)^* \succeq 0$ in the sense of positive semi-definiteness of an operator (resp., $\|f(z)\| \leq 1$) for all $z \in \mathbb{D}$. Let us recall some known characterizations of the classes C_ρ.

Theorem 2.1. *Let $A \in L(\mathcal{X})$ and $\rho > 0$. The following statements are equivalent:*

(i) *$A \in C_\rho$;*

(ii) *the function $k^A_\rho(z,w) := \rho I_{\mathcal{X}} - (\rho - 1)((zA + (wA)^*) + (\rho - 2)(wA)^* zA$ satisfies $k^A_\rho(z,z) \succeq 0$ for all $z \in \operatorname{clos}(\mathbb{D})$;*

(iii) *the function $\psi^A_\rho(z) := (1 - \frac{2}{\rho})I_{\mathcal{X}} + \frac{2}{\rho}(I_{\mathcal{X}} - zA)^{-1}$ belongs to $\mathcal{H}(\mathcal{X})$;*

(iv) *the function $\varphi^A_\rho(z) := zA\left((\varrho - 1)zA - \rho I_{\mathcal{X}}\right)^{-1}$ belongs to $\mathcal{S}(\mathcal{X})$.*

Conditions (ii) and (iii) of Theorem 2.1 each characterizing the class C_ρ appear in [32], while condition (iv) is due to C. Davis [11].

Corollary 2.2. *Condition* (ii) *in Theorem* 2.1 *can be replaced by*

(ii′) $k_\rho^A(C,C) := \rho I_{\mathcal{X}} \otimes I_{\mathcal{H}_C} - (\rho-1)(A\otimes C + (A\otimes C)^*)$
$+ (\rho-2)(A\otimes C)^*(A\otimes C) \succeq 0$
for any contraction C *on a Hilbert space* $\mathcal{H}_C$.

Proof. Indeed, (ii′)⇒(ii), hence (ii′)⇒(i). Conversely, if $A \in C_\rho \cap L(\mathcal{X})$ then for any contraction C on $\mathcal{H}_C$ one has $A\otimes C \in C_\rho$ because, by [31], C admits a unitary dilation $\widetilde{C}$, and A admits a unitary ρ-dilation $\widetilde{A}$, thus $\widetilde{A}\otimes\widetilde{C}$ is a unitary ρ-dilation of $A\otimes C$:

$$\begin{aligned}(A\otimes C)^n &= A^n \otimes C^n = (\rho P_{\mathcal{X}} \widetilde{A}^n|\mathcal{X}) \otimes (P_{\mathcal{H}_C}\widetilde{C}^n|\mathcal{H}_C)\\ &= \rho P_{\mathcal{X}\otimes\mathcal{H}_C}(\widetilde{A}^n \otimes \widetilde{C}^n)|\mathcal{X}\otimes\mathcal{H}_C\\ &= \rho P_{\mathcal{X}\otimes\mathcal{H}_C}(\widetilde{A}\otimes\widetilde{C})^n|\mathcal{X}\otimes\mathcal{H}_C, \quad n\in\mathbb{N}.\end{aligned}$$

Therefore, $k_\rho^A(C,C) = k_\rho^{A\otimes C}(1,1) \succeq 0$, i.e., (ii′) is valid. □

Corollary 2.3. *Condition*

(v): $A\otimes C \in C_\rho$ *for any contraction* C *on a Hilbert space,*

is equivalent to each of conditions (i)–(iv) *of Theorem* 2.1.

Proof. See the proof of Corollary 2.2. □

Any operator $A \in C_\rho$ is *power-bounded*:

$$\|A^n\| \le \rho, \quad n\in\mathbb{N}, \tag{2.2}$$

moreover, its *spectral radius*

$$\nu(A) = \lim_{n\to+\infty} \|A^n\|^{\frac{1}{n}} \tag{2.3}$$

is at most one. In [32] an example of a power-bounded operator which is not contained in any of the classes C_ρ, $\rho>0$, is given. However, J.A.R. Holbrook [15] showed that any bounded linear operator A with $\nu(A)\le 1$ can be approximated in the operator norm topology by elements of the classes C_ρ. More precisely, if C_∞ denotes the class of bounded linear operators with spectral radius at most one, and $\mathcal{X}$ is a Hilbert space, then

$$C_\infty \cap L(\mathcal{X}) = \operatorname{clos}\left\{ \bigcup_{0<\rho<\infty} C_\rho \cap L(\mathcal{X}) \right\}. \tag{2.4}$$

For a fixed Hilbert space $\mathcal{X}$, the class C_ρ as a function of ρ increases [32]:

$$C_\rho \subset C_{\rho'} \text{ for } \rho<\rho'. \tag{2.5}$$

Moreover, it was shown by E. Durszt [13] that C_ρ increases strictly for $\dim\mathcal{X}\ge 2$:

$$C_\rho \ne C_{\rho'} \text{ for } \rho\ne\rho'.$$

Proposition 2.4. *For $\mathcal{X} = \mathbb{C}$, the classes C_ρ coincide for all $\rho \geq 1$, and strictly increase for $0 < \rho < 1$:*

$$C_\rho \subsetneq C_{\rho'} \text{ for } 0 < \rho < \rho' \leq 1.$$

Proof. If $a \in \mathbb{C} \cong L(\mathbb{C})$ belongs to C_ρ then $\|a\| = |a| = \nu(a) \leq 1$. Hence $C_\rho \subset C_1$ for any $\rho > 0$. Since (2.5) implies $C_\rho \supset C_1$ for $\rho \geq 1$, we get $C_\rho = C_1$ for this case, that proves the first part of this proposition.

For the proof of the second part, we will show that for any $\varepsilon, \rho : 0 < \varepsilon < \rho < 1$, one has

$$a := \frac{\rho}{2-\rho} \in C_\rho \backslash C_{\rho-\varepsilon}. \tag{2.6}$$

If $0 \leq \varepsilon < \rho$ then, by condition (ii) in Theorem 2.1, the inclusion $a \in C_{\rho-\varepsilon}$ is equivalent to

$$\rho - \varepsilon - (\rho - \varepsilon - 1)(az + \bar{a}\bar{z}) + (\rho - \varepsilon - 2)|az|^2 \geq 0, \quad z \in \operatorname{clos}(\mathbb{D}),$$

which for $a = \frac{\rho}{2-\rho}$ turns into

$$\rho - \varepsilon - 2(\rho - \varepsilon - 1)\frac{\rho}{2-\rho} r \cos\theta + (\rho - \varepsilon - 2)\left(\frac{\rho r}{2-\rho}\right)^2 \geq 0, \quad r \in [0,1], \theta \in [0, 2\pi).$$

Since $\rho - \varepsilon - 1 < 0$, the left-hand side of this inequality, as a function of θ for a fixed r, has a minimum at $\theta = \pi$, so the latter condition turns into

$$\rho - \varepsilon + 2(\rho - \varepsilon - 1)\frac{\rho r}{2-\rho} + (\rho - \varepsilon - 2)\left(\frac{\rho r}{2-\rho}\right)^2 \geq 0, \quad r \in [0,1].$$

The left-hand side attains its minimum at $r = 1$, thus the latter inequality turns into

$$\rho - \varepsilon + 2(\rho - \varepsilon - 1)\frac{\rho}{2-\rho} + (\rho - \varepsilon - 2)\left(\frac{\rho}{2-\rho}\right)^2 = -\frac{4\varepsilon}{(2-\rho)^2} \geq 0,$$

which is possible if and only if $\varepsilon = 0$. Thus, (2.6) is true. □

The properties of the classes C_ρ become more clear due to the following numerical characteristics of operators. J.A.R. Holbrook [15] and J.P. Williams [35], independently, introduced for any $A \in L(\mathcal{X})$ the *operator radii*

$$w_\rho(A) := \inf\{u > 0 : \frac{1}{u} A \in C_\rho\}. \tag{2.7}$$

Theorem 2.5. *$w_\rho(\cdot)$ has the following properties:*

(i) *$w_\rho(A) < \infty$;*
(ii) *$w_\rho(A) > 0$ unless $A = 0$, moreover, $w_\rho(A) \geq \frac{1}{\rho}\|A\|$;*
(iii) *$\forall \mu \in \mathbb{C}, \quad w_\rho(\mu A) = |\mu| w_\rho(A)$;*
(iv) *$w_\rho(A) \leq 1$ if and only if $A \in C_\rho$;*
(v) *$w_\rho(\cdot)$ is a norm on $L(\mathcal{X})$ for any $\rho : \ 0 < \rho \leq 2$, and not a norm on $L(\mathcal{X})$, $\dim \mathcal{X} \geq 2$, for any $\rho > 2$;*
(vi) *$w_1(A) = \|A\|$ (of course, here $\|\cdot\|$ is the operator norm on $L(\mathcal{X})$ with respect to the Hilbert-space metric on $\mathcal{X}$);*

(vii) $w_2(A) = w(A)$;

(viii) $w_\infty(A) := \lim_{\rho\to+\infty} w_\rho(A) = \nu(A)$;

(ix) $w_\rho(I_{\mathcal{X}}) = \begin{cases} 1 & \textit{for } \rho \geq 1, \\ \frac{2}{\rho} - 1 & \textit{for } 0 < \rho < 1; \end{cases}$

(x) *if* $0 < \rho < \rho'$ *then* $w_{\rho'}(A) \leq w_\rho(A) \leq \left(\frac{2\rho'}{\rho} - 1\right) w_{\rho'}(A)$, *thus* $w_\rho(A)$ *is continuous in* ρ *and non-increasing as* ρ *increases;*

(xi) *if* $\|A\| = 1$ *and* $A^2 = 0$ *then, for any* $\rho > 0$, $w_\rho(A) = \frac{1}{\rho}$;

(xii) *if for some* ρ_0 *one has* $w_{\rho_0}(A) > w_\infty(A)$ $(= \nu(A))$ *then for any* $\rho > \rho_0$ *one has* $w_{\rho_0}(A) > w_\rho(A)$;

(xiii) $\lg w_\rho(A)$ *is a convex function in* ρ, $0 < \rho < +\infty$;

(xiv) $w_\rho(A)$ *is a convex function in* ρ, $0 < \rho < +\infty$;

(xv) *the function* $h_A(\rho) := \rho w_\rho(A)$ *is non-decreasing on* $[1, +\infty)$, *and non-increasing on* $(0, 1)$;

(xvi) *for any* ρ *such that* $0 < \rho < 2$ *one has* $\rho w_\rho(A) = (2 - \rho) w_{2-\rho}(A)$, *and* $\lim_{\rho\downarrow 0} \frac{\rho}{2} w_\rho(A) = w_2(A)$ $(= w(A))$;

(xvii) $\forall \rho : \; 0 < \rho \leq 1, \quad w_\rho(A) \geq \left(\frac{2}{\rho} - 1\right) w_2(A)$;

(xviii) $\forall A, B \in L(\mathcal{X})$, $\forall \rho \geq 1, \quad w_\rho(AB) \leq \rho^2 w_\rho(A) w_\rho(B)$, *moreover,* ρ^2 *is the best constant in this inequality for the case* $\dim \mathcal{X} \geq 2$;

(xix) $\forall A, B \in L(\mathcal{X})$, $\forall \rho : \; 0 < \rho < 1, \quad w_\rho(AB) \leq (2 - \rho)\rho w_\rho(A) w_\rho(B)$, *moreover,* $(2-\rho)\rho$ *is the best constant in this inequality for the case* $\dim \mathcal{X} \geq 2$;

(xx) $\forall \rho > 0$, $\forall n \in \mathbb{N}, \quad w_\rho(A^n) \leq w_\rho(A)^n$.

Properties (i)–(xii), (xviii), and (xx) were proved by J.A.R. Holbrook [15], properties (xiii)–(xvi) were discovered by T. Ando and K. Nishio [4]. Property (xix) was shown by K. Okubo and T. Ando [26], and follows also from (xvi) and (xviii). Finally, property (xvii) easily follows from (x) and (xvi). Indeed, for $0 < \rho \leq 1$ one has $w_{2-\rho}(A) \geq w_2(A)$, hence $\rho w_\rho(A) = (2 - \rho) w_{2-\rho}(A) \geq (2 - \rho) w_2(A)$, which implies (xvii).

We have listed in Theorem 2.5 only the most important, as it seems to us, properties of operator radii $w_\rho(\cdot)$. Other properties of $w_\rho(\cdot)$ can be found in [15, 16, 14, 4, 26, 5] and elsewhere.

Let us note that properties of the classes C_ρ discussed before Theorem 2.5, including Proposition 2.4, can be deduced from properties (iv), (vi)–(x) in Theorem 2.5. Due to property (iv) in Theorem 2.5, operators from the classes C_ρ are called *ρ-contractions.*

Any $A \in C_\rho$ satisfies the following *generalized von Neumann inequality* [32]: for any polynomial p of one variable

$$\|p(A)\| \leq \max_{|z|\leq 1} |\rho p(z) + (1 - \rho) p(0)|. \tag{2.8}$$

Let $A \in L(\mathcal{X}), B \in L(\mathcal{Y})$. Then A is said to be *similar* to B if there exists a bounded invertible operator $S \in L(\mathcal{X}, \mathcal{Y})$ such that

$$A = S^{-1}BS. \tag{2.9}$$

B. Sz.-Nagy and C. Foiaş proved in [33] (see also [34]) that any $A \in C_\rho$ is similar to some $T \in C_1$, i.e., any ρ-contraction is similar to a contraction.

To conclude this section, let us remark that the classes C_ρ are of continuous interest, e.g., see recent works [12, 10, 5, 24, 27]. In [30] the classes C_ρ were extended to a multivariable setting; we shall discuss this generalization in Section 5.

3. The classes $C_{\rho,N}$

Let $\rho > 0$. We will say that an N-tuple of operators $\widetilde{\mathbf{A}} = (\widetilde{A}_1, \dots, \widetilde{A}_N) \in L(\widetilde{\mathcal{X}})^N$ is a *ρ-dilation of an N-tuple of operators* $\mathbf{A} = (A_1, \dots, A_N) \in L(\mathcal{X})^N$ if $\widetilde{\mathcal{X}} \supset \mathcal{X}$, and for any $z = (z_1, \dots, z_N) \in \mathbb{C}^N$ the operator $z\widetilde{\mathbf{A}} = \sum_{k=1}^N z_k \widetilde{A}_k$ is a ρ-dilation, in the sense of [32], of the operator $z\mathbf{A} = \sum_{k=1}^N z_k A_k$, i.e.,

$$(z\mathbf{A})^n = \rho P_{\mathcal{X}} (z\widetilde{\mathbf{A}})^n | \mathcal{X}, \quad z \in \mathbb{C}^N, \ n \in \mathbb{N}. \tag{3.1}$$

These relations are equivalent to

$$\mathbf{A}^t = \rho P_{\mathcal{X}} \widetilde{\mathbf{A}}^t | \mathcal{X}, \quad t \in \mathbb{Z}_+^N := \{\tau \in \mathbb{Z}^N : \tau_k \geq 0, \ k = 1, \dots, N\}, \tag{3.2}$$

where $\mathbf{A}^t$, $t \in \mathbb{Z}_+^N$, are *symmetrized multi-powers of* $\mathbf{A}$:

$$\mathbf{A}^t := \frac{t!}{|t|!} \sum_\sigma A_{[\sigma(1)]} \cdots A_{[\sigma(|t|)]},$$

and analogously for $\widetilde{\mathbf{A}}$. Here for a multi-index $t = (t_1, \dots, t_N)$, $t! := t_1! \cdots t_N!$ and $|t| := t_1 + \cdots + t_N$; σ runs over the set of all permutations with repetitions in a string of $|t|$ numbers from the set $\{1, \dots, N\}$ such that the κth number $[\kappa] \in \{1, \dots, N\}$ appears in this string $t_{[\kappa]}$ times. Say, if $t = (1, 2, 0, \dots, 0)$ then

$$\mathbf{A}^t = \frac{A_1 A_2^2 + A_2 A_1 A_2 + A_2^2 A_1}{3}.$$

In the case of a commutative N-tuple $\mathbf{A}$ one has $\mathbf{A}^t = A_1^{t_1} \cdots A_N^{t_N}$, i.e., a usual multi-power.

Note 3.1. Compare (3.1) and (3.2) with (2.1).

In the case $\rho = 1$ the notion of ρ-dilation of an N-tuple of operators $\mathbf{A} = (A_1, \dots, A_N)$ coincides with the notion of dilation of $\mathbf{A}$ (or corresponding linear pencil $z\mathbf{A}$) as defined in [18].

We will call $\widetilde{\mathbf{A}} \in L(\widetilde{\mathcal{X}})^N$ a *unitary ρ-dilation of* $\mathbf{A} \in L(\mathcal{X})^N$ if $\widetilde{\mathbf{A}}$ is a ρ-dilation of $\mathbf{A}$ and for any $\zeta \in \mathbb{T}^N$ the operator $\zeta\widetilde{\mathbf{A}} = \sum_{k=1}^N \zeta_k \widetilde{A}_k$ is unitary. The class of operator N-tuples which admit a unitary ρ-dilation will be denoted by $C_{\rho,N}$.

Let $\mathcal{C}^N$ denote the family of all N-tuples $\mathbf{C} = (C_1, \ldots, C_N)$ of *commuting strict contractions* on a common Hilbert space $\mathcal{H}_{\mathbf{C}}$, i.e., $C_k C_j = C_j C_k$ and $\|C_k\| < 1$ for all $k, j \in \{1, \ldots, N\}$. An $L(\mathcal{X})$-valued function

$$k(z, w) = \sum_{(t,s)\in\mathbb{Z}_+^N\times\mathbb{Z}_+^N} \hat{k}(t, s)\bar{w}^s z^t, \quad (z, w) \in \mathbb{D}^N \times \mathbb{D}^N,$$

which is holomorphic in $z \in \mathbb{D}^N$ and anti-holomorphic in $w \in \mathbb{D}^N$, will be called an *Agler kernel* if

$$k(\mathbf{C}, \mathbf{C}) := \sum_{(t,s)\in\mathbb{Z}_+^N\times\mathbb{Z}_+^N} \hat{k}(t, s) \otimes \mathbf{C}^{*s}\mathbf{C}^t \succeq 0, \quad \mathbf{C} \in \mathcal{C}^N, \tag{3.3}$$

where the series converges in the operator norm topology on $L(\mathcal{X} \otimes \mathcal{H}_{\mathbf{C}})$. The *Agler–Herglotz class* $\mathcal{AH}_N(\mathcal{X})$ (resp., the *Agler–Schur class* $\mathcal{AS}_N(\mathcal{X})$) is the class of all $L(\mathcal{X})$-valued functions f holomorphic on $\mathbb{D}^N$ for which $k(z, w) = f(z)+f(w)^*$ (resp., $k(z, w) = I_{\mathcal{X}} - f(w)^* f(z)$) is an Agler kernel. Agler kernels, as well as the classes $\mathcal{AH}_N(\mathcal{X})$ and $\mathcal{AS}_N(\mathcal{X})$, were defined and studied by J. Agler in [1]. The von Neumann inequality [25] implies that $\mathcal{AS}_1(\mathcal{X}) = \mathcal{S}(\mathcal{X})$ and $\mathcal{AH}_1(\mathcal{X}) = \mathcal{H}(\mathcal{X})$.

Remark 3.2. The function $k_\rho^A(z, w)$ from condition (ii) in Theorem 2.1, due to Corollary 2.2, is an Agler kernel ($N = 1$).

Theorem 3.3. *Let* $\mathbf{A} \in L(\mathcal{X})^N$, $\rho > 0$. *The following conditions are equivalent:*

(i) $\mathbf{A} \in C_{\rho,N}$;

(ii) *the function* $k_{\rho,N}^{\mathbf{A}}(z, w) := \rho I_{\mathcal{X}} - (\rho - 1)\left((z\mathbf{A} + (w\mathbf{A})^*\right) + (\rho - 2)(w\mathbf{A})^* z\mathbf{A}$ *is an Agler kernel on* $\mathbb{D}^N \times \mathbb{D}^N$;

(iii) *the function* $\psi_{\rho,N}^{\mathbf{A}}(z) := (1 - \frac{2}{\rho})I_{\mathcal{X}} + \frac{2}{\rho}(I_{\mathcal{X}} - z\mathbf{A})^{-1}$ *belongs to* $\mathcal{AH}_N(\mathcal{X})$;

(iv) *the function* $\varphi_{\rho,N}^{\mathbf{A}}(z) := z\mathbf{A}\left((\varrho - 1)z\mathbf{A} - \rho I_{\mathcal{X}}\right)^{-1}$ *belongs to* $\mathcal{AS}_N(\mathcal{X})$;

(v) $\mathbf{A} \otimes \mathbf{C} := \sum_{k=1}^N A_k \otimes C_k \in C_\rho = C_{\rho,1}$ *for all* $\mathbf{C} \in \mathcal{C}^N$.

Remark 3.4. This theorem generalizes Theorem 2.1 with condition (ii) replaced by condition (ii′) from Corollary 2.2, and added condition (v) from Corollary 2.3.

Proof of Theorem 3.3. (i)⇔(iii). The proof of this part combines the idea of B. Sz.-Nagy and C. Foiaş [32] for the proof of the equivalence (i)⇔(iii) in Theorem 2.1 (see Remark 3.4) with Agler's representation of functions from $\mathcal{AH}_N(\mathcal{X})$ [1]. Let $\mathbf{A} = (A_1, \ldots, A_N) \in C_{\rho,N} \cap L(\mathcal{X})^N$, and $\widetilde{\mathbf{A}} = (\widetilde{A}_1, \ldots, \widetilde{A}_N) \in L(\widetilde{\mathcal{X}})^N$ be a unitary ρ-dilation of $\mathbf{A}$. By Corollary 4.3 in [18], the linear function $L_{\widetilde{\mathbf{A}}}(z) = z\widetilde{\mathbf{A}}$ belongs to the class $\mathcal{AS}_N(\widetilde{\mathcal{X}})$. Since for any $\mathbf{C} \in \mathcal{C}^N$ one has $(1+\varepsilon)\mathbf{C} \in \mathcal{C}^N$ for a sufficiently small $\varepsilon > 0$, the operator $\widetilde{\mathbf{A}} \otimes \mathbf{C}$, as well as $\widetilde{\mathbf{A}} \otimes (1 + \varepsilon)\mathbf{C}$, is contractive. Thus, $\widetilde{\mathbf{A}} \otimes \mathbf{C}$ is a strict contraction, and the series

$$I_{\widetilde{\mathcal{X}}\otimes\mathcal{H}_{\mathbf{C}}} + 2\sum_{n=1}^{\infty}(\widetilde{\mathbf{A}} \otimes \mathbf{C})^n$$

converges in the $L(\widetilde{\mathcal{X}} \otimes \mathcal{H}_{\mathbf{C}})$-norm to

$$(I_{\widetilde{\mathcal{X}} \otimes \mathcal{H}_{\mathbf{C}}} + \widetilde{\mathbf{A}} \otimes \mathbf{C})(I_{\widetilde{\mathcal{X}} \otimes \mathcal{H}_{\mathbf{C}}} - \widetilde{\mathbf{A}} \otimes \mathbf{C})^{-1}.$$

Moreover,

$$\operatorname{Re}[(I_{\widetilde{\mathcal{X}} \otimes \mathcal{H}_{\mathbf{C}}} + \widetilde{\mathbf{A}} \otimes \mathbf{C})(I_{\widetilde{\mathcal{X}} \otimes \mathcal{H}_{\mathbf{C}}} - \widetilde{\mathbf{A}} \otimes \mathbf{C})^{-1}] \succeq 0. \tag{3.4}$$

Therefore,

$$\begin{aligned}
&P_{\mathcal{X} \otimes \mathcal{H}_{\mathbf{C}}}(I_{\widetilde{\mathcal{X}} \otimes \mathcal{H}_{\mathbf{C}}} + \widetilde{\mathbf{A}} \otimes \mathbf{C})(I_{\widetilde{\mathcal{X}} \otimes \mathcal{H}_{\mathbf{C}}} - \widetilde{\mathbf{A}} \otimes \mathbf{C})^{-1}|\mathcal{X} \otimes \mathcal{H}_{\mathbf{C}} \\
&= P_{\mathcal{X} \otimes \mathcal{H}_{\mathbf{C}}}\left(I_{\widetilde{\mathcal{X}} \otimes \mathcal{H}_{\mathbf{C}}} + 2\sum_{n=1}^{\infty}(\widetilde{\mathbf{A}} \otimes \mathbf{C})^n\right)\Bigg| \mathcal{X} \otimes \mathcal{H}_{\mathbf{C}} \\
&= I_{\mathcal{X} \otimes \mathcal{H}_{\mathbf{C}}} + 2\sum_{n=1}^{\infty}\sum_{|t|=n}\frac{n!}{t!}(P_{\mathcal{X}} \otimes I_{\mathcal{H}_{\mathbf{C}}})(\widetilde{\mathbf{A}}^t \otimes \mathbf{C}^t)|\mathcal{X} \otimes \mathcal{H}_{\mathbf{C}} \\
&= I_{\mathcal{X} \otimes \mathcal{H}_{\mathbf{C}}} + \frac{2}{\rho}\sum_{n=1}^{\infty}\sum_{|t|=n}\frac{n!}{t!}\mathbf{A}^t \otimes \mathbf{C}^t = I_{\mathcal{X} \otimes \mathcal{H}_{\mathbf{C}}} + \frac{2}{\rho}\sum_{n=1}^{\infty}(\mathbf{A} \otimes \mathbf{C})^n \\
&= (1 - \frac{2}{\rho})I_{\mathcal{X} \otimes \mathcal{H}} + \frac{2}{\rho}(I_{\mathcal{X} \otimes \mathcal{H}} - \mathbf{A} \otimes \mathbf{C})^{-1} = \psi^{\mathbf{A}}_{\rho,N}(\mathbf{C}),
\end{aligned}$$

and (3.4) implies $\operatorname{Re}\psi^{\mathbf{A}}_{\rho,N}(\mathbf{C}) \succeq 0$. Since the function $(I_{\widetilde{\mathcal{X}}} + z\widetilde{\mathbf{A}})(I_{\widetilde{\mathcal{X}}} - z\widetilde{\mathbf{A}})^{-1}$ is well defined and holomorphic on $\mathbb{D}^N$, so is

$$\psi^{\mathbf{A}}_{\rho,N}(z) = P_{\mathcal{X}}(I_{\widetilde{\mathcal{X}}} + z\widetilde{\mathbf{A}})(I_{\widetilde{\mathcal{X}}} - z\widetilde{\mathbf{A}})^{-1}|\mathcal{X}, \quad z \in \mathbb{D}^N, \tag{3.5}$$

and we obtain $\psi^{\mathbf{A}}_{\rho,N} \in \mathcal{AH}_N(\mathcal{X})$.

Conversely, let $\psi^{\mathbf{A}}_{\rho,N} \in \mathcal{AH}_N(\mathcal{X})$. Since $\psi^{\mathbf{A}}_{\rho,N}(0) = I_{\mathcal{X}}$, according to [1], there exist a Hilbert space $\widetilde{\mathcal{X}} \supset \mathcal{X}$, its subspaces $\widetilde{\mathcal{X}}_1, \ldots, \widetilde{\mathcal{X}}_N$ satisfying $\widetilde{\mathcal{X}} = \bigoplus_{k=1}^N \widetilde{\mathcal{X}}_k$, and a unitary operator $U \in L(\widetilde{\mathcal{X}})$ such that

$$\psi^{\mathbf{A}}_{\rho,N}(z) = P_{\mathcal{X}}(I_{\widetilde{\mathcal{X}}} + U(z\mathbf{P}))(I_{\widetilde{\mathcal{X}}} - U(z\mathbf{P}))^{-1}|\mathcal{X}, \quad z \in \mathbb{D}^N, \tag{3.6}$$

where $z\mathbf{P} := \sum_{k=1}^N z_k P_{\widetilde{\mathcal{X}}_k}$, i.e., we get (3.5) with $\widetilde{A}_k = UP_{\widetilde{\mathcal{X}}_k}$, $k = 1, \ldots, N$. Note that for each $\zeta \in \mathbb{T}^N$ the operator $\zeta\widetilde{\mathbf{A}}$ is unitary. Developing both parts of (3.6) into the series in homogeneous polynomials convergent in the operator norm, we get

$$I_{\mathcal{X}} + \frac{2}{\rho}\sum_{n=1}^{\infty}(z\mathbf{A})^n = I_{\mathcal{X}} + 2\sum_{n=1}^{\infty}P_{\mathcal{X}}(z\widetilde{\mathbf{A}})^n|\mathcal{X}, \quad z \in \mathbb{D}^N,$$

that implies the relations

$$(z\mathbf{A})^n = \rho P_{\mathcal{X}}(z\widetilde{\mathbf{A}})^n|\mathcal{X}, \quad n \in \mathbb{N},$$

for all $z \in \mathbb{D}^N$, and hence for all $z \in \mathbb{C}^N$. Thus, $\widetilde{\mathbf{A}}$ is a unitary ρ-dilation of $\mathbf{A}$, and $\mathbf{A} \in C_{\rho,N}$. The equivalence (i)⇔(iii) is proved.

Note that in this proof we have established that each Agler representation (3.6) of $\psi^{\mathbf{A}}_{\rho,N}$ gives rise to a unitary ρ-dilation $\widetilde{\mathbf{A}}$ of $\mathbf{A}$, and vice versa. Indeed, we already showed that (3.6) determines $\widetilde{\mathbf{A}}$. Conversely, if $\widetilde{\mathbf{A}} \in L(\widetilde{\mathcal{X}})^N$ is a unitary ρ-dilation of $\mathbf{A}$, then (3.5) holds. Set $U := \sum_{k=1}^N \widetilde{A}_k \in L(\widetilde{\mathcal{X}})$ and $\widetilde{\mathcal{X}}_k := \widetilde{A}_k^* \widetilde{\mathcal{X}}$, $k = 1, \dots, N$. Then U is unitary, $\widetilde{\mathcal{X}}_k$ is a closed subspace in $\widetilde{\mathcal{X}}$ for each $k = 1, \dots, N$, the subspaces $\widetilde{\mathcal{X}}_k$ are pairwise orthogonal, and $\widetilde{\mathcal{X}} = \bigoplus_{k=1}^N \widetilde{\mathcal{X}}_k$ (see Proposition 2.4 in [17]). Thus, (3.5) turns into (3.6).

(v)⇔(iv). Let (v) be true. By Theorem 2.1 applied for $\mathbf{A} \otimes \mathbf{C}$ with a $\mathbf{C} \in \mathcal{C}^N$, one has $\varphi_\rho^{\mathbf{A}\otimes\mathbf{C}} \in \mathcal{S}(\mathcal{X} \otimes \mathcal{H}_\mathbf{C})$. For $\varepsilon > 0$ small enough, $(1+\varepsilon)\mathbf{C} \in \mathcal{C}^N$, hence $\mathbf{A} \otimes (1+\varepsilon)\mathbf{C} \in C_\rho$, and $\varphi_\rho^{\mathbf{A}\otimes(1+\varepsilon)\mathbf{C}} \in \mathcal{S}(\mathcal{X} \otimes \mathcal{H}_\mathbf{C})$. Thus,

$$\varphi_{\rho,N}^{\mathbf{A}}(\mathbf{C}) = \mathbf{A} \otimes \mathbf{C}((\rho - 1)\mathbf{A} \otimes \mathbf{C} - \rho I_{\mathcal{X}\otimes\mathcal{H}_\mathbf{C}})^{-1} = \varphi_\rho^{\mathbf{A}\otimes(1+\varepsilon)\mathbf{C}}\left(\frac{1}{1+\varepsilon}\right)$$

is a contraction on $\mathcal{X} \otimes \mathcal{H}_\mathbf{C}$. In particular, $\varphi_{\rho,N}^{\mathbf{A}}(z)$ is well defined, holomorphic and contractive on $\mathbb{D}^N$. Finally, $\varphi_{\rho,N}^{\mathbf{A}} \in \mathcal{AS}_N(\mathcal{X})$.

Conversely, if (iv) is true then for any $\mathbf{C} \in \mathcal{C}^N$:

$$\varphi_\rho^{\mathbf{A}\otimes\mathbf{C}}(\lambda) = \lambda\mathbf{A} \otimes \mathbf{C}((\rho - 1)\lambda\mathbf{A} \otimes \mathbf{C} - \rho I_{\mathcal{X}\otimes\mathcal{H}_\mathbf{C}})^{-1} = \varphi_{\rho,N}^{\mathbf{A}}(\lambda\mathbf{C})$$

is well defined, holomorphic and contractive for $\lambda \in \mathbb{D}$. Thus, $\varphi_\rho^{\mathbf{A}\otimes\mathbf{C}} \in \mathcal{S}(\mathcal{X} \otimes \mathcal{H}_\mathbf{C})$, and by Theorem 2.1, $\mathbf{A} \otimes \mathbf{C} \in C_\rho$.

(v)⇔(iii) and (v)⇔(ii) are proved analogously, using the following relations for $\mathbf{C} \in \mathcal{C}^N$, $\lambda \in \mathbb{D}$:

$$\psi_{\rho,N}^{\mathbf{A}}(\mathbf{C}) = \psi_\rho^{\mathbf{A}\otimes(1+\varepsilon)\mathbf{C}}\left(\frac{1}{1+\varepsilon}\right), \quad \psi_\rho^{\mathbf{A}\otimes\mathbf{C}}(\lambda) = \psi_{\rho,N}^{\mathbf{A}}(\lambda\mathbf{C}),$$
$$k_{\rho,N}^{\mathbf{A}}(\mathbf{C},\mathbf{C}) = k_\rho^{\mathbf{A}\otimes\mathbf{C}}(1,1), \quad k_\rho^{\mathbf{A}\otimes\mathbf{C}}(\lambda,\lambda) = k_{\rho,N}^{\mathbf{A}}(\lambda\mathbf{C},\lambda\mathbf{C}).$$

The proof is complete. □

Remark 3.5. For the case $\rho = 1$ each of conditions (ii)–(v) in Theorem 3.3 means that for any $\mathbf{C} \in \mathcal{C}^N$ the operator $\mathbf{A} \otimes \mathbf{C}$ is a contraction. In other words,

$$\mathbf{A} \in C_{1,N} \cap L(\mathcal{X})^N \iff L_\mathbf{A} \in \mathcal{AS}_N(\mathcal{X}),$$

that coincides with in [18, Corollary 4.3] (here $L_\mathbf{A}(z) := z\mathbf{A}$, $z \in \mathbb{C}^N$).

Let us also note that using [18, Corollary 4.3] one can deduce (v) from (i) directly. Indeed, if $\widetilde{\mathbf{A}} \in L(\widetilde{\mathcal{X}})^N$ is a unitary ρ-dilation of $\mathbf{A} \in L(\mathcal{X})^N$ then for any $\mathbf{C} \in \mathcal{C}^N$ by [18, Corollary 4.3] the operator $\widetilde{\mathbf{A}} \otimes \mathbf{C}$ is a contraction. Therefore, due to [31], $\widetilde{\mathbf{A}} \otimes \mathbf{C} \in L(\widetilde{\mathcal{X}} \otimes \mathcal{H}_\mathbf{C})$ has a unitary dilation $U \in L(\mathcal{K})$, $\mathcal{K} \supset \widetilde{\mathcal{X}} \otimes \mathcal{H}_\mathbf{C}$. Then for any $n \in \mathbb{N}$:

$$\begin{aligned}(\mathbf{A} \otimes \mathbf{C})^n &= \rho P_{\mathcal{X}\otimes\mathcal{H}_\mathbf{C}}(\widetilde{\mathbf{A}} \otimes \mathbf{C})^n|\mathcal{X} \otimes \mathcal{H}_\mathbf{C} \\ &= \rho P_{\mathcal{X}\otimes\mathcal{H}_\mathbf{C}}(P_{\widetilde{\mathcal{X}}\otimes\mathcal{H}_\mathbf{C}} U^n|\widetilde{\mathcal{X}} \otimes \mathcal{H}_\mathbf{C})|\mathcal{X} \otimes \mathcal{H}_\mathbf{C} \\ &= \rho P_{\mathcal{X}\otimes\mathcal{H}_\mathbf{C}} U^n|\mathcal{X} \otimes \mathcal{H}_\mathbf{C},\end{aligned}$$

i.e., U is a unitary ρ-dilation of the operator $\mathbf{A} \otimes \mathbf{C}$. Thus, $\mathbf{A} \otimes \mathbf{C} \in C_\rho$.

Let us define the *numerical radius of an N-tuple of operators* $\mathbf{A} \in L(\mathcal{X})^N$ as

$$w^{(N)}(\mathbf{A}) := \sup_{\mathbf{C}\in\mathcal{C}^N} w(\mathbf{A} \otimes \mathbf{C}). \tag{3.7}$$

For $N = 1$, $w^{(1)}(A) = w(A)$. Indeed,

$$\begin{aligned} w^{(1)}(A) &= \sup_{\|C\|<1} w(A \otimes C) \geq \sup_{0<\varepsilon<1} w(A \otimes (1-\varepsilon)I_{\mathcal{H}_C}) = \sup_{0<\varepsilon<1} (1-\varepsilon)w(A) \\ &= w(A); \\ w^{(1)}(A) &= \sup_{\|C\|<1} w(A \otimes C) \leq \sup_{\|C\|<1} w(A)\|C\| = w(A). \end{aligned}$$

Here we used the properties $w(A \otimes I_{\mathcal{H}}) = w(A)$ and $w(A \otimes B) \leq w(A)\|B\|$ valid for any $A \in L(\mathcal{X}), B \in L(\mathcal{H})$ (see, e.g., [14]).

Proposition 3.6. $\mathbf{A} \in C_{2,N} \iff w^{(N)}(\mathbf{A}) \leq 1$.

Proof. By Theorem 3.3, $\mathbf{A} \in C_{2,N}$ if and only if $\mathbf{A} \otimes \mathbf{C} \in C_2 = C_{2,1}$ for any $\mathbf{C} \in \mathcal{C}^N$. This, in turn, means that $w(\mathbf{A} \otimes \mathbf{C}) \leq 1$ for any $\mathbf{C} \in \mathcal{C}^N$ (by Berger's result mentioned in Section 2), i.e., $w^{(N)}(\mathbf{A}) \leq 1$. □

Theorem 3.7. *If $\mathbf{A} \in C_{\rho,N} \cap L(\mathcal{X})^N$ for a $\rho > 0$, then $L_{\mathbf{A}} \in \rho\mathcal{AS}_N(\mathcal{X})$. For any $\rho > 0$ such that $\rho \neq 1$, there exists an $\mathbf{A} \in L(\mathcal{X})^N$ such that $L_{\mathbf{A}} \in \rho\mathcal{AS}_N(\mathcal{X})$ and $\mathbf{A} \notin C_{\rho,N}$.*

Proof. Let $\mathbf{A} \in C_{\rho,N} \cap L(\mathcal{X})^N$ for some $\rho > 0$, and $\mathbf{C} \in \mathcal{C}^N$. Then $\mathbf{A}$ has a unitary ρ-dilation $\widetilde{\mathbf{A}} \in L(\widetilde{\mathcal{X}})^N$, and

$$\begin{aligned} \|\mathbf{A} \otimes \mathbf{C}\| &= \left\| \sum_{k=1}^N A_k \otimes C_k \right\| = \left\| \rho(P_{\mathcal{X}} \otimes I_{\mathcal{H}_\mathbf{C}}) \left(\sum_{k=1}^N \widetilde{A}_k \otimes C_k \right) \middle| \mathcal{X} \otimes \mathcal{H}_\mathbf{C} \right\| \\ &\leq \rho \left\| \sum_{k=1}^N \widetilde{A}_k \otimes C_k \right\| = \rho \left\| \widetilde{\mathbf{A}} \otimes \mathbf{C} \right\| \leq \rho \end{aligned}$$

(here we used again Corollary 4.3 in [18]). Thus, $L_{\mathbf{A}} \in \rho\mathcal{AS}_N(\mathcal{X})$.

Now, let $0 < \rho \neq 1$, and $\mathbf{A} \in L(\mathcal{X})^N$ be such that $\frac{1}{\rho} L_{\mathbf{A}}(\zeta) = \frac{1}{\rho}\zeta\mathbf{A}$ is a unitary operator for each $\zeta \in \mathbb{T}^N$. Then, again by Corollary 4.3 in [18], $L_{\mathbf{A}} \in \rho\mathcal{AS}_N(\mathcal{X})$. Suppose there exists a unitary ρ-dilation $\widetilde{\mathbf{A}} \in L(\widetilde{\mathcal{X}})^N$ of $\mathbf{A}$. Then for any $\zeta \in \mathbb{T}^N$, $L_{\mathbf{A}}(\zeta) = \zeta\mathbf{A} = \rho P_{\mathcal{X}}(\zeta\widetilde{\mathbf{A}})|\mathcal{X}$. Hence, for any $\zeta \in \mathbb{T}^N$ and $x \in \mathcal{X}$,

$$\|\zeta\widetilde{\mathbf{A}}x\| = \|x\| = \|\frac{1}{\rho}\zeta\mathbf{A}x\| = \|P_{\mathcal{X}}(\zeta\widetilde{\mathbf{A}})x\|,$$

that is possible only if $\zeta\widetilde{\mathbf{A}}x \in \mathcal{X}$ for all $\zeta \in \mathbb{T}^N$ and $x \in \mathcal{X}$. Therefore, for $n > 1$,

$$\rho^n\|x\| = \|(\zeta\mathbf{A})^n x\| = \|\rho P_{\mathcal{X}}(\zeta\widetilde{\mathbf{A}})^n x\| = \rho\|(\zeta\widetilde{\mathbf{A}})^n x\| = \rho\|x\|,$$

that is impossible for $x \neq 0$. Thus, $\mathbf{A} \notin C_{\rho,N}$. □

Note 3.8. Compare Theorem 3.7 with Remark 3.5.

The same argument as in the proof of the first part of Theorem 3.7 shows that, for $\mathbf{A} \in C_{\rho,N}$,

$$\|(\mathbf{A} \otimes \mathbf{C})^n\| \leq \rho, \quad n \in \mathbb{N}, \ \mathbf{C} \in \mathcal{C}^N. \tag{3.8}$$

Note 3.9. Compare (3.8) with (2.2).

This uniform (in $\mathbf{C} \in \mathcal{C}^N$) power-boundedness of an N-tuple of operators $\mathbf{A}$ is, in our setting, a generalization of power-boundedness of a single operator. Let us define the *spectral radius of an N-tuple of operators* $\mathbf{A} \in L(\mathcal{X})^N$ as

$$\nu^{(N)}(\mathbf{A}) := \lim_{n \to +\infty} \left(\sup_{\mathbf{C} \in \mathcal{C}^N} \|(\mathbf{A} \otimes \mathbf{C})^n\| \right)^{\frac{1}{n}}. \tag{3.9}$$

Note 3.10. Compare (3.9) with (2.3).

In other words, $\nu^{(N)}(\mathbf{A}) = \nu^{(N,\infty)}(L_{\mathbf{A}})$, where $\nu^{(N,\infty)}(f)$ is the *spectral radius of an element f of the Banach algebra $H_N^\infty(\mathcal{X})$* consisting of holomorphic $L(\mathcal{X})$-valued functions f on $\mathbb{D}^N$ which satisfy

$$\|f\|_{\infty,N} := \sup_{\mathbf{C} \in \mathcal{C}^N} \|f(\mathbf{C})\| < \infty$$

(this algebra was introduced in [1]). Here $f(\mathbf{C})$ is defined in the same manner as $k(\mathbf{C}, \mathbf{C})$ in (3.3), i.e., for

$$\begin{aligned} f(z) &= \sum_{t \in \mathbb{Z}_+^N} \hat{f}_t z^t, \quad z \in \mathbb{D}^N, \\ f(\mathbf{C}) &:= \sum_{t \in \mathbb{Z}_+^N} \hat{f}_t \otimes \mathbf{C}^t, \quad \mathbf{C} \in \mathcal{C}^N, \end{aligned}$$

where the latter series converges in the $L(\mathcal{X} \otimes \mathcal{H}_{\mathbf{C}})$-norm. For $N = 1$, $\nu^{(1)}(A) = \nu(A)$. Indeed,

$$\begin{aligned} \nu^{(1)}(A) &= \lim_{n \to +\infty} \left(\sup_{\|C\|<1} \|(A \otimes C)^n\| \right)^{\frac{1}{n}} = \lim_{n \to +\infty} \left(\sup_{\|C\|<1} \|A^n \otimes C^n\| \right)^{\frac{1}{n}} \\ &= \lim_{n \to +\infty} \left(\|A^n\| \sup_{\|C\|<1} \|C^n\| \right)^{\frac{1}{n}} = \lim_{n \to +\infty} \|A^n\|^{\frac{1}{n}} = \nu(A). \end{aligned}$$

Remark 3.11. For any $\mathbf{A} \in C_{\rho,N}$, by virtue of (3.8), $\nu^{(N)}(\mathbf{A}) \leq 1$.

Theorem 3.12. *For a fixed Hilbert space $\mathcal{X}$ and any $N \geq 1$ the class $C_{\rho,N}$ increases as a function of ρ:*

$$C_{\rho,N} \subset C_{\rho',N} \ \textit{for} \ \rho < \rho'.$$

Moreover, for $\dim \mathcal{X} \geq 2$, *$C_{\rho,N}$ increases strictly:*

$$C_{\rho,N} \neq C_{\rho',N} \ \textit{for} \ \rho \neq \rho'.$$

For $\dim \mathcal{X} = 1$ *the classes $C_{\rho,N}$ coincide for all $\rho \geq 1$, and strictly increase for $0 < \rho < 1$.*

Proof. For $N = 1$ this theorem is true (see Section 2). For $N > 1$ it follows from the equivalence (i)$\Leftrightarrow$(v) in Theorem 3.3. □

Theorem 3.13. *For any* $\mathbf{A} \in C_{\rho,N}$, $\mathbf{C} \in \mathcal{C}^N$, *and a polynomial* p *of one variable,*

$$\|p(\mathbf{A} \otimes \mathbf{C})\| \le \max_{|z|\le 1} |\rho p(z) + (1-\rho)p(0)|.$$

Proof. This result follows from the generalized von Neumann inequality (2.8) and the equivalence (i)$\Leftrightarrow$(v) in Theorem 3.3. □

Let us remark that results of this section on N-tuples of operators from the classes $C_{\rho,N}$ can be extended to elements of $H_N^\infty(\mathcal{X})$, though the notion of unitary ρ-dilation no longer makes sense for this case. Define $C_{\rho,N}^{(\infty)}$ as a class of functions $f \in H_N^\infty(\mathcal{X})$ such that $f(\mathbf{C}) \in C_\rho = C_{\rho,1}$ for any $\mathbf{C} \in \mathcal{C}^N$. Then, in particular, Theorem 3.3 implies that $\mathbf{A} \in C_{\rho,N}$ if and only if $L_{\mathbf{A}} \in C_{\rho,N}^{(\infty)}$. The following analogue of Theorem 3.3 is easily obtained.

Theorem 3.14. *Let* $f \in H_N^\infty(\mathcal{X})$ *and* $\rho > 0$. *The following conditions are equivalent:*

(i) $f \in C_{\rho,N}^{(\infty)}$;

(ii) *the function* $k_{\rho,N}^f(z,w) := \rho I_{\mathcal{X}} - (\rho-1)(f(z) + (f(w)^*) + (\rho-2)f(w)^* f(z)$ *is an Agler kernel on* $\mathbb{D}^N \times \mathbb{D}^N$;

(iii) *the function* $\psi_{\rho,N}^f(z) := (1 - \frac{2}{\rho})I_{\mathcal{X}} + \frac{2}{\rho}(I_{\mathcal{X}} - f(z))^{-1}$ *belongs to* $\mathcal{AH}_N(\mathcal{X})$;

(iv) *the function* $\varphi_{\rho,N}^f(z) := f(z)((\varrho-1)f(z) - \rho I_{\mathcal{X}})^{-1}$ *belongs to* $\mathcal{AS}_N(\mathcal{X})$.

Clearly, $H_N^\infty(\mathcal{X}) \cap C_{1,N}^{(\infty)} = \mathcal{AS}_N(\mathcal{X})$. Set

$$w^{(N,\infty)}(f) := \sup_{\mathbf{C}\in\mathcal{C}^N} w(f(\mathbf{C})). \tag{3.10}$$

Note 3.15. Compare (3.10) with (3.7).

Remark 3.16. Proposition 3.6 extends directly to the class $C_{2,N}^{(\infty)}$, with $f \in H_N^\infty(\mathcal{X})$ in the place of $\mathbf{A} \in L(\mathcal{X})^N$, and $w^{(N,\infty)}(f)$ in the place of $w^{(N)}(\mathbf{A})$. Remark 3.11 extends directly to $f \in H_N^\infty(\mathcal{X})$ in the place of $\mathbf{A} \in L(\mathcal{X})^N$, and $\nu^{(N,\infty)}(f)$ in the place of $\nu^{(N)}(\mathbf{A})$. Also, Theorems 3.12 and 3.13 extend to the classes $C_{\rho,N}^{(\infty)}$.

4. Multivariable operator and operator-function radii

In this section we extend the notion of operator radii w_ρ, $0 < \rho \le \infty$, to the multivariable case, i.e., to N-tuples of bounded linear operators and to elements of the Banach algebra $H_N^\infty(\mathcal{X})$. Let $0 < \rho < \infty$ and $f \in H_N^\infty(\mathcal{X})$. Set

$$w_{\rho,N}^{(\infty)}(f) := \inf\{u > 0 : \frac{1}{u} f \in C_{\rho,N}^{(\infty)}\},$$

and for $\mathbf{A} \in L(\mathcal{X})^N$, define

$$w_{\rho,N}(\mathbf{A}) := w_{\rho,N}^{(\infty)}(L_{\mathbf{A}}).$$

Due to our remark preceding to Theorem 3.14,

$$w_{\rho,N}(\mathbf{A}) = \inf\{u > 0 : \frac{1}{u}\mathbf{A} \in C_{\rho,N}\}. \tag{4.1}$$

Note 4.1. Compare (4.1) with (2.7).

Clearly, for $N = 1$ and $A \in L(\mathcal{X})$, $w_{\rho,1}(A) = w_\rho(A)$.

Lemma 4.2. *For* $f \in H_N^\infty(\mathcal{X})$, $\mathbf{A} \in L(\mathcal{X})^N$,

$$w_{\rho,N}^{(\infty)}(f) = \sup_{\mathbf{C}\in\mathcal{C}^N} w_\rho(f(\mathbf{C})), \tag{4.2}$$

$$w_{\rho,N}(\mathbf{A}) = \sup_{\mathbf{C}\in\mathcal{C}^N} w_\rho(\mathbf{A}\otimes\mathbf{C}). \tag{4.3}$$

Proof. Let $f \in H_N^\infty(\mathcal{X})$. Then for $u > 0$, $\frac{1}{u}f \in C_{\rho,N}^{(\infty)}$ if and only if for any $\mathbf{C} \in \mathcal{C}^N$ one has $\frac{1}{u}f(\mathbf{C}) \in C_\rho$. Therefore,

$$\begin{aligned} w_{\rho,N}^{(\infty)}(f) &= \inf\{u > 0 : \frac{1}{u}f \in C_{\rho,N}^{(\infty)}\} = \inf\{u > 0 : \forall\mathbf{C} \in \mathcal{C}^N,\ \frac{1}{u}f(\mathbf{C}) \in C_\rho\} \\ &= \sup_{\mathbf{C}\in\mathcal{C}^N} \inf\{u > 0 : \frac{1}{u}f(\mathbf{C}) \in C_\rho\} = \sup_{\mathbf{C}\in\mathcal{C}^N} w_\rho(f(\mathbf{C})), \end{aligned}$$

i.e., (4.2) is true. Now, (4.3) follows from (4.2) and the definition of $w_{\rho,N}(\mathbf{A})$. □

Theorem 4.3. 1. *All properties* (i)–(xx) *listed in Theorem* 2.5 *are satisfied for* $w_{\rho,N}^{(\infty)}(\cdot)$ *in the place of* $w_\rho(\cdot)$; $f, g \in H_N^\infty(\mathcal{X})$ *in the place of* $A, B \in L(\mathcal{X})$; $w^{(N,\infty)}(\cdot)$ *in the place of* $w(\cdot)$; *and* $\nu^{(N,\infty)}(\cdot)$ *in the place of* $\nu(\cdot)$.

2. *Properties* (i)–(xvii) *listed in Theorem* 2.5 *are satisfied for* $w_{\rho,N}(\cdot)$ *in the place of* $w_\rho(\cdot)$; $\mathbf{A} \in L(\mathcal{X})^N$ *in the place of* $A \in L(\mathcal{X})$; $w^{(N)}(\cdot)$ *in the place of* $w(\cdot)$; *and* $\nu^{(N)}(\cdot)$ *in the place of* $\nu(\cdot)$.

Proof. **1.** Let $f \in H_N^\infty(\mathcal{X})$. Then $\|f\|_{\infty,N} = \sup_{\mathbf{C}\in\mathcal{C}^N}\|f(\mathbf{C})\| < \infty$. By properties (vi) and (x) in Theorem 2.5, and Lemma 4.2, if $0 < \rho \le 1$ then

$$\begin{aligned} w_{\rho,N}^{(\infty)}(f) &= \sup_{\mathbf{C}\in\mathcal{C}^N} w_\rho(f(\mathbf{C})) \le \left(\frac{2}{\rho} - 1\right) \sup_{\mathbf{C}\in\mathcal{C}^N} w_1(f(\mathbf{C})) \\ &= \left(\frac{2}{\rho} - 1\right) \sup_{\mathbf{C}\in\mathcal{C}^N} \|f(\mathbf{C})\| < \infty, \end{aligned}$$

and if $\rho > 1$ then

$$w_{\rho,N}^{(\infty)}(f) = \sup_{\mathbf{C}\in\mathcal{C}^N} w_\rho(f(\mathbf{C})) \le \sup_{\mathbf{C}\in\mathcal{C}^N} w_1(f(\mathbf{C})) = \sup_{\mathbf{C}\in\mathcal{C}^N} \|f(\mathbf{C})\| < \infty.$$

Thus, property (i) is fulfilled.

Properties (ii)–(vii), (ix)–(xi), (xiii)–(xv), and (xvii)–(xx) easily follow from the properties in Theorem 2.5 with the same numbers, and Lemma 4.2.

The proof of property (viii) is an adaptation of the proof of Theorem 5.1 in [15] to our case. First of all, let us remark that property (iv) implies that if

$u > w_{\rho,N}^{(\infty)}(f)$ then $\frac{1}{u}f \in C_{\rho,N}^{(\infty)}$, and for any $\mathbf{C} \in \mathcal{C}^N$ one has $\frac{1}{u}f(\mathbf{C}) \in C_\rho$. In particular,

$$\sup_{\mathbf{C}\in\mathcal{C}^N} \left\| \left(\frac{f(\mathbf{C})}{u} \right)^n \right\| \le \rho, \quad n \in \mathbb{N}.$$

Therefore, $\nu^{(N,\infty)}\left(\frac{f}{u}\right) \le 1$, i.e., $\nu^{(N,\infty)}(f) \le u$. Thus, for any $\rho > 0$, $\nu^{(N,\infty)}(f) \le w_{\rho,N}^{(\infty)}(f)$, moreover,

$$\nu^{(N,\infty)}(f) \le \lim_{\rho\to+\infty} w_{\rho,N}^{(\infty)}(f)$$

(note, that due to property (x), $w_{\rho,N}^{(\infty)}(f)$ is a non-increasing and bounded from below function of ρ, hence it has a limit as $\rho \to +\infty$).

For the proof of the opposite inequality, let us first show that if $\nu^{(N,\infty)}(g) < 1$ for some $g \in H_N^\infty(\mathcal{X})$ then beginning with some $\rho_0 > 0$ (i.e., for all $\rho \ge \rho_0$) one has $g \in C_{\rho,N}^{(\infty)}$. Indeed, in this case there exists an $s > 1$ such that $\nu^{(N,\infty)}(sg) < 1$. Then there exists a $B > 0$ such that

$$s^n \sup_{\mathbf{C}\in\mathcal{C}^N} \|g(\mathbf{C})^n\| \le B, \quad n \in \mathbb{N}.$$

Hence, for any $\mathbf{C} \in \mathcal{C}^N$,

$$\begin{aligned}
\operatorname{Re} \psi_{\rho,N}^g(\mathbf{C}) &= \left(1 - \frac{2}{\rho}\right) I_{\mathcal{X}\otimes\mathcal{H}_\mathbf{C}} + \frac{2}{\rho} \operatorname{Re}(I_{\mathcal{X}\otimes\mathcal{H}_\mathbf{C}} - g(\mathbf{C}))^{-1} \\
&= I_{\mathcal{X}\otimes\mathcal{H}_\mathbf{C}} + \frac{2}{\rho} \operatorname{Re} \sum_{n=1}^\infty g(\mathbf{C})^n \succeq \left(1 - \frac{2}{\rho} \sum_{n=1}^\infty \|g(\mathbf{C})^n\|\right) I_{\mathcal{X}\otimes\mathcal{H}_\mathbf{C}} \\
&\succeq \left(1 - \frac{2}{\rho} \sum_{n=1}^\infty \frac{B}{s^n}\right) I_{\mathcal{X}\otimes\mathcal{H}_\mathbf{C}} = \left(1 - \frac{2B}{\rho(s-1)}\right) I_{\mathcal{X}\otimes\mathcal{H}_\mathbf{C}} \succeq 0
\end{aligned}$$

as soon as $\rho \ge \frac{2B}{s-1}$. Thus, by Theorem 3.14, $g \in C_{\rho,N}^{(\infty)}$ for any $\rho \ge \frac{2B}{s-1}$.

Now, if $\nu^{(N,\infty)}(f) = 0$ then for any $k \in \mathbb{N}$, $\nu^{(N,\infty)}(kf) = 0$. Hence, for $\rho \ge \rho_0$ we have $kf \in C_{\rho,N}^{(\infty)}$, and by property (iv), $w_{\rho,N}^{(\infty)}(kf) \le 1$. Thus,

$$\lim_{\rho\to+\infty} w_{\rho,N}^{(\infty)}(f) \le \frac{1}{k}$$

for any $k \in \mathbb{N}$, and

$$\lim_{\rho\to+\infty} w_{\rho,N}^{(\infty)}(f) = 0 = \nu^{(N,\infty)}(f),$$

as required.

If $\nu^{(N,\infty)}(f) > 0$ then for any $\varepsilon > 0$,

$$\nu^{(N,\infty)}\left(\frac{f}{(1+\varepsilon)\nu^{(N,\infty)}(f)}\right) = \frac{1}{1+\varepsilon} < 1.$$

Then for $\rho \ge \rho_0$,

$$w_{\rho,N}^{(\infty)}\left(\frac{f}{(1+\varepsilon)\nu^{(N,\infty)}(f)}\right) \le 1,$$

hence $w_{\rho,N}^{(\infty)}(f) \leq (1+\varepsilon)\nu^{(N,\infty)}(f)$. Passing to the limit as $\rho \to +\infty$, and then as $\varepsilon \downarrow 0$, we get

$$\lim_{\rho\to+\infty} w_{\rho,N}^{(\infty)}(f) \leq \nu^{(N,\infty)}(f),$$

as required. Thus, property (viii) is proved.

For the proof of property (xii), it is enough to suppose, by virtue of positive homogeneity of $w_{\rho,N}^{(\infty)}(\cdot)$ and $\nu^{(N,\infty)}(\cdot)$, that for $f \in H_N^\infty(\mathcal{X})$ one has $w_{\rho_0,N}^{(\infty)}(f) = 1$, $\nu^{(N,\infty)}(f) < 1$, and prove that for any $\rho > \rho_0$, $w_{\rho,N}^{(\infty)}(f) < 1$. By Theorem 3.14 and property (iv) in the present theorem,

$$\frac{\rho_0}{2}\psi_{\rho_0,N}^f(z) = \left(\frac{\rho_0}{2} - 1\right) I_{\mathcal{X}} + (I_{\mathcal{X}} - f(z))^{-1} \in \mathcal{AH}_N(\mathcal{X}),$$

i.e., for any $\mathbf{C} \in \mathcal{C}^N$,

$$\operatorname{Re}\left[\frac{\rho_0}{2}\psi_{\rho_0,N}^f(\mathbf{C})\right] = \operatorname{Re}\left[\left(\frac{\rho_0}{2} - 1\right) I_{\mathcal{X}\otimes\mathcal{H}_{\mathbf{C}}} + (I_{\mathcal{X}\otimes\mathcal{H}_{\mathbf{C}}} - f(\mathbf{C}))^{-1}\right] \succeq 0,$$

and for any $\rho > \rho_0$,

$$\operatorname{Re}\left[\frac{\rho}{2}\psi_{\rho,N}^f(\mathbf{C})\right] = \operatorname{Re}\left[\left(\frac{\rho}{2} - 1\right) I_{\mathcal{X}\otimes\mathcal{H}_{\mathbf{C}}} + (I_{\mathcal{X}\otimes\mathcal{H}_{\mathbf{C}}} - f(\mathbf{C}))^{-1}\right] \succeq \frac{\rho-\rho_0}{2} I_{\mathcal{X}\otimes\mathcal{H}_{\mathbf{C}}}.$$

Since the resolvent $R_f(\lambda) := (\lambda I_{\mathcal{X}} - f)^{-1}$ is continuous in the $H_N^\infty(\mathcal{X})$-norm on the resolvent set of f, and $\nu^{(N,\infty)}(f) < 1$, for $\varepsilon > 0$ small enough, one has $\nu^{(N,\infty)}((1+\varepsilon)f) < 1$, and

$$\operatorname{Re}\left[\frac{\rho}{2}\psi_{\rho,N}^{(1+\varepsilon)f}(\mathbf{C})\right] = \operatorname{Re}\left[\left(\frac{\rho}{2} - 1\right) I_{\mathcal{X}\otimes\mathcal{H}_{\mathbf{C}}} + (I_{\mathcal{X}\otimes\mathcal{H}_{\mathbf{C}}} - (1+\varepsilon)f(\mathbf{C}))^{-1}\right] \succeq 0$$

for any $\mathbf{C} \in \mathcal{C}^N$, i.e., $\frac{\rho}{2}\psi_{\rho,N}^{(1+\varepsilon)f} \in \mathcal{AH}_N(\mathcal{X})$, and $\psi_{\rho,N}^{(1+\varepsilon)f} \in \mathcal{AH}_N(\mathcal{X})$. Hence, by Theorem 3.14, $(1+\varepsilon)f \in C_{\rho,N}^{(\infty)}$ which means, by property (iv), that $w_{\rho,N}^{(\infty)}((1+\varepsilon)f) \leq 1$. Thus, $w_{\rho,N}^{(\infty)}(f) \leq \frac{1}{1+\varepsilon} < 1$, as required.

The first part of property (xvi) in this theorem follows from property (xvi) in Theorem 2.5, and Lemma 4.2. For the proof of the second part of (xvi), we use properties (xv) and (xvi) from Theorem 2.5, property (xv) in the present theorem, and Lemma 4.2:

$$\begin{aligned}
\lim_{\rho\downarrow 0}\frac{\rho}{2} w_{\rho,N}^{(\infty)}(f) &= \sup_{0<\rho<1}\left\{\frac{\rho}{2} w_{\rho,N}^{(\infty)}(f)\right\} = \sup_{0<\rho<1}\sup_{\mathbf{C}\in\mathcal{C}^N}\left\{\frac{\rho}{2} w_\rho(f(\mathbf{C}))\right\} \\
&= \sup_{\mathbf{C}\in\mathcal{C}^N}\sup_{0<\rho<1}\left\{\frac{\rho}{2} w_\rho(f(\mathbf{C}))\right\} = \sup_{\mathbf{C}\in\mathcal{C}^N}\left\{\lim_{\rho\downarrow 0}\frac{\rho}{2} w_\rho(f(\mathbf{C}))\right\} \\
&= \sup_{\mathbf{C}\in\mathcal{C}^N} w_2(f(\mathbf{C})) = w_{2,N}^{(\infty)}(f).
\end{aligned}$$

The proof of property (xvi), as well as part **1** of this theorem, is complete.

Part **2** follows from part **1**. □

Denote by $C_{\infty,N}^{(\infty)}$ (resp., $C_{\infty,N}$) the class of $\mathcal{C}^N$-bounded holomorphic operator valued functions on $\mathbb{D}^N$ (resp., N-tuples of bounded linear operators on a common Hilbert space) with spectral radius at most one.

Theorem 4.4. *Let $\mathcal{X}$ be a Hilbert space. Then*

$$C^{(\infty)}_{\infty,N} \cap H^\infty_N(\mathcal{X}) = \operatorname{clos}\left\{ \bigcup_{0<\rho<\infty} (C^{(\infty)}_{\rho,N} \cap H^\infty_N(\mathcal{X})) \right\}; \tag{4.4}$$

$$C_{\infty,N} \cap L(\mathcal{X})^N = \operatorname{clos}\left\{ \bigcup_{0<\rho<\infty} (C_{\rho,N} \cap L(\mathcal{X})^N) \right\}. \tag{4.5}$$

Note 4.5. Compare (4.4) and (4.5) with (2.4).

Proof of Theorem 4.4. The inclusion "$\supset$" in (4.4) and (4.5) follows from Remarks 3.11 and 3.16, and the fact that the set of $\mathcal{C}^N$-bounded holomorphic operator-valued functions on $\mathbb{D}^N$ (resp., N-tuples of bounded operators) with spectral radius at most one is closed in $H^\infty_N(\mathcal{X})$ (resp., $L(\mathcal{X})^N$).

To show the inclusion "$\subset$" in (4.4), observe that for $f \in C^{(\infty)}_{\infty,N} \cap H^\infty_N(\mathcal{X})$ and $0 < r < 1$, $\nu^{(N,\infty)}(rf) \leq r < 1$. By property (viii) from Theorem 4.3, for $\rho_0 > 0$ big enough, $w^{(\infty)}_{\rho_0,N}(rf) < 1$, and by property (iv) from the same theorem,

$$rf \in C^{(\infty)}_{\rho_0,N} \cap H^\infty_N(\mathcal{X}) \subset \operatorname{clos}\left\{ \bigcup_{0<\rho<\infty} (C^{(\infty)}_{\rho,N} \cap H^\infty_N(\mathcal{X})) \right\}.$$

Passing to the limit as $r \uparrow 1$, we get

$$f \in \operatorname{clos}\left\{ \bigcup_{0<\rho<\infty} (C_{\rho,N} \cap H^\infty_N(\mathcal{X})) \right\},$$

and the inclusion "$\subset$" in (4.4) follows. Analogously for the inclusion "$\subset$" in (4.5). □

In view of property (iv) in Theorem 4.3, let us call the elements of the class $C_{\rho,N}$ *(N-variable) ρ-contractions.*

5. On similarity of ρ-contractions to 1-contractions in several variables

An N-tuple of operators $\mathbf{A} = (A_1, \ldots, A_N) \in L(\mathcal{X})^N$ is said to be *simultaneously similar* to an N-tuple of operators $\mathbf{B} = (B_1, \ldots, B_N) \in L(\mathcal{Y})^N$ if there exists a boundedly invertible operator $S \in L(\mathcal{X}, \mathcal{Y})$ such that

$$A_k = S^{-1} B_k S, \quad k = 1, \ldots, N, \tag{5.1}$$

or equivalently,

$$z\mathbf{A} = S^{-1}(z\mathbf{B})S, \quad z \in \mathbb{C}^N. \tag{5.2}$$

Note 5.1. Compare (5.1) and (5.2) with (2.9).

Theorem 5.2. *For any $\rho > 1$ and $N > 1$, there exists an $\mathbf{A} = (A_1, \ldots, A_N) \in C_{\rho,N}$ which is not simultaneously similar to any $\mathbf{T} = (T_1, \ldots, T_N) \in C_{1,N}$.*

Proof. Let $N = 2$, and for any $\varepsilon \geq 0$ set $\mathbf{A}^{(\varepsilon)} = (A_1^{(\varepsilon)}, A_2^{(\varepsilon)}) \in L(\mathbb{C}^3)^2$, where

$$A_1^{(\varepsilon)} := \begin{bmatrix} 0 & \frac{1+\varepsilon}{\sqrt{2}} & 0 \\ 0 & 0 & 0 \\ -\frac{1+\varepsilon}{\sqrt{2}} & 0 & 0 \end{bmatrix}, \quad A_2^{(\varepsilon)} := \begin{bmatrix} 0 & 0 & \frac{1+\varepsilon}{\sqrt{2}} \\ \frac{1+\varepsilon}{\sqrt{2}} & 0 & 0 \\ 0 & 0 & 0 \end{bmatrix}.$$

Then for any $\varepsilon \geq 0$ and $z \in \mathbb{C}^2$,

$$\begin{aligned} z\mathbf{A}^{(\varepsilon)} &= \begin{bmatrix} 0 & \frac{1+\varepsilon}{\sqrt{2}} z_1 & \frac{1+\varepsilon}{\sqrt{2}} z_2 \\ \frac{1+\varepsilon}{\sqrt{2}} z_2 & 0 & 0 \\ -\frac{1+\varepsilon}{\sqrt{2}} z_1 & 0 & 0 \end{bmatrix}, \\ (z\mathbf{A}^{(\varepsilon)})^2 &= \begin{bmatrix} 0 & 0 & 0 \\ 0 & \frac{(1+\varepsilon)^2}{2} z_1 z_2 & \frac{(1+\varepsilon)^2}{2} z_2^2 \\ 0 & -\frac{(1+\varepsilon)^2}{2} z_1^2 & -\frac{(1+\varepsilon)^2}{2} z_1 z_2 \end{bmatrix}, \\ (z\mathbf{A}^{(\varepsilon)})^3 &= (z\mathbf{A}^{(\varepsilon)})^4 = \ldots = 0, \end{aligned}$$

i.e., $z\mathbf{A}^{(\varepsilon)}$ is a nilpotent operator of degree 3. Hence, for any $\rho > 1$ and $z \in \mathbb{D}^2$,

$$\begin{aligned} \|\varphi_{\rho,2}^{\mathbf{A}^{(0)}}(z)\| &= \left\| z\mathbf{A}^{(0)}((\rho-1)z\mathbf{A}^{(0)} - \rho I)^{-1} \right\| = \left\| \frac{z\mathbf{A}^{(0)}}{\rho} \left(I - \frac{\rho-1}{\rho} z\mathbf{A}^{(0)} \right)^{-1} \right\| \\ &= \left\| \frac{z\mathbf{A}^{(0)}}{\rho} + (\rho-1) \left(\frac{z\mathbf{A}^{(0)}}{\rho} \right)^2 \right\| \leq \frac{1}{\rho} \|z\mathbf{A}^{(0)}\| + \frac{\rho-1}{\rho^2} \|(z\mathbf{A}^{(0)})^2\| \\ &= \frac{1}{\rho} \left\| \begin{bmatrix} 0 & \frac{z_1}{\sqrt{2}} & \frac{z_2}{\sqrt{2}} \\ \frac{z_2}{\sqrt{2}} & 0 & 0 \\ -\frac{z_1}{\sqrt{2}} & 0 & 0 \end{bmatrix} \right\| + \frac{\rho-1}{\rho^2} \left\| \begin{bmatrix} 0 \\ \frac{z_2}{\sqrt{2}} \\ -\frac{z_1}{\sqrt{2}} \end{bmatrix} \begin{bmatrix} 0 \\ \frac{z_1}{\sqrt{2}} \\ \frac{z_2}{\sqrt{2}} \end{bmatrix}^T \right\| \\ &\leq \frac{1}{\rho} + \frac{\rho-1}{\rho^2} = \frac{2\rho-1}{\rho^2} < 1. \end{aligned}$$

Then, due to the von Neumann inequality in two variables [3], one has

$$\|\varphi_{\rho,2}^{\mathbf{A}^{(0)}}(\mathbf{C})\| \leq \frac{2\rho-1}{\rho^2} < 1, \quad \mathbf{C} \in \mathcal{C}^2,$$

i.e., $\varphi_{\rho,2}^{\mathbf{A}^{(0)}} \in \mathcal{AS}_2(\mathbb{C}^3)$. Analogously, for $\varepsilon > 0$ small enough (the choice of ε depends on ρ), one has

$$\sup_{\mathbf{C} \in \mathcal{C}^2} \|\varphi_{\rho,2}^{\mathbf{A}^{(\varepsilon)}}(\mathbf{C})\| = \sup_{z \in \mathbb{D}^2} \|\varphi_{\rho,2}^{\mathbf{A}^{(\varepsilon)}}(\mathbf{z})\| \leq \frac{1+\varepsilon}{\rho} + (\rho-1)\left(\frac{1+\varepsilon}{\rho}\right)^2 < 1,$$

i.e., $\varphi_{\rho,2}^{\mathbf{A}^{(\varepsilon)}} \in \mathcal{AS}_2(\mathbb{C}^3)$, and by Theorem 3.3, $\mathbf{A}^{(\varepsilon)} \in C_{\rho,2}$.

Let us show now that for any $\varepsilon > 0$ the pair $\mathbf{A}^{(\varepsilon)} = (A_1^{(\varepsilon)}, A_2^{(\varepsilon)})$ is not simultaneously similar to any pair $\mathbf{T} = (T_1, T_2) \in C_{1,2}$. Observe that

$$\begin{aligned}(A_1^{(\varepsilon)} + A_2^{(\varepsilon)})(A_1^{(\varepsilon)} - A_2^{(\varepsilon)}) &= \begin{bmatrix} 0 & \frac{1+\varepsilon}{\sqrt{2}} & \frac{1+\varepsilon}{\sqrt{2}} \\ \frac{1+\varepsilon}{\sqrt{2}} & 0 & 0 \\ -\frac{1+\varepsilon}{\sqrt{2}} & 0 & 0 \end{bmatrix} \begin{bmatrix} 0 & \frac{1+\varepsilon}{\sqrt{2}} & -\frac{1+\varepsilon}{\sqrt{2}} \\ -\frac{1+\varepsilon}{\sqrt{2}} & 0 & 0 \\ -\frac{1+\varepsilon}{\sqrt{2}} & 0 & 0 \end{bmatrix} \\ &= \begin{bmatrix} -(1+\varepsilon)^2 & 0 & 0 \\ 0 & \frac{(1+\varepsilon)^2}{2} & -\frac{(1+\varepsilon)^2}{2} \\ 0 & -\frac{(1+\varepsilon)^2}{2} & \frac{(1+\varepsilon)^2}{2} \end{bmatrix}.\end{aligned}$$

Then

$$\lim_{n\to+\infty} \|[(A_1^{(\varepsilon)} + A_2^{(\varepsilon)})(A_1^{(\varepsilon)} - A_2^{(\varepsilon)})]^n\| = \infty. \tag{5.3}$$

On the other hand, if $\mathbf{A}^{(\varepsilon)} = (A_1^{(\varepsilon)}, A_2^{(\varepsilon)})$ is simultaneously similar to some $\mathbf{T} = (T_1, T_2) \in C_{1,2}$ then for any $n \in \mathbb{N}$ one would have

$$\|[(A_1^{(\varepsilon)} + A_2^{(\varepsilon)})(A_1^{(\varepsilon)} - A_2^{(\varepsilon)})]^n\| = \|S^{-1}[(T_1+T_2)(T_1-T_2)]^n S\| \le \|S\|\|S^{-1}\| < \infty,$$

since $\|T_1 \pm T_2\| \le 1$. We get a contradiction with (5.3).

Examples of N-tuples of operators from $C_{\rho,N}$, $\rho > 1$, which are not simultaneously similar to any $\mathbf{T} \in C_{1,N}$ for the case $N > 2$ can be obtained from the examples of pairs $\mathbf{A} = (A_1^{(\varepsilon)}, A_2^{(\varepsilon)})$ above, for sufficiently small $\varepsilon > 0$, by setting zeros for the rest of operators in these N-tuples: $\widetilde{\mathbf{A}}^{(\varepsilon)} := (A_1^{(\varepsilon)}, A_2^{(\varepsilon)}, 0, \dots, 0)$. □

Let $\mathbf{A} = (A_1, \dots, A_N) \in L(\mathcal{X})^N$. Then $\widetilde{\mathbf{A}} = (\widetilde{A}_1, \dots, \widetilde{A}_N) \in L(\widetilde{\mathcal{X}})^N$ is called a *uniform ρ-dilation* of $\mathbf{A}$ if $\widetilde{\mathcal{X}} \supset \mathcal{X}$ and

$$\forall n \in \mathbb{N},\ \forall i_1, \dots, i_n \in \{1, \dots, N\}, \quad A_{i_1} \cdots A_{i_n} = \rho P_{\mathcal{X}} \widetilde{A}_{i_1} \cdots \widetilde{A}_{i_n} | \mathcal{X}, \tag{5.4}$$

or equivalently,

$$\forall n \in \mathbb{N},\ \forall z^{(1)}, \dots, z^{(n)} \in \mathbb{C}^N, \quad z^{(1)}\mathbf{A} \cdots z^{(n)}\mathbf{A} = \rho P_{\mathcal{X}} z^{(1)}\widetilde{\mathbf{A}} \cdots z^{(n)}\widetilde{\mathbf{A}} | \mathcal{X}. \tag{5.5}$$

Note 5.3. Compare (5.4) and (5.5) with (3.1) and (3.2).

Clearly, a uniform ρ-dilation is a ρ-dilation. If $\widetilde{\mathbf{A}} \in L(\widetilde{\mathcal{X}})^N$ is a uniform ρ-dilation of $\mathbf{A} \in L(\mathcal{X})$, and for any $\zeta \in \mathbb{T}^N$, $\zeta\widetilde{\mathbf{A}}$ is a unitary operator, then $\widetilde{\mathbf{A}}$ is called a *uniform unitary ρ-dilation* of $\mathbf{A}$. Denote by $C^u_{\rho,N}$ the class of N-tuples of operators $\mathbf{A} = (A_1, \dots, A_N)$ on a common Hilbert space which admit a uniform unitary ρ-dilation. Clearly, $C^u_{\rho,N} \subset C_{\rho,N}$.

Theorem 5.4. *Any $\mathbf{A} = (A_1, \dots, A_N) \in C^u_{\rho,N}$ is simultaneously similar to some $\mathbf{T} = (T_1, \dots, T_N) \in C^u_{1,N}$.*

Proof. Let $\mathbf{A} = (A_1, \dots, A_N) \in C^u_{\rho,N} \cap L(\mathcal{X})^N$, and $\mathbf{U} = (U_1, \dots, U_N) \in L(\widetilde{\mathcal{X}})^N$ be a uniform unitary ρ-dilation of $\mathbf{A}$. Let $\mathcal{A} \subset L(\widetilde{\mathcal{X}})$ be the minimal C^*-algebra which contains the operators $I_{\widetilde{\mathcal{X}}}, U_1, \dots, U_N$, and $\mathcal{B} \subset L(\widetilde{\mathcal{X}})$ be the minimal algebra over

$\mathbb{C}$ which contains the operators $U_1, \dots, U_N$. Clearly, $\mathcal{B} \subset \mathcal{A}$. Let $\varphi : \mathcal{B} \longrightarrow L(\mathcal{X})$ be a homomorphism defined on the generators as

$$\varphi : U_k \longmapsto A_k, \ k = 1, \dots, N.$$

The algebra $\mathcal{B}$ consists of operators of the form

$$p(\mathbf{U}) = \sum_{1 \le k \le m, \ i_1, \dots, i_k \in \{1, \dots, N\}} \alpha_{i_1, \dots, i_k} U_{i_1} \cdots U_{i_k},$$

where $\alpha_{i_1, \dots, i_k} \in \mathbb{C}$ for all $i_1, \dots, i_k \in \{1, \dots, N\}$. Then

$$\begin{aligned} \varphi(p(\mathbf{U})) &= \varphi(\sum \alpha_{i_1, \dots, i_k} U_{i_1} \cdots U_{i_k}) = \sum \alpha_{i_1, \dots, i_k} A_{i_1} \cdots A_{i_k} \\ &= p(\mathbf{A}) = \rho P_{\mathcal{X}} p(\mathbf{U})|\mathcal{X}. \end{aligned}$$

Therefore, if $p(\mathbf{U}) = 0$ then $\varphi(p(\mathbf{U})) = 0$, and φ is correctly defined. The homomorphism φ is *completely bounded*, i.e.,

$$\|\varphi\|_{cb} := \sup_{n \in \mathbb{N}} \| \operatorname{id}_n \otimes \varphi \| < \infty,$$

where id_n is the identical map of the matrix algebra $\mathcal{M}_n(\mathbb{C})$ onto itself. Moreover, $\|\varphi\|_{cb} \le \rho$. Indeed, for any $n \in \mathbb{N}$ and a polynomial $n \times n$ matrix of N noncommuting variables,

$$P(\mathbf{X}) = [p_{ij}(\mathbf{X})]_{i,j=1}^n = \left[\sum_{1 \le k \le m, \ i_1, \dots, i_k \in \{1, \dots, N\}} \alpha^{(ij)}_{i_1, \dots, i_k} X_{i_1} \cdots X_{i_k} \right]_{i,j=1}^n,$$

$$\begin{aligned} \|(\operatorname{id}_n \otimes \varphi)(P(\mathbf{U}))\| &= \|(\operatorname{id}_n \otimes \varphi) \left([p_{ij}(\mathbf{U})]_{i,j=1}^n \right) \| = \| [\varphi(p_{ij}(\mathbf{U}))]_{i,j=1}^n \| \\ &= \| [p_{ij}(\mathbf{A})]_{i,j=1}^n \| = \| [\rho P_{\mathcal{X}} p_{ij}(\mathbf{U})|\mathcal{X}]_{i,j=1}^n \| \\ &= \rho \| (I_{\mathbb{C}^n} \otimes P_{\mathcal{X}}) [p_{ij}(\mathbf{U})]_{i,j=1}^n | \mathbb{C}^n \otimes \mathcal{X} \| \\ &\le \rho \| [p_{ij}(\mathbf{U})]_{i,j=1}^n \| = \rho \| P(\mathbf{U}) \|. \end{aligned}$$

Then, by Theorem 3.1 in [28], there exist a Hilbert space $\mathcal{N}$, a *completely contractive* homomorphism $\gamma : \mathcal{B} \longrightarrow L(\mathcal{N})$ (i.e., such that $\|\gamma\|_{cb} \le 1$), and a boundedly invertible operator $S \in L(\mathcal{X}, \mathcal{N})$ such that

$$\varphi(b) = S^{-1} \gamma(b) S, \quad b \in \mathcal{B}.$$

Moreover, as was shown in the proof of Theorem 3.1 in [28], γ can be chosen in the form

$$\gamma(b) = P_{\mathcal{N}} \pi(b) | \mathcal{N}, \quad b \in \mathcal{B},$$

where $\pi : \mathcal{A} \longrightarrow L(\mathcal{K})$ is a $*$-homomorphism, for some Hilbert space $\mathcal{K} \supset \mathcal{N}$. In addition, it follows from Theorem 2.7 and the proof of Theorem 2.8 in [28] that one can choose $\mathcal{K} = \mathcal{K}_1 \oplus \mathcal{K}_1$, for some Hilbert space $\mathcal{K}_1$, and

$$\pi(a) = \pi_1(a) \oplus 0, \quad a \in \mathcal{A},$$

where $\pi_1 : \mathcal{A} \longrightarrow L(\mathcal{K}_1)$ is a unital $*$-homomorphism. Set

$$T_k := \gamma(U_k) \in L(\mathcal{N}), \quad k = 1, \dots, N.$$

Then

$$A_k = \varphi(U_k) = S^{-1}T_kS, \quad k = 1, \dots, N.$$

It remains to show that $\mathbf{T} = (T_1, \dots, T_N) \in C_{1,N}^u$. Set

$$W_k := \pi(U_k) \in L(\mathcal{K}), \quad k = 1, \dots, N.$$

Since for any $n \in \mathbb{N}$ and $i_1, \dots, i_n \in \{1, \dots, N\}$ one has

$$\begin{aligned} T_{i_1} \cdots T_{i_n} &= \gamma(U_{i_1} \cdots U_{i_n}) = P_{\mathcal{N}}\pi(U_{i_1} \cdots U_{i_n})|\mathcal{N} \\ &= P_{\mathcal{N}} W_{i_1} \cdots W_{i_n}|\mathcal{N}, \end{aligned}$$

$\mathbf{W} = (W_1, \dots, W_N)$ is a uniform 1-dilation of $\mathbf{T} = (T_1, \dots, T_N)$, however, still not unitary. Actually,

$$W_k = \pi(U_k) = \pi_1(U_k) \oplus 0 \ (=: W_k^{(1)} \oplus 0), \quad k = 1, \dots, N.$$

Since π_1 is a unital $*$-homomorphism, and for any $\zeta \in \mathbb{T}^N$,

$$(\zeta\mathbf{U})^*\zeta\mathbf{U} = I_{\widetilde{\mathcal{X}}} = \zeta\mathbf{U}(\zeta\mathbf{U})^*,$$

one has, for any $\zeta \in \mathbb{T}^N$,

$$(\zeta\mathbf{W}^{(1)})^*\zeta\mathbf{W}^{(1)} = I_{\mathcal{K}_1} = \zeta\mathbf{W}^{(1)}(\zeta\mathbf{W}^{(1)})^*,$$

where $\mathbf{W}^{(1)} = (W_1^{(1)}, \dots, W_N^{(1)})$. Set

$$\widetilde{W}_k := W_k^{(1)} \oplus \delta_{1k}V \in L(\mathcal{K}_1 \oplus \mathcal{R}), \quad k = 1, \dots, N,$$

where δ_{ij} is the Kronecker symbol, and V is a unitary dilation of the zero operator on $\mathcal{K}_1$, e.g., the two-sided shift on the space $\mathcal{R} := l^2(\mathcal{K}_1) = \bigoplus_{-\infty}^{+\infty} \mathcal{K}_1$ (here we identify the space $\mathcal{K}_1$ with the subspace $\ldots \oplus \{0\} \oplus \{0\} \oplus \mathcal{K}_1 \oplus \{0\} \oplus \{0\} \oplus \ldots$ in $\mathcal{R}$). Then, for any $\zeta \in \mathbb{T}^N$,

$$(\zeta\widetilde{\mathbf{W}})^*\zeta\widetilde{\mathbf{W}} = I_{\mathcal{K}_1 \oplus \mathcal{R}} = \zeta\widetilde{\mathbf{W}}(\zeta\widetilde{\mathbf{W}})^*,$$

and $\widetilde{\mathbf{W}} = (\widetilde{W}_1, \dots, \widetilde{W}_N) \in L(\mathcal{K}_1 \oplus \mathcal{R})$ is a uniform unitary 1-dilation of $\mathbf{W} = (W_1, \dots, W_N)$, and therefore, of $\mathbf{T} = (T_1, \dots, T_N)$. Thus, $\mathbf{T} \in C_{1,N}^u$, as required. □

Theorem 5.4 is similar to the result of G. Popescu [30] on simultaneous similarity of ρ-contractions to 1-contractions in several variables, however his notion of multivariable ρ-contractions is different. Let us clarify the relation between these two results. Denote by $C_{\rho,N}^P$ (we use here this notation instead of just C_ρ, as in [30]) the *Popescu class* of all N-tuples $\mathbf{A} = (A_1, \dots, A_N)$ of bounded linear operators on a common Hilbert space, say $\mathcal{X}$, which have a *uniform isometric ρ-dilation*, i.e., such an N-tuple of operators $\mathbf{V} = (V_1, \dots, V_N) \in L(\widetilde{\mathcal{X}})^N$, $\widetilde{\mathcal{X}} \supset \mathcal{X}$, for which

(1) $V_k^*V_k = I_{\widetilde{\mathcal{X}}}, \quad k = 1, \dots, N;$

(2) $V_k^*V_j = 0, \quad k \neq j;$

(3) $\forall n \in \mathbb{N}, \ \forall i_1, \dots, i_n, \quad A_{i_1} \cdots A_{i_n} = \rho P_{\mathcal{X}} V_{i_1} \cdots V_{i_n}|\mathcal{X}.$

Condition (2) in this definition can be replaced by

(2′) $\sum_{k=1}^{N} V_k V_k^* \preceq I_{\widetilde{\mathcal{X}}}$,

since (1)&(2)$\Longleftrightarrow$(1)&(2'). According to [29], the class $C^P_{1,N}$ coincides with the class of N-tuples of operators $\mathbf{A} = (A_1, \ldots, A_N)$ ($\in L(\mathcal{X})^N$, for some Hilbert space $\mathcal{X}$) such that

$$\sum_{k=1}^{N} A_k A_k^* \preceq I_{\mathcal{X}}.$$

By Theorem 4.5 in [30], any $\mathbf{A} = (A_1, \ldots, A_N) \in C^P_{\rho,N}$, $\rho > 0$, is simultaneously similar to some $\mathbf{T} = (T_1, \ldots, T_N) \in C^P_{1,N}$. This is a generalization of the theorem of Sz.-Nagy and Foiaş [33] to several variables. Theorem 5.4 of the present paper is a different generalization of the same result, since our classes $C^u_{\rho,N}$ are different from Popescu's classes $C^P_{\rho,N}$ for $N > 1$. More precisely, the following is true.

Theorem 5.5. *For any $N > 1$ and $\rho > 0$, $C^u_{\rho,N} \subsetneq C^P_{\rho,N}$.*

Proof. Let $\mathbf{A} \in C^u_{\rho,N} \cap L(\mathcal{X})^N$, $N > 1$, and $\mathbf{U} \in L(\widetilde{\mathcal{X}})^N$ be a uniform unitary ρ-dilation of $\mathbf{A}$. Since for any $\zeta \in \mathbb{T}^N$ the operator $\zeta\mathbf{U}$ is unitary, it follows that

$$\zeta\mathbf{U}(\zeta\mathbf{U})^* = I_{\widetilde{\mathcal{X}}}, \quad \zeta \in \mathbb{T}^N,$$

which implies

$$\sum_{k=1}^{N} U_k U_k^* = I_{\widetilde{\mathcal{X}}}.$$

Thus, by [29], $\mathbf{U} \in C^P_{1,N}$. Let $\mathbf{V} \in L(\hat{\mathcal{X}})^N$ be a uniform isometric 1-dilation of $\mathbf{U}$ in the sense of Popescu. Then for any $n \in \mathbb{N}$ and $i_1, \ldots, i_n \in \{1, \ldots, N\}$,

$$\begin{aligned} A_{i_1} \cdots A_{i_n} &= \rho P_{\mathcal{X}} U_{i_1} \cdots U_{i_n}|\mathcal{X} = \rho P_{\mathcal{X}}(P_{\widetilde{\mathcal{X}}} V_{i_1} \cdots V_{i_n}|\widetilde{\mathcal{X}})|\mathcal{X} \\ &= \rho P_{\mathcal{X}} V_{i_1} \cdots V_{i_n}|\mathcal{X}, \end{aligned}$$

i.e., $\mathbf{V}$ is a uniform isometric ρ-dilation of $\mathbf{A}$ in the sense of Popescu. Thus, $\mathbf{A} \in C^P_{\rho,N}$. This proves the inclusion $C^u_{\rho,N} \subset C^P_{\rho,N}$.

Let us prove that this inclusion is proper for any $N > 1$ and $\rho > 0$. Firstly, consider the case $N = 2$. Let $B \in L(\mathcal{X}_0)$ be any operator of the class C_ρ with $\|B\| = \rho$. For example,

$$B := \begin{bmatrix} 0 & \rho \\ 0 & 0 \end{bmatrix} \in L(\mathbb{C}^2)$$

satisfies $B^2 = 0$ and $\|B\| = \rho$, therefore by properties (iii) and (xi) in Theorem 2.5, $w_\rho(B) = 1$, and by property (iv) in the same theorem, $B \in C_\rho$. Set $\mathcal{X} := \mathcal{X}_0 \oplus \mathcal{X}_0$,

$$A_1 := \begin{bmatrix} B & 0 \\ 0 & 0 \end{bmatrix} \in L(\mathcal{X}), \quad A_2 := \begin{bmatrix} 0 & 0 \\ B & 0 \end{bmatrix} \in L(\mathcal{X}). \tag{5.6}$$

Let $U \in L(\widetilde{\mathcal{X}}_0)$ be a unitary ρ-dilation of B. Set $\widetilde{\mathcal{X}} := \widetilde{\mathcal{X}}_0 \oplus \widetilde{\mathcal{X}}_0 \oplus \ldots$, and identify $\mathcal{X} = \mathcal{X}_0 \oplus \mathcal{X}_0$ with the subspace $\mathcal{X}_0 \oplus \mathcal{X}_0 \oplus \{0\} \oplus \{0\} \oplus \ldots$ in $\widetilde{\mathcal{X}}$. Set

$$V_1 := \begin{bmatrix} \boxed{\begin{matrix} U \\ 0 \end{matrix}} & & \\ & \boxed{\begin{matrix} U \\ 0 \end{matrix}} & \\ & & \ddots \end{bmatrix} \in L(\widetilde{\mathcal{X}}), \quad V_2 := \begin{bmatrix} \boxed{\begin{matrix} 0 \\ U \end{matrix}} & & \\ & \boxed{\begin{matrix} 0 \\ U \end{matrix}} & \\ & & \ddots \end{bmatrix} \in L(\widetilde{\mathcal{X}}),$$

i.e., the operators V_1 and V_2 are introduced here as infinite block-diagonal matrices with equal operator blocks $\begin{bmatrix} U \\ 0 \end{bmatrix} \in L(\widetilde{\mathcal{X}}_0, \widetilde{\mathcal{X}}_0 \oplus \widetilde{\mathcal{X}}_0)$ (resp., $\begin{bmatrix} 0 \\ U \end{bmatrix} \in L(\widetilde{\mathcal{X}}_0, \widetilde{\mathcal{X}}_0 \oplus \widetilde{\mathcal{X}}_0)$) on the main diagonal. We will show that the pair $\mathbf{V} = (V_1, V_2)$ is a uniform isometric ρ-dilation of the pair $\mathbf{A} = (A_1, A_2)$ in the sense of Popescu. First of all, observe that

$$V_1^* V_1 = I_{\widetilde{\mathcal{X}}} = V_2^* V_2, \quad V_1^* V_2 = V_2^* V_1 = 0.$$

Next, the following relations hold:

$$\begin{aligned} \forall k \in \mathbb{N},\ A_1^k = \begin{bmatrix} B^k & 0 \\ 0 & 0 \end{bmatrix} = \begin{bmatrix} \rho P_{\mathcal{X}_0} U^k | \mathcal{X}_0 & 0 \\ 0 & 0 \end{bmatrix} &= \rho P_{\mathcal{X}} V_1^k | \mathcal{X}; \\ \forall k, n \in \mathbb{N},\ \forall i_1, \ldots, i_n \in \{1, 2\},\ A_1^k A_2 A_{i_1} \cdots A_{i_n} &= 0 \\ &= \rho P_{\mathcal{X}} V_1^k V_2 V_{i_1} \cdots V_{i_n} | \mathcal{X} \end{aligned}$$

(since $A_1^k A_2 = 0,\ P_{\widetilde{\mathcal{X}}_0 \oplus \widetilde{\mathcal{X}}_0 \oplus \{0\} \oplus \{0\} \oplus \ldots} V_1^k V_2 = 0$);

$$\begin{aligned} A_2 = \begin{bmatrix} 0 & 0 \\ B & 0 \end{bmatrix} = \begin{bmatrix} 0 & 0 \\ \rho P_{\mathcal{X}_0} U | \mathcal{X}_0 & 0 \end{bmatrix} &= \rho P_{\mathcal{X}} V_2 | \mathcal{X}; \\ \forall k, n \in \mathbb{N},\ \forall i_1, \ldots, i_n \in \{1, 2\},\ A_2^{k+1} A_{i_1} \cdots A_{i_n} &= 0 \\ &= \rho P_{\mathcal{X}} V_2^{k+1} V_{i_1} \cdots V_{i_n} | \mathcal{X} \end{aligned}$$

(since $A_2^2 = 0,\ P_{\widetilde{\mathcal{X}}_0 \oplus \widetilde{\mathcal{X}}_0 \oplus \{0\} \oplus \{0\} \oplus \ldots} V_2^2 = 0$);

$$\begin{aligned} \forall k \in \mathbb{N},\ A_2 A_1^k = \begin{bmatrix} 0 & 0 \\ B^{k+1} & 0 \end{bmatrix} &= \begin{bmatrix} 0 & 0 \\ \rho P_{\mathcal{X}_0} U^{k+1} | \mathcal{X}_0 & 0 \end{bmatrix} = \rho P_{\mathcal{X}} V_2 V_1^k | \mathcal{X}; \\ \forall k \in \mathbb{N},\ A_2 A_1^k A_2 = 0 &= P_{\mathcal{X}} V_2 V_1^k V_2 | \mathcal{X}; \\ \forall k, n \in \mathbb{N},\ \forall i_1, \ldots, i_n \in \{1, 2\}, &\quad A_2 A_1^k A_2 A_{i_1} \cdots A_{i_n} = 0 \\ &= \rho P_{\mathcal{X}} V_2 V_1^k V_2 V_{i_1} \cdots V_{i_n} | \mathcal{X} \end{aligned}$$

(since $A_1^k A_2 = 0,\ P_{\widetilde{\mathcal{X}}_0 \oplus \widetilde{\mathcal{X}}_0 \oplus \{0\} \oplus \{0\} \oplus \ldots} V_2 V_1^k V_2 = 0$). Finally, we get

$$\forall n \in \mathbb{N},\ \forall i_1, \ldots, i_n \in \{1, 2\},\ A_{i_1} \cdots A_{i_n} = \rho P_{\mathcal{X}} V_{i_1} \cdots V_{i_n} | \mathcal{X}.$$

Thus, $\mathbf{V}$ is a uniform isometric ρ-dilation of $\mathbf{A}$ in the sense of Popescu. However, for any $\zeta \in \mathbb{T}^N$,

$$\|\zeta\mathbf{A}\| = \left\| \begin{bmatrix} \zeta_1 B & 0 \\ \zeta_2 B & 0 \end{bmatrix} \right\| = \sqrt{2}\|B\| = \sqrt{2}\rho > \rho.$$

Therefore, $\zeta\mathbf{A} \notin C_\rho$ for all $\zeta \in \mathbb{T}^N$. We obtain $\mathbf{A} \in C^P_{\rho,2} \backslash C^u_{\rho,2}$ (moreover, $\mathbf{A} \notin C_{\rho,2}$).

For the case $N > 2$ (and any $\rho > 0$) an analogous example of $\widetilde{\mathbf{A}} \in C^P_{\rho,N} \backslash C^u_{\rho,N}$ is easily obtained from the previous one, by setting zeros for the rest of operators in the N-tuple, i.e., $\widetilde{\mathbf{A}} := (A_1, A_2, 0, \dots, 0)$, where A_1 and A_2 are defined in (5.6). In this case the construction of a uniform isometric ρ-dilation of $\mathbf{A}$ in the sense of Popescu should be slightly changed (we leave this to a reader as an easy exercise). □

Remark 5.6. The pair $\mathbf{A}^{(\varepsilon)} = (A_1^{(\varepsilon)}, A_2^{(\varepsilon)})$ constructed in Theorem 5.2 doesn't belong to the class $C^u_{\rho,2}$ for any $\varepsilon > 0$ and $\rho > 1$. Indeed, we have shown in Theorem 5.2 that $\mathbf{A}^{(\varepsilon)}$ is not simultaneously similar to any $\mathbf{T} = (T_1, T_2) \in C_{1,2}$, not speaking of $\mathbf{T} \in C^u_{1,2}$. Thus, by Theorem 5.5, $\mathbf{A}^{(\varepsilon)} \notin C^u_{\rho,2}$. This can be shown also by the following estimate: if $\mathbf{A}^{(\varepsilon)} \in C^u_{\rho,2}$ for some $\varepsilon > 0$ and $\rho > 1$, then there exists a uniform unitary ρ-dilation $\mathbf{U}^{(\varepsilon)} = (U_1^{(\varepsilon)}, U_2^{(\varepsilon)})$ of $\mathbf{A}^{(\varepsilon)} = (A_1^{(\varepsilon)}, A_2^{(\varepsilon)})$, and for any $n \in \mathbb{N}$,

$$\begin{aligned} \|[(A_1^{(\varepsilon)} + A_2^{(\varepsilon)})(A_1^{(\varepsilon)} - A_2^{(\varepsilon)})]^n\| &= \|\rho P_{\mathcal{X}}[(U_1^{(\varepsilon)} + U_2^{(\varepsilon)})(U_1^{(\varepsilon)} - U_2^{(\varepsilon)})]^n|\mathcal{X}\| \\ &\leq \rho\|[(U_1^{(\varepsilon)} + U_2^{(\varepsilon)})(U_1^{(\varepsilon)} - U_2^{(\varepsilon)})]^n\| = \rho < \infty. \end{aligned}$$

This contradicts to (5.3). Thus, for each $\rho > 1$ we obtain for $\varepsilon > 0$ small enough, $\mathbf{A}^{(\varepsilon)} = (A_1^{(\varepsilon)}, A_2^{(\varepsilon)}) \in C_{\rho,2} \backslash C^u_{\rho,2}$, as well as $\widetilde{\mathbf{A}} := (A_1, A_2, 0, \dots, 0) \in C_{\rho,N} \backslash C^u_{\rho,N}$.

Acknowledgements

I am grateful for the hospitality of the Universities of Leeds and Newcastle upon Tyne where a part of this work was carried out during my visits under the International Short Visit Scheme of the LMS (grant no. 5620). I wish to thank also Dr. Michael Dritschel from the University of Newcastle upon Tyne for useful discussions.

References

[1] J. Agler, 'On the representation of certain holomorphic functions defined on a polydisc', in Topics in Operator Theory: Ernst D. Hellinger Memorial Volume (L. de Branges, I. Gohberg, and J. Rovnyak, eds.), *Oper. Theory Adv. Appl.* 48 (1990) 47–66 (Birkhäuser Verlag, Basel).

[2] J. Agler and J.E. McCarthy, 'Nevanlinna–Pick interpolation on the bidisk', *J. Reine Angew. Math.* 506 (1999) 191–204.

[3] T. Ando, 'On a pair of commutative contractions', *Acta Sci. Math. (Szeged)* 24 (1963) 88–90.

[4] T. Ando and K. Nishio, 'Convexity properties of operator radii associated with unitary ρ-dilations', *Michigan Math. J.* 20 (1973) 303–307.

[5] C. Badea and G. Cassier, 'Constrained von Neumann inequalities', *Adv. Math.* 166 no. 2 (2002) 260–297.

[6] J.A. Ball, W.S. Li, D. Timotin and T.T. Trent, 'A commutant lifting theorem on the polydisc', *Indiana Univ. Math. J.* 48 no. 2 (1999) 653–675.

[7] J.A. Ball, C. Sadosky and V. Vinnikov, 'Conservative input-state-output systems with evolution on a multidimensional integer lattice', Multidimens. Syst. Signal Process., to appear.

[8] J.A. Ball and T.T. Trent, 'Unitary colligations, reproducing kernel Hilbert spaces, and Nevanlinna-Pick interpolation in several variables', *J. Funct. Anal.* 157 no. 1 (1998) 1–61.

[9] C.A. Berger, 'A strange dilation theorem', *Notices Amer. Math. Soc.* 12 (1965) 590.

[10] G. Cassier and T. Fack, 'Contractions in von Neumann algebras', *J. Funct. Anal.* 135 no. 2 (1996) 297–338.

[11] C. Davis, 'The shell of a Hilbert-space operator', *Acta Sci. Math. (Szeged)* 29 (1968) 69–86.

[12] M.A. Dritschel, S. McCullough and H.J. Woerdeman, 'Model theory for ρ-contractions, $\rho \leq 2$', *J. Operator Theory* 41 no. 2 (1999) 321–350.

[13] E. Durszt, 'On unitary ρ-dilations of operators', *Acta Sci. Math. (Szeged)* 27 (1966) 247–250.

[14] C.K. Fong and J.A.R. Holbrook, 'Unitarily invariant operator norms', *Canad. J. Math.* 35 no. 2 (1983) 274–299.

[15] J.A.R. Holbrook, 'On the power-bounded operators of Sz.-Nagy and Foiaş', *Acta Sci. Math. (Szeged)* 29 (1968) 299–310.

[16] J.A.R. Holbrook, 'Inequalities governing the operator radii associated with unitary ρ-dilations', *Michigan Math. J.* 18 (1971) 149–159.

[17] D.S. Kalyuzhniy, 'Multiparametric dissipative linear stationary dynamical scattering systems: discrete case', *J. Operator Theory* 43 no. 2 (2000) 427–460.

[18] D.S. Kalyuzhniy, 'Multiparametric dissipative linear stationary dynamical scattering systems: discrete case. II. Existence of conservative dilations', *Integral Equations Operator Theory* 36 no. 1 (2000) 107–120.

[19] D.S. Kalyuzhniy, 'On the notions of dilation, controllability, observability, and minimality in the theory of dissipative scattering linear nD systems', in Proceedings CD of the Fourteenth International Symposium of Mathematical Theory of Networks and Systems (MTNS), June 19–23, 2000, Perpignan, France (A. El Jai and M. Fliess, Eds.), or `http://www.univ-perp.fr/mtns2000/articles/SI13_3.pdf/`.

[20] D.S. Kalyuzhniy-Verbovetzky, 'Cascade connections of linear systems and factorizations of holomorphic operator functions around a multiple zero in several variables', *Math. Rep. (Bucur.)* 3(53) no. 4 (2001) 323–332.

[21] D.S. Kalyuzhniy-Verbovetzky, 'On J-conservative scattering system realizations in several variables', *Integral Equations Operator Theory* 43 no. 4 (2002) 450–465.

[22] D.S. Kalyuzhnyĭ, 'The von Neumann inequality for linear matrix functions of several variables', *Mat. Zametki* 64 no. 2 (1998) 218–223 (Russian). English transl. in *Math. Notes* 64 no. 1–2 (1998) 186–189 (1999).

[23] D.S. Kalyuzhnyĭ-Verbovetskiĭ, 'Cascade connections of multiparameter linear systems and the conservative realization of a decomposable inner operator function on the bidisk', *Mat. Stud.* 15 no. 1 (2001) 65–76 (Russian).

[24] T. Nakazi and K. Okubo, 'ρ-contraction and 2×2 matrix', *Linear Algebra Appl.* 283 no. 1-3 (1998) 165–169.

[25] J. von Neumann, 'Eine Spektraltheorie für allgemeine Operatoren eines unitären Raumes', *Math. Nachr.* 4 (1951) 258–281 (German).

[26] K. Okubo and T. Ando, 'Operator radii of commuting products', *Proc. Amer. Math. Soc.* 56 no. 1 (1976) 203–210.

[27] K. Okubo and I. Spitkovsky, 'On the characterization of 2×2 ρ-contraction matrices', *Linear Algebra Appl.* 325 no. 1-3 (2001) 177–189.

[28] V.I. Paulsen, 'Every completely polynomially bounded operator is similar to a contraction', *J. Funct. Anal.* 55 no. 1 (1984) 1–17.

[29] G. Popescu, 'Isometric dilations for infinite sequences of noncommuting operators', *Trans. Amer. Math. Soc.* 316 no. 2 (1989) 523–536.

[30] G. Popescu, 'Positive-definite functions on free semigroups', *Canad. J. Math.* 48 no. 4 (1996) 887–896.

[31] B. Sz.-Nagy, 'Sur les contractions de l'espace de Hilbert', *Acta Sci. Math. (Szeged)* 15 (1953) 87–92 (French).

[32] B. Sz.-Nagy and C. Foiaş, 'On certain classes of power-bounded operators in Hilbert space', *Acta Sci. Math. (Szeged)* 27 (1966) 17–25.

[33] B. Sz.-Nagy and C. Foiaş, 'Similitude des opérateurs de class $\mathcal{C}_\rho$ à des contractions', *C. R. Acad. Sci. Paris Sér. A-B* 264 (1967) A1063–A1065 (French).

[34] B. Sz.-Nagy and C. Foiaş, *Harmonic analysis of operators on Hilbert space* (North-Holland, Amsterdam–London, 1970).

[35] J.P. Williams, 'Schwarz norms for operators', *Pacific J. Math.* 24 (1968) 181–188.

Dmitry S. Kalyuzhnyĭ-Verbovetzkiĭ
Department of Mathematics
Ben-Gurion University of the Negev
P.O. Box 653
Beer-Sheva 84105, Israel
e-mail: `dmitryk@wisdom.weizmann.ac.il`

Operator Theory:
Advances and Applications, Vol. 160, 299–309

The Singularly Continuous Spectrum and Non-Closed Invariant Subspaces

Vadim Kostrykin and Konstantin A. Makarov

Dedicated to Israel Gohberg on the occasion of his 75th birthday

Abstract. Let $\mathbf{A}$ be a bounded self-adjoint operator on a separable Hilbert space $\mathfrak{H}$ and $\mathfrak{H}_0 \subset \mathfrak{H}$ a closed invariant subspace of $\mathbf{A}$. Assuming that $\mathfrak{H}_0$ is of codimension 1, we study the variation of the invariant subspace $\mathfrak{H}_0$ under bounded self-adjoint perturbations $\mathbf{V}$ of $\mathbf{A}$ that are off-diagonal with respect to the decomposition $\mathfrak{H} = \mathfrak{H}_0 \oplus \mathfrak{H}_1$. In particular, we prove the existence of a one-parameter family of dense non-closed invariant subspaces of the operator $\mathbf{A} + \mathbf{V}$ provided that this operator has a nonempty singularly continuous spectrum. We show that such subspaces are related to non-closable densely defined solutions of the operator Riccati equation associated with generalized eigenfunctions corresponding to the singularly continuous spectrum of $\mathbf{B}$.

Mathematics Subject Classification (2000). Primary 47A55, 47A15; Secondary 47B15.

Keywords. Invariant subspaces, operator Riccati equation, singular spectrum.

1. Introduction

In the present article we address the problem of a perturbation of invariant subspaces of self-adjoint operators on a separable Hilbert space $\mathfrak{H}$ and related questions on the existence of solutions to the operator Riccati equation.

Given a self-adjoint operator $\mathbf{A}$ and a closed invariant subspace $\mathfrak{H}_0 \subset \mathfrak{H}$ of $\mathbf{A}$ we set $A_i = \mathbf{A}|_{\mathfrak{H}_i}$, $i = 0, 1$, with $\mathfrak{H}_1 = \mathfrak{H} \ominus \mathfrak{H}_0$. Assuming that the perturbation $\mathbf{V}$ is off-diagonal with respect to the orthogonal decomposition $\mathfrak{H} = \mathfrak{H}_0 \oplus \mathfrak{H}_1$ consider the self-adjoint operator

$$\mathbf{B} = \mathbf{A} + \mathbf{V} = \begin{pmatrix} A_0 & V \\ V^* & A_1 \end{pmatrix} \quad \text{with} \quad \mathbf{V} = \begin{pmatrix} 0 & V \\ V^* & 0 \end{pmatrix},$$

where V is a linear operator from $\mathfrak{H}_1$ to $\mathfrak{H}_0$. It is well known (see, e.g., [7]) that the Riccati equation

$$A_1X - XA_0 - XVX + V^* = 0 \tag{1}$$

has a closed (possibly unbounded) solution $X : \mathfrak{H}_0 \to \mathfrak{H}_1$ if and only if its graph

$$\mathcal{G}(\mathfrak{H}_0, X) := \{x \in \mathfrak{H} \,|\, x = x_0 \oplus Xx_0,\ x_0 \in \mathrm{Dom}(X) \subset \mathfrak{H}_0\} \tag{2}$$

is an invariant closed subspace for the operator $\mathbf{B}$.

Sufficient conditions guaranteeing the existence of a solution to equation (1) require in general the assumption that the spectra of the operators A_0 and A_1 are separated,

$$d := \mathrm{dist}(\mathrm{spec}(A_0), \mathrm{spec}(A_1)) > 0, \tag{3}$$

and hence $\mathfrak{H}_0$ and $\mathfrak{H}_1$ are necessarily spectral invariant subspaces of the operator $\mathbf{A}$. In particular (see [9]), if

$$\|V\| < c_\pi d \quad \text{with} \quad c_\pi = \frac{3\pi - \sqrt{\pi^2 + 32}}{\pi^2 - 4} = 0.503288\ldots, \tag{4}$$

then the Riccati equation (1) has a bounded solution X satisfying the bound

$$\frac{\|X\|}{\sqrt{1 + \|X\|^2}} \le \frac{\pi}{2} \frac{\|V\|}{d - \delta_V} < 1$$

with

$$\delta_V = \|V\| \tan\left(\frac{1}{2}\arctan\frac{2\|V\|}{d}\right).$$

It is plausible to conjecture that condition (4) can be relaxed by the weaker requirement $\|V\| < \sqrt{3}d/2$ (see [9] for details). However, no proof of that is available as yet.

In general, without additional assumptions, neither condition (3) nor a smallness assumption like (4) on the magnitude of the perturbation V can be dropped. However, if the spectra of A_0 and A_1 are subordinated in the sense that

$$\sup \mathrm{spec}(A_0) \le \inf \mathrm{spec}(A_1),$$

then for any V with arbitrary large norm the Riccati equation (1) has a contractive solution [8] (see also [1]). Note that in this case the invariant subspaces $\mathfrak{H}_0$ and $\mathfrak{H}_1$ are not necessarily supposed to be spectral invariant subspaces of $\mathbf{A}$.

In the present work we prove new existence results for the Riccati equation under the assumption that the subspace $\mathfrak{H}_1$ is *one-dimensional.* In particular, these results imply the existence of a one-parameter family of non-closed invariant subspaces of the self-adjoint operator $\mathbf{B}$, provided that $\mathbf{B}$ has nonempty singularly continuous spectrum.

The main result of our paper is presented by the following theorem.

Theorem 1. *Assume that* $\dim \mathfrak{H}_1 = 1$ *and suppose that* $\mathfrak{H}_0$ *is a cyclic subspace for the operator* A_0 *generated by the one-dimensional subspace* $\mathrm{Ran}\, V$. *Let* S_{pp} *denote the set of all eigenvalues of the operator* $\mathbf{B}$.

Then there exists a minimal support S_s of the singular part of the spectral measure of the operator $\mathbf{B}$ *such that:*

(i) *For any* $\lambda \in S_{sc} = S_s \setminus S_{pp}$ *the subspace* $\Psi(\lambda) = \mathcal{G}(\mathfrak{H}_0, X_\lambda) \subset \mathfrak{H}$ *is a dense non-closed graph subspace with* $X_\lambda : \mathfrak{H}_0 \to \mathfrak{H}_1$ *a non-closed densely defined operator solving the Riccati equation* (1) *in the sense of Definition* 2.3 *below.*
(ii) *For any* $\lambda \in S_{pp} \subset S_s$ *the subspace* $\Psi(\lambda) = \mathcal{G}(\mathfrak{H}_0, X_\lambda) \subset \mathfrak{H}$ *is a closed graph subspace of codimension* 1 *with* $X_\lambda : \mathfrak{H}_0 \to \mathfrak{H}_1$ *a bounded operator solving the Riccati equation* (1). *Moreover, the operator* X_λ *is an isolated point (in the operator norm topology) of the set of all bounded solutions to the Riccati equation.*

The mapping Ψ *from* S_s *to the set* $\mathcal{M}(\mathbf{B})$ *of all (not necessarily closed) subspaces of* $\mathfrak{H}$ *invariant with respect to the operator* $\mathbf{B}$ *is injective.*

The article is organized as follows. In Section 2 we establish a link between non-closable densely defined solutions to the Riccati equation (1) and the associated non-closed invariant subspaces of the operator $\mathbf{B}$. In Section 3 accommodating the Simon-Wolff theory [10] to rank two off-diagonal perturbations we perform the spectral analysis of this operator under the assumption that $\dim \mathfrak{H}_1 = 1$. The main result of this section is Theorem 3.4. Theorem 1 will be proven in Section 4.

Throughout the whole work the Hilbert space $\mathfrak{H}$ will assumed to be separable. The notation $\mathcal{B}(\mathfrak{M}, \mathfrak{N})$ is used for the set of bounded linear operators from the Hilbert space $\mathfrak{M}$ to the Hilbert space $\mathfrak{N}$. We will write $\mathcal{B}(\mathfrak{N})$ instead of $\mathcal{B}(\mathfrak{N}, \mathfrak{N})$.

2. Non-closed graph subspaces

Let $\mathfrak{H}_0$ be a closed subspace of a Hilbert space $\mathfrak{H}$ and X a densely defined (possibly unbounded and not necessarily closed) operator from $\mathfrak{H}_0$ to $\mathfrak{H}_1 = \mathfrak{H}_0^\perp := \mathfrak{H} \ominus \mathfrak{H}_0$ with domain $\mathrm{Dom}(X)$. A linear subspace

$$\mathcal{G}(\mathfrak{H}_0, X) := \{x \in \mathfrak{H} \,|\, x = x_0 \oplus X x_0,\ x_0 \in \mathrm{Dom}(X) \subset \mathfrak{H}_0\}$$

is called the graph subspace of $\mathfrak{H}$ associated with the pair $(\mathfrak{H}_0, X)$ or, in short, the graph of X.

Recalling general facts on densely defined closable operators (see, e.g., [6]) we mention the following: If $X : \mathfrak{H}_0 \to \mathfrak{H}_1$ is a densely defined non-closable operator, then $\mathcal{G}(\mathfrak{H}_0, X)$ is a non-closed subspace of $\mathfrak{H}$. Its closure is not a graph subspace, i.e., there is no closed operator Y such that

$$\overline{\mathcal{G}(\mathfrak{H}_0, X)} = \mathcal{G}(\mathfrak{H}_0, Y).$$

Proposition 2.1. *Let* $X : \mathfrak{H}_0 \to \mathfrak{H}_1$ *be a densely defined non-closable operator. Then the closed subspace* $\overline{\mathcal{G}(\mathfrak{H}_0, X)}$ *contains an element orthogonal to* $\mathfrak{H}_0$.

Proof. First, for $X : \mathfrak{H}_0 \to \mathfrak{H}_1$ being a densely defined non-closable operator we prove the following alternative: either the closed subspace $\overline{\mathcal{G}(\mathfrak{H}_0, X)}$ contains an element orthogonal to $\mathfrak{H}_0$ or the subspace $\mathfrak{H}_0$ contains an element orthogonal to $\overline{\mathcal{G}(\mathfrak{H}_0, X)}$. Indeed, assume on the contrary that neither the closed subspace $\overline{\mathcal{G}(\mathfrak{H}_0, X)}$ contains an element orthogonal to $\mathfrak{H}_0$ nor the subspace $\mathfrak{H}_0$ contains an element orthogonal to $\overline{\mathcal{G}(\mathfrak{H}_0, X)}$. Then by Theorem 3.2 in [7] there is a closed densely defined operator $Y : \mathfrak{H}_0 \to \mathfrak{H}_1$ such that $\overline{\mathcal{G}(\mathfrak{H}_0, X)} = \mathcal{G}(\mathfrak{H}_0, Y)$, which is a contradiction.

Now assume that the subspace $\mathfrak{H}_0$ contains an element x_0 orthogonal to $\overline{\mathcal{G}(\mathfrak{H}_0, X)}$. Obviously, this element is orthogonal to $\mathcal{G}(\mathfrak{H}_0, X)$, that is, $\langle x_0 \oplus 0, x_0 \oplus Xx_0\rangle = 0$, and hence $x_0 = 0$. Then, by the alternative proven above the subspace $\overline{\mathcal{G}(\mathfrak{H}_0, X)}$ contains an element orthogonal to $\mathfrak{H}_0$, completing the proof. □

For notational setup assume the following hypothesis.

Hypothesis 2.2. *Let* $\mathbf{B}$ *be a self-adjoint operator represented with respect to the decomposition* $\mathfrak{H} = \mathfrak{H}_0 \oplus \mathfrak{H}_1$ *as a* 2×2 *operator block matrix*

$$\mathbf{B} = \begin{pmatrix} A_0 & V \\ V^* & A_1 \end{pmatrix}, \tag{5}$$

where $A_i \in \mathcal{B}(\mathfrak{H}_i)$, $i = 0, 1$, *are bounded self-adjoint operators in* $\mathfrak{H}_i$ *while* $V \in \mathcal{B}(\mathfrak{H}_1, \mathfrak{H}_0)$ *is a bounded operator from* $\mathfrak{H}_1$ *to* $\mathfrak{H}_0$. *More explicitly,* $\mathbf{B} = \mathbf{A} + \mathbf{V}$, *where* $\mathbf{A}$ *is the bounded diagonal self-adjoint operator,*

$$\mathbf{A} = \begin{pmatrix} A_0 & 0 \\ 0 & A_1 \end{pmatrix}, \tag{6}$$

and the operator $\mathbf{V} = \mathbf{V}^*$ *is an off-diagonal bounded operator*

$$\mathbf{V} = \begin{pmatrix} 0 & V \\ V^* & 0 \end{pmatrix}. \tag{7}$$

Definition 2.3. *A densely defined (possibly unbounded and not necessarily closable) operator* X *from* $\mathfrak{H}_0$ *to* $\mathfrak{H}_1$ *with domain* $\mathrm{Dom}(X)$ *is called a* strong solution *to the Riccati equation*

$$A_1X - XA_0 - XVX + V^* = 0 \tag{8}$$

if

$$\mathrm{Ran}(A_0 + VX)|_{\mathrm{Dom}(X)} \subset \mathrm{Dom}(X)$$

and

$$A_1Xx - X(A_0 + VX)x + V^*x = 0 \quad \textit{for any} \quad x \in \mathrm{Dom}(X).$$

Theorem 2.4. *Assume Hypothesis* 2.2. *A densely defined (possibly unbounded and not necessarily closed) operator* X *from* $\mathfrak{H}_0$ *to* $\mathfrak{H}_1$ *with domain* $\mathrm{Dom}(X)$ *is a strong solution to the Riccati equation* (8) *if and only if the graph subspace* $\mathcal{G}(\mathfrak{H}_0, X)$ *is invariant for the operator* $\mathbf{B}$.

Proof. First, assume that $\mathcal{G}(\mathfrak{H}_0, X)$ is invariant for $\mathbf{B}$. Then

$$\mathbf{B}(x \oplus Xx) = (A_0x + VXx) \oplus (A_1Xx + V^*x) \in \mathcal{G}(\mathfrak{H}_0, X)$$

for any $x \in \mathrm{Dom}(X)$. In particular, $A_0x + VXx \in \mathrm{Dom}(X)$ and

$$A_1Xx + V^*x = X(A_0x + VXx) \text{ for all } x \in \mathrm{Dom}(X),$$

which proves that X is a strong solution to the Riccati equation (21).

To prove the converse statement assume that X is a strong solution to the Riccati equation (8), that is,

$$A_0x + VXx \in \mathrm{Dom}(X)$$

and

$$A_1Xx + V^*x = X(A_0x + VXx), \qquad x \in \mathrm{Dom}(X),$$

which proves that the graph subspace $\mathcal{G}(\mathfrak{H}_0, X)$ is $\mathbf{B}$-invariant. □

Remark 2.5. *By Lemma* 4.3 *in* [7] *a closed densely defined operator* $X : \mathfrak{H}_0 \to \mathfrak{H}_1$ *is a strong solution to the Riccati equation* (8) *if and only if it is a weak solution to* (8).

3. The singular spectrum of the operator B

Assume the following hypothesis.

Hypothesis 3.1. *Assume Hypothesis* 2.2. *Assume in addition that the Hilbert space* $\mathfrak{H}_1$ *is one-dimensional,*

$$\mathfrak{H}_1 = \mathbb{C},$$

and the Hilbert space $\mathfrak{H}_0$ *is the cyclic subspace generated by* $\mathrm{Ran}\, V$.

Note that under Hypothesis 3.1 the Hilbert space $\mathfrak{H}_0$ can be realized as a space of square integrable functions with respect to a Borel probability measure m with compact support,

$$\mathfrak{H}_0 = L^2(\mathbb{R}; m)$$

such that the bounded operator A_0 acts on $L^2(\mathbb{R}, m)$ as the multiplication operator

$$(A_0x_0)(\lambda) = \lambda x_0(\lambda), \qquad x_0 \in L^2(\mathbb{R}, m),$$

A_1 is the multiplication by a real number a_1 and, finally, the linear bounded map

$$V^* : \mathfrak{H}_0 \to \mathfrak{H}_1$$

is given by

$$V^*x_0 = \langle v, x_0 \rangle_{\mathfrak{H}_0}, \qquad x_0 \in \mathfrak{H}_0$$

for some $v \in \mathfrak{H}_0$.

Lemma 3.2. *Assume Hypothesis* 3.1. *Then the element* $0 \oplus 1 \in \mathfrak{H} = \mathfrak{H}_0 \oplus \mathfrak{H}_1$ *is cyclic for the operator* $\mathbf{B}$ *given by* (5) – (7) *and, hence,* $\mathbf{B}$ *has a simple spectrum.*

Proof. By hypothesis (in the above notations) the element $v \in \mathfrak{H}_0$ is cyclic for the operator A_0. Therefore, the cyclic subspace with respect to the operator $\mathbf{B}$ generated by the elements $v \oplus 0 \in \mathfrak{H}$ and $0 \oplus 1 \in \mathfrak{H}$ is the whole $\mathfrak{H}$. Without loss of generality we may assume that $a_1 = 0$. Observing that $\mathbf{B}(0 \oplus 1) = v \oplus 0$ proves the claim. □

Theorem 3.3. *Assume Hypothesis 3.1. Then the Herglotz function*

$$\phi(z) = \frac{1 + (a_1 - z)\langle v, (A_0 - z)^{-1} v\rangle_{\mathfrak{H}_0}}{(a_1 - z) - \langle v, (A_0 - z)^{-1} v\rangle_{\mathfrak{H}_0}} \tag{9}$$

admits the representation

$$\phi(z) = \int \frac{d\omega(\lambda)}{\lambda - z},$$

where ω is a probability measure on $\mathbb{R}$ with compact support. Moreover, the operator $\mathbf{B}$ *is unitarily equivalent to the multiplication operator by the independent variable on* $L^2(\mathbb{R}, \omega)$.

Proof. Introduce the Borel measure Ω with values in the set of non-negative operators on $\mathfrak{H}_1 \oplus \mathfrak{H}_1$ by

$$\Omega(\delta) = \begin{pmatrix} V & 0 \\ 0 & 1 \end{pmatrix}^* \mathsf{E}_{\mathbf{B}}(\delta) \begin{pmatrix} V & 0 \\ 0 & 1 \end{pmatrix},$$

where $\begin{pmatrix} V & 0 \\ 0 & 1 \end{pmatrix}$ is the linear map from $\mathfrak{H}_1 \oplus \mathfrak{H}_1$ to $\mathfrak{H}_0 \oplus \mathfrak{H}_1$ and let

$$\omega(\delta) = \operatorname{tr} \Omega(\delta), \qquad \delta \subset \mathbb{R} \text{ a Borel set.}$$

Clearly, the measure ω vanishes on all Borel sets δ such that $\mathsf{E}_{\mathbf{B}}(\delta) = 0$. In fact, these measures have the same families of Borel sets, on which they vanish. Indeed, assuming $\omega(\delta) = 0$ yields

$$\langle v \oplus 0, \mathsf{E}_{\mathbf{B}}(\delta)\, v \oplus 0\rangle_{\mathfrak{H}} + \langle 0 \oplus 1, \mathsf{E}_{\mathbf{B}}(\delta)\, 0 \oplus 1\rangle_{\mathfrak{H}} = 0$$

and, hence, in particular,

$$\langle 0 \oplus 1, \mathsf{E}_{\mathbf{B}}(\delta)\, 0 \oplus 1\rangle_{\mathfrak{H}} = 0, \tag{10}$$

which implies $\mathsf{E}_{\mathbf{B}}(\delta) = 0$.

Introducing the $\mathcal{B}(\mathfrak{H}_1 \oplus \mathfrak{H}_1)$-valued Herglotz function

$$M(z) = \begin{pmatrix} V & 0 \\ 0 & 1 \end{pmatrix}^* (\mathbf{B} - z)^{-1} \begin{pmatrix} V & 0 \\ 0 & 1 \end{pmatrix} \tag{11}$$

one concludes that the Herglotz function $M(z)$ admits the representation

$$M(z) = \int_{\mathbb{R}} \frac{d\Omega(\lambda)}{\lambda - z},$$

and hence

$$\operatorname{tr} M(z) = \int_{\mathbb{R}} \frac{d\omega(\lambda)}{\lambda - z}.$$

Straightforward computations show that the operator-valued function (11) with respect to the orthogonal decomposition $\mathfrak{H} = \mathfrak{H}_0 \oplus \mathfrak{H}_1$ can be represented as the 2×2 matrix

$$M(z) = \begin{pmatrix} M_{00}(z) & M_{01}(z) \\ M_{10}(z) & M_{11}(z) \end{pmatrix}$$

with the entries given by

$$\begin{aligned}
M_{00}(z) &= (a_1 - z)\langle v, (A_0 - z)^{-1} v\rangle [a_1 - z - \langle v, (A_0 - z)^{-1} v\rangle]^{-1}, \\
M_{11}(z) &= [a_1 - z - \langle v, (A_0 - z)^{-1} v\rangle]^{-1}, \\
M_{01}(z) &= -(a_1 - z)^{-1} M_{00}(z), \\
M_{10}(z) &= -(a_1 - z)^{-1} M_{00}(z).
\end{aligned}$$

Taking the trace of $M(z)$ yields representation (9).

Since by Lemma 3.2 the element $0 \oplus 1$ is cyclic and the measure ω and the spectral measure $\mathsf{E}_{\mathbf{B}}$ have the same families of Borel sets, on which they vanish, one concludes (see, e.g., [3]) that the operator $\mathbf{B}$ is unitarily equivalent to the multiplication operator by the independent variable on $L^2(\mathbb{R}, \omega)$, completing the proof. □

Recall that a measurable not necessarily closed set $S \subset \mathbb{R}$ is a support of a measure ν if $\nu(\mathbb{R} \setminus S) = 0$. A support S is said to be minimal if any measurable subset $S' \subset S$ with $\nu(S') = 0$ has Lebesgue measure zero.

Theorem 3.4. *The sets*

$$S_{\mathrm{s}} := \left\{ \lambda \in \mathbb{R} \,\middle|\, a_1 - \lambda = \int \frac{|v(\mu)|^2 dm(\mu)}{\mu - \lambda - \mathrm{i}0} \right\} \tag{12}$$

and

$$S_{\mathrm{sc}} := \left\{ \lambda \in \mathbb{R} \,\middle|\, a_1 - \lambda = \int \frac{|v(\mu)|^2 dm(\mu)}{\mu - \lambda - \mathrm{i}0}, \quad \int \frac{|v(\mu)|^2 dm(\mu)}{|\mu - \lambda|^2} = \infty \right\} \tag{13}$$

are minimal supports of the singular part ω_{s} and the singularly continuous part ω_{sc} of the measure ω, respectively. The set

$$S_{\mathrm{pp}} := \left\{ \lambda \in \mathbb{R} \,\middle|\, a_1 - \lambda = \int \frac{|v(\mu)|^2 dm(\mu)}{\mu - \lambda}, \quad \int \frac{|v(\mu)|^2 dm(\mu)}{|\mu - \lambda|^2} < \infty \right\} \tag{14}$$

coincides with the set of all atoms of the measure ω.

Proof. The fact that (12) is a minimal support of ω_{s} follows from Lemma 3.5 in [4], where one sets $m_a^+(z) = (a_1 - z)$ and

$$m_b^+(z) = \langle v, (A_0 - z)^{-1} v\rangle_{\mathfrak{H}_0} = \int \frac{|v(\mu)|^2 dm(\mu)}{\mu - z}, \quad \operatorname{Im} z \neq 0.$$

It is not hard to see (cf., e.g., Example 1 in [2]) that the set S_{pp} coincides with the set of all eigenvalues of the operator $\mathbf{B}$. Hence, by Theorem 3.3 one proves that S_{pp} coincides with the set of all atoms of the measure ω. Therefore, to prove that (13) is a minimal support of ω_{sc} it suffices to check the inclusion

$$S_{\mathrm{pp}} \subset S_{\mathrm{s}}. \tag{15}$$

Assume that $\lambda \in S_{\mathrm{pp}}$, that is,

$$a_1 - \lambda = \int \frac{|v(\mu)|^2 dm(\mu)}{\mu - \lambda} \tag{16}$$

and

$$\int \frac{|v(\mu)|^2 dm(\mu)}{|\mu-\lambda|^2} < \infty.$$

Since

$$\int \frac{|v(\mu)|^2 dm(\mu)}{|\mu-\lambda|} \le \left(\int \frac{|v(\mu)|^2 dm(\mu)}{|\mu-\lambda|^2} \right)^{1/2} \|v\|_{L^2(\mathbb{R};m)},$$

the dominated convergence theorem yields

$$\int \frac{|v(\mu)|^2 dm(\mu)}{\mu-\lambda-\mathrm{i}0} \equiv \lim_{\varepsilon\to+0} \int \frac{|v(\mu)|^2 dm(\mu)}{\mu-\lambda-\mathrm{i}\varepsilon} = \int \frac{|v(\mu)|^2 dm(\mu)}{\mu-\lambda},$$

which together with (16) proves inclusion (15). The proof is complete. □

Remark 3.5. *By Lemma* 5 *in* [5] *from Theorem* 3.3 *it follows that there exist minimal supports of the absolutely continuous part* ω_{ac}*, the singular part* ω_{s}*, and the singularly continuous part* ω_{sc} *of the measure* ω *such that their closures coincide with the absolute continuous part* $\mathrm{spec}_{\mathrm{ac}}(\mathbf{B})$*, the singular part* $\mathrm{spec}_{\mathrm{s}}(\mathbf{B})$*, and the singularly continuous part* $\mathrm{spec}_{\mathrm{sc}}(\mathbf{B})$ *of the spectrum, respectively.*

4. Riccati equation

Given $\lambda \in \mathbb{R}$, introduce the operator (linear functional)

$$X_\lambda : L^2(\mathbb{R};m) \to \mathfrak{H}_1 = \mathbb{C}$$

on

$$\mathrm{Dom}(X_\lambda) = \left\{ \varphi \in L^2(\mathbb{R};m) \,\middle|\, \lim_{\varepsilon\to+0} \int \frac{\overline{v(\mu)}\varphi(\mu)}{\mu-\lambda-\mathrm{i}\varepsilon} dm(\mu) \text{ exists finitely} \right\}$$

by

$$X_\lambda \varphi = \lim_{\varepsilon\to+0} \int \frac{\overline{v(\mu)}\varphi(\mu)}{\mu-\lambda-\mathrm{i}\varepsilon} dm(\mu), \qquad \varphi \in \mathrm{Dom}(X_\lambda). \tag{17}$$

Lemma 4.1. *If* $\lambda \in S_{\mathrm{s}}$*, then the operator* X_λ *is densely defined.*

Proof. Since the element $v \in L^2(\mathbb{R};m)$ is generating for the operator A_0, the set

$$D = \{\varphi \mid \varphi(\mu) = v(\mu)\psi(\mu),\ \psi \text{ is continuously differentiable on } \mathbb{R}\}$$

is dense in $L^2(\mathbb{R};m)$. For $\varphi \in D$ and $\varepsilon > 0$ one obtains

$$\int \frac{\overline{v(\mu)}\varphi(\mu)}{\mu-\lambda-\mathrm{i}\varepsilon} dm(\mu) = \psi(\lambda) \int \frac{|v(\mu)|^2}{\mu-\lambda-\mathrm{i}\varepsilon} dm(\mu) \tag{18}$$

$$+ \int \frac{|v(\mu)|^2(\psi(\mu)-\psi(\lambda))}{\mu-\lambda-\mathrm{i}\varepsilon} dm(\mu). \tag{19}$$

Since $\lambda \in S_{\mathrm{s}}$, by Theorem 3.4 the limit

$$\lim_{\varepsilon\to+0} \int \frac{|v(\mu)|^2 dm(\mu)}{\mu-\lambda-\mathrm{i}\varepsilon} = \int \frac{|v(\mu)|^2 dm(\mu)}{\mu-\lambda-\mathrm{i}0}$$

exists finitely. The integral (19) also has a limit as $\varepsilon \to +0$ since ψ is a continuously differentiable which proves that the left-hand side of (18) has a finite limit as $\varepsilon \to +0$. Therefore, $D \subset \mathrm{Dom}(X_\lambda)$, that is, X_λ is densely defined. □

Remark 4.2. *Note that by the Riesz representation theorem X_λ is bounded whenever the condition*

$$\int \frac{|v(\mu)|^2}{|\lambda - \mu|^2} dm(\mu) < \infty \tag{20}$$

holds true. The converse is also true: If X_λ is bounded, then (20) holds. Indeed, by the uniform boundedness principle from definition (17) it follows that

$$\sup_{\varepsilon \in (0,1]} \int \frac{|v(\mu)|^2}{(\mu - \lambda)^2 + \varepsilon^2} dm(\mu) < \infty,$$

proving (20) by the monotone convergence theorem.

Theorem 4.3. *Let $\lambda \in S_{\mathrm{s}}$. Then the operator X_λ is a strong solution to the Riccati equation*

$$A_1 X - X A_0 - X V X + V^* = 0. \tag{21}$$

Moreover, if $\lambda \in S_{\mathrm{pp}}$, the solution X_λ is bounded and if $\lambda \in S_{\mathrm{sc}} = S_{\mathrm{s}} \setminus S_{\mathrm{pp}}$, the operator X_λ is non-closable.

Proof. Note that $A_0 \,\mathrm{Dom}(X_\lambda) \subset \mathrm{Dom}(X_\lambda)$. If $\lambda \in S_{\mathrm{s}}$, then by Theorem 3.4

$$a_1 - \lambda = \int \frac{|v(\mu)|^2 dm(\mu)}{\mu - \lambda - \mathrm{i}0}.$$

In particular, $v \in \mathrm{Dom}(X_\lambda)$ and

$$\begin{aligned} X_\lambda V X_\lambda \varphi &= \int \frac{|v(\mu)|^2}{\mu - \lambda - \mathrm{i}0} dm(\mu) \cdot X_\lambda \varphi \\ &= (a_1 - \lambda) \int \frac{\overline{v(\mu)} \varphi(\mu)}{\mu - \lambda - \mathrm{i}0} dm(\mu), \qquad \varphi \in \mathrm{Dom}(X_\lambda). \end{aligned}$$

Therefore, for an arbitrary $\varphi \in \mathrm{Dom}(X_\lambda)$ one gets

$$\begin{aligned} & A_1 X_\lambda \varphi - X_\lambda A_0 \varphi - X_\lambda V X_\lambda \varphi \\ &= \int \frac{\overline{v(\mu)} \varphi(\mu)(a_1 - \mu)}{\mu - \lambda - \mathrm{i}0} dm(\mu) - (a_1 - \lambda) \int \frac{\overline{v(\mu)} \varphi(\mu)}{\mu - \lambda - \mathrm{i}0} dm(\mu) \\ &= \int \frac{\overline{v(\mu)} \varphi(\mu)(\lambda - \mu)}{\mu - \lambda - \mathrm{i}0} dm(\mu) = - \int \overline{v(\mu)} \varphi(\mu) dm(\mu) = -V^* \varphi, \end{aligned}$$

which proves that the operator X_λ is a strong solution to the Riccati equation (21).

If $\lambda \in S_{\mathrm{pp}}$, then (20) holds, in which case X_λ is bounded. If $\lambda \in S_{\mathrm{sc}} = S_{\mathrm{s}} \setminus S_{\mathrm{pp}}$, then X_λ is an unbounded densely defined operator (functional) (cf. Remark 4.2). Since every closed finite-rank operator is bounded [6], it follows that for $\lambda \in S_{\mathrm{sc}}$ the unbounded solution X_λ is non-closable. □

Proof of Theorem 1. Introduce the mapping

$$\Psi(\lambda) = \mathcal{G}(\mathfrak{H}_0, X_\lambda), \quad \lambda \in S_{\mathrm{s}}, \tag{22}$$

where X_λ is the strong solution to the Riccati equation referred to in Theorem 4.3. By Theorem 2.4 the subspace $\Psi(\lambda)$, $\lambda \in S_{\mathrm{s}}$ is invariant with respect to $\mathbf{B}$. To prove the injectivity of the mapping Ψ, assume that $\Psi(\lambda_1) = \Psi(\lambda_2)$ for some $\lambda_1, \lambda_2 \in S_{\mathrm{s}}$. Due to (22), $X_{\lambda_1} = X_{\lambda_2}$ which by (17) implies $\lambda_1 = \lambda_2$.

(i) Let $\lambda \in S_{\mathrm{sc}}$. By Theorem 4.3 the functional X_λ is non-closable. Since X_λ is densely defined, the closure $\overline{\mathcal{G}(\mathfrak{H}_0, X_\lambda)}$ of the subspace $\mathcal{G}(\mathfrak{H}_0, X_\lambda)$ contains the subspace $\mathfrak{H}_0$. By Proposition 2.1, the subspace $\overline{\mathcal{G}(\mathfrak{H}_0, X_\lambda)}$ contains an element orthogonal to $\mathfrak{H}_0$. Since $\mathfrak{H}_0 \subset \mathfrak{H}$ is of codimension 1, one concludes that $\overline{\mathcal{G}(\mathfrak{H}_0, X_\lambda)} = \mathfrak{H}_0 \oplus \mathfrak{H}_1 = \mathfrak{H}$.

(ii) Let $\lambda \in S_{\mathrm{pp}}$. By Theorem 5.3 in [7] the solution X_λ is an isolated point (in the operator norm topology) of the set of all bounded solutions to the Riccati equation (21) if and only if the subspace $\mathcal{G}(\mathfrak{H}_0, X_\lambda)$ is spectral, that is, there is a Borel set $\Delta \subset \mathbb{R}$ such that

$$\mathcal{G}(\mathfrak{H}_0, X_\lambda) = \operatorname{Ran} \mathsf{E}_{\mathbf{B}}(\Delta).$$

Observe that the one-dimensional graph subspace $\mathcal{G}(\mathfrak{H}_1, -X_\lambda^*)$ is invariant with respect to the operator $\mathbf{B}$. This subspace is spectral since by Lemma 3.2 λ is a simple eigenvalue of the operator $\mathbf{B}$. Thus, $\mathcal{G}(\mathfrak{H}_0, X_\lambda) = \mathcal{G}(\mathfrak{H}_1, -X_\lambda^*)^\perp$ is also a spectral subspace of the operator $\mathbf{B}$. □

Acknowledgments

The authors are grateful to C. van der Mee for useful suggestions. K.A. Makarov is indebted to the Graduiertenkolleg "Hierarchie und Symmetrie in mathematischen Modellen" for its kind hospitality during his stay at the RWTH Aachen in the Summer of 2003.

References

[1] V. Adamyan, H. Langer, and C. Tretter, *Existence and uniqueness of contractive solutions of some Riccati equations*, J. Funct. Anal. **179** (2001), 448–473.

[2] S. Albeverio, K.A. Makarov, and A.K. Motovilov, *Graph subspaces and the spectral shift function*, Canad. J. Math. **55** (2003), 449–503. `arXiv:math.SP/0105142`

[3] M.S. Birman and M.Z. Solomyak, *Spectral Theory of Self-Adjoint Operators in Hilbert Space*, D. Reidel, Dordrecht, 1987.

[4] D.J. Gilbert, *On subordinacy and analysis of the spectrum of Schrödinger operators with two singular endpoints*, Proc. Royal Soc. Edinburgh **112A** (1989), 213–229.

[5] D.J. Gilbert and D.B. Pearson, *On subordinacy and analysis of the spectrum of one-dimensional Schrödinger operators*, J. Math. Anal. Appl. **128** (1987), 30–56.

[6] T. Kato, *Perturbation Theory for Linear Operators*, Springer-Verlag, Berlin, 1966.

[7] V. Kostrykin, K.A. Makarov, and A.K. Motovilov, *Existence and uniqueness of solutions to the operator Riccati equation. A geometric approach*, in Yu. Karpeshina, G. Stolz, R. Weikard, Y. Zeng (Eds.), *Advances in Differential Equations and Mathematical Physics*, Contemporary Mathematics **327**, Amer. Math. Soc., 2003, pp. 181–198. arXiv:math.SP/0207125

[8] V. Kostrykin, K.A. Makarov, and A.K. Motovilov, *A generalization of the* tan 2Θ *theorem*, in J.A. Ball, M. Klaus, J.W. Helton, and L. Rodman (Eds.), *Current Trends in Operator Theory and Its Applications.* Operator Theory: Advances and Applications **149**, Birkhäuser, Basel, 2004, pp. 349–372. arXiv:math.SP/0302020

[9] V. Kostrykin, K.A. Makarov, and A.K. Motovilov, *Perturbation of spectra and spectral subspaces*, Trans. Amer. Math. Soc. (to appear), arXiv:math.SP/0306025

[10] B. Simon and T. Wolff, *Singular continuous spectrum under rank one perturbations and localization for random Hamiltonians*, Comm. Pure Appl. Math. **39** (1986), 75–90.

Vadim Kostrykin
Fraunhofer-Institut für Lasertechnik
Steinbachstraße 15
D-52074 Aachen, Germany
e-mail: kostrykin@ilt.fraunhofer.de, kostrykin@t-online.de
URL: http://home.t-online.de/home/kostrykin

Konstantin A. Makarov
Department of Mathematics
University of Missouri
Columbia, MO 65211, USA
e-mail: makarov@math.missouri.edu
URL: http://www.math.missouri.edu/people/kmakarov.html

Operator Theory:
Advances and Applications, Vol. 160, 311–336

Numerical Methods for Cauchy Singular Integral Equations in Spaces of Weighted Continuous Functions

G. Mastroianni, M.G. Russo and W. Themistoclakis

Dedicated to Professor Israel Gohberg on the occasion of his 75th birthday

Abstract. Some convergent and stable numerical procedures for Cauchy singular integral equations are given. The proposed approach consists of solving the regularized equation and is based on the weighted polynomial interpolation. The convergence estimates are sharp and the obtained linear systems are well conditioned.

Mathematics Subject Classification (2000). Primary 65R20; Secondary 45E05.

Keywords. Cauchy singular integral equation, projection method, Lagrange interpolation.

1. Introduction

We consider the Cauchy singular integral equation (CSIE)

$$(D + \nu K)f = g \tag{1.1}$$

where g is a known function on $(-1, 1)$, f is the unknown, $\nu \in \mathbb{R}$ and the operators D and K are defined as follows

$$Df(y) = \cos \pi\alpha f(y)v^{\alpha,-\alpha}(y) - \frac{\sin \pi\alpha}{\pi} \int_{-1}^{1} \frac{f(x)}{x-y} v^{\alpha,-\alpha}(x)dx, \tag{1.2}$$

$$Kf(y) = \int_{-1}^{1} k(x, y)f(x)v^{\alpha,-\alpha}(x)dx, \tag{1.3}$$

where $0 < \alpha < 1$ and $v^{\alpha,-\alpha}(x) = (1-x)^{\alpha}(1+x)^{-\alpha}$ is a Jacobi weight.

Work supported by INDAM-GNCS project 2003 "Metodi numerici per equazioni integrali".

In the last decades the idea of approximating the solution of (1.1) by polynomials has been presented in several papers (we mention for instance [26, 30, 1, 3, 4, 5, 10, 11, 12, 13, 14, 18, 25, 27] and the references therein).

Such procedures, usually called "collocation" and "discrete collocation" methods, project the equation onto the subspace of polynomials, replacing the integral by a quadrature formula. By collocation on a suitable set of nodes, one can construct a linear system, the solution of which gives the coefficients of the polynomial approximating the exact solution. The convergence and stability of the method is usually studied by considering a finite-dimensional equation equivalent to the system.

Anyway this approach does not take into account the condition number of the matrix of the system. Indeed if, for an approximation method applied to an operator equation, convergence and stability are proved, then immediately it follows that the norms of the discrete operators and the norms of their inverses are uniformly bounded, and, consequently, the condition numbers of these operators are also uniformly bounded. However infinite linear systems exist, depending on the choice of the polynomial base, that are equivalent to the given finite-dimensional equation and some of these systems can be ill conditioned (see, e.g., [15, 8]). For example if we use as a polynomial basis the fundamental Lagrange polynomials based on equispaced points, we get a linear system that is strongly ill conditioned. As a consequence we get a not reliable numerical procedure for evaluating the coefficients of the approximating polynomial.

In this paper, following an idea in [28], we assume Kf and g sufficiently smooth and solve the equivalent regularized equation

$$(I + \nu\widehat{D}K)f = \widehat{D}g \tag{1.4}$$

where

$$\widehat{D}f(y) = \cos\pi\alpha v^{-\alpha,\alpha}(y)f(y) + \frac{\sin\pi\alpha}{\pi}\int_{-1}^{1}\frac{f(x)}{x-y}v^{-\alpha,\alpha}(x)dx.$$

Hence we will construct two polynomial sequences $\{K_m f\}_m$ and $\{G_m\}_m$ which are convergent to $\widehat{D}Kf$ and $\widehat{D}g$, respectively, like the best approximation in some suitable spaces. Hence we consider the finite-dimensional equation

$$(I + \nu K_m)f_m = G_m \tag{1.5}$$

where f_m is the unknown polynomial. Via standard arguments, it follows that (1.5) has a unique solution f_m which converges to f (if f is the solution of (1.4)). The convergence estimates proved here are sharp. Moreover the condition number of $I + \nu K_m$ is uniformly bounded. Then expanding both sides of (1.5) in a suitable basis, we get a linear system that is equivalent to (1.5) and whose matrix is well conditioned (except for some log factor). Moreover in the case when K has a smooth kernel the entries of the matrix can be easily computed. Using suitable polynomial bases the exposed procedure includes the "collocation" and "discrete collocation" methods.

We remark that we are not able to use the above mentioned procedure when the index of equation (1.1) is ± 1, since at the moment, the behavior of the corresponding operator D in the Zygmund type spaces is not clear. Anyway the authors believe that the main results of this paper can be extended to the case of the Cauchy singular integral equations of index ± 1.

The paper is structured as follows. Section 2 collects basic tools of the approximation theory, the mapping properties of the operators D and K and some results on special interpolation processes. By using such processes some sequences of operators convergent in norm are constructed (see Lemma 2.3 and Lemma 2.4). In Section 3 some numerical methods are shown and the related theorems about the convergence, the stability and the behavior of the condition number of the linear systems are given. In Section 4 some weakly singular perturbation operators (frequently appearing in the literature) are considered. Several numerical tests are given in Section 5. Finally Section 6 and the Appendix are devoted to the proofs of the main results and to other technical details.

2. Preliminary results

2.1. Functional spaces

In order to introduce some functional spaces we will denote by $C(-1,1)$ the set of all continuous functions on the open interval $(-1,1)$. Let $v^{\gamma,\delta}(x) = (1-x)^\gamma(1+x)^\delta$ be the Jacobi weight with exponents $\gamma, \delta > -1$. Let us define $C_{v^{\gamma,\delta}}$, $\gamma, \delta > 0$, as

$$C_{v^{\gamma,\delta}} = \{f \in C(-1,1) : \lim_{x\to\pm 1} f(x)v^{\gamma,\delta}(x) = 0\}.$$

Further in the case $\gamma = 0$ (respectively $\delta = 0$), $C_{v^{\gamma,\delta}}$ consists of all functions which are continuous on $(-1,1]$ (respectively on $[-1,1)$) such that $\lim_{x\to -1}(fv^{\gamma,\delta})(x) = 0$ (respectively $\lim_{x\to 1}(fv^{\gamma,\delta})(x) = 0$). Moreover if $\gamma = \delta = 0$ we set $C_{v^{0,0}} = C[-1,1]$. The norm of a function $f \in C_{v^{\gamma,\delta}}$ is defined as $\|f\|_{C_{v^{\gamma,\delta}}} := \sup_{|x|\le 1}|f(x)v^{\gamma,\delta}(x)| = \|fv^{\gamma,\delta}\|_\infty$. Somewhere for brevity we will write $\|G\|_A = \sup_{x\in A}|G(x)|$, $A \subseteq [-1,1]$. To deal with smoother functions we define the Sobolev type space

$$W_r = W_r(v^{\gamma,\delta}) = \{f \in C_{v^{\gamma,\delta}} : f^{(r-1)} \in AC(-1,1) \text{ and } \|f^{(r)}\varphi^r v^{\gamma,\delta}\|_\infty < \infty\}$$

where $\varphi(x) = \sqrt{1-x^2}$, $r \ge 1$ is an integer and $AC(-1,1)$ is the set of absolutely continuous functions on $(-1,1)$. The norm in W_r is $\|f\|_{W_r} := \|fv^{\gamma,\delta}\|_\infty + \|f^{(r)}\varphi^r v^{\gamma,\delta}\|_\infty$.

Now let us introduce some suitable moduli of smoothness. Following Ditzian and Totik [7] $\forall f \in C_{v^{\gamma,\delta}}$ we define

$$\Omega_\varphi^k(f,\tau)_{v^{\gamma,\delta}} = \sup_{0<h\le\tau}\|v^{\gamma,\delta}\Delta_{h\varphi}^k f\|_{I_{hk}} \tag{2.1}$$

where $\Delta_{h\varphi}^k f(x) = \sum_{i=0}^k (-1)^i \binom{k}{i} f(x + (k/2 - i)h\varphi(x))$, $0 < k \in \mathbb{N}$, $I_{hk} = [-1 + 4h^2k^2, 1 - 4h^2k^2]$.

Denoting by $\mathbb{P}_m$ the set of all algebraic polynomials of degree at most m, we define the complete modulus as

$$\omega_\varphi^k(f,\tau)_{v^{\gamma,\delta}} = \Omega_\varphi^k(f,\tau)_{v^{\gamma,\delta}} + \inf_{q_1\in\mathbb{P}_{k-1}} \|v^{\gamma,\delta}(f-q_1)\|_{[-1,-1+4k^2\tau^2]} \quad (2.2)$$
$$+ \inf_{q_2\in\mathbb{P}_{k-1}} \|v^{\gamma,\delta}(f-q_2)\|_{[1-4k^2\tau^2,1]}.$$

For the sake of simplicity we will also write $\Omega_\varphi = \Omega_\varphi^1$, $\omega_\varphi = \omega_\varphi^1$, $\Omega_\varphi^k(f,\tau) = \Omega_\varphi^k(f,\tau)_{v^{0,0}}$, $\omega_\varphi^k(f,\tau) = \omega_\varphi^k(f,\tau)_{v^{0,0}}$, $k>0$.

Now denoting by

$$E_m(f)_{v^{\gamma,\delta}} = \inf_{P\in\mathbb{P}_m} \|(f-P)v^{\gamma,\delta}\|_\infty$$

the error of best approximation in $C_{v^{\gamma,\delta}}$ ($E_m(f)\equiv E_m(f)_{v^{0,0}}$), the following inequalities hold

$$E_m(f)_{v^{\gamma,\delta}} \le C\omega_\varphi^k(f,1/m)_{v^{\gamma,\delta}} \quad (2.3)$$

and

$$\omega_\varphi^k(f,\tau)_{v^{\gamma,\delta}} \le C\tau^k \sum_{0\le i\le \frac{1}{\tau}} (1+i)^{k-1} E_i(f)_{v^{\gamma,\delta}} \quad (2.4)$$

where C is a positive constant independent of f,m and τ. Estimates (2.3)–(2.4) can be deduced from [7], but they both explicitly appeared in [19].

By definition $\Omega_\varphi^k(f,\tau)_{v^{\gamma,\delta}} \le \omega_\varphi^k(f,\tau)_{v^{\gamma,\delta}}$. On the other hand by (2.3)–(2.4) $\Omega_\varphi^k(f,\tau)_{v^{\gamma,\delta}} \sim \tau^\beta$, $0<\beta<k$ implies $\omega_\varphi^k(f,\tau)_{v^{\gamma,\delta}} \sim \tau^\beta$. Therefore, by using $\Omega_\varphi^k(f,\tau)_{v^{\gamma,\delta}}$ in place of $\omega_\varphi^k(f,\tau)_{v^{\gamma,\delta}}$, we can define the Zygmund space

$$Z_r(v^{\gamma,\delta}) = \left\{ f\in C_{v^{\gamma,\delta}} : \|f\|_{Z_r(v^{\gamma,\delta})} = \|fv^{\gamma,\delta}\|_\infty + \sup_{\tau>0} \frac{\Omega_\varphi^k(f,\tau)_{v^{\gamma,\delta}}}{\tau^r} < \infty \right\}$$

where $0<r<k\in\mathbb{N}$ and we set $Z_r\equiv Z_r(v^{0,0})$. In conclusion we remark that for differentiable functions in $(-1,1)$ we can estimate Ω_φ^k by means of the inequality

$$\Omega_\varphi^k(f,\tau)_{v^{\gamma,\delta}} \le C \sup_{0<h\le\tau} h^k \|f^{(k)}\varphi^k v^{\gamma,\delta}\|_{I_{hk}}, \quad C\ne C(f,\tau) \quad (2.5)$$

where here and in the sequel $C\ne C(a,b,c,\dots)$ means that C is a positive constant independent of the parameters a, b, c, ….

Finally the following Lemma could be useful in several contexts.

Lemma 2.1. *Let $f\in C_{v^{\gamma,\delta}}$ and $P_m\in\mathbb{P}_m$ be such that*

$$\|(f-P_m)v^{\gamma,\delta}\|_\infty \le cE_m(f)_{v^{\gamma,\delta}}$$

holds with some positive constant $c\ne c(m,f)$. Then

$$\int_0^{\frac{1}{m}} \frac{\omega_\varphi(f-P_m,t)_{v^{\gamma,\delta}}}{t}dt \le C\int_0^{\frac{1}{m}} \frac{\omega_\varphi^k(f,t)_{v^{\gamma,\delta}}}{t}dt, \quad k<m, \quad (2.6)$$

where C is a positive constant independent of f and m.

2.2. Mapping properties of D and assumptions on K

The properties of the operator

$$Df(y) = \cos \pi\alpha f(y) v^{\alpha,-\alpha}(y) - \frac{\sin \pi\alpha}{\pi} \int_{-1}^{1} \frac{f(x)}{x-y} v^{\alpha,-\alpha}(x)dx, \quad 0 < \alpha < 1$$

in weighted L^2 spaces are well known (see, e.g., [26, 30]). In this paper we consider D as a mapping from $Z_r(v^{\alpha,0})$ into $Z_r(v^{0,\alpha})$, $r > 0$. In [24] the authors extensively studied this operator. For the convenience of the reader we recall here the following results

- $D : Z_r(v^{\alpha,0}) \longrightarrow Z_r(v^{0,\alpha})$ is bounded and invertible, $\forall r > 0$;
- the inverse (and bounded) operator, $\widehat{D} : Z_r(v^{0,\alpha}) \longrightarrow Z_r(v^{\alpha,0})$ is given by

$$\widehat{D}f(y) = \cos \pi\alpha v^{-\alpha,\alpha}(y) f(y) + \frac{\sin \pi\alpha}{\pi} \int_{-1}^{1} \frac{f(x)}{x-y} v^{-\alpha,\alpha}(x)dx.$$

With respect to the operator

$$Kf(y) = \int_{-1}^{1} k(x,y) f(x) v^{\alpha,-\alpha}(x)dx \tag{2.7}$$

we assume

$$\sup_{\tau>0} \frac{\Omega_\varphi^k(Kf,\tau)_{v^{0,\alpha}}}{\tau^r} \leq C\|fv^{\alpha,0}\|_\infty \tag{2.8}$$

with $k > r > 0$ and C independent of f. Therefore the operator $K : C_{v^{\alpha,0}} \longrightarrow C_{v^{0,\alpha}}$ is compact.

Indeed (2.8) and (2.3) imply that $\lim_n \sup_{f\in S} E_n(Kf)_{v^{0,\alpha}} = 0$, where $S = \{f \in C_{v^{\alpha,0}} : \|fv^{\alpha,0}\| = 1\}$ and this is equivalent to the compactness of K [32]. Moreover by (2.8) Kf belongs to $Z_r(v^{0,\alpha})$, $r > 0$.

In conclusion we remark that (2.8) is satisfied, for instance, if the kernel of K is of the type $k(x,y) = |x-y|^\mu$, $\mu > -1, \mu \neq 0$, with $r = \mu + 1$ (see Section 4) or if $k(x,y) = k_x(y)$ satisfies

$$\sup_{|x|\leq 1} \sup_{\tau>0} \frac{\Omega_\varphi^k(k_x,\tau)_{v^{0,\alpha}}}{\tau^r} < \infty, \quad k \geq r > 0,$$

(see Lemma 2.4).

2.3. Some interpolation processes

Let $\{p_m(v^{\alpha,-\alpha})\}_m = \{p_m^{\alpha,-\alpha}\}_m$ be the sequence of the orthonormal Jacobi polynomials with respect to $v^{\alpha,-\alpha}$ (α is the same parameter appearing in the definition of the operators D and K) and having positive leading coefficients. Denote by $t_1 < \cdots < t_m$ the zeros of $p_m^{\alpha,-\alpha}$ and by $L_m(v^{\alpha,-\alpha}, F)$, $F \in C(-1,1)$, the Lagrange polynomial interpolating F on the nodes $t_1, \ldots, t_m$. Moreover let $L_{m,1,1}(v^{\alpha,-\alpha}, F)$ denote the Lagrange polynomial interpolating $F \in C(-1,1)$ on the knots $\frac{t_1-1}{2} < t_1 < \ldots < t_m < -\frac{t_1-1}{2}$ (we choose symmetric additional points only in order to simplify the computations). Sometimes for the sake of brevity we will use

$$\mathcal{L}_m^{\alpha,-\alpha} = \begin{cases} L_m(v^{\alpha,-\alpha}), & \text{if} \quad \alpha \geq 1/2 \\ L_{m,1,1}(v^{\alpha,-\alpha}), & \text{if} \quad 0 < \alpha < 1/2 \,. \end{cases}$$

As a consequence of [23, Th. 2.1, 2.2, 2.4] and of [21, Th. 3.1] the following estimates hold for any $F \in C_{v^{\alpha,0}}$

$$\|[F - \mathcal{L}_m^{\alpha,-\alpha} F] v^{\alpha,0}\|_\infty \leq C E_{m-1}(F)_{v^{\alpha,0}} \log m \tag{2.9}$$

$$\|[F - \mathcal{L}_m^{\alpha,-\alpha} F] v^{0,-\alpha}\|_1 \leq C E_{m-1}(F) \tag{2.10}$$

where in both cases C is a positive constant independent of F and m and $\|\cdot\|_1$ denotes the L^1-norm.

With an analogous meaning of the symbols we denote by $L_m(v^{-\alpha,\alpha}, F)$, $F \in C(-1,1)$, the Lagrange polynomial interpolating F on the zeros $x_1, \ldots, x_m$, of $p_m^{-\alpha,\alpha}$ and by $L_{m,1,1}(v^{-\alpha,\alpha}, F)$ the Lagrange polynomial interpolating $F \in C(-1,1)$ on the knots $-\dfrac{x_m+1}{2} < x_1 < \ldots < x_m < \dfrac{x_m+1}{2}$. Also in this case sometimes we will use the notation

$$\mathcal{L}_m^{-\alpha,\alpha} = \begin{cases} L_m(v^{-\alpha,\alpha}), & \text{if} \quad \alpha \geq 1/2 \\ L_{m,1,1}(v^{-\alpha,\alpha}), & \text{if} \quad 0 < \alpha < 1/2 \,. \end{cases} \tag{2.11}$$

Recalling the definition of $\widehat{D}$ we can state the following theorem.

Theorem 2.2. *Let $\phi \in Z_r(v^{0,\alpha})$, $r > 0$, $0 < \alpha < 1$. Then*

$$\|\widehat{D}[\phi - \mathcal{L}_m^{-\alpha,\alpha}\phi] v^{\alpha,0}\|_\infty \leq C \frac{\log m}{m^r} \|\phi\|_{Z_r(v^{0,\alpha})} \tag{2.12}$$

where C is a positive constant independent of ϕ and m.

2.4. Some operator sequences

As we already remarked, the assumption (2.8) we are making on K, allows the kernel $k(x,y)$ to be also singular. In this case we can define the following operator sequence

$$K_m f(y) = \widehat{D} \mathcal{L}_m^{-\alpha,\alpha}(Kf, y). \tag{2.13}$$

Due to the invariance property of $\widehat{D}$ on polynomials, K_m maps $C_{v^{\alpha,0}}$ into $\mathbb{P}_{m+1}$. Moreover the following Lemma holds.

Lemma 2.3. *Let $0 < \alpha < 1$. If the operator K satisfies the assumption (2.8) then*

$$\|\widehat{D}K - K_m\|_{C_{v^{\alpha,0}} \to C_{v^{\alpha,0}}} = \mathcal{O}\left(\frac{\log m}{m^r}\right), \qquad r > 0 \tag{2.14}$$

where the constant in "$\mathcal{O}$" is independent of m.

When the kernel $k(x,y)$ is smooth we introduce

$$K^* f(y) = \int_{-1}^{1} \mathcal{L}_m^{\alpha,-\alpha}(k_y, x) f(x) v^{\alpha,-\alpha}(x) dx, \tag{2.15}$$

where $k_y(x) = k(x,y)$ and the interpolation is made with respect to x.

Hence we define the sequence $\{K_m^*\}_m$ as

$$K_m^* f(y) = \widehat{D}\mathcal{L}_m^{-\alpha,\alpha}(K^* f)(y). \tag{2.16}$$

K_m^* maps $C_{v^{\alpha,0}}$ into $\mathbb{P}_{m+1}$. Moreover, if $k_x(y) = k(x,y)$, we can state the following Lemma.

Lemma 2.4. *Let $0 < \alpha < 1$. Assume that*

$$\sup_{t>0} \left\{ \sup_{|y|\le 1} (1-y)^\alpha \Omega_\varphi^k(k_y, t) \right\} \frac{1}{t^r} < \infty \tag{2.17}$$

and

$$\sup_{t>0} \left\{ \sup_{|x|\le 1} \Omega_\varphi^k(k_x, t)_{v^{\alpha,0}} \right\} \frac{1}{t^r} < \infty \tag{2.18}$$

with $r > 0$ and $k > r$. Then

$$\|\widehat{D}K - K_m^*\|_{C_{v^{\alpha,0}} \to C_{v^{\alpha,0}}} = \mathcal{O}\left(\frac{\log m}{m^r}\right) \tag{2.19}$$

where the constant in "$\mathcal{O}$" is independent of m.

3. Numerical methods

Now go back to the equation $(D + \nu K)f = g$. Assume $g \in Z_r(v^{0,\alpha})$, $r > 0$ and that operator K satisfies condition (2.8):

$$\sup_{\tau>0} \frac{\Omega_\varphi^k(Kf, \tau)_{v^{0,\alpha}}}{\tau^r} \le C\|f v^{\alpha,0}\|_\infty.$$

Under these assumptions we solve the equivalent equation

$$(I + \nu\widehat{D}K)f = G, \qquad G := \widehat{D}g. \tag{3.1}$$

By the mapping properties of $\widehat{D}$ it follows that $G \in Z_r(v^{\alpha,0}) \subset C_{v^{\alpha,0}}$ and $\widehat{D}K : C_{v^{\alpha,0}} \longrightarrow C_{v^{\alpha,0}}$ is compact (see, e.g., Lemma 2.3). Thus $I + \nu\widehat{D}K$ is invertible in $C_{v^{\alpha,0}}$. From now on we will assume that (3.1) has a unique solution in $C_{v^{\alpha,0}}$ for any fixed $g \in Z_r(v^{0,\alpha})$ and we denote by $\bar{f}$ the solution of (3.1). In order to construct an approximate solution of (3.1), we solve the finite-dimensional equation

$$(I + \nu K_m) f_m = G_m, \tag{3.2}$$

where $G_m = \widehat{D}\mathcal{L}_m^{-\alpha,\alpha} g$, $f_m \in \mathbb{P}_{m+1}$ is the unknown polynomial and K_m was defined in (2.13).

Theorem 3.1. *Let $0 < \alpha < 1$. If the previous assumptions on K and g hold true, then for m sufficiently large, there exists a unique polynomial $\bar{f}_m \in \mathbb{P}_{m+1}$, which is the solution of (3.2). Moreover the following estimate holds*

$$\|(\bar{f} - \bar{f}_m)v^{\alpha,0}\|_\infty \le C \frac{\log m}{m^r} \|g\|_{Z_r(v^{0,\alpha})} \tag{3.3}$$

where C is a positive constant independent of $\bar{f}$ and m.

Consequently $\bar{f} \in Z_{r-\epsilon}(v^{\alpha,0})$, *with* $\epsilon > 0$ *arbitrarily small. Moreover*

$$|cond(I + \nu K_m) - cond(I + \nu \widehat{D}K)| = \mathcal{O}\left(\frac{\log m}{m^r}\right) \tag{3.4}$$

where $cond(B) = \|B\|\|B^{-1}\|$ *is the condition number of the bounded and invertible operator* B *in* $C_{v^{\alpha,0}}$ *and the constant in* $\mathcal{O}$ *is independent of* m.

In order to compute the coefficients of $\bar{f}_m$ we construct a linear system which is equivalent to equation (3.2) and whose coefficient matrix has a bounded (up to some log factor) condition number w.r.t. the ℓ_∞ norm. This goal can be reached in several ways, also taking into account the complexity in building the matrix and solving the linear system.

Here we propose the following procedure for $\alpha \geq 1/2$. We note that by definition we have

$$\begin{aligned}(K_m f)(x) &= (\widehat{D}\mathcal{L}_m^{-\alpha,\alpha} Kf)(x) = (\widehat{D}L_m^{-\alpha,\alpha} Kf)(x) \\ &= \sum_{i=1}^{m} \frac{\widehat{D}l_i^{-\alpha,\alpha}(x)}{v^{0,\alpha}(x_i)} (Kf)(x_i) v^{0,\alpha}(x_i)\end{aligned} \tag{3.5}$$

where x_i denote the zeros of $p_m^{-\alpha,\alpha}$, while $l_i^{-\alpha,\alpha}$ are the fundamental Lagrange polynomials defined on the same zeros. Analogously

$$\begin{aligned}G_m(x) &= (\widehat{D}\mathcal{L}_m^{-\alpha,\alpha} g)(x) = (\widehat{D}L_m^{-\alpha,\alpha} g)(x) \\ &= \sum_{i=1}^{m} \frac{\widehat{D}l_i^{-\alpha,\alpha}(x)}{v^{0,\alpha}(x_i)} g(x_i) v^{0,\alpha}(x_i).\end{aligned} \tag{3.6}$$

Hence both the polynomials $K_m f$ and G_m are expanded in the basis

$$\left\{\varphi_i(x) := \frac{\widehat{D}l_i^{-\alpha,\alpha}(x)}{v^{0,\alpha}(x_i)}\right\}_{i=1,\ldots,m}. \tag{3.7}$$

Therefore we express also the unknown $\bar{f}_m$ in the same basis, i.e., we set $\bar{f}_m = \sum_{j=1}^m a_j \varphi_j$. Now by substituting (3.5), (3.6) and the expression of $\bar{f}_m$ in (3.2) and making equal the corresponding coefficients in the basis we get

$$a_i + \nu(K\bar{f}_m)(x_i) v^{0,\alpha}(x_i) = g(x_i) v^{0,\alpha}(x_i), \qquad i = 1, \ldots, m. \tag{3.8}$$

Taking into account that $l_j^{-\alpha,\alpha}(x) = \lambda_j^{-\alpha,\alpha} \sum_{k=0}^{m-1} p_k^{-\alpha,\alpha}(x) p_k^{-\alpha,\alpha}(x_j)$, where

$$\lambda_j^{-\alpha,\alpha} \equiv \lambda_m(v^{-\alpha,\alpha}, x_j) = \left[\sum_{i=0}^{m-1} p_i^2(v^{-\alpha,\alpha}, x_j)\right]^{-1},$$

denotes the jth Christoffel number, we get

$$\begin{aligned} K\bar{f}_m(x_i) &= \sum_{j=1}^{m} a_j \int_{-1}^{1} k(x,x_i)\varphi_j(x)v^{\alpha,-\alpha}(x)dx \\ &= \sum_{j=1}^{m} \frac{a_j}{v^{0,\alpha}(x_j)} \int_{-1}^{1} k(x,x_i)\lambda_j^{-\alpha,\alpha} \sum_{k=0}^{m-1} (\widehat{D}p_k^{-\alpha,\alpha})(x)p_k^{-\alpha,\alpha}(x_j)v^{\alpha,-\alpha}(x)dx \\ &= \sum_{j=1}^{m} \frac{a_j}{v^{0,\alpha}(x_j)}\lambda_j^{-\alpha,\alpha} \sum_{k=0}^{m-1} p_k^{-\alpha,\alpha}(x_j) \int_{-1}^{1} k(x,x_i)p_k^{\alpha,-\alpha}(x)v^{\alpha,-\alpha}(x)dx. \end{aligned}$$

Using this expression in (3.8) we finally have the linear system

$$a_i + \nu \sum_{j=1}^{m} a_j \frac{v^{0,\alpha}(x_i)}{v^{0,\alpha}(x_j)}\lambda_j^{-\alpha,\alpha} \sum_{k=0}^{m-1} p_k^{-\alpha,\alpha}(x_j)m_k(x_i) = b_i, \quad i=1,\dots,m \tag{3.9}$$

where $b_i = g(x_i)v^{0,\alpha}(x_i)$ and

$$m_k(y) := \int_{-1}^{1} k(x,y)p_k^{\alpha,-\alpha}(x)v^{\alpha,-\alpha}(x)dx.$$

The obtained system can be rewritten in a more significant matrix form. Indeed, if we put

$$\mathbf{D}_m = (p_j^{-\alpha,\alpha}(x_i))_{i=1,\dots,m,j=0,\dots,m-1},$$

then

$$\mathbf{D}_m^{-1} = (\lambda_j(v^{-\alpha,\alpha})p_k^{-\alpha,\alpha}(x_j))_{k=0,\dots,m-1,j=1,\dots,m}$$

(see, e.g., [9]).

Hence, setting

$$\mathbf{M}_m = (m_k(x_i))_{i=1,\dots,m,k=0,\dots,m-1}, \quad \mathbf{\Lambda}_m = diag((v^{0,\alpha}(x_j))_{j=1,m})$$

and denoting by $\mathbf{I}_m$ the identity matrix of order m, system (3.9) becomes

$$(\mathbf{I}_m + \nu\mathbf{\Lambda}_m^{-1}\mathbf{M}_m\mathbf{D}_m^{-1}\mathbf{\Lambda}_m)\mathbf{a_m} = \mathbf{b}_m \tag{3.10}$$

where $\mathbf{a_m} = (a_i)_{i=1,\dots,m}^T$ and $\mathbf{b}_m = (g(x_1)v^{0,\alpha}(x_1),\dots,g(x_m)v^{0,\alpha}(x_m))^T$.

For the sake of simplicity set $\mathbf{C}_m = \mathbf{I}_m + \nu\mathbf{\Lambda}_m^{-1}\mathbf{M}_m\mathbf{D}_m^{-1}\mathbf{\Lambda}_m$. Since in the case when $\alpha \geq 1/2$, $\mathbf{C}_m$ is the matrix representation of the operator $I + \nu K_m$ in the basis $\{\varphi_i\}_i$, for arbitrary polynomials $f_m = \sum_{i=1}^m a_i\varphi_i$ and $G_m = \sum_{i=1}^m b_i\varphi_i$ there holds that $\mathbf{C}_m\mathbf{a}_m = \mathbf{b}_m \Leftrightarrow (I+\nu K_m)f_m = G_m$, i.e., equation (3.10) is equivalent to (3.2). Now denote by *cond*($\mathbf{C}_m$) the condition number of $\mathbf{C}_m$ considered as a linear operator in $\mathbb{R}^m$ equipped with the ℓ_∞ norm.

Proposition 3.2. *Under the assumptions of Theorem* 3.1 *and with* $\alpha \geq 1/2$ *we have*

$$\sup_m \frac{cond(\mathbf{C}_m)}{\log^4 m} < \infty. \tag{3.11}$$

During the construction of system (3.10) we supposed $\alpha \geq 1/2$. The case $0 < \alpha < 1/2$ only needs some additional computations.

The described numerical method can be used when the kernel $k(x,t)$ has some weak singularities. The main effort consists of the computation of the so called "modified moments" $m_k(x_i)$, which, sometimes, satisfy stable recurrent relations, as we will show in some significant examples later on.

Assume now that the kernel $k(x,t)$ is sufficiently smooth. We propose a different approach, avoiding the computation of the $m_j(x_k)$.

Indeed in this case instead of (3.2) we consider the equation

$$(I + \nu K_m^*) f_m = G_m \tag{3.12}$$

where f_m is the unknown polynomial, G_m is the same as before and K_m^* was defined in (2.16).

Theorem 3.3. *Let $0 < \alpha < 1$. Under the assumptions of Lemma 2.4, if $g \in Z_r(v^{0,\alpha})$, then for any sufficiently large m, there exists a unique polynomial $\bar{f}_m \in \mathbb{P}_{m+1}$, which is the solution of (3.12). Moreover the following estimate holds*

$$\|(\bar{f} - \bar{f}_m)v^{\alpha,0}\|_\infty \leq C \frac{\log m}{m^r} \|g\|_{Z_r(v^{0,\alpha})} \tag{3.13}$$

where C is a positive constant independent of $\bar{f}$ and m.

Consequently $\bar{f} \in Z_{r-\epsilon}(v^{\alpha,0})$, with $\epsilon > 0$ arbitrarily small. Moreover

$$|cond(I + \nu K_m^*) - cond(I + \nu \widehat{D} K)| = \mathcal{O}\left(\frac{\log m}{m^r}\right), \tag{3.14}$$

where the constant in $\mathcal{O}$ is independent of m.

Also in this case in order to compute the coefficients of $\bar{f}_m$ assume $\alpha \geq 1/2$ and set $\bar{f}_m = \sum_{i=1}^m a_i \varphi_i$ and $G_m = \sum_{i=1}^m b_i \varphi_i$. Using the same argument as before, i.e., expanding both sides of (3.12) in the basis $\{\varphi_i\}_i$ and making equal the corresponding coefficients we get

$$a_i + \nu (K^* \bar{f}_m)(x_i) v^{0,\alpha}(x_i) = b_i, \qquad i = 1, \dots, m \tag{3.15}$$

where K^* was defined in (2.15). So we have to evaluate the quantities $(K^* \bar{f}_m)(x_i)$. Using the gaussian quadrature formula we have

$$\begin{aligned} (K^* \bar{f}_m)(x_i) &= \int_{-1}^{1} L_m^{\alpha,-\alpha}(k(\cdot, x_i), x) \bar{f}_m(x) v^{\alpha,-\alpha}(x) dx \\ &= \sum_{k=1}^{m} \lambda_k^{\alpha,-\alpha} k(t_k, x_i) \sum_{j=1}^{m} a_j \varphi_j(t_k), \end{aligned}$$

where t_k denote the zeros of $p_m^{\alpha,-\alpha}$ and $\lambda_k^{\alpha,-\alpha}$ the Christoffel numbers with respect to the weight $v^{\alpha,-\alpha}$.

By (3.7), using $l_j^{-\alpha,\alpha}(x) = p_m^{-\alpha,\alpha}(x)/[(x - x_j)p_m'(v^{-\alpha,\alpha}, x_j)]$, since

$$\widehat{D}\left[\frac{p_m^{-\alpha,\alpha}}{\cdot - x_j}\right](x) = \frac{p_m^{\alpha,-\alpha}(x) - p_m^{\alpha,-\alpha}(x_j)}{x - x_j}$$

and [22]

$$\frac{p_m^{\alpha,-\alpha}(x_j)}{p'_m(v^{-\alpha,\alpha},x_j)} = \frac{\sin\pi\alpha}{\pi}\lambda_j^{-\alpha,\alpha},$$

we get

$$(K^*\bar{f}_m)(x_i) = \frac{\sin\pi\alpha}{\pi}\sum_{k=1}^{m}\lambda_k^{\alpha,-\alpha}k(t_k,x_i)\sum_{j=1}^{m}\frac{a_j}{v^{0,\alpha}(x_j)}\frac{\lambda_j^{-\alpha,\alpha}}{x_j-t_k}.$$

Using this expression in (3.15) we finally get the system

$$\begin{aligned} a_i + \nu\frac{\sin\pi\alpha}{\pi}\sum_{j=1}^{m}a_j\frac{v^{0,\alpha}(x_i)}{v^{0,\alpha}(x_j)}\lambda_j^{-\alpha,\alpha}\sum_{k=1}^{m}\lambda_k^{\alpha,-\alpha}\frac{k(t_k,x_i)}{x_j-t_k} &= b_i, \\ i &= 1,\dots,m \end{aligned} \tag{3.16}$$

where $b_i = g(x_i)v^{0,\alpha}(x_i)$. We remark that, since [26] $\min_{j,k}|\vartheta_j - \tau_k| \sim m^{-1}$, where $x_j = \cos\vartheta_j$ and $t_k = \cos\tau_k$, the last term at the left-hand side in (3.16) always makes sense.

Now denote by $\mathbf{B}_m$ the matrix of system (3.16) and by cond$\,(\mathbf{B}_m)$ its condition number in the ℓ_∞ norm. We have

Proposition 3.4. *Under the assumptions of Theorem* 3.3 *and with* $\alpha \geq 1/2$ *we have*

$$\sup_m \frac{\text{cond}\,(\mathbf{B}_m)}{\log^4 m} < \infty. \tag{3.17}$$

Finally the case $0 < \alpha < 1/2$ can be handled in a similar way but with some additional computations.

Remark 3.5. Looking at (3.16) and (3.9) we underline that, even if in some cases the computational efforts may be comparable, the entries of the matrix in (3.16) can be always computed, while sometimes the computation of the modified moments in (3.9) can be very hard.

4. Some special cases

In this section we will consider some special cases of the operator K. Consider $Kf(y) = K^\mu f(y) := \int_{-1}^{1} k^\mu(x,y)f(x)v^{\alpha,-\alpha}(x)dx$ where we set

$$k^\mu(x,y) := \begin{cases} |x-y|^\mu, & \mu > -1, \mu \neq 0 \\ \log|x-y|, & \mu = 0\,. \end{cases} \tag{4.1}$$

We have the following result.

Lemma 4.1. *Let $f \in C_{v^{\alpha,0}}$, $0 < \alpha < 1$ and $\mu > -1$. Then*

$$\sup_{\tau>0} \frac{\Omega_\varphi^k(K^\mu f,\tau)_{v^{0,\alpha}}}{\tau^{1+\mu}} \leq C\|fv^{\alpha,0}\|_\infty, \quad k > 1+\mu, \quad \mu \neq 0, \tag{4.2}$$

$$\sup_{\tau>0} \frac{\Omega_\varphi(K^\mu f,\tau)_{v^{0,\alpha}}}{\tau \log \tau^{-1}} \leq C\|fv^{\alpha,0}\|_\infty, \quad \mu = 0 \tag{4.3}$$

where in both cases C is a positive constant independent of f.

The proof of the previous lemma is technical and we refer the reader to the Appendix for a sketch of it.

The previous result assures that assumption (2.8) is satisfied with $r = 1+\mu$ in the case $\mu \neq 0$, and $0 < r < 1$, for $\mu = 0$. Hence in the case $-1 < \mu \leq 0$ we will solve the linear system (3.9), since Theorem 3.1 holds true. In the case $\mu > 0$ both of the Theorems 3.1 and 3.3 are true and so we can solve the linear systems (3.9) or (3.16).

In the cases when we have to solve (3.9) it is necessary to compute the integrals

$$m_j(y) = \int_{-1}^{1} k^\mu(x,y)p_j^{\alpha,-\alpha}(x)v^{\alpha,-\alpha}(x)dx.$$

The quantities $m_j(y)$ can be computed by means of suitable recurrence relations. In the Appendix we will give such recurrence relations in the cases $-1 < \mu < 0$ and $\mu = 0$, $\alpha = 1/2$.

We note that in the particular case $\alpha = 1/2$, $\mu = 0$, since $\frac{d}{dy}K^\mu f(y) = -\pi Df(y)$, using the boundedness of $D : Z_s(v^{\alpha,0}) \longrightarrow Z_s(v^{0,\alpha})$ we immediately get $\|K^\mu f\|_{Z_{s+1}(v^{0,\alpha})} \leq C\|f\|_{Z_s(v^{\alpha,0})}$ for any $f \in Z_s(v^{\alpha,0})$, $s > 0$. For the numerical method (3.2) this means that the rate of convergence will only depend on the smoothness of the right-hand side of (3.1). For instance if $g \in Z_s(v^{0,\alpha})$, by Theorem 3.1 we get that the rate of convergence will be $\mathcal{O}(\log m/m^s)$.

Finally we remark that the previous results can be generalized for operators of the type $Kf(y) = \mathcal{K}^\mu f(y) := \int_{-1}^{1} q(x,y)k^\mu(x,y)f(x)v^{\alpha,-\alpha}(x)dx$ where q is a smooth function (assume for instance that q is many times differentiable with respect to both variables). Anyway from the computational point of view the problem is how to compute efficiently the corresponding integrals of the type $m_j(y)$.

5. Numerical examples

In this section we give some numerical tests for the proposed methods. All the computations were performed in 16-digits arithmetic. In every example we give the values of the weighted approximating polynomials in two internal points of $[-1,1]$, the condition number of the matrix in ℓ_∞ norm (denoted by κ_∞) and the graph of one weighted approximating polynomial of suitable degree.

Example 1. Consider the equation

$$\frac{1}{2}f(y)v^{\frac{2}{3},-\frac{2}{3}}(y) + \frac{\sqrt{3}}{2\pi}\int_{-1}^{1}\frac{f(x)}{x-y}v^{\frac{2}{3},-\frac{2}{3}}(x)dx + \int_{-1}^{1}|x-y|^{-\frac{1}{8}}f(x)v^{\frac{2}{3},-\frac{2}{3}}(x)dx = 1+y^2.$$

In this case the right-hand side is smooth while the kernel of the perturbation is weakly singular. We apply the method (3.2), i.e., we solve system (3.9). According to estimate (3.3) we expect an order of convergence $\mathcal{O}(\log m/m^{7/8})$. Anyway in the internal points of $[-1,1]$ the convergence is faster.

m	$y=.5$	$y=.9$	κ_∞
16	1.0009	.6346	
32	1.000993	.6346	12.253
64	1.0009932	.634659	12.611
128	1.0009932	.63465960	12.817
256	1.00099326	.63465960	12.935
512	1.0009932693	.634659602	13.002

The graph of the weighted approximating polynomial $f_{512}(y)v^{\frac{2}{3},0}(y)$ is given in Fig. 1.

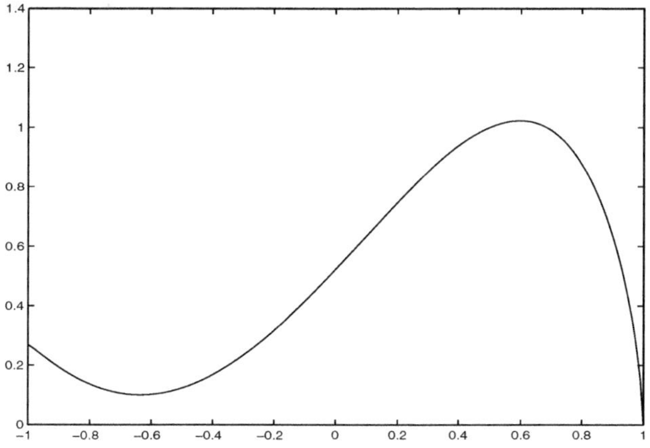

FIGURE 1. $f_{512}(y)v^{\frac{2}{3},0}(y)$

Example 2. Consider the equation

$$\cos(.7\pi)f(y)v^{.7,-.7}(y) - \frac{\sin(.7\pi)}{\pi}\int_{-1}^{1}\frac{f(x)}{x-y}v^{.7,-.7}(x)dx + \frac{1}{2}\int_{-1}^{1}e^{(x+y)}f(x)v^{.7,-.7}(x)dx = \sin(1+y)$$

In this case both the right-hand side and the kernel are analytic functions. We apply method (3.12), i.e., linear system (3.16). According to (3.13) a very fast convergence is expected. Indeed we get the machine precision with $m = 32$.

m	$y = -.8$	$y = -.4$	κ_∞
8	.524545	.481560	
16	.52454584772060	.48156018319029	8.984526522194360
32	.52454584772060	.481560183190292	10.24891532148221

The graph of the approximation $f_{32}(y)v^{0.7,0}(y)$ is given in Fig. 2.

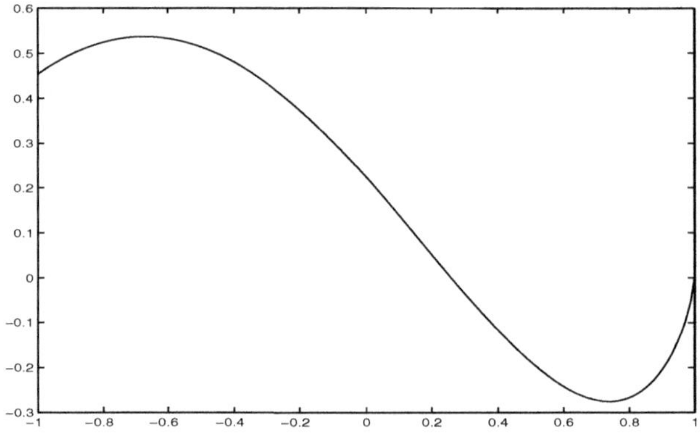

FIGURE 2. $f_{32}(y)v^{0.7,0}(y)$

Example 3. Consider the equation

$$\frac{1}{2}f(y)v^{\frac{2}{3},-\frac{2}{3}}(y) + \frac{\sqrt{3}}{2\pi}\int_{-1}^{1}\frac{f(x)}{x-y}v^{\frac{2}{3},-\frac{2}{3}}(x)dx - \frac{1}{2\pi}\int_{-1}^{1}|x-y|^{3.5}f(x)v^{\frac{2}{3},-\frac{2}{3}}(x)dx = \cos(1+y)$$

In this case the kernel is smooth but is of the type (4.1). So we can solve system (3.16) or (3.9). In both cases the rate of convergence expected is $\mathcal{O}(\log m/m^{3.5})$. The numerical test shows essentially the same results, using one system or the other. There is only a slightly better behavior in the case of (3.9) due to the exact computation of the modified moments.

m	$y = .5$	$y = .9$	κ_∞
16	-.97619	-.4845	35.69328
32	-.9761992	-.484502	44.05081
64	-.97619925	-.4845025	48.33887
128	-.976199252	-.48450252	50.99769
256	-.97619925225	-.48450252058	52.46197

The graph of the weighted approximation $f_{512}(y)v^{\frac{2}{3},0}(y)$ is given in Fig. 3.

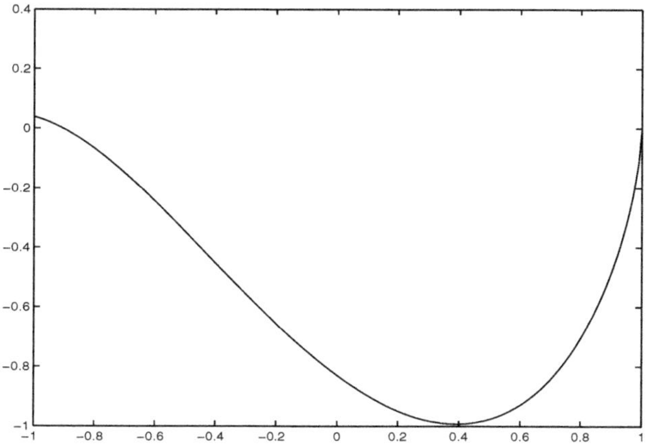

FIGURE 3. $f_{512}(y)v^{\frac{2}{3},0}(y)$

Example 4. The last test is devoted to the case of the so called "generalized airfoil equation"

$$-\frac{1}{\pi}\int_{-1}^{1}\frac{f(x)}{x-y}\sqrt{\frac{1-x}{1+x}}dx - \frac{1}{2}\int_{-1}^{1}\log|x-y|f(x)\sqrt{\frac{1-x}{1+x}}dx = e^{3x}$$

As remarked at the end of Section 4 in this case the rate of convergence depends only on the smoothness of the right-hand side. The numerical evidence confirms this expectation.

m	$y = .5$	$y = .9$	κ_∞
8	5.09	6.37	
16	5.097410331	6.372656359	8.68451
32	5.09741033161161	6.37265635964955	9.92191

The graph of the weighted polynomial $f_{32}(y)\sqrt{1-y}$ is given in Fig. 4

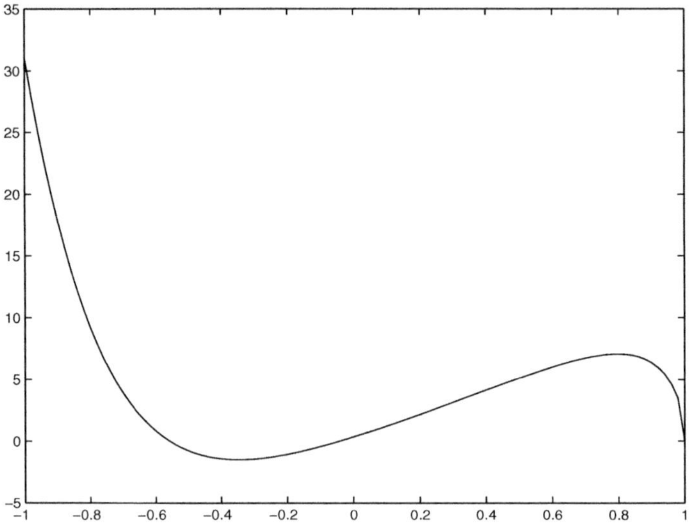

FIGURE 4. $f_{32}(y)\sqrt{1-y}$

6. Proofs of the main results

Proof of Lemma 2.1.

Let $P_m \in \mathbb{P}_m$ as in the assumption. By (2.4) we have

$$\int_0^{\frac{1}{m}} \frac{\omega_\varphi(f-P_m,t)_{v^{\gamma,\delta}}}{t}dt = \sum_{j\geq m}\int_{\frac{1}{j+1}}^{\frac{1}{j}} \frac{\omega_\varphi(f-P_m,t)_{v^{\gamma,\delta}}}{t}dt$$

$$\leq \sum_{j\geq m}\omega_\varphi\left(f-P_m,\frac{1}{j}\right)_{v^{\gamma,\delta}}\log\left(1+\frac{1}{j}\right) \leq \sum_{j\geq m}\frac{\omega_\varphi(f-P_m,1/j)_{v^{\gamma,\delta}}}{j}$$

$$\leq C\sum_{j\geq m}\frac{1}{j^2}\sum_{i=0}^{j}E_i(f-P_m)_{v^{\gamma,\delta}}$$

$$\leq C\sum_{j\geq m}\frac{1}{j^2}\left(mE_m(f)_{v^{\gamma,\delta}}+\sum_{i=m}^{j}E_i(f)_{v^{\gamma,\delta}}\right)$$

$$\leq CE_m(f)_{v^{\gamma,\delta}}+C\sum_{j\geq m}\frac{1}{j^2}\sum_{i=m}^{j}E_i(f)_{v^{\gamma,\delta}}$$

$$\leq \quad CE_m(f)_{v^{\gamma,\delta}} + C\sum_{i=m}^{\infty}\frac{E_i(f)_{v^{\gamma,\delta}}}{i}$$

$$\leq \quad C\int_0^{\frac{1}{m}}\frac{\omega_\varphi^k(f,t)_{v^{\gamma,\delta}}}{t}dt + C\sum_{i=m}^{\infty}\frac{E_i(f)_{v^{\gamma,\delta}}}{i}$$

having used (2.3). Now using again (2.3) we have that the sum on the right-hand side can be estimated as

$$\sum_{i=m}^{\infty}\frac{E_i(f)_{v^{\gamma,\delta}}}{i} \leq C\sum_{i=m}^{\infty}\frac{1}{i}\omega_\varphi^k(f,1/i)_{v^{\gamma,\delta}}$$

$$\leq \quad C\sum_{i=m}^{\infty}\int_{\frac{1}{i+1}}^{\frac{1}{i}}\frac{\omega_\varphi^k(f,t)_{v^{\gamma,\delta}}}{t}dt \leq C\int_0^{\frac{1}{m}}\frac{\omega_\varphi^k(f,t)_{v^{\gamma,\delta}}}{t}dt$$

and then the Lemma follows. □

In order to prove Theorem 2.2 we first need the following result.

Proposition 6.1. *Let $\mathcal{L}_m^{-\alpha,\alpha}$, $0<\alpha<1$, be the Lagrange operator defined in* (2.11)*. Then for every $\phi \in C_{v^{0,\alpha}}$ we get*

$$\|(\widehat{D}\mathcal{L}_m^{-\alpha,\alpha}\phi)v^{\alpha,0}\|_\infty \leq C\log m\|\phi v^{0,\alpha}\|_\infty \tag{6.1}$$

where C is a positive constant independent of m and ϕ.

Proof. First consider the case $\alpha \geq 1/2$. Using the property $\widehat{D}p_m^{-\alpha,\alpha} = p_m^{\alpha,-\alpha}$, we get $\forall x \in [-1,1]$

$$\widehat{D}\left[\frac{p_m^{-\alpha,\alpha}}{\cdot - x_k}\right](x) = \frac{p_m^{\alpha,-\alpha}(x) - p_m^{\alpha,-\alpha}(x_k)}{x-x_k} \tag{6.2}$$

where x_k, $k=1,\ldots,m$, denote the zeros of $p_m^{-\alpha,\alpha}$. Hence denoting by x_d the closest node to x, i.e., $|x_d - x| \leq |x_k - x|, k=1,2,\ldots,m$, we get

$$\begin{aligned}
&\left|(1-x)^\alpha\widehat{D}\mathcal{L}_m^{-\alpha,\alpha}\phi(x)\right| \qquad (6.3)\\
\leq\ &\left|(1-x)^\alpha\frac{\phi(x_d)}{p_m'(v^{-\alpha,\alpha},x_d)}\frac{p_m^{\alpha,-\alpha}(x)-p_m^{\alpha,-\alpha}(x_d)}{x-x_d}\right|\\
+\ &\left|(1-x)^\alpha p_m^{\alpha,-\alpha}(x)\sum_{\substack{k=1\\k\neq d}}^{m}\frac{\phi(x_k)}{x-x_k}\frac{1}{p_m'(v^{-\alpha,\alpha},x_k)}\right|\\
+\ &\left|(1-x)^\alpha\sum_{\substack{k=1\\k\neq d}}^{m}\frac{\phi(x_k)}{x-x_k}\frac{p_m^{\alpha,-\alpha}(x_k)}{p_m'(v^{-\alpha,\alpha},x_k)}\right| =: A_1+A_2+A_3
\end{aligned}$$

Let us estimate the quantities A_i separately. Since [22, (3.1), p. 124],

$$\frac{p_m^{\alpha,-\alpha}(x_k)}{p_m'(v^{-\alpha,\alpha},x_k)} = \frac{\sin\pi\alpha}{\pi}\lambda_k^{-\alpha,\alpha} \tag{6.4}$$

for A_3 we get (see, e.g., [6],[23])

$$A_3 \le \frac{\sin \pi\alpha}{\pi}(1-x)^\alpha \sum_{\substack{k=1\\k\ne d}}^m \frac{|\phi(x_k)|}{|x-x_k|}\lambda_k^{-\alpha,\alpha} \le C\log m\|\phi v^{0,\alpha}\|_\infty. \tag{6.5}$$

In order to estimate A_2 we recall the following inequalities [29]

$$\left(\sqrt{1-x}+\frac{1}{m}\right)^{\gamma+\frac12}\left(\sqrt{1+x}+\frac{1}{m}\right)^{\delta+\frac12}|p_m(v^{\gamma,\delta},x)| \le C, \quad \forall x\in[-1,1], \tag{6.6}$$

$$|p'_m(v^{\gamma,\delta},y_k)|^{-1} \sim \varphi(y_k)v^{\gamma/2+1/4,\delta/2+1/4}(y_k)m^{-1}. \tag{6.7}$$

Here $\gamma,\delta>-1$, $\varphi(x)=\sqrt{1-x^2}$, $\{y_k\}_{k=1,\dots,m}$ are the zeros of $p_m^{\gamma,\delta}$ and C and the constants in $\sim$ are independent of m and k. We need also the following estimate [20, Lemma 4.1]

$$\sum_{\substack{k=1\\k\ne d}}^m \frac{v^{\gamma,\delta}(y_k)}{m|x-y_k|} \le C\log m\left(\sqrt{1-x}+\frac{1}{m}\right)^{2\gamma-1}\left(\sqrt{1+x}+\frac{1}{m}\right)^{2\delta-1} \tag{6.8}$$

where $-1/2\le\gamma,\delta\le 1/2$ and y_d is the zero of $p_m^{\gamma,\delta}$ nearest to x.

Therefore by (6.7) we get

$$\begin{aligned} A_2 &\le \|\phi v^{0,\alpha}\|_\infty|(1-x)^\alpha p_m^{\alpha,-\alpha}(x)|\sum_{\substack{k=1\\k\ne d}}^m \frac{(1+x_k)^{-\alpha}}{|p'_m(v^{-\alpha,\alpha},x_k)|}\frac{1}{|x-x_k|} \\ &\le C\|\phi v^{0,\alpha}\|_\infty|(1-x)^\alpha p_m^{\alpha,-\alpha}(x)|\sum_{\substack{k=1\\k\ne d}}^m \frac{\varphi(x_k)}{m}\frac{v^{-\alpha/2+1/4,-\alpha/2+1/4}(x_k)}{|x-x_k|}. \end{aligned}$$

Since $\alpha\ge 1/2$, by (6.8) and (6.6), we have

$$\begin{aligned} A_2 &\le C\log m\|\phi v^{0,\alpha}\|_\infty|(1-x)^\alpha p_m^{\alpha,-\alpha}(x)| \\ &\cdot \left(\sqrt{1-x}+\frac{1}{m}\right)^{-\alpha+\frac12}\left(\sqrt{1+x}+\frac{1}{m}\right)^{-\alpha+\frac12} \le C\log m\|\phi v^{0,\alpha}\|_\infty. \end{aligned} \tag{6.9}$$

Only the estimation of A_1 is left. Using the mean value theorem we have

$$A_1 \le \|\phi v^{0,\alpha}\|_\infty(1-x)^\alpha|p'_m(v^{\alpha,-\alpha},\xi)|\frac{(1+x_d)^{-\alpha}}{|p'_m(v^{-\alpha,\alpha},x_d)|}$$

where $|\xi-x|\le|x_d-x|$. Since x_d is one of the nodes closest to x, $1\pm x_d\sim 1\pm\xi$ and, by (6.7), we have

$$\begin{aligned} A_1 &\le \frac{C}{m}\|\phi v^{0,\alpha}\|_\infty|p'_m(v^{\alpha,-\alpha},\xi)|v^{\alpha/2+3/4,-\alpha/2+3/4}(\xi) \\ &\sim C\|\phi v^{0,\alpha}\|_\infty|p_{m-1}^{\alpha+1,-\alpha+1}(\xi)v^{\alpha/2+3/4,-\alpha/2+3/4}(\xi)| \le C\|\phi v^{0,\alpha}\|_\infty \end{aligned} \tag{6.10}$$

where we used [31, (4.5.5), p. 72] and (6.6).

Hence (6.1), in the case $\alpha \geq 1/2$, follows using estimates (6.5), (6.9) and (6.10) in (6.3) and then taking the maximum with respect to x.

The proof in the case $\alpha < 1/2$ is exactly the same. Indeed if we set $x_{m+1} = (x_m + 1)/2$, $x_0 = -x_{m+1}$ and $q_{-\alpha,\alpha}(x) = (x - x_0)(x - x_{m+1})p_m^{-\alpha,\alpha}(x)$, we get

$$q'_{-\alpha,\alpha}(x_k) = \left[\begin{array}{ll} -2x_{m+1}p_m^{-\alpha,\alpha}(-x_{m+1}), & k = 0 \\ (x_k + x_{m+1})(x_k - x_{m+1})p'_m(v^{-\alpha,\alpha}, x_k) & k = 1, \dots, m \\ 2x_{m+1}p_m^{-\alpha,\alpha}(x_{m+1}), & k = m + 1. \end{array} \right. \tag{6.11}$$

Identities (6.2), (6.4) still hold true with $q_{-\alpha,\alpha}$ playing the role of $p_m^{-\alpha,\alpha}$. Thus, taking also into account that [29]

$$p_m^{-\alpha,\alpha}(x_{m+1}) \sim m^{-\alpha+1/2}, \qquad p_m^{-\alpha,\alpha}(-x_{m+1}) \sim m^{\alpha+1/2}$$

the previous proof can be repeated word by word with $q_{-\alpha,\alpha}$ in place of $p_m^{-\alpha,\alpha}$. □

Proof of Theorem 2.2. Let P be an arbitrary polynomial of degree $m - 1$, $m > 1$. Since there holds [24]

$$\|(\widehat{D}\phi)v^{\alpha,0}\|_\infty \leq C\left[\|\phi v^{0,\alpha}\|_\infty + \int_0^1 \omega_\varphi(\phi, t)_{v^{0,\alpha}} \frac{dt}{t}\right] \tag{6.12}$$

and $\widehat{D}P \in \mathbb{P}_{m-1}$, we have

$$\|\widehat{D}[\phi - \mathcal{L}_m^{-\alpha,\alpha}\phi]v^{\alpha,0}\|_\infty \leq \|\widehat{D}[\phi - P]v^{\alpha,0}\|_\infty + \|\widehat{D}[P - \mathcal{L}_m^{-\alpha,\alpha}\phi]v^{\alpha,0}\|_\infty$$

$$\leq C\|(\phi - P)v^{0,\alpha}\|_\infty + C\int_0^1 \frac{\omega_\varphi(\phi - P, t)_{v^{0,\alpha}}}{t}dt + \|\widehat{D}[\mathcal{L}_m^{-\alpha,\alpha}(P - \phi)]v^{\alpha,0}\|_\infty.$$

Now by applying Proposition 6.1 we get

$$\begin{aligned} \|\widehat{D}[\phi - \mathcal{L}_m^{-\alpha,\alpha}\phi]v^{\alpha,0}\|_\infty &\leq C\log m\|(\phi - P)v^{0,\alpha}\|_\infty \\ &+ C\int_0^{\frac{1}{m}} \frac{\omega_\varphi(\phi - P, t)_{v^{0,\alpha}}}{t}dt \end{aligned}$$

where C is a positive constant independent of m and ϕ. Now choosing P as the best approximation of ϕ in $C_{v^{0,\alpha}}$ and using Lemma 2.1 we get

$$\|\widehat{D}[\phi - \mathcal{L}_m^{-\alpha,\alpha}\phi]v^{\alpha,0}\|_\infty \leq C\log m E_m(\phi)_{v^{0,\alpha}} + C\int_0^{\frac{1}{m}} \frac{\omega_\varphi^k(\phi, t)_{v^{0,\alpha}}}{t}dt.$$

Therefore (2.12) follows by inequality (2.3) and recalling that if $\omega_\varphi^k(\phi, t) \sim t^r$ then $\omega_\varphi^k(\phi, t) \sim \Omega_\varphi^k(\phi, t)$. □

Proof of Lemma 2.3. The lemma follows by Theorem 2.2 and assumption (2.8). □

Proof of Lemma 2.4. First we note that, by definition, (2.18) implies (2.8). Moreover we get

$$\begin{aligned} \|[(\widehat{D}K - K_m^*)f]v^{\alpha,0}\|_\infty &\leq \|[(\widehat{D}K - K_m)f]v^{\alpha,0}\|_\infty \\ &+ \|[\widehat{D}\mathcal{L}_m^{-\alpha,\alpha}(K^* - K)f]v^{\alpha,0}\|_\infty \end{aligned}$$

where K_m is the operator defined in (2.13). Since (2.8) holds true, by Lemma 2.3 and Proposition 6.1 it follows

$$\|[(\widehat{D}K - K_m^*)f]v^{\alpha,0}\|_\infty \le C\frac{\log m}{m^r}\|fv^{\alpha,0}\|_\infty + C\log m\|(K^* - K)fv^{0,\alpha}\|_\infty.$$

So we have to evaluate the second term on the right-hand side. By the definition of K^* and using (2.10) we have

$$\begin{aligned}
&\|(K^* - K)fv^{0,\alpha}\|_\infty \\
= \;& \max_{|y|\le 1}(1+y)^\alpha \left|\int_{-1}^{1}[k_y - \mathcal{L}_m^{-\alpha,\alpha}(k_y)](x)f(x)v^{\alpha,-\alpha}(x)dx\right| \\
\le \;& \|fv^{\alpha,0}\|_\infty \max_{|y|\le 1}(1+y)^\alpha \int_{-1}^{1}\left|k_y(x) - \mathcal{L}_m^{-\alpha,\alpha}(k_y)(x)\right| v^{0,-\alpha}(x)dx \\
\le \;& C\|fv^{\alpha,0}\|_\infty \max_{|y|\le 1}(1+y)^\alpha E_m(k_y) \\
\le \;& C\|fv^{\alpha,0}\|_\infty \max_{|y|\le 1}(1+y)^\alpha \omega_\varphi^k(k_y, m^{-1}).
\end{aligned}$$

Since by (2.17) it follows that $\omega_\varphi^k(k_y, m^{-1}) \sim \Omega_\varphi^k(k_y, m^{-1})$, again by (2.17) we finally get

$$\|(K^* - K)fv^{0,\alpha}\|_\infty \le \frac{C}{m^r}\|fv^{\alpha,0}\|_\infty$$

and the Lemma follows. □

Proof of Theorem 3.1. By Lemma 2.3 we immediately get that $I + \nu\widehat{D}K_m$ is invertible and uniformly bounded in $C_{v^{\alpha,0}}$. Therefore by the identity

$$\bar{f} - \bar{f}_m = (I + \nu\widehat{D}K_m)^{-1}\left[(\widehat{D}g - \widehat{D}\mathcal{L}_m^{-\alpha,\alpha}g) + \nu(\widehat{D}K - K_m)\bar{f}\right]$$

it follows

$$\begin{aligned}
\|(\bar{f} - \bar{f}_m)v^{\alpha,0}\|_\infty \;&\le\; C\|(\widehat{D}g - \widehat{D}\mathcal{L}_m^{-\alpha,\alpha}g)v^{\alpha,0}\|_\infty \\
&+\; C\|(\widehat{D}K - K_m)(\bar{f})v^{\alpha,0}\|_\infty
\end{aligned}$$

where C is a positive constant independent of m and $\bar{f}$. Hence (3.3) can be obtained by applying Theorem 2.2 and Lemma 2.3. Finally working as in [15] estimate (3.4) can be deduced by Lemma 2.3. □

Proof of Proposition 3.2. To prove the proposition we need some preliminary results. First of all since [24]

$$\|(D\phi)v^{0,\alpha}\|_\infty \le C\left[\|\phi v^{\alpha,0}\|_\infty + \int_0^1 \omega_\varphi(\phi, t)_{v^{\alpha,0}}\frac{dt}{t}\right], \tag{6.13}$$

by (2.5) we get, for any $\phi \in \mathbb{P}_{m-1}$,

$$\begin{aligned} \|(D\phi)v^{0,\alpha}\|_\infty &\le C\left[\|\phi v^{\alpha,0}\|_\infty + \frac{1}{m}\|\phi'\varphi v^{\alpha,0}\|_\infty\right. \\ &+ \left.\int_{\frac{1}{m}}^1 \omega_\varphi(\phi,t)_{v^{\alpha,0}}\frac{dt}{t}\right] \le C\log m\|\phi v^{\alpha,0}\|_\infty \end{aligned} \tag{6.14}$$

having used the Bernstein inequality and the definition of ω_φ and where $C \neq C(\phi, m)$.

On the other hand we remark that if $\phi \in \mathbb{P}_{m-1}$ and $\phi = \sum_{i=1}^m c_i\varphi_i$, with φ_i defined in (3.7), then $c_i = (D\phi)(x_i)v^{0,\alpha}(x_i)$, where x_i are the zeros of $p_m^{-\alpha,\alpha}$. Moreover by Proposition 6.1 we have

$$\begin{aligned} \|\phi v^{\alpha,0}\|_\infty &\le \|\mathbf{c}\|_{\ell_\infty}\max_{|x|\le 1} v^{\alpha,0}(x)\sum_{i=1}^m \frac{|\widehat{D}l_i^{-\alpha,\alpha}(x)|}{v^{0,\alpha}(x_i)} \\ &= \|\mathbf{c}\|_{\ell_\infty}\|\widehat{D}\mathcal{L}_m^{-\alpha,\alpha}\|_{C_{v^{\alpha,0}}\longrightarrow C_{v^{0,\alpha}}} \le C\log m\|\mathbf{c}\|_{\ell_\infty} \end{aligned} \tag{6.15}$$

where $C \neq C(m)$ and $\mathbf{c} = (c_1, \ldots, c_m)^T$.

Now let $\mathbf{a} = (a_0, \ldots, a_{m-1})^T \in \mathbb{R}^m$, $\mathbf{b} = (b_0, \ldots, b_{m-1})^T \in \mathbb{R}^m$ and set

$$f_m = \sum_{i=0}^{m-1} a_i\varphi_i, \qquad G_m = \sum_{i=0}^{m-1} b_i\varphi_i.$$

Since $\mathbf{C}_m$ represents the operator $I + \nu K_m$ in the basis $\{\varphi_i\}_i$ we have

$$(I + \nu K_m)f_m = G_m \Longleftrightarrow \mathbf{C}_m\mathbf{a} = \mathbf{b}.$$

By (6.14) and for any $\mathbf{a} \in \mathbb{R}^m$ we get

$$\begin{aligned} \|\mathbf{C}_m\mathbf{a}\|_{\ell_\infty} &= \|\mathbf{b}\|_{\ell_\infty} = \max_{1\le i\le m}|(DG_m)(x_i)v^{0,\alpha}(x_i)| \\ &\le \max_{|x|\le 1}|(DG_m)(x)v^{0,\alpha}(x)| \le C\log m\|G_m v^{\alpha,0}\|_\infty \\ &= C\log m\|[(I+\nu K_m)f_m]v^{\alpha,0}\|_\infty \\ &\le C\log m\|f_m v^{\alpha,0}\|_\infty\|I+\nu K_m\|_{C_{v^{\alpha,0}}\longrightarrow C_{v^{\alpha,0}}} \\ &\le C\log^2 m\|\mathbf{a}\|_{\ell_\infty}\|I+\nu\widehat{D}K\|_{C_{v^{\alpha,0}}\longrightarrow C_{v^{\alpha,0}}} \end{aligned}$$

where $C \neq C(m)$ and we used (6.15).

Analogously for any $\mathbf{b} \in \mathbb{R}^m$, if $\mathbf{a} = \mathbf{C}_m^{-1}\mathbf{b}$ we get

$$\begin{aligned} \|\mathbf{C}_m^{-1}\mathbf{b}\|_{\ell_\infty} &= \|\mathbf{a}\|_{\ell_\infty} = \max_{1\le i\le m}|(Df_m)(x_i)v^{0,\alpha}(x_i)| \\ &\le \max_{|x|\le 1}|(Df_m)(x)v^{0,\alpha}(x)| \le C\log m\|f_m v^{\alpha,0}\|_\infty \\ &= C\log m\|[(I+\nu K_m)^{-1}G_m]v^{\alpha,0}\|_\infty \\ &\le C\log m\|G_m v^{\alpha,0}\|_\infty\|(I+\nu K_m)^{-1}\|_{C_{v^{\alpha,0}}\longrightarrow C_{v^{\alpha,0}}} \\ &\le C\log^2 m\|\mathbf{b}\|_{\ell_\infty}\|(I+\nu\widehat{D}K)^{-1}\|_{C_{v^{\alpha,0}}\longrightarrow C_{v^{\alpha,0}}}, \quad C \neq C(m) \end{aligned}$$

having used the same arguments as before. Thus the proof is complete. □

Proof of Theorem 3.3. The proof is similar to that of Theorem 3.1. Indeed we can repeat the same arguments using Lemma 2.4 instead of Lemma 2.3. □

Proof of Proposition 3.4. We can repeat word by word the proof of Proposition 3.2 using operator K_m^*, defined in (2.16), instead of K_m. □

7. Appendix

Proof of Lemma 4.1. Let $\mu > -1$, k be an integer s.t. $k > 1+\mu$ and $|y| \leq 1-4h^2k^2$, $0 < h \leq \tau$. Taking into account the definition of Ω_φ^k we have to estimate

$$
\begin{aligned}
|\Delta_{h\varphi(y)}^k(K^\mu f)(y)(1+y)^\alpha| &\leq \left| \int_{-1}^1 \Delta_{h\varphi(y)}^k (k_x^\mu(y)) f(x)(1-x)^\alpha(1+x)^{-\alpha}dx \right| (1+y)^\alpha \\
&\leq (1+y)^\alpha \|fv^{\alpha,0}\|_\infty \int_{-1}^1 |\Delta_{h\varphi(y)}^k k_x^\mu(y)|(1+x)^{-\alpha}dx. \qquad (7.1)
\end{aligned}
$$

We split the integration interval on the right-hand side as follows

$$
\int_{-1}^1 |\Delta_{h\varphi(y)}^k k_x^\mu(y)|(1+x)^{-\alpha}dx = \left\{ \int_{-1}^{y-2kh\varphi(y)} + \int_{y-2kh\varphi(y)}^{y+2kh\varphi(y)} \right. \qquad (7.2)
$$

$$
+ \left. \int_{y+2kh\varphi(y)}^1 \right\} |\Delta_{h\varphi(y)}^k k_x^\mu(y)|(1+x)^{-\alpha}dx := \sum_{i=1}^3 G_i(y,h).
$$

We remark that in the case of G_3 we can use that [7, p. 21]

$$
|\Delta_{h\varphi(y)}^k k_x^\mu(y)| \leq C\,(2kh\varphi(y))^k\,(x-\xi)^{\mu-k} \leq Ch^k(x-\xi)^{\mu-k},
$$

where $\xi \in \left[y - \frac{k}{2}h\varphi(y), y + \frac{k}{2}h\varphi(y) \right]$. Since $k > 1+\mu$ and $\mu > -1$ (also $\mu = 0$) we get

$$
\begin{aligned}
G_3(y,h) &\leq Ch^k \int_{y+2kh\varphi(y)}^1 (x-y-kh\varphi(y))^{\mu-k}(1+x)^{-\alpha}dx \\
&\leq Ch^k(1+y)^{-\alpha} \int_{kh\varphi(y)}^{1-(y+kh)} u^{\mu-k}du \\
&\leq Ch^k(1+y)^{-\alpha} \int_{kh\varphi(y)}^{\infty} u^{\mu-k}du \leq C(1+y)^{-\alpha}h^{\mu+1}.
\end{aligned}
$$

In order to estimate $G_1(y,h)$ it is sufficient to split the integration interval in $[-1,-1+(1+y)/2]$ and $[-1+(1+y)/2, y-2hk\varphi(y)]$.

Finally estimate $G_2(y,h)$. We note that in this case $1+y \sim 1+x$. Hence we can write

$$
G_2(y,h) \leq C(1+y)^{-\alpha} \sum_{j=0}^k \int_{y-2kh\varphi(y)}^{y+2kh\varphi(y)} \left| k^\mu \left(x, y + \left(\frac{k}{2} - j \right) h\varphi(y) \right) \right| dx.
$$

By means of basic computations it is easy to see that each term in the sum has order $h^{\mu+1}$ if $\mu \neq 0$ and $h \log h^{-1}$ if $\mu = 0$.

Therefore using the obtained estimates for $G_i(y,h)$, $i = 1,2,3$ in (7.1)–(7.2) and taking the sup on $0 < h \leq \tau$ the Lemma immediately follows. □

7.1. The recurrence relations for the modified moments.

For the convenience of the reader and for further references, we collect here some recurrence relations performing the modified moments of the type

$$m_j(y) = \int_{-1}^{1} k^\mu(x,y) p_j^{\alpha,\beta}(x) v^{\alpha,\beta}(x)dx, \quad \alpha, \beta > -1$$

where k^μ is the kernel defined in (4.1).

The case $\mu \neq 0$. Using [31, (4.5.5), p. 72] and the Rodrigues' formula [31, (4.10.1), p. 94]

$$v^{\alpha,\beta}(x)p_n^{\alpha,\beta}(x) = -\frac{1}{\sqrt{n(n+\alpha+\beta+1)}}\frac{d}{dx}v^{\alpha+1,\beta+1}(x)p_{n-1}^{\alpha+1,\beta+1}(x)$$

it is possible to deduce, for $\alpha + \beta \neq -1$, the recurrence relation

$$\delta_{j+1}\left(1 + \frac{\mu+1}{j+\alpha+\beta+1}\right) m_{j+1}(y) = (y+\gamma_j)\, m_j(y)$$
$$+ \quad \delta_j \left(\frac{\mu+1}{j} - 1\right) m_{j-1}(y), \qquad j = 1,2,\ldots$$

where

$$\delta_j = \sqrt{\frac{4j(j+\alpha)(j+\beta)(j+\alpha+\beta)}{(2j+\alpha+\beta-1)(2j+\alpha+\beta)^2(2j+\alpha+\beta+1)}}$$
$$\gamma_j = \frac{(\alpha-\beta)(\alpha+\beta+2\mu+2)}{(2j+\alpha+\beta)(2j+\alpha+\beta+2)}.$$

The starting moments are defined as follows

$$m_0(y) = \frac{1}{\gamma(\alpha,\beta)}[m_0^-(y,\mu) + m_0^+(y,\mu)]$$

with

$$\gamma(\alpha,\beta) = \sqrt{2^{\alpha+\beta+1}B(1+\alpha,1+\beta)}$$

$$m_0^-(y,\mu) := 2^\alpha(1+y)^{\beta+\mu+1}B(1+\mu,1+\beta)\; {}_2F_1\left(-\alpha, 1+\beta, \mu+\beta+2, \frac{1+y}{2}\right)$$
$$m_0^+(y,\mu) := 2^\beta(1-y)^{\alpha+\mu+1}B(1+\mu,1+\alpha)\; {}_2F_1\left(-\beta, 1+\alpha, \mu+\alpha+2, \frac{1-y}{2}\right)$$

where B and ${}_2F_1$ denote respectively the Beta and the hypergeometric function, and

$$\begin{aligned} m_1(y) &= \frac{1}{2}\sqrt{\frac{\alpha+\beta+3}{(\alpha+1)(\beta+1)}}\{[(\alpha+\beta+2)y+(\alpha-\beta)]m_0(y) \\ &+ \frac{(\alpha+\beta+2)}{\gamma(\alpha,\beta)}(m_0^+(y,\mu+1)-m_0^-(y,\mu+1))\Big\} \end{aligned}$$

with $\gamma(\alpha,\beta)$, m_0^- and m_0^+ defined above. In the case $\alpha+\beta=-1$ the recurrence relation still holds but with $j=2,3,\dots$ and hence it is necessary to compute separately also the starting moment $m_2(y)$.

The case $\mu=0$, $\alpha=\frac{1}{2}$. In [2, 17] can be found a recurrence relation for the modified moments

$$m_j(y)=\int_{-1}^{1}\log|x-y|p_j^{\frac{1}{2},-\frac{1}{2}}(x)\sqrt{\frac{1-x}{1+x}}dx$$

expressed by using only the polynomials $p_j^{-\frac{1}{2},\frac{1}{2}}$. The relation is

$$\begin{aligned} m_0(y) &= \sqrt{\pi}(t-\log 2), \\ m_j(y) &= \frac{\pi}{2}\left[\frac{1}{j+1}p_{j+1}^{-\frac{1}{2},\frac{1}{2}}(y)-\frac{1}{j(j+1)}p_j^{-\frac{1}{2},\frac{1}{2}}(y)-\frac{1}{j}p_{j-1}^{-\frac{1}{2},\frac{1}{2}}(y)\right], \quad j=1,2,\dots \end{aligned}$$

and it can be computed by means of the formula

$$\begin{cases} p_0^{-\frac{1}{2},\frac{1}{2}}(x) &= \frac{1}{\sqrt{\pi}}, \quad p_1^{-\frac{1}{2},\frac{1}{2}}(x)=\frac{1}{\sqrt{\pi}}(2x-1) \\ p_j^{-\frac{1}{2},\frac{1}{2}}(x) &= 2xp_{j-1}^{-\frac{1}{2},\frac{1}{2}}(x)-p_{j-2}^{-\frac{1}{2},\frac{1}{2}}(x), \quad j=2,3,\dots \end{cases}$$

Acknowledgments

The authors are grateful to the referees for the accurate reading of the paper and their pertinent remarks.

References

[1] Berthold D., Hoppe W., Silbermann B., *A fast algorithm for solving the generalized airfoil equation*, J. Comp. Appl. Math. **43** (1992), 185–219.

[2] Berthold D., Hoppe W., Silbermann B., *The numerical solution of the generalized airfoil equation*, J. Integr. Eq. Appl. **4** (1992), 309–336.

[3] Capobianco M.R., *The stability and the convergence of a collocation method for a class of Cauchy singular integral equation*, Math. Nachr. **162** (1993), 45–58.

[4] Capobianco M.R., Russo M.G., *Uniform convergence estimates for a collocation method for the Cauchy Singular integral equation*, J. Integral Equations Appl. **9**, (1997), no. 1, 21–45.

[5] Capobianco M.R., Junghanns P., Luther U., Mastroianni G., *Weighted uniform convergence of the quadrature method for Cauchy singular integral equations*, Singular integral operators and related topics (Tel Aviv, 1995) 153–181, Oper. Theory Adv. Appl. **90**, Birkhäuser, Basel, 1996.

[6] Criscuolo G., Mastroianni G., *On the uniform convergence of Gaussian quadrature rules for Cauchy principal value integrals*, Numer. Math., 54 (1989), no. 4, 445–461.

[7] Ditzian Z., Totik V., *Moduli of smoothness*, SCMG Springer-Verlag, New York Berlin Heidelberg London Paris Tokyo, 1987.

[8] Frammartino C., Russo M.G., *Numerical remarks on the condition numbers and the eigenvalues of matrices arising from integral equations*, Advanced special functions and integration methods (Melfi, 2000), 291–310, Proc. Melfi Sch. Adv. Top. Math. Phys., 2, Aracne, Rome, 2001.

[9] Gautschi W., *The condition of Vandermonde-like matrices involving orthogonal polynomials*, Linear Algebra and Appl. **52**, (1983) 293–300.

[10] Junghanns P., Luther U., *Cauchy singular integral equations in spaces of continuous functions and methods for their numerical solution*, ROLLS Symposium (Leipzig, 1996). J. Comput. Appl. Math. **77** (1997), no. 1–2, 201–237.

[11] Junghanns P., Luther U., *Uniform convergence of the quadrature method for Cauchy singular integral equations with weakly singular perturbation kernels*, Proceedings of the Third International Conference on Functional Analysis and Approximation Theory, Vol. II (Acquafredda di Maratea, 1996). Rend. Circ. Mat. Palermo (2) Suppl. No. **52**, Vol. II (1998), 551–566.

[12] Junghanns P., Luther U., *Uniform convergence of a fast algorithm for a Cauchy singular integral equations*, Proceedings of the Sixth Conference of the International Linear Algebra Society (Chemnitz, 1996), Linear Algebra and Appl. 275/276 (1998), 327–347.

[13] Junghanns P., Silbermann B., *Zur Theorie der Näherungsverfahren für singuläre Integralgleichungen auf Intervallen*, Math. Nachr. **103** (1981), 199–244.

[14] Junghanns P., Silbermann B., *The numerical treatment of singular integral equations by means of polynomial approximations*, Preprint, P–MAT–35/86, AdW der DDR, Karl Weierstrass Institut für Mathematik, Berlin (1986).

[15] Laurita C., Mastroianni G. *Revisiting a quadrature method for Cauchy singular integral equations with a weakly singular perturbation kernel*, Problems and methods in mathematical plysics (Chemnitz, 1999), 307–326, Operator Theory: Advances and Applications, 121, Birkhäuser, Basel, 2001.

[16] Laurita C., Mastroianni G., Russo M. G., *Revisiting CSIE in L^2: condition numbers and inverse theorems*, Integral and Integrodifferential Equations, 159–184, Ser. Math. Anal. Appl., 2, Gordon and Breach, Amsterdam, 2000.

[17] Laurita C., Occorsio D., *Numerical solution of the generalized airfoil equation*, Advanced special functions and applications (Melfi, 1999), 211–226, Proc. Melfi Sch. Adv. Top. Math. Phys., 1, Aracne, Rome, 2000.

[18] Luther U., *Generalized Besov spaces and CSIE*, Ph.D. Dissertation, 1998.

[19] Luther U., Russo M.G., *Boundedness of the Hilbert transformation in some Besov type spaces*, Integr. Equ. Oper. Theory **36** (2000), no.2, 220–240.

[20] Mastroianni G., *Uniform convergence of derivatives of Lagrange interpolation*, J. Comput. Appl. Math. **43** (1992), 37–51.

[21] Mastroianni G., Nevai P., *Mean convergence of derivatives of Lagrange interpolation*, J. Comput. Appl. Math. **34** (1991), no. 3, 385–396.

[22] Mastroianni G., Prössdorf S., *Some nodes matrices appearing in the numerical analysis for singular integral equations*, BIT **34** (1994), no. 1, 120–128.

[23] Mastroianni G., Russo M.G., *Lagrange interpolation in some weighted uniform spaces*, Facta Universitatis, Ser. Math. Inform. **12** (1997), 185–201.

[24] Mastroianni G., Russo M.G., Themistoclakis W., *The boundedness of the Cauchy singular integral operator in weighted Besov type spaces with uniform norms*, Integr. Equ. Oper. Theory **42**, (2002), no.1, 57–89.

[25] Mastroianni G., Themistoclakis W., *A numerical method for the generalized airfoil equation based on the de la Vallée Poussin interpolation*, J. Comput. Appl. Math. 180(1), 71–105 (2005).

[26] Mikhlin S.G., Prössdorf S., *Singular Integral Operators*, Akademie-Verlag, Berlin, 1986.

[27] Monegato G., Prössdorf S., *Uniform convergence estimates for a collocation and discrete collocation method for the generalized airfoil equation*, Contributions to Numerical Mathematics (A.G. Agarval, ed.), World Scientific Publishing Company 1993, 285–299 (see also the errata corrige in the Internal Reprint No. 14 (1993) Dip. Mat. Politecnico di Torino).

[28] Muskhelishvili N.I., *Singular Integral Equations*, Noordhoff, Groningen, 1953.

[29] Nevai P., *Mean convergence of Lagrange interpolation III*, Trans. Amer. Math. Soc. **282** (1984), 669–698.

[30] Prössdorf S., Silbermann B., *Numerical Analysis for Integral and related Operator Equations*, Akademie-Verlag, Berlin 1991 and Birkhäuser Verlag, Basel-Boston-Stuttgart 1991.

[31] Szegő G., *Orthogonal Polynomials*, AMS, Providence, Rhode Island, 1939.

[32] Timan A.F., *Theory of approximation of functions of a real variable*, Pergamonn Press, Oxford, England, 1963.

G. Mastroianni and M.G. Russo
Dipartimento di Matematica
Università degli Studi della Basilicata
Campus Macchia Romana
I-85100 Potenza, Italy
e-mail: `mastroianni@unibas.it`
e-mail: `russo@unibas.it`

W. Themistoclakis
CNR, Istituto per le Applicazioni del Calcolo "Mauro Picone"
Sezione di Napoli
Via P. Castellino 111
I-80131 Napoli, Italy
e-mail: `wt@na.iac.cnr.it`

Operator Theory:
Advances and Applications, Vol. 160, 337–356

On a Gevrey-Nonsolvable Partial Differential Operator

Alessandro Oliaro

Abstract. We consider an operator whose principal part is the mth power of a Mizohata hypoelliptic operator. We assume that the lower order term vanishes at a small rate with respect to the principal part, and we prove the local nonsolvability in Gevrey classes, for large Gevrey index.

Mathematics Subject Classification (2000). 35A07, 35A20, 35D05.

Keywords. Gevrey classes, local solvability, Mizohata operator.

1. Introduction

In this paper we study the nonsolvability in Gevrey classes of the operator

$$P = (D_t + iat^{2k}D_x)^m + ct^\ell D_x^n, \quad (t,x) \in \mathbb{R}^2. \tag{1.1}$$

We recall that, given a real number $s > 1$ and an open set $\Omega \subset \mathbb{R}^N$ the Gevrey space $G^s(\Omega)$ is the set of all C^∞ functions f such that for every compact set $K \subset \Omega$ there exists $C_K > 0$ satisfying $\sup_{x \in K} |\partial^\alpha f(x)| \leq C_K^{|\alpha|+1}\alpha!^s$, for every $\alpha \in \mathbb{Z}_+$.

A differential operator Q is said to be G^s locally solvable at the point x_0 if there exists a neighborhood Ω of x_0 such that for any compactly supported function $f \in G^s$ there is a solution $u \in \mathcal{D}'_s$ of the equation $Qu = f$ in Ω, $\mathcal{D}'_s$ being the ultradistribution space, topological dual of $G_0^s := G^s \cap C_0^\infty$. Since

$$\mathcal{A}(\Omega) = G^1(\Omega) \subset \cdots \subset G^{s_1}(\Omega) \subset G^{s_2}(\Omega) \subset \cdots \subset C^\infty(\Omega) \ for\ 1 < s_1 \leq s_2,$$

then any operator Q which is G^{s_2} locally solvable is also G^{s_1} locally solvable for $s_1 < s_2$. Therefore, dealing with non C^∞ locally solvable operators we can look for the bigger s up to which they remain G^s solvable, finding a critical index for the local solvability of Q.

Work supported by NATO grant PST.CLG.979347.

Operators connected to P, cf. (1.1), have been considered in many papers. It is well known that the operator

$$D_t + it^h D_x,$$

h odd, is not C^∞ locally solvable, neither G^s locally solvable at the origin for any $1 < s < \infty$; the same result holds for the operator

$$(D_t + it^h D_x)^m + \text{lower order terms}; \tag{1.2}$$

the C^∞ nonlocal solvability of (1.2) was proved by Cardoso-Trèves [3] for $m = 2$; Goldman [10] proved the C^∞ nonlocal solvability of (1.2) for arbitrary m but under conditions on the lower order terms; finally Cicognani-Zanghirati [4] and Gramchev [12] showed that (1.2) is not G^s locally solvable for $1 < s < \infty$, without any assumption on the lower order terms. Further generalizations of this results are given in Georgiev-Popivanov [9] and Marcolongo-Rodino [17], in which the case of infinite order vanishing coefficients is treated. Regarding the operator P, cf. (1.1), it follows from general results that P is G^s locally solvable for $s < \frac{m}{m-1}$, cf. Gramchev-Rodino [14], Marcolongo-Oliaro [16], Spagnolo [24], De Donno-Oliaro [7]. Okaji [19], [20] studied P in the case $m = 2$, $n = 1$ proving that, for $\ell = 0$, P is G^s hypoelliptic and G^s locally solvable for $1 \leq s < \frac{4k}{2k-1}$; moreover, P is C^∞ locally solvable if $\ell \geq 2k - 1$. The case $m = 2$, $n = 1$, $\ell < 2k - 1$ has been studied by Oliaro-Popivanov-Rodino [21], who proved the G^s nonsolvability at the origin for $s > \frac{4k-\ell}{2k-\ell-1}$, and by Calvo-Popivanov [2], in which the G^s local solvability is proved for $s < \frac{4k-\ell}{2k-\ell-1}$; if $m = 2$, $n = 1$ and $\ell < 2k - 1$ the critical index for the Gevrey local solvability of P is then found. Regarding the more general case $m \geq 2$ we have a result due to Gramchev [11], who proved the G^s local solvability of

$$(D_t + it^{2k} D_x)^m + \text{lower order terms}$$

for $s < \frac{2km}{2km-2k-1}$, under nonvanishing conditions on the lower order terms. In particular, the operator P was studied by Popivanov [22] in the case $2km - \ell > 0$, $m + \ell < n(2k+1)$, proving its C^∞ nonlocal solvability at the origin; in the present paper, under an additional condition, cf. (2.3) below, we prove that the operator P is not G^s locally solvable at the origin for $s > s_{\text{cr}}$, where $s_{\text{cr}} = \frac{2km-\ell}{2km-\ell-(2k+1)(m-n)}$; this result have been already conjectured by Popivanov [22]. The additional condition (2.3) is technical, and we think that the result is true also when it is not satisfied, but with the technique used in this paper we cannot avoid it. On the other hand we observe that our result is sharp; indeed, at least in the case $m = 2$, $n = 1$, the index s_{cr} that we find in this paper coincides with the critical one, in the sense that we have solvability for $s < s_{\text{cr}}$, cf. Calvo-Popivanov [2].

We finally want to give some references where background material on Partial Differential Equations can be found: many results on Gevrey classes and (non)solvability of Partial Differential Equations in Gevrey and other functional spaces are proved for example in the books of Gramchev-Popivanov [13] and Rodino [23], see also the references therein.

2. The main theorem and the fundamental tool

Let us consider the operator (1.1), where a and c are constants, $c \in \mathbb{C} \setminus \{0\}$, $a \in \mathbb{R}$, $a < 0$; we suppose that $1 \le n \le m-1$, $\ell, k \in \mathbb{N}$, $\ell \ge 0$, and moreover we shall assume that the next conditions hold:

(a) If m is odd, then $c = i^m d$, $d > 0$;
(b) If m is even, $c = -i^m d$, $d > 0$.

We then have the following result.

Theorem 2.1. *Assume that the previous conditions are satisfied, and moreover*

$$2km - \ell > 0 \tag{2.1}$$

$$m + \ell < n(2k+1) \tag{2.2}$$

$$\begin{cases} m \ge 2n & \textit{or otherwise} \\ m < 2n & \textit{and} \ \ \frac{m}{n}(2k+1)(m-n) \ge 2km - \ell. \end{cases} \tag{2.3}$$

Then the transposed ${}^tP$ of the operator P is G^s-non locally solvable at the origin for $s > s_{\mathrm{cr}}$, where

$$s_{\mathrm{cr}} = 1 + \frac{(2k+1)(m-n)}{2kn + n - m - \ell}.$$

Remark 2.2. The conditions (2.1) and (2.2) mean, roughly speaking, that ℓ cannot be too large with respect to k, m and n; on the other hand, if $m < 2n$ the hypothesis (2.3) prevents ℓ from being too small. For example, if $n = m-1$ and $k = m$ we are requiring

$$2m^2 - \frac{m}{m-1}(2m+1) \le \ell < 2m^2 - 2m - 1;$$

for instance, when $k = m = 4$, $n = 3$, we must assume $20 \le \ell < 23$.

In the sequel we use the following notation:

$$\varepsilon = \frac{m-n}{2km - \ell}; \tag{2.4}$$

then we can write

$$s_{\mathrm{cr}} = 1 + \frac{(2k+1)(m-n)}{2kn + n - m - \ell} = \frac{1}{1-(2k+1)\varepsilon}. \tag{2.5}$$

The tool that we shall use to prove our theorem is a necessary condition for the G^s local solvability, proved by Corli [5]. First of all we can introduce a topology in $G_0^s(\Omega)$ and $G^s(\Omega)$ in the following way. Let us fix $K \subset\subset \Omega$ and $C > 0$; we define $G_0^s(\Omega, K, C)$ as the set of all functions $f \in C^\infty(\Omega)$ such that $\operatorname{supp} f \subset K$ and

$$\|f\|_{s,K,C} := \sup_{\alpha \in \mathbb{Z}_+^n} \left(C^{-|\alpha|} (\alpha!)^{-s} \sup_{x \in K} |\partial^\alpha f(x)| \right) < \infty. \tag{2.6}$$

Analogously we write $G^s(\Omega, K, C)$ for the space of all functions $f \in G^s(\Omega)$ for which the norm (2.6) of the restriction of f to K is finite. A topology in $G_0^s(\Omega)$

and $G^s(\Omega)$ is then introduced as follows:

$$G_0^s(\Omega) = \operatorname*{ind\,lim}_{K\nearrow\Omega}\ \lim_{C\nearrow\infty} G_0^s(\Omega, K, C)$$

$$G^s(\Omega) = \operatorname*{proj\,lim}_{K\nearrow\Omega}\ \operatorname*{ind\,lim}_{C\nearrow\infty} G^s(\Omega, K, C).$$

The following result is due to Corli [5], who extended to the Gevrey frame the well-known necessary condition of Hörmander [15] for the local solvability in the Schwartz distribution space $\mathcal{D}'$.

Theorem 2.3 (Necessary condition). *Let us fix $s > 1$ and let P be a linear partial differential operator with G^s coefficients; we suppose that P is G^s solvable in Ω, (i.e., for every $f \in G_0^s(\Omega)$ there exists $u \in \mathcal{D}_s'(\Omega)$, solution of $Pu = f$). Then for every compact set $K \subset \Omega$, for every $\eta > \epsilon > 0$, there exists a constant $C > 0$ such that*

$$\|u\|^2_{L^\infty(K)} \le C\|u\|_{s,K,\frac{1}{\eta-\epsilon}} \|\,{}^tPu\|_{s,K,\frac{1}{\eta-\epsilon}} \tag{2.7}$$

for every $u \in G_0^s(\Omega, K, \frac{1}{\eta})$.

3. Construction of a suitable function violating Corli's inequality

In this section we shall construct a function $u_\lambda(t,x)$, depending on a (large) parameter λ, that shall contradict the condition (2.7) for $\lambda \to +\infty$. This construction is the same as in Popivanov [22], so we give here only some lines, referring to the above mentioned paper for a more precise treatment.

- First of all, up to a nonvanishing factor, we have that

$${}^tP = (D_t + iat^{2k}D_x)^m + (-1)^{m+n}ct^\ell D_x^n.$$

- By making a partial Fourier transform with respect to x in the equation $P(t, D_x, D_t)u = 0$ we obtain

$$P(t, \xi, D_t)\hat{u}(\xi, t) = 0, \tag{3.1}$$

  where $\hat{u}(\xi,t) = \int e^{-ix\xi}u(t,x)\,dx$.
- If we choose $\hat{u}(\xi,t) = e^{a\frac{t^{2k+1}}{2k+1}}\hat{v}(\xi,t)$ in (3.1) we obtain that $\hat{u}(\xi,t)$ can be written in the form

$$\hat{u}(\xi,t) = e^{a\frac{t^{2k+1}}{2k+1}} w(t\xi^{\frac{n}{m+\ell}}), \tag{3.2}$$

  ξ being considered as a positive (large) parameter; $w(s)$ is then solution of the equation

$$(\partial_s^m - A^m s^\ell)w(s) = 0, \tag{3.3}$$

  where $A^m = -ci^m$, so $A \in \mathbb{R}$, $A > 0$, according to (a)–(b).

From known results of asymptotical analysis, see, e.g., Fedoryuk [8, Chapter 5], Turrittin [25], Braaksma [1], we have that

$$
\begin{aligned}
w(s) &\sim C_0 s^{\ell\frac{1-m}{2m}} e^{A\frac{s^{\frac{\ell}{m}+1}}{\frac{\ell}{m}+1}}, \quad s\to\infty \\
w^{(r)}(s) &\sim C_r s^{\ell\frac{1-m}{2m}+r\frac{\ell}{m}} e^{A\frac{s^{\frac{\ell}{m}+1}}{\frac{\ell}{m}+1}}, \quad s\to\infty,\ r\geq 1.
\end{aligned}
\tag{3.4}
$$

where C_r are positive constants, see also Popivanov [22]. These asymptotical expansions are not suitable for our purposes, because they are not uniform with respect to r: indeed they mean that for every $\epsilon > 0$

$$
\left| \frac{w^{(r)}(s)}{C_r s^{\ell\frac{1-m}{2m}+r\frac{\ell}{m}} e^{A\frac{s^{\frac{\ell}{m}+1}}{\frac{\ell}{m}+1}}} - 1 \right| < \epsilon
$$

for $s > s_0$, but s_0 depends on r. Since the Gevrey seminorms, cf. (2.6), involve all the derivatives of the function, we need some kind of uniform estimates with respect to r.

Lemma 3.1. *Let $w(s)$ be solution of (3.3); then there exists a positive constant D such that for $s > S_0 > 1$ and for every $r \geq 1$ we have*

$$
|w(s)| \leq D s^{\ell\frac{1-m}{2m}} e^{A\frac{s^{\frac{\ell}{m}+1}}{\frac{\ell}{m}+1}} \tag{3.5}
$$

$$
|w^{(r)}(s)| \leq D^{r+1} r!\, s^{\ell\frac{1-m}{2m}+r\frac{\ell}{m}} e^{A\frac{s^{\frac{\ell}{m}+1}}{\frac{\ell}{m}+1}}, \tag{3.6}
$$

where S_0 does not depend on r.

Proof. For $r \leq m$ the thesis follows immediately from (3.4), by taking D sufficiently large in (3.5)–(3.6). For $r > m$ we proceed by induction: let us suppose that the estimate (3.6) holds for every $r < m+h$, $h \in \mathbb{N}$ fixed, and let us estimate $|w^{(m+h)}(s)|$. Remembering that $w^{(m)}(s) = A^m s^\ell w(s)$, cf. (3.3), we have

$$
\begin{aligned}
|w^{(m+h)}(s)| &= |\partial_s^h (A^m s^\ell w(s))| \\
&\leq A^m \sum_{\mu=0}^{\min\{\ell,h\}} \left| \binom{h}{\mu} \ell \ldots (\ell-\mu+1) s^{\ell-\mu} w^{(h-\mu)}(s) \right|;
\end{aligned}
$$

by the inductive hypothesis and since $\ell\ldots(\ell-\mu+1) \leq \ell!$, for $s > S_0$ we then have:

$$
\begin{aligned}
|w^{(m+h)}(s)| &\leq A^m \sum_{\mu=0}^{\min\{\ell,h\}} \frac{h!\,\ell!}{\mu!(h-\mu)!} s^{\ell-\mu} D^{h-\mu+1} (h-\mu)!\, s^{\ell\frac{1-m}{2m}+(h-\mu)\frac{\ell}{m}} e^{A\frac{s^{\frac{\ell}{m}+1}}{\frac{\ell}{m}+1}} \\
&\leq D^{h+m+1} (h+m)!\, s^{\ell\frac{1-m}{2m}+(h+m)\frac{\ell}{m}} e^{A\frac{s^{\frac{\ell}{m}+1}}{\frac{\ell}{m}+1}},
\end{aligned}
$$

by taking $D \geq l! A^m$. □

Let us consider now the cut-off functions $\varphi(x), \psi(\rho), g_1(t) \in G_0^{s'}(\mathbb{R})$, $1 < s' < s_{\mathrm{cr}}$, cf. (2.5), in such a way that:

- $\varphi \equiv 1$ for $|x| \ll 1$, $\varphi \equiv 0$ for $|x| > 1$, $0 \le \varphi(x) \le 1$ for every $x \in \mathbb{R}$;
- $\operatorname{supp} \psi = [1, 1 + \mu_0]$, $\psi(\rho) > 0$ for $\rho \in (1, 1 + \mu_0)$, $\int_{-\infty}^{+\infty} \psi(\rho)\, d\rho = 1$;
- $g_1 \equiv 1$ for $|t - 1| \le \epsilon_1$, $g_1 \equiv 0$ for $|t - 1| \ge 2\epsilon_1$, $0 \le g_1(t) \le 1$ for every $t \in \mathbb{R}$;

Now (3.4) gives an asymptotic behavior of $\hat{u}(\xi, t)$, cf. (3.2), for $\xi \to +\infty$ and $t \ge \delta_0 > 0$:

$$\hat{u}(\xi, t) \sim C_0 (t\xi^{\frac{n}{m+\ell}})^{\ell \frac{1-m}{2m}} e^{f(\xi, t)},$$

$f(\xi, t)$ being given by

$$f(\xi, t) = a \frac{t^{2k+1}}{2k+1} \xi + A \frac{t^{\frac{\ell}{m}+1}}{\frac{\ell}{m}+1} \xi^{n/m}, \tag{3.7}$$

where $a < 0$ and $A > 0$. We can easily prove the following assertions, cf. Popivanov [22]:

(i) $f(\xi, t)$ has a maximum for $t = t_\xi$, where

$$t_\xi = c_0 \xi^{-\varepsilon}, \quad c_0 = \left(-\frac{A}{a}\right)^{\frac{m}{2km-\ell}} > 0, \tag{3.8}$$

ε being given by (2.4), and moreover

$$f(\xi, t_\xi) = \alpha_0 \xi^{1-(2k+1)\varepsilon}, \quad \alpha_0 > 0. \tag{3.9}$$

(ii) There exists a positive constant ϵ_0, sufficiently small, such that for $|t - t_\xi| \le \epsilon_0 |t_\xi|$ we can write

$$f(\xi, t) = \alpha_0 \xi^{1-(2k+1)\varepsilon} - e_0 (t - t_\xi)^2 \xi^{1-(2k-1)\varepsilon} \tag{3.10}$$

where α_0 is a positive constant and $0 < \mathrm{const}_1 \le e_0 \le \mathrm{const}_2$.

Let us fix now a cut-off function

$$g_{\lambda\rho}(t, x) = \varphi(x) g_1\left(\frac{t}{t_{\lambda\rho}}\right), \tag{3.11}$$

$1 \le \rho \le 1 + \mu_0$, λ being a (large) parameter, and define

$$u_\lambda(t, x) = \int_{-\infty}^{+\infty} \psi(\rho) e^{-\alpha_0 (\lambda\rho)^{1-(2k+1)\varepsilon}} e^{ix\lambda\rho + a\frac{t^{2k+1}}{2k+1}\lambda\rho} w(t(\lambda\rho)^{\frac{n}{m+\ell}}) g_{\lambda\rho}(t, x)\, d\rho. \tag{3.12}$$

Remark 3.2. The function $u_\lambda(t, x)$ is compactly supported, and its support satisfies $\operatorname{supp} u_\lambda \subset \{(1 - \epsilon_2)t_\lambda \le t \le (1 + \epsilon_2)t_\lambda\} \times \{|x| \le 1\}$ for a suitable constant ϵ_2, with $0 < \epsilon_2 < 1$. Observe in particular that $\operatorname{supp} u_\lambda$ does not contain any point with $t = 0$, for every $\lambda > 0$.

4. Fundamental estimates

In this section we shall estimate the left- and right-hand sides of Corli's inequality for the function $u_\lambda(t,x)$. The compact K in (2.7) can contain the point $(0,0)$, but since $\operatorname{supp} u_\lambda$ does not contain $(0,0)$, we shall fix in the following

$$K_\lambda = \{(1-\epsilon_2)t_\lambda \leq t \leq (1+\epsilon_2)t_\lambda\} \times \{|x| \leq 1\}, \tag{4.1}$$

cf. Remark 3.2.

4.1. Some preliminary results

We state here several technical lemmas, that we shall use in the following. To start with, we recall the next known result.

Lemma 4.1. *For every compact K and for all $\eta > \epsilon > 0$ there exists a constant C such that*

$$\|fg\|_{s,K,\frac{1}{\eta-\epsilon}} \leq C\|f\|_{s,K,\frac{1}{\eta}}\|g\|_{s,K,\frac{1}{\eta}}$$

for all $f,g \in G^s(\Omega,K,\frac{1}{\eta})$ (or $f,g \in G_0^s(\Omega,K,\frac{1}{\eta})$).

For the proof of the previous lemma, see for example Corli [5, 6].

In the following we shall use the Faà di Bruno formula: if $f,g : \mathbb{R} \to \mathbb{R}$, $f,g \in C^n(\mathbb{R})$, then for every $\nu = 1,\ldots,n$, we have:

$$\frac{d^\nu}{dx^\nu}(f(g(x))) = \nu! \sum_{k=1}^{\nu} f^{(k)}(g(x)) \sum_{\substack{k_1+\cdots+k_\nu=k \\ k_1+2k_2+\cdots+\nu k_\nu=\nu}} \prod_{j=1}^{\nu} \frac{1}{k_j!}\left(\frac{1}{j!}g^{(j)}(x)\right)^{k_j}. \tag{4.2}$$

Let us recall now the definition of Bell polynomials: if $\{x_j\}_{j\in\mathbb{Z}_+}$ is a sequence of real numbers, $\mu \in \mathbb{Z}_+$ and h is a positive integer, we define

$$B_{\mu,h}(\{x_j\}) = \mu! \sum_{\sum_{j=1}^{\infty} h_j=h,\ \sum_{j=1}^{\infty} jh_j=\mu} \prod_{j=1}^{\infty} \frac{1}{h_j!}\left(\frac{1}{j!}x_j\right)^{h_j}, \tag{4.3}$$

where h_j are non-negative integers. The following identity holds:

$$\frac{1}{h!}\left(\sum_{j=1}^{\infty} x_j \frac{z^j}{j!}\right)^h = \sum_{\mu=h}^{\infty} B_{\mu,h}(\{x_j\})\frac{z^\mu}{\mu!}, \tag{4.4}$$

cf. Mascarello-Rodino [18, Section 5.5]. We now want to prove the following proposition.

Proposition 4.2. *Let $w(s)$ be the function of Lemma 3.1, solution of (3.3). Let $K \subset \mathbb{R}^2$ be a compact set satisfying the following condition: for every $(t_0,x_0) \in K$*

$$\lim_{\lambda\to+\infty} t_0(\lambda\rho)^{\frac{n}{m+\ell}} = +\infty \tag{4.5}$$

for every $\rho \in [1, 1+\mu_0]$*. Then for every* $s > 1$*,* $C > 0$*, there exist positive constants* C_1 *and* d *such that for all* $\rho \in [1, 1+\mu_0]$ *and* $h \in \mathbb{Z}_+$ *we have:*

$$\begin{aligned} &\|e^{ix\lambda\rho + a\frac{t^{2k+1}}{2k+1}\lambda\rho} w^{(h)}(t(\lambda\rho)^{\frac{n}{m+\ell}})\|_{s,K,C} \\ &\leq C_1^{h+1} h!\, \lambda^{h\frac{n\ell}{m(m+\ell)}} \\ &\quad \times e^{m(\lambda\rho) + s(d\lambda)^{1/s} + (s-1)(d\lambda^{n/m})^{\frac{1}{s-1}}}, \end{aligned} \tag{4.6}$$

for λ *sufficiently large, where*

$$m(\lambda\rho) = \sup_{(t,x)\in K} \Big(a\frac{t^{2k+1}}{2k+1}\lambda\rho + A\frac{t^{\frac{\ell}{m}+1}}{\frac{\ell}{m}+1}(\lambda\rho)^{n/m} \Big).$$

Proof. For every $\alpha, \beta \in \mathbb{Z}_+$ and for fixed h, by formula (4.2) and Lemma 3.1 we have:

$$\begin{aligned} &|\partial_x^\alpha \partial_t^\beta (e^{ix\lambda\rho + a\frac{t^{2k+1}}{2k+1}\lambda\rho} w^{(h)}(t(\lambda\rho)^{\frac{n}{m+\ell}}))| \\ &= \Big| (i\lambda\rho)^\alpha e^{ix\lambda\rho} \sum_{\mu=0}^{\beta} \binom{\beta}{\mu} \partial_t^\mu \big(e^{a\frac{t^{2k+1}}{2k+1}\lambda\rho} \big) (\lambda\rho)^{(\beta-\mu)\frac{n}{m+\ell}} w^{(h+\beta-\mu)}(t(\lambda\rho)^{\frac{n}{m+\ell}}) \Big| \\ &\leq (\lambda\rho)^\alpha \sum_{\mu=0}^{\beta} \binom{\beta}{\mu} \sum_{0\leq q\leq\mu} e^{a\frac{t^{2k+1}}{2k+1}\lambda\rho} \mu! \sum_{\substack{q_1+\cdots+q_\mu=q \\ q_1+2q_2+\cdots+\mu q_\mu=\mu}} \prod_{j=1}^{\mu} \frac{1}{q_j!} \Big(\frac{|\partial_t^j (a\frac{t^{2k+1}}{2k+1}\lambda\rho)|}{j!} \Big)^{q_j} \\ &\quad \times (\lambda\rho)^{(\beta-\mu)\frac{n}{m+\ell}} D^{h+\beta-\mu+1} (h+\beta-\mu)! \big(t(\lambda\rho)^{\frac{n}{m+\ell}} \big)^{\ell\frac{1-m}{2m} + (h+\beta-\mu)\frac{\ell}{m}} \\ &\quad \times e^{A\frac{t^{\frac{\ell}{m}+1}}{\frac{\ell}{m}+1}(\lambda\rho)^{n/m}}, \end{aligned}$$

for $\lambda > \lambda_0$, λ_0 being independent on α, β. Now the condition (4.5) implies that there exists $\lambda_1 > 0$ such that $t(\lambda\rho)^{\frac{n}{m+\ell}} \geq 1$ for every $(t,x) \in K$, $\lambda > \lambda_1$; then we can find $\tilde{C} > 0$ satisfying $\big(t(\lambda\rho)^{\frac{n}{m+\ell}}\big)^{\ell\frac{1-m}{2m} + (h+\beta-\mu)\frac{\ell}{m}} \leq \tilde{C}^{(h+\beta-\mu)\frac{\ell}{m}} \lambda^{\frac{n}{m+\ell}(h+\beta-\mu)\frac{\ell}{m}}$ for all $(t,x) \in K$, $\rho \in [1, 1+\mu_0]$ and $\lambda > \lambda_1$. From this fact and the estimate $(h+\beta-\mu)! \leq 2^{h+\beta} h! (\beta-\mu)!$ we have that for $\lambda > \tilde{\lambda}_0 = \max\{\lambda_0, \lambda_1\}$

$$\begin{aligned} &|\partial_x^\alpha \partial_t^\beta (e^{ix\lambda\rho + a\frac{t^{2k+1}}{2k+1}\lambda\rho} w^{(h)}(t(\lambda\rho)^{\frac{n}{m+\ell}}))| \\ &\leq C_0^{h+1+\alpha+\beta} \lambda^\alpha h! \lambda^{h\frac{n\ell}{m(m+\ell)}} \\ &\times e^{a\frac{t^{2k+1}}{2k+1}\lambda\rho + A\frac{t^{\frac{\ell}{m}+1}}{\frac{\ell}{m}+1}(\lambda\rho)^{n/m}} \sum_{\mu=0}^{\beta} \lambda^{(\beta-\mu)\frac{n}{m}} (\beta-\mu)! \sum_{0\leq q\leq\mu} B_{\mu,q}\Big(\Big\{ \big|\partial_t^j \big(a\frac{t^{2k+1}}{2k+1}\lambda\rho\big)\big| \Big\} \Big), \end{aligned}$$

where $B_{\mu,q}\big(\{|\partial_t^j (a\frac{t^{2k+1}}{2k+1}\lambda\rho)|\}\big)$ is a Bell Polynomial, cf. (4.3).

Now, using the identity (4.4) with $z = \mu$ and $x_j = \left|\partial_t^j\left(a\frac{t^{2k+1}}{2k+1}\lambda\rho\right)\right|$ for every j, we obtain:

$$\begin{aligned}
B_{\mu,q}\Big(\Big\{\big|\partial_t^j\big(a\frac{t^{2k+1}}{2k+1}\lambda\rho\big)\big|\Big\}\Big) &\le B_{\mu,q}\Big(\Big\{\big|\partial_t^j\big(a\frac{t^{2k+1}}{2k+1}\lambda\rho\big)\big|\Big\}\Big)\frac{\mu^\mu}{\mu!}\\
&\le \frac{1}{q!}\Big(\sum_{j=1}^{\infty}\big|\partial_t^j\big(a\frac{t^{2k+1}}{2k+1}\lambda\rho\big)\big|\frac{\mu^j}{j!}\Big)^q\\
&\le \Big(\sum_{j=1}^{2k+1}\big[a\lambda\rho 2k(2k-1)\dots(2k-j+2)t^{2k+1-j}\big]\frac{\mu^j}{j!}\Big)^q\\
&\le C^q\lambda^q e^{q\mu} \le \tilde{C}^\beta\lambda^\mu.
\end{aligned}$$

We then get:

$$\begin{aligned}
\big|\partial_x^\alpha\partial_t^\beta\big(e^{ix\lambda\rho+a\frac{t^{2k+1}}{2k+1}\lambda\rho}w^{(h)}(t(\lambda\rho)^{\frac{n}{m+\ell}})\big)\big| &\le C_1^{h+1}h!\,\lambda^{h\frac{n\ell}{m(m+\ell)}}C_1^{\alpha+\beta}\lambda^\alpha e^{m(\lambda\rho)}\\
&\quad\times\sum_{\mu=0}^{\beta}\lambda^{(\beta-\mu)\frac{n}{m}}(\beta-\mu)!\,\lambda^\mu
\end{aligned}$$

for every $(t,x)\in K$ and $\lambda > \tilde{\lambda}_0$. Since $\beta!^{-s} \le \mu!^{-s}(\beta-\mu)!^{-s}$ for every $\mu = 0,\dots,\beta$, we finally obtain:

$$\begin{aligned}
&C^{-\alpha-\beta}(\alpha!\,\beta!)^{-s}\sup_{(t,x)\in K}\big|\partial_x^\alpha\partial_t^\beta\big(e^{ix\lambda\rho+a\frac{t^{2k+1}}{2k+1}\lambda\rho}w^{(h)}(t(\lambda\rho)^{\frac{n}{m+\ell}})\big)\big|\\
&\le C_1^{h+1}h!\,\lambda^{h\frac{n\ell}{m(m+\ell)}}e^{m(\lambda\rho)}\frac{(C_1C^{-1}\lambda)^\alpha}{\alpha!^s}\sum_{\mu=0}^{\beta}\frac{1}{2^\beta}\frac{(2C_1C^{-1}\lambda)^\mu}{\mu!^s}\frac{(2C_1C^{-1}\lambda^{n/m})^{\beta-\mu}}{(\beta-\mu)!^{s-1}}\\
&\le C_1^{h+1}h!\,\lambda^{h\frac{n\ell}{m(m+\ell)}}e^{m(\lambda\rho)+s(d\lambda)^{1/s}+(s-1)(d\lambda^{n/m})^{\frac{1}{s-1}}}
\end{aligned}\tag{4.7}$$

where $d = 3C_1C^{-1}$, since $\frac{(C_1C^{-1}\lambda)^\alpha}{\alpha!^s} \le \left(\frac{\left((C_1C^{-1}\lambda)^{1/s}\right)^\alpha}{\alpha!}\right)^s \le e^{s(d\lambda)^{\frac{1}{s}}}$, and similar estimates hold for $\frac{(2C_1C^{-1}\lambda)^\mu}{\mu!^s}$ and $\frac{(2C_1C^{-1}\lambda^{n/m})^{\beta-\mu}}{(\beta-\mu)!^{s-1}}$. Since (4.7) is valid for $\lambda > \tilde{\lambda}_0$ and $\tilde{\lambda}_0$ is independent of $\alpha,\beta\in\mathbb{Z}_+$, taking the $\sup\limits_{\alpha,\beta\in\mathbb{Z}_+}$ in the left-hand side of (4.7) we have that (4.6) holds for every $\lambda > \tilde{\lambda}_0$. □

Remark 4.3. The compact (4.1) satisfies the condition (4.5): indeed it is sufficient to prove (4.5) for $\rho = 1$ and $t_0 = (1-\epsilon_2)t_\lambda = (1-\epsilon_2)c_0\lambda^{-\varepsilon}$. We then have to prove that

$$\lim_{\lambda\to+\infty}(1-\epsilon_2)c_0\lambda^{-\epsilon+\frac{n}{m+\ell}} = +\infty,$$

that is true since $\frac{n}{m+\ell} - \varepsilon > 0$, as we can deduce from (2.4), (2.1) and (2.2).

Lemma 4.4. *Let us consider the cut-off function $g_{\lambda\rho}(t,x)$, cf. (3.11). There exist positive constants D and G_0 such that*

$$\left\|(D^\alpha g_1)\Big(\frac{t}{t_{\lambda\rho}}\Big)D^\beta\varphi(x)\right\|_{s,K,\frac{1}{\eta}} \le D^{\alpha+\beta+1}(\alpha!\,\beta!)^{s'}e^{G_0\lambda^{\frac{\varepsilon}{s-s'}}}, \tag{4.8}$$

for every $\alpha,\beta\in\mathbb{Z}_+$ and $s>s'$, s' being the Gevrey order of the functions g_1 and φ. The constants D and G_0 are independent of α, β, λ and $\rho\in[1,1+\mu_0]$.

Proof. First of all we observe that $D_t^\delta\big[D^\alpha g_1\big(\frac{t}{t_{\lambda\rho}}\big)\big]=(t_{\lambda\rho})^{-\delta}(D^{\alpha+\delta}g_1)\big(\frac{t}{t_{\lambda\rho}}\big)$. So we have:

$$\begin{aligned}
&\left\|(D^\alpha g_1)\Big(\frac{t}{t_{\lambda\rho}}\Big)D^\beta\varphi(x)\right\|_{s,K,\frac{1}{\eta}}\\
&\quad\le \sup_{\delta,\gamma\in\mathbb{Z}_+}\left\{\Big(\frac{1}{\eta}\Big)^{-\delta-\gamma}(\delta!\,\gamma!)^{-s}\sup_{(t,x)\in\mathbb{R}^2}\left|D_t^\delta\Big[(D^\alpha g_1\Big(\frac{t}{t_{\lambda\rho}}\Big)\Big]D_x^\gamma(D^\beta\varphi)(x)\right|\right\}\\
&\quad\le \sup_{\gamma\in\mathbb{Z}_+}\left[\Big(\frac{1}{\eta}\Big)^{-\gamma}\gamma!^{-s}\sup_{x\in\mathbb{R}}\big|\partial^{\beta+\gamma}\varphi(x)\big|\right]\\
&\quad\quad\times\sup_{\delta\in\mathbb{Z}_+}\left[\Big(\frac{1}{\eta}\Big)^{-\delta}\delta!^{-s}\sup_{t\in\mathbb{R}}\left|(t_{\lambda\rho})^{-\delta}(\partial^{\alpha+\delta}g_1)\Big(\frac{t}{t_{\lambda\rho}}\Big)\right|\right].
\end{aligned}$$

Taking into account that φ and g_1 are compactly supported Gevrey functions of order s' we have: $\sup_{x\in\mathbb{R}}|\partial^{\beta+\gamma}\varphi(x)|\le C^{\beta+\gamma+1}(\beta+\gamma)!^{s'}\le C_1^{\beta+1}\beta!^{s'}\,C_1^\gamma\gamma!^{s'}$, for every $\gamma\in\mathbb{Z}_+$, since $(\beta+\gamma)!\le 2^{\beta+\gamma}\beta!\,\gamma!$; a similar estimate holds for g_1. Then recalling the expression of $t_{\lambda\rho}$, cf. (3.8), we obtain:

$$\begin{aligned}
\left\|(D^\alpha g_1)\Big(\frac{t}{t_{\lambda\rho}}\Big)D^\beta\varphi(x)\right\|_{s,K,\frac{1}{\eta}} &\le C_1^{\beta+1}\beta!^{s'}\sup_{\gamma\in\mathbb{Z}_+}\left[\Big(\frac{1}{C_1\eta}\Big)^{-\gamma}\gamma!^{-(s-s')}\right]\\
&\quad\times C_2^{\alpha+1}\alpha!^{s'}\sup_{\delta\in\mathbb{Z}_+}\left[\Big(\frac{c_0}{\lambda^\varepsilon\rho^\varepsilon C_2\eta}\Big)^{-\delta}\delta!^{-(s-s')}\right].
\end{aligned}\tag{4.9}$$

Now we observe that

$$\sup_{\gamma\in\mathbb{Z}_+}\left[\Big(\frac{1}{C_1\eta}\Big)^{-\gamma}\gamma!^{-(s-s')}\right]=\sup_{\gamma\in\mathbb{Z}_+}\left[\frac{\big[(C_1\eta)^{\frac{1}{s-s'}}\big]^\gamma}{\gamma!}\right]^{s-s'}\le e^{(s-s')(C_1\eta)^{\frac{1}{s-s'}}};$$

in the same way we deduce that

$$\sup_{\gamma\in\mathbb{Z}_+}\left[\Big(\frac{c_0}{\lambda^\varepsilon\rho^\varepsilon C_2\eta}\Big)^{-\delta}\delta!^{-(s-s')}\right]\le e^{(s-s')\left((1+\mu_0)^\varepsilon C_2\eta c_0^{-1}\right)^{\frac{1}{s-s'}}\lambda^{\frac{\varepsilon}{s-s'}}}.$$

Applying the last two estimates in (4.9) we obtain (4.8). □

Lemma 4.5. *Let R be a real number; then for every $s>1$ and $C>0$ there exist positive constants b and c such that:*

$$\|t^R\|_{s,K_\lambda,C}\le b\lambda^{-\varepsilon R}e^{c\lambda^{\frac{\varepsilon}{s-1}}} \tag{4.10}$$

K_λ being the compact (4.1); the constants b and c are independent of λ.

Proof. By the definition of Gevrey seminorms we have:

$$\|t^R\|_{s,K_\lambda,C} = \sup_{\alpha\in\mathbb{Z}_+} \left[C^{-\alpha}\alpha!^{-s} \sup_{(1-\epsilon_2)t_\lambda\le t\le(1+\epsilon_2)t_\lambda} |R(R-1)\dots(R-\alpha+1)t^{R-\alpha}|\right];$$

now we observe that $|R(R-1)\dots(R-\alpha+1)| \le |R|(|R|+1)\dots(|R|+\alpha-1) = \binom{|R|+\alpha-1}{|R|-1}\alpha! \le 2^{|R|+\alpha-1}\alpha!$ Moreover, $|t^{R-\alpha}| \le ((1+\mathrm{sign}(R-\alpha)\epsilon_2)t_\lambda)^{R-\alpha}$. Then we have:

$$\|t^R\|_{s,K_\lambda,C} \le 2^{|R|-1}((1+\epsilon_2)c_0)^{|R|}\lambda^{-\varepsilon R} \sup_{\alpha\in\mathbb{Z}_+}\left[\left(\frac{C(1-\epsilon_2)c_0}{2}\lambda^{-\varepsilon}\right)^{-\alpha}\alpha!^{-(s-1)}\right].$$

By the same technique used to estimate the right-hand side of (4.9) we obtain that

$$\sup_{\alpha\in\mathbb{Z}_+}\left[\left(\frac{C(1-\epsilon_2)c_0}{2}\lambda^{-\varepsilon}\right)^{-\alpha}\alpha!^{-(s-1)}\right] \le e^{(s-1)\left(2(C(1-\epsilon_2)c_0)^{-1}\right)^{\frac{1}{s-1}}\lambda^{\frac{\varepsilon}{s-1}}}.$$

We then have proved the estimate (4.10) with $c = (s-1)\left(2(C(1-\epsilon_2)c_0)^{-1}\right)^{\frac{1}{s-1}}$ and $b = 2^{|R|-1}((1+\epsilon_2)c_0)^{|R|}$. □

4.2. Lower bound for $\|u_\lambda\|_{L^\infty(K_\lambda)}$

Let us define $K_1 = \{t\in\mathbb{R} : |t| \le \frac{1}{2}\}$, and observe that $K_1 \supset \mathrm{supp}\big(g_1(\frac{t}{t_{\lambda\rho}})\big)$ for $\lambda \gg 0$. Since $\mathrm{meas}(K_1) = 1$, for $\lambda \gg 0$ we have:

$$\|u_\lambda\|_{L^\infty(K_\lambda)} \ge \|u_\lambda(t,0)\|_{L^\infty(K_1)} \ge \|u_\lambda(t,0)\|_{L^1(K_1)},$$

where K_λ is the set (4.1); then, recalling the definition of $u_\lambda(t,x)$, cf. (3.12), we have:

$$\|u_\lambda\|_{L^\infty(K_\lambda)} \ge \left|\iint \psi(\rho)e^{-\alpha_0(\lambda\rho)^{1-(2k+1)\varepsilon}}e^{a\frac{t^{2k+1}}{2k+1}\lambda\rho}w(t(\lambda\rho)^{\frac{n}{m+\ell}})g_1\left(\frac{t}{t_{\lambda\rho}}\right)d\rho\,dt\right|. \tag{4.11}$$

Now we can apply the same computations as in Popivanov [22] and conclude that there exists a constant $E_0 > 0$ such that

$$\begin{aligned}\iint \psi(\rho)e^{-\alpha_0(\lambda\rho)^{1-(2k+1)\varepsilon}}e^{a\frac{t^{2k+1}}{2k+1}\lambda\rho}w(t(\lambda\rho)^{\frac{n}{m+\ell}})g_1\left(\frac{t}{t_{\lambda\rho}}\right)d\rho\,dt = \\ = E_0\lambda^{-p+\ell\frac{1-m}{2m}(\frac{n}{m+\ell}-\varepsilon)}(1+o(1)) \qquad \text{as } \lambda\to\infty,\end{aligned} \tag{4.12}$$

where $p = \frac{1-(2k+1)\varepsilon}{2}$ and $\varepsilon = \frac{m-n}{2km-\ell}$. Formula (4.12) is obtained by (3.4), by making the change of variable $y_1 = \lambda^p(t-t_{\lambda\rho})$ in the integral over $\mathbb{R}_t$ and then by applying the Lebesgue Dominated Convergence Theorem for $\lambda\to\infty$. By (4.11) and (4.12) we can conclude that there exists a positive constant E satisfying

$$\|u_\lambda\|_{L^\infty(K_\lambda)} \ge E\lambda^{-p+\ell\frac{1-m}{2m}(\frac{n}{m+\ell}-\varepsilon)}, \tag{4.13}$$

for $\lambda \gg 0$.

4.3. Upper bound for $\|u_\lambda\|_{s,K_\lambda,\frac{1}{\eta-\epsilon}}$

By Lemma 4.1 we have:

$$\|u_\lambda\|_{s,K_\lambda,\frac{1}{\eta-\epsilon}} \leq \int_{-\infty}^{+\infty} \psi(\rho)e^{-\alpha_0(\lambda\rho)^{1-(2k+1)\varepsilon}} \times \|e^{ix\lambda\rho+a\frac{t^{2k+1}}{2k+1}\lambda\rho}w(t(\lambda\rho)^{\frac{n}{m+\ell}})\|_{s,K_\lambda,\frac{1}{\eta}}\|g_{\lambda\rho}(t,x)\|_{s,K_\lambda,\frac{1}{\eta}}\,d\rho;$$

then applying Proposition 4.2 with $h=0$ and Lemma 4.4 we obtain:

$$\|u_\lambda\|_{s,K_\lambda,\frac{1}{\eta-\epsilon}} \leq \leq C\int_{-\infty}^{+\infty} \psi(\rho)e^{-\alpha_0(\lambda\rho)^{1-(2k+1)\varepsilon}} e^{m(\lambda\rho)+s(d\lambda)^{1/s}+(s-1)(d\lambda^{n/m})^{\frac{1}{s-1}}} e^{G_0\lambda^{\frac{\varepsilon}{s-s'}}}\,d\rho.$$

We already know that $m(\lambda\rho) = \sup\limits_{(t,x)\in K_\lambda} f(\lambda\rho,t) = \alpha_0(\lambda\rho)^{1-(2k+1)\varepsilon}$, cf. (3.9); so we can conclude that

$$\|u_\lambda\|_{s,K_\lambda,\frac{1}{\eta,\epsilon}} \leq Ce^{s(d\lambda)^{1/s}+(s-1)(d\lambda^{n/m})^{\frac{1}{s-1}}+G_0\lambda^{\frac{\varepsilon}{s-s'}}}. \tag{4.14}$$

4.4. Upper bound for $\|Pu_\lambda\|_{s,K_\lambda,\frac{1}{\eta-\epsilon}}$

Observe at first that, since by construction

$$P(t,D_t,D_x)\Big(e^{ix\lambda\rho+a\frac{t^{2k+1}}{2k+1}\lambda\rho}w(t(\lambda\rho)^{\frac{n}{m+\ell}})\Big)=0,$$

cf. Popivanov [22], we have:

$$P(t,D_t,D_x)u_\lambda(t,x) = \sum_{\substack{\alpha_1+\alpha_2\leq m\\ \beta_1+\beta_2\leq m\\ (\alpha_2,\beta_2)\neq(0,0)}} P_{\alpha\beta}(t)\int_{-\infty}^{+\infty}\psi(\rho)e^{-\alpha_0(\lambda\rho)^{1-(2k+1)\varepsilon}} \times D_t^{\alpha_1}D_x^{\beta_1}\Big(e^{ix\lambda\rho+a\frac{t^{2k+1}}{2k+1}\lambda\rho}w(t(\lambda\rho)^{\frac{n}{m+\ell}})\Big)D_t^{\alpha_2}D_x^{\beta_2}\big(g_{\lambda\rho}(t,x)\big)\,d\rho,$$

where $\alpha=(\alpha_1,\alpha_2)$, $\beta=(\beta_1,\beta_2)$ and $P_{\alpha\beta}(t)$ is a polynomial in t with constant coefficients of degree $\leq 2km$, $P_{\alpha\beta}(t)=\sum\limits_{r=0}^{2km} c_rt^r$, $c_r=c_r(\alpha,\beta)\in\mathbb{C}$. Observe that for $\alpha_3\in\mathbb{Z}_+$ we can write $\partial_t^{\alpha_3}e^{a\frac{t^{2k+1}}{2k+1}\lambda\rho} = \sum\limits_{\substack{0\leq j\leq 2k\alpha_3\\ 0\leq q\leq\alpha_3}} c_{jq}t^j(\lambda\rho)^q\,e^{a\frac{t^{2k+1}}{2k+1}\lambda\rho}$ for suitable $c_{jq}\in\mathbb{C}$; then, using Leibnitz rule and recalling that $t_{\lambda\rho}=c_0(\lambda\rho)^{-\varepsilon}$, cf. (3.8), we have:

$$
\begin{aligned}
P(t,D_t,D_x)u_\lambda(t,x) = & \sum_{\substack{\alpha_1+\alpha_2\le m\\ \beta_1+\beta_2\le m\\ (\alpha_2,\beta_2)\neq(0,0)}} \sum_{\alpha_3+\alpha_4=\alpha_1} \sum_{\substack{0\le j\le 2k\alpha_3\\ 0\le q\le\alpha_3}} \sum_{r=0}^{2km} C t^{r+j}\lambda^{\beta_1+q+\varepsilon\alpha_2+\frac{n}{m+\ell}\alpha_4} \\
& \times \int_{-\infty}^{+\infty} \psi(\rho) e^{-\alpha_0(\lambda\rho)^{1-(2k+1)\varepsilon}} \rho^{\beta_1+q+\varepsilon\alpha_2+\frac{n}{m+\ell}\alpha_4} e^{ix\lambda\rho+a\frac{t^{2k+1}}{2k+1}\lambda\rho} \\
& \times w^{(\alpha_4)}(t(\lambda\rho)^{\frac{n}{m+\ell}}) g_1^{(\alpha_2)}\Big(\frac{t}{t_{\lambda\rho}}\Big)\varphi^{(\beta_2)}(x)\,d\rho \\
= & \sum_{\substack{\alpha_1+\alpha_2\le m\\ \beta_1+\beta_2\le m\\ (\alpha_2,\beta_2)\neq(0,0)}} J_{\alpha\beta}(t,x,\lambda),
\end{aligned}
\tag{4.15}
$$

where $C = C(\alpha_1,\alpha_2,\beta_1,\beta_2,\alpha_3,j,q,r) = \binom{\alpha_1}{\alpha_3} c_0^{-\alpha_2} c_r c_{jq} (-i)^{\alpha_1+\alpha_2+\beta_2}$. We can write

$$
P(t,D_t,D_x)u_\lambda(t,x) = \sum_{\substack{\alpha_1+\alpha_2\le m\\ \beta_1+\beta_2\le m\\ \alpha_2\neq 0}} J_{\alpha\beta}(t,x,\lambda) + \sum_{\substack{\alpha_1+\alpha_2\le m\\ \beta_1+\beta_2\le m\\ \alpha_2=0,\ \beta_2\neq 0}} J_{\alpha\beta}(t,x,\lambda) := I_1 + I_2.
\tag{4.16}
$$

Now let us analyze separately I_1 and I_2.

Regarding I_1, since $\alpha_2\neq 0$, in the t variable we can limit ourselves to $\operatorname{supp}\big(g_1'\big(\frac{t}{t_{\lambda\rho}}\big)\big)$, and so we have:

$$
\|I_1\|_{s,K_\lambda,\frac{1}{\eta-\epsilon}} = \|I_1\|_{s,\tilde K,\frac{1}{\eta-\epsilon}},
$$

where $\tilde K = \{(t,x)\in\mathbb{R}^2 : \frac{t}{t_\lambda}\in[1-\tilde\epsilon_2, 1-\tilde\epsilon_1]\cup[1+\tilde\epsilon_1,1+\tilde\epsilon_2],\ |x|\le 1\}$ for suitable constants $0<\tilde\epsilon_1<\tilde\epsilon_2\ll 1$. Then we can write:

$$
\begin{aligned}
\|I_1\|_{s,K_\lambda,\frac{1}{\eta-\epsilon}} \le & \sum_{\substack{\alpha_1+\alpha_2\le m\\ \beta_1+\beta_2\le m\\ \alpha_2\neq 0}} \sum_{\alpha_3+\alpha_4=\alpha_1} \sum_{\substack{0\le j\le 2k\alpha_3\\ 0\le q\le\alpha_3}} \sum_{r=0}^{2km} C\|t^{r+j}\|_{s,\tilde K,\frac{1}{\eta}} \lambda^{\beta_1+q+\varepsilon\alpha_2+\frac{n}{m+\ell}\alpha_4} \\
& \times \int_1^{1+\mu_0} \psi(\rho) e^{-\alpha_0(\lambda\rho)^{1-(2k+1)\varepsilon}} \rho^{\beta_1+q+\varepsilon\alpha_2+\frac{n}{m+\ell}\alpha_4} \\
& \times \|e^{ix\lambda\rho+a\frac{t^{2k+1}}{2k+1}\lambda\rho} w^{(\alpha_4)}(t(\lambda\rho)^{\frac{n}{m+\ell}})\|_{s,\tilde K,\frac{1}{\eta+\epsilon}} \\
& \times \|g_1^{(\alpha_2)}\Big(\frac{t}{t_{\lambda\rho}}\Big)\varphi^{(\beta_2)}(x)\|_{s,\tilde K,\frac{1}{\eta+\epsilon}}\,d\rho.
\end{aligned}
$$

In order to apply Proposition 4.2 we observe that $m(\lambda\rho) = \sup_{(t,x)\in\tilde K}\big(a\frac{t^{2k+1}}{2k+1}\lambda\rho + A\frac{t^{\frac{\ell}{m}+1}}{\frac{\ell}{m}+1}(\lambda\rho)^{n/m}\big) \le (\alpha_0 - L_0)(\lambda\rho)^{1-(2k+1)\varepsilon}$ where $L_0 = e_0\tilde\epsilon_1^2 c_0^2 > 0$, as we can deduce using the expression (3.10). So we can apply Lemma 4.5, Proposition 4.2 and Lemma 4.4 and deduce that

$$\|I_1\|_{s,K_\lambda,\frac{1}{\eta-\epsilon}} \le C\lambda^{2m+\varepsilon m} e^{-L_0\lambda^{1-(2k+1)\varepsilon}} e^{s(d\lambda)^{1/s}+(s-1)(d\lambda^{n/m})^{\frac{1}{s-1}}+G_0\lambda^{\frac{\varepsilon}{s-s'}}+c\lambda^{\frac{\varepsilon}{s-1}}}, \tag{4.17}$$

for $\lambda \gg 0$.

In order to estimate $\|I_2\|_{s,K_\lambda,\frac{1}{\eta-\epsilon}}$ we observe that for every $M \in \mathbb{N}$ we have:

$$e^{ix\lambda\rho+a\frac{t^{2k+1}}{2k+1}\lambda\rho} = \lambda^{-M}\Big(ix + a\frac{t^{2k+1}}{2k+1}\Big)^{-M} \frac{\partial^M}{\partial\rho^M}\big(e^{ix\lambda\rho+a\frac{t^{2k+1}}{2k+1}\lambda\rho}\big);$$

then integration by parts in (4.15) gives us:

$$\begin{aligned} I_2 = &\sum_{\substack{\alpha_1+\alpha_2\le m\\ \beta_1+\beta_2\le m\\ \beta_2\neq 0}} \sum_{\alpha_3+\alpha_4=\alpha_1} \sum_{\substack{0\le j\le 2k\alpha_3\\ 0\le q\le\alpha_3}} \sum_{r=0}^{2km} (-1)^M C t^{r+j}\lambda^{-M}\lambda^{\beta_1+q+\varepsilon\alpha_2+\frac{n}{m+\ell}\alpha_4}\\ &\times \Big(ix + a\frac{t^{2k+1}}{2k+1}\Big)^{-M} \int_1^{1+\mu_0} e^{ix\lambda\rho+a\frac{t^{2k+1}}{2k+1}\lambda\rho} \frac{\partial^M}{\partial\rho^M}\Big[\psi(\rho)e^{-\alpha_0(\lambda\rho)^{1-(2k+1)\varepsilon}}\\ &\times \rho^{\beta_1+q+\varepsilon\alpha_2+\frac{n}{m+\ell}\alpha_4} w^{(\alpha_4)}(t(\lambda\rho)^{\frac{n}{m+\ell}}) g_1\Big(\frac{t}{t_{\lambda\rho}}\Big)\Big]\varphi^{(\beta_2)}(x)\,d\rho. \end{aligned}$$

We can compute $\frac{\partial^M}{\partial\rho^M}[\dots]$ in the previous integral via Leibnitz and Faà di Bruno formulas, cf. (4.2); moreover, since now $\beta_2 \neq 0$ we can limit our attention to $\mathrm{supp}(\varphi'(x))$, and so

$$\|I_2\|_{s,K_\lambda,\frac{1}{\eta-\epsilon}} = \|I_2\|_{s,K',\frac{1}{\eta-\epsilon}}; \tag{4.18}$$

where $K' = \{(t,x) \in \mathbb{R}^2 : (1-\epsilon_2)t_\lambda \le t \le (1+\epsilon_2)t_\lambda,\ x \in [-1,-\tilde{\epsilon}] \cup [\tilde{\epsilon},1]\}$, for a fixed $0 < \tilde{\epsilon} < 1$. Thus we have:

$$\begin{aligned} &\|I_2\|_{s,K_\lambda,\frac{1}{\eta-\epsilon}}\\ &\le \sum C^{M+1}\binom{M}{M_1}\binom{K_1}{M_2}\binom{K_2}{M_3}\binom{K_3}{M_4}\lambda^{-M}\lambda^R\lambda^{M_2(1-(2k+1)\varepsilon)+M_4\frac{n}{m+\ell}+M_5\varepsilon}\\ &\times \sum_{h=1}^{M_5}\sum_{p=1}^{M_4} \|t^{S+p+h}\|_{s,K',\frac{1}{\eta}} \Big\|\Big(ix + a\frac{t^{2k+1}}{2k+1}\Big)^{-M}\Big\|_{s,K',\frac{1}{\eta+\epsilon}} F_{M_2}F_pF_hC_{M_3}\\ &\times \int_1^{1+\mu_0} |\psi^{(M_1)}(\rho)| e^{-\alpha_0(\lambda\rho)^{1-(2k+1)\varepsilon}} \|e^{ix\lambda\rho+a\frac{t^{2k+1}}{2k+1}\lambda\rho} w^{(\alpha_4+p)}(t(\lambda\rho)^{\frac{n}{m+\ell}})\|_{s,K',\frac{1}{\eta+2\epsilon}}\\ &\times \Big\|g_1^{(h)}\Big(\frac{t}{t_{\lambda\rho}}\Big)\varphi^{(\beta_2)}(x)\Big\|_{s,K',\frac{1}{\eta+2\epsilon}} \rho^V\rho^{M_2(1-(2k+1)\varepsilon)+M_4\frac{n}{m+\ell}+M_5\varepsilon}\,d\rho, \end{aligned} \tag{4.19}$$

where $S, R, V \in \mathbb{R}$ are independent of M and the first sum in the right-hand side is over $\alpha_1+\alpha_2 \le m$, $\beta_1+\beta_2 \le m$, $\beta_2 \neq 0$, $\alpha_3+\alpha_4 = \alpha_1$, $0 \le j \le 2k\alpha_3$, $0 \le q \le \alpha_3$, $r = 0,\dots,2km$, $K_1+M_1 = M$, $K_2+M_2 = K_1$, $K_3+M_3 = K_2$, $M_4+M_5 = K_3$;

moreover the quantities F_{M_2}, F_p, F_h (coming from the Faà di Bruno formula) and C_{M_3} have the following expression:

– $$F_{M_2} = M_2! \sum_{i=1}^{M_2} \sum_{\substack{\gamma_1+\dots+\gamma_{M_2}=i \\ \gamma_1+2\gamma_2+\dots+M_2\gamma_{M_2}=M_2}} \prod_{\nu=1}^{M_2} \frac{1}{\gamma_\nu!}\Big(\frac{C_\nu}{\nu!}\Big)^{\gamma_\nu},$$

where $C_\nu = |\alpha_0(1-(2k+1)\varepsilon)\dots(1+(2k+1)\varepsilon-\nu+1)|$;

– $$F_p = M_4! \sum_{\substack{\delta_1+\dots+\delta_{M_4}=p \\ \delta_1+2\delta_2+\dots+M_4\delta_{M_4}=M_4}} \prod_{\mu=1}^{M_4} \frac{1}{\delta_\mu!}\Big(\frac{C_\mu}{\mu!}\Big)^{\delta_\mu},$$

where $C_\mu = \Big|\frac{n}{m+\ell}\dots\Big(\frac{n}{m+\ell}-\mu+1\Big)\Big|$;

– $$F_h = M_5! \sum_{\substack{\sigma_1+\dots+\sigma_{M_5}=h \\ \sigma_1+2\sigma_2+\dots+M_5\sigma_{M_5}=M_5}} \prod_{\tau=1}^{M_5} \frac{1}{\sigma_\tau!}\Big(\frac{C_\tau}{\tau!}\Big)^{\sigma_\tau},$$

where $C_\tau = |\varepsilon(\varepsilon-1)\dots(\varepsilon-\tau+1)|$;

– $C_{M_3} = \big|(\beta_1+q+\varepsilon\alpha_2+\frac{n}{m+\ell}\alpha_4)\dots(\beta_1+q+\varepsilon\alpha_2+\frac{n}{m+\ell}\alpha_4-M_3+1)\big|$.

So we have now to estimate the various quantities appearing in the right-hand side of (4.19). To start with we give the following lemma.

Lemma 4.6. *Let K' be as before, cf.* (4.18). *Then*

$$\Big\|\Big(ix+a\frac{t^{2k+1}}{2k+1}\Big)^{-M}\Big\|_{s,K',\frac{1}{\eta+\epsilon}} \le C\Big(\frac{2}{\tilde\epsilon}\Big)^M.$$

Proof. By Faà di Bruno formula we have:

$$\Big|\partial_x^\alpha\partial_t^\beta\Big(\Big(ix+a\frac{t^{2k+1}}{2k+1}\Big)^{-M}\Big)\Big| \le \sum_{0<h\le\beta} M(M+1)\dots(M+\alpha+h-1)$$

$$\times\Big|\Big(ix+a\frac{t^{2k+1}}{2k+1}\Big)^{-M-\alpha-h}\Big|\beta! \sum_{\substack{h_1+\dots+h_\beta=h \\ h_1+2h_2+\dots+\beta h_\beta=\beta}} \prod_{j=1}^{\beta}\frac{1}{h_j!}\Big(\frac{1}{j!}\Big)^{h_j}\Big|\partial_t^j\Big(a\frac{t^{2k+1}}{2k+1}\Big)\Big|^{h_j}.$$

Now since $(1-\epsilon_2)t_\lambda \le t \le (1+\epsilon_2)t_\lambda$ there exists a constant $D>0$ such that for every $\lambda \ge 1$ and $j=1,\dots,\beta$

$$\Big|\partial_t^j\Big(a\frac{t^{2k+1}}{2k+1}\Big)\Big|^{h_j} \le D^\beta;$$

moreover the condition $|x| \ge \tilde\epsilon$, $0<\tilde\epsilon<1$, gives us

$$\Big|\Big(ix+a\frac{t^{2k+1}}{2k+1}\Big)^{-M-\alpha-h}\Big| \le \tilde\epsilon^{-M-\alpha-\beta}.$$

Using the identity (4.4) with $z = 1$ we can see that

$$\beta! \sum_{\substack{h_1+\cdots+h_\beta=h \\ h_1+2h_2+\cdots+\beta h_\beta=\beta}} \prod_{j=1}^{\beta} \frac{1}{h_j!}\Big(\frac{1}{j!}\Big)^{h_j} \leq \beta! \sum_{\beta=h}^{\infty} B_{\beta,h}(\{1\})\frac{1}{\beta!} = \frac{\beta!}{h!}\Big(\sum_{j=1}^{\infty}\frac{1}{j!}\Big)^h \leq \frac{\beta!}{h!}e^\beta;$$

finally

$$M(M+1)\dots(M+\alpha+h-1) = \binom{M+\alpha+h-1}{M-1}(\alpha+h)! \leq 2^M 4^{\alpha+\beta}\alpha!\, h!,$$

since $(\alpha+h)! \leq 2^{\alpha+h}\alpha!\, h!$ and $\binom{M+\alpha+h-1}{M-1} \leq 2^{M+\alpha+h-1} \leq 2^{M+\alpha+\beta}$. We then have

$$\left|\partial_x^\alpha \partial_t^\beta\left(\Big(ix + a\frac{t^{2k+1}}{2k+1}\Big)^{-M}\right)\right| \leq \Big(\frac{2}{\tilde{\epsilon}}\Big)^M C^{\alpha+\beta}\alpha!\,\beta!$$

for a constant C independent of α, β and M. Now remembering the definition of Gevrey seminorms, cf. (2.6), we obtain:

$$\begin{aligned}\Big\|\Big(ix + a\frac{t^{2k+1}}{2k+1}\Big)^{-M}\Big\|_{s,K',\frac{1}{\eta+\epsilon}} &\leq \Big(\frac{2}{\tilde{\epsilon}}\Big)^M \sup_{\alpha,\beta\in\mathbb{Z}_+}\Big(\big(C(\eta+\epsilon)\big)^{\alpha+\beta}(\alpha!\,\beta!)^{-(s-1)}\Big) \\ &= C\Big(\frac{2}{\tilde{\epsilon}}\Big)^M,\end{aligned}$$

since $s > 1$. □

Taking into account that $\psi \in G_0^{s'}(\mathbb{R})$ we have

$$|\psi^{(M_1)}(\rho)| \leq \tilde{C}^{M_1+1}M_1!^{s'}. \tag{4.20}$$

Regarding C_{M_3}, if we denote by N_0 the smallest integer satisfying $N_0 \geq \beta_1 + q + \varepsilon\alpha_2 + \frac{n}{m+\ell}\alpha_4$ for every β_1, q, α_2, α_4 (observe that N_0 depends only on n, m, k and ℓ) we have:

$$\begin{aligned}C_{M_3} &\leq N_0\dots(N_0+M_3-1) = \binom{N_0+M_3-1}{N_0-1}M_3! \\ &\leq 2^{N_0+M_3-1}M_3! \leq C_3^{M_3+1}M_3!\end{aligned} \tag{4.21}$$

We have now to analyze the quantities F_{M_2}, F_p and F_h. As an example let us consider F_h. Since $C_\tau \leq \tau!$ we have:

$$\begin{aligned}F_h &\leq M_5! \sum_{\substack{\sigma_1+\cdots+\sigma_{M_5}=h \\ \sigma_1+2\sigma_2+\cdots+M_5\sigma_{M_5}=M_5}} \prod_{\tau=1}^{M_5}\frac{1}{\sigma_\tau!} \\ &\leq B_{M_5,h}(\{\tau!\}) \\ &\leq 2^{M_5}M_5!\,B_{M_5,h}(\{\tau!\})\frac{1}{M_5!}\Big(\frac{1}{2}\Big)^{M_5} \leq 2^{M_5}M_5!\frac{1}{h!},\end{aligned} \tag{4.22}$$

as we can deduce by (4.4) with $z = \frac{1}{2}$. The quantities F_{M_2} and F_p can be treated in a similar way: we estimate C_ν and C_μ as in (4.21) and then we use the same procedure as in (4.22), obtaining:

$$F_{M_2} \le C_1^{M_2+1} M_2! \tag{4.23}$$

$$F_p \le C_2^{M_4+1} M_4! \, \frac{1}{p!} \tag{4.24}$$

Now we can estimate $\|I_2\|_{s,K_\lambda,\frac{1}{\eta-\epsilon}}$ by considering (4.19) and applying Proposition 4.2, Lemma 4.4, Lemma 4.5, Lemma 4.6, (4.20), (4.21), (4.22), (4.23) and (4.24); we then have:

$$\|I_2\|_{s,K_\lambda,\frac{1}{\eta-\epsilon}} \le C^{M+1}\lambda^{-M}\lambda^{M\max\{1-(2k+1)\varepsilon,\frac{n}{m},\varepsilon\}} M!^{s'} \lambda^{\tilde{R}} \times e^{s(d\lambda)^{1/s}+(s-1)(d\lambda^{n/m})^{\frac{1}{s-1}}} e^{c\lambda^{\frac{\varepsilon}{s-1}}} e^{G_0\lambda^{\frac{\varepsilon}{s-s'}}},$$

for every $M \in \mathbb{Z}_+$ and $\lambda \gg 0$, where $\tilde{R}$ does not depend on M. By the Stirling formula we have: $M! \le C_0 M^M e^{-M}\sqrt{M}$; we fix now $\lambda = M^h$, with $h > \frac{s'}{1-\max\{1-(2k+1)\varepsilon,\frac{n}{m},\varepsilon\}}$ (observe that if h is fixed then $\lambda = M^h \to +\infty \Leftrightarrow M \to +\infty$, h being positive); in this way we obtain

$$C^M\left(\sqrt{M}M^M\right)^{s'}\lambda^{-M}\lambda^{M\max\{1-(2k+1)\varepsilon,\frac{n}{m},\varepsilon\}} \le C_0.$$

Moreover e^{-M} in Stirling formula gives rise to the term $e^{-Ms'} = e^{-s'\lambda^{1/h}}$, so the following estimate holds:

$$\|I_2\|_{s,K_\lambda,\frac{1}{\eta-\epsilon}} \le C\lambda^{\overline{R}} e^{-s'\lambda^{1/h}} e^{s(d\lambda)^{1/s}+(s-1)(d\lambda^{n/m})^{\frac{1}{s-1}}} e^{c\lambda^{\frac{\varepsilon}{s-1}}} e^{G_0\lambda^{\frac{\epsilon}{s-s'}}}, \tag{4.25}$$

for $\lambda \gg 0$.

From (4.16) we have that $\|Pu_\lambda\|_{s,K_\lambda,\frac{1}{\eta-\epsilon}} \le \|I_1\|_{s,K_\lambda,\frac{1}{\eta-\epsilon}} + \|I_2\|_{s,K_\lambda,\frac{1}{\eta-\epsilon}}$, and so by (4.17) and (4.25) the following estimate holds:

$$\|Pu_\lambda\|_{s,K_\lambda,\frac{1}{\eta-\epsilon}} \le C\lambda^{R_0} e^{s(d\lambda)^{1/s}+(s-1)(d\lambda^{n/m})^{\frac{1}{s-1}}+G_0\lambda^{\frac{\varepsilon}{s-s'}}+c\lambda^{\frac{\varepsilon}{s-1}}} \times \left\{e^{-L_0\lambda^{1-(2k+1)\varepsilon}} + e^{-s'\lambda^{1/h}}\right\}, \tag{4.26}$$

for $\lambda \gg 0$ where R_0, L_0 are positive constants, $h > \frac{s'}{1-\max\{1-(2k+1)\varepsilon,\frac{n}{m},\varepsilon\}}$ and $s' > 1$ is arbitrarily fixed.

5. Proof of Theorem 2.1

Let us fix $s > s_{\mathrm{cr}}$ and suppose that ${}^tP$ is G^s locally solvable at the origin. Then the condition (2.7) holds in particular for $u = u_\lambda(t,x)$, for every λ. Applying (4.13), (4.14) and (4.26) the following estimate must be true for every $\lambda \gg 0$:

$$E^2\lambda^{-2p+2\ell\frac{1-m}{2m}\left(\frac{n}{m+\ell}-\varepsilon\right)} \le C\lambda^{R_0} e^{2s(d\lambda)^{1/s}+2(s-1)(d\lambda^{n/m})^{\frac{1}{s-1}}+2G_0\lambda^{\frac{\varepsilon}{s-s'}}+2c\lambda^{\frac{\varepsilon}{s-1}}} \times \left\{e^{-L_0\lambda^{1-(2k+1)\varepsilon}} + e^{-s'\lambda^{1/h}}\right\}, \tag{5.1}$$

where R_0, L_0 are positive constants, $h > \frac{s'}{1-\max\{1-(2k+1)\varepsilon, \frac{n}{m}, \varepsilon\}}$ and $s' > 1$ is arbitrarily fixed. Now we want to show that, for a suitable choice of h and s' the following condition is satisfied:

$$\min\left\{1-(2k+1)\varepsilon, \frac{1}{h}\right\} > \max\left\{\frac{1}{s}, \frac{n}{m}\frac{1}{s-1}, \frac{\varepsilon}{s-s'}, \frac{\varepsilon}{s-1}\right\}. \tag{5.2}$$

We observe at first that $\frac{1}{h} < \frac{1-\max\{1-(2k+1)\varepsilon, \frac{n}{m}, \varepsilon\}}{s'} < 1-\max\{1-(2k+1)\varepsilon, \frac{n}{m}, \varepsilon\}$; then if we prove that

$$\begin{aligned}\min\left\{1-(2k+1)\varepsilon, 1-\max\left\{1-(2k+1)\varepsilon, \frac{n}{m}, \varepsilon\right\}\right\} \\ > \max\left\{\frac{1}{s}, \frac{n}{m}\frac{1}{s-1}, \frac{\varepsilon}{s-s'}, \frac{\varepsilon}{s-1}\right\}\end{aligned} \tag{5.3}$$

we can choose h and s' in such a way that (5.2) is satisfied, since we can always fix $s' > 1$ and h such that $1-\max\{1-(2k+1)\varepsilon, \frac{n}{m}, \varepsilon\} - h < \nu$ for an arbitrary $\nu > 0$. Now, since $s > s_{\mathrm{cr}}$, the following estimates hold:

$$\frac{n}{m}\frac{1}{s_{\mathrm{cr}}-1} > \frac{n}{m}\frac{1}{s-1},$$
$$\frac{\varepsilon}{s_{\mathrm{cr}}-1} > \frac{\varepsilon}{s-1};$$

moreover, choosing $1 < s' < s - s_{\mathrm{cr}} + 1$ we have:

$$\frac{\varepsilon}{s_{\mathrm{cr}}-1} > \frac{\varepsilon}{s-s'}.$$

It is then sufficient to prove that

$$\begin{aligned}\min\left\{1-(2k+1)\varepsilon, 1-\max\left\{1-(2k+1)\varepsilon, \frac{n}{m}, \varepsilon\right\}\right\} \\ \geq \max\left\{\frac{1}{s_{\mathrm{cr}}}, \frac{n}{m}\frac{1}{s_{\mathrm{cr}}-1}, \frac{\varepsilon}{s_{\mathrm{cr}}-1}\right\}.\end{aligned} \tag{5.4}$$

Now the condition (2.2) implies that $\frac{n}{m} > \varepsilon$, so we have only to show that

$$\min\left\{1-(2k+1)\varepsilon, (2k+1)\varepsilon, 1-\frac{n}{m}\right\} \geq \max\left\{\frac{1}{s_{\mathrm{cr}}}, \frac{n}{m}\frac{1}{s_{\mathrm{cr}}-1}\right\}. \tag{5.5}$$

The condition (2.3) implies that $(2k+1)\varepsilon \geq 1-(2k+1)\varepsilon$, and so, recalling the expression of s_{cr}, cf. (2.5), (5.5) is equivalent to

$$\min\left\{1-(2k+1)\varepsilon, 1-\frac{n}{m}\right\} \geq \max\left\{1-(2k+1)\varepsilon, \frac{n}{m}\frac{1-(2k+1)\varepsilon}{(2k+1)\varepsilon}\right\}; \tag{5.6}$$

finally, since $(2k+1)\varepsilon \geq \frac{n}{m}$, cf. (2.3), we have that $1-(2k+1)\varepsilon \leq 1-\frac{n}{m}$ and $1-(2k+1)\varepsilon \geq \frac{n}{m}\frac{1-(2k+1)\varepsilon}{(2k+1)\varepsilon}$; then (5.6) is satisfied, and so (5.2) holds.

Now the condition (5.2) assures us that, when $\lambda \to \infty$, for suitable choices of h and s' and for $s > s_{\mathrm{cr}}$ the leading terms of the two summands in the right-hand side of (5.1) are respectively $e^{-L_0\lambda^{1-(2k+1)\varepsilon}}$ and $e^{-s'\lambda^{1/h}}$. Then (5.1) cannot be true for $\lambda \to \infty$, and so ${}^tP$ is not G^s locally solvable at the origin for $s > s_{\mathrm{cr}}$.

Acknowledgment

The author expresses his deep gratitude to Prof. Popivanov for the very instructive and pleasant discussions on the subject of this paper during his visit in Sofia.

References

[1] B.L.J. Braaksma, *Asymptotic analysis of a differential equation of Turrittin*, SIAM J. Math. Anal. **2** (1971), 1–16.

[2] D. Calvo and P. Popivanov, *Solvability in Gevrey classes for second powers of the Mizohata operator*, C. R. Acad. Bulgare Sci. **57** (2004), n. 6, 11–18.

[3] F. Cardoso and F. Trèves, *A necessary condition of local solvability for pseudo differential equations with double characteristics*, Ann. Inst. Fourier, Grenoble **24** (1974), 225–292.

[4] M. Cicognani and L. Zanghirati, *On a class of unsolvable operators*, Ann. Scuola Norm. Sup. Pisa **20** (1993), 357–369.

[5] A. Corli, *On local solvability in Gevrey classes of linear partial differential operators with multiple characteristics*, Comm. Partial Differential Equations **14** (1989), 1–25.

[6] A. Corli, *On local solvability of linear partial differential operators with multiple characteristics*, J. Differential Equations **81** (1989), 275–293.

[7] G. De Donno and A. Oliaro, *Local solvability and hypoellipticity for semilinear anisotropic partial differential equations*, Trans. Amer. Math. Soc. **355**, 8 (2003), 3405–3432.

[8] M. Fedoryuk, *Asymptotic Analysis*, Springer, Berlin, 1993.

[9] Ch. Georgiev and P. Popivanov, *A necessary condition for the local solvability of a class of operators having double characteristics*, Annuaire Univ. Sofia, Fac. Math. Mec, I **75** (1981), 57–71.

[10] R. Goldman, *A necessary condition for the local solvability of a pseudodifferential equation having multiple characteristics*, J. Differential Equations **19** (1975), 176–200.

[11] T. Gramchev, *Powers of Mizohata type operators in Gevrey classes*, Boll. Un. Mat. Ital. B (7) **5** (1991), 135–156.

[12] T. Gramchev, *Nonsolvability for analytic partial differential operators with multiple characteristics*, J. Math. Kyoto Univ. **33** (1993), 989–1002.

[13] T. Gramchev and P. Popivanov, *Partial differential equations: approximate solutions in scales of functional spaces*, Math. Res., **108**, Wiley-VCH Verlag, Berlin, 2000.

[14] T. Gramchev and L. Rodino, *Gevrey solvability for semilinear partial differential equations with multiple characteristics*, Boll. Un. Mat. Ital. B (8) **2** (1999), 65–120.

[15] L. Hörmander, *Linear Partial Differential Operators*, Springer, Berlin, 1963.

[16] P. Marcolongo and A. Oliaro, *Local Solvability for Semilinear Anisotropic Partial Differential Equations*, Ann. Mat. Pura Appl. (4) **179** (2001), 229–262.

[17] P. Marcolongo and L. Rodino, *Nonsolvability for an operator with multiple complex characteristics*, Progress in Analysis, Vol. I,II (Berlin, 2001), World Sci. Publishing, River Edge, NJ, 2003, 1057–1065.

[18] M. Mascarello and L. Rodino, *Partial differential equations with multiple characteristics*, Wiley-Akademie Verlag, Berlin, 1997.

[19] T. Okaji, *The local solvability of partial differential operator with multiple characteristics in two independent variables*, J. Math. Kyoto Univ. **20**, 1 (1980), 125–140.

[20] T. Okaji, *Gevrey-hypoelliptic operators which are not C^∞-hypoelliptic*, J. Math. Kyoto Univ. **28**, 2 (1988), 311–322.

[21] A. Oliaro, P. Popivanov, and L. Rodino, *Local solvability for partial differential equations with multiple characteristics*, Proceedings of Abstract and Applied Analysis conference 2002 (Hanoi) (Kluwer, ed.), 2002, pp. 143–157.

[22] P. Popivanov, *On a nonsolvable partial differential operator*, Ann. Univ. Ferrara, Sez. VII, Mat. **49** (2003), 197–208.

[23] L. Rodino, *Linear partial differential operators in Gevrey spaces*, World Scientific, Singapore, 1993.

[24] S. Spagnolo, *Local and semi-global solvability for systems of non-principal type*, Comm. Partial Differential Equations **25** (2000), 1115–1141.

[25] H.L. Turrittin, *Stokes multipliers for asymptotic solutions of a certain differential equation*, Trans. Amer. Math. Soc. **68** (1950), 304–329.

Alessandro Oliaro
Dip. Matematica
Università di Torino
Via Carlo Alberto, 10
I-10123 Torino, Italy
e-mail: oliaro@dm.unito.it

Operator Theory:
Advances and Applications, Vol. 160, 357–366

Optimal Prediction of Generalized Stationary Processes

Vadim Olshevsky and Lev Sakhnovich

To Israel Gohberg on the occasion of his 75th anniversary with appreciation and friendship

Abstract. Methods for solving optimal filtering and prediction problems for the *classical* stationary processes are well known since the late forties. Practice often gives rise to what is called *generalized* stationary processes [GV61], e.g., to white noise and to many other examples. Hence it is of interest to carry over optimal prediction and filtering methods to them. For arbitrary generalized stochastic processes this could be a challenging problem. It was shown recently [OS04] that the generalized matched filtering problem can be efficiently solved for a rather general class of S_J*-generalized stationary* processes introduced in [S96]. Here it is observed that the optimal prediction problem admits an efficient solution for a slightly narrower class of T_J*-generalized stationary* processes. Examples indicate that the latter class is wide enough to include white noise, positive frequencies white noise, as well as generalized processes occurring when the smoothing effect gives rise to a situation in which the distribution of probabilities may not exist at some time instances. One advantage of the suggested approach is that it connects solving the optimal prediction problem with inverting the corresponding integral operators S_J. The methods for the latter, e.g., those using the Gohberg-Semençul formula, can be found in the extensive literature, and we include an illustrative example where a computationally efficient solution is feasible.

Mathematics Subject Classification (2000). Primary 60G20, 93E11 ; Secondary 60G25.

Keywords. Generalized Stationary Processes, Prediction, Filtering, Integral Equations, Gohberg-Semençul Formula.

1. Introduction

1.1. Optimal prediction of classical stationary processes

A complex-valued stochastic process $X(t)$ is called *stationary in the wide sense* (see, e.g., [D53]), if its expectation is a constant,

$$E[X(t)] = \text{const}, \qquad -\infty < t < \infty$$

and the correlation function depends only on the difference $(t-s)$, i.e.,

$$K_X(t,s) = E[X(t)\overline{X(s)}] = K_X(t-s).$$

We assume that $E[|X(t)|^2] < \infty$. Let us consider a system with the memory depth ω that maps the input stochastic process $X(t)$ into the output stochastic process $Y(t)$ in accordance with the following rule:

$$Y(t) = \int_{t-\omega}^{t} X(s)g(t-s)ds, \qquad g(x) \in L(0,w). \tag{1.1}$$

In the optimal prediction problem one needs to find a filter $g(t)$ in

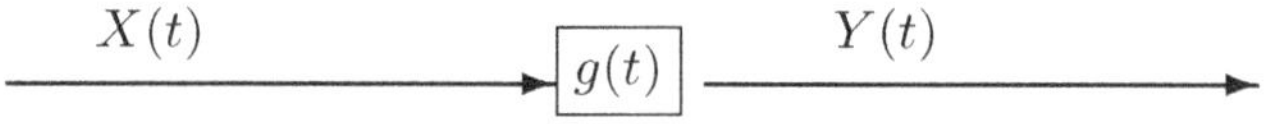

Figure 1. Classical Optimal Filter.

so that the output process $Y(t)$ is as close as possible to the true process $X(t+\tau)$ where $\tau > 0$ is a given constant. The measure of closeness is understood in the sense of minimizing the quantity

$$E[|X(t+\tau) - Y(t)|^2].$$

Wiener's seminal monograph [W49] solves the above problem for the case $\omega = \infty$ in (1.1). His results were extended to the case $\omega < \infty$ in [ZR50].

1.2. Generalized stationary processes. Motivation

White noise $X(t)$ (having equal intensity at all frequencies within a broad band) is not a stochastic process in the *classical sense* as defined above. In fact, white noise can be thought of as the *derivative of a Brownian motion*, which is a continuous stationary stochastic process $W(t)$. It can be shown that $W(t)$ is nowhere differentiable, a fact explaining the highly irregular motions that Robert Brown observed. This means that white noise $\frac{dW(t)}{dt}$ does not exist in the ordinary sense. In fact, it is is a *generalized stochastic process* whose definition is stated in [GV61].

Generally, any receiving device has a certain "inertia" and hence instead of actually measuring the classical stochastic process $X(t)$ it measures its averaged value

$$\Phi(\varphi) = \int \varphi(t)X(t)dt, \tag{1.2}$$

where $\varphi(t)$ is a certain function characterizing the device. Small changes in φ yield small changes in $\Phi(\varphi)$ (small changes in the receiving devices yield closer measurements), hence Φ is a continuous linear functional (see (1.2)), i.e., a generalized stochastic process whose definition was given in [GV61].

Hence it is very natural and important to solve the optimal filtering and prediction problems in the case of generalized stochastic processes. This correspondence is a sequel to [OS04] where we solved the problem of constructing the matched filtering problem for generalized stationary processes. Here, formulas for solving the optimal prediction problem are given.

1.3. The main result and the structure of the correspondence

In the next Section 2 we recall the definition [GV61] of generalized stationary processes and describe the system action on it. Then in Section 3 we introduce a class of T_J-generalized processes for which we will be solving the optimal prediction problem in Section 4. Finally, in Section 5 we consider a new model of colored noise. It is shown how our general solution to the optimal prediction problem can be a basis to provide a computationally efficient solution in this important example.

2. Generalized stationary processes. Auxiliary results

2.1. The definition of [GV61]

Let $\mathcal{K}$ denotes the set of all infinitely differentiable finite functions. Let a stochastic functional Φ (i.e., a functional assigning to any $\varphi(t) \in \mathcal{K}$ a stochastic value $\Phi(\varphi)$) be linear, i.e.,

$$\Phi(\alpha\varphi + \beta\psi) = \alpha\Phi(\varphi) + \beta\Phi(\psi).$$

Let us further assume that all the stochastic values $\Phi(\varphi)$ have expectations given by

$$m(\varphi) = E[\Phi(\varphi)] = \int_{-\infty}^{\infty} x dF(x), \qquad \text{where} \qquad F(x) = P[\Phi(\varphi) \leq x].$$

Notice that $m(\varphi)$ is a linear functional acting in the space $\mathcal{K}$ that depends continuously on φ. The bilinear functional

$$B(\varphi, \psi) = E[\Phi(\varphi)\overline{\Phi(\psi)}]$$

is the correlation functional of a stochastic process. It is supposed that $B(\varphi, \psi)$ is continuously dependent on either of the arguments.

The stochastic process Φ is called *generalized stationary in the wide sense* [GV61], [S97] if for any functions $\varphi(t)$ and $\psi(t)$ from $\mathcal{K}$ and for any number h the equalities

$$m[\varphi(t)] = m[\varphi(t+h)], \tag{2.1}$$

$$B[\varphi(t), \psi(t)] = B[\varphi(t+h), \psi(t+h)] \tag{2.2}$$

hold true.

2.2. System action on generalized stationary processes

If the processes $X(t)$ and $Y(t)$ were classical, then it is standard to assume that the system action shown in Figure 1 obeys

$$Y(t) = \int_0^T X(t-\tau)g(\tau)d\tau, \qquad \text{(where } T = w \text{ is the memory depth).} \tag{2.3}$$

Let us now consider a more general situation when the system shown in Figure 2.

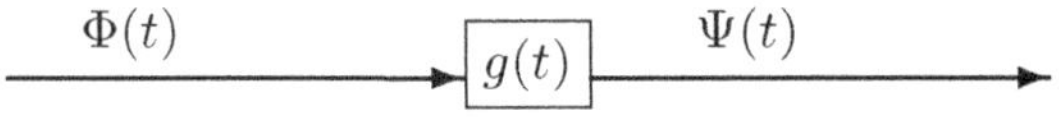

Figure 2. System action on generalized stationary processes

receives the generalized stationary signal Φ that we assume to be zero-mean. Then we define the system action as follows:

$$\Psi(\varphi) = \Phi[\int_0^T g(\tau)\varphi(t+\tau)d\tau], \tag{2.4}$$

The motivation for the latter definition is that if $X(t)$ and $Y(t)$ were the classical stationary processes then the Formula (2.4) for the corresponding functionals

$$\Phi(\varphi) = \int_{-\infty}^{\infty} X(t)\varphi(t)dt, \qquad \Psi(\varphi) = \int_{-\infty}^{\infty} Y(t)\varphi(t)dt \tag{2.5}$$

is equivalent to the classical relation (2.3), so that the former is a natural generalization of the latter.

3. S_J-generalized and T_J-generalized stationary processes

3.1. Definitions

Let us denote by $\mathcal{K}_J$ the set of functions in $\mathcal{K}$ such that $\varphi(t) = 0$ when $t \notin J = [a, b]$. The correlation functional $B_J(\varphi, \psi)$ is called a *segment* of the correlation functional $B(\varphi, \psi)$ if

$$B_J(\varphi, \psi) = B(\varphi, \psi), \qquad \varphi, \psi \in \mathcal{K}_J. \tag{3.1}$$

Definition 3.1. *Generalized stationary processes are called* S_J-generalized processes *if their segments satisfy*

$$B_J(\varphi, \psi) = (S_J\varphi, \psi)_{L^2}, \tag{3.2}$$

where S_J is a bounded nonnegative operator acting in $L^2(a, b)$ having the form

$$S_J\varphi = \frac{d}{dt}\int_a^b \varphi(u)s(t-u)du. \tag{3.3}$$

Here $(\cdot, \cdot)_{L^2}$ is the inner product in the space $L^2(a, b)$.

Formulas for *matched filters* for the above S_J-generalized processes have been recently derived in [OS04]. Here we consider a different problem of *optimal prediction* that is shown to have an efficient solution under the additional assumption captured by the next definition.

Definition 3.2. *An S_J-generalized process is referred to as a T_J-generalized process if in addition to (3.2) and (3.3) the kernel $s(t)$ of S_J in (3.3) has a continuous derivative (for $t \neq 0$) that we denote by $k(t)$, and moreover*

$$s'(t) = k(t) \qquad (t \neq 0), \qquad\qquad k(0) = \infty. \tag{3.4}$$

3.2. Examples

Before solving the optimal prediction problem we provide some illustrative examples.

Example 3.3. **White noise.** It is well known that white noise W (which is the derivative of a nowhere differentiable Brownian motion) is not a continuous stochastic process. In fact, it is a generalized stationary process whose correlation functional is known [L68] to be

$$B'(\varphi, \psi) = \int_{-\infty}^{\infty}\int_{-\infty}^{\infty} \delta(t-s)\varphi(t)\overline{\psi(s)}ds.$$

Thus, in this case we have $B'(\varphi, \psi) = (\varphi, \psi)_{L_2}$ and hence (3.2) implies that white noise Φ is a very special S_J-generalized stationary process with

$$S_J = I. \tag{3.5}$$

It means that the corresponding kernel function $s(t)$ has the form

$$s(t) = \begin{cases} \frac{1}{2} & t > 0 \\ -\frac{1}{2} & t < 0. \end{cases}$$

In accordance with (3.4) the white noise Φ is a T_J-generalized stationary process.

Example 3.4. **Positive frequencies white noise (PF-white noise).** Observe that the operator

$$S_J f = fD + \frac{j}{\pi}⨍_0^T \frac{f(t)}{x-t}dt, \quad f \in L_2[0,T], \tag{3.6}$$

(where $⨍_0^T$ is the Cauchy Principal Value integral, and $D \geq 1$) defines an S_J-generalized process. When $D > 1$ then S_J is invertible (see example 4.1), and if $D = 1$ then S_J is noninvertible. Notice that if $D = 1$ then the kernel of S_J is the Fourier transform of $f_{PW}(z) = 1$ having equal intensity at all *positive frequencies* and the zero intensity at the negative frequencies (hence the name *PF-white noise*). Observe that the process is, in fact, T_J-generalized. Indeed, it can be shown that the kernel function of S_J has the form

$$s(t) = \begin{cases} \frac{D}{2} + \frac{i}{\pi}\ln t & t > 0 \\ -\frac{D}{2} + \frac{i}{\pi}\ln|t| & t < 0 \end{cases}$$

which implies

$$k(t) = s'(t) = \frac{1}{t} \qquad (t \neq 0), \qquad k(0) = \infty.$$

In accordance with (3.4) the PF-white noise is a T_J-generalized stationary process.

4. Solution to the optimal prediction problem

As is well known [L68], [M65], in the classical case the solution $g(t)$ of the optimal prediction problem can be found by solving

$$\int_0^w g(u)k_x(u-v)dv = k_x(u+\tau). \tag{4.1}$$

In the generalized case the solution g to the optimal prediction problem can be found by solving a generalization of (4.1),

$$S_J g = k_x(u+\tau) = s'(u+\tau). \tag{4.2}$$

If S_J is invertible then

$$g = S_J^{-1} s'(u+\tau). \tag{4.3}$$

Example 4.1. **PF-white noise revisited.** Here we return to the case considered in example 3.4. It can be shown that if $D > 1$ then S_J in (3.6) is positive definite and invertible with

$$S_J^{-1} f = f(x)D_1 - \beta\frac{j}{\pi}\int_0^T \left(\frac{t}{T-y}\right)^{j\alpha}\left(\frac{x}{T-x}\right)^{-j\alpha}\frac{f(t)}{x-t}dt, \tag{4.4}$$

where

$$D_1 = \frac{D}{D^2-1}, \qquad \beta = \frac{1}{D^2-1}, \tag{4.5}$$

and the number α is obtained from

$$\cosh\alpha\pi = D\sinh\alpha\pi. \tag{4.6}$$

Clearly, (4.4) and (4.3) solve the optimal prediction problem in this case.

5. Some practical consequences. A connection to the Gohberg-Semençul formula

The main focus of the sections 2–4 had mostly a theoretical nature. In this section we indicate that the Formula (4.3) offers a novel technique allowing one to work out practical problems. Specifically:

- Filtering problems for *classical* stationary processes typically lead to non-invertible operators S_J, and to find the solution $g(t)$ to the optimal prediction problem one needs to solve (4.1). In the case of *generalized* stationary processes the operator S_J is often invertible and hence there is a better Formula (4.3).

- Secondly, the operator S_J in (3.2) and (3.3) can be seen as a new way of modeling colored noise. This model is useful since the existing integral equations literature already describes many particular examples on inverting S_J, either explicitly or numerically. Hence (4.3) solves the corresponding optimal prediction problem.

Before providing one such example let us rewrite the operator S_J of (3.3),

$$S_J f = \frac{d}{dx}\int_a^b f(t)s(x-t)dt, \qquad f(x) \in L^2(a,b)$$

in a more familiar form.

Proposition 5.1. *Let*

$$a = 0, \qquad b = T, \qquad s(0) - s(-0) = 1, \qquad s'(t) = k(t) \quad (t \neq 0).$$

With these settings S_J can be rewritten as

$$S_J f = f(x) + \int_0^T f(t)k(x-t)dt. \tag{5.1}$$

If $k(t)$ is continuous for $t \neq 0$ then the corresponding process is T_J-generalized.

The integral equations literature (see, e.g., [GF74], [M77], [S96]) contains results on the inversion of the operator S_J of the form (5.1). The following theorem is well known.

Theorem 5.2 (Gohberg-Semençul [GS72] (see also [GF74]).). *Let the operator S_J have the form (5.1) with $k(x) \in L(-w,w)$. If there are two functions $\gamma_\pm(x) \in L(0,w)$ such that*

$$S_J\gamma_+(x) = k(x), \qquad S_J\gamma_-(x) = k(x-w) \tag{5.2}$$

then $S_{[0,w]}$ is invertible in $L^p(0,w)$ $(p \geq 1)$ and

$$S_J^{-1} f = f(x) + \int_0^w f(t)\gamma(x,t)dt, \tag{5.3}$$

where $\gamma(x,t)$ is given by (5.4).

$$\gamma(x,t) = \begin{cases} -\gamma_+(x-t) - \int_t^{w+t-x} \big[\gamma_-(w-s)\gamma_+(s+x-t) \\ \qquad\qquad -\gamma_+(w-s)\gamma_-(s+x-t)\big]ds, & x > t, \\ -\gamma_-(x-t) - \int_t^{w} \big[\gamma_-(w-s)\gamma_+(s+x-t) \\ \qquad\qquad -\gamma_+(w-s)\gamma_-(s+x-t)\big]ds, & x < t \end{cases} \tag{5.4}$$

The latter result leads to a number of interesting special cases when the operator S_J can be explicitly inverted and hence the Formula (4.3) solves the optimal prediction problem in these cases. For example, the processes corresponding to $k(x) = |x|^{-h}$, with $0 < h < 1$, or to $k(x) = -\log|x-t|$ are of interest. We elaborate the details for another example next.

Example 5.3. **Colored noise approximated by rational functions. The exponential kernel.** Let us consider *colored noise* approximated by a combination of rational functions with the fixed poles $\{\pm i\alpha_m\}$,

$$f(t) = \sum_{m=1}^{N} \gamma_m \frac{1}{t^2+\alpha_m^2}, \qquad \alpha_m > 0, \qquad \gamma_m > 0.$$

Let us use its Fourier transform

$$k(x) = \sum_{m=1}^{N} \beta_m e^{-\alpha_m |x|}, \qquad \beta_j = \frac{\pi}{\alpha_m}\gamma_m \tag{5.5}$$

to define the operator S_J via (5.1).

Solution to the filtering problem. The situation is exactly the one captured by theorem 5.2 where the operator (5.1) has the special kernel (5.5). A procedure to solve (5.2), and hence to find the inverse of S_J is obtained next.

Theorem 5.4 (Computational Procedure). *Let S_J be given by* (5.1) *and its kernel $k(x)$ have the special form* (5.5). *Then S_J^{-1} is given by the Formulas* (5.3), (5.4), (5.2), *where*

$$\gamma_+(x) = -\gamma(x,0), \qquad \gamma_-(x) = -\gamma(w-x,0). \tag{5.6}$$

Here

$$\gamma(x,0) = G(x)\begin{bmatrix} F_1 \\ F_2 \end{bmatrix}^{-1} B, \tag{5.7}$$

where the $1\times 2N$ row $G(x)$, the $N\times 2N$ matrices F_1, F_2, and the $2N\times 1$ column B are defined by

$$G(x) = \begin{bmatrix} e^{\nu_1 x} & e^{\nu_2 x} & \cdots & e^{\nu_{2N} x} \end{bmatrix}, \qquad F_1 = \begin{bmatrix} \frac{1}{\alpha_i+\nu_k} \end{bmatrix}_{1\le i\le N, 1\le k\le 2N},$$

$$F_2 = \begin{bmatrix} \frac{-e^{\nu_k w}}{\alpha_i-\nu_k} \end{bmatrix}_{1\le i\le N, 1\le k\le 2N}, \qquad B = \begin{bmatrix} \underbrace{1 \ \cdots \ 1}_{N} & \underbrace{0 \ \cdots \ 0}_{N} \end{bmatrix}.$$

The numbers $\{\nu_k\}$ are the roots (we assume them to be pairwise different) of the polynomial

$$Q(z) = P(z) - 2\sum_{m=1}^{N} \delta_m \sum_{s=1}^{m} z^{2(m-s)} \sum_{k=1}^{N} \alpha_k^{2s-1}\beta_k, \tag{5.8}$$

where the numbers $\{\delta_k\}$ are the coefficients of the polynomial

$$P(z) = \prod_{m=1}^{N} (z^2 - \alpha_m^2) = \sum_{m=1}^{N} \delta_m z^{2m}. \tag{5.9}$$

Hence, in this important case, the Formula (4.3) allows us to find the explicit solution $g(t)$ to the optimal prediction problem by plugging in it the Formulas (5.3), (5.4) together with (5.6) (5.7). Notice that there are fast and superfast algorithms to solve the Cauchy-like linear system in (5.7), see, e.g., [O03] and the references therein.

Remark 5.5. Note that the problem in example 5.3 can also be solved by using the Kalman filtering method, see, e.g., [K61, KSH00]. Hence, in such simplest cases when we have a classical stationary process corrupted by white noise our paper suggests an alternative way to derive the solution via solving integral equations.

However, the Kalman filtering method has not yet been carried over to generalized processes, and in the case of T_J-generalized stationary processes our technique is currently the only one available.

Problem 5.6. We would like to conclude this paper with the interesting open problem of extending the Kalman filtering method to generalized stationary processes.

Acknowledgment

This work was supported in part by the NSF contracts 0242518 and 0098222. We would also like to thank the editor Cornelis Van der Mee and the anonymous referee for the very careful reading of the manuscript, and for a number of helpful suggestions.

References

[D53] J.L. Doob, *Stochastic processes*, Wiley, 1953.

[GF74] I.C. Gohberg and I.A. Feldman, *Convolution equations and projection methods for their solution,* Transl. Math. Monographs, **v. 41**, AMS Publications, Providence, Rhode Island, 1974.

[GS72] I. Gohberg and A. Semençul, *On the inversion of finite Toeplitz matrices and their continual analogues,* Math. Issled. 7:2, 201–223. (In Russian.)

[GV61] I.M. Gelfand and N.Ya. Vilenkin, *Generalized Functions, No. 4. Some Applications of Harmonic Analysis. Equipped Hilbert Spaces*, Gosud. Izdat. Fiz.-Mat. Lit., Moscow, 1961 (Russian);
translated as: Generalized Functions. Vol. 4: Applications of Harmonic Analysis, Academic Press, 1964.

[K61] T. Kailath, *Lectures on Wiener and Kalman filtering*, Springer Verlag, 1961.

[KSH00] T. Kailath, A.S. Sayed, and B. Hassibi, *Linear Estimation*, Prentice Hall, 2000.

[L68] B.R. Levin, *Theoretical Foundations of Statistical Radio Engineering*, Moskow, 1968.

[M65] D. Middleton, *Topics in Communication Theory,* McGraw-Hill, 1965.

[M77] N.I. Muskhelishvili, *Singular Integral Equations,* Aspen Publishers Inc, 1977.

[O03] V. Olshevsky, *Pivoting for structured matrices and rational tangential interpolation*, in Fast Algorithms for Structured Matrices: Theory and Applications, CONM/323, p. 1–75, AMS publications, May 2003.

[OS04] V. Olshevsky and L. Sakhnovich, *Matched filtering for generalized Stationary Processes,* 2004, submitted.

[S96] L.A. Sakhnovich, *Integral Equations with Difference Kernels on Finite Intervals,* Operator Theory Series, v. 84, Birkhäuser Verlag, 1996.

[S97] L.A. Sakhnovich, *Interpolation Theory and its Applications,* Kluwer Academic Publications, 1997.

[W49] N. Wiener, *Extrapolation, Interpolation and Smoothing of Stationary Time Series*, Wiley, 1949.

[ZR50] L.A. Zadeh and J.R. Ragazzini, *An extension of Wiener's Theory of Prediction*, J.Appl.Physics, 1950, 21:7, 645–655.

Vadim Olshevsky
Department of Mathematics
University of Connecticut
Storrs, CT 06269, USA
e-mail: olshevsky@math.uconn.edu

Lev Sakhnovich
Department of Mathematics
University of Connecticut
Storrs, CT 06269, USA
e-mail: lev.sakhnovich@verizon.net

Operator Theory:
Advances and Applications, Vol. 160, 367–382

Symmetries of 2D Discrete-Time Linear Systems

Paula Rocha, Paolo Vettori and Jan C. Willems

Abstract. Static symmetries of linear shift invariant 1D systems had been thoroughly investigated, and the resulting theory is now a well-established topic. Nevertheless, only partial results were available for the multidimensional case, since the extension of the theory for 1D systems proved not to be straightforward, as usual. Actually, a non trivial regularity assumption and also restrictions on the set of allowed symmetries had to be imposed. In this paper we show how it is possible to overcome these difficulties using a particular canonical form for 2D discrete linear systems.

Mathematics Subject Classification (2000). Primary 93C55; Secondary 47B39.

Keywords. Discrete-time linear multidimensional systems, symmetry, behavioral approach.

1. Introduction

After Noether's Theorem, the importance of the analysis of symmetries in the study of dynamical systems became more than evident. Indeed, the study of symmetries is a very important tool to comprehend and clarify many intrinsic properties of physical systems. In fact, the knowledge of the symmetries of a dynamical system often leads to a simplification of its mathematical description.

A basic but illuminating example is a dynamical system given by a collection of equal particles. As is well known, the behavior of this kind of system can be

The research of the first two authors is partially supported by the *Unidade de Investigação Matemática e Aplicações* (UIMA), University of Aveiro, Portugal, through the *Programa Operacional "Ciência, Tecnologia, Inovação"* (POCTI) of the *Fundação para a Ciência e Tecnologia* (FCT), co-financed by the European Community fund FEDER.
The research of the third author is supported by the Belgian Federal Government under the DWTC program *Interuniversity Attraction Poles, Phase V, 2002–2006, Dynamical Systems and Control: Computation, Identification and Modelling,* by the KUL Concerted Research Action (GOA) MEFISTO–666, and by several grants en projects from IWT-Flanders and the Flemish Fund for Scientific Research.

described by the dynamics of their center of mass. The same simplified representation can be achieved without any consideration of its physical nature, just by analyzing the symmetry law which is here given by the invariance with respect to the exchange of particles.

The first and essential aim of our research is precisely to characterize through canonical forms the special structures of system representations that are induced by the underlying symmetries.

As regards linear systems, in the framework of the behavioral approach, the first results were presented by Fagnani and Willems [2, 3, 4] for regular behaviors. However, while in the 1D case every behavior is regular, this condition is not necessarily fulfilled by generic multidimensional systems. Moreover, only a restricted class of symmetries was allowed in the study of real-valued trajectories.

In Section 2 we introduce the class of dynamical discrete systems we deal with in this paper: the *behaviors*, sets of trajectories which can be described by a finite number of linear (ordinary or partial) difference equations. To introduce the notion of symmetric behavior, we recall the necessary concepts regarding symmetries and their representations in Section 3.

In Section 4, we provide a canonical way to write the equations which define a behavior. This canonical form is then used in Section 5 to show how the results about symmetric 1D behaviors can be extended to the 2D case without the need of any further assumption.

We conclude by showing how each 2D behavior that exhibits some symmetry, can be described by a set of equations that reflect the intrinsic structure of its symmetry.

2. Behavioral systems

We briefly recall the notion of dynamical system in the behavioral approach [12, 13] and some results that we will need later on. According to this approach a dynamical system is defined as a triple

$$\Sigma = (T, W, \mathcal{B}),$$

where T denotes the domain, W the signal space, and $\mathcal{B}$, which is a subset of $W^T = \{w : T \to W\}$, represents the set of trajectories which are allowed to occur by the definition of the system. This is called the behavior of the system. We will consider only discrete, linear, complete and shift-invariant systems. This amounts to say that: the domain is $T = \mathbb{Z}^n$, $n = 1, 2$, i.e., for 1D and 2D systems, respectively; W and $\mathcal{B}$ are vector spaces over $\mathbb{K}$ (the real or the complex field); the dimension of W is finite and $\mathcal{B}$ is closed in the topology of pointwise convergence; for any trajectory $w \in \mathcal{B}$ and $\tau \in T$ we have $\sigma^\tau w \in \mathcal{B}$ where σ^τ is the shift operator such that $(\sigma^\tau w)(t) = w(t + \tau)$.

Remark 2.1. The multi-index notation is here used to handle 1D and 2D systems in a unified way. So, if $\tau = (\tau_1, \tau_2) \in \mathbb{Z}^2$, then $\sigma^\tau = \sigma_1^{\tau_1}\sigma_2^{\tau_2}$ where σ_i are the (commuting) partial shift operators on the i-th component. Analogous is the notation for monomials: $s^\tau = s_1^{\tau_1} s_2^{\tau_2}$. In the following sections we shall be more explicit to distinguish between 1D and 2D systems. In particular, note that for 2D systems,

$$\sigma^\tau w(t) = \sigma_1^{\tau_1}\sigma_2^{\tau_2} w(t_1, t_2) = w(t_1 + \tau_1, t_2 + \tau_2).$$

This way of defining the dynamics leads to a representation free theory, i.e., to a theory which does not require a specific model for the system equations, as for example the input/state/output model of classical systems theory. Indeed, it is possible to characterize the trajectories of a dynamical system in many ways. Those studied most are the kernel and image representations, which define the behavior as the kernel and, respectively, as the image of a suitable operator.

To be more precise, we introduce operators on trajectories $w \in W^T$ of the form $\sum_{i\in\mathcal{I}} R_i w(t+i)$, where $\mathcal{I} \subseteq T$ is a finite subset of the domain and R_i are constant matrices with suitable dimensions. We may write

$$\sum_{i\in\mathcal{I}} R_i w(t+i) = \sum_{i\in\mathcal{I}} R_i \sigma^i w(t) = R(\sigma, \sigma^{-1}) w(t), \tag{2.1}$$

where $R(s, s^{-1})$ is a univariate or bivariate Laurent polynomial matrix.

Using this notation, a behavior $\mathcal{B} \subseteq (\mathbb{K}^q)^T$ is defined by a kernel representation if there exists $R(s, s^{-1}) \in \mathbb{K}[s, s^{-1}]^{p\times q}$, for some $p \in \mathbb{N}$, such that

$$\mathcal{B} = \ker R(\sigma, \sigma^{-1}) = \{w \in (\mathbb{K}^q)^T : R(\sigma, \sigma^{-1})w = 0\},$$

i.e., if $\mathcal{B}$ is the set of solutions of a matrix difference equation represented by the difference operator $R(\sigma, \sigma^{-1})$. Note that, by shift-invariance of the system, we may always suppose that the summation in (2.1) is made over indices with non-negative components and therefore we assume without loss of generality that kernel representations are polynomial matrices $R(s) \in \mathbb{K}[s]^{p\times q}$.

This representation is very general as the following theorem [8] states.

Theorem 2.2. *Every nD behavior admits a kernel representation.*

Moreover, it is possible to establish a deep relation between a behavior and any matrix providing its kernel representation that leads, for instance, to the following fundamental result about inclusion of behaviors [8, Thm. 2.61].

Theorem 2.3. *The condition $\ker R_1(\sigma) \subseteq \ker R_2(\sigma)$ holds if and only if there exists a Laurent polynomial matrix $X(s, s^{-1})$ such that $X(s, s^{-1})R_1(s) = R_2(s)$.*

In this theorem, $X(s, s^{-1})$ has to be a Laurent polynomial matrix. Consider, for instance, the case $R_1(s) = s$ and $R_2(s) = 1$.

If $X(s, s^{-1})$ in Theorem 2.3 is unimodular, i.e., it has a polynomial inverse, say $Y(s, s^{-1})$, then $Y(s, s^{-1})R_2(s) = R_1(s)$ and so $\ker R_1(\sigma) = \ker R_2(\sigma)$. We say that in this case R_1 and R_2 are equivalent representations. Additional hypotheses are needed to prove the converse statement.

Corollary 2.4. *If $\ker R_1(\sigma) = \ker R_2(\sigma)$ and both $R_1(s)$ and $R_2(s)$ have full row rank, then $X(s, s^{-1})R_1(s) = R_2(s)$ with $X(s, s^{-1})$ Laurent unimodular matrix.*

Note that while every 1D behavior that can be defined by a kernel representation admits a full row rank kernel representation too, called *minimal*, this does not hold for 2D behaviors. Behaviors that have a full row rank kernel representation are called *regular*.

3. Symmetries and their representations

To define in a proper way the class of symmetries we will be dealing with, we first have to introduce some notions of representation theory. For more details we refer the reader to [10].

Given a finite dimensional vector space $\mathcal{W}$ over the field $\mathbb{K}$, we will denote by $\mathrm{GL}(\mathcal{W})$ the group of $\mathbb{K}$-isomorphisms of $\mathcal{W}$.

Definition 3.1. A representation of the group G on the vector space $\mathcal{W}$ is a group homomorphism

$$\rho : G \to \mathrm{GL}(\mathcal{W}),\ g \mapsto \rho_g.$$

The degree of a representation is defined by $\deg \rho = \dim \mathcal{W}$.

Remark 3.2. We suppose in this paper that G is equipped with a topology that makes it into a Hausdorff compact topological group. Even if not mentioned, every representation will be assumed to be continuous. Thus, if G is finite, the discrete topology is employed, which trivially ensures continuity.

Note that, according to the definition, ρ_g is an isomorphism of $\mathcal{W}$ onto itself for every g. With a little abuse of notation we may implicitly assume that some basis of $\mathcal{W}$ has already been fixed and therefore we will always identify ρ_g with its matrix representation.

Definition 3.3. Given a representation ρ on $\mathcal{W}$, a subspace $\mathcal{U} \subseteq \mathcal{W}$ is ρ-symmetric if $\rho\mathcal{U} \subseteq \mathcal{U}$, i.e., $\rho_g\mathcal{U} \subseteq \mathcal{U}$ for any $g \in G$.

Note that when $\mathcal{U}$ is a ρ-symmetric subspace of $\mathcal{W}$, the restrictions of ρ_g to $\mathcal{U}$ are isomorphisms of $\mathcal{U}$ and thus $\rho|_{\mathcal{U}}$ is itself a representation which is called subrepresentation of ρ. It can be proved that in the case of finite-degree representations there exists another ρ-symmetric subspace $\mathcal{V}$ such that $\mathcal{W} = \mathcal{U} \oplus \mathcal{V}$. We write also $\rho = \rho|_{\mathcal{U}} \oplus \rho|_{\mathcal{V}}$. A representation which does not admit proper symmetric subspaces, that is to say subrepresentations, is called irreducible. The decomposition of $\mathcal{W}$ into minimal symmetric subspaces gives then rise to a decomposition of ρ into irreducible subrepresentations. This decomposition becomes unique only if we identify different irreducible representations which are isomorphic — η^1 is isomorphic to η^2, or also $\eta^1 \cong \eta^2$, if there exists an isomorphism $\pi : \mathcal{W} \to \mathcal{W}$ such that $\pi\eta^1_g = \eta^2_g\pi$ for every g.

Eventually, the standard way to write such a decomposition of ρ into subrepresentations is

$$\rho = m_1\eta^1 \oplus m_2\eta^2 \oplus \cdots \oplus m_r\eta^r, \tag{3.1}$$

where the notation $m_i\eta^i$ stands for the direct sum of m_i copies of subrepresentations isomorphic to η^i.

Example. Consider the symmetric group $S_3 = \{e,(12),(13),(23),(123),(132)\}$, i.e., the group of all permutations of three elements. The most natural representation of S_3 is given by the 3×3 matrix group generated by

$$A = \rho_{(12)} = \begin{bmatrix} 0 & 1 & 0 \\ 1 & 0 & 0 \\ 0 & 0 & 1 \end{bmatrix} \text{ and } B = \rho_{(123)} = \begin{bmatrix} 0 & 1 & 0 \\ 0 & 0 & 1 \\ 1 & 0 & 0 \end{bmatrix}.$$

Note that $\rho_e = I = A^2 = B^3$, $\rho_{(13)} = AB$, $\rho_{(23)} = BA$, and $\rho_{(132)} = B^2$.

As it can be easily verified, the subspace $\mathcal{U} \subseteq \mathbb{R}^3$ generated by $\begin{bmatrix} 1 & 1 & 1 \end{bmatrix}^\top$ is ρ-symmetric and $\rho|_{\mathcal{U}} = 1$ is the corresponding (trivial) subrepresentation. On the other hand, for any direct summand $\mathcal{V}$ of $\mathcal{U}$, $\rho = \rho|_{\mathcal{U}} \oplus \rho|_{\mathcal{V}}$ is a decomposition of ρ into irreducible subrepresentations.

To write the representation $\eta = \rho|_{\mathcal{V}}$ in matrix form, it is necessary to fix a basis of $\mathcal{V}$. If, e.g. , $\mathcal{V} = \left\langle \left[\begin{smallmatrix} 1 \\ 0 \\ -1 \end{smallmatrix}\right], \left[\begin{smallmatrix} 0 \\ 1 \\ -1 \end{smallmatrix}\right] \right\rangle$, it follows that η is generated by

$$\eta_{(12)} = \begin{bmatrix} 0 & 1 \\ 1 & 0 \end{bmatrix} \text{ and } \eta_{(123)} = \begin{bmatrix} 0 & 1 \\ -1 & -1 \end{bmatrix}. \tag{3.2}$$

The theory that was just exposed can be used to define and analyze symmetries of dynamical systems. A quite general approach would be to choose $\mathcal{W} = W^T$, the set of trajectories. Nevertheless, in this paper we deal with a simpler class of symmetries, called *static symmetries*, since they act on the coordinates of the trajectories, i.e., $\mathcal{W} = W$, the signal space of the behavior. Therefore, the representations are homomorphism of the type

$$\rho : G \to \mathrm{GL}(W).$$

In this case, if $W = \mathbb{K}^q$, a representation of G is a family of invertible matrices $\rho_g \in \mathbb{K}^{q\times q}$ that act on trajectories as one would expect. Actually, for any $w \in W^T$, $\rho_g w \in W^T$ is such that

$$\rho_g w : T \to W,\ t \mapsto \rho_g(w(t))\ \forall g \in G.$$

Definition 3.4. Given a representation ρ on W, a behavior $\mathcal{B} \subseteq W^T$ is ρ-symmetric if

$$\rho\mathcal{B} \subseteq \mathcal{B}.$$

Note that, since ρ_g are isomorphisms, with $\rho_g^{-1} = \rho_{g^{-1}}$, a behavior is ρ-symmetric if and only if $\rho\mathcal{B} = \mathcal{B}$

Remark 3.5. The class of symmetries we are dealing with is not the most general. There are many simple symmetries that are group actions [5] but not representations. One example is given by time-reversal symmetries (if $w(t) \in \mathcal{B}$ then also $w(-t) \in \mathcal{B}$), studied for 1D systems in [2, 11]. Also shift-invariance is a symmetry. Indeed, the behaviors we are considering are symmetric with respect to the action of the group $\mathbb{Z}$ given by the shift operator σ^τ.

4. A canonical form for 2D behaviors

Our aim is to provide a canonical structure for kernel representations of 2D behaviors, that will be useful in the analysis of their symmetries. This canonical form was first introduced as *forward computational representation* in [9] for the analysis of 2D linear discrete systems.

In this paper we illustrate some of the already investigated properties along with new ones that are necessary to our purposes.

The idea is to obtain a representation of a behavior $\mathcal{B}$ as a product of two factors which depend only on one indeterminate. As we will see, this permits to derive properties of $\mathcal{B}$ using theorems about 1D behaviors (namely, Corollary 2.4).

The factors of canonical representations mentioned above are constructed recursively by computing minimal representations of properly defined 1D behaviors. However, just to clarify the whole process and to give some first important definitions, we begin by assuming that a kernel representation of $\mathcal{B}$ is already given.

Let $\mathcal{B} = \ker R(\sigma_1, \sigma_2)$, where $R(s_1, s_2) \in \mathbb{K}[s_1, s_2]^{p\times q}$ is a polynomial matrix in two indeterminates, and let $N-1$ be its degree in s_2. We may write

$$R(s_1, s_2) = \sum_{j=0}^{N-1} R_j(s_1)s_2^j = \begin{bmatrix} R_0(s_1) & R_1(s_1) & \cdots & R_{N-1}(s_1) \end{bmatrix} \begin{bmatrix} I \\ Is_2 \\ \vdots \\ Is_2^{N-1} \end{bmatrix}, \quad (4.1)$$

where I are $q \times q$ identity matrices. If we call $R^N(s) = \begin{bmatrix} R_0(s) & \cdots & R_{N-1}(s) \end{bmatrix}$ and $\Phi^N(s) = \begin{bmatrix} I & \cdots & Is^{N-1} \end{bmatrix}^\top$, then (4.1) is a factorization of $R(s_1, s_2)$ into polynomial factors in one indeterminate

$$R(s_1, s_2) = R^N(s_1)\Phi^N(s_2), \quad (4.2)$$

where $R^N(s) \in \mathbb{K}[s]^{p\times Nq}$ is uniquely determined by $R(s_1, s_2)$.

We will show how to build a canonical representation $C^N(s_1)\Phi^N(s_2)$ of $\mathcal{B}$ where the matrix $C^N(s)$, that is equivalent to $R^N(s)$, has a special nested structure and therefore can be defined recursively. In order to do this, it is necessary to shed some light on the relation between $\mathcal{B}$ and the 1D behavior $\ker R^N(\sigma)$.

First of all note that $w \in \mathcal{B}$ if and only if

$$R(\sigma_1, \sigma_2)w(t_1, t_2) = R^N(\sigma_1)\Phi^N(\sigma_2)w(t_1, t_2) = R^N(\sigma_1) \begin{bmatrix} w(t_1, t_2) \\ w(t_1, t_2 + 1) \\ \vdots \\ w(t_1, t_2 + N - 1) \end{bmatrix} = 0.$$

So, if we partition the trajectories of $\ker R^N(\sigma)$ into N blocks, $w^j : \mathbb{Z} \to \mathbb{K}^q$, $j = 0, \ldots, N-1$, then by the shift invariance of $\mathcal{B}$ we may write that

$$w \in \mathcal{B} \Rightarrow R^N(\sigma) \begin{bmatrix} w^0 \\ \vdots \\ w^{N-1} \end{bmatrix} = 0 \text{ where } w^j(t) = w(t, j)\, \forall j = 0, \ldots, N-1. \quad (4.3)$$

This fact tells us that, roughly speaking, trajectories in the restriction of $\mathcal{B}$ to the domain $\mathbb{Z}\times\{0,1,\dots,N-1\}$ (i.e., made up of N consecutive *horizontal lines* of $\mathcal{B}$), belong to $\ker R^N(\sigma)$. This shows the opportunity of the following definition.

Definition 4.1. Given a 2D behavior $\mathcal{B}$, we define

$$\mathcal{B}^i = \left\{ \begin{bmatrix} w^0 \\ \vdots \\ w^{i-1} \end{bmatrix} : \exists w \in \mathcal{B} \text{ such that } w^j(t) = w(t,j)\,\forall j = 0,\dots,i-1 \right\}.$$

Theorem 4.2. *The sets $\mathcal{B}^i$ are 1D behaviors* [7].

Remark 4.3. Note that in general $\mathcal{B}^N \subseteq \ker R^N(\sigma)$ is not an equality. Indeed, let $R(s_1,s_2) = s_2$. So, $\mathcal{B} = \{0\}$, hence $\mathcal{B}^2 = \{0\}$ too, but $\ker R^2(s) = \ker \begin{bmatrix} 0 & 1 \end{bmatrix} \neq \{0\}$.

In Section 4.1 we show how it is possible to take advantage of kernel representations of $\mathcal{B}^i$ in the construction of a particular kernel representation of $\mathcal{B}^{i+1}$. These *canonical* representations of the behaviors $\mathcal{B}^i$ will be the main tool for defining canonical representations of $\mathcal{B}$ in Section 4.2.

4.1. Iterative construction of representations of $\mathcal{B}^i$

In this section we show how to compute minimal representations of $\mathcal{B}^i$ in an efficient way. More precisely, the structure of a canonical form relative to $\mathcal{B}^i$ will be defined by induction. Actually, in the following proposition, we begin by showing how to construct a kernel representations of $\mathcal{B}^{i+1}$ that is based on a representation of $\mathcal{B}^i$.

Proposition 4.4. $\mathcal{B}^i = \ker C^i(\sigma)$ *if and only if there exist a full row rank matrix* $C_{i+1}(s)$ *and a matrix* $T_{i+1}(s)$ *such that* $\mathcal{B}^{i+1} = \ker C^{i+1}(\sigma)$ *with*

$$C^{i+1}(s) = \begin{bmatrix} C^i(s) & 0 \\ -T_{i+1}(s) & C_{i+1}(s) \end{bmatrix}. \tag{4.4}$$

Proof. Assume first that $\mathcal{B}^i = \ker C^i(\sigma)$. Given any kernel representation $\tilde{C}^{i+1}(s)$ of $\mathcal{B}^{i+1}$, it is always possible to put its last q columns in Hermite form just using elementary operations on the rows. In other words, there exists a unimodular matrix $U(s)$ such that

$$U(s)\tilde{C}^{i+1}(s) = \begin{bmatrix} \tilde{C}^i(s) & 0 \\ -T_{i+1}(s) & C_{i+1}(s) \end{bmatrix}, \tag{4.5}$$

where $C_{i+1}(s)$ has q columns and full row rank.

Moreover, it follows from Definition 4.1 that the trajectories of $\mathcal{B}^i$ are restrictions (to the first iq components) of some trajectory of $\mathcal{B}^{i+1}$. As a consequence, $\mathcal{B}^{i+1} \subseteq \mathcal{B}^i \times (\mathbb{K}^q)^{\mathbb{Z}}$ and thus $\mathcal{B}^{i+1} \subseteq \ker \begin{bmatrix} C^i(\sigma) & 0 \end{bmatrix}$. So, since (4.5) is still a kernel representation of $\mathcal{B}^{i+1}$, we can write

$$\mathcal{B}^{i+1} = \ker \begin{bmatrix} C^i(\sigma) & 0 \\ \tilde{C}^i(\sigma) & 0 \\ -T_{i+1}(\sigma) & C_{i+1}(\sigma) \end{bmatrix}.$$

However, since $\tilde{C}^i(\sigma)w = 0$ for every $w \in \mathcal{B}^i$, the operator $\tilde{C}^i(\sigma)$ does not impose any further restriction on the trajectories of $\mathcal{B}^{i+1}$. Hence it can be deleted from the kernel representation to obtain the equivalent representation (4.4).

Let us suppose now that (4.4) holds, with $C_{i+1}(s)$ having full row rank. Clearly $\mathcal{B}^i \subseteq \ker C^i(\sigma)$. Indeed,

$$w \in \mathcal{B}^i \;\Leftrightarrow\; \exists w^* : \begin{bmatrix} w \\ w^* \end{bmatrix} \in \mathcal{B}^{i+1} \;\Rightarrow\; C^i(\sigma)w = 0.$$

On the other hand, let $w \in \ker C^i(\sigma)$. Since $C_{i+1}(s)$ has full row rank, $C_{i+1}(\sigma)$ is a surjective operator and there exists a w^* such that $T_{i+1}(\sigma)w = C_{i+1}(\sigma)w^*$. Hence $\left[\begin{smallmatrix} w \\ w^* \end{smallmatrix}\right] \in \mathcal{B}^{i+1}$ and so $w \in \mathcal{B}^i$. Thus $\mathcal{B}^i = \ker C^i(\sigma)$. □

In Proposition 4.4 nothing is said about the minimality of $C^i(s)$. However, the statement assures that, once $C^i(s)$ is minimal, $C^{i+1}(s)$ is minimal too. So, the recursive scheme assures that $C^i(s)$ is minimal for any i if $C_1(s) = C^1(s)$ is a minimal representation of $\mathcal{B}^1$, the restriction of $\mathcal{B}$ to one horizontal line (the axis, for example).

Definition 4.5. We call canonical representation of $\mathcal{B}^i$ a lower triangular block matrix (each block having q columns)

$$C^i(s) = \begin{bmatrix} C_1(s) & & 0 \\ & \ddots & \\ * & & C_i(s) \end{bmatrix} \tag{4.6}$$

such that $\mathcal{B}^i = \ker C^i(\sigma)$ is a minimal representation.

To state the properties of this canonical form, a very important role will be played by the family of behaviors defined as follows:

$$\mathcal{B}_i = \left\{ w \in (\mathbb{K}^q)^{\mathbb{Z}} : \begin{bmatrix} 0 \\ w \end{bmatrix} \in \mathcal{B}^i \right\}. \tag{4.7}$$

These can be thought of as the sets of 'line-values' that in $\mathcal{B}$ follow $i-1$ null 'lines'.

Note that $\mathcal{B}_1 = \ker C_1(\sigma)$ by definition. However, this is a general fact: given a canonical representation (4.6) of $\mathcal{B}^i$, the behavior $\mathcal{B}_j$ has minimal representation $C_j(s)$ for any $j = 1, \ldots, i$. Indeed, by the form of $C^j(s)$ given in (4.4),

$$w \in \mathcal{B}_j \;\Leftrightarrow\; C^j(\sigma)\begin{bmatrix} 0 \\ w \end{bmatrix} = \begin{bmatrix} 0 \\ C_j(\sigma)w \end{bmatrix} = 0 \;\Leftrightarrow\; w \in \ker C_j(\sigma). \tag{4.8}$$

Also the converse result holds, as the following statement shows.

Lemma 4.6. *The matrix $C^i(s)$ defined by (4.6) is a canonical representation of $\mathcal{B}^i$ if and only if $C_j(s)$ is a minimal representation of $\mathcal{B}_j$ for any $j = 1, \ldots, i$.*

Proof. We already showed that the sufficiency holds. To prove the necessary condition we proceed by induction.

If we let $C^1(s) = C_1(s)$, we only have to prove that if $\mathcal{B}^j = \ker C^j(\sigma)$ and $\mathcal{B}_{j+1} = \ker C_{j+1}(\sigma)$, then there exists $T_{j+1}(s)$ such that $\mathcal{B}^{j+1} = \ker C^{j+1}(\sigma)$, with the matrix $C^{j+1}(s)$ as in (4.4).

By Proposition 4.4, there exist matrices $\tilde{T}_{j+1}(s)$ and $\tilde{C}_{j+1}(s)$ with full row rank such that

$$\mathcal{B}^{j+1} = \ker \begin{bmatrix} C^j(\sigma) & 0 \\ -\tilde{T}_{j+1}(\sigma) & \tilde{C}_{j+1}(\sigma) \end{bmatrix}.$$

By (4.8), $\mathcal{B}_{j+1} = \ker \tilde{C}_{j+1}(\sigma)$. Hence there must exist a unimodular $U(s)$ such that $C_{j+1}(s) = U(s)\tilde{C}_{j+1}(s)$. So, if we put $T_{j+1}(s) = U(s)\tilde{T}_{j+1}(s)$, we get that

$$C^{j+1}(s) = \begin{bmatrix} C^j(s) & 0 \\ -T_{j+1}(s) & C_{j+1}(s) \end{bmatrix} = \begin{bmatrix} I & 0 \\ 0 & U(s) \end{bmatrix} \begin{bmatrix} C^j(\sigma) & 0 \\ -\tilde{T}_{j+1}(\sigma) & \tilde{C}_{j+1}(\sigma) \end{bmatrix},$$

which provides the claimed canonical representation of $\mathcal{B}^{j+1}$, $j = 1, \dots, i$. □

This fact permits us to state the following important theorem that, restricting to a particular class of unimodular matrices, specializes Corollary 2.4 to the case of canonical representations.

Definition 4.7. Given $\mathcal{B}^i$ or, equivalently, a representation (4.6), the set of lower triangular unimodular $i \times i$ block matrices with block sizes $n_l \times n_k$, where n_j is the number of rows of any minimal representations of $\mathcal{B}_j$, is denoted by $\mathcal{U}^i$.

Theorem 4.8. *Let $C^i(s)$ be one canonical representation of $\mathcal{B}^i$. Then $\tilde{C}^i(s)$ is a canonical representation of $\mathcal{B}^i$ if and only if $\tilde{C}^i(s) = U^i(s)C^i(s)$ for some $U^i(s) \in \mathcal{U}^i$.*

Proof. On the diagonal of any matrix $U^i(s) \in \mathcal{U}^i$ there are necessarily unimodular matrices $U_j(s)$, $j = 1, \dots, i$. Therefore, if $C^i(s)$ has the form (4.6), the diagonal blocks of $U^i(s)C^i(s)$, i.e., $U_j(s)C_j(s)$, still have full row rank and so $U^i(s)C^i(s)$ is a canonical representation of $\mathcal{B}^i$, by Definition 4.5.

Suppose now that $\tilde{C}^i(s)$ is a canonical representation of $\mathcal{B}^i$, with diagonal blocks $\tilde{C}_j(s)$, $j = 1, \dots, i$. By Lemma 4.6, both the matrices $C_j(s)$ and $\tilde{C}_j(s)$ are minimal representations of the behavior $\mathcal{B}_j$, hence they have the same dimensions and, by Corollary 2.4, there exist unimodular matrices $U_j(s)$ such that $\tilde{C}_j(s) = U_j(s)C_j(s)$. The question is whether there exists a matrix $U^i(s) \in \mathcal{U}^i$ with diagonal blocks $U_j(s)$ such that $\tilde{C}^i(s) = U^i(s)C^i(s)$. We prove it by induction on j. The fact is trivially true for $j = 1$. So let us suppose that $\tilde{C}^j(s) = U^j(s)C^j(s)$ with $U^j(s) \in \mathcal{U}^j$. We want to find $V_{j+1}(s)$ such that $U^{j+1}(s) = \begin{bmatrix} U^j(s) & 0 \\ -V_{j+1}(s) & U_{j+1}(s) \end{bmatrix}$. Explicitly,

$$\tilde{C}^{j+1}(s) = \begin{bmatrix} \tilde{C}^j(s) & 0 \\ -\tilde{T}_{j+1}(s) & \tilde{C}_{j+1}(s) \end{bmatrix} = \begin{bmatrix} U^j(s) & 0 \\ -V_{j+1}(s) & U_{j+1}(s) \end{bmatrix} \begin{bmatrix} C^j(s) & 0 \\ -T_{j+1}(s) & C_{j+1}(s) \end{bmatrix}.$$

This condition, since all the other equations are satisfied, is equivalent to $\tilde{T}_{j+1}(s) = V_{j+1}(s)C^j(s) + U_{j+1}(s)T_{j+1}(s)$. Now, for every $w \in \mathcal{B}^j$ there exist w^* such that $\left[\begin{smallmatrix} w \\ w^* \end{smallmatrix}\right] \in \mathcal{B}^{j+1}$. Therefore,

$$\tilde{T}_{j+1}(\sigma)w = \tilde{C}_{j+1}(\sigma)w^* = U_{j+1}(\sigma)C_{j+1}(\sigma)w^* = U_{j+1}(\sigma)T_{j+1}(\sigma)w.$$

This means that $(\tilde{T}_{j+1}(\sigma) - U_{j+1}(\sigma)T_{j+1}(\sigma))w = 0$, i.e., that

$$\ker C^j(\sigma) = \mathcal{B}^j \subseteq \ker(\tilde{T}_{j+1}(\sigma) - U_{j+1}(\sigma)T_{j+1}(\sigma)),$$

which, by Theorem 2.3, implies that there exists a matrix $V_{j+1}(s)$ such that $\tilde{T}_{j+1}(s) - U_{j+1}(s)T_{j+1}(s) = V_{j+1}(s)C^j(s)$, thus proving the theorem. □

4.2. Canonical representations of a 2D behavior

We use now the canonical form introduced in Definition 4.5 to give a canonical form for 2D systems.

Definition 4.9. The representation $\mathcal{B} = \ker C(\sigma_1, \sigma_2)$ of a 2D behavior is canonical (with respect to the second variable) if the degree N in s_2 of $C(s_1, s_2)$ is minimal over all kernel representations of $\mathcal{B}$ and in the factorization (4.2),

$$C(s_1, s_2) = C^N(s_1)\Phi^N(s_2), \tag{4.9}$$

the matrix $C^N(s)$ is a canonical representation (4.6).

Remark 4.10. A similar construction would lead to the definition of a canonical form with respect to the first variable but we will not use it in this paper.

Note that, by its definition, when $\mathcal{B}$ is given by a canonical representation $C^N(s_1)\Phi^N(s_2)$, then $\mathcal{B}^N = \ker C^N(\sigma)$. The canonical representation of $\ker \sigma_2$, given as example in Remark 4.3, is simply $C(s_1, s_2) = 1$.

Theorem 4.11. *Every 2D behavior $\mathcal{B}$ admits a canonical representation.*

Proof. By Theorem 2.2, there exist a kernel representations of $\mathcal{B}$. Let $R(s_1, s_2)$ have minimal degree N in the second variable and factorize it as in Equation (4.1), $R(s_1, s_2) = R^N(s_1)\Phi^N(s_2)$. Then, given minimal representations of the 1D behaviors $\mathcal{B}_i$, Proposition 4.4 and Lemma 4.6 show how to construct canonical representations $C^i(s)$ of $\mathcal{B}^i$ for any $i = 1, \ldots, N$.

We now show that $C(s_1, s_2) = C^N(s_1)\Phi^N(s_2)$ is a canonical representation of $\mathcal{B}$. Indeed, as we saw in Remark 4.3, $\ker C^N(\sigma) = \mathcal{B}^N \subseteq \ker R^N(\sigma)$. So,

$$\ker C^N(\sigma_1)\Phi^N(\sigma_2) \subseteq \ker R^N(\sigma_1)\Phi^N(\sigma_2) = \mathcal{B}.$$

On the other hand, if $w \in \mathcal{B}$ then $\tilde{w}_k(h) = \Phi^N(\sigma_2)w(h, k)$ belongs to $\mathcal{B}^N$ for any $k \in \mathbb{Z}$ and thus

$$\mathcal{B} \subseteq \ker C(\sigma_1, \sigma_2) = \ker C^N(\sigma_1)\Phi^N(\sigma_2).$$

Hence, since $\mathcal{B} = \ker C(\sigma_1, \sigma_2)$, the proof is concluded. □

Another fundamental fact regards the relation between two canonical representations, which is clarified by the following theorem.

Theorem 4.12. *Given a canonical representation $C(s_1, s_2) = C^N(s_1)\Phi^N(s_2)$ of the 2D behavior $\mathcal{B}$, any other canonical representation is equal to $U(s_1)C(s_1, s_2)$ for some $U(s) \in \mathcal{U}^N$, where $\mathcal{U}^N$ depends on $C^N(s)$ as in Definition 4.7.*

Note that using this class of canonical representations, we overcome the two problems occurring in the equivalence of 2D behaviors. Indeed, two matrices that are representations of the same behavior are generally not unimodularly equivalent. However, by Theorem 4.12, this holds always true for canonical representations of 2D systems and, moreover, the unimodular matrix is a polynomial matrix in one indeterminate.

These facts will be very useful in the proof of the main result about symmetries in Section 5.

Example. Let the 2D behavior $\mathcal{B}$ be defined by the kernel representation

$$R(s_1, s_2) = \begin{bmatrix} s_2 - s_1 & s_2 - 1 \\ 1 - s_1 & s_2 - s_1 \end{bmatrix}. \tag{4.10}$$

Since $R(s_1, s_2)$ is a first order polynomial matrix in s_2, i.e., $N = 2$, it is possible to write $R(s_1, s_2) = R_0(s_1) + R_1(s_1)s_2$ where

$$R_0(s) = \begin{bmatrix} -s & -1 \\ 1 - s & -s \end{bmatrix} \text{ and } R_1(s) = \begin{bmatrix} 1 & 1 \\ 0 & 1 \end{bmatrix}.$$

Note that both $R_0(s)$ and $R_1(s)$ are full row rank matrices and therefore $R_0(\sigma)$ and $R_1(\sigma)$ are surjective operators.

First of all, we show that $\mathcal{B}^1 = (\mathbb{R}^2)^{\mathbb{Z}}$, i.e., the restriction of $\mathcal{B}$ to one (horizontal) line is *free*. In other words, we prove that any sequence $w(t, 0)$ can be extended to a 2D sequence $w(t, \tau) \in \mathcal{B}$.

Let us fix $w(t, 0)$. By using the kernel representation of $\mathcal{B}$ we have that

$$R_0(\sigma)w(t, 0) + R_1(\sigma)w(t, 1) = 0. \tag{4.11}$$

Since $R_1(\sigma)$ is surjective, there exists a sequence $w(t, 1)$ such that (4.11) holds true. This shows that any $w(t, 0)$ can be recursively extended to a 2D sequence $w(t, \tau)$ which satisfies the defining equation of $\mathcal{B}$ for any positive τ.

The same reasoning can be done for negative τ, in this case by using the equation $R_0(\sigma)w(t, -1) + R_1(\sigma)w(t, 0) = 0$ and the fact that $R_0(\sigma)$ is surjective. We conclude that $\mathcal{B}^1 = \ker C^1(\sigma)$, with $C^1(s) = 0$.

To find a kernel representation of $\mathcal{B}^2$ we use Proposition 4.4 and Lemma 4.6, i.e., we look for a kernel representation of $\mathcal{B}_2$. By definition, $w \in \mathcal{B}_2$ if and only if $\left[\begin{smallmatrix} 0 \\ w \end{smallmatrix}\right] \in \mathcal{B}^2$. So, by shift-invariance, $\mathcal{B}_2 = \{w(t, 1) : w \in \mathcal{B} \text{ and } w(t, 0) = 0\}$. However, by equation (4.11), if $w(t, 0) = 0$ then $R_1(\sigma)w(t, 1) = 0$ and, being $R_1(s)$ clearly invertible, $w(t, 1) = 0$. Hence, $\mathcal{B}_2 = \ker C_2(\sigma)$ with $C_2(s) = I$.

To find $T_2(s)$ in equation (4.4), observe that $-T_2(\sigma)w(t, 0) + C_2(\sigma)w(t, 1) = 0$ must hold, i.e., $w(t, 1) = T_2(\sigma)w(t, 0)$. By substituting it into equation (4.11), we get $[R_0(\sigma) + R_1(\sigma)T_2(\sigma)]w(t, 0) = 0$. By the freeness of $w(t, 0)$ and the invertibility of $R_1(\sigma)$, it follows that

$$-T_2(s) = R_1(s)^{-1}R_0(s) = \begin{bmatrix} 1 & -1 \\ 0 & 1 \end{bmatrix} \begin{bmatrix} -s & -1 \\ 1-s & -s \end{bmatrix} = \begin{bmatrix} -1 & s-1 \\ 1-s & -s \end{bmatrix}.$$

Finally, since $C^1(s)$ is null, the canonical representation of $\mathcal{B}^2$ is

$$C^2(s) = \begin{bmatrix} -T_2(s) & C_2(s) \end{bmatrix} = \left[\begin{array}{cc|cc} -1 & s-1 & 1 & 0 \\ 1-s & -s & 0 & 1 \end{array}\right] \tag{4.12}$$

and a canonical representation of $\mathcal{B}$ is

$$C(s_1, s_2) = C^2(s_1)\Phi^2(s_2) = \begin{bmatrix} s_2 - 1 & s_1 - 1 \\ 1 - s_1 & s_2 - s_1 \end{bmatrix}.$$

Notice that $U(s_1)C(s_1, s_2) = R(s_1, s_2)$ where $U(s) = R_1(s)$. This means, by Theorem 4.12, that $R(s_1, s_2)$ was already a 2D canonical form.

5. Static symmetries of dynamical systems

The aim of this section is to show the effect of static symmetries of $\mathcal{B}$, corresponding to the condition $\rho\mathcal{B} \subseteq \mathcal{B}$, on its kernel representations.

For 1D systems the following theorem holds [2].

Theorem 5.1. *Given a representation ρ of G on $W = \mathbb{K}^q$, the behavior $\mathcal{B} \subseteq W^{\mathbb{Z}}$ is ρ-symmetric if and only if it admits a minimal representation provided by $R(s) \in \mathbb{K}[s]^{p \times q}$ such that*

$$R(s)\rho = \rho' R(s)$$

where ρ' is a subrepresentation of ρ.

This theorem has been extended to nD systems [1] but under the restrictive assumption of behavior regularity. Moreover, in the real case $\mathbb{K} = \mathbb{R}$, another hypothesis has to be added: the representation cannot have irreducible quaternionic components (see Section 5.2). However, we will prove that for 2D systems a complete extension of Theorem 5.1 is possible.

First of all, note that if $\mathcal{B} = \ker R(s_1, s_2)$ with $R(s_1, s_2) \in \mathbb{K}[s_1, s_2]^{p \times q}$, then for any representation ρ' of G on $\mathbb{K}^p$, the condition

$$R(s_1, s_2)\rho = \rho' R(s_1, s_2) \tag{5.1}$$

is sufficient for $\mathcal{B}$ to be ρ-symmetric. Indeed, for any $w \in \mathcal{B}$ we have

$$R(s_1, s_2)\rho w = \rho' R(s_1, s_2) w = 0 \Rightarrow \rho w \in \mathcal{B}.$$

We are going to show that this condition is also necessary in the main theorem of this paper.

Theorem 5.2. *Given a representation ρ of G on $W = \mathbb{K}^q$, the behavior $\mathcal{B} \subseteq W^{\mathbb{Z}^2}$ is ρ-symmetric if and only if it admits a kernel representation $R(s_1, s_2) \in \mathbb{K}[s_1, s_2]^{p \times q}$ such that for a suitable representation ρ' of G*

$$R(s_1, s_2)\rho = \rho' R(s_1, s_2). \tag{5.2}$$

Moreover, if $\rho = m_1\eta^1 \oplus \cdots \oplus m_r\eta^r$*, then* $\rho' = m'_1\eta^1 \oplus \cdots \oplus m'_r\eta^r$ *for some* m'_i*,* $i = 1, \dots, r$*.*

Proof. We already showed that if equation (5.2) holds, $\mathcal{B}$ is ρ-symmetric. We prove now the converse and the condition on ρ'.

Let $C(s_1, s_2)$ be a canonical representation of $\mathcal{B}$. The behavior is ρ-symmetric if and only if $\rho\mathcal{B} = \mathcal{B}$, which is equivalent to $\ker C(\sigma_1, \sigma_2) = \ker C(\sigma_1, \sigma_2)\rho$. Note that $\Phi^N(s)\rho = \rho^N\Phi(s)^N$ where

$$\rho^N = \mathrm{diag}(\overbrace{\rho, \dots, \rho}^{N}). \tag{5.3}$$

Therefore, if we decompose $C(s_1, s_2)$ as in the factorization (4.9), we obtain that

$$C(s_1, s_2)\rho = C^N(s_1)\Phi^N(s_2)\rho = C^N(s_1)\rho^N\Phi^N(s_2). \tag{5.4}$$

The matrix $C^N(s)\rho^N$ is still a canonical representation of $\mathcal{B}^N$. Indeed it is a lower triangular matrix with diagonal blocks $C_j(s)\rho$, where $C_j(s)$ are the corresponding blocks of $C^N(s)$. These blocks are clearly full row rank matrices, hence the conditions of Definition 4.5 are met and $C(s_1, s_2)\rho$ is a canonical representation of $\mathcal{B}$. Therefore, by Theorem 4.12, a family of unimodular matrices $U_g(s_1) \in \mathcal{U}^N$ is uniquely determined such that

$$C(s_1, s_2)\rho_g = U_g(s_1)C(s_1, s_2),\ \forall g \in G. \tag{5.5}$$

The rest of the proof is analogous to the 1D case. We just give a sketch without the details that can be found in [2].

It is possible to prove that $U_g(s)$ is a continuous *polynomial* representation of G. However, by a theorem proved in [6], it is also isomorphic to a constant representation ρ', i.e., there exists a unimodular matrix $V(s)$ such that $V(s)U_g(s) = \rho'_g V(s)$ for any $g \in G$. This means that, if we let $R(s_1, s_2) = V(s_1)C(s_1, s_2)$, then $\mathcal{B} = \ker R(\sigma_1, \sigma_2)$ and, by (5.5),

$$\rho'_g R(s_1, s_2) = \rho'_g V(s_1)C(s_1, s_2) = V(s_1)C(s_1, s_2)\rho_g = R(s_1, s_2)\rho_g,\ \forall g \in G.$$

To prove that the decomposition of ρ' uses the irreducible subrepresentations of ρ, note that, as we already said, matrix (5.4) is a canonical representation, $\mathcal{B}^N = \ker C^N(\sigma)\rho^N$, and thus $\rho^N\mathcal{B}^N \subseteq \mathcal{B}^N$. Using the matrix $V(s)$ that we have just found, we see that $\rho'V(s)C^N(s) = V(s)C^N(s)\rho^N$ and so, by Theorem 5.1, ρ' is a subrepresentation of ρ^N. This shows that $m'_i \le Nm_i$. □

One could wonder whether it is possible to take a canonical representation $R(s_1, s_2)$ in formula (5.2) of Theorem 5.2. The answer, in general, is negative. The only result which is possible to state is the following.

Corollary 5.3. *If* $\mathcal{B}$ *is* ρ*-symmetric, then also the* 1*D behaviors* $\mathcal{B}_j$ *are* ρ*-symmetric. Moreover, let* $C_j(s)$ *be the minimal kernel representations of these behaviors such that* $\rho'^j C_j(s) = C_j(s)\rho$*, where* ρ'^j *are subrepresentations of* ρ *(as stated by Theorem* 5.1*). Then there exist* $U_g(s) \in \mathcal{U}^N$ *such that*

$$U_g(s_1)C(s_1, s_2) = C(s_1, s_2)\rho_g,\ \forall g \in G,$$

where $U_g(s) \in \mathcal{U}^N$ *has diagonal blocks* ρ'^j_g*.*

Proof. This is a straightforward consequence of the proof of Theorem 5.2. Indeed, by the properties of $U_g(s)$ therein stated, by the structure (4.6) of $C^N(s)$, and by minimality of $C_j(s)$, the statement follows. □

If we take into account also the decomposition into irreducible components of the representation ρ inducing the symmetry, it is possible to characterize $R(s_1, s_2)$ in a more detailed way. This leads to the definition of a canonical structure for kernel representations of symmetric systems.

However, the decomposition of representations depends highly on the base field, and so we have to distinguish the real and complex cases. We will only expose the main ideas, referring the reader to [2] for further details.

5.1. Symmetric representations of complex behaviors

Suppose that the basis of W is such that the decomposition (3.1), corresponds to orthogonal subspaces $U_1, \ldots, U_r$. If $n_i = \deg \eta^i$, then $\dim U_i = m_i n_i$. The matrices ρ_g have therefore a block diagonal structure. We will suppose without loss of generality that also each ρ'_g is a block diagonal matrix. We can then partition $R(s_1, s_2)$ into blocks $R_{ij}(s_1, s_2)$ according to ρ_g as regards the columns and to ρ'_g for the rows – so that equation (5.2) can be written blockwise

$$m'_i \eta^i R_{ij}(s_1, s_2) = R_{ij}(s_1, s_2) m_j \eta^j.$$

As a consequence of a fundamental result in representation theory, the Schur Lemma [10], we obtain that $R_{ij}(s_1, s_2) = 0$ for $i \neq j$ and that $R_{ii}(s_1, s_2)$, a block of dimension $m'_i n_i \times m_i n_i$, can be expressed as

$$R_{ii}(s_1, s_2) = \Lambda_i(s_1, s_2) \otimes I_{n_i}, \tag{5.6}$$

where $\Lambda_i(s_1, s_2) \in \mathbb{C}[s_1, s_2]^{m'_i \times m_i}$ is a suitable polynomial matrix, I_{n_i} is the $n_i \times n_i$ identity matrix and $\otimes$ is the Kronecker product, such that $[a_{ij}] \otimes B = [a_{ij} B]$.

Therefore, by partitioning also trajectories $w \in \mathcal{B}$ into r block vectors according to ρ, the kernel representation of $\mathcal{B}$ can be written as

$$\mathcal{B} = \ker(\Lambda_1(s_1, s_2) \otimes I_{n_1}) \oplus \cdots \oplus \ker(\Lambda_r(s_1, s_2) \otimes I_{n_r}).$$

5.2. Symmetric representations of real behaviors

When $\mathbb{K}$ is the field of real numbers, the general structure presented in the previous section for a complex representation still holds, but the matrices $\Lambda_i(s_1, s_2)$ in equation 5.6 are not so simple anymore. Indeed, three different types of irreducible real representations (*real*, *complex* and *quaternionic*, as defined in [5, §3.5]) give rise to different kinds of blocks $R_{ii}(s_1, s_2)$. If ρ_i is a real irreducible representation, the corresponding block looks like

$$A_i(s_1, s_2) \otimes I_{n_i}$$

where $A_i(s_1, s_2) \in \mathbb{R}[s_1, s_2]^{m'_i \times m_i}$.

For a complex irreducible representation ρ_i the typical block looks like

$$\begin{bmatrix} A_i(s_1,s_2) & B_i(s_1,s_2) \\ -B_i(s_1,s_2) & A_i(s_1,s_2) \end{bmatrix} \otimes I_{n_i}, \tag{5.7}$$

where both $A_i(s_1,s_2), B_i(s_1,s_2) \in \mathbb{R}[s_1,s_2]^{m_i' \times m_i}$.

Finally, if ρ_i is a quaternionic irreducible representation, the block $R_{ii}(s_1,s_2)$ is equal to

$$\begin{bmatrix} A_i(s_1,s_2) & B_i(s_1,s_2) & C_i(s_1,s_2) & D_i(s_1,s_2) \\ -B_i(s_1,s_2) & A_i(s_1,s_2) & -D_i(s_1,s_2) & C_i(s_1,s_2) \\ -C_i(s_1,s_2) & D_i(s_1,s_2) & A_i(s_1,s_2) & -B_i(s_1,s_2) \\ -D_i(s_1,s_2) & -C_i(s_1,s_2) & B_i(s_1,s_2) & A_i(s_1,s_2) \end{bmatrix} \otimes I_{n_i},$$

where $A_i(s_1,s_2), B_i(s_1,s_2), C_i(s_1,s_2), D_i(s_1,s_2) \in \mathbb{R}[s_1,s_2]^{m_i' \times m_i}$.

Note that the size of $R_{ii}(s_1,s_2)$ is divisible by 2 or by 4 when the corresponding subrepresentation is complex or quaternionic.

Example. Consider again the 2D behavior $\mathcal{B}$ defined by the kernel representation (4.10). We want to show that it is ρ-symmetric, where ρ is the representation of the group of permutations $G = \{e, (123), (132)\}$ defined in (3.2), i.e.,

$$\rho_{(123)} = \begin{bmatrix} 0 & 1 \\ -1 & -1 \end{bmatrix} \text{ and } \rho_{(132)} = \rho_{(123)}^{-1} = \begin{bmatrix} -1 & -1 \\ 1 & 0 \end{bmatrix}.$$

By equations (5.4) and (5.5), the behavior is symmetric if and only if there exist a family of unimodular matrices $U_g(s)$ such that

$$C^2(s)\rho_g^2 = U_g(s)C^2(s),\ \forall g \in G,$$

where $C^2(s) = \begin{bmatrix} -T_2(s) & C_2(s) \end{bmatrix}$ was found in equation (4.12). However, this means that, in particular, $C_2(s)\rho_g = U_g(s)C_2(s)$ and, being $C_2(s) = I$, it follows that $\rho_g = U_g(s)$.

So, we can affirm that $\mathcal{B}$ is ρ-symmetric once we check that also $T_2(s)\rho_g = U_g(s)T_2(s)$ holds true, i.e., if and only if

$$\begin{bmatrix} -1 & s-1 \\ 1-s & -s \end{bmatrix} \rho_g = \rho_g \begin{bmatrix} -1 & s-1 \\ 1-s & -s \end{bmatrix},\ \forall g \in G.$$

As a last remark, note that ρ is a complex irreducible representation: in complex form it is just a cubic root of unity (e.g, $-\frac{1}{2} + i\frac{\sqrt{3}}{2}$). So, by changing the base of $\mathcal{B}$, its kernel representation and the representation of G can be written

$$R(s_1,s_2) = \begin{bmatrix} 2s_2 - s_1 - 1 & \sqrt{3}(s_1-1) \\ -\sqrt{3}(s_1-1) & 2s_2 - s_1 - 1 \end{bmatrix} \text{ and } \rho_{(123)} = \begin{bmatrix} -\frac{1}{2} & \frac{\sqrt{3}}{2} \\ -\frac{\sqrt{3}}{2} & -\frac{1}{2} \end{bmatrix},$$

respectively, according to the structure showed in equation (5.7).

References

[1] C. De Concini and F. Fagnani. Symmetries of differential behaviors and finite group actions on free modules over a polynomial ring. *Math. Control Signals Systems*, 6(4):307–321, 1993.

[2] F. Fagnani and J.C. Willems. Representations of symmetric linear dynamical systems. *SIAM J. Control Optim.*, 31(5):1267–1293, 1993.

[3] F. Fagnani and J.C. Willems. Interconnections and symmetries of linear differential systems. *Math. Control Signals Systems*, 7(2):167–186, 1994.

[4] F. Fagnani and J.C. Willems. Symmetries of differential systems. In *Differential Equations, Dynamical Systems, and Control Science*, pages 491–504. Marcel Dekker Inc., New York, 1994.

[5] W. Fulton and J. Harris. *Representation Theory. A First Course.* Springer-Verlag, New York, 1991.

[6] V.G. Kac and D.H. Peterson. On geometric invariant theory for infinite-dimensional groups. In *Algebraic Groups, Utrecht 1986*, pages 109–142. Springer-Verlag, Berlin, 1987.

[7] J. Komorník, P. Rocha, and J.C. Willems. Closed subspaces, polynomial operators in the shift, and ARMA representations. *Appl. Math. Lett.*, 4(3):15–19, 1991.

[8] U. Oberst. Multidimensional constant linear systems. *Acta Appl. Math.*, 20(1–2):1–175, 1990.

[9] P. Rocha and J.C. Willems. Canonical Computational Forms for AR 2-D Systems. *Multidimens. Systems Signal Process.*, 1:251–278, 1990.

[10] J.-P. Serre. *Linear Representations of Finite Groups.* Springer-Verlag, Berlin, 1977.

[11] P. Vettori. Symmetric controllable behaviors. In *Proceedings of the 5th Portuguese Conference on Automatic Control, Controlo 2002*, pages 552–557, Aveiro, Portugal, 2002.

[12] J.C. Willems. Models for dynamics. In *Dynamics Reported*, volume 2, pages 171–269. John Wiley & Sons Ltd., Chichester, 1989.

[13] J.C. Willems. Paradigms and puzzles in the theory of dynamical systems. *IEEE Trans. Automat. Control*, 36(3):259–294, Mar. 1991.

Paula Rocha and Paolo Vettori
Departamento de Matemática
Universidade de Aveiro,
Campus de Santiago
3810-193, Aveiro, Portugal
e-mail: `procha@mat.ua.pt`
e-mail: `pvettori@mat.ua.pt`

Jan C. Willems
K.U. Leuven
ESAT/SCD (SISTA),
Kasteelpark Arenberg 10
B-3001 Leuven-Heverlee, Belgium
e-mail: `Jan.Willems@esat.kuleuven.ac.be`

Operator Theory:
Advances and Applications, Vol. 160, 383–401

An Algorithm for Solving Toeplitz Systems by Embedding in Infinite Systems

G. Rodriguez, S. Seatzu and D. Theis

Dedicated to Israel Gohberg on the occasion of his 75th birthday

Abstract. In this paper we propose a new algorithm to solve large Toeplitz systems. It consists of two steps. First, we embed the system of order N into a semi-infinite Toeplitz system and compute the first N components of its solution by an algorithm of complexity $O(N \log_2 N)$. Then we check the accuracy of the approximate solution, by an *a posteriori* criterion, and update the inaccurate components by solving a small Toeplitz system. The numerical performance of the method is then compared with the conjugate gradient method, for 3 different preconditioners. It turns out that our method is compatible with the best PCG methods concerning the accuracy and superior concerning the execution time.

Mathematics Subject Classification (2000). Primary 65F05; Secondary 47B35.

Keywords. Toeplitz linear systems, infinite linear systems, Wiener-Hopf factorization, projection method.

1. Introduction

One of the most widespread approaches to the solution of a finite Toeplitz linear system

$$A^N \mathbf{x}^N = \mathbf{b}^N, \qquad (A^N)_{ij} = a^N_{i-j}, \ i,j = 0,1,\ldots,N-1, \tag{1.1}$$

of large order N, in order to exploit the Toeplitz structure to optimize execution time and storage, is to employ an iterative method preconditioned by a circulant matrix. If the matrix A^N is symmetric and positive definite the preconditioned conjugate gradient (PCG) is the most commonly used method [7]. This technique

---

Research partially supported by MIUR under COFIN grants No. 2002014121 and 2004015437.

consists of determining a nonsingular circulant matrix C^N such that the preconditioned linear system

$$(C^N)^{-1}A^N\mathbf{x}^N = (C^N)^{-1}\mathbf{b}^N \tag{1.2}$$

can be solved by a small number of iterations. Within this family of preconditioning techniques we can distinguish those usually attributed to G. Strang [8], T. Chan [6], and E. Tyrtyshnikov [22] (see also [21]), characterized by different strategies to compute C^N.

For a review of fast direct methods for the solution of Toeplitz systems, see also [13, 14, 15].

We recall that for the solution of a semi-infinite Toeplitz system

$$A\mathbf{x} = \mathbf{b}, \qquad (A)_{ij} = a_{i-j},\ i,j = 0,1,2,\ldots, \tag{1.3}$$

there exists the classical Wiener-Hopf factorization method [9, 10]. Let us now assume that the symbol

$$\hat{A}(z) = \sum_{j\in\mathbb{Z}} a_j z^j, \quad |z| = 1, \tag{1.4}$$

does not vanish on the unit circle and its winding number is zero. Under this hypothesis, using a (canonical) Wiener-Hopf factorization of the inverse $\hat{A}(z)^{-1}$ of the symbol for semi-infinite Toeplitz matrix of Wiener class, i.e., for those satisfying the Wiener condition $\sum_{j\in\mathbb{Z}} |a_j| < \infty$, the solution of Eq. (1.3) can be given in terms of Krein's resolvent formula [5, 9, 16]. Various numerical techniques exist for computing a Wiener-Hopf factorization, in particular for symbols of positive definite semi-infinite Toeplitz matrices [1, 2, 3, 20, 23] (see also [12] where a comparative analysis of these techniques was made). Here we employ Krein's method (also called the cepstral method in signal processing [3, 20]) which combines FFT techniques with Krein's resolvent formula to solve Eq. (1.3) numerically.

In this article, assuming A^N can be extended to a semi-infinite Toeplitz matrix whose symbol does not vanish and has winding number zero, we propose a numerical method to solve large Toeplitz systems of the form (1.1). Embedding Eq. (1.1) into a semi-infinite Toeplitz system, solving the semi-infinite system by Krein's method and truncating its solution to its first N components, we obtain a first approximation of the solution of the original Eq. (1.1). The crucial observation is that for large enough N there exists a comparatively small positive integer n such that the first $N - n$ components of the solution vector have an acceptable accuracy whereas the remaining n components are to be recomputed by solving a finite Toeplitz system of order n. Since $n \ll N$, this can be done by any of the established techniques for solving Toeplitz systems. The numerical results obtained are quite satisfactory if compared with those obtained by the conjugate gradient method preconditioned by the techniques of Strang, Chan and Tyrtyshnikov.

Let us now discuss the contents of this paper. In Section 2 we shall explain the algorithm used to solve Eq. (1.1) and also analyze the pointwise error generated in the first step of the method in a simple case. Section 3 is devoted to the various test matrices of Toeplitz type used in the calculations and Section 4 to the numerical results. Throughout the paper $\|\cdot\|$ denotes the Euclidean vector norm $\|\mathbf{x}\| = \left[|x_0|^2 + \cdots + |x_{N-1}|^2\right]^{1/2}$ for $\mathbf{x} = (x_0, \ldots, x_{N-1})^T$.

2. The algorithm

Given the linear Toeplitz system (1.1), we embed it in the semi-infinite system

$$A^{\infty}\mathbf{x}^{\infty} = \mathbf{b}^{\infty}, \tag{2.1}$$

where

$$a_{ij}^{\infty} = \begin{cases} a_{i-j}^{N}, & |i-j| \leq N-1 \\ 0, & |i-j| \geq N, \end{cases}$$

and

$$b_i^{\infty} = \begin{cases} b_i^{N}, & i = 0, 1, \ldots, N-1 \\ 0, & i \geq N. \end{cases}$$

We assume that A^N and A^{∞} are invertible, the symbol $\hat{A}(z)$ does not vanish and its winding number is zero.

The method that we propose is based on the following two steps:

1. computation of the first N components of the solution $\mathbf{x}^{\infty}$ of the semi-infinite system by an algorithm whose computational complexity has order $N \log_2 N$;
2. correction of the components of $\mathbf{x}^{\infty}$ that eventually do not satisfy an *a posteriori* error criterion, by solving a small Toeplitz system.

Let us preliminarily analyze how large the pointwise error $|x_i^{\infty} - x_i^N|$, $i = 0, 1, \ldots, N-1$, is in a simple case. More precisely, let us consider the $N \times N$ tridiagonal real Toeplitz system

$$T^N \mathbf{y}^N = \mathbf{b}^N$$

given by

$$\begin{bmatrix} \alpha & \beta & & 0 \\ \beta & \ddots & \ddots & \\ & \ddots & \ddots & \beta \\ 0 & & \beta & \alpha \end{bmatrix} \begin{bmatrix} y_0 \\ \vdots \\ \vdots \\ y_{N-1} \end{bmatrix} = \begin{bmatrix} b_0 \\ \vdots \\ \vdots \\ b_{N-1} \end{bmatrix} \tag{2.2}$$

where $\alpha > 2|\beta| > 0$. Letting $\mathbf{b}^N = (b_i)_{i=0}^{N-1}$ for $\mathbf{b}^{\infty} = (b_i)_{i=0}^{\infty} \in l^2$ and letting $\mathbf{y}^{\infty} = (y_i)_{i=0}^{\infty}$ be the unique solution of the corresponding semi-infinite Toeplitz system

$$T^{\infty}\mathbf{y}^{\infty} = \mathbf{b}^{\infty},$$

the vector $(\mathbf{y}^{\infty})^N = (y_i^{\infty})_{i=0}^{N-1}$ is easily seen to satisfy

$$T^N(\mathbf{y}^{\infty})^N = \mathbf{b}^N - \beta\, y_N^{\infty}\mathbf{e}^N$$

where $\mathbf{e}_i^N = \delta_{i,N-1}$, $i = 0, 1, \ldots, N-1$ and $(\mathbf{y}^\infty)^N = (y_0^\infty, \ldots, y_{N-1}^\infty)^T$. Hence the solution $\mathbf{y}^N$ of (2.2) is given by

$$\mathbf{y}^N = (\mathbf{y}^\infty)^N + \beta\, y_N^\infty (\mathbf{x}^N)^\updownarrow ,$$

where $(\mathbf{x}^N)^\updownarrow = (x_{N-1-i}^N)_{i=0}^{N-1}$ satisfies $T^N(\mathbf{x}^N)^\updownarrow = \mathbf{e}^N$. Since the unique solution of $T^\infty \mathbf{x}^\infty = (\delta_{i,0})_{i=0}^\infty$ equals $\mathbf{x}^\infty = (\delta c^i)_{i=0}^\infty$ where $c = [-\alpha + \sqrt{\alpha^2 - 4\beta^2}]/2\beta$ and $\delta = 1/(\alpha + \beta c) = -c/\beta$, we easily find

$$\begin{aligned} y_i^N - y_i^\infty &= \beta y_N^\infty x_{N-1-i}^N \\ &= \frac{\beta\delta y_N^\infty}{1 - \beta^2\delta^2 c^{2N}} \left(c^{N-1-i} + \beta\delta c^{N+i}\right) \\ &= -\frac{y_N^\infty}{1 - c^{2N+2}} c^{N-i} \left(1 - c^{2i+2}\right), \end{aligned}$$

for $i = 0, 1, \ldots, N-1$.

It is well known that y_N^∞ is exponentially decaying as $N \to \infty$ whenever b_N is, in particular when $b_i = 0$ for i large enough. Since

$$\frac{y_{i+1}^N - y_{i+1}^\infty}{y_i^N - y_i^\infty} = \frac{1}{c} \frac{1 - c^{2i+4}}{1 - c^{2i+2}},$$

which exceeds 1 in absolute value, the deviation of y_i^N from the corresponding expression y_i^∞ for the semi-infinite system increases monotonically as i increases from 0 to $N-1$. As a result, the maximum absolute value of the error occurs in the last components of the vector, that is

$$|y_{N-1}^N - y_{N-1}^\infty| = \max_{i=0,\ldots,N-1} |y_i^N - y_i^\infty|.$$

Furthermore, as

$$\max_{i=0,\ldots,N-1} |y_i^N - y_i^\infty| \simeq |c\, y_N^\infty|,$$

the error goes to zero exponentially fast as $N \to \infty$.

Let us now illustrate the first part of the algorithm. It is well known [9, 10] that the invertibility of a semi-infinite Toeplitz matrix A^∞ is equivalent to the symbol (1.4), associated to the bi-infinite matrix $A_{ij} = a_{i-j}$, $i, j \in \mathbb{Z}$, having a (canonical) Wiener-Hopf factorization

$$\hat{A}(z) = \hat{A}_+(z)\hat{A}_-(z), \quad |z| = 1,$$

where $\hat{A}_+(z)$ and $\left(\hat{A}_+(z)\right)^{-1}$ are continuous for $|z| \leq 1$ and analytic for $|z| < 1$ and $\hat{A}_-(z)$ and $\left(\hat{A}_-(z)\right)^{-1}$ are continuous for $|z| \geq 1$ and analytic for $|z| > 1$ and at infinity. A Wiener-Hopf factorization of A exists, in particular, if A is positive definite.

Let

$$\left(\hat{A}(z)\right)^{-1} = \hat{\Gamma}_+(z)\hat{\Gamma}_-(z), \quad |z| = 1, \tag{2.3}$$

where

$$\hat{\Gamma}_+(z) = \sum_{j\in\mathbb{Z}_+} \Gamma_j^{(1)} z^j, \quad \hat{\Gamma}_-(z) = \sum_{j\in\mathbb{Z}_+} \Gamma_{-j}^{(2)} z^{-j},$$

are its Wiener-Hopf factors. Such a factorization exists if and only if A^∞ is invertible on l^2.

With this notation, we have the following resolvent formula [5, 9, 16] for the semi-infinite system (2.1)

$$x_i^\infty = \sum_{j\in\mathbb{Z}_+} \Gamma_{ij} b_j^\infty, \quad i \in \mathbb{Z}_+,$$

with

$$\Gamma_{ij} = \sum_{0\le h\le \min(i,j)} \Gamma_{i-h}^{(1)} \Gamma_{h-j}^{(2)}.$$

We note that $b_j^\infty = 0$ for $j \ge N$ and that, as the matrix A^∞ is banded, the coefficients $(A^\infty)_j^{-1}$ of its inverse decay exponentially as well as the elements $\Gamma_j^{(1)}$ and $\Gamma_{-j}^{(2)}$ of the Wiener-Hopf factors, as can be shown by using Gelfand theory and Sec. XXX.4(v) of [10].

According to Krein's method [5, 16], the Wiener-Hopf factorization (2.3) can be obtained by decomposing the Fourier series of the negative logarithm of $\hat{A}(z)$ additively in Fourier series in nonnegative and nonpositive powers of z and exponentiating the terms obtained. More precisely, we compute

$$-\log \hat{A}(z) = \gamma_+(z) + \gamma_-(z),$$

where

$$\gamma_+(z) = \frac{\gamma_0}{2} + \sum_{j=1}^{\infty} \gamma_j z^j,$$

$$\gamma_-(z) = \frac{\gamma_0}{2} + \sum_{j=1}^{\infty} \gamma_{-j} z^{-j},$$

and then expand in Fourier series the functions

$$\hat{\Gamma}_+(z) = e^{\gamma_+(z)} \quad \text{and} \quad \hat{\Gamma}_-(z) = e^{\gamma_-(z)}.$$

Numerical computations confirm that, as the matrix A^∞ is banded, the coefficients $\{\gamma_j\}$ and $\{\gamma_{-j}\}$ decay exponentially, within machine precision, as to be expected for theoretical reasons ([10], Sections XXX.1(v) and XXX.4(v)). This fact justifies the approximation of $\log \hat{A}(z)$ by a polynomial whose degree is comparatively small with respect to N. The typical decays of these coefficients are depicted in Figures 1 and 2, where the coefficients $\{\gamma_j\}$ and $\{\Gamma_j^{(1)}\}$ correspond to a positive definite matrix generated by a Gaussian decaying function (see Section 3). In these figures, as throughout in the paper, $\mu = \text{cond}(A^\infty)$, that is $\mu = \lim_{N\to\infty} \text{cond}(A^N)$ (see Section 3). Though the absolute values of $\{\gamma_j\}$ and $\{\gamma_{-j}\}$, as those of $\{\Gamma_j^{(1)}\}$ and $\{\Gamma_{-j}^{(2)}\}$, should decay exponentially fast, starting from an index $m \ll N$, for

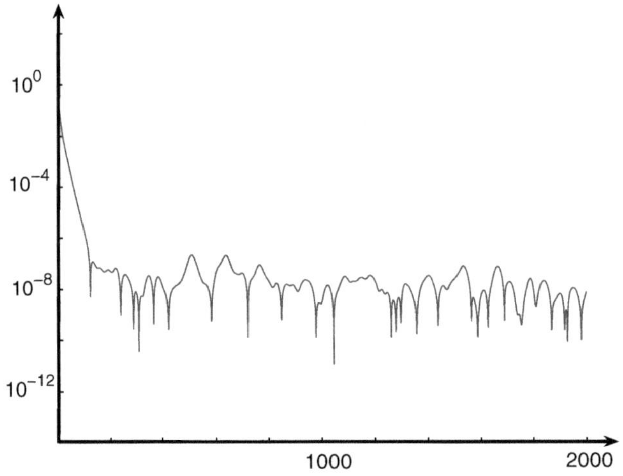

FIGURE 1. Decay of $\{|\gamma_j|\}$ (Gaussian decay, $\sigma = 0.17$, $\mu \simeq 2{\cdot}10^{12}$, $N = 2^{20}$)

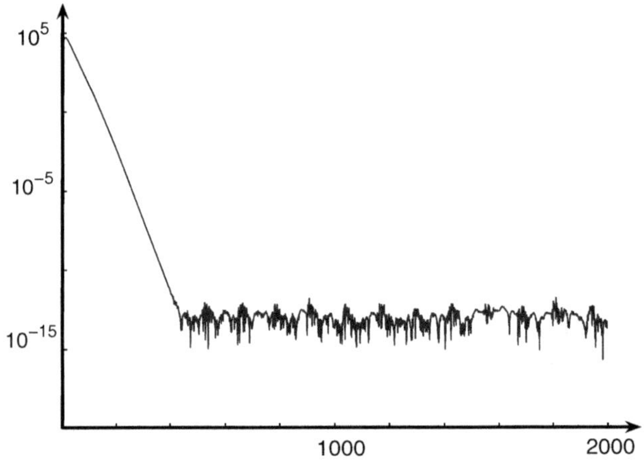

FIGURE 2. Decay of $\{|\Gamma_j^{(1)}|\}$ (Gaussian decay, $\sigma = 0.17$, $\mu \simeq 2 \cdot 10^{12}$, $N = 2^{20}$)

numerical reasons they become stagnant. This observation induced us to set to zero all coefficients with $j > m$, determining in this way the polynomials that approximate $\gamma_+(z)$ and $\gamma_-(z)$.

It is worthwhile to remark that an accurate approximation of the vector $(x_i^\infty)_{i=0}^{N-1}$ can be computed in *FFT*-time, since the Wiener-Hopf factors $\Gamma_+ =$

$(\Gamma^{(1)}_{i-j})$ and $\Gamma_- = (\Gamma^{(2)}_{i-j})$ of $(A^\infty)^{-1}$ are lower and upper semi-infinite Toeplitz matrices, respectively. More precisely, we need 2 FFT's for computing the polynomial approximation of the Fourier series of $\log \hat{A}(z)$, 2 for the coefficients $\{\gamma_j\}$ and $\{\gamma_{-j}\}$ and 2 more FFT's for the coefficients $\{\Gamma^{(1)}_j\}$ and $\{\Gamma^{(2)}_{-j}\}$. If the matrix A^∞ is symmetric positive definite we just need 4 FFT's, as $\gamma_{-j} = \gamma_j$, $j \in \mathbb{N}$, and $\Gamma^{(2)}_{-j} = \Gamma^{(1)}_j$, $j \in \mathbb{Z}_+$. The vector $(\mathbf{x}^\infty)^N$ of the first N components of $\mathbf{x}^\infty$ can then be computed as

$$(\mathbf{x}^\infty)^N = \Gamma^{(1)}\Gamma^{(2)}\mathbf{b}^N,$$

i.e., by 4 additional FFT's (3 in the positive definite case).

Let us now discuss the asymptotic behavior of the pointwise error as $N \to \infty$. To this end we consider a generalization to weighted spaces of well-known results on the projection method [4, 11].

More precisely, let us assume A^∞ is a semi-infinite Toeplitz matrix whose symbol $\hat{A}(z)$ does not vanish in an annulus about the unit circle and has winding number zero, and let $\mathbf{b}^\infty$ be a vector whose elements decay exponentially fast. Furthermore, let $A^N\mathbf{x}^N = \mathbf{b}^N$ be the system of order N with

$$\begin{aligned} a^N_{ij} &= a^\infty_{ij}\,, \qquad i,j = 0,1,\ldots,N-1 \qquad \text{and} \\ b^N_i &= b^\infty_i\,, \qquad i = 0,1,\ldots,N-1\,. \end{aligned}$$

We assume that A^N is non singular for each N value. Following [4, 11] it is then straightforward to prove

Theorem 2.1. *Assuming A^N, A^∞ and $\mathbf{b}^\infty$ as before specified, the finite Toeplitz system $A^N\mathbf{x}^N = \mathbf{b}^N$ has a unique solution for each N and, for fixed $\rho > 1$ such that $\hat{A}(z) \neq 0$ for $\frac{1}{\rho} \leq |z| \leq \rho$, the sequence*

$$\alpha_N(\rho) := \max_{i=0,1,\ldots,N-1} \rho^i |x^N_i - x^\infty_i| \to 0\,, \quad \text{as } N \to \infty\,.$$

A direct consequence is that $\rho^{N-1}|x^N_{N-1} - x^\infty_{N-1}|$ converges to zero as $N \to \infty$. Furthermore, our numerical results highlight that, as in the tridiagonal case, the error

$$\|\mathbf{x}_N - P_N\mathbf{x}^\infty\|_\infty := \max_{i=0,\ldots,N-1} |x^N_i - x^\infty_i|$$

decays exponentially fast. This property is illustrated in Figure 3, where $\|\mathbf{x}_N - P_N\mathbf{x}^\infty\|_\infty$ is depicted in log-log scale in the Gaussian case (see Section 3). A similar decay holds true if we consider both the exponential and the algebraic decaying matrices A^N considered in Section 3.

The second step of the algorithm is based on the following observation: our numerical experiments highlight that the pointwise error

$$|x^N_i - x^\infty_i|$$

is generally very small for the first $N - n$ components and large for the last n components, with n comparatively very small with respect to N. More precisely, for large enough N, the value of n does not depend on N, so that the larger N is,

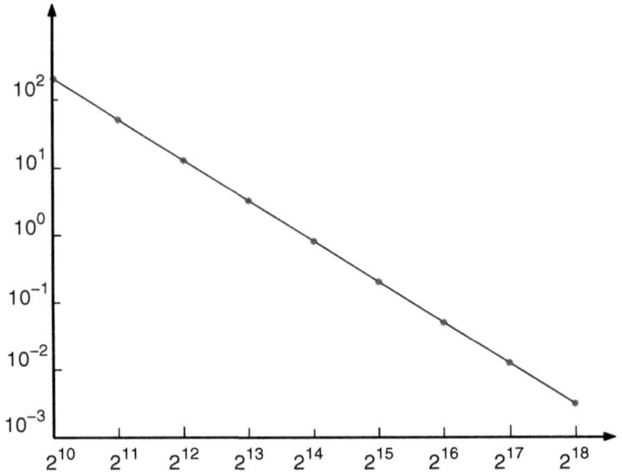

FIGURE 3. Decay of $\|\mathbf{x}_N - P_N \mathbf{x}^\infty\|_\infty$ (Gaussian decay, $\sigma = 0.2$, $\mu \simeq 3 \cdot 10^{10}$, $x_i^N = 1/(i+1)^2$)

the comparatively smaller is n, that is the number of components to be updated. This result is based on numerical computations for matrices whose elements have a Gaussian, exponential or algebraic decay (see Section 3). Unfortunately, we cannot explain this phenomenon for non tridiagonal Toeplitz systems. Such a typical situation is depicted in Figure 4.

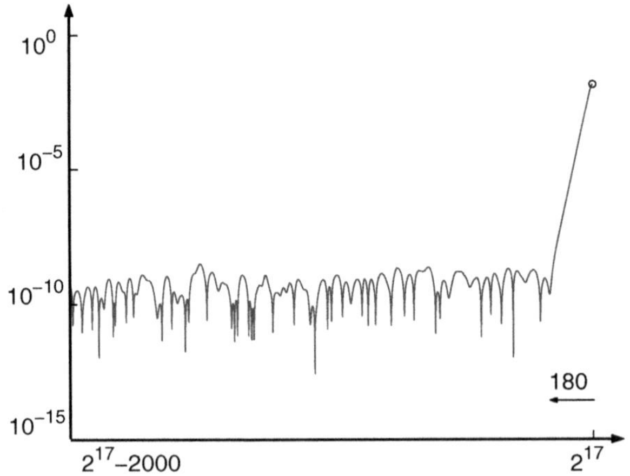

FIGURE 4. Pointwise errors on the last components of the solution (Gaussian decay, $\sigma = 0.2$, $\mu \simeq 3 \cdot 10^{10}$; $N = 2^{17}$, $x_i^N = 1/(i+1)^2$)

The observation of the above phenomenon suggested us to treat n as a *recovery parameter*, meaning that we consider the first $N - n$ components of $(\mathbf{x}^\infty)^N$

to be substantially correct and the last n to be recomputed. Fixing the value of this parameter, by a criterion we shall make precise later, we partition the finite system (1.1) as follows

$$\begin{bmatrix} A_{EE} & A_{EF} \\ A_{FE} & A_{FF} \end{bmatrix} \begin{bmatrix} \mathbf{x}_E \\ \mathbf{x}_F \end{bmatrix} = \begin{bmatrix} \mathbf{b}_E \\ \mathbf{b}_F \end{bmatrix},$$

where $\mathbf{x}_E = (x_0^N, x_1^N, \ldots, x_{N-n-1}^N)^T$, $\mathbf{b}_E = (b_0^N, b_1^N, \ldots, b_{N-n-1}^N)^T$, $\mathbf{x}_F = (x_{N-n}^N, x_{N-n+1}^N, \ldots, x_{N-1}^N)^T$ and $\mathbf{b}_F = (b_{N-n}^N, b_{N-n+1}^N, \ldots, b_{N-1}^N)^T$. Hence, if A_{FF} is nonsingular, we can improve the approximation coming from the infinite system (2.1) by setting $\mathbf{x}_E = (x_0^\infty, x_1^\infty, \ldots, x_{N-n-1}^\infty)^T$ and then taking $\mathbf{x}_F$ as the solution of the small Toeplitz system

$$A_{FF}\mathbf{x}_F = \mathbf{b}_F - A_{FE}\mathbf{x}_E. \tag{2.4}$$

This step takes 3 FFT's for computing the product $A_{FE}\mathbf{x}_E$, so that the total number of FFT's required by the algorithm is 13 (10 if the matrix is positive definite). Finally, we take the vector $\begin{bmatrix} \mathbf{x}_E \\ \mathbf{x}_F \end{bmatrix}$ as an acceptable approximation of $\mathbf{x}^N$.

To estimate the recovery parameter n we adopted the following heuristic criterion. Let $(\mathbf{x}^N)^{(n)}$ denote the approximation of the solution obtained by taking the first $N-n$ components of the solution of the semi-infinite system (2.1) and recomputing the last n entries by solving system (2.4). Then, choosing an increment k, we compare the last computed correction $\|(\mathbf{x}^N)^{(n+k)} - (\mathbf{x}^N)^{(n)}\|$ with the previous one. More precisely, fixing $0 < c_1 < c_2$, we look for the smallest value of n such that

$$c_1 < \frac{\|(\mathbf{x}^N)^{(n+k)} - (\mathbf{x}^N)^{(n)}\|}{\|(\mathbf{x}^N)^{(n)} - (\mathbf{x}^N)^{(n-k)}\|} < c_2. \tag{2.5}$$

Our numerical experiments suggest that $c_1 = 0.9$ and $c_2 = 1.1$ are suitable values for these two parameters. Furthermore, we found out that $n = 40$ and $k = 20$ are good choices for the recovery parameter and the increment. We iterate the correction process, by setting $n = n + k$ and correspondingly updating the solution until condition (2.5) is verified, and we take the last computed vector as the approximation of the solution.

Our experiments show that the ratio (2.5) oscillates for low values of n and stabilizes around 1, as shown in Figures 5 and 6. In these figures, the value of n selected by the adopted criterion is marked by a circle.

3. Test matrices

In this section we illustrate the method adopted to generate the matrices we used in our numerical experiments. This method, recently proposed in [18] and extended in various directions in [19], allowed us to generate several bi-infinite positive definite Toeplitz matrices, each characterized by a parameter, whose condition number is a known function of this parameter.

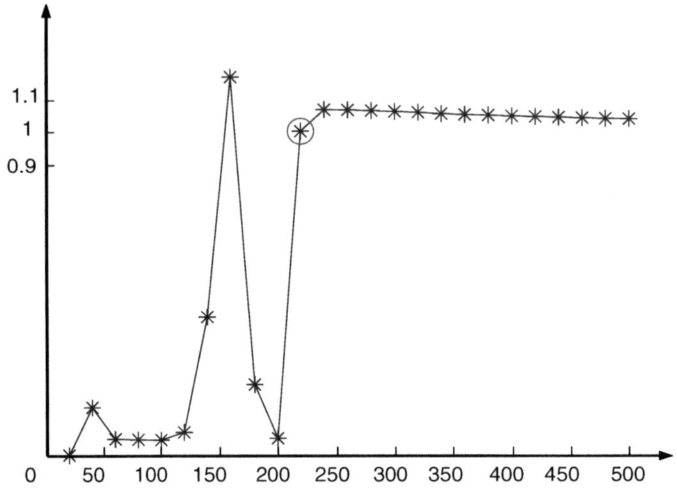

FIGURE 5. Ratio (2.5) vs. recovery parameter n (Algebraic decay, $\sigma = 5.5$, $\mu \simeq 3 \cdot 10^{12}$, $N = 2^{17}$, $x_i^N = \sin \frac{\pi(i+1)}{N+1}$)

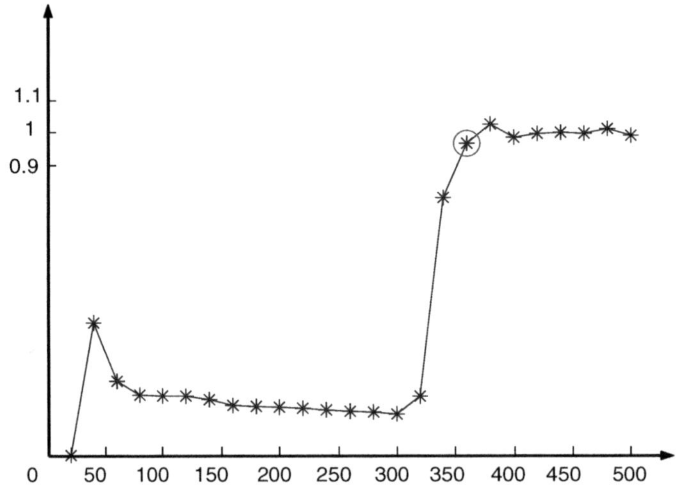

FIGURE 6. Ratio (2.5) vs. recovery parameter n (Gaussian decay, $\sigma = 0.17$, $\mu \simeq 2 \cdot 10^{12}$, $N = 2^{17}$, $x_i^N = 1$)

The following, basically well-known, result is also of great relevance in our numerical experiments.

Theorem 3.1. *If A^∞ is a bi-infinite positive definite Toeplitz matrix, then the condition number of the semi-infinite matrix A_+^∞, whose elements are $(A_+^\infty)_{ij} = a_{ij}$, $i, j \in \mathbb{Z}_+$, equals the condition number of A^∞.*

As a consequence, the condition number of a bi-infinite matrix turns out to be the limit of the condition numbers of finite projections of the corresponding semi-infinite matrix. Let us now introduce the method.

Let ϕ be a real function in $L^1(\mathbb{R}) \cap L^2(\mathbb{R})$ satisfying the two following properties:

(a) there exists a number $\gamma > 1$ such that

$$\int_{\mathbb{R}} (1 + |t|^2)^\gamma \, \phi(t)^2 \, dt < \infty;$$

(b) the Fourier transform $\hat{\phi}$ does not have real zeros.

Every function satisfying these two properties can be used to generate a bi-infinite positive matrix, as the following theorem claims.

Theorem 3.2. *Let the function ϕ satisfy the above properties and consider the sampling points $t_j = \alpha j$, $j \in \mathbb{Z}$, for some $\alpha > 0$. Then the Gram matrix with elements*

$$(A_\phi^\infty)_{ij} = k(t_i, t_j) = \int_{\mathbb{R}} \phi(t - t_i) \, \phi(t - t_j) \, dt, \quad i, j \in \mathbb{Z},$$

is a positive definite Toeplitz matrix which is bounded on $\ell^2(\mathbb{Z})$. Further, its symbol

$$\hat{A}_\phi(z; \alpha) = \sum_{j \in \mathbb{Z}} z^j \int_{\mathbb{R}} \phi(t) \, \phi(t + \alpha j) \, dt = \sum_{j \in \mathbb{Z}} z^j \kappa(\alpha j), \quad |z| = 1,$$

satisfies the Wiener condition

$$\sum_{j \in \mathbb{Z}} |\kappa(\alpha j)| < \infty,$$

and the condition number of the Toeplitz matrix equals

$$\frac{\max_{|z|=1} \hat{A}_\phi(z; \alpha)}{\min_{|z|=1} \hat{A}_\phi(z; \alpha)}.$$

The proof of this theorem can be found in [18, Theorems 3.1 and 3.2].

Let us now give some examples of bi-infinite positive definite Toeplitz matrices generated by functions having Gaussian, exponential and algebraic decay, respectively. We consider $\alpha = 1$, that is $t_i = i$, $i \in \mathbb{Z}$.

Gaussian decay. For $\phi_\sigma(t) = \mathrm{e}^{-\sigma t^2}$, $\sigma > 0$, the Gram matrix generated as above specified is the following positive definite Toeplitz matrix:

$$(A_\phi^\infty)_{ij} = \sqrt{\frac{\pi}{2\sigma}} \, \mathrm{e}^{-\frac{\sigma}{2}(i-j)^2}, \quad i, j \in \mathbb{Z},$$

whose corresponding symbol is

$$\hat{A}_\phi(z) = c_\sigma \prod_{j=1}^{\infty} \left[\left(1 + \mathrm{e}^{-(j-\frac{1}{2})\sigma} z\right) \left(1 + \mathrm{e}^{-(j-\frac{1}{2})\sigma} z^{-1}\right) \right],$$

where $z = \mathrm{e}^{\mathrm{i}\theta}$ and

$$c_\sigma = \sqrt{\frac{\pi}{2\sigma}} \prod_{j=1}^{\infty} \left(1 - \mathrm{e}^{-j\sigma}\right).$$

The condition number of A_ϕ^∞ is then

$$\mathrm{cond}(A_\phi^\infty) = \frac{\hat{A}_\phi(1)}{\hat{A}_\phi(-1)} = \left(\prod_{j=1}^{\infty} \frac{1 + \mathrm{e}^{-(j-\frac{1}{2})\sigma}}{1 - \mathrm{e}^{-(j-\frac{1}{2})\sigma}} \right)^2,$$

which is strictly decreasing from ∞ to 1 as σ goes from zero to ∞, so that for any chosen $\mu > 1$ there is a unique value of σ for which $\mathrm{cond}(A_\phi^\infty) = \mu$.

Exponential decay. Let $\phi_\sigma(t) = \mathrm{e}^{-\sigma|t|}$. Then we have

$$(A_\phi^\infty)_{ij} = \frac{1 + \sigma|i-j|}{\sigma} \mathrm{e}^{-\sigma|i-j|}, \qquad i, j \in \mathbb{Z}$$

and

$$\hat{A}_\phi(z) = \frac{p(\sigma) + q(\sigma)(z + z^{-1})}{(1 - z\mathrm{e}^{-\sigma})^2 (1 - z^{-1}\mathrm{e}^{-\sigma})^2},$$

where $z = \mathrm{e}^{\mathrm{i}\theta}$ and

$$\begin{cases} p(\sigma) = \frac{1}{\sigma}\left(1 - \mathrm{e}^{-4\sigma}\right) - 4\mathrm{e}^{-2\sigma} \\ q(\sigma) = \left(1 + \frac{1}{\sigma}\right)\mathrm{e}^{-3\sigma} + \left(1 - \frac{1}{\sigma}\right)\mathrm{e}^{-\sigma}. \end{cases}$$

The condition number is

$$\mathrm{cond}(A_\phi^\infty) = \frac{\hat{A}_\phi(1)}{\hat{A}_\phi(-1)} = \frac{p(\sigma) + 2q(\sigma)}{p(\sigma) - 2q(\sigma)} \left(\frac{1 + \mathrm{e}^{-\sigma}}{1 - \mathrm{e}^{-\sigma}}\right)^4.$$

As in the Gaussian case, $\mathrm{cond}(A_\phi^\infty)$ strictly decreases from ∞ to 1 as σ increases from zero to ∞.

Algebraic decay. Let $\phi_\sigma(t) = \frac{1}{(\sigma^2+t^2)^2}$ for $\sigma > 0$. Then

$$(A_\phi^\infty)_{ij} = \frac{\pi}{8\sigma^3} \left[\frac{1}{(4\sigma^2 + |i-j|^2)^2} + \frac{4\sigma^2}{(4\sigma^2 + |i-j|^2)^3} \right], \qquad i, j \in \mathbb{Z}$$

$$\hat{A}_\phi(\mathrm{e}^{\mathrm{i}\theta}) = \left(\frac{\pi}{8\sigma^3 \sinh(2\pi\sigma)} \right)^2 \left[F_2(\sigma, \theta) + \frac{1}{4\sinh(2\pi\sigma)} F_3(\sigma, \theta) \right]$$

and

$$\mathrm{cond}(A_\phi^\infty) = \frac{F_2(\sigma, 0) + \frac{F_3(\sigma,0)}{4\sinh(2\pi\sigma)}}{F_2(\sigma, \pi) + \frac{F_3(\sigma,\pi)}{4\sinh(2\pi\sigma)}},$$

where

$$\begin{aligned} F_2(\sigma,\theta) =& \cosh(2(\pi-\theta)\sigma)\sinh(2\pi\sigma) + 2\sigma\big[\pi\cosh(2\sigma\theta)+ \\ & \theta\sinh(2(\pi-\theta)\sigma)\sinh(2\pi\sigma)\big], \\ F_3(\sigma,\theta) =& \big[3-4\theta(\pi-\theta)\sigma^2\big]\cosh(2(\pi-\theta)\sigma)\sinh^2(2\pi\sigma) \\ & + 4\pi\sigma\cosh(2\sigma\theta)\sinh(2\pi\sigma) \\ & + 2(3\theta-\pi)\sigma\sinh(2(\pi-\theta)\sigma)\sinh^2(2\pi\sigma) \\ & + 2\pi\sigma\cosh(2(\pi-\theta)\sigma)\cosh(2\pi\sigma)\sinh(2\pi\sigma) \\ & + 8\pi^2\sigma^2\cosh(2\sigma\theta)\cosh(2\pi\sigma) \\ & + 4\pi^2\sigma^2\theta\sinh(2(\pi-\theta)\sigma)\cosh(2\pi\sigma)\sinh(2\pi\sigma) \\ & - 4\pi\sigma^2\theta\sinh(2\sigma\theta)\sinh(2\pi\sigma). \end{aligned}$$

In this case, $\mathrm{cond}(A_\phi^\infty)$ strictly increases from 1 to ∞ as σ increases from zero to ∞.

4. Numerical results

In order to assess the effectiveness of the method proposed, we carried out an extensive experimentation by using matrices generated by the truncation of the semi-infinite matrices introduced in the previous section. In every case $\mathrm{cond}(A_\phi^\infty)$ is the upper bound of the condition numbers of these truncations. Further, we note that our experiments concern matrices whose entries exhibit a Gaussian, exponential or algebraic decay, away from the main diagonal.

In our numerical experiments, for each test matrix A_ϕ^N, we consider a set of sample solutions $\mathbf{x}^N$ and generate the corresponding data vector $\mathbf{b}^N = A_\phi^N \mathbf{x}^N$. The error values quoted in each table are the relative errors

$$E_r(N) = \frac{\|\mathbf{x}^N - (\mathbf{x}^N)^{(n)}\|}{\|\mathbf{x}^N\|},$$

where n is the recovery parameter, chosen by the heuristic criterion described in Section 2.

The results quoted in Tables 1–6 have been computed by our embedding method (EM) and by the preconditioned conjugate gradient (PCG) method with the Chan (PCG/Chan) [6], Strang (PCG/Strang) [8] and Tyrtyshnikov (PCG/Tyrt) [22] preconditioners. The solutions pertaining the PCG method have been obtained by performing 100 iterations.

The computations were performed in double precision with Matlab vers. 6.5 [17] running under Linux on an AMD Athlon 64 3200+ processor, with 1.5 Gbyte RAM. We remark that while the total computing time for solving a system of order 2^{17} by our method is $\approx$ 10 sec., 100 iterations of the preconditioned conjugate gradient method usually take $\approx$ 110 sec. In either algorithm all Toeplitz matrix products and inversions of circulant preconditioners have been implemented, as usual, by means of the Fast Fourier Transform (FFT) in order to optimize their computational complexity.

In each table, as throughout the paper, σ denotes the parameter that identifies the function ϕ and the corresponding matrix A_ϕ^∞, and μ is the upper bound of $\text{cond}(A_\phi^N)$.

TABLE 1. ALGEBRAIC DECAY ($N = 2^{20}$; $\sigma = 4.5$, $\mu \simeq 6.6 \cdot 10^9$)

	EM		E_r for PCG		
x_i^N	n	E_r	Chan	Strang	Tyrt
1	340	$3.7 \cdot 10^{-3}$	$1.2 \cdot 10^{-3}$	$3.2 \cdot 10^{-7}$	$4.5 \cdot 10^{-5}$
$\sin \frac{\pi(i+1)}{N+1}$	200	$5.4 \cdot 10^{-7}$	$2.1 \cdot 10^{-7}$	$2.2 \cdot 10^{-7}$	$1.1 \cdot 10^{-9}$
$1/(i+1)^2$	320	$9.6 \cdot 10^{-8}$	$2.6 \cdot 10^{-2}$	$7.5 \cdot 10^{-8}$	$6.1 \cdot 10^{-1}$
$(-1)^{(i+1)}/(i+1)$	340	$2.5 \cdot 10^{-7}$	$5.2 \cdot 10^{-3}$	$3.0 \cdot 10^{-8}$	$1.0 \cdot 10^{0}$
$(1-10^{-4})^{(i+1)}$	40	$3.0 \cdot 10^{-5}$	$8.3 \cdot 10^{-3}$	$3.0 \cdot 10^{-7}$	$5.5 \cdot 10^{-3}$
$(1-10^{-1})^{(i+1)}$	60	$2.2 \cdot 10^{-7}$	$1.3 \cdot 10^{-1}$	$2.0 \cdot 10^{-7}$	$1.8 \cdot 10^{-1}$

TABLE 2. ALGEBRAIC DECAY ($N = 2^{20}$; $\sigma = 5$, $\mu \simeq 10^{11}$)

	EM		E_r for PCG		
x_i^N	n	E_r	Chan	Strang	Tyrt
1	240	$3.9 \cdot 10^{-3}$	$2.2 \cdot 10^{-2}$	$4.6 \cdot 10^{-6}$	$4.9 \cdot 10^{-5}$
$\sin \frac{\pi(i+1)}{N+1}$	200	$9.3 \cdot 10^{-5}$	$1.2 \cdot 10^{-6}$	$4.7 \cdot 10^{-6}$	$1.4 \cdot 10^{-9}$
$1/(i+1)^2$	160	$6.8 \cdot 10^{-5}$	$4.4 \cdot 10^{-1}$	$1.4 \cdot 10^{-6}$	$7.1 \cdot 10^{-1}$
$(-1)^{(i+1)}/(i+1)$	300	$3.4 \cdot 10^{-6}$	$2.1 \cdot 10^{-1}$	$4.8 \cdot 10^{-7}$	$1.1 \cdot 10^{0}$
$(1-10^{-4})^{(i+1)}$	200	$2.7 \cdot 10^{-2}$	$2.1 \cdot 10^{-1}$	$4.9 \cdot 10^{-6}$	$6.5 \cdot 10^{-3}$
$(1-10^{-1})^{(i+1)}$	120	$1.9 \cdot 10^{-5}$	$2.9 \cdot 10^{0}$	$3.5 \cdot 10^{-6}$	$2.1 \cdot 10^{-1}$

TABLE 3. EXPONENTIAL DECAY ($N = 2^{20}$; $\sigma = 0.025$, $\mu \simeq 10^8$)

	EM		E_r for PCG		
x_i^N	n	E_r	Chan	Strang	Tyrt
1	240	$7.5 \cdot 10^{-6}$	$8.6 \cdot 10^{-9}$	$8.0 \cdot 10^{-9}$	$1.3 \cdot 10^{-4}$
$\sin \frac{\pi(i+1)}{N+1}$	120	$7.6 \cdot 10^{-8}$	$9.5 \cdot 10^{-9}$	$9.1 \cdot 10^{-9}$	$4.0 \cdot 10^{-8}$
$1/(i+1)^2$	80	$6.7 \cdot 10^{-9}$	$9.3 \cdot 10^{-10}$	$9.2 \cdot 10^{-10}$	$6.9 \cdot 10^{-1}$
$(-1)^{(i+1)}/(i+1)$	100	$2.6 \cdot 10^{-9}$	$3.4 \cdot 10^{-10}$	$3.3 \cdot 10^{-10}$	$9.6 \cdot 10^{-1}$
$(1-10^{-4})^{(i+1)}$	140	$1.7 \cdot 10^{-6}$	$1.0 \cdot 10^{-8}$	$1.0 \cdot 10^{-8}$	$6.0 \cdot 10^{-3}$
$(1-10^{-1})^{(i+1)}$	180	$1.7 \cdot 10^{-8}$	$2.6 \cdot 10^{-9}$	$2.7 \cdot 10^{-9}$	$2.0 \cdot 10^{-1}$

The numerical results shown in each table have been obtained by considering the matrices introduced in Section 3, for different values of the parameter σ on

TABLE 4. EXPONENTIAL DECAY ($N = 2^{20}$; $\sigma = 0.005$, $\mu \simeq 8 \cdot 10^{10}$)

x_i^N	EM n	EM E_r	E_r for PCG Chan	E_r for PCG Strang	E_r for PCG Tyrt
1	100	$1.0 \cdot 10^{0}$	$7.7 \cdot 10^{-6}$	$5.4 \cdot 10^{-6}$	$1.9 \cdot 10^{-4}$
$\sin \frac{\pi(i+1)}{N+1}$	100	$1.3 \cdot 10^{-3}$	$6.3 \cdot 10^{-6}$	$6.3 \cdot 10^{-6}$	$2.7 \cdot 10^{-7}$
$1/(i+1)^2$	60	$4.0 \cdot 10^{-7}$	$2.8 \cdot 10^{-7}$	$2.7 \cdot 10^{-7}$	$9.6 \cdot 10^{-1}$
$(-1)^{(i+1)}/(i+1)$	100	$1.1 \cdot 10^{-6}$	$2.3 \cdot 10^{-7}$	$9.5 \cdot 10^{-8}$	$1.0 \cdot 10^{0}$
$(1-10^{-4})^{(i+1)}$	40	$1.8 \cdot 10^{-3}$	$2.4 \cdot 10^{-5}$	$1.0 \cdot 10^{-6}$	$1.6 \cdot 10^{-2}$
$(1-10^{-1})^{(i+1)}$	260	$1.0 \cdot 10^{-6}$	$1.0 \cdot 10^{-6}$	$8.0 \cdot 10^{-7}$	$5.6 \cdot 10^{-1}$

TABLE 5. GAUSSIAN DECAY ($N = 2^{20}$; $\sigma = 0.2$, $\mu \simeq 3 \cdot 10^{10}$)

x_i^N	EM n	EM E_r	E_r for PCG Chan	E_r for PCG Strang	E_r for PCG Tyrt
1	360	$1.1 \cdot 10^{-6}$	$7.9 \cdot 10^{-3}$	$8.5 \cdot 10^{-7}$	$3.1 \cdot 10^{-5}$
$\sin \frac{\pi(i+1)}{N+1}$	280	$8.6 \cdot 10^{-7}$	$3.3 \cdot 10^{-7}$	$7.5 \cdot 10^{-7}$	$3.6 \cdot 10^{-10}$
$1/(i+1)^2$	320	$4.2 \cdot 10^{-7}$	$5.9 \cdot 10^{-1}$	$3.1 \cdot 10^{-7}$	$5.1 \cdot 10^{-1}$
$(-1)^{(i+1)}/(i+1)$	380	$4.8 \cdot 10^{-7}$	$1.4 \cdot 10^{-1}$	$1.2 \cdot 10^{-7}$	$1.0 \cdot 10^{0}$
$(1-10^{-4})^{(i+1)}$	320	$1.4 \cdot 10^{-6}$	$7.2 \cdot 10^{-2}$	$1.1 \cdot 10^{-6}$	$4.5 \cdot 10^{-3}$
$(1-10^{-1})^{(i+1)}$	120	$6.1 \cdot 10^{-7}$	$1.5 \cdot 10^{0}$	$6.2 \cdot 10^{-7}$	$1.5 \cdot 10^{-1}$

TABLE 6. GAUSSIAN DECAY ($N = 2^{20}$; $\sigma = 0.17$, $\mu \simeq 2 \cdot 10^{12}$)

x_i^N	EM n	EM E_r	E_r for PCG Chan	E_r for PCG Strang	E_r for PCG Tyrt
1	360	$1.6 \cdot 10^{-4}$	$1.9 \cdot 10^{-2}$	$4.8 \cdot 10^{-5}$	$3.3 \cdot 10^{-5}$
$\sin \frac{\pi(i+1)}{N+1}$	280	$5.5 \cdot 10^{-5}$	$6.4 \cdot 10^{-7}$	$6.0 \cdot 10^{-5}$	$4.4 \cdot 10^{-10}$
$1/(i+1)^2$	240	$3.1 \cdot 10^{-5}$	$1.2 \cdot 10^{0}$	$3.6 \cdot 10^{-4}$	$5.9 \cdot 10^{-1}$
$(-1)^{(i+1)}/(i+1)$	320	$3.5 \cdot 10^{-5}$	$2.6 \cdot 10^{-1}$	$8.2 \cdot 10^{-6}$	$1.0 \cdot 10^{0}$
$(1-10^{-4})^{(i+1)}$	380	$8.4 \cdot 10^{-4}$	$2.0 \cdot 10^{-1}$	$8.7 \cdot 10^{-5}$	$5.3 \cdot 10^{-3}$
$(1-10^{-1})^{(i+1)}$	260	$5.0 \cdot 10^{-5}$	$4.2 \cdot 10^{0}$	$9.9 \cdot 10^{-5}$	$1.7 \cdot 10^{-1}$

which they depend. For each matrix, the following sample solutions have been considered:

$$x_i^N = 1, \qquad x_i^N = \sin \frac{\pi(i+1)}{N+1},$$

$$x_i^N = \frac{1}{(i+1)^2}, \qquad x_i^N = \frac{(-1)^{i+1}}{(i+1)},$$

$$x_i^N = (1-10^{-4})^{i+1}, \qquad x_i^N = (1-10^{-1})^{i+1},$$

with $i = 0, 1, \ldots, N-1$.

In each table we have reported the relative error $E_r(N)$, with respect to the variation of the dimension N of the linear system, together with the value of the parameter σ and the asymptotic condition number μ of the corresponding matrix.

For the sake of clarity, we make some specific comments pertaining to each class of matrices.

1. **Algebraic decay.** In this case both our method and the Strang preconditioning technique give reasonable accuracy, also for large values of the condition number of the matrix (Tables 1 and 2). The performance of the Chan preconditioner is not as good, especially in Table 2. Furthermore, it totally fails in one case, as does the Tyrtyshnikov preconditioner. On the whole, both these two preconditioners produce a limited performance on this test problem.
2. **Exponential decay.** If the condition number is moderately large, $\mu \simeq 10^8$ say, our results, as well as those obtained with the Strang and Chan preconditioners are very good whereas those generated by the Tyrtyshnikov preconditioner are not always as good (Table 3). We note that to reach a good accuracy in Table 3 by using the Tyrtyshnikov preconditioner we need about 1000 iterations of the PCG method, which take ≈ 200 times the computation time of our method.

   If the condition number is $\approx 10^{10}$ our method fails in the first example and gives good results in all the other examples, while we have very good results using both the Strang and Chan preconditioners (Table 4). The Tyrtyshnikov preconditioner fails in one case and gives poor results in the other two cases. In these cases we need more than 1000 iterations to obtain acceptable results. In particular, the relative errors decrease to $\approx 10^{-3}$ in the last example when 5000 iterations are considered.
3. **Gaussian decay.** All of the methods are quite effective if the condition number is moderately large, $\mu \leq 10^8$ say, though our method is very effective and the Strang and Chan preconditioners perform better than the Tyrtyshnikov preconditioner (Table 5).

   If the condition number is much larger (Table 6) our method still gives very good results as does the Strang preconditioner, whereas the Chan and Tyrtyshnikov preconditioners sometimes fail or give poor results. When they fail, we did not obtain acceptable results even when considering 10^4 iterations.

The results shown in Tables 1–6 may suggest that the performance of the Tyrtyshnikov preconditioner is always unsatisfactory. However, our numerical experiments show that there are situations in which the Tyrtyshnikov preconditioner proves to be compatible with the other two preconditioning techniques on the test problems considered. This holds true, in particular, when the dimension of the linear system is considerably smaller than in the previous tables; a typical example is displayed in Table 7, where $N = 2^{10}$. We feel that the poorer performance of this preconditioner for larger dimensions is due to rounding errors propagation, caused by the larger complexity of the algorithm for its computation with respect to Strang and Chan preconditioners.

TABLE 7. EXPONENTIAL DECAY ($N = 2^{10}$; $\sigma = 0.005$, $\mu \approx 7.6 \cdot 10^{10}$)

x_i^N	E_r for PCG Chan	Strang	Tyrt
1	$6.1 \cdot 10^{-2}$	$1.1 \cdot 10^{0}$	$4.0 \cdot 10^{-5}$
$\sin \frac{\pi(i+1)}{N+1}$	$6.4 \cdot 10^{-2}$	$1.0 \cdot 10^{0}$	$1.8 \cdot 10^{-5}$
$1/(i+1)^2$	$7.8 \cdot 10^{-2}$	$7.2 \cdot 10^{-1}$	$6.7 \cdot 10^{-1}$
$(-1)^{(i+1)}/(i+1)$	$3.7 \cdot 10^{-1}$	$9.6 \cdot 10^{-1}$	$9.4 \cdot 10^{-1}$
$(1-10^{-4})^{(i+1)}$	$7.7 \cdot 10^{-2}$	$1.4 \cdot 10^{0}$	$6.4 \cdot 10^{-5}$
$(1-10^{-1})^{(i+1)}$	$1.0 \cdot 10^{-1}$	$1.5 \cdot 10^{-1}$	$6.0 \cdot 10^{-2}$

Conclusions

Our numerical results show that our method is reliable if the order of the system is large enough. Furthermore, we generally obtain good results also for moderately ill-conditioned matrices, though the computational effort of our method is lower with respect to preconditioned conjugate gradient method.

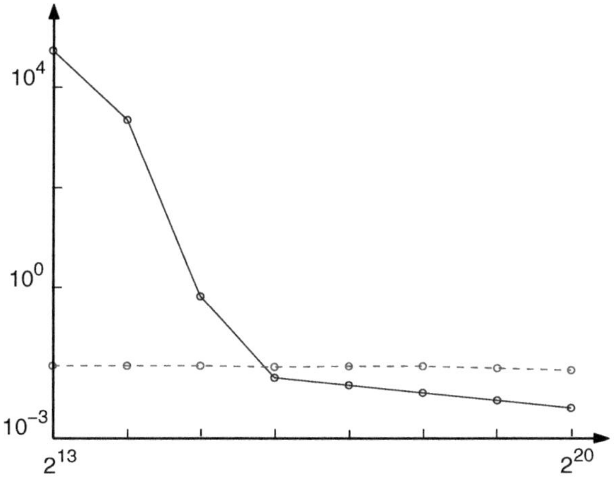

FIGURE 7. Relative error $E_r(N)$ (EM (solid line), PCG/Chan (dashed line); Algebraic decay $\sigma = 5$, $\mu \simeq 10^{11}$, $x_i^N = 1$)

As our method is generated by the resolvent formula for semi-infinite systems, the results improve as N increases, whereas this conclusion certainly does not hold for the iterative methods, even when preconditioned. More precisely, in our method $E_r(N)$ decreases as N increases, whereas this does not always happen for the other methods we tested. In order to give a geometrical idea of this fact, in Figure 7 we plotted $E_r(N)$ for $2^{13} \leq N \leq 2^{20}$, obtained applying our method (solid line) and the PCG/Chan method (dashed line) to one of the test problems considered (see Section 3).

In conclusion, for large Toeplitz systems our method is compatible with the best PCG methods concerning accuracy and superior concerning the computation time.

Acknowledgments
The authors are greatly indebted to their friend C.V.M. van der Mee for his help and advice in writing the paper. They also express their gratitude to the referees for their useful comments that allowed us to improve the paper.

References

[1] F.L. Bauer. Ein direktes Iterations Verfahren zur Hurwitz-Zerlegung eines Polynoms. *Arch. Elektr. Übertragung*, 9:285–290, 1955.

[2] F.L. Bauer. Beiträge zur Entwicklung numerischer Verfahren für programmgesteuerte Rechenanlagen, ii. Direkte Faktorisierung eines Polynoms. *Sitz. Ber. Bayer. Akad. Wiss.*, pp. 163–203, 1956.

[3] B.P. Bogert, M.J.R. Healy, and J.W. Tukey. The frequency analysis of time series for echoes: cepstrum pseudo-autocovariance, cross-cepstrum and saphe cracking. In: M. Rosenblatt (ed.), *Proc. Symposium Time Series Analysis*, John Wiley, New York, 1963, pp. 209–243.

[4] A. Böttcher and B. Silbermann. Operator-valued Szegő-Widom theorems. In: E.L. Basor and I. Gohberg (Eds.), *Toeplitz Operators and Related Topics*, volume 71 of *Operator Theory: Advances and Applications*. Birkhäuser, Basel-Boston, 1994, pp. 33–53.

[5] A. Calderón, F. Spitzer, and H. Widom. Inversion of Toeplitz matrices. *Illinois J. Math.*, 3:490–498, 1959.

[6] T.F. Chan. An optimal circulant preconditioner for Toeplitz systems. *SIAM J. Sci. Stat. Comput.*, 9(4):766–771, 1988.

[7] R.H. Chan and M.K. Ng. Conjugate Gradient Methods for Toeplitz Systems. *SIAM Review*, 38(3):297–386, 1996.

[8] R. Chan and G. Strang. Toeplitz equations by conjugate gradients with circulant preconditioner. *SIAM J. Sci. Stat. Comput.*, 10(1):104–119, 1989.

[9] I.C. Gohberg and I.A. Feldman. *Convolution Equations and Projection Methods for their Solution*, volume 41 of *Transl. Math. Monographs*. Amer. Math. Soc., Providence, RI, 1974.

[10] I. Gohberg, S. Goldberg, and M.A. Kaashoek. *Classes of Linear Operators, Vol. II*, volume 63 of *Operator Theory: Advances and Applications*. Birkhäuser, Basel-Boston, 1993.

[11] I. Gohberg and M.A. Kaashoek. Projection method for block Toeplitz operators with operator-valued symbols. In: E.L. Basor and I. Gohberg (Eds.), *Toeplitz Operators and Related Topics*, volume 71 of *Operator Theory: Advances and Applications*. Birkhäuser, Basel-Boston, 1994, pp. 79–104.

[12] T.N.T. Goodman, C.A. Micchelli, G. Rodriguez, and S. Seatzu. Spectral factorization of Laurent polynomials. *Adv. Comput. Math.*, 7:429–454, 1997.

[13] G. Heinig and K. Rost, *Algebraic Methods for Toeplitz-like Matrices and Operators*, volume 13 of *Operator Theory: Advances and Applications.* Birkhäuser, Basel-Boston, 1984.

[14] T. Kailath and A.H. Sayed. Displacement structure: theory and applications. *SIAM Review*, 37(3):297–386, 1996.

[15] T. Kailath and A.H. Sayed. Fast reliable algorithms for matrices with structure. SIAM, Philadelphia, PA, 1999.

[16] M.G. Krein. Integral equations on the half-line with kernel depending upon the difference of the arguments. *Uspehi Mat. Nauk.*, 13(5):3–120, 1958. (Russian, translated in *AMS Translations*, 22:163–288, 1962).

[17] The MathWorks Inc. *Matlab ver. 6.5.* Natick, MA, 2002.

[18] C.V.M. van der Mee, M.Z. Nashed, and S. Seatzu. Sampling expansions and interpolation in unitarily translation invariant reproducing kernel Hilbert spaces. *Adv. Comput. Math.*, 19(4):355–372, 2003.

[19] C.V.M. van der Mee and S. Seatzu. A method for generating infinite positive definite self-adjoint testmatrices and Riesz bases. *SIAM J. Matrix Anal. Appl.* (to appear).

[20] A.V. Oppenheim and R.W. Schafer. *Discrete-Time Signal Processing.* Prentice Hall Signal Processing Series. Prentice Hall, Englewood Cliffs, NJ, 1989.

[21] M. Tismenetsky. A Decomposition of Toeplitz Matrices and Optimal Circulant Preconditioning. Linear Algebra Appl., 154–156:105–121, 1991.

[22] E.E. Tyrtyshnikov. Optimal and superoptimal circulant preconditioners. *SIAM J. Matrix Anal. Appl.*, 13:459–473, 1992.

[23] G. Wilson. Factorization of the covariance generating function of a pure moving average process. *SIAM J. Numer. Anal.*, 6(1):1–7, 1969.

G. Rodriguez, S. Seatzu and D. Theis
Dipartimento di Matematica e Informatica
Università degli Studi di Cagliari
viale Merello 92
I-09123 Cagliari, Italy
e-mail: `rodriguez@unica.it, seatzu@unica.it, theis@bugs.unica.it`

Operator Theory:
Advances and Applications, Vol. 160, 403–411

Fredholm Theory and Numerical Linear Algebra

Bernd Silbermann

Dedicated to Israel Gohberg on the occasion of his seventyfifth birthday

Abstract. It is shown that for every bounded linear operator A on a separable Hilbert space, the finite sections of A^*A reflect perfectly the Fredholm properties of this operator. A few applications are briefly discussed.

Mathematics Subject Classification (2000). 47B35.

1. Introduction

Given a Hilbert space H we denote by $\mathcal{B}(H)$ the C^*-algebra of all bounded linear operators acting on H and let $\mathcal{K}(H) \subset \mathcal{B}(H)$ be the ideal of all compact operators. The strong limit of a strongly converging sequence $(A_n) \subset \mathcal{B}(H)$ will be denoted by s-lim A_n. We shall use without further comments, that strong convergence on compact subsets is uniform and that strongly convergent sequences $(A_n) \subset \mathcal{B}(H)$ are uniformly bounded (Banach-Steinhaus Theorem). Suppose we are given a sequence $(P_n) \subset \mathcal{B}(H)$ of finite rank orthogonal projections such that (P_n) converges strongly to the identity operator I (thus, H is separable, and for every separable Hilbert space there exist such sequences) and consider the sequence (P_nAP_n) where $A \subset \mathcal{B}(H)$ is fixed and P_nAP_n : im $P_n \to$ im P_n. Clearly, one can identify P_nAP_n with an $l_n \times l_n$-matrix, where $l_n =$ dim im P_n. The paper is organized as follows.

Section 2 is devoted to the study of the asymptotic behavior of the singular values associated to the operators (matrices) P_nAP_n. It will be shown that for so-called Fredholm sequences the singular values have remarkable behavior: they are subject to the finite splitting property. The result of Section 2 will then be used in Section 3 to show that the Fredholm properties of an operator $A \subset \mathcal{B}(H)$ are perfectly reflected in the asymptotic behavior of the sequences $(P_nA^*AP_n)$ and $(P_nAA^*P_n)$. A few examples are mentioned in Section 4.

2. The finite splitting property

Let $H, (P_n)$ be as in the Introduction. Form the collection $\mathcal{E}$ of all bounded sequences (A_n) with $A_n : \text{ im } P_n \to \text{ im } P_n$ where both sequences $(A_n P_n)$, $(A_n^* P_n)$ converge strongly (this implies $(s\text{-lim } A_n P_n)^* = s\text{-lim } A_n^* P_n$). By defining the algebraic operations and the involution componentwise, and the norm by

$$||(A_n)|| := \sup ||A_n P_n|| ,$$

$\mathcal{E}$ actually becomes a unital C^*-algebra containing the closed and two-sided ideals

$$\begin{aligned} G &:= \{(A_n) \in \mathcal{E} : ||A_n P_n|| \to 0 \text{ as } u \to \infty\} , \\ J &:= \{(A_n) \in \mathcal{E} : A_n = P_n k P_n + C_n , \, k \in \mathcal{K}(H) , \, (C_n) \in G\} . \end{aligned}$$

Definition 1. A sequence $(A_n) \in \mathcal{E}$ is said to be Fredholm if the coset $(A_n)+J$ is invertible in $\mathcal{E}/J$. This definition is justified by the circumstance that the Fredholmness of (A_n) implies the Fredholmness of s-lim $A_n P_n$. Moreover, the Fredholmness of a sequence is stable under perturbations which are small or belong to J.

Notice that there are more general and involved notions of Fredholm sequences (see [7], Chapter 6). Theorem 1 below indicates that Fredholm sequences in the sense of Definition 1 are also Fredholm in more general contents. The reverse is however not true. An instructive example will be presented later on. For technical reasons we need

Definition 2. A sequence $(A_n) \in \mathcal{E}$ is called stable if the operators A_n are invertible for n large enough, say for $n \geq n_0$, and if $\sup\limits_{n \geq n_0} ||A_n^{-1} P_n|| < \infty$.

Notice that $(A_n) \in \mathcal{E}$ is stable if and only if the coset $(A_n)+G$ is invertible in $\mathcal{E}/G$. If $(A_n) \in \mathcal{E}$ is stable, then $s\text{-}\lim\limits_{n\to\infty} A_n P_n =: A$ is invertible and $s\text{-}\lim\limits_{k\to\infty} A_n^{-1} P_n$ exists and equals A^{-1}. The following proposition is well known (see [7], Theorem 1.20).

Proposition 1. If $(A_n) \in \mathcal{E}$ is Fredholm and s-lim $A_n P_n$ is invertible then (A_n) is stable.

Let $(A_n) \in \mathcal{E}$ be arbitrary. We order the singular values of A_n as follows:

$$0 \leq s_1(A_n) \leq s_2(A_n) \leq \cdots \leq s_{l_n}(A_n) .$$

For the sake of convenience, let us also put $s_0(A_n) = 0$

Definition 3. We say that $(A_n) \in \mathcal{E}$ owns the finite splitting property (k-splitting property) if there is a k such that

$$\lim_{n\to\infty} s_k(A_n) = 0 ,$$

while the remaining $l_n - k$ singular values stay away from zero, that is

$$s_{k+1}(A_n) \geq \delta > 0$$

for n large enough. k is also called the splitting number.

Remark 1. If $(A_n) \in \mathcal{E}$ is subject to the finite splitting property and $k = 0$ then (A_n) is stable.

The following theorem is probably new; for so-called standard models related results were proved in [11] (see also [7]).

Theorem 1. Let $(A_n) \in \mathcal{E}$ be Fredholm. Then (A_n) is subject to the finite splitting property, and the splitting number k equals

$$k = \dim\ker\left(s - \lim A_n P_n\right) .$$

Proof. We shall make use of the following alternative description of the singular values (see [6], Theorem 2.1):

$$s_j(A_n) := \min_{\mathcal{B}\in\mathcal{F}^{l_n}_{l_n-j}} ||A_n - B|| ,$$

where $\mathcal{F}^{l_n}_m$ denotes the collection of all $l_n \times l_n$-matrices of rank at most m. Let R_n be the orthoprojection onto im $P_n P_{\ker A} P_n$. From Lemma 6.21 in [7] and its proof it follows that

$$\text{im } R_n = \text{ im } P_n P_{\ker A} P_n ,$$

rank $R_n =$ rank $P_n P_{\ker A} P_n =$ rank $P_{\ker A} = \dim\ker A =: k$ for n large enough, and

$$||R_n - P_n P_{\ker A} P_n|| \to 0 \text{ as } n \to \infty .$$

Consequently, $||A_n R_n|| \to 0$ as $n \to \infty$, and $(A_n R_n) \in G$. Consider the sequence $(B_n) \in \mathcal{E}$, $B_n := A_n^* A_n (P_n - R_n) + P_n P_{\ker A} P_n$. Obviously, this sequence is also Fredholm and s-lim $B_n P_n = A^* A + P_{\ker A}$ is invertible. Then (B_n) is stable by Proposition 1. Since rank $(P_n - R_n) = l_n - k$ we get for n large enough

$$\begin{aligned} s_k(A) &\leq ||\left(A_n - A_n A_n^* A_n (P_n - R_n) B_n^{-1}\right) P_n|| \\ &\leq ||\left(A_n B_n - A_n A_n^* A_n (P_n - R_n)\right) P_n||\, ||B_n^{-1} P_n|| \\ &\leq ||B_n^{-1} P_n||\, ||A_n P_n P_{\ker A} P_n|| . \end{aligned}$$

Since (B_n) is stable, there exists for n large enough a constant C with $||B_n^{-1} P_n|| \leq C$. Thus we have

$$s_k(A_n) \leq C ||A_n P_n P_{\ker A} P_n|| \to 0 \text{ as } n \to \infty .$$

Now consider $s_{k+1}(A_n)$. By using the well-known inequality $s_{k+1}(A_n^* A_n) \leq s_{k+1}(A_n) ||A_n^*||$ and that $||A_n^*||$ is bounded away from zero for n large enough (recall that $A_n^* P_n$ converges strongly to $A^* \neq 0$) it has to be shown that $s_{k+1}(A^* A_n)$ is bounded away from zero (n large enough). We have

$$\begin{aligned} s_{k+1}\left(A_n^* A_n\right) &= \min_{B\in\mathcal{F}^{l_n}_{l_n-k-1}} ||\left(A_n^* A_n - B\right) P_n|| = \\ &= \min_{B\in\mathcal{F}^{l_n}_{l_n-k-1}} ||\left(\left(A_n^* A_n + P_n P_{\ker A} P_n\right) - B - P_n P_{\ker A} P_n\right) P_n|| \\ &\geq \min_{B\in\mathcal{F}^{l_n}_{l_n-1}} ||\left(\left(A_n^* A_n + P_n P_{\ker A} P_n\right) - B\right) P_n|| = \\ &= s_1\left(A_n^* A_n + P_n P_{\ker A} P_n\right) \geq \delta > 0 \end{aligned}$$

for n large enough since $(A_n^* A_n + P_n P_{\ker A} P_n)$ is stable, and we are done. □

Theorem 1 has remarkable consequences. One of them should be mentioned here and a further one in the next section. Recall that an element a belonging to a C^*-algebra $\mathcal{A}$ is called Moore-Penrose invertible if there exists an element $b \in \mathcal{A}$ such that

$$aba = a\,,\; bab = b\,,\; (ab)^* = ab\,,\; (ba)^* = ba\,.$$

If a is Moore-Penrose invertible then there exists only one element b with the properties cited above. This element is called Moore-Penrose inverse to a and is often denoted by a^+. It is well known that $A \in \mathcal{B}(H)$ is Moore-Penrose invertible if and only if A is normally solvable, that is im $A = \overline{\text{im } A}$.

Theorem 2. Let $(A_n) \in \mathcal{E}$ be Fredholm and $A = s - \lim A_n P_n$. Then $A_n^+ P_n \to A^+$ if and only if

$$\dim \ker A_n = \dim \ker A$$

for all n large enough.

Proof. Recall that $B \in \mathcal{B}(H)$ is Moore-Penrose invertible if and only if $d := \inf(\text{sp } B^*B) \setminus \{0\} > 0$ (Theorem 2.5 in [7]). In this case $d^{-1} = ||B^+||$. Let now A be $A := s - \lim A_n P_n$. This operator is Fredholm by assumption and Moore-Penrose invertible as well as the operators A_n.

If $A_n^+ P_n \to A^+$ strongly, then dim ker $A_n = \dim \ker A$ for n large enough. Otherwise there would be a sequence (n_k) such that dim ker $A_{n_k} < \dim \ker A_n$ (by Theorem 1 again) and this would imply that $||A_{n_k}^+|| \to \infty$. Conversely, if dim ker $A_n = \dim \ker A$ for n large enough then $s_{k+1} = ||A_n^+||^{-1}$ is bounded away from zero by Theorem 1. Hence, $(||A_n^+||^{-1})$ is bounded and therefore s-lim $A_n^+ P_n = A^+$ (by Theorem 2.12 in [7]). □

Remark 2. The well-known results about the continuity of the Moore-Penrose inversion for matrices are an immediate corollary to the last theorem.

Remark 3. If $(A_n) \in \mathcal{E}$ is Fredholm then ind $(s - \lim A_n P_n) = 0$.

This result follows immediately from [7]. It is easy to prove it directly. One only has to use that the matrices $A_n^* A_n$ and $A_n A_n^*$ are unitarily equivalent. Indeed, this gives that the splitting numbers of (A_n) and (A_n^*) coincide; as a consequence we get the claim.

Example 1. Consider the shift operator $V : l^2(\mathbb{N}) \to l^2(\mathbb{N})$ given by $Ve_n = e_{n+1}$, where $(e_n)_{n\in\mathbb{N}}$ denotes the standard orthonormal basis in $l^2(\mathbb{N})$. Let (P_n) denote the sequence of the orthoprojections that map $l^2(\mathbb{N})$ onto span $(e_1, \ldots, e_n)$. Then $(P_n V P_n)$ is Fredholm in the sense of Section 6.1.4 or even of Section 6.3.1 in [7], but not in the sense of Definition 1 because ind$V = -1$. The reason for this unpleasant fact is the circumstance that the ideal J is too small. However, the Fredholmness of a sequence $(A) \in \mathcal{E}$ implies in any case the Fredholmness in more general situations discussed in [7].

3. Classes of normally solvable operators and the finite splitting property

The collection of all normally solvable linear and bounded operators acting on H with $\dim\ker A < \infty$ ($\dim\ker A^* < \infty$) will be denoted by $\mathcal{F}_+(H)(\mathcal{F}_-(H))$. The elements in $\mathcal{F}_+(H)$ are called semi-Fredholm operators (and the intersection $\mathcal{F}(H) := \mathcal{F}_+(H) \cap \mathcal{F}_-(H)$ consists exactly of the set of all Fredholm operators, and for $A \in \mathcal{F}(H)$ the number $\operatorname{ind} A = \dim\ker A - \dim\ker A^*$ is well defined).

Theorem 3. $A \in \mathcal{F}_+(H)\,(A \in \mathcal{F}_-(H))$ if and only if the sequence $(P_nA^*AP_n)$ $((P_nAA^*P_n))$ is subject to the finite splitting property, and the splitting number k equals $\dim\ker A$ ($\dim\ker A^*$).

Proof. Suppose that $A \in \mathcal{F}_+(H)$. Then it follows that $A^*A \in \mathcal{F}(H)$ and $A^*A + P_{\ker A}$ is an invertible and positive operator. Then $(P_n(A^*A + P_{\ker A})P_n)$ is stable (see [5], Chapter II, § 2 or [7], Theorem 1.10), and $P_n(A^*A)P_n$ is Fredholm since $(P_nP_{\ker A}P_n) \in J$. By Theorem 2 the if-part is proved. Conversely, if $A \notin \mathcal{F}_+(H)$ then A^*A is not Fredholm and Theorem 6.67 in [7] applies what gives

$$s_l(A_n) \to 0 \text{ as } n \to \infty$$

for any $l \in \mathbb{N}$, and we are done. □

The last theorem leads to an alternative description of the classes $\mathcal{F}_\pm(H)$ and $\mathcal{F}(H)$ including the determination of dim ker A for $A \in \mathcal{F}_+(H)$ (dim ker A^* for $A \in \mathcal{F}_-(H)$). Moreover, it allows (at least theoretically) to answer the question whether a given complex number belongs to the spectrum, essential spectrum of A, or whether it is an eigenvalue of finite multiplicity. Notice that these results are in sharp contrast to some investigations of Ben-Artzi [1], where selfadjoint band operators with at least 5 nonzero main diagonals were considered.

Example 2. Toeplitz operator with rational generating matrix-function. Consider the familiar Toeplitz operators $T(a) : l_2^2(\mathbb{N}) \to l^2(\mathbb{N})$ on the $\mathbb{C}^2$-valued l^2 space $l_2^2(\mathbb{N})$ with the following generating functions

$$\begin{aligned} a_1(t) &= \begin{pmatrix} 2t^2 + 7t + 3 + \frac{1}{2}t^{-1} & \frac{1}{2}t^{-2} \\ t + 3 + t^{-1} & t^{-2} \end{pmatrix}, \\ a_2(t) &= \begin{pmatrix} t^3 + 2t & 2t^{-2} + t \\ 3t^2 + 1 - 5t^{-1} & t^2 - 4 \end{pmatrix}. \end{aligned}$$

We have $\det a_1(t) \neq 0$ for all $t \in \mathbb{T}$ whereas $\det a_2(-1) = 0$. Then $T(a_1)$ is Fredholm and $T(a_2)$ is even not normally solvable (see [5], Chapter VIII). Below there are plotted the 10 smallest singular values for $n = 50k$, $k = 1, \dots, 10$, for each of the matrices $P_nT^*(a_i)T(a_i)P_n$, $i = 1, 2$:

It is seen that for $i = 1$ the finite splitting property is in force with $k = 2$. For $i = 2$ the finite splitting property cannot be observed which agrees with Theorem 3.

$i=1$	50	100	150	200	250	300	350	400	450	500
s_{10}	0.7572	0.7453	0.7431	0.7423	0.7420	0.7418	0.7417	0.7416	0.7415	0.7415
s_9	0.7535	0.7444	0.7427	0.7421	0.7418	0.7417	0.7416	0.7415	0.7415	0.7415
s_8	0.7503	0.7436	0.7423	0.7419	0.7417	0.7416	0.7415	0.7415	0.7415	0.7414
s_7	0.7475	0.7429	0.7420	0.7417	0.7416	0.7415	0.7415	0.7414	0.7414	0.7414
s_6	0.7453	0.7423	0.7418	0.7416	0.7415	0.7415	0.7414	0.7414	0.7414	0.7414
s_5	0.7436	0.7419	0.7416	0.7415	0.7414	0.7414	0.7414	0.7414	0.7414	0.7414
s_4	0.7423	0.7416	0.7415	0.7414	0.7414	0.7414	0.7414	0.7414	0.7414	0.7414
s_3	0.7416	0.7414	0.7414	0.7414	0.7414	0.7414	0.7414	0.7414	0.7414	0.7414
s_2	0.0000	0.0000	0.0000	0.0000	0.0000	0.0000	0.0000	0.0000	0.0000	0.0000
s_1	0.0000	0.0000	0.0000	0.0000	0.0000	0.0000	0.0000	0.0000	0.0000	0.0000

$i=2$	50	100	150	200	250	300	350	400	450	500
s_{10}	0.8610	0.6781	0.6202	0.6001	0.5470	0.4595	0.3957	0.3474	0.3095	0.2790
s_9	0.8603	0.6493	0.6029	0.5951	0.4837	0.4058	0.3493	0.3066	0.2731	0.2462
s_8	0.7726	0.6448	0.6028	0.5192	0.4199	0.3520	0.3029	0.2658	0.2367	0.2134
s_7	0.7720	0.6107	0.5753	0.4405	0.3556	0.2980	0.2564	0.2249	0.2004	0.1806
s_6	0.6957	0.6098	0.4739	0.3611	0.2912	0.2439	0.2098	0.1841	0.1639	0.1478
s_5	0.6518	0.5365	0.3699	0.2812	0.2266	0.1898	0.1632	0.1432	0.1275	0.1150
s_4	0.6417	0.3867	0.2648	0.2010	0.1620	0.1356	0.1166	0.1023	0.0911	0.0821
s_3	0.4316	0.2331	0.1591	0.1207	0.0972	0.0814	0.0700	0.0614	0.0547	0.0493
s_2	0.1461	0.0778	0.0531	0.0402	0.0324	0.0271	0.0233	0.0205	0.0182	0.0164
s_1	0.0009	0.0000	0.0000	0.0000	0.0000	0.0000	0.0000	0.0000	0.0000	0.0000

Quasidiagonal operators and their finite sections

The algebra $\mathcal{E}$ contains interesting subalgebras. We call a C^*-subalgebra $\mathcal{E}_0$ of $\mathcal{E}$ standard (or a standard model) if $J \subset \mathcal{E}_0$ and if the invertibility of s-lim $A_nP_n(A_n) \in \mathcal{E}_0$ already implies the stability of (A_n) (for more general situation, see [7]). Every quasidiagonal operator gives raise for introducing a standard subalgebra of $\mathcal{E}$.

Recall that a bounded linear operator T acting on a separable (complex) Hilbert space is said to be quasidiagonal if there exists a sequence $(P_n)_{n\in\mathbb{N}}$ of finite rank orthogonal projections such that s-lim $P_n = I$ and which asymptotically commute with T, that is

$$||[T, P_n]|| := ||TP_n - P_nT|| \to 0 \text{ as } n \to \infty .$$

In particular, every selfadjoint or even normal operator is quasidiagonal as well as their perturbations by compact operators. However it is by no means trivial to single out a related sequence (P_n). For instance, for multiplication operators in periodic Sobolev spaces H^λ related sequences can explicitly be given: these are orthogonal projections on some spline spaces (see [10], Section 2.12.).

Let T be quasidiagonal with respect to $(P_n) = (P_n)_{n\in\mathbb{N}}$ as given above. Let $C_{(P_n)}(T)$ denote the smallest closed C^*-subalgebra containing the sequence (P_nTP_n) and the ideal J.

Proposition 2. $C_{(P_n)}(T)$ actually forms a standard model. Moreover, $C_{(P_n)}(T)/G$ is isometrically isomorphic to the smallest C^*-subalgebra $C(T)$ of $\mathcal{B}(H)$ containing T and all compact operators.

Sketch of proof. Suppose s-lim $A_nP_n =: A$ is invertible ($(A_n) \in C_{(P_n)}(T)$). Then $A^{-1} \in C(T)$, and since every element in $C(T)$ is quasidiagonal, A^{-1} owns this property, and

$$||P_n - P_nAP_nA^{-1}P_n|| = ||P_nAA^{-1}P_n - P_nAP_nA^{-1}P_n|| = \\ = ||P_n\,(P_nA - AP_n)\,A^{-1}P_n|| \to 0\,.$$

Hence, (P_nAP_n) is stable. Now it is sufficient to prove that $(P_nAP_n) - (A_n) \in G$. For it is sufficient to show this for the special case $A_n = (P_nB_1P_n)(P_nB_2P_n)$. We have

$$||P_nB_1B_2P_n - P_nB_1P_nB_2P_n|| = ||P_n(P_nB_1 - B_1P_n)B_2P_n|| \to 0$$

as $n \to \infty$, and the stability of (A_n) is proved. The converse is immediate. The second claim in the theorem is obvious. □

An immediate consequence of Remark 3 and Proposition 2 is the well-known fact that the index of a Fredholm quasidiagonal operator is necessarily equal to zero. Now it is obvious that the theory of spectral approximations explained in [7] completely applies to the case at hand.

In the papers [3], [4] Nathaniel Brown proposed further refinements into two directions: speed of convergence and how to choose the sequence (P_n) of orthoprojections in some special cases such as quasidiagonal unilateral band operators, bilateral band operators or operators in irrational rotation algebras.

We will now take up one problem considered already in [3], namely the weighted shift operator $T : l^2(\mathbb{N}) \to l^2(\mathbb{N})$, acting by the rule $Te_n = \alpha_n e_{n+1}$ where $(\alpha_n)_{n\in\mathbb{N}}$ is a bounded sequence of complex numbers (the so-called weight sequence). In [3] there was assumed that the sequence (α_n) possesses an infinite subsequence (α_{n_k}) tending to zero. This condition is equivalent to the quasidiagonality of T. It is known that if T is any weighted shift and $r(T)$ is the spectral radius of T then sp $T = \{\lambda \in \mathbb{C} : |\lambda| \le \rho(T)\}$, that is, the spectrum of T is as large as possible. In [3] there was proposed a simple proof of this result under the condition that T is quasidiagonal.

We will give a simple proof in the general case; this proof is not based on the material before, but it is in its spirit. Note that the spectral radius of any weighted shift T is given by (see [9])

$$r(T) = \lim_{l\to\infty} ||T^l||^{\frac{1}{e}} = \lim_{l\to\infty} \left(\sup_m |\alpha_m\alpha_{m+1}\dots\alpha_{m+l-1}|\right)^{\frac{1}{l}}.$$

Proposition 3. Let T be any weighted shift with weight sequence (α_n). Then $\operatorname{sp} T = \{x \in \mathbb{C} : |\lambda| \le r(T)\}$.

Proof. Suppose $\lambda \notin \operatorname{sp} T$ and $0 < |\lambda| < r(T)$. Let $P_n \in \mathbb{C}(l^2(\mathbb{N}))$ be the orthoprojection onto span $\{e_1, \dots, e_n\}$, where $(e_n)_{n\in\mathbb{N}}$ is the orthonormal standard basis in $l^2(\mathbb{N})$. Because of

$$C_n := P_n\,(\lambda I - T)P_n = P_n(\lambda I - T)$$

we get $P_n(\lambda I - T)P_n(\lambda I - T)^{-1}P_n = P_n$ and therefore the stability of $(P_n(\lambda I - T)P_n)$. Consider

$$C_n^{-1} = \begin{pmatrix} \frac{1}{\lambda} & & & & \\ \frac{\alpha_1}{\lambda} & \frac{1}{\lambda} & & & 0 \\ \frac{\alpha_1\alpha_2}{\lambda^2} & \frac{\alpha_2}{\lambda} & \frac{1}{\lambda} & & \\ \dots & \dots & \dots & \ddots & \\ \frac{\Pi_1^{n-1}\alpha_i}{\lambda^n} & \frac{\Pi_2^{n-1}\alpha_i}{\lambda^{n-1}} & \dots & \frac{\alpha_{n-1}}{\lambda^2} & \frac{1}{\lambda} \end{pmatrix}.$$

It is easy to show (using arguments from [3]) that the sequence (C_n^{-1}) is not uniformly bounded: Let $\delta := r(T) - |\lambda|$ and l_0 be such that

$$\left| r(t) - \sup_m |\alpha_m\alpha_{m+1}\dots\alpha_{m+l-1}|^{\frac{1}{l}} \right| < \frac{\delta}{4} \text{ for all } l \geq l_0\,.$$

Hence, $(|\lambda| + \frac{\delta}{4}) < \sup_m |\alpha_m\alpha_{m+l}\dots\alpha_{m+l-1}|^{\frac{1}{l}}$ for all $l \geq l_0$, whence follows the existence of an $m_0 = m_0(l)$ such that

$$\left(1 + \frac{\delta}{4|\lambda|}\right)^l < \frac{|\alpha_{m_0}\alpha_{m_0+l-1}\dots\alpha_{m_0+l-1}|}{|\lambda|^l}\,.$$

This estimate shows that

$$\frac{|\alpha_{m_0}\alpha_{m_0+l-1}\dots\alpha_{m_0+l-1}|}{|\alpha|^l} \longrightarrow \infty \text{ as } l \to \infty\,.$$

However, the numbers $\frac{\alpha_{m_0}\alpha_{m_0+l-1}\dots\alpha_{m_0+l-1}}{\lambda^l}$ are entries in all of the matrices above for sufficiently large n and therefore (C_n) cannot be stable. The obtained contradiction proves the claim. □

Notice that this idea can also be used to determine the spectrum of unilateral block weighted shifts. A more refined analysis for such problems where both the weight sequence and their inverse are uniformly bounded is contained in [2].

References

[1] A. Ben-Artzi: On approximation spectrum of bounded selfadjoint operators. *Integral Equations Operator Theory* **9** (1986), no. 2, 266–274.

[2] A. Ben-Artzi, I. Gohberg: Dichotomy, Discrete Bohl Exponents, and Spectrum of Block weighted shifts. *Integral Equations Operator Theory* **14** (1991), 615–677.

[3] N.P. Brown: Quasidiagonality and the finite section method. Preprint, Department of Mathematics, Penn State University (2003).

[4] N.P. Brown: An Embedding and the Numerical Computation of Spectra in Irrational Rotation Algebras. Preprint, *Department of Mathematics*, Penn State University (2003).

[5] I. Gohberg, I. Feldman: *Convolution Equations and Projection Method for Their Solution.* Nauka, Moskva 1971 (Russian; English translat.: Am. Math. Soc. Transl. of Math. Monographs 41, Providence, R. I. 1974).

[6] I. Gohberg, M. Krein: *Introduction to the theory of linear nonselfadjoint operators.* Nauka, Moskva (Russian; Engl. translat.: Am. Math. Soc. Transl. of Math. Monographs 18, Providence, R. I. 1969).

[7] R. Hagen, S. Roch, B. Silbermann: *C^*-Algebras and Numerical Analysis.* Marcel Dekker, Inc., New York, Basel (2001).

[8] S. Prössdorf, B. Silbermann: *Numerical Analysis for Integral and Related Operator Equations.* Akademie Verlag, Berlin (1991), and Birkhäuser Verlag, Basel, Boston, Stuttgart (1991).

[9] P. Halmos: *A Hilbert space problem book.* D. Van Nostrand Company, Inc., Toronto, London (1967).

[10] S. Prössdorf, B. Silbermann: *Numerical Analysis for Integral and Related Operator equations.* Akademie Verlag, Berlin, 1991, and Birkhäuser Verlag, Basel, Boston, Stuttgart, (1991).

[11] S. Roch, B. Silbermann: Index calculus for approximation methods and singular value decomposition. *J. Math. Anal. Appl.* **225** (1998), 401–426.

Bernd Silbermann
Technical University Chemnitz
Department of Mathematics
D-09107 Chemnitz, Germany

Operator Theory:
Advances and Applications, Vol. 160, 413–424

Additive and Multiplicative Perturbations of Exponentially Dichotomous Operators on General Banach Spaces

Cornelis V.M. van der Mee and André C.M. Ran

To Israel Gohberg on the occasion of his 75th *birthday*

Abstract. Recent perturbation results for exponentially dichotomous operators are generalized, in part by replacing compactness conditions on the perturbation by resolvent compactness. Both additive and multiplicative perturbations are considered.

Mathematics Subject Classification (2000). Primary 47D06; Secondary 47A55.

Keywords. block operator, exponential dichotomy, semigroup perturbation, Riccati equation.

1. Introduction

In [14] perturbation results for exponentially dichotomous operators on Banach spaces were discussed. In this paper we continue the investigation started in that paper.

Recall that an exponentially dichotomous operator is a direct sum $A_0\dot{+}(-A_1)$, in which A_0 and A_1 are generators of exponentially decaying C_0-semigroups. Such operators were introduced in [2, 3] in connection with convolution equations on the half-line. Operators of this type also occur in various other applications, see, e.g., [7, 8, 9, 10, 11].

Perturbation results for exponentially dichotomous operators were already studied in [2], where additive perturbations were considered. Results in this direction were later obtained for more particular operators on Hilbert spaces in [8, 9, 10, 11]. In [14] the authors considered additive perturbations for exponentially dichotomous operators on Banach spaces. Multiplicative perturbations were studied in [7, 13].

The research leading to this article was supported in part by MIUR under the COFIN grants 2002014121 and 2004015437, and by INdAM.

In this article we accomplish the following two tasks. First we generalize the main results on additive and bounded perturbations of exponentially dichotomous operators derived in [14] by requiring the corresponding bisemigroup multiplied from the right by a bounded additive perturbation to be continuous in the operator norm, except possibly as $t \to 0^{\pm}$. This greatly simplifies the treatment in [14], where it is assumed that either the corresponding bisemigroup itself is continuous in the operator norm (except possibly as $t \to 0^{\pm}$) or the perturbation is a compact operator. We shall prove a lemma that will allow us to use the same proofs as in [14] and to refer to [14] for these proofs. Secondly, for exponentially dichotomous operators having bounded analytic constituent semigroups, we study perturbations obtained by multiplying the given exponentially dichotomous operator from the right by a compact perturbation of the identity. We shall prove that the newly obtained operator is exponentially dichotomous and has bounded analytic constituent semigroups. We thus generalize results obtained before in [7] in a Hilbert space setting. All of our results will be derived in general complex Banach spaces, including those on Riccati equations.

The main body of this paper consists of two sections. In Section 2 we indicate how one of the main results of [14] can be generalized to the present setting without changing its proof and discuss the consequences of this result for canonical factorization and for block operators. In Section 3 we study perturbations of analytic bisemigroup generators. We refer to the introduction of [14] for a more comprehensive discussion of the existing literature.

Let us introduce some notations. We let $\mathbb{R}^{\pm}$ stand for the right (left, resp.) half-line, including the point at zero. For two complex Banach spaces $\mathcal{X}$ and $\mathcal{Y}$, we let $\mathcal{L}(\mathcal{X}, \mathcal{Y})$ stand for the Banach spaces of all bounded linear operators from $\mathcal{X}$ into $\mathcal{Y}$. We write $\mathcal{L}(\mathcal{X})$ instead of $\mathcal{L}(\mathcal{X}, \mathcal{X})$.

Let $\mathcal{X}$ be a complex Banach space and E an interval of the real line $\mathbb{R}$. Then $L^p(E; \mathcal{X})$ denotes the Banach space of all strongly measurable functions $\phi : E \to \mathcal{X}$ such that $\|\phi(\cdot)\|_{\mathcal{X}} \in L^p(E)$, endowed with the L^p-norm, and $C_0(E; \mathcal{X})$ stands for the Banach space of all bounded continuous functions $\phi : E \to \mathcal{X}$ which vanish at infinity if E is unbounded, endowed with the supremum norm. In particular, $C_0(\mathbb{R}^-; \mathcal{X}) \dot{+} C_0(\mathbb{R}^+; \mathcal{X})$ is the Banach space of all bounded continuous functions $\phi : \mathbb{R} \to \mathcal{X}$ which vanish at $\pm\infty$ and may have a jump discontinuity in zero.

2. Bisemigroups and their perturbations

2.1. Bisemigroup perturbation results

A C_0-semigroup $(T(t))_{t \geq 0}$ on a complex Banach space $\mathcal{X}$ is called *uniformly exponentially stable* if

$$\|T(t)\| \leq M e^{-\varepsilon t}, \qquad t \geq 0, \tag{2.1}$$

for certain $M, \varepsilon > 0$.

A closed and densely defined linear operator $-S$ on a Banach space $\mathcal{X}$ is called *exponentially dichotomous* [2] if for some projection P commuting with S, the restrictions of S to $\operatorname{Im} P$ and of $-S$ to $\operatorname{Ker} P$ are the infinitesimal generators of exponentially decaying C_0-semigroups. We then define the *bisemigroup* generated by $-S$ as

$$E(t;-S) = \begin{cases} e^{-tS}(I-P), & t>0 \\ -e^{-tS}P, & t<0. \end{cases}$$

Its *separating projection* P is given by $P = -E(0^-;-S) = I_{\mathcal{X}} - E(0^+;-S)$. One easily verifies the existence of $\varepsilon > 0$ such that $\{\lambda \in \mathbb{C} : |\operatorname{Re}\lambda| \leq \varepsilon\}$ is contained in the resolvent set $\rho(S)$ of S and for every $x \in \mathcal{X}$

$$(\lambda - S)^{-1}x = -\int_{-\infty}^{\infty} e^{\lambda t} E(t;-S)x\,dt, \qquad |\operatorname{Re}\lambda| \leq \varepsilon. \tag{2.2}$$

As a result, for every $x \in \mathcal{X}$ we have $\|(\lambda - S)^{-1}x\| \to 0$ as $\lambda \to \infty$ in $\{\lambda \in \mathbb{C} : |\operatorname{Re}\lambda| \leq \varepsilon'\}$ for some $\varepsilon' \in (0,\varepsilon]$. We call the restrictions of e^{-tS} to $\operatorname{Ker} P$ and of e^{tS} to $\operatorname{Im} P$ the *constituent semigroups* of the exponentially dichotomous operator $-S$. Observe that $\{x \in \mathcal{X} : (\lambda - S)^{-1}x \text{ is analytic for } \operatorname{Re}\lambda < 0\} = \operatorname{Ker} P$, and $\{x \in \mathcal{X} : (\lambda - S)^{-1}x \text{ is analytic for } \operatorname{Re}\lambda > 0\} = \operatorname{Im} P$.

Before deriving our main perturbation result, we prove the following lemma. Note that Theorem 3 of [14] is an immediate consequence of this lemma.

Lemma 2.1. *Let $-S_0$ be exponentially dichotomous, Γ a bounded operator such that $E(t;-S_0)\Gamma$ is norm continuous in $0 \neq t \in \mathbb{R}$, and $-S = -S_0 + \Gamma$, where $\mathcal{D}(S) = \mathcal{D}(S_0)$. Suppose the strip $\{\lambda \in \mathbb{C} : |Re\,\lambda| < \varepsilon\}$ is contained in the resolvent set of S for some $\varepsilon > 0$. Then $-S$ is exponentially dichotomous. Moreover, $E(t;-S)\Gamma$ is norm continuous in $0 \neq t \in \mathbb{R}$ with norm continuous limits as $t \to 0^{\pm}$.*

Proof. There exists $\varepsilon > 0$ such that

$$\int_{-\infty}^{\infty} e^{\varepsilon|t|}\|E(t;-S_0)\|\,dt < \infty. \tag{2.3}$$

Using the resolvent identity

$$(\lambda - S)^{-1} - (\lambda - S_0)^{-1} = -(\lambda - S_0)^{-1}\Gamma(\lambda - S)^{-1}, \quad |\operatorname{Re}\lambda| \leq \varepsilon, \tag{2.4}$$

for some $\varepsilon > 0$, we obtain the convolution integral equation

$$E(t;-S)x - \int_{-\infty}^{\infty} E(t-\tau;-S_0)\Gamma E(\tau;-S)x\,d\tau = E(t;-S_0)x, \tag{2.5}$$

where $x \in \mathcal{H}$ and $0 \neq t \in \mathbb{R}$. By assumption, in (2.5) the convolution kernel $E(\cdot;-S_0)\Gamma$ is continuous in the norm except for a jump discontinuity in $t = 0$. Further, (2.3) implies that $e^{\varepsilon|\cdot|}E(\cdot;-S_0)\Gamma$ is Bochner integrable.

The symbol of the convolution integral equation (2.5), which equals $I_{\mathcal{H}} + (\lambda - S_0)^{-1}\Gamma = (\lambda - S_0)^{-1}(\lambda - S)$, tends to $I_{\mathcal{H}}$ in the norm as $\lambda \to \infty$ in the strip $|\operatorname{Re}\lambda| \leq \varepsilon$, because of the Riemann-Lebesgue lemma. Thus there exists $\varepsilon_0 \in (0,\varepsilon]$ such that the symbol only takes invertible values on the strip $|\operatorname{Re}\lambda| \leq \varepsilon_0$. By the Bochner-Phillips theorem ([4], also [6]), the convolution equation (2.5) has a unique solution $u(\cdot;x) = E(\cdot;-S)x$ with the following properties:

1) $E(\cdot;-S)$ is strongly continuous, except for a jump discontinuity at $t=0$,
2) $\int_{-\infty}^{\infty} e^{\varepsilon_0|t|}\|E(t;-S)\|\,dt < \infty$; hence $E(\cdot;-S)$ is exponentially decaying,
3) the identity (2.2) holds.

As a result [2], $-S$ is exponentially dichotomous. □

We now present our main perturbation result. Note that Theorem 2 of [14] is an immediate consequence of this result.

Theorem 2.2. *Let $-S_0$ be exponentially dichotomous, Γ a bounded operator such that the operator $(\lambda - S_0)^{-1}\Gamma$ is compact for imaginary λ, and $-S = -S_0 + \Gamma$, where $\mathcal{D}(S) = \mathcal{D}(S_0)$. Suppose the imaginary axis is contained in the resolvent set of S. Then $-S$ is exponentially dichotomous. Moreover, $E(t;-S) - E(t;-S_0)$ is a compact operator, also in the limits as $t \to 0^{\pm}$.*

Proof. It suffices to prove that $E(t;-S_0)\Gamma$ is a compact operator for $0 \neq t \in \mathbb{R}$. This would imply that (1) $E(t;-S_0)\Gamma$ is norm continuous in $0 \neq t \in \mathbb{R}$ with norm continuous limits as $t \to 0^{\pm}$, and (2) the symbol $I_{\mathcal{H}} + (\lambda - S_0)^{-1}\Gamma = (\lambda - S_0)^{-1}(\lambda - S)$ of the convolution integral equation (2.5) tends to $I_{\mathcal{H}}$ in the norm as $\lambda \to \infty$ in the strip $|\mathrm{Re}\,\lambda| \le \varepsilon$. In combination with the absence of imaginary spectrum of S, the latter would imply that the strip $\{\lambda \in \mathbb{C} : |\mathrm{Re}\,\lambda| < \varepsilon_0\}$ is contained in the resolvent set of S for some $\varepsilon_0 > 0$. Theorem 2.2 would then be immediate from Lemma 2.1.

By analytic continuation, we easily prove that $(\lambda - S_0)^{-1}\Gamma$ is a compact operator on a strip $\{\lambda \in \mathbb{C} : |\mathrm{Re}\,\lambda| \le \varepsilon\}$ for some $\varepsilon > 0$. Thus $(\lambda - S_0)^{-1}E(0^+,-S_0)\Gamma$ is analytic and compact operator valued for $\mathrm{Re}\,\lambda < \varepsilon$, while $(\lambda - S_0)^{-1}E(0^-,-S_0)\Gamma$ is analytic and compact operator valued for $\mathrm{Re}\,\lambda > -\varepsilon$.

Now it is well known ([5], Corollary III 5.5) that

$$E(t;-S_0)x = \begin{cases} \lim\limits_{n\to\infty} (I + \dfrac{t}{n}S_0)^{-n}E(0^+;-S_0)x, & t > 0, \\ \lim\limits_{n\to\infty} (I + \dfrac{t}{n}S_0)^{-n}E(0^-;-S_0)x, & t < 0. \end{cases}$$

uniformly in x on relatively compact sets. Since for every $0 \neq t \in \mathbb{R}$ we have that $(I + \frac{t}{n}S_0)^{-1}\Gamma$ is compact for sufficiently large $n \in \mathbb{N}$, it follows that

$$E(t;-S_0)\Gamma = \begin{cases} \lim\limits_{n\to\infty} (I + \dfrac{t}{n}S_0)^{-n}E(0^+;-S_0)\Gamma, & t > 0, \\ \lim\limits_{n\to\infty} (I + \dfrac{t}{n}S_0)^{-n}E(0^-;-S_0)\Gamma, & t < 0, \end{cases}$$

in the operator norm. Since $(I + \frac{t}{n}S_0)^{-n}E(0^{\pm};-S_0)\Gamma$ is compact for $(\pm t) > 0$, it follows that $E(t;-S_0)\Gamma$ is compact for $0 \neq t \in \mathbb{R}$, which completes the proof. □

2.2. Canonical factorization and matching of subspaces

Let $-S_0$ be exponentially dichotomous and Γ a bounded operator on a complex Banach space $\mathcal{X}$, and let $-S = -S_0 + \Gamma$, where $\mathcal{D}(S) = \mathcal{D}(S_0)$ and $\{\lambda \in \mathbb{C} : |\mathrm{Re}\,\lambda| \leq \varepsilon\} \subset \rho(S)$ for some $\varepsilon > 0$. Then $-S$ is exponentially dichotomous if $E(t; -S_0)\Gamma$ is continuous in $0 \neq t \in \mathbb{R}$ in the operator norm. In this section we consider the analogous vector-valued Wiener-Hopf integral equation

$$\phi(t) - \int_0^\infty E(t-\tau; -S_0)\Gamma\phi(\tau)\,d\tau = g(t) \tag{2.6}$$

where $t > 0$.

Suppose W is a continuous function from the extended imaginary axis $i(\mathbb{R} \cup \{\infty\})$ into $\mathcal{L}(\mathcal{X})$. Then by a *left canonical (Wiener-Hopf) factorization* of W we mean a representation of W of the form

$$W(\lambda) = W_+(\lambda)W_-(\lambda), \qquad \mathrm{Re}\,\lambda = 0, \tag{2.7}$$

in which $W_\pm(\pm\lambda)$ is continuous on the closed right half-plane (the point at ∞ included), is analytic on the open right half-plane, and takes only invertible values for λ in the closed right half-plane (the point at infinity included). Obviously, such an operator function only takes invertible values on the extended imaginary axis. By a *right canonical (Wiener-Hopf) factorization* we mean a representation of W of the form

$$W(\lambda) = W_-(\lambda)W_+(\lambda), \qquad \mathrm{Re}\,\lambda = 0, \tag{2.8}$$

where $W_\pm(\lambda)$ are as above.

Theorems 6 and 7 and Corollary 8 of [14] can now easily be generalized with exactly the same proofs. We now require $-S_0$ to be an exponentially dichotomous and Γ a bounded operator on a complex Banach space $\mathcal{X}$ (Hilbert space when generalizing Corollary 8 of [14]) such that $E(t; -S_0)\Gamma$ is continuous in $0 \neq t \in \mathbb{R}$ in the operator norm, instead of requiring that either (i) $E(t; -S_0)$ itself is continuous in the operator norm for $0 \neq t \in \mathbb{R}$ or (ii) Γ is a compact operator. Lemma 2.1 then enables us to apply Theorems 6 and 7 and Corollary 8 of [14] in the case $(\lambda - S_0)^{-1}\Gamma$ is a compact operator for imaginary λ.

The above generalizations of Theorems 6 and 7 of [14] yield results on the equivalence of (i) left (resp., right) canonical Wiener-Hopf factorizability of $W(\lambda) = (\lambda - S_0)^{-1}(\lambda - S)$, (ii) the complementarity in $\mathcal{X}$ of the range of one of the separating projections P_0 and P and the kernel of the other, and (iii) the unique solvability of the vector-valued convolution equation on the positive (resp. negative) half-line with convolution kernel $E(\cdot; -S_0)\Gamma$. The above generalization of Corollary 8 of [14] yields left and right canonical factorizability if the symbol $W(\lambda) = (\lambda - S_0)^{-1}(\lambda - S)$ of the half-line convolution equation involved is either close to the identity operator or has a strictly positive definite real part. Similar results in various different contexts exist in the finite-dimensional case [1], for equations with symbols analytic in a strip and at infinity [3], for extended Pritchard-Salamon realizations [8], and for abstract kinetic equations [7].

2.3. Block operators

Suppose $-S_0$ is exponentially dichotomous and Γ is a bounded linear operator on a complex Banach space $\mathcal{X}$. Define S by $-S = -S_0 + \Gamma$, and put

$$\mathcal{X}^{\pm} = \operatorname{Im} E(0^{\pm}; -S_0),$$

i.e., $\mathcal{X}^+ = \operatorname{Im}(I - P_0) = \operatorname{Ker} P_0$ and $X^- = \operatorname{Im} P_0$. Assuming that $\Gamma[\mathcal{X}^{\pm}] \subset \mathcal{X}^{\mp}$, we have the following block decompositions of S_0 and S with respect to the direct sum $\mathcal{X} = \mathcal{X}^+ \dot{+} \mathcal{X}^-$:

$$S_0 = \begin{pmatrix} A_0 & 0 \\ 0 & -A_1 \end{pmatrix}, \qquad S = \begin{pmatrix} A_0 & -D \\ -Q & -A_1 \end{pmatrix}, \tag{2.9}$$

where $-A_0$ and $-A_1$ are the generators of uniformly exponentially stable C_0-semigroups and $Q : \mathcal{X}^+ \to \mathcal{X}^-$ and $D : \mathcal{X}^- \to \mathcal{X}^+$ are bounded. Then we call S written in the form (2.9) a *block operator*, which is in line with the definition used in [14]. In the literature the notion of a block operator is also used in a wider sense (e.g., without assumptions about semigroup generators).

Theorem 9 of [14] can now be generalized in the same way without changing its proof. We now require $-S_0$ to be an exponentially dichotomous operator and Γ a bounded operator on a complex Banach space $\mathcal{X}$ satisfying $\Gamma[\mathcal{X}^{\pm}] \subset \mathcal{X}^{\mp}$, instead of requiring that either (i) $E(t; -S_0)$ itself is continuous in the operator norm for $0 \neq t \in \mathbb{R}$ or (ii) Γ is a compact operator. Here $\mathcal{X}^+$ and $\mathcal{X}^-$ are the kernel and range of the separating projection of $-S_0$, respectively.

The above generalization of Theorem 9 of [14] states that there exists a bounded linear operator Π_+ from $\mathcal{X}^-$ into $\mathcal{X}^+$ which maps $\mathcal{D}(A_1)$ into $\mathcal{D}(A_0)$, has the property that $B_1 = A_1 + Q\Pi_+$ generates an exponentially stable semigroup on $\mathcal{X}^-$, and satisfies the Riccati equation

$$A_0\Pi_+ x + \Pi_+ A_1 x - Dx + \Pi_+ Q \Pi_+ x = 0, \quad x \in \mathcal{D}(A_1), \tag{2.10}$$

if and only if the equivalent statements (a)–(e) of Theorem 7 of [14] are true. Analogously, it states that there exists a bounded linear operator Π_- from $\mathcal{X}^+$ into $\mathcal{X}^-$ which maps $\mathcal{D}(A_0)$ into $\mathcal{D}(A_1)$, has the property that $B_0 = A_0 - D\Pi_-$ generates an exponentially stable semigroup on $\mathcal{X}^+$, and satisfies the Riccati equation

$$\Pi_- A_0 x + A_1 \Pi_- x - \Pi_- D \Pi_- x + Qx = 0, \qquad x \in \mathcal{D}(A_0). \tag{2.11}$$

if and only if the equivalent statements (a)–(e) of Theorem 8 of [14] are true. Similar results are valid in the finite-dimensional case [1] and for extended Pritchard-Salamon realizations [8].

3. Analytic bisemigroups and unbounded perturbations

3.1. Preliminaries on analytic semigroups

As in [5] (but in contrast to the definition given in [12]), a closed linear operator A densely defined on a complex Banach space $\mathcal{X}$ is called *sectorial* if there exists a δ with $0 < \delta \leq (\pi/2)$ such that the sector

$$\Sigma_{\frac{\pi}{2}+\delta} = \{\lambda \in \mathbb{C} : |\arg \lambda| < \frac{\pi}{2} + \delta\} \setminus \{0\}$$

is contained in the resolvent set of A, and if for each $\zeta \in (0, \delta)$ there exists $M_\zeta \geq 1$ such that

$$\|(\lambda - A)^{-1}\| \leq \frac{M_\zeta}{|\lambda|}, \qquad \lambda \in \overline{\Sigma_{\frac{\pi}{2}+\delta-\zeta}} \setminus \{0\}.$$

According to [5], Theorem II 4.6, the sectorial operators are exactly the generators of bounded analytic semigroups. Thus A is the generator of a uniformly exponentially stable analytic semigroup if and only if there exist δ and γ with $0 < \delta \leq (\pi/2)$ and $\gamma > 0$ such that (1) the sector

$$-\gamma + \Sigma_{\frac{\pi}{2}+\delta} = \{\lambda \in \mathbb{C} : |\arg(\lambda + \gamma)| < \frac{\pi}{2} + \delta\} \setminus \{-\gamma\} \tag{3.1}$$

is contained in the resolvent set of A, and (2) for each $\zeta \in (0, \delta)$ there exists $M_\zeta \geq 1$ such that

$$\|(\lambda - A)^{-1}\| \leq \frac{M_\zeta}{|\lambda + \gamma|}, \qquad \lambda \in -\gamma + \left[\overline{\Sigma_{\frac{\pi}{2}+\delta-\zeta}} \setminus \{-\gamma\}\right]. \tag{3.2}$$

3.2. Perturbation results for analytic bisemigroups

A bisemigroup is called *analytic* if its constituent semigroups are analytic. Writing $-S$ for its generator and P for its separating projection, we can define

$$H(t; -S) = \begin{cases} Se^{-tS}(I - P), & t > 0 \\ -Se^{-tS}P, & t < 0, \end{cases}$$

for the derivative of $E(t; -S)$ with respect to $0 \neq t \in \mathbb{R}$.

Next, we note that the generator $-S$ has the following two properties (cf. (3.1)–(3.2)):

1. there exist δ and γ with $0 < \delta \leq (\pi/2)$ and $\gamma > 0$ such that the set

$$\Omega_{\delta,\gamma} = \left\{\lambda \in \mathbb{C} : \left|\frac{\pi}{2} - \arg \lambda\right| < \delta \text{ or } |\mathrm{Re}\, \lambda| < \gamma\right\} \tag{3.3}$$

   is contained in the resolvent set of S, and
2. for each $\zeta \in (0, \delta)$ there exists $N_\zeta \geq 1$ such that

$$\|(\lambda - S)^{-1}\| \leq N_\zeta \left(\frac{1}{|\lambda + \gamma|} + \frac{1}{|\lambda - \gamma|}\right), \qquad \lambda \in \overline{\Omega_{\zeta,\gamma}} \setminus \{\gamma, -\gamma\}. \tag{3.4}$$

It is not clear if a closed and densely defined linear operator $-S$ on $\mathcal{X}$ having the properties (3.3)–(3.4) generates an analytic bisemigroup.

Starting from an exponentially dichotomous operator $-S_0$ on a complex Banach space $\mathcal{X}$ generating an analytic bisemigroup and a bounded linear operator Δ on $\mathcal{X}$, we now study sufficient conditions under which the unbounded perturbation $-S = -S_0 + \Gamma$ of $-S_0$ for which $\Gamma = S_0\Delta$, is a generator of an analytic bisemigroup. We will always assume that $1 \notin \sigma(\Delta)$ and define $-S$ by

$$\mathcal{D}(S) = (I - \Delta)^{-1}[\mathcal{D}(S_0)], \qquad -S = -S_0(I - \Delta).$$

Before deriving our main perturbation result, we prove the following lemma.

Lemma 3.1. *Let $-S_0$ be the generator of an analytic bisemigroup and Δ a bounded linear operator such that $1 \notin \sigma(\Delta)$. Suppose that*

1. *there exist δ and γ with $0 < \delta \leq (\pi/2)$ and $\gamma > 0$ such that the set $\Omega_{\delta,\gamma}$ defined by (3.3) is contained in the resolvent set of $S = S_0(I - \Delta)$, and*
2. $\int_{-\infty}^{\infty} \|H(t; -S_0)\Delta\| \, dt < \infty$.

Then $-S$ is the generator of an analytic bisemigroup.

Proof. There exists $\varepsilon > 0$ such that (2.3) is true. Using the resolvent identity (2.4), we obtain the convolution integral equation

$$E(t; -S)x - \int_{-\infty}^{\infty} H(t - \tau; -S_0)\Delta E(\tau; -S)x \, d\tau = E(t; -S_0)x, \qquad (3.5)$$

where $x \in \mathcal{H}$ and $0 \neq t \in \mathbb{R}$. By assumption, in (3.5) the convolution kernel $H(\cdot; -S_0)\Delta$ is continuous in the norm except for a jump discontinuity in $t = 0$ and satisfies $\int_{-\infty}^{\infty} e^{\varepsilon|t|} \, \|H(t; -S_0)\Delta\| \, dt < \infty$. Indeed, the integral is an improper integral at 0 and at $\pm\infty$. Convergence at $t = 0$ is guaranteed by the second assumption. Convergence at $\pm\infty$ follows from (2.3) and a line of argument as on page 103 (bottom) of [5], which together prove that $H(t; -S_0)$ is exponentially decaying. Thus $e^{\varepsilon|\cdot|}H(\cdot; -S_0)\Delta$ is Bochner integrable.

The symbol of the convolution integral equation (3.5), which equals $I - \Delta + \lambda(\lambda - S_0)^{-1}\Delta = (\lambda - S_0)^{-1}(\lambda - S)$, tends to I in the norm as $\lambda \to \infty$ in the strip $|\mathrm{Re}\,\lambda| < \varepsilon$, because of the Riemann-Lebesgue lemma. Thus there exists $\varepsilon_0 \in (0, \min(\varepsilon, \gamma)]$ such that the symbol only takes invertible values on the strip $|\mathrm{Re}\,\lambda| \leq \varepsilon_0$. By the Bochner-Phillips theorem [4], the convolution equation (3.5) has a unique solution $u(\cdot; x) = E(\cdot; -S)x$ with the following properties:

1) $E(\cdot; -S)$ is strongly continuous, except for a jump discontinuity at $t = 0$,
2) $\int_{-\infty}^{\infty} e^{\varepsilon_0|t|}\|E(t; -S)\| \, dt < \infty$; hence $E(\cdot; -S)$ is exponentially decaying,
3) the identity (2.2) holds.

As a result [2], $-S$ is exponentially dichotomous. □

The following result has been established in [7] for the case in which S_0 is the inverse of a bounded and injective selfadjoint operator on a Hilbert space. In [7] it has been sketched how the arguments used to prove the Hilbert space case can also be applied to prove the Banach space case, without rendering details.

Theorem 3.2. *Let $-S_0$ be the generator of an analytic bisemigroup and Δ a compact operator such that $1 \notin \sigma(\Delta)$ and $S = S_0(I - \Delta)$ does not have purely imaginary eigenvalues. Suppose that*

$$\int_{-\infty}^{\infty} \|H(t; -S_0)\Delta\| \, dt < \infty. \qquad (3.6)$$

Then $-S$ is the generator of an analytic bisemigroup.

Proof. It suffices to prove the first condition of Lemma 3.1. Indeed, since the symbol of the convolution equation (3.5) is a compact perturbation of the identity and is invertible on a strip $|\mathrm{Re}\,\lambda| \le \varepsilon_0$ about the imaginary axis while it has invertible limits in the operator norm as $\lambda \to 0$ and $\lambda \to \pm i\infty$, the spectrum of S in this strip must consist of finitely many normal eigenvalues. Thus the first condition of Lemma 3.1 amounts to requiring the absence of purely imaginary eigenvalues of S, as assumed. □

It is clear from the proof that the hypotheses of Theorem 3.2 can be replaced by the hypotheses that $-S_0$ is the generator of an analytic bisemigroup, $(\lambda - S_0)^{-1}\Delta$ is compact for purely imaginary λ, $1 \notin \sigma(\Delta)$, $S = S_0(I - \Delta)$ does not have purely imaginary eigenvalues, and (3.6) holds. It is not necessary to have Δ itself compact.

It is well known that sectorial operators have fractional powers [12]. Thus generators $-S = (-A_0)\dot{+}A_1$ of analytic bisemigroups, where $-A_0$ and $-A_1$ are generators of uniformly exponentially stable analytic semigroups, have fractional powers defined by $|S|^\alpha \stackrel{\text{def}}{=} (-A_0)^\alpha \dot{+} (-A_1)^\alpha$ for any $\alpha \in \mathbb{R}$. Moreover,

$$\||S|^\alpha E(t;-S)\| = O(|t|^{-\alpha}), \qquad t \to 0^\pm; \tag{3.7}$$

$$\exists c > 0 : \||S|^\alpha E(t;-S)\| = O(|t|^{-\alpha}\, e^{-c|t|}), \qquad t \to \pm\infty. \tag{3.8}$$

As a result of (3.7)–(3.8) we have

$$\left\||S|^{-\alpha} H(t;-S)\right\| = O(|t|^{\alpha-1}), \qquad t \to 0^\pm;$$

$$\exists c > 0 : \left\||S|^{-\alpha} H(t;-S)\right\| = O(|t|^{\alpha-1}\, e^{-c|t|}), \qquad t \to \pm\infty.$$

The following corollary is now clear.

Corollary 3.3. *Let $-S_0$ be the generator of an analytic bisemigroup and Δ a compact operator such that $1 \notin \sigma(\Delta)$, $S = S_0(I - \Delta)$ does not have purely imaginary eigenvalues, and $Im\,\Delta \subset \mathcal{D}(|S_0|^\alpha)$ for some $\alpha > 0$. Then $-S$ is the generator of an analytic bisemigroup.*

3.3. Canonical factorization and matching of subspaces

The following results can all be found in [7] for the case in which S_0 is the inverse of a bounded and injective selfadjoint operator on a Hilbert space.

Theorem 3.4. *Suppose $\mathcal{X}$ is a complex Banach space. Let $-S_0$ be the generator of an analytic bisemigroup and Δ a bounded operator with $1 \notin \sigma(\Delta)$ such that $(\lambda - S_0)^{-1}\Delta$ is compact for purely imaginary λ, $S_0(I - \Delta)$ does not have purely imaginary eigenvalues, and (3.6) is true. Let P_0 and P stand for the separating projections of $-S_0$ and $-S$, respectively. Then the following statements are equivalent:*

(a) *The operator function*

$$W(\lambda) = (\lambda - S_0)^{-1}(\lambda - S) = I_{\mathcal{X}} - \Delta + \lambda(\lambda - S_0)^{-1}\Delta, \quad |Re\,\lambda| \le \varepsilon, \tag{3.9}$$

has a left canonical factorization with respect to the imaginary axis.

(b) *We have the decomposition*

$$Ker\,P\dot{+}Im\,P_0 = \mathcal{X}. \tag{3.10}$$

(c) *For some (and hence every) $E(\mathbb{R}^+;\mathcal{X})$, the vector-valued Wiener-Hopf equation*

$$\phi(t) - \int_0^\infty H(t-\tau;-S_0)\Delta\phi(\tau)\,d\tau = g(t), \qquad t>0, \tag{3.11}$$

is uniquely solvable in $E(\mathbb{R}^+;\mathcal{X})$ for any $g \in E(\mathbb{R}^+;\mathcal{X})$.

Theorem 3.5. *Suppose $\mathcal{X}$ is a complex Banach space. Let $-S_0$ be the generator of an analytic bisemigroup and Δ a bounded operator with $1 \notin \sigma(\Delta)$ such that $(\lambda - S_0)^{-1}\Delta$ is compact for purely imaginary λ, $S_0(I-\Delta)$ does not have purely imaginary eigenvalues, and (3.6) is true. Let P_0 and P stand for the separating projections of $-S_0$ and $-S$, respectively. Then the following statements are equivalent:*

(a) *The operator function*

$$W(\lambda) = (\lambda - S_0)^{-1}(\lambda - S) = I_{\mathcal{X}} - \Delta + \lambda(\lambda - S_0)^{-1}\Delta, \quad |Re\,\lambda| \le \varepsilon,$$

has a right canonical factorization with respect to the imaginary axis.

(b) *We have the decomposition*

$$Ker\,P_0\dot{+}Im\,P = \mathcal{X}. \tag{3.12}$$

(c) *For some (and hence every) $E(\mathbb{R}^-;\mathcal{X})$, the vector-valued Wiener-Hopf equation*

$$\phi(t) - \int_{-\infty}^0 H(t-\tau;-S_0)\Delta\phi(\tau)\,d\tau = g(t), \qquad t<0, \tag{3.13}$$

is uniquely solvable in $E(\mathbb{R}^-;\mathcal{X})$ for any $g \in E(\mathbb{R}^-;\mathcal{X})$.

Corollary 3.6. *Suppose $\mathcal{H}$ is a complex Hilbert space. Let $-S_0$ be the generator of an analytic bisemigroup and Δ a bounded operator with $1 \notin \sigma(\Delta)$ such that $(\lambda - S_0)^{-1}\Delta$ is compact for purely imaginary λ, $S_0(I-\Delta)$ does not have purely imaginary eigenvalues, and (3.6) is true. Let P_0 and P be the separating projections of $-S_0$ and $-S$, respectively. Suppose*

$$\sup_{Re\,\lambda=0} \|-\Delta + \lambda(\lambda - S_0)^{-1}\Delta\| < 1.$$

Then all of the following statements are true:

(a) *The operator function $W(\cdot)$ in (3.9) has a left and a right canonical factorization with respect to the imaginary axis.*

(b) *We have the decompositions (3.10) and (3.12).*

(c) *For some (and hence every) $E(\mathbb{R}^\pm;\mathcal{H})$, the vector-valued Wiener-Hopf equation (3.11) [(3.13), respectively] is uniquely solvable in $E(\mathbb{R}^\pm;\mathcal{H})$ for any $g \in E(\mathbb{R}^\pm;\mathcal{H})$.*

Acknowledgment

The authors are greatly indebted to Prof. Karl-Heinz Förster for a question suggesting a generalization of previous results on bisemigroup perturbation.

References

[1] H. Bart, I. Gohberg, and M.A. Kaashoek, *Minimal Factorization of Matrix and Operator Functions.* Birkhäuser OT **1**, Basel and Boston, 1979.

[2] H. Bart, I. Gohberg, and M.A. Kaashoek, *Wiener-Hopf factorization, inverse Fourier transforms and exponentially dichotomous operators.* J. Funct. Anal. **68** (1986), 1–42.

[3] H. Bart, I. Gohberg, and M.A. Kaashoek, *Wiener-Hopf equations with symbols analytic in a strip.* In: I. Gohberg and M.A. Kaashoek, eds., *Constructive Methods of Wiener-Hopf Factorization.* Birkhäuser OT **21**, Basel, 1986, pp. 39–74.

[4] S. Bochner and R.S. Phillips, *Absolutely convergent Fourier expansions for non-commutative normed rings.* Ann. Math. **43** (1942), 409–418.

[5] K.-J. Engel and R. Nagel, *One-parameter Semigroups for Linear Evolution Equations.* Springer GTM **194**, Berlin, 2000.

[6] I.C. Gohberg and J. Leiterer, *Factorization of operator functions with respect to a contour. II. Canonical factorization of operator functions close to the identity.* Math. Nachrichten **54** (1972), 41–74 [Russian].

[7] W. Greenberg, C.V.M. van der Mee, and V. Protopopescu, *Boundary Value Problems in Abstract Kinetic Theory.* Birkhäuser OT **23**, Basel and Boston, 1987.

[8] M.A. Kaashoek, C.V.M. van der Mee, and A.C.M. Ran, *Wiener-Hopf factorization of transfer functions of extended Pritchard-Salamon realizations.* Math. Nachrichten **196** (1998), 71–102.

[9] H. Langer, A.C.M. Ran, and B.A. van de Rotten, *Invariant subspaces of infinite-dimensional Hamiltonians and solutions of the corresponding Riccati equations.* In: I. Gohberg and H. Langer, eds., *Linear Operators and Matrices.* Birkhäuser OT **130**, Basel and Boston, 2001, pp. 235–254.

[10] H. Langer and C. Tretter, *Spectral decomposition of some nonselfadjoint block operator matrices.* J. Operator Theory **39** (1998), 339–359.

[11] H. Langer and C. Tretter, *Diagonalization of certain block operator matrices and applications to Dirac operators.* In: H. Bart, I. Gohberg, and A.C.M. Ran, eds., *Operator Theory and Analysis.* Birkhäuser OT **122**, Basel and Boston, 2001, pp. 331–358.

[12] A. Lunardi, *Analytic Semigroups and Optimal Regularity in Parabolic Problems,* Birkhäuser PNDEA **16**, Basel and Boston, 1995.

[13] C.V.M. van der Mee, *Transport theory in L_p-spaces.* Integral Equations and Operator Theory **6** (1983), 405–443.

[14] A.C.M. Ran and C. van der Mee, *Perturbation results for exponentially dichotomous operators on general Banach spaces.* J. Func. Anal. 210 (2004), 193–213.

Cornelis V.M. van der Mee
Dipartimento di Matematica e Informatica
Università di Cagliari
Viale Merello 92
I-09123 Cagliari, Italy
e-mail: `cornelis@bugs.unica.it`

André C.M. Ran
Afdeling Wiskunde, FEW
Vrije Universiteit
De Boelelaan 1081a
NL-1081 HV Amsterdam, The Netherlands
e-mail: `ran@cs.vu.nl`

Operator Theory:
Advances and Applications, Vol. 160, 425–439

Factorization of Block Triangular Matrix Functions with Off-diagonal Binomials

Cornelis V.M. van der Mee, Leiba Rodman and Ilya M. Spitkovsky

Dedicated to Israel Gohberg on the occasion of his 75th birthday

Abstract. Factorizations of Wiener–Hopf type are considered in the abstract framework of Wiener algebras of matrix-valued functions on connected compact abelian groups, with a non-archimedean linear order on the dual group. A criterion for factorizability is established for 2×2 block triangular matrix functions with elementary functions on the main diagonal and a binomial expression in the off-diagonal block.

Mathematics Subject Classification (2000). Primary 47A68. Secondary 43A17.

Keywords. Wiener–Hopf factorization, Wiener algebras, linearly ordered groups.

1. Introduction and the main result

Let G be a (multiplicative) connected compact abelian group and let Γ be its (additive) character group. Recall that Γ consists of continuous homomorphisms of G into the group of unimodular complex numbers. Since G is compact, Γ is discrete. In applications, often Γ is an additive subgroup of $\mathbb{R}$, the group of real numbers, or of $\mathbb{R}^k$, and G is the Bohr compactification of Γ. The group G can be also thought of as the character group of Γ, an observation that will be often used.

The group G has a unique invariant measure ν satisfying $\nu(G) = 1$, while Γ is equipped with the discrete topology and the (translation invariant) counting measure. It is well known [17] that, because G is connected, Γ can be made into a linearly ordered group. So let $\preceq$ be a linear order such that $(\Gamma, \preceq)$ is an ordered group, i. e., if $x, y, z \in \Gamma$ and $x \preceq y$, then $x + z \preceq y + z$. *Throughout the paper it will*

The research of van der Mee leading to this article was supported by INdAM and MIUR under grants No. 2002014121 and 2004015437. The research of Rodman and Spitkovsky was partially supported by NSF grant DMS-9988579.

be assumed that Γ *is ordered with a fixed linear order* $\preceq$. The notations $\prec$, $\succeq$, $\succ$, max, min (with obvious meaning) will also be used. We put $\Gamma_+ = \{x \in \Gamma : x \succeq 0\}$ and $\Gamma_- = \{x \in \Gamma : x \preceq 0\}$.

For any nonempty set M, let $\ell^1(M)$ stand for the complex Banach space of all complex-valued M-indexed sequences $\boldsymbol{x} = \{x_j\}_{j \in M}$ having at most countably many nonzero terms that are finite with respect to the norm

$$\|\boldsymbol{x}\|_1 = \sum_{j \in M} |x_j|.$$

Then $\ell^1(\Gamma)$ is a commutative Banach algebra with unit element with respect to the convolution product $(\boldsymbol{x} * \boldsymbol{y})_j = \sum_{k \in \Gamma} x_k \, y_{j-k}$. Further, $\ell^1(\Gamma_+)$ and $\ell^1(\Gamma_-)$ are closed subalgebras of $\ell^1(\Gamma)$ containing the unit element.

Given $a = \{a_j\}_{j \in \Gamma} \in \ell^1(\Gamma)$, by the *symbol* of a we mean the complex-valued continuous function $\hat{a}$ on G defined by

$$\hat{a}(g) = \sum_{j \in \Gamma} a_j \langle j, g \rangle, \qquad g \in G, \tag{1}$$

where $\langle j, g \rangle$ stands for the action of the character $j \in \Gamma$ on the group element $g \in G$ (thus, $\langle j, g \rangle$ is a unimodular complex number), or, by Pontryagin duality, of the character $g \in G$ on the group element $j \in \Gamma$. The set

$$\sigma(\hat{a}) := \{j \in \Gamma \, : \, a_j \neq 0\}$$

will be called the *Fourier spectrum* of $\hat{a}$ given by (1). Since Γ is written additively and G multiplicatively, we have

$$\langle \alpha + \beta, g \rangle = \langle \alpha, g \rangle \cdot \langle \beta, g \rangle, \qquad \alpha, \beta \in \Gamma, \quad g \in G,$$
$$\langle \alpha, gh \rangle = \langle \alpha, g \rangle \cdot \langle \alpha, h \rangle, \qquad \alpha \in \Gamma, \quad g, h \in G.$$

We will use the shorthand notation e_α for the function $e_\alpha(g) = \langle \alpha, g \rangle$, $g \in G$. Thus, $e_{\alpha+\beta} = e_\alpha e_\beta$, $\alpha, \beta \in \Gamma$.

The set of all symbols of elements $a \in \ell^1(\Gamma)$ forms an algebra $W(G)$ of continuous functions on G. The algebra $W(G)$ (with pointwise multiplication and addition) is isomorphic to $\ell^1(\Gamma)$. Denote by $W(G)_+$ (resp., $W(G)_-$) the algebra of symbols of elements in $\ell^1(\Gamma_+)$ (resp., $\ell^1(\Gamma_-)$).

We have the following result. For every unital Banach algebra $\mathcal{A}$ we denote its group of invertible elements by $\mathcal{G}(\mathcal{A})$.

Theorem 1. *Let* G *be a compact abelian group with character group* Γ, *and let* $W(G)^{n \times n}$ *be the corresponding Wiener algebra of* $n \times n$ *matrix functions. Then* $\hat{A} \in \mathcal{G}(W(G)^{n \times n})$ *if and only if* $\hat{A}(g) \in \mathcal{G}(\mathbb{C}^{n \times n})$ *for every* $g \in G$.

This is an immediate consequence of Theorem A.1 in [8] (also proved in [1], and see [14]).

We now consider the discrete abelian subgroup Γ' of Γ and denote its character group by G'. Then we introduce the annihilator

$$\Lambda = \{g \in G : \langle j, g \rangle = 1 \text{ for all } j \in \Gamma'\}, \tag{2}$$

which is a closed subgroup of G and hence a compact group. According to Theorem 2.1.2 in [17], we have $G' \simeq (G/\Lambda)$.

Let us now introduce the natural projection $\pi : G \to G/\Lambda$. We observe that the above theorem also applies to $W(G')^{n\times n}$. Given $A \in \ell^1(\Gamma)^{n\times n}$ with its Fourier spectrum restricted to Γ' (i.e., $A_j = 0$ for $j \in \Gamma \setminus \Gamma'$), we have two symbol definitions:

$$\hat{A}_\Gamma(g) = \sum_{j\in\Gamma'} A_j\langle j, g\rangle, \qquad g \in G,$$

$$\hat{A}_{\Gamma'}(g) = \sum_{j\in\Gamma'} A_j\langle j, g\rangle, \qquad g \in G',$$

where we have taken into account that $A_j = 0$ for $j \in \Gamma \setminus \Gamma'$. The latter can be replaced by

$$\hat{A}_{\Gamma'}([g]) = \sum_{j\in\Gamma'} A_j\langle j, g\rangle, \qquad [g] \in (G/\Lambda),$$

where $[g] = \pi(g)$ for $g \in G$. Obviously, $\langle j, g\rangle$ only depends on $[g] = \pi(g)$ if $j \in \Gamma'$. (If $[g_1] = [g_2]$, then $g_1g_2^{-1} \in \Lambda$ and hence $\langle j, g_1g_2^{-1}\rangle = 1$ for all $j \in \Gamma'$, which implies the statement.) Thus the two symbol definitions are equivalent in the sense that the value of "the" symbol $\hat{A}$ on $g \in G$ only depends on $[g] = \pi(g)$.

Theorem 2. *Let Γ' be a subgroup of the discrete abelian group Γ, let G and G' be the character groups of Γ and Γ', respectively, and let Λ be defined by (2). If $\hat{A} \in W(G)^{n\times n}$ is an element which has all of its Fourier spectrum within Γ', then $\hat{A} \in \mathcal{G}(W(G')^{n\times n})$ if and only if $\hat{A}(g) \in \mathcal{G}(\mathbb{C}^{n\times n})$ for every $g \in G$.*

For the proof see [14].

We now consider factorizations. A *(left) factorization* of $A \in (W(G))^{n\times n}$ is a representation of the form

$$A(g) = A_+(g)\,(\mathrm{diag}\,(e_{j_1}(g), \dots, e_{j_n}(g)))\,A_-(g), \quad g \in G, \tag{3}$$

where $A_+ \in \mathcal{G}((W(G)_+)^{n\times n})$, $A_- \in \mathcal{G}((W(G)_-)^{n\times n})$, and $j_1, \dots, j_n \in \Gamma$. Here and elsewhere we use $\mathrm{diag}\,(x_1, \dots, x_n)$ to denote the $n \times n$ diagonal matrix with $x_1, \dots, x_n$ on the main diagonal, in that order. The elements j_k are uniquely defined (if ordered $j_1 \preceq j_2 \preceq \cdots \preceq j_n$); this can be proved by a standard argument (see [9, Theorem VIII.1.1]). The elements $j_1, \dots, j_n$ in (3) are called the *(left) factorization indices* of A.

If all factorization indices coincide with the zero element of Γ, the factorization is called *canonical.* If a factorization of A exists, the function A is called *factorizable.* For $\Gamma = \mathbb{Z}$ and G the unit circle, the definitions and the results are classical [10], [9], [4]; many results have been generalized to $\Gamma = \mathbb{R}^k$ (see [2] and references there), and Γ a subgroup of $\mathbb{R}^k$ (see [15],[16]). The notion of factorization in the abstract abelian group setting was introduced and studied, in particular, for block triangular matrices, in [14]. The present paper can be thought of as a follow up of [14].

In this paper we prove the following result.

Theorem 3. *Let A have the form*

$$A(g) = \begin{bmatrix} e_{\lambda}(g)I_p & 0 \\ c_1 e_{\sigma}(g) - c_2 e_{\mu}(g) & e_{-\lambda}(g)I_q \end{bmatrix}, \qquad g \in G, \tag{4}$$

and assume that $\lambda \succ 0$, $\mu \succ \sigma$, and

$$n\mu \prec \lambda, \qquad n\sigma \prec \lambda \quad \textit{for all integers } n. \tag{5}$$

Then A admits a factorization if and only if

$$\begin{aligned} \operatorname{rank}(\lambda_1 c_1 - \lambda_2 c_2) \quad &= \quad \max\{\operatorname{rank}(z_1 c_1 - z_2 c_2) \, : \, z_1, z_2 \in \mathbb{C}\} \\ &\textit{for every } \lambda_1, \lambda_2 \in \mathbb{C} \textit{ satisfying } |\lambda_1| = |\lambda_2| = 1. \end{aligned} \tag{6}$$

Moreover, in case a factorization exists, the factorization indices of A belong to the set

$$\{\pm\sigma, \pm\mu, \pm\lambda, \lambda - (\mu - \sigma), \ldots, \lambda - \min\{p, q\}(\mu - \sigma)\}.$$

We emphasize that the setting of Theorem 3 is a *non-archimedean* linearly ordered abelian group $(\Gamma, \succeq)$, in contrast with the archimedean linear order of $\mathbb{R}$ and its subgroups. The setting of non-archimedean, as well as archimedean, linearly ordered abelian subgroups was studied in [14].

2. Preliminary results on factorization

Theorem 4. *If A admits a factorization* (3), *and if the Fourier spectrum $\sigma(A)$ is bounded:*

$$\lambda_{\min} \preceq \sigma(A) \preceq \lambda_{\max},$$

for some $\lambda_{\min}, \lambda_{\max} \in \Gamma$, then the factorization indices are also bounded with the same bounds

$$\lambda_{\min} \preceq j_k \preceq \lambda_{\max}, \quad k = 1, 2, \ldots, n, \tag{7}$$

and moreover,

$$\sigma(A_-) \subseteq \{j \in \Gamma \, : \, -\lambda_{\max} + \lambda_{\min} \preceq j \preceq 0\}, \tag{8}$$

and

$$\sigma(A_+) \subseteq \{j \in \Gamma \, : \, 0 \preceq j \preceq \lambda_{\max} - \lambda_{\min}\}. \tag{9}$$

Proof. We follow well-known arguments. Rewrite (3) in the form

$$A_+^{-1}(e_{-\lambda_{\min}} A) = e_{-\lambda_{\min}} \Lambda A_-.$$

Since the left-hand side is in $W(G)_+^{n \times n}$, so is the right-hand side, and we have $j_k \succeq \lambda_{\min}$ for all $k = 1, \ldots, n$ (otherwise, A_- would contain a zero row, which is impossible because A_- is invertible). Analogously the second inequality in (7) is proved. Now

$$e_{\lambda_{\max} - \lambda_{\min}} A_- = \left(e_{\lambda_{\max}} \Lambda^{-1}\right) A_+^{-1} \left(e_{-\lambda_{\min}} A\right)$$

is a product of three matrix functions in $W(G)_+^{n\times n}$, and therefore also

$$e_{\lambda_{\max}-\lambda_{\min}} A_- \in W(G)_+^{n\times n}.$$

This proves (8); (9) is proved analogously. □

It follows from the proof that "one-sided" bounds are valid for the factorization indices:

$$\lambda_{\min} \preceq \sigma(A) \quad \Longrightarrow \quad \lambda_{\min} \preceq j_k, \quad \text{for} \quad k = 1, 2, \ldots, n;$$

$$\sigma(A) \preceq \lambda_{\max} \quad \Longrightarrow \quad j_k \preceq \lambda_{\max} \quad \text{for} \quad k = 1, 2, \ldots, n.$$

For future use we record the next corollary of Theorem 4. On $\Gamma_+ \setminus \{0\}$ we consider the equivalence relation (cf. [6])

$$i \sim j \Longleftrightarrow \exists n, m \in \mathbb{N} : (ni \succ j \text{ and } mj \succ i).$$

Here $\mathbb{N}$ is the set of positive integers. Any such i, j are called *archimedeally equivalent* (with respect to $(\Gamma, \preceq)$). The set $\mathrm{Arch}(\Gamma, \preceq)$ of archimedean equivalence classes, which are additive semigroups (in the sense that they are closed under addition), can be linearly ordered in a natural way. Given $\boldsymbol{J} \in \mathrm{Arch}(\Gamma, \preceq)$, it is easily seen that

$$\Gamma_{\boldsymbol{J}} := \{i - j : i, j \in \boldsymbol{J}\}$$

is the smallest additive subgroup of Γ containing $\boldsymbol{J}$ and that $\Gamma_{\boldsymbol{J}}$ in fact contains all archimedean components $\preceq \boldsymbol{J}$ in $\mathrm{Arch}(\Gamma, \preceq)$.

Before proceeding we first discuss some illustrative examples.

a. If $\mathbb{Z}^k$ is ordered lexicographically, in increasing order the archimedean components are as follows: $J_0 = (0)$, $J_1 = (0)_{k-1} \times \mathbb{N}$, $J_2 = (0)_{k-2} \times \mathbb{N} \times \mathbb{Z}$, $J_3 = (0)_{k-3} \times \mathbb{N} \times \mathbb{Z}^2$, …, $J_{k-1} = (0)_1 \times \mathbb{N} \times \mathbb{Z}^{k-2}$, and $J_k = \mathbb{N} \times \mathbb{Z}^{k-1}$.
b. $\mathbb{Z}^2$ with linear order $(i_1, i_2) \succ (0,0)$ whenever $i_1 + i_2\sqrt{5} > 0$. Then the ordered group is archimedean and in increasing order the archimedean components are $J_0 = (0)$ and $J_1 = \{i \in \mathbb{Z}^2 : i \succ 0\}$.
c. Let $(i_1, i_2) \succ (0,0)$ whenever $i_1 + i_2 > 0$. Then in increasing order the archimedean components are $J_0 = (0)$, $J_1 = \{(j, -j) : j \in \mathbb{N}\}$, and $J_2 = \{(i_1, i_2) : i_1 + i_2 > 0\}$.

We now have the following corollary.

Corollary 5. *If A admits a factorization* (3), *and if the Fourier spectrum $\sigma(A)$ is contained in $\Gamma_{\boldsymbol{J}}$ for some $\boldsymbol{J} \in \mathrm{Arch}(\Gamma, \preceq)$, then*

$$j_k \in \Gamma_{\boldsymbol{J}}, \qquad k = 1, \ldots, n,$$

and

$$\sigma(A_-^{\pm 1}) \in \Gamma_{\boldsymbol{J}}, \qquad \sigma(A_+^{\pm 1}) \in \Gamma_{\boldsymbol{J}}.$$

Indeed, in addition to using Theorem 4 we need only to observe that if $X \in \mathcal{G}(W(G)^{n\times n})$ is such that $\sigma(X) \subseteq \Gamma'$ for some subgroup $\Gamma' \subseteq \Gamma$, then $\sigma(X^{-1}) \subseteq \Gamma'$, and apply this observation for $X = A_\pm$ (the observation follows easily from Theorem 2).

Corollary 5 may be considered as asserting the hereditary property of Fourier spectra for additive subgroups of Γ of the form Γ_j. We say that a subgroup Γ' of Γ has the *hereditary property* if for each matrix function A that admits a factorization (3), and the Fourier spectrum of A is contained in Γ', we have that the factorization indices as well as the Fourier spectra of $A_\pm$ and of $A_\pm^{-1}$ are also contained in Γ'. This notion was introduced in [16] for Γ the additive group $\mathbb{R}^k$; the hereditary property of certain subgroups of $\mathbb{R}^k$ was proved there as well. It is an open question whether or not the hereditary property holds for every subgroup of the character group of every connected compact abelian group.

A factorization (3) will be called *finitely generated* if the Fourier spectra of A_+ and of A_- are contained in some finitely generated subgroup of Γ. Clearly, a necessary condition for existence of a finitely generated factorization of A is that the Fourier spectrum of A is contained in a finitely generated subgroup of Γ. We shall prove below that this condition is also sufficient.

In the proof of the following theorem we make use of a natural projection: If $B \in W(G)^{n\times n}$ is given by the series

$$B(g) = \sum_{j\in\Gamma} B_j \langle j, g\rangle, \qquad g \in G,$$

and if Ω is a subset of Γ, we define B_Ω by

$$B_\Omega(g) = \sum_{j\in\Omega} B_j \langle j, g\rangle, \qquad g \in G.$$

Clearly, $B_\Omega \in W(G)^{n\times n}$ and the Fourier spectrum of B_Ω is contained in Ω.

Theorem 6. *If $A \in W(G)^{n\times n}$ is factorizable, and if the Fourier spectrum of A is contained in a finitely generated subgroup of Γ, then A admits a finitely generated factorization.*

Proof. Let $\tilde{\Gamma}$ be a finitely generated subgroup of Γ that contains the Fourier spectrum of A. Let (3) be a factorization of A. Since $(W(G)_\pm)^{n\times n}$ are unital Banach algebras, the set of invertible elements $\mathcal{G}((W(G)_\pm)^{n\times n})$ is open in $(W(G)_\pm)^{n\times n}$. Thus, there exists a finitely generated subgroup $\breve{\Gamma}$ of Γ with the following properties:

(a) $\breve{\Gamma}$ contains $\tilde{\Gamma}$;
(b) $\breve{\Gamma}$ contains the elements $j_1, \dots, j_n$;
(c) $(A_-)_\Omega$ and $(A_+^{-1})_\Omega$ are invertible in $(W(G)_\pm)^{n\times n}$ for every set $\Omega \supseteq \breve{\Gamma}$.

For verification of (c), note the following estimate:

$$\begin{aligned} \|A_- - (A_-)_\Omega\|_{(W(G)_\pm)^{n\times n}} &= \sum_{j\in\Gamma\setminus\Omega} \|(A_-)_j\| \le \sum_{j\in\Gamma\setminus\breve{\Gamma}} \|(A_-)_j\| \\ &= \|A_- - (A_-)_{\breve{\Gamma}}\|_{(W(G)_\pm)^{n\times n}}. \end{aligned}$$

Letting $\breve{G}$ be the dual group of $\breve{\Gamma}$, by Theorem 2 and (3) we have

$$(A_+^{-1})_{\breve{\Gamma}}, (A_-)_{\breve{\Gamma}} \in \mathcal{G}((W(\breve{G})_\pm)^{n\times n}).$$

Rewrite the equality (3) in the form

$$(A_+(g))^{-1}A(g) = (\mathrm{diag}\,(e_{j_1}(g),\dots,e_{j_n}(g)))\,A_-(g).$$

Write also (omitting the argument $g \in G$ in the formulas)

$$\begin{aligned}&\left((A_+^{-1})_{\breve{\Gamma}} + ((A_+)^{-1})_{\Gamma_+\setminus\breve{\Gamma}}\right)A\\ &= (\mathrm{diag}\,(e_{j_1},\dots,e_{j_n}))\left((A_-)_{\breve{\Gamma}} + (A_-)_{\Gamma_-\setminus\breve{\Gamma}}\right).\end{aligned} \tag{10}$$

Since $j_1,\dots,j_n \in \breve{\Gamma}$ and the Fourier spectrum of A is contained in $\breve{\Gamma}$, (10) implies

$$(A_+^{-1})_{\breve{\Gamma}}A = (\mathrm{diag}\,(e_{j_1},\dots,e_{j_n}))\,(A_-)_{\breve{\Gamma}}.$$

Rewriting this equality in the form

$$A = \left((A_+^{-1})_{\breve{\Gamma}}\right)^{-1}(\mathrm{diag}\,(e_{j_1},\dots,e_{j_n}))\,(A_-)_{\breve{\Gamma}},$$

we obtain a finitely generated factorization of A. □

Theorem 7. *Let A be given as in Theorem 3 with $p = q$, and assume that (5) holds. If the matrix c_1 is invertible and the spectrum of $c_1^{-1}c_2$ does not intersect the unit circle, or if c_2 is invertible and the spectrum of $c_2^{-1}c_1$ does not intersect the unit circle, then A admits a finitely generated factorization. Moreover, the factorization indices belong to the set $\{\pm\sigma, \pm\mu\}$.*

For the proof see [14]. In fact, the proof of Theorem 7 shows more detailed information about the factorization indices:

Theorem 8. *Under the hypotheses of Theorem 7, assume that c_1 is invertible and the spectrum of $c_1^{-1}c_2$ does not intersect the unit circle, and let r be the dimension of the spectral subspace of $c_1^{-1}c_2$ corresponding to the eigenvalues inside the unit circle. Then the factorization indices of A are σ (r times), $-\sigma$ (r times), μ ($p-r$ times), and $-\mu$ ($p-r$ times).*

If c_2 is invertible and the spectrum of $c_2^{-1}c_1$ does not intersect the unit circle, then the factorization indices of A are μ (r times), $-\mu$ (r times), σ ($p-r$ times), and $-\sigma$ ($p-r$ times), where r be the dimension of the spectral subspace of $c_2^{-1}c_1$ corresponding to the eigenvalues inside the unit circle.

Finally, we present a result concerning linearly ordered groups that will be used in the next section.

Proposition 9. *Let $(\Gamma, \preceq)$ be a finitely generated additive ordered abelian group. Let Γ_0 stand for the additive subgroup of Γ generated by all archimedean equivalence classes preceding the archimedean equivalence class E. Then there exists an additive subgroup Γ_1 of Γ such that the direct sum decomposition*

$$\Gamma = \Gamma_0 \dot{+} \Gamma_1 \tag{11}$$

holds and the coordinate projection $\Gamma \to \Gamma_1$ is $\preceq$-order preserving.

Proof. With no loss of generality we assume that $\Gamma = \mathbb{Z}^k$ and that the order $\preceq$ on $\mathbb{Z}^k$ has been extended to a so-called term order on $\mathbb{R}^k$. That is, if $x \preceq y$ in $\mathbb{R}^k$, $z \in \mathbb{R}^k$ and $c \geq 0$, then $x+z \preceq y+z$ and $cx \preceq cy$. Such an extension is always possible but is often nonunique [3]. There now exists an orthonormal basis $\{e_1, \ldots, e_k\}$ of $\mathbb{R}^k$ and a decreasing sequence $\{H_0, H_1, \ldots, H_k\}$ of linear subspaces of $\mathbb{R}^k$ with $\dim H_r = k - r$ $(r = 0, 1, \ldots, k)$ such that $e_r \succ 0$, $e_r \in H_{r-1}$ and $e_r \perp H_r$ $(r = 1, \ldots, k)$ (cf. [5]). Here we note that the orthonormal basis is completely determined by the term order $\preceq$ on $\mathbb{R}^k$, with the one-to-one correspondence between term order (on $\mathbb{R}^k$) and orthonormal basis given by

$$x = (x_1, \ldots, x_k) \succ (0, \ldots, 0) \Leftrightarrow \begin{cases} x_1 > 0 \text{ or} \\ x_1 = 0 \text{ and } x_2 > 0, \text{ or} \\ \vdots \\ x_1 = \cdots = x_{k-1} = 0 \text{ and } x_k > 0. \end{cases}$$

Indeed, put $H_0 = \mathbb{R}^k$ and let H_1 stand for the set of those points in $\mathbb{R}^k$ all of whose neighborhoods contain elements of both Γ_+ and Γ_-. Then H_1 is a linear subspace of $\mathbb{R}^k$ of dimension $k - 1$ [5]. We now let e_1 be the unique unit vector in $\mathbb{R}^k$ that is $\preceq$-positive and orthogonal to H_1 and restrict the term order to H_1. We now repeat the same construction in H_1 and find a linear subspace H_2 of H_1 of dimension $k - 2$ and a unique $\preceq$-positive unit vector e_2 in H_1 orthogonal to H_2. After finitely many such constructions we arrive at the sequence of linear subspaces $\mathbb{R}^k = H_0 \supset H_1 \supset \cdots \supset H_{k-1} \supset H_k = \{0\}$ and the orthonormal basis $e_1, \ldots, e_k$ of $\mathbb{R}^k$ as indicated above.

Next, let $\tilde{H}_r$ be the smallest linear subspace of $\mathbb{R}^k$ spanned by $H_r \cap \mathbb{Z}^k$ $(r = 0, 1, \ldots, k)$. From this nonincreasing set of linear subspaces of $\mathbb{R}^k$ we select a maximal strictly decreasing set of nontrivial linear subspaces $\mathbb{R}^k = L_0 \supset L_1 \supset \cdots \supset L_{\mu-1} \neq \{0\}$. Also let ν be the largest among the integers $s \in \{1, \ldots, k\}$ such that $L_{\mu-1}$ is spanned by $H_{s-1} \cap \mathbb{Z}^k$; then $H_\nu \cap \mathbb{Z}^k = \{0\}$. If $\mu = 1$, we have $H_1 \cap \mathbb{Z}^k = \{0\}$, so that the ordered group $(\mathbb{Z}^k, \preceq)$ is archimedean; in that case $i \mapsto \xi_1(i) \stackrel{\text{def}}{=} (i, e_1)$ (i.e., the signed distance from i to H_1) is an order preserving group homomorphism from $(\mathbb{Z}^k, \preceq)$ into $\mathbb{R}$. On the other hand, if $\mu \geq 2$, we let (i) $\xi_1(i)$ stand for the signed distance from i to H_1 and $p_1(i)$ for the orthogonal projection of i onto L_1, (ii) $\xi_r(i)$ for the signed distance from $p_{r-1}(i)$ to H_q for $q = \min\{s : L_r = \operatorname{span}(H_s \cap \mathbb{Z}^k)\}$ and $p_r(i)$ for the orthogonal projection of $p_{r-1}(i)$ onto L_r $(r = 2, \ldots, \mu-1)$, and finally (iii) $\xi_\mu(i)$ as the signed distance from $p_{\mu-1}(i)$ to H_ν. In this way

$$i \stackrel{\varphi}{\mapsto} (\xi_1(i), \ldots, \xi_\mu(i))$$

is an order preserving group homomorphism from $(\mathbb{Z}^k, \preceq)$ into $\mathbb{R}^\mu$ with lexicographical order. It then appears that μ is the number of nontrivial (i.e., different from $\{0\}$) archimedean components. Moreover, in increasing order the archimedean components of $(\Gamma, \preceq)$ are now as follows:

$$\boldsymbol{J}_0 = \{0\}, \qquad \boldsymbol{J}_r = [L_{\mu-r} \cap \Gamma_+] \setminus \cup_{s=0}^{r-1} \boldsymbol{J}_s \ (r = 1, \ldots, \mu). \tag{12}$$

The additive subgroups of $\mathbb{Z}^k$ generated by the smallest archimedean components are as follows:

$$\Gamma_{J_0} = \{0\}, \qquad \Gamma_{J_r} = L_{\mu-r} \cap \Gamma \ (r = 1, \dots, \mu). \tag{13}$$

Let us now define the group homomorphisms π_r on Γ with image Γ_{J_r} and q_r with kernel Γ_{J_r} by $\pi_0 = 0$, q_0 equal the identity, and

$$\begin{cases} \pi_r i = \varphi^{-1}(0, \dots, 0, \xi_{\mu-r+1}(i), \dots, \xi_\mu(i)), \\ q_r i = \varphi^{-1}(\xi_1(i), \dots, \xi_{\mu-r}(i), 0, \dots, 0). \end{cases} \tag{14}$$

Then the fact that the linear order on $\varphi[\mathbb{Z}^k] \subset \mathbb{R}^\mu$ is lexicographical, implies that the additive group homomorphisms $q_0, q_1, \dots, q_\mu$ are order preserving, but $\pi_1, \dots, \pi_{\mu-1}$ are not. Putting $\Gamma'_{J_r} = q_r[\Gamma]$ we obtain the direct sum decomposition

$$\Gamma = \Gamma_{J_r} \dot{+} \Gamma'_{J_r}, \qquad r = 0, 1, \dots, \mu,$$

which completes the proof. □

3. Proof of Theorem 3

Using Theorem 6 we can assume without loss of generality that Γ is finitely generated, and furthermore assume that $\Gamma = \mathbb{Z}^k$ for some positive integer k.

Consider the part "if". Applying the transformation

$$c_1 \mapsto Sc_1T, \qquad c_2 \mapsto Sc_2T,$$

for suitable invertible matrices S and T, we may assume that the pair (c_1, c_2) is in the Kronecker normal form (see, e.g, [7]); in other words, c_1 and c_2 are direct sums of blocks of the following types:

(a) c_1 and c_2 are of size $k \times (k+1)$ of the form

$$c_1 = \begin{bmatrix} I_k & 0_{k\times 1} \end{bmatrix}, \quad c_2 = \begin{bmatrix} 0_{k\times 1} & I_k \end{bmatrix}.$$

(b) c_1 and c_2 are of size $(k+1) \times k$ of the form

$$c_1 = \begin{bmatrix} I_k \\ 0_{1\times k} \end{bmatrix}, \quad c_2 = \begin{bmatrix} 0_{1\times k} \\ I_k \end{bmatrix}.$$

(c) c_1 is the $k \times k$ upper triangular nilpotent Jordan block, denoted by V_k, and $c_2 = I_k$.

(d) $c_1 = I_k$, and $c_2 = V_k$.

(e) c_1 and c_2 are both invertible of the same size.

(f) c_1 and c_2 are both zero matrices of the same size.

Note that if c_1 (resp., c_2) is invertible, then condition (6) is equivalent to the condition that the spectrum of $c_1^{-1}c_2$ (resp., of $c_1c_2^{-1}$) does not intersect the unit circle. Thus, by Theorem 7 we are done in cases (c), (d), and (e), as well as in the trivial case (f).

Consider the cases (a) and (b), where the condition (6) is obviously satisfied. We follow arguments similar to those presented in [13], and also in the proof of [14, Theorem 7].

Let J_k be the $k\times k$ matrix with 1's along the top-right to the left-bottom diagonal and zeros in all other positions. If $A(g) = [a_{i,j}(g)]_{i,j=1}^n \in (W(G))^{n\times n}$, then A^* will denote the matrix function defined by $[\overline{a_{j,i}(g)}]_{i,j=1}^n$; clearly, $A^* \in (W(G))^{n\times n}$, and if $A \in (W(G)_\pm)^{n\times n}$, then $A^* \in (W(G)_\mp)^{n\times n}$. The transformation

$$A \mapsto \begin{bmatrix} 0 & J_{k+1} \\ J_k & 0 \end{bmatrix} A^* \begin{bmatrix} 0 & J_k \\ J_{k+1} & 0 \end{bmatrix}$$

transforms the case (b) to the case (a). Thus, it will suffice to consider the case (a):

$$A = \begin{bmatrix} e_\lambda I_k & 0 & 0 \\ 0 & e_\lambda & 0 \\ e_\sigma I_k - e_\mu V_k & h & e_{-\lambda} I_k \end{bmatrix}, \quad \text{where} \quad h = \begin{bmatrix} 0_{(k-1)\times 1} \\ -e_\mu \end{bmatrix}.$$

Let

$$B_+ = \begin{bmatrix} I_k - e_{\mu-\sigma} V_k & b & -e_{\lambda-\sigma} I_k \\ 0 & 1 & 0 \\ 0 & 0 & \sum_{j=0}^{k-1} e_{j(\mu-\sigma)} V_k^j \end{bmatrix}, \quad \text{where} \quad b = \begin{bmatrix} 0_{(k-1)\times 1} \\ -e_{\mu-\sigma} \end{bmatrix},$$

$$B_- = \begin{bmatrix} \sum_{j=0}^{k-1} e_{j(\mu-\sigma)-\lambda-\sigma} V_k^j & 0 & I_k \\ 0 & 1 & 0 \\ -I_k & 0 & 0 \end{bmatrix}.$$

Clearly,

$$B_+ \in \mathcal{G}((W(G)_+)^{(2k+1)\times(2k+1)}) \quad \text{and} \quad B_- \in \mathcal{G}((W(G)_-)^{(2k+1)\times(2k+1)})$$

(the latter inclusion follows from (5) and from $\mu \succeq \sigma$). A direct computation shows that

$$\Phi_0 := B_+ A B_- = \begin{bmatrix} e_{-\sigma} I_k & 0 & 0 \\ 0 & e_\lambda & 0 \\ 0 & h_k & e_\sigma I_k \end{bmatrix},$$

where

$$(h_k)^T = \begin{bmatrix} -e_{(k-1)(\mu-\sigma)+\mu} & \cdots & -e_{(\mu-\sigma)+\mu} & -e_\mu \end{bmatrix}. \tag{15}$$

Define for $j = 0, 1, \ldots, k-1$ the auxiliary matrices

$$R_{+,k-j} = \begin{bmatrix} 1 & 0 & e_{\lambda-\mu-j(\mu-\sigma)} \\ 0 & I_{k-j-1} & h_{k-j-1} e_{-\sigma} \\ 0 & 0 & 1 \end{bmatrix},$$

$$R_{-,k-j} = \begin{bmatrix} e_{\sigma-\mu} & 0 & -1 \\ 0 & I_{k-j-1} & 0 \\ 1 & 0 & 0 \end{bmatrix}, \quad R_{k-j} = \begin{bmatrix} e_{\lambda-j(\mu-\sigma)} & 0 \\ h_{k-j} & e_\sigma I_{k-j} \end{bmatrix}.$$

Clearly, $R_{-,k-j} \in \mathcal{G}((W(G)_-)^{(k-j+1)\times(k-j+1)})$, and in view of (5),

$$R_{+,k-j} \in \mathcal{G}((W(G)_+)^{(k-j+1)\times(k-j+1)}).$$

We also have the recurrence relations

$$R_{+,k-j}R_{k-j}R_{-,k-j} = \begin{bmatrix} R_{k-j-1} & 0 \\ 0 & e_{\mu} \end{bmatrix}, \quad R_0 = e_{\lambda-k(\mu-\sigma)}, \tag{16}$$

for $j = 0,\dots,k-1$. Note that $\Phi_0 = \operatorname{diag}(e_{-\sigma}I_k, R_k)$. Applying consecutively (16) for $j = 0,\dots,k-1$, we obtain a factorization $A = A_+\Lambda A_-$ with $\Lambda = \operatorname{diag}(e_{-\sigma}I_k, e_{\lambda-k(\mu-\sigma)}, e_{\mu}I_k)$. This completes the proof of the "if" part of the theorem.

For the part "only if", we make use of the archimedean structure on $\Gamma = (\mathbb{Z}^k, \preceq)$ (see the previous section). Let Γ_0 be the subgroup of Γ generated by all archimedean classes of Γ that are $\prec \lambda$. Condition (5) guarantees that $0 \neq \mu-\sigma \in \Gamma_0$ and hence that $\Gamma_0 \neq \{0\}$. Since

$$\alpha \in \Gamma, \quad n\alpha \in \Gamma_0 \quad \text{for some} \quad n \in \mathbb{N} \qquad \Longrightarrow \qquad \alpha \in \Gamma_0,$$

it follows that

$$\Gamma = \Gamma_0 \dot{+} \Gamma_1, \tag{17}$$

a direct sum, for some subgroup Γ_1 of $\Gamma = \mathbb{Z}^k$, where the coordinate projection onto Γ_1 along Γ_0 is order preserving (Proposition 9). Also, by [11, Theorem 23.18], we may assume

$$G = G_0 \times G_1, \tag{18}$$

where G_j is the character group of Γ_j, $j = 0, 1$. We write

$$\lambda = \lambda_0 + \lambda_1, \quad \mu = \mu_0 + \mu_1, \quad \sigma = \sigma_0 + \sigma_1,$$

in accordance with (17). By construction of Γ_0, we have $\lambda_1 \succ 0$, and by (5) $\mu, \sigma \in \Gamma_0$, and so $\mu_1 = \sigma_1 = 0$.

Assume that A has a factorization

$$A(g) = A_+(g)\left(\operatorname{diag}(e_{j_1}(g),\dots,e_{j_n}(g))\right)A_-(g), \quad g \in G. \tag{19}$$

In accordance with (17) and (18) write

$$j_k = j_{k,0} + j_{k,1}, \quad j_{k,0} \in \Gamma_0, \quad j_{k,1} \in \Gamma_1, \quad k = 1,\dots,n,$$

$$g = g_0 g_1, \qquad g_0 \in G_0, \quad g_1 \in G_1,$$

and consider the equation (19) in which g_0 is kept fixed, whereas g_1 is kept variable. To emphasize this interpretation, we write (19) in the form

$$\begin{aligned} A_{g_0}(g_1) \;&=\; A_{+,g_0}(g_1)\left(\operatorname{diag}(e_{j_{1,0}}(g_0),\dots,e_{j_{n,0}}(g_0))\right) \\ &\cdot\; \left(\operatorname{diag}(e_{j_{1,1}}(g_1),\dots,e_{j_{n,1}}(g_1))\right)A_{-,g_0}(g_1). \end{aligned} \tag{20}$$

We consider Γ_1 with the linear order induced by $(\Gamma, \preceq)$. Since the property that $\alpha = \alpha_0 + \alpha_1 \in \Gamma_{\pm}$, where $\alpha_j \in \Gamma_j$, $j = 0, 1$, implies that $\alpha_1 \in (\Gamma_1)_{\pm}$, we obtain

$$A_{\pm,g_0} \in \mathcal{G}((W(G_1)_{\pm})^{n\times n})$$

for every $g_0 \in \Gamma_0$. Thus, (20) is in fact a factorization of $A_{g_0}(g_1)$ whose factorization indices are $j_{1,1},\dots,j_{n,1}$, and moreover we have the following property:

(ℵ) the factorization indices of $A_{g_0}(g_1)$ are independent of $g_0 \in G_0$.

Arguing by contradiction, we assume that

$$\operatorname{rank}(\lambda_1 c_1 - \lambda_2 c_2) \quad < \quad \max\{\operatorname{rank}(z_1 c_1 - z_2 c_2) \, : \, z_1, z_2 \in \mathbb{C}\}$$
$$\text{for some } \lambda_1, \lambda_2 \in \mathbb{C} \text{ satisfying } |\lambda_1| = |\lambda_2| = 1. \quad (21)$$

A contradiction will be obtained with Property ($\aleph$). We can assume, using the Kronecker normal from (see, e.g., [7]), that c_1 and c_2 have the form

$$c_1 = \operatorname{diag}(c_{1,1}, \dots, c_{1,s}), \quad c_2 = \operatorname{diag}(c_{2,1}, \dots, c_{2,s}),$$

where each pair of blocks $(c_{1,w}, c_{2,w})$ has one of the forms (a) - (f). After a permutation transformation, we obtain (keeping the same notation for the transformed $A_{g_0}(g_1)$):

$$A_{g_0}(g_1) = \operatorname{diag}(A_{g_0,1}(g_1), \dots, A_{g_0,s}(g_1)),$$

where

$$A_{g_0,w}(g_1) = \begin{bmatrix} e_{\lambda_1}(g_1) I_{p_w} & 0 \\ c_{1,w} e_\beta(g_0) - c_{2,w} e_\kappa(g_0) & e_{-\lambda_1}(g_1) I_{q_w} \end{bmatrix} Q, \quad w = 1, \dots, s,$$

with $\beta, \kappa \in \Gamma_0$ independent of w, and Q is a diagonal matrix (also independent of w) with terms of the form $e_\alpha(g_0)$, $\alpha \in \Gamma_0$ on the main diagonal. Note that $\beta \neq \kappa$ (otherwise we would have $\mu = \sigma$, which is excluded by the hypotheses of the theorem). The "if" part of the theorem shows that $A_{g_0,w}(g_1)$ is factorable with indices independent of g_0 if the pair $(c_{1,w}, c_{2,w})$ has one of the forms (a), (b), (c), (d), and (f).

Suppose that the pair $(c_{1,w}, c_{2,w})$ is of the form (e). Then we may further assume that $c_{1,w} = I$ and $c_{2,w}$ is in the Jordan form:

$$c_{2,w} = J_{\tau_1}(\rho_1) \oplus \cdots \oplus J_{\tau_u}(\rho_u),$$

where $J_{\tau_j}(\rho_j)$ is the upper triangular $\tau_j \times \tau_j$ Jordan block with the eigenvalue ρ_j (for notational simplicity, we suppress the dependence of ρ_j, τ_j, and u on w in the notation used). Accordingly, after a permutation transformation we have $A_{g_0,w}(g_1) Q^{-1}$ in the following form:

$$\tilde{A}_{g_0,w}(g_1) = \operatorname{diag}(\tilde{A}_{g_0,w,1}(g_1), \dots, \tilde{A}_{g_0,w,u}(g_1)),$$

where

$$\tilde{A}_{g_0,w,j}(g_1) = \begin{bmatrix} e_{\lambda_1}(g_1) I_{\tau_j} & 0 \\ e_\beta(g_0) I_{\tau_j} - e_\kappa(g_0) J_{\tau_j}(\rho_j) & e_{-\lambda_1}(g_1) I_{\tau_j} \end{bmatrix}, \; j = 1, \dots, u.$$

If $|\rho_j| \neq 1$, then by the "if" part of the theorem, the factorization indices of $\tilde{A}_{g_0,w,j}(g_1)$ are independent of g_0 (this can be also checked directly). Assume $|\rho_j| = 1$; then the factorization indices of $\tilde{A}_{g_0,w,j}(g_1)$ equal zero if

$$e_\beta(g_0) - \rho_j e_\kappa(g_0) \neq 0. \tag{22}$$

Indeed,

$$\begin{bmatrix} e_{\lambda_1} I & 0 \\ S & e_{-\lambda_1} I \end{bmatrix} = \begin{bmatrix} I & e_{\lambda_1} S^{-1} \\ 0 & I \end{bmatrix} \begin{bmatrix} 0 & -S^{-1} \\ S & e_{-\lambda_1} I \end{bmatrix},$$

where

$$S = e_\beta(g_0)I_{\tau_j} - e_\kappa(g_0)J_{\tau_j}(\rho_j).$$

If $|\rho_j| = 1$ and

$$e_\beta(g_0) - \rho_j e_\kappa(g_0) = 0, \tag{23}$$

then the factorization indices of $\tilde{A}_{g_0,w,j}(g_1)$ are zeros ($2(\tau_j - 1)$ times) and $\pm\lambda_1$. It follows that for the values of g_0 such that

$$e_\beta(g_0) - \rho e_\kappa(g_0) \neq 0 \tag{24}$$

for any eigenvalue ρ of $c_{2,w}$ the factorization indices of $A_{g_0,w}(g_1)$ are all zeros, whereas in case the equality

$$e_\beta(g_0) - \rho e_\kappa(g_0) = 0 \tag{25}$$

holds for some eigenvalue ρ of $c_{2,w}$ not all factorization indices of $A_{g_0,w}(t_1)$ are zeros. Since $\kappa \neq \beta$, the range of the function

$$e_{\beta-\kappa}(g_0) = e_\beta(g_0)\left(e_\kappa(g_0)\right)^{-1}$$

coincides with the unit circle (since G is connected and the characters are continuous), and therefore by hypothesis (21) there do exist eigenvalues ρ of $c_{2,w}$ for which (25) holds. We obtain a contradiction with Property $\aleph$.

This completes the proof of Theorem 3.

4. Invertibility vs factorizability

The following conjecture was stated in [14].

Conjecture 10. *Every function $A \in \mathcal{G}(W(G)^{n\times n})$ admits a factorization if and only if Γ (as an abstract group without regard to $\succeq$) is isomorphic to a subgroup of the additive group of rational numbers $\mathbb{Q}$.*

Regarding this conjecture we quote a result from [14]:

Theorem 11. *If Γ is not isomorphic to a subgroup of $\mathbb{Q}$, then there exists a 2×2 matrix function of the form*

$$A(g) = \begin{bmatrix} e_\lambda(g) & 0 \\ c_1 e_{\alpha_1}(g) + c_2 + c_3 e_{\alpha_3}(g) & e_{-\lambda}(g) \end{bmatrix}, \qquad g \in G, \tag{26}$$

where $\lambda, \alpha_1, \alpha_2, \alpha_3 \in \Gamma$, and $c_1, c_2, c_3 \in \mathbb{C}$, which does not admit a factorization with the factors $A_\pm$ and their inverses $A_\pm^{-1}$ having finite Fourier spectrum.

We improve on Theorem 11:

Theorem 12. *If Γ is not isomorphic (as an abstract group) to a subgroup of $\mathbb{Q}$, then there exists a 2×2 matrix function of the form (26) which is not factorable.*

Proof. Consider two cases: (1) Γ is archimedean. Then $(\Gamma, \succeq)$ is isomorphic to a subgroup of the additive group of real numbers (Hölder's theorem, see, e.g., [6]) and since Γ is not isomorphic to a subgroup of $\mathbb{Q}$, there exist non-commensurable elements $x, y \in \Gamma \setminus \{0\}$. Using x and y, a known construction (see [12], also [2, Section 8.5]) may be used to produce a 2×2 matrix function of the required form.

(2) Γ is not archimedean. Then there exist $\sigma = 0 \prec \mu \prec \lambda \in \Gamma$ such that (5) holds. Theorem 3 now implies that the function

$$\begin{bmatrix} e_\lambda(g) & 0 \\ 1 - e_\mu(g) & e_{-\lambda}(g) \end{bmatrix}, \qquad g \in G,$$

is not factorable. □

Theorem 12 and its proof show that if Γ is not isomorphic to a subgroup of $\mathbb{Q}$, then there exists a 2×2 matrix function of the form

$$A(g) = \sum_{j=1}^{k} c_j e_{\alpha_j}(g), \qquad \det A(g) \equiv 1,$$

which is not factorable, with $k = 5$ if Γ is archimedean, and $k = 4$ if Γ is not archimedean. On the other hand, for every linearly ordered group Γ, every $n \times n$ matrix function of the form

$$A(g) = c_1 e_{\alpha_1}(g) + c_2 e_{\alpha_2}(g)$$

with $\det A(g) \neq 0$, $g \in G$, is factorable. Indeed, this follows easily from the Kronecker form of the pair of matrices (c_1, c_2). This leaves the following problem open:

Problem 13. *Assume that Γ is not isomorphic to a subgroup of $\mathbb{Q}$.*

(a) *If the subgroup generated by $\alpha_1, \alpha_2, \alpha_3 \in \Gamma$ is not archimedean, prove or disprove that every $n \times n$ matrix function of the form*

$$A(g) = c_1 e_{\alpha_1}(g) + c_2 e_{\alpha_2}(g) + c_3 e_{\alpha_3}(g)$$

with $\det A(g) \neq 0$, $g \in G$, is factorable.

(b) *If Γ is archimedean, prove or disprove that every $n \times n$ matrix function of the form $A(g) = \sum_{j=1}^{k} c_j e_{\alpha_j}(g)$ with $\det A(g) \neq 0$, $g \in G$, and with $k = 3$ or $k = 4$, is factorable.*

References

[1] G.R. Allan, *One-sided inverses in Banach algebras of holomorphic vector-valued functions*, J. London Math. Soc. **42**, 463–470 (1967).

[2] A. Böttcher, Yu.I. Karlovich, and I.M. Spitkovsky, *Convolution Operators and Factorization of Almost Periodic Matrix Functions*, Birkhäuser OT 131, Basel and Boston, 2002.

[3] L. Cerlienco and M. Mureddu, *Rappresentazione matriciale degli ordini l.c. su $\mathbb{R}^n$ e su $\mathbb{N}^n$*, Rend. Sem. Fac. Sc. Univ. Cagliari 66, 49–68 (1996).

[4] K.F. Clancey and I. Gohberg, *Factorization of Matrix Functions and Singular Integral Operators*, Birkhäuser OT 3, Basel and Boston, 1981.

[5] J. Erdős, *On the structure of ordered real vector spaces*, Publ. Math. Debrecen 4, 334–343 (1956).

[6] L. Fuchs, *Partially Ordered Algebraic Systems*, Pergamon Press, Oxford, 1963.

[7] F.R. Gantmacher, *Applications of the Theory of Matrices*, Interscience Publishers, New York, 1959. (Translation from Russian.)

[8] I.C. Gohberg and Yu. Leiterer, *Factorization of operator functions with respect to a contour. II. Canonical factorization of operator functions close to the identity*, Math. Nachr. **54**, 41–74 (1972). (Russian)

[9] I.C. Gohberg and I.A. Feldman, *Convolution Equations and Projection Methods for their Solution*, Transl. Math. Monographs 41, Amer. Math. Soc., Providence, R. I., 1974.

[10] I.C. Gohberg and M.G. Krein, *Systems of integral equations on a half line with kernels depending on the difference of arguments*, Amer. Math. Soc. Transl. (2)14, 217–287 (1960).

[11] E. Hewitt and K.A. Ross, *Abstract Harmonic Analysis I*, 2nd edition, Springer-Verlag, Berlin, Heidelberg, New York, 1979.

[12] Yu.I. Karlovich and I.M. Spitkovsky, *On the Noether property for certain singular integral operators with matrix coefficients of the class SAP and the systems of convolution equations on a finite interval connected with them*, Soviet Math. Doklady 27, 358–363 (1983).

[13] Yu.I. Karlovich and I.M. Spitkovsky, *Factorization of almost periodic matrix functions and (semi)-Fredholmness of some convolution type equations*, No. 4421–85 dep., VINITI, Moscow, 1985. (Russian).

[14] C.V.M. van der Mee, L. Rodman, I.M. Spitkovsky, and H. J. Woerdeman, *Factorization of block triangular matrix functions in Wiener algebras on ordered abelian groups*. In: J.A. Ball, J.W. Helton, M. Klaus, and L. Rodman (eds.), *Current Trends in Operator Theory and its Applications*, Birkhäuser OT 149, Basel and Boston, 2004, pp. 441–465.

[15] L. Rodman, I.M. Spitkovsky, and H.J. Woerdeman, *Carathéodory–Toeplitz and Nehari problems for matrix-valued almost periodic functions*, Trans. Amer. Math. Soc. 350, 2185–2227 (1998).

[16] L. Rodman, I.M. Spitkovsky, and H.J. Woerdeman, *Noncanonical factorizations of almost periodic multivariable matrix functions*, Operator Theory: Advances and Applications 142, 311–344 (2003).

[17] W. Rudin, *Fourier Analysis on Groups*, John Wiley, New York, 1962.

Cornelis V.M. van der Mee
Dipartimento di Matematica e Informatica
Università di Cagliari
Viale Merello 92
I-09123 Cagliari, Italy
e-mail: `cornelis@bugs.unica.it`

Leiba Rodman and Ilya M. Spitkovsky
Department of Mathematics
The College of William and Mary
Williamsburg, VA 23187-8795, USA
e-mail: `lxrodm@math.wm.edu`
e-mail: `ilya@math.wm.edu`

Operator Theory:
Advances and Applications, Vol. 160, 441–468

Closely Connected Unitary Realizations of the Solutions to the Basic Interpolation Problem for Generalized Schur Functions

Gerald Wanjala

Abstract. A generalized Schur function which is holomorphic at $z = 0$ can be written as the characteristic function of a closely connected unitary colligation with a Pontryagin state space. We describe the closely connected unitary colligation of a solution $s(z)$ of the basic interpolation problem for generalized Schur functions (studied in [3]) in terms of the interpolation data and the canonical unitary colligation of the parameter function $s_1(z)$ appearing in the formula for $s(z)$. In particular, we consider the case where the interpolation data and the Taylor coefficients of $s_1(z)$ at $z = 0$ are real. We also show that the canonical unitary colligation of $s_1(z)$ can be recovered from that of $s(z)$.

Mathematics Subject Classification (2000). Primary 47A48, 47B32, 47B50.

Keywords. Schur transform, generalized Schur function, reproducing kernel Pontryagin space, J-unitary colligation, closely connected colligation, realization.

1. Introduction

We recall that a *Schur* function is a holomorphic function $s(z)$ defined on the open unit disc $\mathbb{D}$ with the property that $|s(z)| \leq 1$, $z \in \mathbb{D}$, and that a *generalized Schur function $s(z)$ with κ negative squares* is a meromorphic function on $\mathbb{D}$ of the form

$$s(z) = \prod_{j=1}^{\kappa} \frac{1 - \alpha_j^* z}{z - \alpha_j} s_0(z), \tag{1.1}$$

where $\alpha_j \in \mathbb{D}$ and $s_0(z)$ is a Schur function with $s_0(\alpha_j) \neq 0$, $j = 1, 2, \ldots, \kappa$. Here κ is a nonnegative integer; evidently, a generalized Schur function with zero negative squares is a Schur function.

A generalized Schur function $s(z)$ which is holomorphic at $z = 0$ determines and is determined by a *realization* of the form

$$s(z) = s(0) + z\langle(1 - zA)^{-1}u, v\rangle_{\mathcal{P}}, \tag{1.2}$$

where A is an operator on some Pontryagin space $\mathcal{P}$, $u, v \in \mathcal{P}$, and the colligation

$$U = \begin{pmatrix} A & u \\ \langle\,\cdot\,, v\rangle & s(0) \end{pmatrix} : \begin{pmatrix} \mathcal{P} \\ \mathbb{C} \end{pmatrix} \to \begin{pmatrix} \mathcal{P} \\ \mathbb{C} \end{pmatrix}$$

is unitary, that is, $UU^* = U^*U = I$, and *closely connected*, which means

$$\mathcal{P} = \overline{\operatorname{span}}\,\{A^m u, A^{*n} v \mid m, n \geq 0\}.$$

The right-hand side of (1.2) is called the *characteristic function* of the colligation U and is denoted by $s_U(z)$. A unitary closely connected realization of $s(z)$ is uniquely determined up to isomorphism. The negative index of the state space $\mathcal{P}$ of any such realization equals the number of negative squares of $s(z)$. If $\mathcal{P}$ is the reproducing kernel space $\mathcal{D}(s)$ (see Section 2), the realization is unique and called *canonical.*

The basic interpolation problem for generalized Schur functions studied in [3] is as follows.

(BIP): *Given $\sigma_0 \in \mathbb{C}$, determine all generalized Schur functions $s(z)$ which are holomorphic at the origin and are such that $s(0) = \sigma_0$.*

It was shown in [3] that the solutions $s(z)$ are given by fractional linear transformations of the form

$$s(z) = T_{\Theta(z)}s_1(z) = \frac{a(z)s_1(z) + b(z)}{c(z)s_1(z) + d(z)}, \tag{1.3}$$

where

$$\Theta(z) = \begin{pmatrix} a(z) & b(z) \\ c(z) & d(z) \end{pmatrix} \tag{1.4}$$

is a polynomial matrix which depends on whether $|\sigma_0| < 1$, $|\sigma_0| > 1$ or $|\sigma_0| = 1$ and the parameter $s_1(z)$ runs through a set of generalized Schur functions which are holomorphic at $z = 0$, also depending on these three cases. The polynomial matrix $\Theta(z)$ is a generalized Schur function relative to the signature matrix

$$J = \begin{pmatrix} 1 & 0 \\ 0 & -1 \end{pmatrix},$$

and hence $\Theta(z)$ can also be written as the characteristic function of a canonical unitary colligation. For more details see Section 2 which contains the preliminaries about canonical unitary realizations for matrix-valued generalized Schur functions and the basic interpolation problem.

In Section 3 for each of the three cases $|\sigma_0| < 1$, $|\sigma_0| > 1$ or $|\sigma_0| = 1$ we describe the state space $\mathcal{D}(\Theta)$ in the canonical unitary realization of $\Theta(z)$.

In Section 4 we construct a closely connected unitary realization of the solution $s(z)$ from the canonical unitary realizations of the corresponding parameter $s_1(z)$ and the polynomial matrix $\Theta(z)$. The state space $\mathcal{P}$ in the realization of $s(z)$

is a finite-dimensional extension of the state space in the realization of $s_1(z)$, and the main operator A in the realization of $s(z)$ is an extension to $\mathcal{P}$ of a finite-dimensional perturbation of the main operator A_1 in the realization of $s_1(z)$. See Theorems 4.2 and 4.3. We also show how the canonical unitary realization of the parameter can be recovered from the closely connected unitary realization of the solution. See Theorem 4.4. Similar results can be found in [3] where closely outer connected coisometric realizations are considered. The fractional linear transformations mentioned above are related to the Schur algorithm for generalized Schur functions developed in [9], [11], [8], and [10]. The connection between the Schur algorithm for generalized Schur functions and their coisometric and unitary realizations have been investigated in [1, 2] and [4], respectively. In these papers a direct method was used which differs from the approach in this paper (and in [3]), where reproducing kernel Pontryagin spaces are the main tool.

In Section 5 we consider the case where the interpolation data and the Taylor coefficients of $s_1(z)$ at $z = 0$ are real. Then the Taylor coefficients of $s(z)$ at $z = 0$ are also real. According to [5] there exist unique signature operators J_s on $\mathcal{D}(s)$ and J_{s_1} on $\mathcal{D}(s_1)$ such that the main operators A and A_1 are J_s-selfadjoint and J_{s_1}-selfadjoint respectively. In cases $|\sigma_0| \neq 1$ we give explicit formulas relating the two signature operators. In the case $|\sigma_0| = 1$ the connection is rather complicated and we consider a special case. These results are new even in the case where only Schur functions are considered, that is, when $\kappa = 0$ and $|\sigma_0| < 1$.

2. Preliminaries

2.1. Realizations

Let J be an $n \times n$ signature matrix, that is, $J = J^* = J^{-1}$. By $\mathbf{S}_\kappa(\mathbb{C}^n, J)$ we denote the class of generalized Schur functions with κ negative squares. These are the $n \times n$ matrix-valued functions $S(z)$ which are meromorphic on $\mathbb{D}$ and for which the kernel

$$K_S(z,w) := \frac{J - S(z)JS(w)^*}{1 - zw^*} : \mathbb{C}^n \to \mathbb{C}^n$$

has κ negative squares, or, equivalently, the kernel

$$D_S(z,w) = \begin{pmatrix} \dfrac{J - S(z)JS(w)^*}{1 - zw^*} & \dfrac{S(z) - S(w^*)}{z - w^*} \\ \dfrac{\tilde{S}(z) - \tilde{S}(w^*)}{z - w^*} & \dfrac{J - \tilde{S}(z)J\tilde{S}(w)^*}{1 - zw^*} \end{pmatrix} : \begin{pmatrix} \mathbb{C}^n \\ \mathbb{C}^n \end{pmatrix} \to \begin{pmatrix} \mathbb{C}^n \\ \mathbb{C}^n \end{pmatrix}$$

has κ negative squares, where $\tilde{S}(z) = S(z^*)^*$. That these kernels have the same number of negative squares can be shown as in [6]. If $J = I_{n\times n}$, $S(z) \in S_\kappa(\mathbb{C}^n, J)$ if and only if it is a product of the inverse of a Blaschke product and a Schur function. See, for example, [7, Section 4.2]; for the case $n = 1$ the formula for $S(z)$ is given by (1.1).

The set of generalized Schur functions which have κ negative squares and which are holomorphic at the origin will be denoted by $\mathbf{S}^0_\kappa(\mathbb{C}^n, J)$ and we set $\mathbf{S}^0(\mathbb{C}^n, J) = \cup_{\kappa\in\mathbb{N}}\mathbf{S}^0_\kappa(\mathbb{C}^n, J)$. To every $S(z) \in \mathbf{S}^0(\mathbb{C}^n, J)$ we associate two reproducing kernel Pontryagin spaces $\mathcal{H}(S)$ and $\mathcal{D}(S)$. These are the reproducing kernel Pontryagin spaces with reproducing kernels $K_S(z,w)$ and $D_S(z,w)$ respectively. They occur as state spaces in the coisometric and unitary realizations of the matrix function $S(z)$. For an elaborate treatment of these spaces we refer to [7, Section 2.1]. In this paper we focus on the unitary realizations.

If $\mathcal{P}$ is a Pontryagin space, all maximal uniformly negative subspaces have the same finite dimension and this number is called the *negative index* of $\mathcal{P}$ and is denoted by $\operatorname{ind}_-\mathcal{P}$. For $S \in \mathbf{S}^0_\kappa(\mathbb{C}^n, J)$ we have that $\operatorname{ind}_-\mathcal{D}(S) = \kappa$.

Theorem 2.1. *Let $S(z) \in \mathbf{S}^0(\mathbb{C}^n, J)$.*

(i) *The operators*

$$A : \mathcal{D}(S) \to \mathcal{D}(S), \quad A\begin{pmatrix} h \\ k \end{pmatrix}(z) = \begin{pmatrix} \dfrac{h(z)-h(0)}{z} \\ zk(z) - \tilde{S}(z)Jh(0) \end{pmatrix},$$

$$B : \mathbb{C}^n \to \mathcal{D}(S), \qquad Bf(z) = \begin{pmatrix} \dfrac{S(z)-S(0)}{z} \\ J - \tilde{S}(z)JS(0) \end{pmatrix} f,$$

$$C : \mathcal{D}(S) \to \mathbb{C}^n, \qquad C\begin{pmatrix} h \\ k \end{pmatrix} = h(0),$$

are bounded operators.

(ii) *Their adjoints are given by*

$$A^*\begin{pmatrix} h \\ k \end{pmatrix}(z) = \begin{pmatrix} zh(z) - S(z)Jk(0) \\ \dfrac{k(z)-k(0)}{z} \end{pmatrix}, \quad B^*\begin{pmatrix} h \\ k \end{pmatrix}(z) = k(0),$$

and for $c \in \mathbb{C}^n$,

$$(C^*c)(z) = \begin{pmatrix} J - S(z)J\widetilde{S}(0) \\ \dfrac{\widetilde{S}(z)-\widetilde{S}(0)}{z} \end{pmatrix} c.$$

(iii) *The colligation*

$$U = \begin{pmatrix} A & B \\ C & S(0) \end{pmatrix} : \begin{pmatrix} \mathcal{D}(S) \\ \mathbb{C}^n \end{pmatrix} \to \begin{pmatrix} \mathcal{D}(S) \\ \mathbb{C}^n \end{pmatrix} \tag{2.1}$$

is J-unitary, that is,

$$U\begin{pmatrix} I & 0 \\ 0 & J \end{pmatrix}U^* = U^*\begin{pmatrix} I & 0 \\ 0 & J \end{pmatrix}U = \begin{pmatrix} I & 0 \\ 0 & J \end{pmatrix},$$

and closely connected, which means that

$$\overline{\operatorname{span}}\,\{\operatorname{ran} A^m B,\ \operatorname{ran} A^{*n} C^* \,|m, n \geq 0\} = \mathcal{D}(S).$$

(iv) $S(z)$ *has the closely connected unitary realization*

$$S(z) = S(0) + zC(I - zA)^{-1}B. \tag{2.2}$$

The colligation U in the theorem is unique and is called the *canonical J-unitary colligation* for $S(z)$. The function on the right-hand side of (2.2) is called the *characteristic function* of U and is denoted by $S_U(z)$. The realization (2.2) of $S(z)$ is called the *canonical J-unitary realization* of $S(z)$. To indicate the dependence on $S(z)$, we sometimes write U_S, A_S, B_S, etc. instead of U, A, B, etc. Any other J-unitary closely connected colligation whose characteristic function coincides with $S(z)$ is unitarily equivalent to the canonical unitary colligation. By this we mean that if $\mathcal{P}'$ is a Pontryagin space such that

$$U' = \begin{pmatrix} A' & B' \\ C' & D' \end{pmatrix} : \begin{pmatrix} \mathcal{P}' \\ \mathbb{C}^n \end{pmatrix} \to \begin{pmatrix} \mathcal{P}' \\ \mathbb{C}^n \end{pmatrix}$$

is J-unitary:

$$U' \begin{pmatrix} I & 0 \\ 0 & J \end{pmatrix} U^{'*} = U^{'*} \begin{pmatrix} I & 0 \\ 0 & J \end{pmatrix} U' = \begin{pmatrix} I & 0 \\ 0 & J \end{pmatrix}$$

and closely connected:

$$\overline{\operatorname{span}} \left\{ \operatorname{ran} A^{'m} B', \operatorname{ran} A^{'*n} C^{'*}, \,|\, m, n \geq 0 \right\} = \mathcal{P}'$$

and $S_{U'}(z) = D' + zC'(I - zA')^{-1}B' = S(z)$, then there exists an isomorphism $W : \mathcal{D}(S) \to \mathcal{P}'$ such that

$$A' = WA_S W^{-1}, \ B' = WB_S, \ C'W = C_S, \text{ and } D' = S(0).$$

We note that Theorem 2.1 can be proved in a similar way as Theorem 2.3.1 in [7] with some minor modifications.

In the sequel we only consider the cases $n = 1, J = 1$, and $n = 2, J = \begin{pmatrix} 1 & 0 \\ 0 & -1 \end{pmatrix}$. In the first case we write $s(z)$ and $\mathbf{S}^0$ instead of $S(z)$ and $\mathbf{S}^0(\mathbb{C}, 1)$.

If $n = 1$ the colligation U in (2.1) can also be written in the form

$$U = \begin{pmatrix} A & u \\ \langle \,\cdot\,, v \rangle_{\mathcal{D}(s)} & s(0) \end{pmatrix} : \begin{pmatrix} \mathcal{D}(s) \\ \mathbb{C} \end{pmatrix} \to \begin{pmatrix} \mathcal{D}(s) \\ \mathbb{C} \end{pmatrix},$$

where $u, v \in \mathcal{D}(s)$ are given by

$$u(z) = B1(z) = \begin{pmatrix} \dfrac{s(z) - s(0)}{z} \\ 1 - \widetilde{s}(z)s(0) \end{pmatrix}, \quad v(z) = C^*1(z) = D_s(z, 0) \begin{pmatrix} 1 \\ 0 \end{pmatrix}.$$

The last equality follows from the reproducing property of the kernel $D_s(z, w)$.

2.2. The basic interpolation problem

In this subsection we recall from [3] the solutions of the basic interpolation problem (BIP) with a given $\sigma_0 \in \mathbb{C}$. We use the following notation. Given any k complex numbers $s_0 \neq 0, s_1, \ldots, s_{k-1}$ we form the polynomial

$$\begin{aligned} Q(z) &= Q(z; s_0, s_1, \ldots, s_{k-1}) \\ &= c_0 + c_1 z + \cdots + c_{k-1} z^{k-1} - (c_{k-1}^* z^{k+1} + c_{k-2}^* z^{k+2} + \cdots + c_0^* z^{2k}) \end{aligned}$$

of degree $2k$, where the coefficients $c_0, c_1, \ldots, c_{k-1}$ are determined by the relation

$$\begin{pmatrix} c_0 & 0 & \cdots & 0 \\ c_1 & c_0 & \cdots & 0 \\ \vdots & \ddots & \ddots & \vdots \\ c_{k-1} & \cdots & c_1 & c_0 \end{pmatrix} \begin{pmatrix} s_0 & 0 & \cdots & 0 \\ s_1 & s_0 & \cdots & 0 \\ \vdots & \ddots & \ddots & \vdots \\ s_{k-1} & \cdots & s_1 & s_0 \end{pmatrix} = \sigma_0 I_k.$$

Setting

$$p(z) = c_0 + c_1 z + \cdots + c_{k-1} z^{k-1} \tag{2.3}$$

we have that $Q(z) = p(z) - z^{2k} p(z^{-*})^*$, which implies that $-Q(z) = z^{2k} Q(z^{-*})^*$. For $s(z) \in \mathbf{S}^0$ we write its Taylor expansion at $z = 0$ as

$$s(z) = \sum_{n=0}^{\infty} \sigma_n z^n.$$

If $|\sigma_0| \leq 1$, then $s(z) \equiv \sigma_0$ is the constant solution of the problem (BIP). If $|\sigma_0| > 1$, then the function $s(z) \equiv \sigma_0$ does not belong to the class $\mathbf{S}^0$ and hence is not a solution of the problem (BIP). The function $s(z) \in \mathbf{S}^0$ is nonconstant if and only if $s(z) - \sigma_0$ has a zero at $z = 0$ of finite order and the following theorem describes all nonconstant solutions $s(z)$ of the problem (BIP) for which this order equals k.

Theorem 2.2. *Let k be an integer ≥ 1 and if $|\sigma_0| = 1$ let q be an integer ≥ 0, let $s_0 \neq 0, s_1, \ldots, s_{k-1}$ be any k complex numbers, and set $Q(z) = Q(z; s_0, s_1, \ldots, s_{k-1})$. Then the formula*

$$s(z) = \begin{cases} \dfrac{z^k s_1(z) + \sigma_0}{\sigma_0^* z^k s_1(z) + 1} & \text{if } |\sigma_0| < 1, \\[2ex] \dfrac{\sigma_0 s_1(z) + z^k}{s_1(z) + \sigma_0^* z^k} & \text{if } |\sigma_0| > 1, \\[2ex] \dfrac{(Q(z) + z^k) s_1(z) - \sigma_0 Q(z) z^q}{\sigma_0^* Q(z) s_1(z) - (Q(z) - z^k) z^q} & \text{if } |\sigma_0| = 1, \end{cases} \tag{2.4}$$

establishes a one to one correspondence between all nonconstant solutions $s(z) \in \mathbf{S}^0$ of the problem (BIP) *with the property that in all three cases*

$$\sigma_1 = \sigma_2 = \cdots = \sigma_{k-1} = 0 \ \textit{and} \ \sigma_k \neq 0,$$

and in the case $|\sigma_0| = 1$ with the additional property

$$\sigma_j = s_{j-k}, \quad j = k, k+1, \ldots, 2k-1,$$

and all parameters $s_1(z) \in \mathbf{S}^0$ *with*

$$s_1(0) \neq \begin{cases} 0 & \text{if } |\sigma_0| \neq 1, \text{ or } |\sigma_0| = 1 \text{ and } q > 0, \\ \sigma_0 & \text{if } |\sigma_0| = 1 \text{ and } q = 0. \end{cases} \tag{2.5}$$

Consider the case $|\sigma_0| < 1$. If in formula (2.4) we let k vary over all integers ≥ 1 and replace $z^{k-1}s_1(z)$ by $s_2(z)$, then Theorem 2.2 implies that the formula

$$s(z) = \frac{zs_2(z) + \sigma_0}{\sigma_0^* zs_2(z) + 1}$$

gives a one to one correspondence between all solutions $s(z) \in \mathbf{S}^0$ and all parameters $s_2(z) \in \mathbf{S}^0$. The constant solution $s(z) \equiv \sigma_0$ corresponds to the case $s_2(z) \equiv 0$. The function $s(z) - \sigma_0$ has a zero of order k at $z = 0$ if and only if the corresponding parameter $s_2(z)$ has a zero of order $k - 1$ at $z = 0$, that is, $s_2(z) = z^{k-1}s_1(z)$ for some $s_1(z) \in \mathbf{S}^0$ with $s_1(0) \neq 0$. In the case $|\sigma_0| = 1$ one gets the set of all solutions of the problem (BIP) first by describing a nonnegative integer q and k arbitrary complex numbers $s_0 \neq 0, s_1, \ldots, s_{k-1}$ and then by applying a fractional linear transformation with a slightly restricted class of parameters $s_1(z) \in \mathbf{S}^0$. Formally, the constant solution $s(z) \equiv \sigma_0$ can be obtained from (2.4) by replacing $Q(z)$ by ∞.

The expressions in formula (2.4) are fractional linear transformations of the form (1.3), where $\Theta(z)$ in (1.4) is given by

$$\Theta(z) = \Theta_1(z) = \frac{1}{\sqrt{1 - |\sigma_0|^2}} \begin{pmatrix} 1 & \sigma_0 \\ \sigma_0^* & 1 \end{pmatrix} \begin{pmatrix} z^k & 0 \\ 0 & 1 \end{pmatrix}, \text{ if } |\sigma_0| < 1, \tag{2.6}$$

$$\Theta(z) = \Theta_2(z) = \frac{1}{\sqrt{|\sigma_0|^2 - 1}} \begin{pmatrix} \sigma_0 & 1 \\ 1 & \sigma_0^* \end{pmatrix} \begin{pmatrix} 1 & 0 \\ 0 & z^k \end{pmatrix}, \text{ if } |\sigma_0| > 1, \tag{2.7}$$

$$\Theta(z) = \Theta_3(z) = \begin{pmatrix} Q(z) + z^k & -\sigma_0 Q(z) \\ \sigma_0^* Q(z) & -Q(z) + z^k \end{pmatrix} \begin{pmatrix} 1 & 0 \\ 0 & z^q \end{pmatrix}, \text{ if } |\sigma_0| = 1. \tag{2.8}$$

If in (2.8) we have $q > 0$ we write $\Theta_3(z) = \Theta_3^0(z)\Psi_q(z)$, where

$$\Theta_3^0(z) = \begin{pmatrix} Q(z) + z^k & -\sigma_0 Q(z) \\ \sigma_0^* Q(z) & -Q(z) + z^k \end{pmatrix}, \quad \Psi_q(z) = \begin{pmatrix} 1 & 0 \\ 0 & z^q \end{pmatrix}.$$

For a proof of the following theorem we refer to [3, Theorem 3.2].

Theorem 2.3. *The three polynomial matrices* $\Theta_1(z), \Theta_2(z)$, *and* $\Theta_3(z)$ *are* J*-unitary and* $\Theta_1(z) \in \mathbf{S}_0^0(\mathbb{C}^2, J)$, $\Theta_2(z) \in \mathbf{S}_k^0(\mathbb{C}^2, J)$, *and* $\Theta_3(z) \in \mathbf{S}_{k+q}^0(\mathbb{C}^2, J)$.

Recall that here

$$J = \begin{pmatrix} 1 & 0 \\ 0 & -1 \end{pmatrix}$$

and that $\Theta(z)$ is J-unitary means that $\Theta(z)^* J \Theta(z) = J$ for all $z \in \mathbb{C}$ with $|z| = 1$. It follows that $\delta(z) := \det \Theta(z) = cz^\ell$ for some nonnegative integer ℓ and some complex number c with $|c| = 1$; in the cases where $\Theta = \Theta_1$, Θ_2, and Θ_3 we have $\delta(z) = z^k$, z^k, and z^{2k+q} respectively.

Remark: If $s(z)$ is a solution of the basic interpolation problem (BIP) with corresponding parameter function $s_1(z) \in \mathbf{S}^0_{\kappa_1}$ then $s(z) \in \mathbf{S}^0_{\kappa}$ where

$$\kappa = \begin{cases} \kappa_1 & \text{if } |\sigma_0| < 1, \\ \kappa_1 + k & \text{if } |\sigma_0| > 1, \\ \kappa_1 + k + q & \text{if } |\sigma_0| = 1. \end{cases}$$

This follows from the equality $\mathrm{ind}_-\mathcal{D}(s) = \mathrm{ind}_-\mathcal{D}(\Theta) + \mathrm{ind}_-\mathcal{D}(s_1)$ which is proved in Theorem 4.1 below.

3. The spaces $\mathcal{D}(\Theta_1)$, $\mathcal{D}(\Theta_2)$, and $\mathcal{D}(\Theta_3)$

In the following description of these spaces $\mathbf{o}$, $\mathbf{u}$, $\mathbf{e_1}$, $\mathbf{e_2}$ will stand for the vectors

$$\mathbf{o} = \begin{pmatrix} 0 \\ 0 \end{pmatrix}, \quad \mathbf{u} = \begin{pmatrix} 1 \\ \sigma_0^* \end{pmatrix}, \quad \mathbf{e_1} = \begin{pmatrix} 1 \\ 0 \end{pmatrix}, \mathbf{e_2} = \begin{pmatrix} 0 \\ 1 \end{pmatrix}$$

and Δ will be the $k \times k$ matrix

$$\Delta = \begin{pmatrix} 0 & 0 & 0 & 0 & \cdots & 0 \\ c_{k-1} & 0 & 0 & 0 & \cdots & 0 \\ c_{k-2} & c_{k-1} & 0 & 0 & \cdots & 0 \\ \vdots & \ddots & \ddots & \ddots & \ddots & \vdots \\ c_2 & \cdots & c_{k-2} & c_{k-1} & 0 & 0 \\ c_1 & \cdots & c_{k-3} & c_{k-2} & c_{k-1} & 0 \end{pmatrix}.$$

Theorem 3.1. *Let $\Theta_1(z)$, $\Theta_2(z)$ and $\Theta_3(z)$ be defined by (2.6)–(2.8).*

(i) *The space $\mathcal{D}(\Theta_1)$ is a Hilbert space spanned by the orthonormal basis*

$$\left\{ \begin{pmatrix} r z^{n-1} \mathbf{u} \\ z^{k-n} \mathbf{e_1} \end{pmatrix} \right\}_{n=1}^{k}, \quad r = \frac{1}{\sqrt{1 - |\sigma_0|^2}},$$

with Gram matrix $I_{k \times k}$. In particular, the elements of the space $\mathcal{D}(\Theta_1)$ are of the form

$$\begin{pmatrix} r t(z) \mathbf{u} \\ z^{k-1} t(z^{-1}) \mathbf{e_1} \end{pmatrix},$$

where $t(z)$ is a polynomial of degree $\le k-1$.

(ii) *The space $\mathcal{D}(\Theta_2)$ is an anti-Hilbert space spanned by the basis*

$$\left\{ \begin{pmatrix} -r z^{n-1} \mathbf{u} \\ z^{k-n} \mathbf{e_2} \end{pmatrix} \right\}_{n=1}^{k}, \quad r = \frac{1}{\sqrt{|\sigma_0|^2 - 1}},$$

with Gram matrix $-I_{k\times k}$. *In particular, the elements of the space* $\mathcal{D}(\Theta_2)$ *are of the form*

$$\begin{pmatrix} -rt(z)\mathbf{u} \\ z^{k-1}t(z^{-1})\mathbf{e_2} \end{pmatrix},$$

where $t(z)$ *is a polynomial of degree* $\leq k-1$.

(iii) *If* $q=0$, *the space* $\mathcal{D}(\Theta_3)$ *is a Pontryagin space spanned by the basis*

$$\left\{ \begin{pmatrix} z^{n-1}\mathbf{u} \\ z^{k-n}J\mathbf{u} \end{pmatrix}, \begin{pmatrix} z^{n-1}(J\mathbf{u}-2z^k p(z^{-*})^*\mathbf{u}) \\ z^{k-n}(\mathbf{u}-2z^k p(z^{-1})J\mathbf{u}) \end{pmatrix} \right\}_{n=1}^{k},$$

with Gram matrix $2\begin{pmatrix} 0 & I_{k\times k} \\ I_{k\times k} & -2(\Delta+\Delta^*) \end{pmatrix}$. *In particular, the elements of the space* $\mathcal{D}(\Theta_3)$ *are of the form*

$$\begin{pmatrix} t_1(z)\mathbf{u} \\ z^{k-1}t_1(z^{-1})J\mathbf{u} \end{pmatrix} + \begin{pmatrix} t_2(z)(J\mathbf{u}-2z^k p(z^{-*})^*\mathbf{u}) \\ z^{k-1}t_2(z^{-1})(\mathbf{u}-2z^k p(z^{-1})J\mathbf{u}) \end{pmatrix},$$

where $t_1(z)$ *and* $t_2(z)$ *are polynomials of degree* $\leq k-1$.

(iv) *If* $q>0$, *the space* $\mathcal{D}(\Theta_3)$ *can be decomposed as the orthogonal sum*

$$\mathcal{D}(\Theta_3) = \begin{pmatrix} 1 & 0 \\ 0 & \Psi_q \end{pmatrix} \mathcal{D}(\Theta_3^0) \oplus \begin{pmatrix} \Theta_3^0 & 0 \\ 0 & 1 \end{pmatrix} \mathcal{D}(\Psi_q).$$

Here the space $\mathcal{D}(\Theta_3^0)$ *is as described in part* (iii) *and the space* $\mathcal{D}(\Psi_q)$ *is an anti-Hilbert space with basis*

$$\left\{ \begin{pmatrix} -z^{n-1}\mathbf{e_2} \\ z^{k-n}\mathbf{e_2} \end{pmatrix} \right\}_{n=1}^{k}$$

whose Gram matrix equals $-I_{q\times q}$. *Moreover, the map*

$$W : \mathcal{D}(\Theta_3) \ni f \mapsto \begin{pmatrix} f_1 \\ f_2 \end{pmatrix} \in \begin{pmatrix} \mathcal{D}(\Theta_3^0) \\ \mathcal{D}(\Psi_q) \end{pmatrix} \tag{3.1}$$

determined by the decomposition

$$f = \begin{pmatrix} 1 & 0 \\ 0 & \Psi_q \end{pmatrix} f_1 + \begin{pmatrix} \Theta_3^0 & 0 \\ 0 & 1 \end{pmatrix} f_2$$

is unitary.

Proof. (i) We have that

$$D_{\Theta_1}(z,w) = \begin{pmatrix} r^2\dfrac{1-z^k w^{*k}}{1-zw^*}\begin{pmatrix} 1 & \sigma_0 \\ \sigma_0^* & |\sigma_0|^2 \end{pmatrix} & r\dfrac{z^k-w^{*k}}{z-w^*}\begin{pmatrix} 1 & 0 \\ \sigma_0^* & 0 \end{pmatrix} \\ r\dfrac{z^k-w^{*k}}{z-w^*}\begin{pmatrix} 1 & \sigma_0 \\ 0 & 0 \end{pmatrix} & \dfrac{1-z^k w^{*k}}{1-zw^*}\begin{pmatrix} 1 & 0 \\ 0 & 0 \end{pmatrix} \end{pmatrix},$$

from which we derive the equality

$$D_{\Theta_1}(z,w)\begin{pmatrix} a\\ b\\ \mathbf{o}\end{pmatrix} = rw^{*(k-1)}D_{\Theta_1}(z,w^{-1})\begin{pmatrix}\mathbf{o}\\ a+\sigma_0 b\\ d\end{pmatrix}$$

for $a,b,d\in\mathbb{C}$. Since $\mathcal{D}(\Theta_1)$ is spanned by the columns of $D_{\Theta_1}(z,w)$, this means that in fact it is spanned by

$$\frac{1}{r}D_{\Theta_1}(z,w)\begin{pmatrix}\mathbf{e_1}\\ \mathbf{o}\end{pmatrix} = \begin{pmatrix} r(1+zw^*+\cdots+z^{k-1}w^{*(k-1)})\mathbf{u}\\ (z^{k-1}+z^{k-2}w^*+\cdots+zw^{*(k-2)}+w^{*(k-1)})\mathbf{e_1}\end{pmatrix}.$$

We divide this element by $2\pi i w^{*n}$, integrate with respect to w^* over a circle around $w^*=0$ and, by Cauchy's theorem, obtain the basis elements described in part (i) of the theorem. By the reproducing property of the kernel we have

$$\left\langle \frac{1}{r}D_{\Theta_1}(z,w)\begin{pmatrix}\mathbf{e_1}\\ \mathbf{o}\end{pmatrix}, \frac{1}{r}D_{\Theta_1}(z,v)\begin{pmatrix}\mathbf{e_1}\\ \mathbf{o}\end{pmatrix}\right\rangle_{\mathcal{D}(\Theta_1)} = 1+vw^*+\cdots+v^{k-1}w^{*(k-1)}.$$

Dividing both sides by $-4\pi^2 v^m w^{*n}$, $1\le m,n\le k$, and integrating with respect to v and w^* over circles around the origin, we see that the Gram matrix associated with this basis is equal to $I_{k\times k}$.

(ii) The case for $\Theta_2(z)$ can be proved similarly and the proof is omitted.

(iii) For this case we have that

$$D_{\Theta_3}(z,w) = \begin{pmatrix} \dfrac{1-z^k w^{*k}}{1-zw^*}J & \dfrac{z^k-w^{*k}}{z-w^*}I\\ \dfrac{z^k-w^{*k}}{z-w^*}I & \dfrac{1-z^k w^{*k}}{1-zw^*}J\end{pmatrix}$$

$$-\begin{pmatrix} \dfrac{z^kQ(w)^*+w^{*k}Q(z)}{1-zw^*}\begin{pmatrix}1 & \sigma_0\\ \sigma_0^* & 1\end{pmatrix} & \dfrac{Q(z)-Q(w^*)}{z-w^*}\begin{pmatrix}-1 & \sigma_0\\ -\sigma_0^* & 1\end{pmatrix}\\ \dfrac{Q(z^*)^*-Q(w)^*}{z-w^*}\begin{pmatrix}-1 & -\sigma_0\\ \sigma_0^* & 1\end{pmatrix} & \dfrac{z^kQ(w^*)+w^{*k}Q(z^*)^*}{1-zw^*}\begin{pmatrix}1 & -\sigma_0\\ -\sigma_0^* & 1\end{pmatrix}\end{pmatrix}.$$

From this it can be shown that

$$D_{\Theta_3}(z,w)\begin{pmatrix}J\mathbf{u}\\ \mathbf{o}\end{pmatrix} = w^{*(k-1)}D_{\Theta_3}(z,w^{-1})\begin{pmatrix}\mathbf{o}\\ \mathbf{u}\end{pmatrix},$$

and

$$w^{*(k-1)}D_{\Theta_3}(z,w^{-1})\begin{pmatrix}\mathbf{u}\\ \mathbf{o}\end{pmatrix} - D_{\Theta_3}(z,w)\begin{pmatrix}\mathbf{o}\\ J\mathbf{u}\end{pmatrix} = -2\frac{Q(w^*)}{w^{*k}}D_{\Theta_3}(z,w)\begin{pmatrix}\mathbf{o}\\ \mathbf{u}\end{pmatrix}.$$

These equalities imply that

$$\mathcal{D}(\Theta_3) = \operatorname{span}\left\{D_{\Theta_3}(z,w)\begin{pmatrix}J\mathbf{u}\\ \mathbf{o}\end{pmatrix}, D_{\Theta_3}(z,w)\begin{pmatrix}\mathbf{u}\\ \mathbf{o}\end{pmatrix}\right\}. \tag{3.2}$$

Since

$$D_{\Theta_3}(z,w)\begin{pmatrix} J\mathbf{u} \\ \mathbf{o}\end{pmatrix} = \begin{pmatrix} \dfrac{1-z^k w^{*k}}{1-zw^*}\mathbf{u} \\ \dfrac{z^k - w^{*k}}{z-w^*}J\mathbf{u}\end{pmatrix},$$

we obtain, using integration as in the proof of part (i), that

$$\operatorname{span}\{D_{\Theta_3}(z,w)\begin{pmatrix} J\mathbf{u} \\ \mathbf{o}\end{pmatrix}\} = \operatorname{span}\begin{pmatrix} z^{j-1}\mathbf{u} \\ z^{k-j}J\mathbf{u}\end{pmatrix}_{j=1}^{k} \tag{3.3}$$

is a neutral space which accounts for the 0 entry in the left upper corner of the Gram matrix. The elements on the right-hand side are linearly independent and the span coincides with the space of functions of the form

$$\begin{pmatrix} t(z)\mathbf{u} \\ z^{k-1}t(z^{-1})J\mathbf{u}\end{pmatrix},$$

where $t(z)$ is a polynomial of degree $\leq k-1$. From

$$D_{\Theta_3}(z,w)\begin{pmatrix} \mathbf{u} \\ \mathbf{o}\end{pmatrix} = \begin{pmatrix} \dfrac{1-z^k w^{*k}}{1-zw^*}J\mathbf{u} - 2\dfrac{z^k Q(w)^* + w^{*k}Q(z)}{1-zw^*}\mathbf{u} \\ \dfrac{z^k - w^{*k}}{z-w^*}\mathbf{u} + 2\dfrac{Q(z^*)^* - Q(w)^*}{z-w^*}J\mathbf{u}\end{pmatrix}$$

and $Q(z) = p(z) - z^{2k}p(z^{-*})^*$, we get

$$D_{\Theta_3}(z,w)\begin{pmatrix} \mathbf{u} \\ \mathbf{o}\end{pmatrix} = \begin{pmatrix} \dfrac{1-z^k w^{*k}}{1-zw^*}J\mathbf{u} - 2z^k\dfrac{1-z^k w^{*k}}{1-zw^*}p(z^{-*})^*\mathbf{u} \\ \dfrac{z^k - w^{*k}}{z-w^*}\mathbf{u} - 2z^k\dfrac{z^k - w^{*k}}{z-w^*}p(\frac{1}{z})J\mathbf{u}\end{pmatrix} - \begin{pmatrix} t_w(z)\mathbf{u} \\ z^{k-1}t_w(z^{-1})J\mathbf{u}\end{pmatrix},$$

where

$$t_w(z) = 2\frac{z^k p(w)^* - z^k p(z^{-*})^* + w^{*k}p(z) - z^k w^{*2k}p(w^{-*})}{1-zw^*}$$

is a polynomial of degree $\leq k-1$ in z. The span of the second summand is contained in the neutral subspace (3.3) and can be dropped from the formula when calculating the span on the right-hand side of (3.2). The remainder of the proof of part (iii) can be given by integration and using the reproducing property of the kernel as in the proof of part (i) and is omitted.

(iv) The proof of the statements concerning the space $\mathcal{D}(\Psi_q)$ is similar to the proof of (i). The orthogonal decomposition of $\mathcal{D}(\Theta_3)$ and the unitarity of the map follow from (a) the equality

$$\begin{aligned} D_{\Theta_3}(z,w) &= \begin{pmatrix} 1 & 0 \\ 0 & \Psi_q(z)\end{pmatrix} D_{\Theta_3^0}(z,w)\begin{pmatrix} 1 & 0 \\ 0 & \Psi_q(w)^*\end{pmatrix} \\ &\quad + \begin{pmatrix} \Theta_3^0(z) & 0 \\ 0 & 1\end{pmatrix} D_{\Psi_q}(z,w)\begin{pmatrix} \Theta_3^0(w)^* & 0 \\ 0 & 1\end{pmatrix}, \end{aligned} \tag{3.4}$$

(b) the implication that if $f_1 \in \mathcal{D}(\Theta_3^0)$ and $f_2 \in \mathcal{D}(\Psi_q)$ then the identity

$$\begin{pmatrix} 1 & 0 \\ 0 & \Psi_q \end{pmatrix} f_1 + \begin{pmatrix} \Theta_3^0 & 0 \\ 0 & 1 \end{pmatrix} f_2 = 0$$

implies $f_1 = 0$ and $f_2 = 0$, and (c) reproducing kernel methods as in [7, Section 1.5] (see also [3, Theorems 2.1 and 2.2]). The implication in (b) follows from

$$\begin{pmatrix} 1 & 0 & 0 & 0 \\ 0 & 1 & 0 & 0 \\ 0 & 0 & 1 & 0 \\ 0 & 0 & 0 & z^q \end{pmatrix} \mathcal{D}(\Theta_3^0) \cap \begin{pmatrix} Q(z) + z^k & -\sigma_0 & 0 & 0 \\ \sigma_0^* Q(z) & -Q(z) + z^k & 0 & 0 \\ 0 & 0 & 1 & 0 \\ 0 & 0 & 0 & 1 \end{pmatrix} \mathcal{D}(\Psi_q) = \{0\},$$

which can be verified by comparing the degrees of the elements in the two sets. □

4. Solutions in terms of colligations

In this section we construct a closely connected unitary colligation U for the solution

$$s = T_\Theta = \frac{as_1 + b}{cs_1 + d}, \quad \Theta = \begin{pmatrix} a & b \\ c & d \end{pmatrix}, \tag{4.1}$$

of the basic interpolation problem (BIP) in terms of the corresponding parameter function s_1, the entries of the matrix function Θ, and their canonical unitary colligations. We shall use the following notation:

$$\Upsilon = \begin{pmatrix} 1 & -s & 0 & 0 \\ 0 & 0 & \frac{\widetilde{s}_1}{\widetilde{n}} & \frac{1}{\widetilde{n}} \end{pmatrix}, \quad \Phi = \begin{pmatrix} a - cs & 0 \\ 0 & \frac{1}{\widetilde{n}} \end{pmatrix},$$

where $n = cs_1 + d = (a - cs)/\delta$, $\delta := \det \Theta$. Here and elsewhere in the sequel when f is a matrix function on some set in $\mathbb{C}$, we denote by $\widetilde{f}$ the matrix function $\widetilde{f}(z) = f(z^*)^*$.

Theorem 4.1. *Let s in (4.1) be a solution of the interpolation problem (BIP) with parameter function s_1 and $\Theta = \Theta_1, \Theta_2$, and Θ_3.*

(i) *The space $\mathcal{D}(s)$ can be decomposed as the orthogonal sum*

$$\mathcal{D}(s) = \Upsilon\mathcal{D}(\Theta) \oplus \Phi\mathcal{D}(s_1),$$

and $\mathrm{ind}_-\mathcal{D}(s) = \mathrm{ind}_-\mathcal{D}(\Theta) + \mathrm{ind}_-\mathcal{D}(s_1)$.

(ii) *The map*

$$\Lambda : \mathcal{D}(s) \ni h \mapsto \begin{pmatrix} f \\ g \end{pmatrix} \in \begin{pmatrix} \mathcal{D}(\Theta) \\ \mathcal{D}(s_1) \end{pmatrix}$$

determined by the decomposition $h = \Upsilon f + \Phi g$ is unitary.

Proof. We claim that (1)

$$D_s(z,w) = \begin{pmatrix} \Upsilon(z) & \Phi(z) \end{pmatrix} \begin{pmatrix} D_\Theta(z,w) & 0 \\ 0 & D_{s_1}(z,w) \end{pmatrix} \begin{pmatrix} \Upsilon(w)^* \\ \Phi(w)^* \end{pmatrix}, \tag{4.2}$$

which implies that $\mathcal{D}(s) = \Upsilon\mathcal{D}(\Theta) + \Phi\mathcal{D}(s_1)$, and (2) the multiplication map

$$\begin{pmatrix}\Upsilon & \Phi\end{pmatrix} : \mathcal{D}(\Theta) \oplus \mathcal{D}(s_1) \ni \begin{pmatrix} f \\ g \end{pmatrix} \mapsto \Upsilon f + \Phi g \in \mathcal{D}(s)$$

is injective. The theorem follows from these claims by reproducing kernel methods as in [7, Section 1.5] (see also [3, Theorems 2.1 and 2.2]).

Proof of (1). In the proof of the first claim we shall also use the notation

$$\Psi_s = \begin{pmatrix} 1 & -s & 0 & 0 \\ 0 & 0 & 1 & -\widetilde{s} \end{pmatrix}, \quad J_1 = \begin{pmatrix} 0 & 1 \\ -1 & 0 \end{pmatrix}.$$

It is easy to see that $\Theta J_1 \Theta^T = \det \Theta J_1$ and hence $\widetilde{\delta} J_1 \widetilde{\Theta}^{-1} = \widetilde{\Theta}^T J_1$. This will be used in the third equality of the following calculation.

$$\begin{aligned}
D_s(z,w) &= \begin{pmatrix} \dfrac{1 - s(z)s(w)^*}{1 - zw^*} & \dfrac{s(z) - s(w^*)}{z - w^*} \\ \dfrac{\widetilde{s}(z) - \widetilde{s}(w^*)}{z - w^*} & \dfrac{1 - \widetilde{s}(z)\widetilde{s}(w)^*}{1 - zw^*} \end{pmatrix} \\
&= \begin{pmatrix} 1 & -s(z) & 0 & 0 \\ 0 & 0 & 1 & -\widetilde{s}(z) \end{pmatrix} \begin{pmatrix} \dfrac{J}{1 - zw^*} & \dfrac{J_1}{z - w^*} \\ \dfrac{J_1}{z - w^*} & \dfrac{J}{1 - zw^*} \end{pmatrix} \begin{pmatrix} 1 & 0 \\ -s(w)^* & 0 \\ 0 & 1 \\ 0 & -\widetilde{s}(w)^* \end{pmatrix} \\
&= \Psi_s(z) \begin{pmatrix} \dfrac{J - \Theta(z)J\Theta(w)^*}{1 - zw^*} & \dfrac{[\Theta(w^*) - \Theta(z)]\Theta(w^*)^{-1}J_1}{z - w^*} \\ \dfrac{J_1\widetilde{\Theta}(z)^{-1}[\widetilde{\Theta}(z) - \widetilde{\Theta}(w^*)]}{z - w^*} & \dfrac{J - J_1\widetilde{\Theta}(z)^{-1}J\widetilde{\Theta}(w)^{-*}J_1}{1 - zw^*} \end{pmatrix} \Psi_s(w)^* \\
&\quad + \Psi_s(z) \begin{pmatrix} \dfrac{\Theta(z)J\Theta(w)^*}{1 - zw^*} & \dfrac{\Theta(z)J_1\Theta(w^*)^T}{\delta(w^*)(z - w^*)} \\ \dfrac{\widetilde{\Theta}(z)^T J_1 \Theta(w)^*}{\widetilde{\delta}(z)(z - w^*)} & \dfrac{\widetilde{\Theta}(z)^T J_1 J J_1 \Theta(w^*)^T}{\widetilde{\delta}(z)\delta(w^*)(1 - zw^*)} \end{pmatrix} \Psi_s(w)^* \\
&= \Psi_s(z) \begin{pmatrix} I & 0 \\ 0 & J_1\widetilde{\Theta}(z)^{-1} \end{pmatrix} D_\Theta(z,w) \begin{pmatrix} I & 0 \\ 0 & -\Theta(w^*)^{-1}J_1 \end{pmatrix} \Psi_s(w)^* \\
&\quad + \Psi_s(z) \begin{pmatrix} \Theta(z) & 0 \\ 0 & \dfrac{\widetilde{\Theta}(z)^T}{\widetilde{\delta}(z)} \end{pmatrix} \begin{pmatrix} \dfrac{J}{1 - zw^*} & \dfrac{J_1}{z - w^*} \\ \dfrac{J_1}{z - w^*} & \dfrac{J}{1 - zw^*} \end{pmatrix} \begin{pmatrix} \Theta(w)^* & 0 \\ 0 & \dfrac{\Theta(w^*)^T}{\delta(w^*)} \end{pmatrix} \\
&\quad \times \Psi_s(w)^*.
\end{aligned}$$

Using

$$\Psi_s \begin{pmatrix} I & 0 \\ 0 & J_1\widetilde{\Theta}^{-1} \end{pmatrix} = \Upsilon$$

and

$$\Psi_s \begin{pmatrix} \Theta & 0 \\ 0 & \frac{\widetilde{\Theta}^T}{\widetilde{\delta}} \end{pmatrix} = \begin{pmatrix} \begin{pmatrix} 1 & -s \end{pmatrix} \Theta & 0 & & 0 \\ 0 & 0 & \begin{pmatrix} 1 & -\widetilde{s} \end{pmatrix} & \frac{\widetilde{\Theta}^T}{\widetilde{\delta}} \end{pmatrix}$$

$$= \begin{pmatrix} a-c & 0 \\ 0 & \frac{\widetilde{a}-\widetilde{c}\widetilde{s}}{\widetilde{\delta}} \end{pmatrix} \Psi_{s_1} = \Phi \Psi_{s_1},$$

where the second equality follows from the relation

$$\begin{pmatrix} 1 & -s \end{pmatrix} \Theta = (a - sc) \begin{pmatrix} 1 & -s_1 \end{pmatrix}, \tag{4.3}$$

we obtain

$$\begin{aligned} D_s(z,w) &= \Upsilon(z) D_\Theta(z,w) \Upsilon(w)^* \\ &+ \Phi(z)\Psi_{s_1}(z) \begin{pmatrix} \frac{J}{1-zw^*} & \frac{J_1}{z-w^*} \\ \frac{J_1}{z-w^*} & \frac{J}{1-zw^*} \end{pmatrix} \Phi(w)^* \Psi_{s_1}(w)^* \\ &= \Upsilon(z) D_\Theta(z,w) \Upsilon(w)^* + \Phi(z) D_{s_1}(z,w) \Phi(w)^*. \end{aligned}$$

This proves the first claim.

Proof of (2). To prove the second claim, consider

$$f = \begin{pmatrix} f_1 \\ f_2 \end{pmatrix} \in \mathcal{D}(\Theta), \quad g = \begin{pmatrix} g_1 \\ g_2 \end{pmatrix} \in \mathcal{D}(s_1),$$

and assume $\Upsilon f + \Phi g = 0$, that is,

$$\begin{pmatrix} 1 & -s \end{pmatrix} f_1 + (a - cs) g_1 = 0 \tag{4.4}$$

and

$$\begin{pmatrix} \widetilde{s}_1 & 1 \end{pmatrix} f_2 + g_2(z) = 0. \tag{4.5}$$

Then $f_1 \in \mathcal{H}(\Theta)$, $g_1 \in \mathcal{H}(s_1)$, and therefore equation (4.4) implies $f_1 = g_1 = 0$, as already shown in [3]. This means that

$$f = \begin{pmatrix} 0 \\ f_2 \end{pmatrix} \in \mathcal{D}(\Theta).$$

Using Theorem 3.1 we conclude that $f_2 = 0$ in all the three cases $\Theta = \Theta_1$, Θ_2 and Θ_3. Equation (4.5) then implies $g_2 = 0$. Thus multiplication by $\begin{pmatrix} \Upsilon & \Psi \end{pmatrix}$ is injective. □

The following theorem is the main result of this paper.

Theorem 4.2. *Under the unitary mapping*

$$\Lambda_1 = \begin{pmatrix} \Lambda & 0 \\ 0 & 1 \end{pmatrix} : \begin{pmatrix} \mathcal{D}(s) \\ \mathbb{C} \end{pmatrix} \to \begin{pmatrix} \begin{pmatrix} \mathcal{D}(\Theta) \\ \mathcal{D}(s_1) \end{pmatrix} \\ \mathbb{C} \end{pmatrix}$$

the canonical unitary colligation

$$U_s = \begin{pmatrix} A_s & u_s \\ \langle\,\cdot\,,\, v_s\rangle_{\mathcal{D}(s)} & s(0) \end{pmatrix} : \begin{pmatrix} \mathcal{D}(s) \\ \mathbb{C} \end{pmatrix} \to \begin{pmatrix} \mathcal{D}(s) \\ \mathbb{C} \end{pmatrix}$$

for $s(z)$ is transformed into the colligation

$$U = \Lambda_1 U_s \Lambda_1^{-1} = \begin{pmatrix} A & u \\ \langle\,\cdot\,,\, v\rangle_{\mathcal{D}(\Theta)\oplus\mathcal{D}(s_1)} & s(0) \end{pmatrix},$$

where

$$A = \Lambda A_s \Lambda^{-1} = \begin{pmatrix} A_\Theta + \dfrac{B_\Theta}{n(0)} \begin{pmatrix} s_1(0) \\ 1 \end{pmatrix} \begin{pmatrix} 0 & -1 \end{pmatrix} C_\Theta & \dfrac{B_\Theta}{n(0)} \begin{pmatrix} d(0) \\ -c(0) \end{pmatrix} C_{s_1} \\ \dfrac{B_{s_1}}{n(0)} \begin{pmatrix} 0 & -1 \end{pmatrix} C_\Theta & A_{s_1} - \dfrac{B_{s_1}}{n(0)} c(0) C_{s_1} \end{pmatrix},$$

$$u = \Lambda u_s = \frac{1}{n(0)} \begin{pmatrix} B_\Theta \begin{pmatrix} s_1(0) \\ 1 \end{pmatrix} \\ B_{s_1} \end{pmatrix}, \quad v = \Lambda v_s = \begin{pmatrix} D_\Theta(\,\cdot\,,\,0) \begin{pmatrix} 1 \\ -\sigma_0^* \\ 0 \\ 0 \end{pmatrix} \\ 0 \end{pmatrix} \in \begin{pmatrix} \mathcal{D}(\Theta) \\ \mathcal{D}(s_1) \end{pmatrix}.$$

The formula for A shows that it is an extension to $\mathcal{D}(s)$ of the operator

$$B_{s_1} = A_{s_1} - \frac{B_{s_1}}{n(0)} c(0) C_{s_1}$$

in $\mathcal{D}(s_1)$, which is at most a one-dimensional perturbation of A_{s_1}. The theorem implies that with $\Lambda h = \begin{pmatrix} f \\ g \end{pmatrix}$ and $c \in \mathbb{C}$ the following diagram commutes.

$$\begin{array}{ccc} \begin{pmatrix} \mathcal{D}(s) \\ \mathbb{C} \end{pmatrix} \ni \begin{pmatrix} h \\ c \end{pmatrix} & \xrightarrow{\;U_s\;} & \begin{pmatrix} \mathcal{D}(s) \\ \mathbb{C} \end{pmatrix} \\ \Big\downarrow \Lambda_1 & & \Big\downarrow \Lambda_1 \\ \begin{pmatrix} \mathcal{D}(\Theta) \\ \mathcal{D}(s_1) \\ \mathbb{C} \end{pmatrix} \ni \begin{pmatrix} f \\ g \\ c \end{pmatrix} & \xrightarrow{\;U\;} & \begin{pmatrix} \mathcal{D}(\Theta) \\ \mathcal{D}(s_1) \\ \mathbb{C} \end{pmatrix} \end{array}$$

Proof of Theorem 4.2. The formula for v follows from $v_s = D_s(\,\cdot\,,0) \begin{pmatrix} 1 \\ 0 \end{pmatrix}$, formula (4.2), and the fact that

$$a(0) - c(0)\sigma_0 = 0 \tag{4.6}$$

in all three cases $\Theta = \Theta_1$, Θ_2, and Θ_3. Now we derive the formula for u and begin with B_s. Denoting by R_0 the difference-quotient operator:

$$R_0 x(z) = \frac{x(z) - x(0)}{z}, \tag{4.7}$$

where $x(z)$ is any holomorphic function in a neighborhood of $z = 0$, we have

$$B_s = \begin{pmatrix} R_0 s \\ 1 - \widetilde{s}s(0) \end{pmatrix}. \tag{4.8}$$

The entries of B_s in (4.8) can be written in the form

$$R_0 s = \frac{1}{n(0)} \left\{ \begin{pmatrix} 1 & -s \end{pmatrix} R_0 \Theta \begin{pmatrix} s_1(0) \\ 1 \end{pmatrix} + (a - cs) R_0 s_1 \right\} \tag{4.9}$$

and

$$1 - \widetilde{s}s(0) = \frac{1}{n(0)\widetilde{n}} \left\{ \begin{pmatrix} \widetilde{s}_1 & 1 \end{pmatrix} (J - \widetilde{\Theta} J \Theta(0)) \begin{pmatrix} s_1(0) \\ 1 \end{pmatrix} + 1 - \widetilde{s}_1 s_1(0) \right\}. \tag{4.10}$$

For the equality (4.9) we refer to the proof of Theorem 4.1 in [3]. As for the equality (4.10), the right-hand side equals

$$\begin{aligned} & \frac{-1}{n(0)\widetilde{n}} \begin{pmatrix} \widetilde{s}_1 & 1 \end{pmatrix} \begin{pmatrix} \widetilde{a} & -\widetilde{c} \\ \widetilde{b} & -\widetilde{d} \end{pmatrix} \begin{pmatrix} a(0) & b(0) \\ c(0) & d(0) \end{pmatrix} \begin{pmatrix} s_1(0) \\ 1 \end{pmatrix} \\ = {} & \frac{1}{n(0)\widetilde{n}} \left[(\widetilde{c}\widetilde{s}_1 + \widetilde{d})(c(0)s_1(0) + d(0)) - (\widetilde{a}\widetilde{s}_1 + \widetilde{b})(a(0)s_1(0) + b(0)) \right], \end{aligned}$$

which equals the left-hand side. It follows that

$$\begin{aligned} B_s = {} & \frac{1}{n(0)} \begin{pmatrix} 1 & -s & 0 & 0 \\ 0 & 0 & \frac{\widetilde{s}_1}{\widetilde{n}} & \frac{1}{\widetilde{n}} \end{pmatrix} \begin{pmatrix} R_0 \Theta \\ J - \widetilde{\Theta} J \Theta(0) \end{pmatrix} \begin{pmatrix} s_1(0) \\ 1 \end{pmatrix} \\ & + \frac{1}{n(0)} \begin{pmatrix} a - cs & 0 \\ 0 & \frac{1}{\widetilde{n}} \end{pmatrix} \begin{pmatrix} R_0 s_1 \\ 1 - \widetilde{s}_1 s_1(0) \end{pmatrix} = \Upsilon \frac{B_\Theta}{n(0)} \begin{pmatrix} s_1(0) \\ 1 \end{pmatrix} + \Phi \frac{B_{s_1}}{n(0)}, \end{aligned} \tag{4.11}$$

which yields the formula for u in the theorem.

To obtain the expression for A_s, we start from

$$A_s h = A_s \Upsilon f + A_s \Phi g, \tag{4.12}$$

where

$$h \in \mathcal{D}(s), \quad f = \begin{pmatrix} f_1 \\ f_2 \end{pmatrix} \in \mathcal{D}(\Theta), \quad g = \begin{pmatrix} g_1 \\ g_2 \end{pmatrix} \in \mathcal{D}(s_1)$$

are related via $h = \Upsilon f + \Phi g$, that is, $\Lambda h = \begin{pmatrix} f \\ g \end{pmatrix}$. Using the notation

$$\Upsilon = \operatorname{diag}\{\upsilon_1, \upsilon_2\}, \quad \upsilon_1 = \begin{pmatrix} 1 & -s \end{pmatrix}, \ \upsilon_2 = \frac{1}{\widetilde{n}} \begin{pmatrix} \widetilde{s}_1 & 1 \end{pmatrix}$$

and the formula

$$R_0(xy)(z) = x(z) R_0 y(z) + R_0 x(z) y(0), \tag{4.13}$$

which holds for any two functions x, y, which are holomorphic in a neighborhood of $z = 0$, we calculate the first summand on the right-hand side of (4.12):

$$A_s \Upsilon f(z) = A_s \begin{pmatrix} v_1 f_1 \\ v_2 f_2 \end{pmatrix}(z) = \begin{pmatrix} R_0(v_1 f_1)(z) \\ z v_2(z) f_2(z) - \widetilde{s}(z) v_1(0) f_1(0) \end{pmatrix} \tag{4.14}$$

$$= \begin{pmatrix} v_1(z) R_0 f_1(z) \\ v_2(z)[z f_2(z) - \widetilde{\Theta}(z) J f_1(0)] \end{pmatrix} + \begin{pmatrix} R_0 v_1(z) \\ v_2(z)\widetilde{\Theta}(z) J - \widetilde{s}(z) v_1(0) \end{pmatrix} f_1(0)$$

$$= \begin{pmatrix} v_1(z) & 0 \\ 0 & v_2(z) \end{pmatrix} A_\Theta \begin{pmatrix} f_1 \\ f_2 \end{pmatrix}(z) - \begin{pmatrix} 0 & R_0 s(z) \\ 0 & 1 - \widetilde{s}(z) s(0) \end{pmatrix} f_1(0)$$

$$= \Upsilon(z) A_\Theta f(z) + (B_s \begin{pmatrix} 0 & -1 \end{pmatrix} f_1(0))(z).$$

We calculate the second summand on the right-hand side of (4.12) in a similar way. We write

$$\Phi = \operatorname{diag}\{\phi_1, \phi_2\}, \quad \phi_1 = a - cs, \ \phi_2 = \frac{1}{\widetilde{\widetilde{n}}},$$

and using (4.6) we get

$$A_s \Phi g = A_s \begin{pmatrix} \phi_1 g_1 \\ \phi_2 g_2 \end{pmatrix} = \Phi A_{s_1} g + \begin{pmatrix} R_0 \phi_1 \\ \phi_2 \widetilde{s}_1 \end{pmatrix} g_1(0). \tag{4.15}$$

We claim that the two components of the vector function on the right-hand side of (4.15) can be written as

$$R_0 \phi_1 = \begin{pmatrix} 1 & -s \end{pmatrix} \frac{R_0 \Theta}{n(0)} \begin{pmatrix} d(0) \\ -c(0) \end{pmatrix} - (a - cs) R_0 s_1 \frac{c(0)}{n(0)} \tag{4.16}$$

and

$$\phi_2 \widetilde{s}_1 = \frac{1}{n(0)\widetilde{n}} \left\{ \begin{pmatrix} \widetilde{s}_1 & 1 \end{pmatrix} (J - \widetilde{\Theta} J \Theta(0)) \begin{pmatrix} d(0) \\ -c(0) \end{pmatrix} - (1 - \widetilde{s}_1 s_1(0)) c(0) \right\}. \tag{4.17}$$

Assuming these claims are true, we find that the vector function takes the form

$$\begin{pmatrix} R_0 \phi_1 \\ \phi_2 \widetilde{s}_1 \end{pmatrix} = \begin{pmatrix} 1 & -s & 0 & 0 \\ 0 & 0 & \frac{1}{\widetilde{\widetilde{n}}}\widetilde{s}_1 & \frac{1}{\widetilde{\widetilde{n}}} \end{pmatrix} \begin{pmatrix} R_0 \Theta \\ J - \widetilde{\Theta} J \Theta(0) \end{pmatrix} \frac{1}{n(0)} \begin{pmatrix} d(0) \\ -c(0) \end{pmatrix}$$

$$- \begin{pmatrix} a - cs & 0 \\ 0 & \frac{1}{\widetilde{\widetilde{n}}} \end{pmatrix} \begin{pmatrix} R_0 s_1 \\ 1 - \widetilde{s}_1 s_1(0) \end{pmatrix} \frac{c(0)}{n(0)}$$

$$= \Upsilon \frac{B_\Theta}{n(0)} \begin{pmatrix} d(0) \\ -c(0) \end{pmatrix} - \Phi \frac{B_{s_1}}{n(0)} c(0).$$

Substituting this in (4.15), and then substituting (4.15) and (4.14) with B_s replaced by (4.11) into (4.12) we obtain the formula for $A = \Lambda A_s \Lambda^{-1}$ in the theorem:

$$\begin{aligned} A_s h &= \Upsilon \left\{ A_\Theta f + \frac{B_\Theta}{n(0)} \begin{pmatrix} s_1(0) \\ 1 \end{pmatrix} \begin{pmatrix} 0 & -1 \end{pmatrix} f_1(0) + \frac{B_\Theta}{n(0)} \begin{pmatrix} d(0) \\ -c(0) \end{pmatrix} g_1(0) \right\} \\ &\quad + \Phi \left\{ \frac{B_{s_1}}{n(0)} \begin{pmatrix} 0 & -1 \end{pmatrix} f_1(0) + A_{s_1} g - \frac{B_{s_1}}{n(0)} c(0) g_1(0) \right\}. \end{aligned}$$

It remains to prove the claims. Equality (4.17) follows from writing out the right-hand side and using

$$\Theta(0) \begin{pmatrix} d(0) \\ -c(0) \end{pmatrix} = \begin{pmatrix} \delta(0) \\ 0 \end{pmatrix} = 0, \quad \delta(z) = \det \Theta(z). \tag{4.18}$$

For the proof of (4.16) we write

$$n = cs_1 + d = \begin{pmatrix} 1 & -s_1 \end{pmatrix} \begin{pmatrix} d \\ -c \end{pmatrix},$$

use (4.18) and repeatedly (4.3) and (4.13). We obtain the following chain of equalities:

$$\begin{aligned} &(a - sc) \begin{pmatrix} 1 & -s_1 \end{pmatrix} R_0 \begin{pmatrix} d \\ -c \end{pmatrix} + (a - sc) R_0 s_1 c(0) + R_0 (a - sc) n(0) \\ &= R_0 (a - sc) n = R_0 \begin{pmatrix} 1 & -s \end{pmatrix} \Theta \begin{pmatrix} d \\ -c \end{pmatrix} = \begin{pmatrix} 1 & -s \end{pmatrix} R_0 \left[\Theta \begin{pmatrix} d \\ -c \end{pmatrix} \right] \\ &= \begin{pmatrix} 1 & -s \end{pmatrix} R_0 \Theta \begin{pmatrix} d(0) \\ -c(0) \end{pmatrix} + \begin{pmatrix} 1 & -s \end{pmatrix} \Theta R_0 \begin{pmatrix} d \\ -c \end{pmatrix} \\ &= \begin{pmatrix} 1 & -s \end{pmatrix} R_0 \Theta \begin{pmatrix} d(0) \\ -c(0) \end{pmatrix} + (a - sc) \begin{pmatrix} 1 & -s_1 \end{pmatrix} R_0 \begin{pmatrix} d \\ -c \end{pmatrix}. \end{aligned}$$

Comparing both sides we find the equality (4.16). □

For the case $\Theta_3(z) = \Theta_3^0 \Psi_q$ and $q > 0$ there is need to modify Theorem 4.2. To do this we define Λ_q to be the composition of the two unitary maps W in (3.1) and Λ, that is,

$$\Lambda_q := \begin{pmatrix} W & 0 & 0 \\ 0 & I_{\mathcal{D}(s_1)} & 0 \\ 0 & 0 & 1 \end{pmatrix} \Lambda : \begin{pmatrix} \mathcal{D}(s) \\ \mathbb{C} \end{pmatrix} \to \begin{pmatrix} \begin{pmatrix} \mathcal{D}(\Theta_3^0) \\ \mathcal{D}(\Psi_q) \end{pmatrix} \\ \mathcal{D}(s_1) \\ \mathbb{C} \end{pmatrix},$$

and set

$$M = \begin{pmatrix} s_1(0) \\ 1 \end{pmatrix} \begin{pmatrix} 0 & -1 \end{pmatrix}, \quad M_1 = \Psi_q(0) M = \begin{pmatrix} s_1(0) \\ 0 \end{pmatrix} \begin{pmatrix} 0 & -1 \end{pmatrix},$$

$$M_2 = \Psi_q(0) M \Theta_3^0(0), \quad \text{and} \quad M_3 = M \Theta_3^0(0).$$

Theorem 4.3. *With $\Theta(z) = \Theta_3(z)$ and $q > 0$, under the map Λ_q the canonical unitary colligation*

$$U_s = \begin{pmatrix} A_s & u_s \\ \langle \cdot\, v_s\rangle & s(0) \end{pmatrix} : \begin{pmatrix} \mathcal{D}(s) \\ \mathbb{C} \end{pmatrix} \to \begin{pmatrix} \mathcal{D}(s) \\ \mathbb{C} \end{pmatrix}$$

is transformed into the colligation

$$U_q = \Lambda_q U_s \Lambda_q^{-1} = \begin{pmatrix} A_q & u_q \\ \langle \cdot\, v_q\rangle & s(0) \end{pmatrix} : \begin{pmatrix} \begin{pmatrix} \mathcal{D}(\Theta_3^0) \\ \mathcal{D}(\Psi_q) \\ \mathcal{D}(s_1) \end{pmatrix} \\ \mathbb{C} \end{pmatrix} \to \begin{pmatrix} \begin{pmatrix} \mathcal{D}(\Theta_3^0) \\ \mathcal{D}(\Psi_q) \\ \mathcal{D}(s_1) \end{pmatrix} \\ \mathbb{C} \end{pmatrix},$$

where

$$A_q = \begin{pmatrix} A_{\Theta_3^0} + \dfrac{B_{\Theta_3^0}}{n(0)} M_1 C_{\Theta_3^0} & B_{\Theta_3^0} C_{\Psi_q} + \dfrac{B_{\Theta_3^0}}{n(0)} M_2 C_{\Psi_q} & 0 \\ \dfrac{B_{\Psi_q}}{n(0)} M C_{\Theta_3^0} & A_{\Psi_q} + \dfrac{B_{\Psi_q}}{n(0)} M_3 C_{\Psi_q} & \dfrac{B_{\Psi_q}}{n(0)} \begin{pmatrix} 0 \\ -c(0) \end{pmatrix} C_{s_1} \\ \dfrac{B_{s_1}}{n(0)} \begin{pmatrix} 0 & -1 \end{pmatrix} C_{\Theta_3^0} & \dfrac{B_{s_1}}{n(0)} \begin{pmatrix} 0 & -1 \end{pmatrix} \Theta_3^0(0) C_{\Psi_q} & A_{s_1} - \dfrac{B_{s_1}}{n(0)} c(0) C_{s_1} \end{pmatrix},$$

$$u_q = \frac{1}{n(0)} \begin{pmatrix} \Theta_3^0 \Psi_q(0) \begin{pmatrix} s_1(0) \\ 1 \end{pmatrix} \\ \Psi_q \begin{pmatrix} s_1(0) \\ 1 \end{pmatrix} \\ B_{s_1} \end{pmatrix}, \quad v_q = \begin{pmatrix} D_{\Theta_3^0}(\,\cdot\,,\,0) \begin{pmatrix} 1 \\ -\sigma_0^* \\ 0 \\ 0 \end{pmatrix} \\ 0 \\ 0 \end{pmatrix} \in \begin{pmatrix} \mathcal{D}(\Theta_3^0) \\ \mathcal{D}(\Psi_q) \\ \mathcal{D}(s_1) \end{pmatrix}.$$

Proof. Using (4.13) and $d(0) = 0$ (since $q > 0$), we find that

$$WA_\Theta = \begin{pmatrix} A_{\Theta_3^0} & B_{\Theta_3^0} C_{\Psi_q} \\ 0 & A_{\Psi_q} \end{pmatrix}, \; WB_\Theta = \begin{pmatrix} B_{\Theta_3^0} \Psi_q(0) \\ B_{\Psi_q} \end{pmatrix}, C_\Theta = \begin{pmatrix} C_{\Theta_3^0} & \Theta_3^0(0) C_{\Psi_q} \end{pmatrix}.$$

Substitution of these formulas into the formulas of Theorem 4.2 yields the desired result for A_q and u_q. The formula for v_q is obtained by using the decomposition in (3.4). □

Let P be the projection in the space $\mathcal{D}(\Theta) \oplus \mathcal{D}(s_1)$ onto the space $\mathcal{D}(s_1)$. From the operator matrix form of A we see that

$$A_{s_1} = PA|_{\mathcal{D}(s_1)} + \frac{B_{s_1}}{n(0)} c(0) C_{s_1}$$

and

$$u_{s_1} = B_{s_1} 1 = n(0) P u.$$

This observation and the next theorem show that the canonical unitary colligation of the parameter $s_1(z)$ can be recovered from the canonical unitary colligation of the solution $s(z)$.

Theorem 4.4.

$$D_{s_1}(\cdot,0)\begin{pmatrix}1\\0\end{pmatrix}=\begin{cases} n(0)^*PA^{*k}v & \text{if } |\sigma_0|\neq 1,\\ n(0)^*PA^{*(2k)}v & \text{if } |\sigma_0|=1 \text{ and } q=0,\\ n(0)^*PA^{*(2k+q)}v & \text{if } |\sigma_0|=1 \text{ and } q>0.\end{cases}$$

Proof. First we note that
$A^*=$

$$\begin{pmatrix} A_\Theta^*+\left\langle\cdot,\frac{B_\Theta}{n(0)}\begin{pmatrix}s_1(0)\\1\end{pmatrix}\right\rangle C_\Theta^*\begin{pmatrix}0\\-1\end{pmatrix} & \left\langle\cdot,\frac{B_{s_1}}{n(0)}\right\rangle C_\Theta^*\begin{pmatrix}0\\-1\end{pmatrix}\\ \left\langle\cdot,\frac{B_\Theta}{n(0)}\begin{pmatrix}d(0)\\-c(0)\end{pmatrix}\right\rangle D_{s_1}(\cdot,0)\begin{pmatrix}1\\0\end{pmatrix} & A_{s_1}^*-\left\langle\cdot,\frac{B_{s_1}}{n(0)}c(0)\right\rangle D_{s_1}(\cdot,0)\begin{pmatrix}1\\0\end{pmatrix}\end{pmatrix},$$

where A_Θ^* is as given in Theorem 2.1 (ii) and

$$B_\Theta\begin{pmatrix}a\\b\end{pmatrix}=D_\Theta(\,\cdot\,,\,0)\begin{pmatrix}0\\0\\a\\b\end{pmatrix},\quad C_\Theta^*\begin{pmatrix}0\\-1\end{pmatrix}=D_\Theta(\,\cdot\,,0)\begin{pmatrix}0\\-1\\0\\0\end{pmatrix}.$$

For $|\sigma_0|<1$ we set $r=1/\sqrt{1-|\sigma_0|^2}$ and since $d(0)=n(0)=r$ and $c(0)=0$ we obtain

$$A^*=\begin{pmatrix} A_\Theta^*+\frac{1}{n(0)^*}\left\langle\cdot,\,D_\Theta(\,\cdot\,,\,0)\begin{pmatrix}0\\0\\s_1(0)\\1\end{pmatrix}\right\rangle C_\Theta^*\begin{pmatrix}0\\-1\end{pmatrix} & A_{12}^*\\ \frac{1}{n(0)^*}\left\langle\cdot,\,D_\Theta(\,\cdot\,,\,0)\begin{pmatrix}0\\0\\r\\0\end{pmatrix}\right\rangle D_{s_1}(\,\cdot\,,0)\begin{pmatrix}1\\0\end{pmatrix} & A_{22}^*\end{pmatrix},$$

where the operators A_{12}^* and A_{22}^* need not be specified because they play no role in the calculations that follow. From

$$v=\begin{pmatrix}v_\Theta\\0\end{pmatrix},\quad v_\Theta(z)=D_\Theta(z,0)\begin{pmatrix}1\\-\sigma_0^*\\0\\0\end{pmatrix}=\begin{pmatrix}1\\ \sigma_0^*\\ z^{k-1}\begin{pmatrix}1/r\\0\end{pmatrix}\end{pmatrix}$$

and using the reproducing property of the kernel $D_\Theta(z,w)$, we get

$$A^{*j}v=\begin{pmatrix}A_\Theta^{*j}\,v_\Theta\\0\end{pmatrix},\ j=0,1,\ldots,k-1,\ \text{and } A^{*k}v=\begin{pmatrix} * \\ r^{-*}D_{s_1}(\,\cdot\,,0)\begin{pmatrix}1\\0\end{pmatrix}\end{pmatrix},$$

where the entry denoted by $*$ is of no consequence here. We conclude that

$$PA^{*k}v=r^{-*}D_{s_1}(\,\cdot\,,0)\begin{pmatrix}1\\0\end{pmatrix}=\frac{1}{n(0)^*}D_{s_1}(\,\cdot\,,0)\begin{pmatrix}1\\0\end{pmatrix}.$$

The proof of the formula for $|\sigma_0| > 1$ can be given in a similar way and is therefore omitted.

For $|\sigma_0| = 1$ and $q = 0$, $d(0) = -Q(0) = -c_0$ and $c(0) = \sigma_0^* Q(0) = \sigma_0^* c_0$. From

$$v = \begin{pmatrix} v_\Theta \\ 0 \end{pmatrix}, \quad v_\Theta(z) = D_\Theta(z,0) \begin{pmatrix} 1 \\ -\sigma_0^* \\ 0 \\ 0 \end{pmatrix} = \begin{pmatrix} 1 \\ \sigma_0^* \\ z^{k-1} \begin{pmatrix} 1 \\ -\sigma_0^* \end{pmatrix} \end{pmatrix},$$

we see that for $1 \le j \le k-1$,

$$A^{*j} v = \begin{pmatrix} A_\Theta^{*j}\, v_\Theta \\ 0 \end{pmatrix}, \quad \text{and } A^{*k} v = \begin{pmatrix} A^{*k}\, v_\Theta + \dfrac{s_1(0)^* - \sigma_0^*}{n(0)^*} C_\Theta^* \begin{pmatrix} 0 \\ -1 \end{pmatrix} \\ 0 \end{pmatrix}.$$

Using

$$D_\Theta(z,0) \begin{pmatrix} 0 \\ -1 \\ 0 \\ 0 \end{pmatrix} = \begin{pmatrix} 0 \\ 1 \\ z^{k-1} \begin{pmatrix} 0 \\ -1 \end{pmatrix} \end{pmatrix} + \begin{pmatrix} z^k Q(0) \begin{pmatrix} \sigma_0 \\ 1 \end{pmatrix} \\ \dfrac{Q(z^*)^* - Q(0)^*}{z} \begin{pmatrix} \sigma_0 \\ -1 \end{pmatrix} \end{pmatrix}$$

and

$$A_\Theta^{*(k-1)} C_\Theta^* \begin{pmatrix} 0 \\ -1 \end{pmatrix} = \begin{pmatrix} * \\ * \\ 0 \\ -1 \end{pmatrix} + \begin{pmatrix} * \\ * \\ * \begin{pmatrix} \sigma_0 \\ -1 \end{pmatrix} \end{pmatrix},$$

we see that for $0 \le j \le k-1$,

$$A^{*(k+j)} v = \begin{pmatrix} A^{*(k+j)}\, v_\Theta \; + \; p_j(A_\Theta^*) C_\Theta^* \begin{pmatrix} 0 \\ -1 \end{pmatrix} \\ 0 \end{pmatrix},$$

where $p_j(z)$ is a polynomial of degree j with leading coefficient

$$\frac{s_1(0)^* - \sigma_0^*}{n(0)^*} = \frac{1}{c_0^* \sigma_0}.$$

Finally we obtain

$$A^{*2k} v = \begin{pmatrix} * \\ c_0^* \sigma_0 \dfrac{s_1(0)^* - \sigma_0^*}{n(0)^{*2}} D_{s_1}(\,\cdot\,,\ 0) \begin{pmatrix} 1 \\ 0 \end{pmatrix} \end{pmatrix} = \begin{pmatrix} * \\ \dfrac{1}{n(0)^*} D_{s_1}(\,\cdot\,,\ 0) \begin{pmatrix} 1 \\ 0 \end{pmatrix} \end{pmatrix},$$

and conclude that

$$PT^{*2k} v = \frac{1}{n(0)^*} D_{s_1}(\,\cdot\,, 0) \begin{pmatrix} 1 \\ 0 \end{pmatrix}.$$

To prove the formula for the case $|\sigma_0| = 1$ and $q > 0$ we use the decomposition in Theorem 4.3. Setting $N_1 = C^*_{\Theta^0_3}\begin{pmatrix} 0 \\ -1 \end{pmatrix}$ and $N_2 = c_0^* C^*_{\Psi_q}\begin{pmatrix} -\sigma_0 \\ 1 \end{pmatrix}$ we get

$$A_q^* = \begin{pmatrix} A^*_{\Theta^0_3} + \langle \,\cdot\,, \dfrac{B_{\Theta^0_3}}{n(0)}\begin{pmatrix} s_1(0) \\ 0 \end{pmatrix}\rangle N_1 & \langle \,\cdot\,, \dfrac{B_{\Psi_q}}{n(0)}\begin{pmatrix} s_1(0) \\ 1 \end{pmatrix}\rangle N_1 & * \\ C^*_{\Psi_q} B^*_{\Theta^0_3} + \langle \,\cdot\,, \dfrac{B_{\Theta^0_3}}{n(0)}\begin{pmatrix} s_1(0) \\ 0 \end{pmatrix}\rangle N_2 & A^*_{\Psi_q} + \langle \,\cdot\,, \dfrac{B_{\Psi_q}}{n(0)}\begin{pmatrix} s_1(0) \\ 1 \end{pmatrix}\rangle N_2 & * \\ 0 & \langle \,\cdot\,, \dfrac{B_{\Psi_q}}{n(0)}\begin{pmatrix} 0 \\ -c(0) \end{pmatrix}\rangle C^*_{s_1} & * \end{pmatrix}.$$

With

$$v_{\Theta^0_3} = D_{\Theta^0_3}(z,0)\begin{pmatrix} 1 \\ -\sigma_0^* \\ 0 \\ 0 \end{pmatrix} = \begin{pmatrix} 1 \\ \sigma_0^* \\ z^{k-1}\begin{pmatrix} 1 \\ -\sigma_0^* \end{pmatrix} \end{pmatrix},$$

$$w_{\Theta^0_3} := D_{\Theta^0_3}(\,\cdot\,,\,0)\begin{pmatrix} 0 \\ -1 \\ 0 \\ 0 \end{pmatrix}, \quad \text{and} \quad w_{\Psi_q} := D_{\Psi_q}(\,\cdot\,,\,0)\begin{pmatrix} 0 \\ -1 \\ 0 \\ 0 \end{pmatrix}$$

we obtain

$$A_q^{*j} v_q = \begin{pmatrix} A^{*j}_{\Theta^0_3} v_{\Theta^0_3} \\ 0 \\ 0 \end{pmatrix}, \quad A^{*j}_{\Theta^0_3} v_{\Theta^0_3} = \begin{pmatrix} z^j\begin{pmatrix} 1 \\ \sigma_0^* \end{pmatrix} \\ z^{k-(1+j)}\begin{pmatrix} 1 \\ -\sigma_0^* \end{pmatrix} \end{pmatrix}, \quad j = 0, 1, \ldots, k-1,$$

$$A_q^{*k+j} = \begin{pmatrix} A^{*k+j}_{\Theta^0_3} v_{\Theta^0_3} + p_j(A^*_{\Theta^0_3}) w_{\Theta^0_3} \\ 0 \\ 0 \end{pmatrix}, \quad j = 0, 1, \ldots, k-1,$$

where $p_j(z)$ is a polynomial of degree j in z with leading coefficient $s_1(0)^*/n(0)^*$, and finally,

$$A_q^{*2k} = \begin{pmatrix} A^{*2k}_{\Theta^0_3} v_{\Theta^0_3} + p_k(A^*_{\Theta^0_3}) w_{\Theta^0_3} \\ \dfrac{s_1^*(0)}{n(0)^*} w_{\Psi_q} \\ 0 \end{pmatrix}.$$

For $1 \le j \le q-1$ the last formula yields

$$A_q^{*2k+j} = \begin{pmatrix} A^{*2k+j}_{\Theta^0_3} v_{\Theta^0_3} + p_{k+j}(A^*_{\Theta^0_3}) w_{\Theta^0_3} \\ \dfrac{s_1(0)^*}{n(0)^*} A^{*j}_{\Psi_q} w_{\Psi_q} \\ 0 \end{pmatrix},$$

which gives the desired result. □

5. The symmetry condition

In this section we apply [5, Corollary 3.5, Theorem 3.1]:

Theorem 5.1. *Let $s(z) \in \mathbf{S}_\kappa^0$ have the Taylor expansion $s(z) = \sum_{n=0}^{\infty} \sigma_n z^n$ at $z = 0$ and assume $s(z) = s_U(z)$, where*

$$U = \begin{pmatrix} A_s & u_s \\ \langle \,\cdot\,, v_s \rangle & s(0) \end{pmatrix} : \begin{pmatrix} \mathcal{P} \\ \mathbb{C} \end{pmatrix} \to \begin{pmatrix} \mathcal{P} \\ \mathbb{C} \end{pmatrix}$$

is a closely connected unitary colligation. The following are equivalent:

(1) *There exists a $\lambda \in \mathbb{C}$ with $|\lambda| = 1$ such that $\lambda\sigma_n$ is real for all n.*
(2) *A_s is J_s-selfadjoint for some signature operator J_s on $\mathcal{P}$.*

In this case J_s is unique and $J_s v_s = \lambda u_s$.

In the following we may assume without loss of generality that $\lambda = 1$. If in the interpolation problem (BIP) the interpolation data are real and the Taylor expansion at $z = 0$ of the parameter function $s_1(z)$:

$$s_1(z) = \sum_{n=0}^{\infty} \tau_n z^n \tag{5.1}$$

has real coefficients τ_n, then the Taylor coefficients σ_n of the corresponding solution $s(z)$ are also real . So there exist signature operators J_{s_1} on the state space $\mathcal{D}(s_1)$ and J_s on the state space $\mathcal{D}(s) = \begin{pmatrix} \mathcal{D}(\Theta) \\ \mathcal{D}(s_1) \end{pmatrix}$ (see Theorem 4.1) such that A_{s_1} is J_{s_1}-selfadjoint and A_s is J_s-selfadjoint. We express $J_s : \begin{pmatrix} \mathcal{D}(\Theta) \\ \mathcal{D}(s_1) \end{pmatrix} \to \begin{pmatrix} \mathcal{D}(\Theta) \\ \mathcal{D}(s_1) \end{pmatrix}$ in terms of J_{s_1} and the interpolation data. We consider three cases corresponding to $|\sigma_0| < 1$, $|\sigma_0| > 1$ and $|\sigma_0| = 1$. For any function $x(z)$ with Taylor expansion at $z = 0$

$$x(z) = \sum_{n=0}^{\infty} x_n z^n,$$

we denote by $[x]_k(z)$ the polynomial consisting of the first k terms of the series:

$$[x]_k(z) = x_0 + x_1 z + \cdots + x_{k-1} z^{k-1}.$$

Recall that R_0 is the difference-quotient operator defined by (4.7).

Case I: $\sigma_0 \in \mathbb{R}$, $|\sigma_0| < 1$ and $\tau_n \in \mathbb{R}$. Using the notation as in Theorem 3.1 (i) and Theorem 4.1 we have

$$J_s \begin{pmatrix} r t_1(z)\mathbf{u} \\ t_2(z)\mathbf{e}_1 \\ g \end{pmatrix} = \begin{pmatrix} r f_1(z)\mathbf{u} \\ f_2(z)\mathbf{e}_1 \\ h \end{pmatrix},$$

where $t_1(z)$ is a polynomial of degree $\leq k$, $g, h \in \mathcal{D}(s_1)$,

$$\begin{aligned} t_2(z) &= z^{k-1}t_1(1/z), \\ f_1(z) &= [s_1t_2]_k(z) + \langle (1-z^kA_{s_1}^k)(1-zA_{s_1})^{-1}J_{s_1}h, v_{s_1}\rangle \\ f_2(z) &= z^{k-1}f_1(1/z) = [s_1]_k(R_0)t_1(z) + \langle (z^k - A_{s_1}^k)(z-A_{s_1})^{-1}J_{s_1}h, v_{s_1}\rangle, \\ h &= t_1(A_{s_1})u_{s_1} + A_{s_1}^kJ_{s_1}g. \end{aligned}$$

Relative to the basis given in Theorem 3.1 (i), J_s has the matrix representation

$$J_s = \begin{pmatrix} 0 & \cdots & 0 & 0 & 0 & \tau_0 & \langle\,\cdot\,,\, u_{s_1}\rangle \\ 0 & \cdots & 0 & 0 & \tau_0 & \tau_1 & \langle\,\cdot\,,\, A_{s_1}u_{s_1}\rangle \\ 0 & \cdots & 0 & \tau_0 & \tau_1 & \tau_2 & \langle\,\cdot\,,\, A_{s_1}^2u_{s_1}\rangle \\ \vdots & \vdots & & & & \vdots & \vdots \\ 0 & \tau_0 & \tau_1 & \tau_2 & \cdots & \tau_{k-2} & \langle\,\cdot\,,\, A_{s_1}^{k-2}u_{s_1}\rangle \\ \tau_0 & \tau_1 & \tau_2 & \tau_3 & \cdots & \tau_{k-1} & \langle\,\cdot\,,\, A_{s_1}^{k-1}u_{s_1}\rangle \\ u_{s_1} & A_{s_1}u_{s_1} & A_{s_1}^2u_{s_1} & A_{s_1}^3u_{s_1} & \cdots & A_{s_1}^{k-1}u_{s_1} & A_{s_1}^kJ_{s_1} \end{pmatrix}.$$

Case II: $\sigma_0 \in \mathbb{R}$, $|\sigma_0| > 1$. In the basic interpolation problem (BIP) the parameter $s_1(z)$ has the property $s_1(0) \neq 0$. Hence $s_1^{-1}(z) = 1/s_1(z)$ is holomorphic at $z = 0$ and we write its Taylor expansion as

$$s_1^{-1}(z) = \sum_{n=0}^{\infty} \mu_n z^n.$$

Since $s_1(z)$ has real Taylor coefficients, the coefficients μ_n are real also. Using the notation as in Theorem 3.1 (ii) and Theorem 4.1 we have

$$J_s \begin{pmatrix} -rt_1(z)\mathbf{u} \\ t_2(z)\mathbf{e_2} \\ g \end{pmatrix} = \begin{pmatrix} -rf_1(z)\mathbf{u} \\ f_2(z)\mathbf{e_2} \\ h \end{pmatrix},$$

where $t_1(z)$ is a polynomial of degree $\leq k$, $g, h \in \mathcal{D}(s_1)$,

$$\begin{aligned} t_2(z) &= z^{k-1}t_1(1/z), \\ f_1(z) &= [s_1^{-1}t_2]_k(z) + \mu_0\langle (1-z^kB_{s_1}^k)(1-zB_{s_1})^{-1}J_{s_1}h, v_{s_1}\rangle, \\ f_2(z) &= z^{k-1}f_1(1/z) = [s_1^{-1}]_k(R_0)t_1(z) + \mu_0\langle (z^k - B_{s_1}^k)(z-B_{s_1})^{-1}J_{s_1}h, v_{s_1}\rangle, \\ h &= -\mu_0t_1(B_{s_1})u_{s_1} + B_{s_1}^kJ_{s_1}g, \end{aligned}$$

and $B_{s_1} = A_{s_1} - \mu_0\langle\,\cdot\,,\, v_{s_1}\rangle u_{s_1}$.

Relative to the basis given in Theorem 3.1 (ii), J_s has the matrix representation

$$J_s = \begin{pmatrix} 0 & \cdots & 0 & 0 & \mu_0 & \mu_0\langle\,\cdot\,,\, u_{s_1}\rangle \\ 0 & \cdots & 0 & \mu_0 & \mu_1 & \mu_0\langle\,\cdot\,,\, B_{s_1}u_{s_1}\rangle \\ 0 & \cdots & \mu_0 & \mu_1 & \mu_2 & \mu_0\langle\,\cdot\,,\, B_{s_1}^2u_{s_1}\rangle \\ \vdots & \vdots & & & \vdots & \vdots \\ 0 & \mu_0 & & & \mu_{k-2} & \mu_0\langle\,\cdot\,,\, B_{s_1}^{k-2}u_{s_1}\rangle \\ \mu_0 & \mu_1 & \mu_2 & \cdots & \mu_{k-1} & \mu_0\langle\,\cdot\,,\, B_{s_1}^{k-1}u_{s_1}\rangle \\ -\mu_0 u_{s_1} & -\mu_0 B_{s_1}u_{s_1} & -\mu_0 B_{s_1}^2 u_{s_1} & \cdots & -\mu_0 B_{s_1}^{k-1}u_{s_1} & B_{s_1}^k J_{s_1} \end{pmatrix}.$$

We sketch the proof of Case II, that of Case I is similar and therefore omitted. Evidently, J_s is selfadjoint in the space $(\mathbb{C}^k)'\oplus\mathcal{D}(s_1)$ where $(\mathbb{C}^k)'$ is the anti-Hilbert space of $\mathbb{C}^k$, that is, the space $\mathbb{C}^k$ provided with the negative inner product $-y^*x$, $x, y \in \mathbb{C}^k$. Let $\mathcal{B}$ be the basis for the space $\mathcal{D}(\Theta_2)$ given in Theorem 3.1 (ii). Then

$$A_{\Theta_2}\mathcal{B} = \mathcal{B}\begin{pmatrix} 0 & 1 & 0 & 0 & \cdots & 0 \\ 0 & 0 & 1 & 0 & \cdots & 0 \\ 0 & 0 & 0 & 1 & \cdots & 0 \\ \vdots & \vdots & \vdots & \ddots & \ddots & \vdots \\ 0 & 0 & 0 & \cdots & 0 & 1 \\ 0 & 0 & 0 & \cdots & 0 & 0 \end{pmatrix}, \quad B_{\Theta_2}\begin{pmatrix} a \\ b \end{pmatrix} = \mathcal{B}\begin{pmatrix} 0 \\ 0 \\ \vdots \\ 1 \end{pmatrix}\begin{pmatrix} 0 & -1 \end{pmatrix}\begin{pmatrix} a \\ b \end{pmatrix},$$

and $C_{\Theta_2}\mathcal{B} = \begin{pmatrix} -ru & 0 & \cdots & 0 \end{pmatrix}$. It follows that A in Theorem 4.2 with $\Theta = \Theta_2$ has the matrix representation

$$A = \left(\begin{array}{cccccc|c} 0 & 1 & 0 & 0 & \cdots & 0 & 0 \\ 0 & 0 & 1 & 0 & \cdots & 0 & 0 \\ \vdots & \vdots & \ddots & \ddots & \ddots & \vdots & \vdots \\ 0 & 0 & 0 & 0 & \ddots & 0 & 0 \\ 0 & 0 & 0 & 0 & \ddots & 1 & 0 \\ -\mu_0\sigma_o^* & 0 & 0 & 0 & \cdots & 0 & \mu_0\langle\,\cdot\,,\, v_{s_1}\rangle \\ \hline \sigma_0^*\mu_0 u_{s_1} & 0 & 0 & 0 & \cdots & 0 & B_{s_1} \end{array}\right).$$

Using the identities (see [1, Lemma 4.3])

$$B_{s_1}B_{s_1}^* - \mu_0^2\langle\,\cdot\,,\, u_{s_1}\rangle u_{s_1} = I, \quad B_{s_1}^*B_{s_1} - \mu_0^2\langle\,\cdot\,,\, v_{s_1}\rangle v_{s_1} = I,$$

$$B_{s_1}v_{s_1} + \mu_0 u_{s_1} = 0, \qquad B_{s_1}^*u_{s_1} + \mu_0 v_{s_1} = 0,$$

$$\langle v_{s_1}\,,\, v_{s_1}\rangle = 1 - \frac{1}{\mu_0^2}, \qquad \langle u_{s_1}\,,\, u_{s_1}\rangle = 1 - \frac{1}{\mu_0^2},$$

we find after some tedious but straightforward calculations that $J_s^2 = I$ and that J_sA is selfadjoint in $(\mathbb{C}^k)'\oplus\mathcal{D}(s_1)$.

Case III: As an example we consider the case where in Theorem 2.2 $k = 1$, $q = 0$, $|\sigma_0| = 1$ and $\sigma_1 = s_0$. We assume that the interpolation data are real, that is, $\sigma_0 = \pm 1$ and $s_0 \in \mathbb{R}$, and that the Taylor coefficients τ_n of $s_1(z)$ in (5.1) are real. Note that the polynomial $p(z)$ in (2.3) has the form $p(z) = c_0 = \sigma_0/s_0$ and that according to (2.5) $\tau_0 \neq s_0$. Using the notation as in Theorem 3.1 (iii) and Theorem 4.1 we find that

$$J_s \left\{ \begin{pmatrix} t_0 \mathbf{u} \\ t_0 J\mathbf{u} \\ g \end{pmatrix} + \begin{pmatrix} t_1(J\mathbf{u} - 2c_0^*\mathbf{u}) \\ t_1(\mathbf{u} - 2c_0 J\mathbf{u}) \\ 0 \end{pmatrix} \right\} = \begin{pmatrix} f_0 \mathbf{u} \\ f_0 J\mathbf{u} \\ h \end{pmatrix} + \begin{pmatrix} f_1(J\mathbf{u} - 2c_0^*\mathbf{u}) \\ f_1(\mathbf{u} - 2c_0 J\mathbf{u}) \\ 0 \end{pmatrix},$$

where t_0 and t_1 are complex numbers, $g, h \in \mathcal{D}(s_1)$,

$$\begin{aligned} f_0 &= \frac{(\tau_0 + \sigma_0)s_0}{2(\tau_0 - \sigma_0)} t_0 + \frac{(\tau_0 + \sigma_0)^2 s_0^2 + 4s_0\tau_1 + 1}{2(\tau_0 - \sigma_0)^2 s_0} t_1 + \frac{\tau_0 + \sigma_0}{2(\tau_0 - \sigma_0)^2} \langle g, u_{s_1} \rangle \\ &\quad - \frac{\sigma_0}{\tau_0 - \sigma_0} \langle g, B_{s_1} u_{s_1} \rangle, \\ f_1 &= \frac{s_0}{2} \left\{ t_0 + \frac{\tau_0 + \sigma_0}{\tau_0 - \sigma_0} t_1 + \frac{1}{\tau_0 - \sigma_0} \langle g, u_{s_1} \rangle \right\}, \\ h &= \frac{s_0 t_0}{\tau_0 - \sigma_0} u_{s_1} + \frac{(\tau_0 + \sigma_0) t_1}{(\tau_0 - \sigma_0)^2} u_{s_1} - 2\frac{\sigma_0 t_1}{\tau_0 - \sigma_0} B_{s_1} u_{s_1} + \frac{s_0}{(\tau_0 - \sigma_0)^2} \langle g, u_{s_1} \rangle u_{s_1} \\ &\quad + B_{s_1}^2 J_{s_1} g, \end{aligned}$$

and $B_{s_1} = A_{s_1} - \dfrac{\langle\,\cdot\,, v_{s_1}\rangle}{\tau_0 - \sigma_0} u_{s_1}$. Relative to the basis given in Theorem 3.1 (iii), J_s has the matrix representation

$$J_s = \begin{pmatrix} \dfrac{(\tau_0 + \sigma_0)s_0}{2(\tau_0 - \sigma_0)} & \dfrac{(\tau_0 + \sigma_0)^2 s_0^2 + 4s_0\tau_1 + 1}{2(\tau_0 - \sigma_0)^2 s_0} & \dfrac{(\tau_0 + \sigma_0)\langle\,\cdot\,, u_{s_1}\rangle}{2(\tau_0 - \sigma_0)^2} - \dfrac{\sigma_0 \langle\,\cdot\,, B_{s_1} u_{s_1}\rangle}{\tau_0 - \sigma_0} \\ \dfrac{s_0}{2} & \dfrac{s_0(\tau_0 + \sigma_0)}{2(\tau_0 - \sigma_0)} & \dfrac{s_0 \langle\,\cdot\,, u_{s_1}\rangle}{2(\tau_0 - \sigma_0)} \\ \dfrac{s_0}{\tau_0 - \sigma_0} u_{s_1} & \dfrac{(\tau_0 + \sigma_0) u_{s_1}}{(\tau_0 - \sigma_0)^2} - 2\dfrac{\sigma_0 B_{s_1} u_{s_1}}{\tau_0 - \sigma_0} & B_{s_1}^2 J_{s_1} + \dfrac{s_0 \langle\,\cdot\,, u_{s_1}\rangle u_{s_1}}{(\tau_0 - \sigma_0)^2} \end{pmatrix}.$$

To show that J_s is selfadjoint one uses the fact that the Gram matrix G is given by

$$G = \begin{pmatrix} 0 & 2 & 0 \\ 2 & 0 & 0 \\ 0 & 0 & I \end{pmatrix}$$

and that $J^* = G^{-1} J_s^{\times} G$ where $J_s^{\times}$ is the complex conjugate transpose of J_s. The equality $J_s^2 = I$ can be established by using that in this case B_{s_1} is a unitary operator on the space $D(s_1)$. Lastly, that $J_s A$ is selfadjoint in the space $\mathbb{C}^2 \oplus D(s_1)$ with Gram matrix G follows from straightforward calculations and the matrix representation

$$A = \begin{pmatrix} -\dfrac{(\tau_0+\sigma_0)\sigma_0 s_0}{2(\tau_0-\sigma_0)} & \dfrac{(\tau_0+\sigma_0)\sigma_0 s_0}{2(\tau_0-\sigma_0)} - 2\dfrac{\sigma_0}{s_0} & -\dfrac{\sigma_0\langle\,\cdot\,, v_{s_1}\rangle}{\tau_0-\sigma_0} \\ -\dfrac{\sigma_0 s_0}{2} & \dfrac{\sigma_0 s_0}{2} & 0 \\ -\dfrac{\sigma_0 s_0}{\tau_0-\sigma_0} u_{s_1} & \dfrac{\sigma_0 s_0}{\tau_0-\sigma_0} u_{s_1} & B_{s_1} \end{pmatrix}$$

of A with respect to the basis given in Theorem 3.1 (iii).

Acknowledgement

I would like to thank my Ph.D. advisor Professor Aad Dijksma for his valuable contribution towards the writing of this paper. His remarks have been very useful in coming up with this final version.

References

[1] D. Alpay, T.Ya. Azizov, A. Dijksma, and H. Langer, *The Schur algorithm for generalized Schur functions I: Coisometric realizations*, Operator Theory: Adv., Appl., vol. 129, Birkhäuser Verlag, Basel, 2001, 1–36.

[2] D. Alpay, T.Ya. Azizov, A. Dijksma, and H. Langer, *The Schur algorithm for generalized Schur functions II: Jordan chains and transformation of characteristic functions*, Monatsh. Math. **138** (2003), 1–29.

[3] D. Alpay, T.Ya. Azizov, A. Dijksma, H. Langer, and G. Wanjala, *A basic interpolation problem for generalized Schur functions and coisometric realizations*, Operator Theory: Adv. Appl., vol 143, Birkhäuser Verlag, Basel, 2003, 39–76.

[4] D. Alpay, T.Ya. Azizov, A. Dijksma, H. Langer, and G. Wanjala, *The Schur algorithm for generalized Schur functions IV: unitary realizations*, Operator Theory: Adv., Appl., Birkhäuser Verlag, Basel, to appear.

[5] D. Alpay, T.Ya. Azizov, A. Dijksma, and J. Rovnyak, *Colligations in Pontryagin spaces with a symmetric characteristic function*, Operator Theory: Adv., Appl., vol. 130, Birkhäuser Verlag, Basel, 2001, 55–82.

[6] D. Alpay, A. Dijksma, J. van der Ploeg, and H.S.V. de Snoo, *Holomorphic operators between Krein spaces and the number of squares of associated kernels*, Operator Theory: Adv. Appl., vol. 59, Birkhäuser Verlag, Basel, 1992, 11–29.

[7] D. Alpay, A. Dijksma, J. Rovnyak, and H. de Snoo, *Schur functions, operator colligations, and reproducing kernel Pontryagin spaces*, Operator Theory: Adv. Appl., vol. 96, Birkhäuser Verlag, Basel, 1997.

[8] M.J. Bertin, A. Decomps-Guilloux, M. Grandet-Hugot, M. Pathiaux-Delfosse, and J.P. Schreiber, *Pisot and Salem numbers*, Birkhäuser Verlag, Basel, 1992.

[9] C. Chamfy, *Fonctions méromorphes sur le circle unité et leurs séries de Taylor*, Ann. Inst. Fourier **8** (1958), 211–251.

[10] P. Delsarte, Y. Genin, and Y. Kamp, *Pseudo-Carathéodory functions and Hermitian Toeplitz matrices*, Philips J. Res. **41**(1) (1986), 1–54.

[11] J. Dufresnoy, *Le problème des coefficients pour certaines fonctions méromorphes dans le cercle unité*, Ann. Acad. Sc. Fenn. Ser. A.I, **250**,9 (1958), 1–7.

Gerald Wanjala
Department of Mathematics
University of Groningen
P.O. Box 800
NL-9700 AV Groningen, The Netherlands
e-mail: gerald@math.rug.nl

Operator Theory:
Advances and Applications, Vol. 160, 469–478

Trace-Class Weyl Transforms

M.W. Wong

This paper is dedicated to Professor Israel Gohberg on the occasion of his 75th birthday.

Abstract. Criteria for Weyl transforms to be in the trace class are given and the traces of these trace-class Weyl transforms are computed. A characterization of trace-class Weyl transforms is proved and a trace formula for all trace-class Weyl transforms is derived.

Mathematics Subject Classification (2000). Primary 47G30.

Keywords. Weyl transforms, localization operators, Weyl-Heisenberg groups, traces.

1. Introduction

Let $\sigma \in L^1(\mathbb{R}^{2n}) \cup L^2(\mathbb{R}^{2n})$. Then the Weyl transform associated to the symbol σ is the bounded linear operator $W_\sigma : L^2(\mathbb{R}^n) \to L^2(\mathbb{R}^n)$ given by

$$(W_\sigma f, g)_{L^2(\mathbb{R}^n)} = (2\pi)^{-\frac{n}{2}} \int_{\mathbb{R}^n} \int_{\mathbb{R}^n} \sigma(x,\xi) W(f,g)(x,\xi)\, dx\, d\xi$$

for all f and g in $L^2(\mathbb{R}^n)$, where $(\,,\,)_{L^2(\mathbb{R}^n)}$ is the inner product in $L^2(\mathbb{R}^n)$ and $W(f,g)$ is the Wigner transform of f and g defined by

$$W(f,g)(x,\xi) = (2\pi)^{-\frac{n}{2}} \int_{\mathbb{R}^n} e^{-i\xi\cdot p} f\left(x + \frac{p}{2}\right) \overline{g\left(x - \frac{p}{2}\right)}\, dp$$

for all x and ξ in $\mathbb{R}^n$.

It is well known that $W(f,g) \in L^2(\mathbb{R}^{2n})$ for all f and g in $L^2(\mathbb{R}^n)$.

Let X be a complex and separable Hilbert space in which the inner product is denoted by $(\,,\,)$, and let $A : X \to X$ be a compact operator. If we denote by $A^* : X \to X$ the adjoint of $A : X \to X$, then the linear operator $(A^*A)^{\frac{1}{2}} : X \to X$ is positive and compact. Let $\{\psi_k : k = 1, 2, \dots\}$ be an orthonormal basis for X

This research has been partially supported by the Natural Sciences and Engineering Research Council of Canada (NSERC) OGP0008562.

consisting of eigenvectors of $(A^*A)^{\frac{1}{2}} : X \to X$, and let $s_k(A)$ be the eigenvalue corresponding to the eigenvector ψ_k, $k = 1, 2, \ldots$. We call $s_k(A)$, $k = 1, 2, \ldots$, the singular values of $A : X \to X$. If $\sum_{k=1}^{\infty} s_k(A) < \infty$, then the linear operator $A : X \to X$ is said to be in the trace class S_1. It can be shown that S_1 is a Banach space in which the norm $\| \|_{S_1}$ is given by

$$\|A\|_{S_1} = \sum_{k=1}^{\infty} s_k(A), \quad A \in S_1.$$

Let $A : X \to X$ be a linear operator in S_1 and let $\{\varphi_k : k = 1, 2, \ldots\}$ be any orthonormal basis for X. Then it can be shown that the series $\sum_{k=1}^{\infty}(A\varphi_k, \varphi_k)$ is absolutely convergent and the sum is independent of the choice of the orthonormal basis $\{\varphi_k : k = 1, 2, \ldots\}$. Thus, we can define the trace $\operatorname{tr}(A)$ of every linear operator $A : X \to X$ in S_1 by

$$\operatorname{tr}(A) = \sum_{k=1}^{\infty}(A\varphi_k, \varphi_k),$$

where $\{\varphi_k : k = 1, 2, \ldots\}$ is any orthonormal basis for X.

Key problems

1. Which functions σ in $L^1(\mathbb{R}^{2n}) \cup L^2(\mathbb{R}^{2n})$ are such that $W_\sigma : L^2(\mathbb{R}^n) \to L^2(\mathbb{R}^n)$ is in S_1?
2. If $W_\sigma : L^2(\mathbb{R}^n) \to L^2(\mathbb{R}^n)$ is in S_1, what is $\operatorname{tr}(W_\sigma)$?

A sample of known results is surveyed in Section 2. In Section 3, we use the theory of two-wavelet localization operators on the Weyl-Heisenberg group to give another class of symbols σ for which $W_\sigma : L^2(\mathbb{R}^n) \to L^2(\mathbb{R}^n)$ is in S_1. In Sections 4 and 5, we recall, respectively, Hilbert-Schmidt operators and twisted convolutions, which are used in Section 6 to provide a characterization of trace-class Weyl transforms. A trace formula for all trace-class Weyl transforms is given in Section 7.

2. Some known results

Good sufficient conditions on σ to ensure that $W_\sigma : L^2(\mathbb{R}^n) \to L^2(\mathbb{R}^n)$ is in S_1 can be formulated in terms of Sobolev spaces and weighted L^2 spaces. For a positive number s, we let $H^{s,2}$ be the set of functions on $\mathbb{R}^{2n}$ defined by

$$H^{s,2} = \left\{\sigma \in L^2(\mathbb{R}^{2n}) : \int_{\mathbb{R}^{2n}} (1 + |q|^2 + |p|^2)^s |\hat{\sigma}(q,p)|^2 dq\, dp < \infty\right\},$$

where $\hat{\sigma}$ is the Fourier transform of σ defined by

$$\hat{\sigma}(z) = (2\pi)^{-n} \lim_{R\to\infty} \int_{|\zeta|\le R} e^{-iz\cdot\zeta}\sigma(\zeta)\, d\zeta, \quad z \in \mathbb{R}^{2n},$$

and the convergence is understood to take place in $L^2(\mathbb{R}^{2n})$. It is clear that $H^{s,2}$ is the L^2 Sobolev space of order s on $\mathbb{R}^{2n}$. We define the space $L^{s,2}$ on $\mathbb{R}^{2n}$ by

$$L^{s,2} = \left\{ \sigma \in L^2(\mathbb{R}^{2n}) : \int_{\mathbb{R}^{2n}} (1 + |x|^2 + |\xi|^2)^s |\sigma(x,\xi)|^2 dx\, d\xi < \infty \right\}.$$

The following result is due to Daubechies [2] and Hörmander [12].

Theorem 2.1. *Let $\sigma \in H^{s,2} \cap L^{s,2}$, $s > 2n$. Then $W_\sigma : L^2(\mathbb{R}^n) \to L^2(\mathbb{R}^n)$ is in S_1.*

The following improvement of Theorem 2.1 can be found in the paper [11] by Heil, Ramanathan and Topiwala.

Theorem 2.2. *Let $\sigma \in H^{s,2} \cap L^{s,2}$, $s > n$. Then $W_\sigma : L^2(\mathbb{R}^n) \to L^2(\mathbb{R}^n)$ is in S_1.*

The following sufficient condition in terms of the Wigner transform can be found in Section 9 of the book [4] by Dimassi and Sjöstrand, and also in the paper [8] by Gröchenig.

Theorem 2.3. *Let $\sigma \in L^2(\mathbb{R}^{2n})$ be such that $W(\sigma,\sigma) \in L^1(\mathbb{R}^{4n})$. Then $W_\sigma : L^2(\mathbb{R}^n) \to L^2(\mathbb{R}^n)$ is in S_1.*

Using the terminology of modulation spaces in the book [9] by Gröchenig, the symbol σ in the preceding theorem is said to be in the space $M^{1,1}$. It is shown in Chapter 11 of [9] that

$$M^{1,1} \subseteq W(\mathbb{R}^{2n}),$$

where $W(\mathbb{R}^{2n})$ is the Wiener space defined by

$$W(\mathbb{R}^{2n}) = \left\{ f \in L^\infty(\mathbb{R}^{2n}) : \sum_{m \in \mathbb{Z}^{2n}} \|f(\cdot + m)\|_{L^\infty([0,1]^{2n})} < \infty \right\}.$$

Since $W(\mathbb{R}^{2n}) \subseteq L^1(\mathbb{R}^{2n})$, it follows that $M^{1,1} \subseteq L^1(\mathbb{R}^{2n})$. Thus, the symbols in Theorems 2.1–2.3 are in $L^1(\mathbb{R}^{2n})$. The advantage of having L^1 symbols is revealed by the following result in the paper [5] by Du and Wong.

Theorem 2.4. *Let $\sigma \in L^1(\mathbb{R}^{2n})$ be such that the Weyl transform $W_\sigma : L^2(\mathbb{R}^n) \to L^2(\mathbb{R}^n)$ is in S_1. Then*

$$\operatorname{tr}(W_\sigma) = (2\pi)^{-n} \int_{\mathbb{R}^n} \int_{\mathbb{R}^n} \sigma(x,\xi)\, dx\, d\xi.$$

We see in a moment that there are functions $\sigma \in L^1(\mathbb{R}^{2n})$ for which $W_\sigma : L^2(\mathbb{R}^n) \to L^2(\mathbb{R}^n)$ is not in S_1.

3. Two-wavelet localization operators

We begin with a recall of the definition of the Weyl-Heisenberg group. Let $(WH)^n = \mathbb{R}^{2n} \times \mathbb{R}/2\pi\mathbb{Z}$, where $\mathbb{Z}$ is the set of all integers. Then we define the binary operation $\cdot$ on $(WH)^n$ by

$$(q_1, p_1, t_1) \cdot (q_2, p_2, t_2) = (q_1 + q_2, p_1 + p_2, t_1 + t_2 + q_1 \cdot p_2)$$

for all points (q_1, p_1, t_1) and (q_2, p_2, t_2) in $(WH)^n$, where $q_1 \cdot p_2$ is the Euclidean inner product of q_1 and p_2 in $\mathbb{R}^n$; t_1, t_2 and $t_1+t_2+q_1\cdot p_2$ are cosets in the quotient group $\mathbb{R}/2\pi\mathbb{Z}$ in which the group law is addition modulo 2π. With respect to the multiplication $\cdot$, $(WH)^n$ is a non-abelian group in which $(0,0,0)$ is the identity element and the inverse element of (q,p,t) is $(-q,-p,-t+q\cdot p)$ for all (q,p,t) in $(WH)^n$.

To simplify the notation a little bit, we identify $\mathbb{R}^{2n}$ with $\mathbb{C}^n$. Thus, $(WH)^n = \mathbb{C}^n \times \mathbb{R}/2\pi\mathbb{Z}$, which can be identified with $\mathbb{C}^n \times [0, 2\pi] = \mathbb{R}^{2n} \times [0, 2\pi]$. Thus, it is plausible, and indeed the case, that the Lebesgue measure $dq\,dp\,dt$ on $\mathbb{R}^{2n}\times[0,2\pi]$ is the left and right Haar measure on $(WH)^n$. Therefore $(WH)^n$ is a locally compact, Hausdorff and unimodular group, which we call the Weyl-Heisenberg group.

Let $\pi : (WH)^n \to U(L^2(\mathbb{R}^n))$ be the mapping defined by

$$(\pi(q,p,t)f)(x) = e^{i(p\cdot x - q\cdot p + t)} f(x-q), \quad x \in \mathbb{R}^n,$$

for all (q,p,t) in $(WH)^n$ and all f in $L^2(\mathbb{R}^n)$, where $U(L^2(\mathbb{R}^n))$ is the group of all unitary operators on $L^2(\mathbb{R}^n)$. Then it can be shown that π is an irreducible and unitary representation of $(WH)^n$ on $L^2(\mathbb{R}^n)$ such that there exists a function φ in $L^2(\mathbb{R}^n)$ with $\|\varphi\|_{L^2(\mathbb{R}^n)} = 1$ and

$$\int_0^{2\pi}\int_{\mathbb{R}^n}\int_{\mathbb{R}^n} |(\varphi, \pi(q,p,t)\varphi)_{L^2(\mathbb{R}^n)}|^2 dq\,dp\,dt < \infty.$$

In more succinct language, we say that π is a square-integrable representation of $(WH)^n$ on $L^2(\mathbb{R}^n)$, φ is an admissible wavelet for π and the number c_φ defined by

$$c_\varphi = \int_0^{2\pi}\int_{\mathbb{R}^n}\int_{\mathbb{R}^n} |(\varphi, \pi(q,p,t)\varphi)_{L^2(\mathbb{R}^n)}|^2 dq\,dp\,dt$$

is the wavelet constant associated to φ. In fact, it is well known that every function φ in $L^2(\mathbb{R}^n)$ with $\|\varphi\|_{L^2(\mathbb{R}^n)} = 1$ is an admissible wavelet and

$$c_\varphi = (2\pi)^{n+1}.$$

We can now look at localization operators with, say, L^1 symbols on the Weyl-Heisenberg group. To this end, let φ and ψ be two admissible wavelets for π and let $F \in L^1((WH)^n)$. Then the localization operator $L_{F,\varphi,\psi} : L^2(\mathbb{R}^n) \to L^2(\mathbb{R}^n)$ associated to the symbol F and the admissible wavelets φ and ψ is defined by

$$\begin{aligned}&(L_{F,\varphi,\psi}u, v)_{L^2(\mathbb{R}^n)}\\ =\ &\frac{1}{c_\varphi}\int_{(WH)^n} F(g)(u, \pi(g)\varphi)_{L^2(\mathbb{R}^n)}(\pi(g)\psi, v)_{L^2(\mathbb{R}^n)} d\mu(g)\end{aligned}$$

for all u and v in $L^2(\mathbb{R}^n)$, where $d\mu(g)$ is the Haar measure on $(WH)^n$.

To specialize, we let $F \in L^1(\mathbb{R}^{2n})$ and let $F^\sharp \in L^1((WH)^n)$ be defined by

$$F^\sharp(q,p,t) = F(q,p), \quad (q,p,t) \in (WH)^n.$$

Then simple calculations give

$$\begin{aligned}&(L_{F^\sharp,\varphi,\psi}u,v)_{L^2(\mathbb{R}^n)}\\ =\ &(2\pi)^{-n}\int_{\mathbb{R}^n}\int_{\mathbb{R}^n}F(q,p)(u,\varphi_{q,p})_{L^2(\mathbb{R}^n)}(\psi_{q,p},v)_{L^2(\mathbb{R}^n)}dq\,dp\end{aligned}$$

for all u and v in $L^2(\mathbb{R}^n)$, where $\varphi_{q,p}$ is the function defined by

$$\varphi_{q,p}(x)=e^{ip\cdot x}\varphi(x-q),\quad x\in\mathbb{R}^n.$$

It is worth pointing out that the localization operator $L_{F^\sharp,\varphi,\psi}:L^2(\mathbb{R}^n)\to L^2(\mathbb{R}^n)$ is the same as the linear operator $D_{F,\varphi,\psi}:L^2(\mathbb{R}^n)\to L^2(\mathbb{R}^n)$ given by

$$\begin{aligned}&(D_{F,\varphi,\psi}u,v)_{L^2(\mathbb{R}^n)}\\ =\ &(2\pi)^{-n}\int_{\mathbb{R}^n}\int_{\mathbb{R}^n}F(q,p)(u,\varphi_{q,p})_{L^2(\mathbb{R}^n)}(\psi_{q,p},v)_{L^2(\mathbb{R}^n)}dq\,dp\end{aligned}$$

for all u and v in $L^2(\mathbb{R}^n)$. If $\varphi=\psi$, then the linear operator $D_{F,\varphi,\varphi}:L^2(\mathbb{R}^n)\to L^2(\mathbb{R}^n)$ is the localization operator first studied in the paper [3] by Daubechies in the context of signal analysis. It is convenient to call $D_{F,\varphi,\psi}:L^2(\mathbb{R}^n)\to L^2(\mathbb{R}^n)$ the Daubechies operator with symbol F and admissible wavelets φ and ψ.

The connection that is useful to us is the following theorem, which is essentially a consequence of Theorems 16.1 and 17.1 in the book [16] by Wong.

Theorem 3.1. *Let $F\in L^1(\mathbb{R}^{2n})$. Then*

$$D_{F,\varphi,\psi}=W_{F*V(\varphi,\psi)},$$

where $V(\varphi,\psi)$ is the Fourier-Wigner transform of φ and ψ given by

$$V(\varphi,\psi)^\wedge=W(\varphi,\psi).$$

The following theorem, which is an immediate consequence of Theorems 2.4 and 3.1, is a special case of Theorem 16.1 in the book [17] by Wong.

Theorem 3.2. *Let $F\in L^1(\mathbb{R}^{2n})$. Then $D_{F,\varphi,\psi}:L^2(\mathbb{R}^n)\to L^2(\mathbb{R}^n)$ is in S_1 and*

$$\operatorname{tr}(L_{F,\varphi,\psi})=(\psi,\varphi)_{L^2(\mathbb{R}^n)}(2\pi)^{-n}\int_{\mathbb{R}^n}\int_{\mathbb{R}^n}F(q,p)\,dq\,dp.$$

From the preceding two theorems, we have the following sufficient condition for a Weyl transform to be in the trace class.

Theorem 3.3. *Let $\sigma\in L^1(\mathbb{R}^{2n})*\{V(\varphi,\psi):\varphi,\psi\in L^2(\mathbb{R}^n)\}$. If we write*

$$\sigma=F*V(\varphi,\psi),$$

where $F\in L^1(\mathbb{R}^{2n})$ and $\varphi,\psi\in L^2(\mathbb{R}^n)$, then $W_\sigma:L^2(\mathbb{R}^n)\to L^2(\mathbb{R}^n)$ is in S_1 and

$$\operatorname{tr}(W_{F*V(\varphi,\psi)})=\|\varphi\|_{L^2(\mathbb{R}^n)}\|\psi\|_{L^2(\mathbb{R}^n)}(\psi,\varphi)_{L^2(\mathbb{R}^n)}\int_{\mathbb{R}^n}\int_{\mathbb{R}^n}F(x,\xi)\,dx\,d\xi$$

for all F in $L^1(\mathbb{R}^{2n})$, and φ and ψ in $L^2(\mathbb{R}^n)$.

The set $L^1(\mathbb{R}^{2n})*\{V(\varphi,\psi):\varphi,\psi\in L^2(\mathbb{R}^n)\}$ can be traced back to the paper [1] by Cohen and is known as the Cohen class in time-frequency analysis.

Remark 3.4. The sufficient condition given in Theorem 3.3 for a Weyl transform to be in the trace class is also a special case of Corollary 1.12 in the paper [15] by Toft. Indeed, it follows from Corollary 1.12 in [15] that $W_{\mu*\tau} \in S_1$ when μ is a bounded measure and $W_\tau \in S_1$. Now, let $F \in L^1(\mathbb{R}^{2n})$ and let $\tau = V(\varphi, \psi)$, where φ and ψ are functions in $L^2(\mathbb{R}^n)$. Then F is a bounded measure. Since W_τ is an operator of rank one, $W_\tau \in S_1$. Thus, by Corollary 1.12 in [15], $W_{F*\tau} \in S_1$.

4. The Hilbert-Schmidt class

A compact operator $A : X \to X$ from a complex and separable Hilbert space X into X is said to be in the Hilbert-Schmidt class S_2 if its singular values $s_k(A)$, $k = 1, 2, \ldots$, are such that $\sum_{k=1}^\infty s_k(A)^2 < \infty$. It can be shown that S_2 is a Hilbert space in which the inner product $(\,,\,)_{S_2}$ is given by

$$(A, B)_{S_2} = \sum_{k=1}^\infty (A\varphi_k, B\varphi_k), \quad A, B \in S_2,$$

where $\{\varphi_k : k = 1, 2, \ldots\}$ is an orthonormal basis for X, the series is absolutely convergent and the sum is independent of the choice of the orthonormal basis $\{\varphi_k : k = 1, 2, \ldots\}$ for X.

We need the following connections between trace-class operators and Hilbert-Schmidt operators.

Theorem 4.1. $S_1 \subseteq S_2$.

Theorem 4.2. *A linear operator A from a complex and separable Hilbert space X into X is in S_1 if and only if $A = BC$, where $B : X \to X$ and $C : X \to X$ are linear operators in S_2. Furthermore, for all B and C in S_2,*

$$\|BC\|_{S_1} \leq \|B\|_{S_2} \|C\|_{S_2}.$$

The literature on S_1 and S_2 abounds. We just mention here the recent books [7] by Gohberg, Goldberg and Krupnik, and [17] by Wong.

If $X = L^2(\mathbb{R}^n)$, then it is well known that a linear operator $A : L^2(\mathbb{R}^n) \to L^2(\mathbb{R}^n)$ is in S_2 if and only if there exists a function h in $L^2(\mathbb{R}^{2n})$ such that

$$(Af)(x) = \int_{\mathbb{R}^n} h(x, y) f(y)\, dy, \quad x \in \mathbb{R}^n,$$

for all f in $L^2(\mathbb{R}^n)$.

The following two results, due to Pool in [13], play an important role in the solutions of the key problems stated in Section 1, and can be found in Chapters 7 and 6 of the book [16] by Wong, respectively.

Theorem 4.3. *Let σ and τ be in $L^2(\mathbb{R}^{2n})$. Then the Weyl transforms $W_\sigma : L^2(\mathbb{R}^n) \to L^2(\mathbb{R}^n)$ and $W_\tau : L^2(\mathbb{R}^n) \to L^2(\mathbb{R}^n)$ associated to the symbols σ and τ, respectively, are in S_2 and*

$$(W_\sigma, W_\tau)_{S_2} = (2\pi)^{-n} (\sigma, \tau)_{L^2(\mathbb{R}^{2n})},$$

where $(\,,\,)_{L^2(\mathbb{R}^{2n})}$ is the inner product in $L^2(\mathbb{R}^{2n})$.

Theorem 4.4. *Every linear operator from $L^2(\mathbb{R}^n)$ into $L^2(\mathbb{R}^n)$ and in S_2 is a Weyl transform $W_\sigma : L^2(\mathbb{R}^n) \to L^2(\mathbb{R}^n)$ associated to some symbol in $L^2(\mathbb{R}^{2n})$.*

In view of Theorems 2.1 and 2.4, we see that a symbol in $L^1(\mathbb{R}^{2n})$, but not in $L^2(\mathbb{R}^{2n})$, gives a Weyl transform $W_\sigma : L^2(\mathbb{R}^n) \to L^2(\mathbb{R}^n)$ that is not in S_1.

5. Twisted convolutions

Let us begin with identifying $\mathbb{R}^{2n}$ with $\mathbb{C}^n$ and any point (x, ξ) in $\mathbb{R}^{2n}$ with the point $z = x + i\xi$ in $\mathbb{C}^n$. We define the symplectic form $[\,,\,]$ on $\mathbb{C}^n$ by

$$[z, w] = 2\,\mathrm{Im}(z \cdot \overline{w}), \quad z, w \in \mathbb{C}^n,$$

where

$$z = (z_1, z_2, \dots, z_n),$$
$$w = (w_1, w_2, \dots, w_n)$$

and

$$z \cdot \overline{w} = \sum_{j=1}^{n} z_j \overline{w_j}.$$

Let λ be a fixed real number. Then we define the twisted convolution $f *_\lambda g$ of two measurable functions f and g on $\mathbb{C}^n$ by

$$(f *_\lambda g)(z) = \int_{\mathbb{C}^n} f(z - w) g(w) e^{i\lambda[z,w]}\,dw, \quad z \in \mathbb{C}^n,$$

provided that the integral exists.

The following formula for the product of two Weyl transforms associated to symbols in $L^2(\mathbb{R}^{2n})$ in the paper [10] by Grossmann, Loupias and Stein plays a pivotal role in the solutions of the key problems.

Theorem 5.1. *Let σ and τ be functions in $L^2(\mathbb{C}^n)$. Then $W_\sigma W_\tau = W_\omega$, where $\omega \in L^2(\mathbb{C}^n)$ and*

$$\hat{\omega} = (2\pi)^{-n}(\hat{\sigma} *_{\frac{1}{4}} \hat{\tau}).$$

A proof of Theorem 5.1 can be found, for instance, in Chapter 12 of the book [14] by Stein and Chapter 9 of the book [16] by Wong.

Further results on extensions, continuity and applications of the twisted convolution can also be found in the book [6] by Folland and the paper [15] by Toft.

6. A characterization

We give in this section a necessary and sufficient condition on the symbol σ to ensure that the Weyl transform $W_\sigma : L^2(\mathbb{R}^n) \to L^2(\mathbb{R}^n)$ is in S_1. To this end, we introduce the subset W of $L^2(\mathbb{C}^n)$ given by

$$W = \left\{ (2\pi)^{-n}(\hat{a} *_{\frac{1}{4}} \hat{b})^\vee : a, b \in L^2(\mathbb{C}^n) \right\},$$

where $(\cdots)^\vee$ denotes the inverse Fourier transform of $(\cdots)$.

Theorem 6.1. *Let $\sigma \in L^2(\mathbb{C}^n)$. Then $W_\sigma : L^2(\mathbb{R}^n) \to L^2(\mathbb{R}^n)$ is in S_1 if and only if $\sigma \in W$. Furthermore, if $\sigma = (2\pi)^{-n}(\hat{a} *_{\frac{1}{4}} \hat{b})^\vee$, where a and b are in $L^2(\mathbb{C}^n)$, then*

$$\|W_\sigma\|_{S_1} \leq (2\pi)^{-n}\|a\|_{L^2(\mathbb{C}^n)}\|b\|_{L^2(\mathbb{C}^n)}.$$

Proof. Suppose that $\sigma \in W$. Then

$$\hat{\sigma} = (2\pi)^{-n}(\hat{a} *_{\frac{1}{4}} \hat{b}),$$

where a and b are in $L^2(\mathbb{C}^n)$. By Theorem 5.1,

$$W_\sigma = W_a W_b.$$

By Theorem 4.3, W_a and W_b are in S_2. So, by Theorem 4.1, $W_\sigma \in S_1$. Conversely, suppose that $W_\sigma \in S_1$. Then, by Theorem 4.1, $W_\sigma \in S_2$. By Theorem 4.2, W_σ is a product of two linear operators in S_2. By Theorem 4.4, we get

$$W_\sigma = W_a W_b,$$

where a and b are in $L^2(\mathbb{C}^n)$. Thus, by Theorem 5.1,

$$\hat{\sigma} = (2\pi)^{-n}(\hat{a} *_{\frac{1}{4}} \hat{b}),$$

and hence $\sigma \in W$. To prove the trace-class norm inequality, we note that

$$\begin{aligned}\|W_\sigma\|_{S_1} &= \|W_a W_b\|_{S_1} \\ &\leq \|W_a\|_{S^2}\|W_b\|_{S_2} \\ &= (2\pi)^{-n}\|a\|_{L^2(\mathbb{C}^n)}\|b\|_{L^2(\mathbb{C}^n)}.\end{aligned}$$

□

Remark 6.2. Theorem 6.1 is a special case of Proposition 1.9 in the paper [15] by Toft. The more straightforward proof given above is new and interesting in its own right.

7. A trace formula

Theorem 7.1. *Let $\sigma = (2\pi)^{-n}(\hat{a} *_{\frac{1}{4}} \hat{b})^\vee$, where a and b are in $L^2(\mathbb{C}^n)$. Then*

$$\operatorname{tr}(W_\sigma) = (2\pi)^{-n}\int_{\mathbb{C}^n} a(w)b(w)\,dw.$$

Proof. We first prove the theorem for a and b in $\mathcal{S}(\mathbb{C}^n)$. Since

$$(\hat{a} *_{\frac{1}{4}} \hat{b})(z) = \int_{\mathbb{C}^n} \hat{a}(z-w)\hat{b}(w)e^{i\frac{1}{4}[z,w]}dw, \quad z \in \mathbb{C}^n,$$

we see that $\hat{a} *_{\frac{1}{4}} \hat{b} \in \mathcal{S}(\mathbb{C}^n)$ and hence $\sigma \in L^1(\mathbb{C}^n)$. By Theorem 2.4, we get

$$\begin{aligned}
\operatorname{tr}(W_\sigma) &= (2\pi)^{-2n} \int_{\mathbb{C}^n} (\hat{a} *_{\frac{1}{4}} \hat{b})^\vee(z)\, dz \\
&= (2\pi)^{-n} (\hat{a} *_{\frac{1}{4}} \hat{b})(0) \\
&= (2\pi)^{-n} \int_{\mathbb{C}^n} \hat{a}(-w)\hat{b}(w)\, dw \\
&= (2\pi)^{-n} \int_{\mathbb{C}^n} \check{a}(w)\hat{b}(w)\, dw \\
&= (2\pi)^{-n} \int_{\mathbb{C}^n} a(w)b(w)\, dw.
\end{aligned}$$

This proves the result when a and b are in $\mathcal{S}(\mathbb{C}^n)$. Since $\mathcal{S}(\mathbb{C}^n)$ is dense in $L^2(\mathbb{C}^n)$, the full result follows by standard limiting arguments. □

Acknowledgement

The author is grateful to the referees and Professor Cornelis Van der Mee for very constructive comments that lead to a much improved version of the paper.

References

[1] L. Cohen, Generalized phase space distributions, J. Math. Phys. 7 (1966), 781–786.

[2] I. Daubechies, On the distributions corresponding to bounded operators in the Weyl quantization, Comm. Math. Phys. 75 (1980), 229–238.

[3] I. Daubechies, Time-frequency localization operators: a geometric phase space approach, IEEE Trans. Inform. Theory 34 (1988), 605–612.

[4] M. Dimassi and J. Sjöstrand, Spectral Asymptotics in the Semi-Classical Limit, Cambridge University Press, 1999.

[5] J. Du and M.W. Wong, A trace formula for Weyl transforms, Approx. Theory Applic. 16 (2000), 41–45.

[6] G.B. Folland, Harmonic Analysis in Phase Space, Princeton University Press, 1989.

[7] I. Gohberg, S. Goldberg and N. Krupnik, Traces and Determinants of Linear Operators, Birkhäuser, 2001.

[8] K. Gröchenig, An uncertainty principle related to the Poisson summation formula, Studia Math. 121 (1996), 87–104.

[9] K. Gröchenig, Foundations of Time-Frequency Analysis, Birkhäuser, 2001.

[10] A. Grossmann, G. Loupias and E.M. Stein, An algebra of pseudodifferential operators and quantum mechanics in phase space, Ann. Inst. Fourier (Grenoble), 18 (1968), 343–368.

[11] C. Heil, J. Ramanathan and P. Topiwala, Singular values of compact pseudodifferential operators, J. Funct. Anal. 150 (1997), 426–452.

[12] L. Hörmander, The Weyl calculus of pseudodifferential operators, Comm. Pure Appl. Math. (32) (1979), 360–444.

[13] J.C.T. Pool, Mathematical aspects of the Weyl correspondence, J. Math. Phys. 7 (1966), 66–76.

[14] E.M. Stein, Harmonic Analysis: Real-Variable Methods, Orthogonality and Oscillatory Integrals, Princeton University Press, 1993.

[15] J. Toft, Regularizations, decompositions and lower bound problems in the Weyl calculus, Comm. Partial Differential Equations 25 (2000), 1201–1234.

[16] M.W. Wong, Weyl Transforms, Springer-Verlag, 1998.

[17] M.W. Wong, Wavelet Transforms and Localization Operators, Birkhäuser, 2002.

M.W. Wong
Department of Mathematics and Statistics
York University
4700 Keele Street
Toronto, Ontario M3J 1P3, Canada
e-mail: `mwwong@pascal.math.yorku.ca`

GPSR Compliance

The European Union's (EU) General Product Safety Regulation (GPSR) is a set of rules that requires consumer products to be safe and our obligations to ensure this.

If you have any concerns about our products, you can contact us on ProductSafety@springernature.com

In case Publisher is established outside the EU, the EU authorized representative is:

Springer Nature Customer Service Center GmbH
Europaplatz 3
69115 Heidelberg, Germany

Batch number: 10732470

Printed by Printforce, the Netherlands